ANNUAL REVIEW OF EARTH AND PLANETARY SCIENCES

ANNUAL REVIEW OF EARTH AND PLANETARY SCIENCES

VOLUME 20, 1992

GEORGE W. WETHERILL, *Editor*
Carnegie Institution of Washington

ARDEN L. ALBEE, *Associate Editor*
California Institute of Technology

KEVIN C. BURKE, *Associate Editor*
National Research Council

ANNUAL REVIEWS INC 4139 EL CAMINO WAY P.O. BOX 10139 PALO ALTO, CALIFORNIA 94303-0897

ANNUAL REVIEWS INC.
Palo Alto, California, USA

International Standard Serial Number: 0084-6597
International Standard Book Number: 0-8243-2020-4
Library of Congress Catalog Card Number: 72-82137

♾ The paper used in this publication meets the minimum requirements of American National Standard for Information Sciences—Permanence of Paper for Printed Library Materials, ANSI Z39.48-1984.

TYPESET BY BPCC-AUP GLASGOW LTD., SCOTLAND
PRINTED AND BOUND IN THE UNITED STATES OF AMERICA

ERRATUM

Annual Review of Earth & Planetary Sciences, Volume 19 (1991)

The copyright for the article "Nuclear Winter: Physics and Physical Mechanisms" by R. P. Turco, O. B. Toon, T. P. Ackerman, J. B. Pollack, and C. Sagan belongs to the US Government.

SOME RELATED ARTICLES IN OTHER *ANNUAL REVIEWS*

From the *Annual Review of Astronomy and Astrophysics*, Volume 29 (1991):

Flares on the Sun and Other Stars, Berhard M. Haisch, Keith T. Strong, and Marcello Rodonò

The Search for Brown Dwarfs, David J. Stevenson

From the *Annual Review of Fluid Mechanics*, Volume 24 (1992):

Advances in Dynamo Theory, P. H. Roberts and A. M. Soward

Atmospheric Turbulence, John C. Wyngaard

Numerical Models of Mantle Convection, G. Schubert

Annual Review of Earth and Planetary Sciences
Volume 20, 1992

CONTENTS

ANNUAL REVIEWS INC. is a nonprofit scientific publisher established to promote the advancement of the sciences. Beginning in 1932 with the *Annual Review of Biochemistry*, the Company has pursued as its principal function the publication of high quality, reasonably priced *Annual Review* volumes. The volumes are organized by Editors and Editorial Committees who invite qualified authors to contribute critical articles reviewing significant developments within each major discipline. The Editor-in-Chief invites those interested in serving as future Editorial Committee members to communicate directly with him. Annual Reviews Inc. is administered by a Board of Directors, whose members serve without compensation.

ANNUAL REVIEWS OF
Anthropology
Astronomy and Astrophysics
Biochemistry
Biophysics and Biomolecular Structure
Cell Biology
Computer Science
Earth and Planetary Sciences
Ecology and Systematics
Energy and the Environment
Entomology
Fluid Mechanics
Genetics
Immunology
Materials Science
Medicine
Microbiology
Neuroscience
Nuclear and Particle Science
Nutrition
Pharmacology and Toxicology
Physical Chemistry
Physiology
Phytopathology
Plant Physiology and Plant Molecular Biology
Psychology
Public Health
Sociology

SPECIAL PUBLICATIONS

Excitement and Fascination of Science, Vols 1, 2, and 3

Intelligence and Affectivity, by Jean Piaget

For the convenience of readers, a detachable order form/envelope is bound into the back of this volume.

Philip H. Abelson

Annu. Rev. Earth Planet. Sci. 1992. 20:1–17

ADVENTURES IN LIFELONG LEARNING

Philip H. Abelson

Science Advisor, American Association for the Advancement of Science, 1333 H Street, N.W., Washington, DC 20005

KEY WORDS: amino acids, mollusks, kerogen, organic geochemistry, isotopes

In earth science, as in other branches of natural science, opportunities for significant research evolve. A new field is opened up either through a liberating concept or by creation of powerful instrumentation. If the field is sufficiently important, many scientists enter it. In due course the cream is skimmed. Unless a new approach is taken, or more powerful instrumentation becomes available, or new and searching questions are posed, or important applications arise, the field becomes mature, and creative people leave it. Most of the recent advances that have occurred in earth science have been associated with increased instrumental capabilities coupled with improved data handling.

I have been an observer of developments in earth and planetary sciences since 1946. During that interval, what were basically observational sciences have evolved into what are predominantly laboratory activities. Some instrumental capabilities have improved by orders of magnitude. During the past 45 years approaches to earth science have also become far more interdisciplinary. Many scientists trained in physics, chemistry, or biology have addressed interesting problems. Throughout this period, part of the activities were devoted to purely academic or basic research problems. Concurrently much of earth science was highly applied. In future, an increasing share of employment opportunities is likely to be in applied areas, with environmental efforts encompassing a larger factor.

In this first chapter of the current *Annual Review of Earth and Planetary Sciences*, I shall follow precedents established by other such contributors. These authors were given free rein to write what they pleased. Most chose to provide autobiographical material, mention some of their associates,

0084–6597/92/0515–0001$02.00

and describe aspects of the key developments in earth science that they had been part of or had witnessed. In addition to these topics, a little space will be devoted to an estimate of some future trends in earth science.

For me, life has been an interesting series of adventures. I have pursued a policy of lifelong learning. But at times the adventure has been physical in nature. My first view of the great copper ore body at Bingham Canyon came in 1918 when I was five years old. That was a sequel to physical combat. My father was a civil engineer with a smelting company, and we were gentiles living in Mormon country. One day when I had ventured away from home, I was surrounded by a group of young Mormons. They designated one of their members to pick a fight with me. He was bigger than I, but I fought so fiercely that he ran home crying. I was promptly accorded the status of an honorary Mormon and later accompanied the boys on a hike to gaze on the canyon.

After the conclusion of World War I, the family returned to my birthplace on the outskirts of Tacoma, Washington. When I was 14, my father got me a summer job with a professional surveyor. We surveyed city lots, accident scenes, and mining claims. This experience, together with a good training in trigonometry, enabled me (following graduation from high school in January) to be employed as a chainman on a hydroelectric construction project located about 70 miles from Tacoma. When I arrived at the construction camp in February 1930, I was assigned to assist a surveyor (transit man). Our assignment was on a steep hill where huge segments of steel pipe (penstocks) were being installed. These were destined to carry a large stream of water to the powerhouse below. We provided the steelworkers with the information essential to place the pipe correctly. The task had only begun and I had been at the camp about three weeks when on a weekend the transit man went to Seattle, got drunk, and was thrown in jail. The next Monday morning, the engineer in overall charge said to me, "Abelson, go up on the hill and line in those penstocks." I had no assistant, and it was necessary to recruit a steelworker to perform what I had previously done with the rod or plumb bob. The steelworkers tended to cheat. I found it expedient to address them in a language they would respond to. My vocabulary of profanity soon equaled or exceeded theirs. We got along well after that.

The seven months spent at the construction camp as a surveyor was a maturing experience. It made me appreciate the potential value of a college education.

In September 1930 I enrolled as a chemical engineer at the then Washington State College. Marvelous years of learning followed. I was especially fond of the chemistry laboratory. By taking a large load of courses I was able to obtain my bachelor's degree in three years. The depression year of 1933 was not a good time to obtain employment. Thus it was pleasing to be offered a teaching assistantship in physics by the departmental chairman, Paul Anderson. During the next two years, I took further courses in physics and did research leading to a master's thesis. A course in Modern Physics taught by S. T. Stephenson included an assignment to read an article by Ernest Lawrence describing his cyclotron. It was immediately clear that Berkeley was going to be an exciting place. Professors at Washington State were helpful in obtaining for me a teaching assistantship and entree to the Radiation Laboratory. I was the only out-of-state student to receive a teaching assistantship in physics at Berkeley that year.

In the era of 1935–1939, the Radiation Laboratory was the world center of nuclear physics. Visitors included leading scientists of the day and other notables such as Leopold Stokowski. Most of the members of the laboratory were post-docs. As a lowly graduate student I was one of the first to experience the effects of big science on education. Ernest Lawrence was a great scientist. From about 1931 and for the remainder of his life, he was a successful promoter of larger and larger accelerators—big science. But during the prewar era, a graduate student in the Radiation Laboratory spent a large fraction of his time operating or repairing the cyclotron.

Graduate students were expected to make their own experimental equipment. I was a competent machinist and glassblower and was able to build the necessary instruments, including a bent-crystal X-ray spectrograph, to study products of neutron irradiation of uranium. I missed the discovery of fission, but during a few months thereafter was able to identify 15 products of uranium fission (Abelson 1939) and obtain my PhD degree.

In September 1939 I became a staff member at the Department of Terrestrial Magnetism of the Carnegie Institution of Washington, where I was assigned to participate in construction of a cyclotron. The intended purpose of the machine was to supply radioactive tracers to Washington area scientists. Events in Europe modified priorities at DTM. Merle Tuve, Lawrence Hafstad, and Richard Roberts chose to develop a promixity fuze. My interest and efforts moved toward uranium. In May 1940, I carried out chemical procedures which provided unique identification of element 93 (McMillan & Abelson 1940). Then I turned to developing the liquid thermal diffusion method for separating uranium isotopes.

I duplicated some of the earlier work that involved aqueous solutions of a target element. However, it was soon evident that separation of isotopes of uranium would require use of UF_6 as the working fluid. The compound was not available, and only a few grams of it had ever been made. I proceeded to build my own fluorine generator and invented a method for making UF_6. I personally synthesized the first 100 kg of the chemical.

Much of the synthetic work was conducted in a laboratory at the old National Bureau of Standards. Successful partial separation of the isotopes was achieved at the Naval Research Laboratory. A pilot plant was built at the Philadelphia Navy Yard. Ultimately 2100 48-foot-long columns were installed at Oak Ridge, Tennessee (Abelson et al 1946).

Following the end of World War II in August 1945, I stayed on with the Navy for another year. During that time, I participated in testing some experimental nuclear reactors (critical masses of enriched uranium) and was in charge of preparing a feasibility report on building a nuclear submarine. In the meantime I was considering what to do next. Ernest Lawrence asked me to go back to Berkeley. Merle Tuve sketched a more intriguing future in biophysics.

Tuve was slated to succeed John Fleming as Director of the Department of Terrestrial Magnetism. He wished to broaden the scope of the department to emphasize experimental aspects of physics over the older observational tradition of the department. The program was to include geophysics as well as other programs. Tuve suggested that physicists might be capable of making unique contributions to such major outstanding problems as the origin of life. He also noted opportunities created by the availability of radioactive isotopes. In September 1946, I returned to DTM as Chairman of the Section on Biophysics.

Soon I became acutely aware that biological processes are extremely complex and that a person trained in chemistry and physics could not expect to make a quick, important discovery in biology. During the next three years I participated in tracer experiments which led to publications in respectable biological journals. During this period I spent two summers at the Marine Biological Laboratory at Woods Hole. There I made the first observations on the changes in ^{32}P metabolism that occur when sea urchin eggs are fertilized.

Richard Roberts, a member of the Biophysics Section, made a crucial contribution to the group after he returned from taking a phage course at Cold Spring Harbor. He brought back microbiological techniques. Soon the group, consisting of myself, Ellis Bolton, Roy Britten, Dean Cowie, and Roberts, was heavily engaged in studying biosynthesis in *Escherichia coli*. We taught ourselves comparative biochemistry. The only carbon tracer then available was ^{14}C in the form of $CaCO_3$. Using plants, I produced ^{14}C-tagged sugars which were useful in probing mechanisms of synthesis of the various amino acids (Abelson 1954a). Ultimately we worked out most of the synthetic steps in the bacterium and especially the amino acids. We treated *E. coli* as a chemical engineer and accounted for the detailed fate of input carbon, sulfur, and phosphorus. We prepared a book that came to be a bible for the many microbiologists working with

E. coli (Roberts et al 1955). The volume had additional value when *E. coli* was chosen as a key biomaterial for carrying out genetic engineering. The achievements of the group demonstrated that it is possible for five people collaborating well to do world-class science in a new field.

Merle Tuve was an innovative scientist who was broadly aware of potential initiatives. He sensed that a great opportunity existed in the dating of the earth's crust. Accordingly, around 1949 a seminar entitled "Milestones in the Pre-Cambrian" was organized, which I attended. Tuve also brought Thomas Aldrich to DTM in 1950 to initiate a dating program. Aldrich had been a graduate student at Minnesota with Alfred Nier. He quickly organized a rock dating program that in its day was the best. Many of the scientists who became leaders in geochronology spent some time with him at DTM. I was able to be helpful to Aldrich early by suggesting use of ion exchange resins in the processing of rubidium and strontium.

In early 1953 I was beginning to have long thoughts about new learning experiences and a possible change in professional activities. Almost by magic, with no initiative on my part, a major possibility arose. The post of Director of the Geophysical Laboratory of the Carnegie Institution of Washington had become vacant and had been open for about a year. I was asked to consider taking the position. I did not respond immediately. I was aware that the laboratory had established an excellent worldwide reputation. About one fourth of the literature cited in advanced books on petrology included publications of the lab. A multidisciplinary group of scientists were principally engaged in physicochemical studies of oxide systems giving rise to minerals found in the igneous and metamorphic rocks. Experiments under precisely controlled conditions were conducted at high temperatures. The experiments, though important and highly significant to field geologists, had become rather routine. Later, when high pressure equipment was developed, the results were more exciting. But at the time, I had reservations about taking the position, and I knew I did not wish to participate personally in phase equilibrium research. However, I discovered that I could establish a different research agenda for myself, and in the end the attraction of a new intellectual adventure prevailed. The decision was reached in the spring of 1953, with the directorship to begin in September of the same year.

Already in early 1953 I was interacting with some of the leading geologists, including W. W. Rubey. He and colleagues organized a "Conference on Biochemistry, Paleoecology and Evolution" held June 9–11, 1953 (Woodring 1954). It involved 23 of the most creative scientists of the time, among them Sterling Hendricks, George Evelyn Hutchinson, and George Gaylord Simpson. The disciplines represented included biochemistry, biophysics, biology, paleontology, geology, and geochemistry. I attended the symposium and benefited greatly from it. One of the items that caught my attention was a discussion of the role of geochemistry in imposing limits on speculations concerning the origin of life and in evaluating paleoecology. In addition there was optimism that biochemistry could make contributions to geochemistry. I came away from the symposium with the resolve of conducting personal research in organic geochemistry.

The policy of the Carnegie Institution of Washington is to identify creative scientists as staff members and then to give them freedom to make their own decisions concerning details of their research. In the course of a year the director of a department must make decisions regarding personnel and budgets and also supervise preparation of an annual report. However, unless a director creates unnecessary chores, time is available to conduct a personal research program, and a director is expected to do so. In addition, a director makes speeches, serves on visiting committees, engages in activities of the relevant scientific societies, and takes on other tasks.

On the day I took office as director in September 1953, I noted that various scientists were proceeding with their research. I found a few administrative functions dealing with the support staff that needed to be improved, but that was soon taken care of. My attention turned to research with a goal of determining if amino acids might be preserved in fossils. In today's world the cost of initiating a new program in geochemistry is likely to amount to $100,000 or much more, and delays in obtaining funds are likely. The funds necessary to begin my program in organic geochemistry amounted to less than $100. The jars for paper chromatography and the necessary chemicals were procured within a week. As a starter I chose to examine the present-day abundant common clam *Mercenaria mercenaria*, of which there were ample fossil shells in exposures nearby. The present-day shells contained about 0.2% of proteins that included the usual amino acids. Miocene shells contained a lesser number and amount of amino acids. Further studies showed that some of the amino acids were much more stable than others, but all were subject to degradation at elevated temperatures. Time of degradation versus temperatures were run on a number of the amino acids, especially alanine. The degradation was found to follow a first-order Arrhenius equation over a wide range of temperatures. At low ambient temperatures some protected amino acids can endure for hundreds of millions of years. The results of the amino acid studies appeared in our annual report (Abelson 1954b). They were also described in an article published in *Scientific American* (Abelson 1956a). This had the effect of stirring interest in paleobiochemistry and fossil biochemicals. Many copies of reprints of the article were sold by Freeman. At that time I was not the only person doing organic geochemistry. A few chemists were active in petroleum company laboratories. However, most of them were secretive about their activities.

For most of my stay (1953–1971) at the Geophysical Laboratory, I was fortunate in having for colleagues in research Thomas Hoering and Ed Hare. Both are skilled scientists with excellent capabilities in instrumentation. Hoering is a talented experimentalist in many fields including mass spectrometry and gas chromatography. Soon after Hoering arrived at the Geophysical Laboratory he made operational a state-of-the-art mass spectrometer. As new equipment for gas and liquid chromatography became available, they were procured and exploited. Ed Hare developed fast, sensitive amino acid analyzers. The capabilities of his instruments were a great improvement over earlier techniques. They made feasible many detailed studies that I participated in. With his equipment he could measure the amounts of various amino acids in a hydrolyzate in about 20 minutes. The device could detect a picomole of an amino acid. The original Stein and Moore Dowex–50 resin column separation method required about a week. It could detect 50 nanomoles of most of the amino acids. Paper chromatography was performed in two days with a convenient level of amino acids of 0.1 micromoles.

Both Hoering and Hare were very productive in research. Their activities were supplemented by fellows, including Patrick Parker, Richard Mitterer, and Jack Valentine. During the 1960s we had one of the liveliest centers of organic geochemistry. Most of the experiments were designed to illuminate major questions. These included:

Did ancient forms of life employ biochemicals and biosynthetic processes similar to those of today?

How stable is organic matter under natural conditions?

What are the mechanisms in the formation of kerogen, the major form of organic matter in nature?

What are the conditions and mechanisms by which kerogen gives rise to petroleum?

What was the nature of the primitive atmosphere of the Earth and of some chemicals formed abiogenically?

Does life exist elsewhere in the solar system?

Each year during my stay at the Geophysical Laboratory I conducted some research. The activities described in annual reports of the laboratory tended to be scouting expeditions. Brief mention of some of the more significant ventures follows.

Together with Ed Hare I examined the amino acid content of 100 different mollusks, including clams, oysters, snails, and nautilus (Hare & Abelson 1964). Some of the species had evolved relatively recently. Some were closely related to species present during Paleozoic times. The total protein plus chitin content of shells ranged from 5% in ancient forms to as little as 0.01% protein in highly evolved species. The fractions of specific amino acids varied greatly among the recently evolved forms, especially the neogastropods. However, the amino acid distribution and content of the various ancient molluskan forms were strikingly similar, pointing toward a single common ancestor.

The proteins of all living forms consist principally of L-amino acids. Ed Hare and I asked the question, "What is the behavior of L-amino acids in the environment as a function of time and temperature?" The answer is that they become a mixture of D and L isomers. The amino acids in nearby upper Pleistocene shells of *Mercenaria mercenaria* were found to contain substantial amounts of the D-amino acids. The characteristic time for racemization ranged from 10,000 to 1,000,000 years, depending on temperature and the particular amino acid. Amino acids in aqueous solutions held at 160°C were racemized in a week. Any primordial soup probably contained a racemic mixture of amino acids. Subsequently, Ed Hare employed an amino acid analyzer to follow the conversion of isoleucine to alloisoleucine as a function of time and temperature. This research led to a geochronometer that was helpful in correlating Pleistocene sedimentation. Today petroleum geochemists often develop useful information for exploration by measuring the degree of isomerization of other chemicals such as the steranes.

Together with Tom Hoering I made detailed studies of the fractionation of carbon isotopes that occurs in photosynthetic microorganisms (Abelson & Hoering 1960, 1961). Some of the organisms were thought to be closely related to Pre-Cambrian species. The organisms were grown under controlled conditions. After harvesting, their amino acids were separated and carbon isotope ratios of parts of each of the various molecules were determined. We found that CO_2 is fixed by at least three different pathways and that isotope fractionation is similar in the various species.

Much of my efforts was focused on kerogen. This complex, insoluble substance of variable composition constitutes most of the earth's organic carbon. Ancient kerogen was the source of the world's petroleum. Kerogen is formed from the constituents of living matter which include complex carbohydrates and proteins as major components. In an aerobic environment these substances are usually consumed as food. However, scavengers are far less efficient in an anaerobic environment. Following death and lysis, constituent simpler sugars and amino acids or peptides are released by hydrolysis or enzyme actions. Thus reactive aldehyde groups of sugars become available to react with amine groups of amino acids. In the laboratory a similar situation to nature can be studied when solutions of glucose and amino acids are brought together at pH 8. The components react noticeably within a day at 4°C. With time and somewhat higher temperatures they combine to form substances that have properties quite similar to natural humic acids and kerogen. Together with Ed Hare I synthesized a number of these artificial kerogens and compared their behavior to that of natural kerogens (Abelson & Hare 1969, 1970). The two types of kerogen were exposed to solutions of a mixture of amino acids. Results with the two types were closely similar. Cysteine disappeared very rapidly; dibasic amino acids, rapidly; hydrophobics at an intermediate pace; and hydrophilics slowly. With both types of kerogen the disappearance of amino acids was accompanied by the appearance of substantial amounts of NH_3. Both types of kerogen exhibited electron spin resonance which was enhanced by the presence of oxygen.

The disappearance of amino acids and other small organic molecules in organic-rich anaerobic sediments is only partially due to microorganisms. Once kerogen is formed, it reacts to a large degree irreversibly with amino acids and proteins. A substantial body of data on such reactions was presented in the 1968–1969 annual report.

In 1962 knowledge concerning the formation of petroleum was primitive. It was known that there are source rocks containing organic matter and porous reservoir rocks that trap petroleum after it is formed. It was also well known that by heating oil shale to 550°C one could obtain petroleum-like fluids. Hoering and I decided to conduct studies at mild temperatures on the kerogen of the Green River shale and other formations to come closer to duplicating petroleum-forming processes occurring in nature (Hoering & Abelson 1963). The research yielded strong evidence that mild thermal degradation of kerogen is the principal mechanism by which hydrocarbons in natural gas and petroleum are produced. We observed the liberation of a series of saturated straight-chain hydrocarbons which are abundant in natural petroleum, ranging from methane (C_1) through n-dodecane (C_{12}). By conducting our experiments at a series of different temperatures ranging from 25°C to 400°C, we were able to monitor maturation processes. For example, emission of CO_2 was completed early. By incubating kerogen in the presence of D_2O, Hoering was able to demonstrate that the D was incorporated in the formed hydrocarbons. Thermal tests on kerogens ranging in age from recent to Pre-Cambrian times were also performed. The Pre-Cambrian carbonaceous matter yielded mainly small amounts of methane.

A paper I presented on June 25, 1963 at the Sixth World Petroleum Congress (Abelson 1963) included the following:

(*a*) The processes of biosynthesis have remained fundamentally unchanged during the past 500 million years.

(*b*) Thus the study of recent sediments is cogent to the problem of petroleum genesis.

(*c*) Important chemical steps leading from living matter toward petroleum occur rather rapidly following death of the algae. Examination of recent sediments discloses that most of the organic matter is in the form of kerogen—an insoluble complex. Although some nitrogen is present, almost none of the nitrogen can be liberated as free amino acids. Recent work by Abelson and Parker shows that even the most recent sediments contain almost no unsaturated acids. This is in striking contrast to the algae, the fatty acids of which are mostly unsaturated. Another puzzle is the relatively low content of saturated fatty acids in recent sediments. Although the algae contain 5–25% fatty acids, the organic matter in sediments is less than 0.1% fatty acid.

(*d*) The organic contents of recent sediments are already much altered and resemble petroleum more closely than biochemicals.

(*e*) In a note added in proof to the paper, the Hoering-Abelson work on pyrolysis of kerogen was mentioned: "Green River shale was incubated at temperatures ranging from 185°C to 407°C. Products were qualitatively similar to those found in natural gas and petroleum. Only minor amounts of unsaturates were formed. Temperature dependence of rate of evolution of methane was consistent with second order reaction kinetics ($A = 10^9$, $E = 40{,}250$ cal/mol)."

One of my ventures was to conduct experiments on the abiogenic synthesis of chemicals on the primitive earth. The procedures were similar to those employed earlier by Stanley Miller, who had worked with a methane-ammonia atmosphere. I tested a number of different compositions including one with CO, N_2, and H_2. These seemed more likely to approach that which actually existed on the early Earth (Abelson 1956b, 1966). I obtained amino acids. A crucial intermediate is HCN. I also noted that irradiation of a mixture of CO_2, N_2, and H_2O did not give rise to any amino acid. These observations were relevant to the question of whether life had originated or existed on Mars. The planet possesses a trivial amount of H_2O, and its atmosphere consists mainly of a limited amount of CO_2. Nevertheless, at the annual meeting of the National Academy of Sciences in 1965, a committee of the Space Science Board took the position: "The biological exploration of Mars is a scientific undertaking of the greatest validity and significance." I reacted to that statement some months later with an article in the *Proceedings of the National Academy of Sciences* (Abelson 1965) with data and an analysis which conveyed the message that existence of earth-like life on Mars was extremely unlikely. The message was disregarded, and a huge sum was expended in an effort to find life there. No life was found. In future, those people considering the possibility of life elsewhere in the solar system should also note the following.

Virtually all organic chemicals are unstable, degrading at rates that are dependent on temperature. Under favorable conditions, some chemicals can persist for hundreds of millions of years. But in natural environments substances are usually altered by mutual reactions. An example of the consequences of such reactions was provided in experiments in which *Chlorella pyrenoidosa* pellets were incubated at 170°C and 143°C. In 10 days at 170°C, half of the amino acid residues disappeared (Abelson 1961). It was estimated that in a 20-minute exposure at 170°C, on the average, each typical protein molecule would experience degradation of about one constituent residue. A similar degree of degradation is estimated to occur in 35,000 years at 15°C. Thus if organisms are in an environment in which they cannot repair damage they lose viability. These estimates are crude, but they indicate limitations on potential survival time for nonmetabolizing organisms.

Before billions of dollars are spent on the assumption that life may exist on Mars, careful experiments should be made on the long-term survivability of nonmetabolizing living matter. A variety of organisms and spores and a broad range of temperatures should be included. There should also be experiments on exposure at various levels to the forms of radiation that penetrate the surface and subsurface of the planet.

During the days of my tenure as director, the style of operations and research at the Geophysical Laboratory was unique by today's standards. Each staff member was expected to personally conduct his or her own research. There was a good machine shop and good procurement but there were practically no technicians. When a graduate student, or post-doc, or other visitor was accepted to work at the laboratory, a staff mentor was chosen. The duty of the mentor was to hasten the day when the new person could be conducting independent research. Often only a few days were required. The laboratory was supported on endowment funds and contingency money was quickly available to foster unexpected initiatives. Our only chore, and that was a major one for me, was to produce an annual report. Each year the lab produced what amounted to a book, and I edited every manuscript. The discipline of assembling material for the annual report was extremely stimulating to the staff. In the month or so before the deadline the lights in the laboratories burned late.

One of the initiatives I took as director was to increase substantially the number of fellows and guest investigators. There were both pre-doctoral and post-doctoral fellows; their tenure was two to four years. My 1966–1967 report stated, "During the past 15 years, 56 fellows and 25 guest investigators have been in residence. They have represented 15 universities of the United States and 15 foreign countries. Most of our alumni are now associated with major universities."

The American Geophysical Union was the focus of one of my adventures. To a degree I was associated with the Union beginning in 1939 when I came to the Department of Terrestrial Magnetism of the Carnegie Institution as a staff member. Most of the members of the department were geophysicists. I was sufficiently interested in geophysics that in 1947 I joined the AGU. I became active in AGU immediately after I was made Director of the Geophysical Laboratory. A few years later I was appointed Chairman of the AGU Meeting Committee. This brought me in frequent contact with leaders of American geophysics. I especially enjoyed conversations with Harry Hess, Maurice Ewing, and Lloyd Berkner. They were great enthusiasts and fun to be with.

In 1958 a crisis loomed in the matter of adequacy of publications in geophysics. Research activity was increasing and programs related to the International Geophysical Year (I.G.Y.) were projected to lead to many articles. Merle Tuve was editor of the *Journal of Geophysical Research*, which published about 800 pages a year and was subsidized by the Carnegie Institution. He was weary of being editor and wished to dispense with the chore. The AGU published *Transactions of the American Geophysical Union*, but the reputation of the *Transcations* was only fair.

Maurice Ewing, President of AGU, and Lloyd Berkner, President-elect, decided that AGU should accept JGR from Merle Tuve and embark on an expanded publication program.

I was invited to a meeting involving those two, together with Waldo Smith and James Peoples, a geophysicist at the University of Kansas. During the meeting I made a few suggestions about the AGU publications program. These were meant only to be constructive. To my surprise when the meeting ended, James Peoples and I had been asked to serve as co-editors of the *Journal of Geophysical Research.*

When I considered the invitation to become co-editor, I could visualize a process whereby with minimal effort on my part the journal could be expanded while quality was maintained. There were excellent capabilities available for style editing, proofreading, manuscript custody, and reviewing. My principal activities were to identify emerging frontiers of geophysics such as space research and to recruit manuscripts. James Peoples capably handled the half of the journal devoted to solid earth. To cover costs of an expanding flood of manuscripts, I saw to it that page charges were established. We maintained a four-month publication time. In a few years the *Journal of Geophysical Research* was publishing 6000 pages annually, and AGU was prospering.

In early 1962 the position of Editor of *Science* became open. I was Chairman of the AAAS Meetings Committee, but not interested in becoming editor. When first approached by Dael Wolfle, Executive Officer of AAAS, I told him that I did not wish to stop my research in organic geochemistry. Ultimately an arrangement was made in which I stayed on at the Geophysical Laboratory but devoted a part of my time to *Science*. A combination of circumstances led to a favorable outcome. Dael Wolfle was both Executive Officer and Publisher of *Science* and was an excellent manager. The staff of *Science* was competent and helpful. The trends in advertising revenues and membership were positive. Techniques of recruitment of articles that were successful with JGR were also applicable to *Science*. The telephone was a useful device. My principal activities were to recruit and select articles, to write editorials, and to seek to make the magazine more interesting and useful to the readers. The staff took care of the large work load associated with the magazine. I enjoyed the stimulus of seeking to be aware of the important developments affecting the scientific enterprise. For a person who loves to learn, it is a great challenge. In addition, the readership of *Science* is alert and critical. Serving it is a privilege.

In 1971 I became President of the Carnegie Institution of Washington and continued as Editor of *Science*, but it was no longer feasible for me to maintain a research program. However, I continued to have learning experiences. I served on visiting committees to universities and on the Advisory Council of the Jet Propulsion Laboratory. I had a glimpse of the corporate world as a member of the Board of Directors of the Washington Gas Light Company and as a member of the Research Coordination Council of the Gas Research Institute. A fairly comprehensive view of U.S. industrial R&D was obtained during two-day visits to more than 20 major industrial research and development laboratories. Another venture was guided tours of automobile production facilities in the U.S. and Japan. I also visited a number of Japanese and Chinese research laboratories. One of my fun travels took me to the Alaska pipeline during its final construction phases. In early March 1977 I was at Valdez, the Brookes Range, and the production facilities at Prudhoe Bay. Earlier I had been underground in a number of mines. In these various circumstances the differing and often contrasting value systems and modes of operation were particularly interesting. For example, in the industrial world research is closely coupled to development, and industrial R&D is usually conducted by interdisciplinary teams. In general the morale of the team members I encountered was at least as high as that in academia.

One of the characteristics of scientific research is that it is on the forefront of societal change. In turn, the frontiers of research are continually evolving. This is particularly true of earth and planetary sciences, and it will continue to be. Prior to World War II, advances in earth science were largely derived from observation, together with use of the pick and hammer and hand lens. Geophysics was in its beginning phases; geochemistry was largely analytical chemistry of rocks. The most sophisticated aspect of earth science was experimental petrology of the kind conducted at the Geophysical Laboratory. The total number of earth scientists who spent most of their time in the laboratory was probably less than 100.

In the late thirties and following World War II, several men whose basic training was in physics or chemistry made seminal contributions to earth science. One was Alfred Nier, a physicist who with a mass spectrometer he had built made the first dating of Pre-Cambrian rocks. Another was Maurice Ewing, whose background was in physics. He was a great enthusiast who personally explored the oceans and invented many pieces of equipment to facilitate his work and that of colleagues. I remember well an inspiring lecture Ewing gave on exploration of the mid-Atlantic ridge. Merle Tuve organized a series of experiments in explosion seismology which discredited the early picture of the onion-like structure of the interior of the earth. Julian Goldsmith (1991) has written of the stimulus to isotope geology and planetary science provided by Harold Urey, a chemist at the University of Chicago. Students who were there at the time have had a profound effect on earth and planetary sciences—one that continues today.

The I.G.Y. and Sputnik were followed by space exploration and expanded funds for research. Opportunities created by return of the moon rocks were a great stimulus. The Allende and other meteorites provided materials for study of early events in the formation of the solar system. In recent years 8000 meteoric fragments collected in Antarctica have provided additional targets for measurements. Objects originating outside the solar system have been identified. Meteorites produced by impacts of other bodies on the Moon and Mars have also been studied. More of such objects will doubtless become available.

An important factor in the vitality of earth science has been isotope studies made feasible by great improvements in mass spectrometry. These have involved improvement of at least an order of magnitude in the precision of measurements, greatly enhanced rates at which determinations can be made, and a decrease of several orders of magnitude in the weight of samples required for the mass spectrometer. Today the mass spectrometer is probably the most ubiquitous and powerful tool in earth and planetary sciences. Its applications include dating of events throughout the history of the Earth and the solar system, studies of metamorphism, studies of determination of prehistoric ocean temperatures, studies of formation of sulfide ore deposits, identification of plumes originating in the mantle, and applications to organic geochemistry.

Activity in organic geochemistry has expanded greatly since the 1960s. By the early 1970s many fossil biochemicals (now called biomarkers) had been identified. These substances were originally components of living matter—for example sterols, which had been modified to become steranes. Modifications of the original structures are often found to be time and temperature dependent. Thus measurement of biomarkers can provide evidence of the thermal history of kerogen and the oil produced from it. Detailed information about the thousands of components of petroleum can be readily obtained by use of gas chromatography coupled with mass spectrometry and computer analysis.

Information obtained by organic geochemists is incorporated into computer models that are important to successful petroleum exploration. The organic geochemists are now members of the exploration team. Their activities and numbers are now reflected in a number of publications, including journals devoted to their field.

I turn now to speculation about future efforts in earth science. Priorities and opportunities will be heavily affected by societal concerns about the environment and subsequently about energy supplies. The level of research activities will, of course, be affected by availability of funds. Prospects for the funding of the kind of academic research currently in vogue are only fair. During the past several years the number of grant proposals have increased faster than funds. In 1990 only 32% of proposals in earth sciences were funded by the National Science Foundation. That rate is down from 50% several years earlier. There will still be a continuation of exciting results as even more powerful equipment becomes available. However, the important findings will largely be made by those with access to the best equipment.

Some of the students currently on the path to a PhD degree may obtain faculty appointments, but a majority will probably be unable to obtain grant support to carry on research similar to that of their thesis. For many years, the federal government was a relatively reliable source of support for basic research. The overall budget squeeze could lead to erratic effects on funds available to granting agencies. At the same time, there is more congressional interest in achieving faster practical effects from research support. We are now in a different era than that of 1955–1985.

One of the questions that the nation faces is what to do about global change and the greenhouse effect. Other countries are proposing reductions in emissions of CO_2, and they are pressuring the U.S. to comply. Trillions of dollars could be involved, and the knowledge base for decisive action is weak. Hence the federal government will be supporting a broad program to measure global change, present and past.

A much larger potential for employment is in the private sector, where huge amounts of money will be involved in remediation of waste dumps and treatment of their leachates. Efforts to produce some of the more than 300 billion barrels of oil left in place after secondary recovery will merit attention. Improvements in knowledge about methods of inflencing the movement of fluids in porous rocks and soils would have tremendous economic payoffs.

The remediation of waste dumps and treatment of their leachates will require efforts over a period of 30 years or more and expenditures of perhaps $500 billion. The waste sites differ in volume, geometry, and content. Some have only inorganic contaminants and soil; others have soil and complex mixtures of many different organic chemicals of varying toxicity located helter-skelter in the waste site. The various chemicals have different solubilities in water and degrees of adsorption to soil. Some of the chemicals are readily leached and are already in ground-water plumes some distance from the waste site. At the other extreme are chemicals that are virtually immobile. To date the science and technology employed in assessing the situation at sites and remediating them has been costly and primitive. Achievement of effective and efficient remediation will require an interdisciplinary approach that includes earth scientists as key members of the team.

Other problems of contamination of ground water arise from pesticides used in agriculture and by leaky underground tanks containing chemicals or petroleum products. Solution to these problems will require the efforts of hydrologists knowledgeable about the behavior of chemicals in soils.

Ultimately it may be found that some of the phenomena encountered at waste sites have counterparts in the movement of petroleum in soils and rocks. The hydrophobic nature of some of the chemicals is comparable to that of components of petroleum. Indeed, benzene, toluene, and the xylenes are present in petroleum, at many waste sites, and in some leachates.

The habitat of oil is, of course, greatly different from that of a waste site. Nevertheless, one can question whether the technology of petroleum production is more advanced than that of the remediation of waste sites. After all, the technology has left about two thirds of the oil in place. Ultimately a combination of bright ideas, computer modeling, and choice of the best drive mechanisms will extract much of that oil at nominal costs.

The past 50 years have witnessed enormous changes in the conduct of earth science. The next 50 years are likely to bring changes presently unknowable but of comparable magnitude. Professors who have a crucial role in shaping the capabilities of their students should seek to prepare them for an uncertain future. One way is to emphasize acquiring knowledge of the fundamentals in mathematics, physics, chemistry, earth science, and biology as undergraduates. Then as new opportunities arise their students can master the special knowledge needed to tackle the new problems.

Literature Cited

Abelson, P. H. 1939. An investigation of the products of the disintegration of uranium by neutrons. *Phys. Rev.* 56: 1–9

Abelson, P. H. 1954a. Amino acid biosynthesis in *Escherichia coli*: Isotopic competition with ^{14}C. *J. Biol. Chem.* 206: 335–43

Abelson, P. H. 1954b. Organic constituents of fossils. In *Carnegie Institution of Washington Year Book 53*, pp. 97–101

Abelson, P. H. 1956a. Paleobiochemistry. *Sci. Am.* 195: 83–92

Abelson, P. H. 1956b. Paleobiochemistry: Inorganic synthesis of amino acids. In *Carnegie Institution of Washington Year Book 55*, pp. 171–74

Abelson, P. H. 1961. Thermal degradation of algal protein. In *Carnegie Institution of Washington Year Book 60*, pp. 207–8

Abelson, P. H. 1963. Organic geochemistry and the formation of petroleum. *Proc. World Petroleum Congr., 6th, Frankfurt*, pp. 397–407

Abelson, P. H. 1965. Abiogenic synthesis in the Martian environment. *Proc. Natl. Acad. Sci. USA* 54: 1490–94

Abelson, P. H. 1966. Chemical events on the primitive earth. *Proc. Natl. Acad. Sci. USA* 55: 1365–72

Abelson, P. H., Hare, P. E. 1969. Uptake of amino acids by kerogen. In *Carnegie Institution of Washington Year Book 68*, pp. 297–303

Abelson, P. H., Hare, P. E. 1970. Reactions of amino acids with natural and artificial humus and kerogens. In *Carnegie Institution of Washington Year Book 69*, pp. 327–34

Abelson, P. H., Hoering, T. C. 1960. The biogeochemistry of the stable isotopes of carbon. In *Carnegie Institution of Washington Year Book 59*, pp. 158–65

Abelson, P. H., Hoering, T. C. 1961. Thermal degradation of algal protein. In *Carnegie Institution of Washington Year Book 60*, pp. 207–8

Abelson, P. H., Rosen, N., Hoover, J. I. 1946. Liquid thermal diffusion. *NRL Classified Rep. 0-2982*. (Declassified 1957.) Issued 1958 by Atomic Energy Commission under No. TID-5229

Goldsmith, J. R. 1991. Some Chicago georecollections. *Annu. Rev. Earth Planet. Sci.* 19: 1–16

Hare, P. E., Abelson, P. H. 1964. Proteins in mollusk shells. In *Carnegie Institution of Washington Year Book 63*, pp. 267–70

Hoering, T. C., Abelson, P. H. 1963. Hydrocarbons from kerogen. In *Carnegie Institution of Washington Year Book 62*, pp. 229–34

McMillan, E., Abelson, P. H. 1940. Radioactive element No. 93. *Phys. Rev.* 57: 1185–86

Roberts, R. B., Abelson, P. H., Cowie, D. B., Bolton, E. T., Britten, R. J. 1955. *Studies of Biosynthesis in Escherichia coli*. Washington DC: Carnegie Inst. (Publ. No. 607). 521 pp. (Other editions 1957, 1963)

Woodring, W. P. 1954. Conference on biochemistry, paleoecology and evolution. *Proc. Natl. Acad. Sci. USA* 40: 219–24

Annu. Rev. Earth Planet. Sci. 1992. 20:19–43

HOTSPOT VOLCANISM AND MANTLE PLUMES

Norman H. Sleep

Departments of Geology and Geophysics, Stanford University, Stanford, California 94305

KEY WORDS: seamounts, Hawaii, Iceland, convection

INTRODUCTION

Voluminous volcanism occurs both along mid-oceanic ridge axes and along island arcs. Plate tectonics provides a framework for discussing ridge volcanism in terms of upwelling mantle material which forms a diverging lithosphere of oceanic plates. Arc volcanism is more complicated but clearly related to the subduction of oceanic lithosphere beneath the arc. However, midplate volcanism that forms seamount chains, such as the Hawaiian Islands, and segments of the mid-oceanic ridge, such as Iceland, with excessive volcanism are not as obviously related to plate tectonics. Both types of features were called "hotspots" by Morgan (1971, 1972) and attributed to the upwelling of plumes of hot material from the deep mantle beneath the active volcanism.

The explanation for midplate seamount chains is relatively simple. The moving plate passes over an upwelling of hot material from great depths in the mantle (Figure 1). The lithosphere of the plate is heated and volcanism occurs above the upwelling. A sequence of volcanoes are each active when they are over the plume and then die as the plate moves away. Thus, the age of volcanism becomes progressively older as one moves down the chain away from the active volcano. This hypothesis was intended to explain basic features of midplate seamount chains: an active volcano at one end and the formation of atolls and submerged seamounts down the chain by erosion of the volcanic edifice, subsidence of the seafloor, and deposition of carbonate reefs. In addition, the parallelism of several Pacific

0084–6597/92/0515–0019$02.00

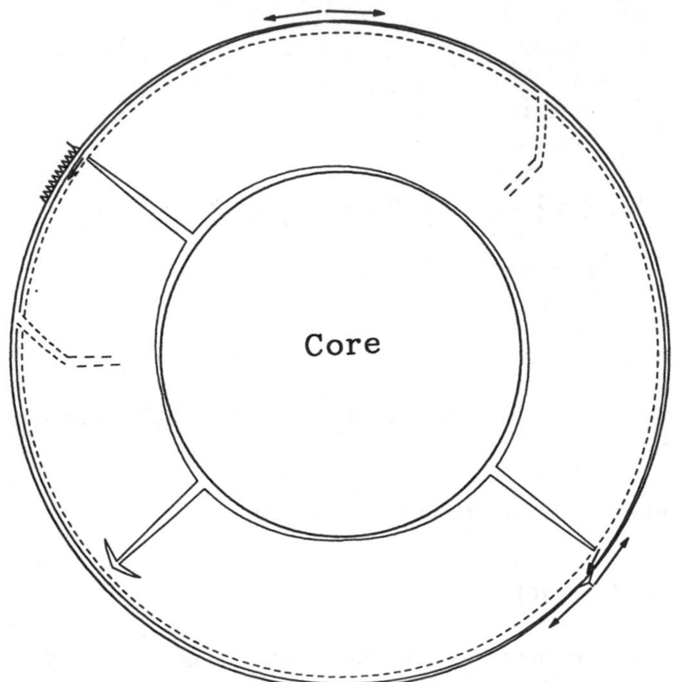

Figure 1 Various aspects of mantle plumes and convection within the Earth are shown schematically more or less to scale. From the bottom up: a thermal boundary layer is produced by heating from the core; plumes arise where the boundary layer is thick (4:30, 7:30, and 10:30); the boundary layer is thinner where slabs impinge (2:00 and 9:00); the cylindrical conduit of the plume tapers upward because viscosity decreases upward; new plumes start with a large head (7:30); hotspot chains are generated as a plate moves over a plume and hot material is entrained from the asthenosphere by the plate (10:30); normal ridges tap the adiabatic interior (12:00); plume material is entrained into hotspot ridges (4:30).

seamount chains in the plumes was presumed to arise from deep regions that are relatively fixed compared with the moving plates.

The history of ideas about the origin of seamount chains with particular emphasis on the Hawaiian Islands is given by Clague & Dalrymple (1987, 1989). A short discussion of this history is relevant to understanding the current controversies about hotspots and mantle plumes. The age progression of the Hawaii chain was known qualitatively in the last century and to some extent before that by native Hawaiians. The idea of a moving heat source also dates from the last century. Explanations of Hawaii and other Pacific seamount chains in terms of mantle plumes were presented even before plate tectonics was well understood and generally accepted

(Wilson 1963, 1965). However, mantle plumes did not receive the general acceptance of the rest of plate tectonics and alternative explanations, again adapting century-old ideas to plate theory, were proposed in terms of propagating cracks in the lithosphere.

Several reasons can be offered for the relatively slow progress toward agreement and understanding on the existence and nature of mantle plumes. First, the physics and kinematics of plumes are complicated. Plumes are supposed to arise from the deep mantle, probably near the core-mantle boundary. That is, the whole Earth, including the core and the deep mantle, is significantly involved. Second, hotspots occur globally and relevant observations involve geochemistry of the lavas, paleontological and radioactive dating methods, sedimentological studies to monitor uplift and subsidence, seismic studies of all depths in the Earth, and geoid and heat flow anomalies. That is, much of geology is involved on a global basis. Thirdly, on a more cultural basis, both Wilson (1963, 1965) and Morgan (1971, 1972, 1981) were vague on how and where plumes would originate at depth from mantle convection. For example, this allowed me (Sleep 1984) to contend that mechanisms involving lithospheric stress and cracks were inherently more testable than those involving plumes. That is, the present author entered the topic of mantle plumes as a skeptic with the working hypothesis that plumes did not exist.

I focus this review on lines of evidence that are useful for showing that mantle plumes really occur and that are thus also useful for demonstrating the physics of plumes and their effects at shallow depths. The concentration is on topics leading to the current level of understanding (Figure 1). That is, plumes are narrow cylindrical conduits of hot low-viscosity material that ascend from near the core-mantle boundary; material enters plumes by horizontal conduit flow along the lowest part of the thermal boundary at the base of the convecting mantle; and hotspots are the result of plume material ponding at the base of the lithosphere or being entrained into mid-oceanic ridge axes. I work from the bottom up starting with the convection in the deep mantle, going to the interaction of hotspots and ridges, and finally returning to midplate hotspot swells, a topic reviewed by Crough (1983). The cleanest evidence comes from isolated hotspots like Hawaii and Iceland. However, many hotspots occur in families including the Pacific superswell hotspots of Caroline, Samoa, Pitcairn, Marqueses, Macdonald, Tahiti, Easter, Juan Fernandez, and San Felix (Bonatti et al 1977, McNutt & Fischer 1987). Such lines of hotspots may form above two-dimensional tabular upwellings at depth (Houseman 1990, Sleep 1990a), but the process is too poorly understood to review usefully. I also concentrate on the physical aspects at the expense of the chemical ones. I do this because the physics is better understood at present (particularly by me) than the chemistry although chemical methods are likely to eventually provide the best constraints.

SOURCE OF MANTLE PLUMES

Several observable features of hotspots are the result of differences between plumes, which are convection heated from below, and the rest of the convection associated with plate tectonics which is heated mainly from within. The differences can be qualitatively understood by considering how heat gets out of and into the mantle. Heat can escape from the interior as surface heat flow only when hot material approaches the surface by convection because the mantle is much thicker than the distance over which heat can be carried by conduction even over the age of the Earth. The heat supplied by radioactive decay and by cooling of the mantle over time is distributed throughout the mantle. Thus the heat can be removed only by cycling all the convecting material near enough to the surface so that conduction can act—that is, through the oceanic lithosphere. Therefore, there are no fixed cores of convection cells—which is expected because ridges migrate with respect to trenches. In addition, the widespread heat sources do not create locally hot areas. In contrast, convection heated from below involves a hot basal boundary layer which is hotter (in terms of "potential" temperature adjusted for adiabatic effects and referenced to a common depth) than the rest of the fluid. Limited volumes of the basal boundary layer, which are heated at any one time and place ascend as isolated upwellings. The upwellings rise until they pond beneath the more rigid lithosphere or are entrained into mid-oceanic ridges.

Detection of local hot upwellings is thus fundamental to establishing the existence of plumes and to identifying the depth of the basal boundary layer. I begin with the latter topic and then discuss evidence that new plumes originate from instabilities on the basal boundary layer. The topic of whether the mantle source region of hotspot volcanism is really hotter than normal mantle is conveniently discussed with regard to the interaction of hotspots with mid-oceanic ridges.

Depth of the Source Region of Plumes

As plumes presumably arise from a basal boundary layer, they are a convenient probe to determine the depth to that layer and hence the depth of convection in the mantle. It is known that the convection in the Earth associated with plate tectonics extends down to at least the depth, 680 km, of the deepest earthquakes associated with downgoing slabs (Frohlich 1989). Thus the basal boundary layer could exist either near that transition between the upper and the lower mantle or just above the core. Both cases imply a compositional boundary in which the lower material has a more dense composition than the upper one so that material does not cross the boundary. Therefore, heat is transferred across the boundary by conduction.

The core-mantle boundary is clearly a compositional boundary between molten iron alloy and solid silicates. The 680 km transition zone is more problematic because most of the difference in density and seismic velocity is associated with a phase change. Laboratory and seismic data are not sufficiently accurate to determine whether an additional compositional density difference exists (cf Jeanloz & Morris 1986, Bina & Silver 1990, Bukowinski & Wolf 1990). Thus a controversy has persisted since the early days of plate tectonics as to whether convection is mantle wide or whether convection systems in the upper and lower mantle are separated by a compositional density difference (see Silver et al 1988).

Davies (1988a) used the properties of convection heated from below to show that plumes are more easily explained if they ascend from near the core-mantle boundary and if the convection associated with plate tectonics is also mantle wide. The relative sizes of the core and mantle are then such that the system behaves as convection heated from within with a modest perturbation from heat from below. Specifically radioactive elements (U, Th, and K) and hence heat generation are probably strongly concentrated in the mantle rather than the core. About 4/5 of the Earth's heat capacity is also in the mantle while 1/5 is in the core (Stacey 1980). As radioactivity and specific heat from cooling are comparable sources at present (Christensen 1985), about 10% of the total heat budget would come from the core if both the mantle and core cooled at equal rates. In contrast the upper mantle is only 18% of the mass of the Earth and 27% of the mass of the mantle. Thus separate upper mantle convection would be essentially convection heated from below.

Continuing with Davies' (1988a) line of reasoning, plate tectonics behaves mainly like convection heated from within. That is, mid-oceanic ridges are passive cracks that tap a relatively adiabatic reservoir from below. The depth-age relationship of oceanic crust is explained by conductive cooling of the oceanic lithosphere as it moves away from ridge axes. In contrast, the temperature contrast across the basal boundary layer associated with convection heated from below is comparable to the temperature contrast across the upper boundary layer—in this case the lithosphere. Ridges would be expected to migrate off upwellings from below. Thus large source temperature anomalies would be expected along the ridge axis depending on how effectively upwellings are locally tapped. Topographic and geoid anomalies comparable to those of ridges would be expected beneath upwellings not associated with ridges.

To be sure, hotspots do exhibit these characteristics of convection heated from below, but the effects are local and minor on a global basis. Using methods reviewed below, Davies (1988a) and Sleep (1990a) determined the heat flux of individual hotspots and summed the total globally. About 6% of the heat flux from the mantle is associated with hotspots. As emphasized by those authors, this is an appropriate amount to be supplied by the core but much too little to be supplied by separate convection beneath the upper mantle. In addition, hotspots tend not to overlie regions where the geometry of mantle-wide convection implies that slabs have collected near the base of the mantle over the last 100 m.y. The cool slabs are expected to make the basal boundary layer locally thin and thus locally suppress plumes (Chase 1979, Chase & Spowl 1983, Richards et al 1988).

Temperature Contrast across the Basal Boundary Layer

The structure at the base of the mantle like the structure of the upper mantle-lower mantle transition is not adequately resolved by seismic studies (see Young & Lay 1987). It is possible that an intrinsically dense layer of dregs exists beneath the basal thermal boundary layer associated with convection in the overlying mantle. This implies that part of the temperature difference between the core and the interior of the mantle occurs across this chemical layer while part occurs across the basal thermal boundary layer. Plumes are a direct probe of the thermal boundary layer and thus yield information whether a chemical layer exists (Jeanloz & Richter 1979, Jeanloz & Morris 1986).

I first consider flow to the plume in the absence of a chemical boundary layer. The highest temperatures occur at the base of the thermal boundary layer in contact with the core. As viscosity is strongly temperature dependent, the lower part of the thermal boundary layer acts as a conduit letting material flow to the plume (see Loper 1984, Stacey & Loper 1984, Loper & Eltayeb 1986, Sleep et al 1988, Davies 1990). The highest flow velocities are at the base of the layer because the core acts as a free-slip boundary. The free-slip nature of the boundary also allows this rapidly flowing material to leave the boundary layer to enter plumes. Thus the initial maximum excess temperature of the upwelling material is the temperature contrast across the thermal boundary layer. The excess temperatures of plumes are thus expected to be globally similar except for the loss of heat to the surrounding mantle on the way up. In contrast, a chemical boundary layer might cause the temperature difference to be much less. Secondarily it might act as a no-slip boundary and retard the flow of the hottest convecting material.

The physics of heat loss from a plume to its surroundings are tolerably understood (Griffiths & Campbell 1991). Heat loss from a vertical plume by conduction to its surrounding is relatively inefficient. However, plumes are tilted by cross currents in the mantle. This causes the hot central region of the plume to underlie cooler surrounding material. Thermal plumes are unstable to convection in the vertical plane perpendicular to the tilt direction. The hot center rises and cooler material from the conductive halo around the plume is entrained. The net effect is to mix the plume with its surroundings and reduce the temperature contrast. Modeling indicates that this effect is important for the weakest plumes, but minor for strong plumes, like Réunion and Hawaii.

As reviewed below, the average excess temperature of hotspot material for strong hotspots relative to normal mid-oceanic ridges is around 250°C implying that the maximum excess temperature is around 300°C. A temperature estimate at the base of the convecting layer is obtained by adjusting the upper mantle plume temperature $\approx$ 1600°C to 2600°C for adiabatic effects down to 2900 km. This temperature is much less than laboratory and thermodynamic estimates of the temperature at the top of the core which are often 1000°C or more higher (cf Jeanloz 1990, Boness & Brown 1990, Boehler et al 1990, Williams et al 1991). Thus an additional boundary—either dregs at the base of the mantle or between separate upper and lower mantle convection systems—is believed to take up some of this difference (Jeanloz & Richter 1979, Jeanloz & Morris 1986). Alternatively, alloying reduces the melting temperature in the core enough so that the temperature contrast above the core is only a few 100°C.

For a stable layer of dregs to persist, it needs to be sufficiently dense so that it is not entrained by the overlying fluid (Sleep 1988, Olson & Kincaid 1991). A more complicated constraint is that the layer needs to be internally stratified or sufficiently viscous so that it does not convect rapidly. Theoretical estimates based on the theory of well developed convection indicate that the temperature dependence of viscosity will make the dense layer sufficiently fluid to convect efficiently enough so that most of the temperature contrast ends up in the overlying thermal boundary layer (Sleep 1988). However in an experiment by Olson & Kincaid (1991), the dense layer convected slowly because it was too thin for well developed convection and only 22% of the total temperature contrast occurred across the overlying boundary layer. It should be noted that their dense layer was intrinsically more viscous by a factor of 3.5 than the upper layer while the additional iron to increase density in the mantle should decrease viscosity. The questions involving convection in the dense layer and entrainment could be resolved by modifying Olson & Kincaid's (1991) experiments so that the intrinsic viscosity difference between the upper and the lower fluid varied independently of density.

Initiation of Plumes: Heads and Tails

It is intuitively expected that new plumes should arise from time to time from instabilities of the basal boundary layer. This topic has received considerable laboratory and theoretical interest and more recently geological interest because of observable effects: massive flood basalts and radial dike swarms. For purposes of this review, the topic is relevant because it demonstrates that plumes consist of lower viscosity material upwelling through the much higher viscosity mantle and because flood basalts and dike swarms are the best evidence of the existence of plumes in the past.

The basic physics describing the upwelling of material which is two or three orders of magnitude less viscous than its surroundings is well understood and similar for either hot plumes or compositionally less dense plumes (Whitehead & Luther 1975, Olson & Singer 1985, Griffiths & Campbell 1990, Loper 1991). The first batch of fluid to rise through the mantle ascends as a mushroom-shaped head which is fed from below by a narrow (less than 100 km diameter) cylindrical conduit (Figure 1). The head ascends by displacing the surrounding mantle at the Stokes' law rate for a sphere of the less dense material moving through the higher viscosity of the mantle. Thus the head spends most of its ascent time and acquires most of its plume material in the deep high-viscosity part of the lower mantle (D. E. Loper, personal communication 1991). Material within the conduit moves as pipe flow at a rate controlled by the lower viscosity of the plume. The head of the plume ponds when it approaches the upper surface and the plume is thereafter a throughgoing (tail) conduit from the basal boundary layer to the uppermost mantle. Normal mantle is entrained into the plume head. Thermal plumes differ from chemical ones mainly because heat diffuses in the head of the plume warming the entrained material and cooling the plume material (Campbell & Griffiths 1990). This causes the head to widen as it ascends.

The geological implications of this process are evident in many plateau basalt provinces (Courtillot et al 1986, Richards et al 1989, Campbell & Griffiths 1990, Duncan & Richards 1991, Richards et al 1991, Hill 1991). The approach of the plume head toward the surface implies the sudden arrival of hot material over a broad area. Thus initial uplift is followed by short-lived voluminous eruption of flood basalts which form a plateau, like the Deccan traps, on land, or a submarine plateau, like Ontong-Java, in the ocean basins. This volcanism occurs over a roughly circular 2000-km diameter region. The 800-km distance of excess volcanism on the Labrador Sea spreading center in the Davis Strait owing to the hotspot location of Vink (1984) is particularly useful for giving a minimum diameter of the Iceland plume head because entrainment of hot mantle rather than possible propagation of dikes away from the source region is involved. This 1600-km diameter is appropriate if the plume head ascends from the base of the mantle but inappropriate for plumes originating at 680 km (Campbell & Griffiths 1990, Hill 1991). Subsequent volcanism is less voluminous and is restricted to the long-lived hotspot track of the plume tail.

An alternative hypothesis by White & McKenzie (1989) is that flood basalts are the result of continental breakup along a hotspot track. Plume material ponds beneath the continental lithosphere. Voluminous volcanism occurs because much plume material ascends to shallow depths along the new ridge axis and beneath highly stretched parts of the continent. As discussed below, entrainment of hot material into ridge axes does produce voluminous volcanism and submarine plateaus, like Iceland. The key difference between the two hypothesis is that lithospheric rifting is not a required side effect of a plume head but rather a fundamental part of the breakup hypothesis.

The case for a plume head is probably best demonstrated for the Deccan traps as reviewed by Duncan & Richards (1991). First, volcanism preceded rifting of the west coast of India (Hooper 1990). Second, the Indian continent was moving rapidly northward at the time making it difficult for the plume to supply a large accumulation of material to the base of the plate. Duncan & Richards (1991) also note that extension is mild to nearly absent in several possible plume head events, particularly the Siberian traps which erupted in a plate interior. Hill (1991) notes that plume head events typically preceed rifting by a few million years when the plate rearrangement is basically a ridge jump and tens of million years for a major rearrangement.

A key feature of plume heads is that they produce a brief widespread episode of volcanism and dike swarms which can be recognized in the geological record. This provides evidence that plumes existed before well developed recent (younger than 200 Ma) tracks on oceanic crust. In particular, the presence of a plume head indicates that the plume was significantly hotter and less viscous than the surrounding mantle. The oldest currently demonstrated plume head event is the MacKenzie dike swarm, which radiated over 1000 km from the associated Muskox intrusion and Coppermine basalts in northern Canada (LeCheminant & Heaman 1989). These events were synchronous at 1267 Ma to within the resolution of the data—a few million years. In this case, continental rifting occurred north of Coppermine, and the hotspot track probably ended up in the new ocean basin where it was not preserved.

Also, plume head events and large submarine plateaus are empirically negatively correlated with the frequency of magnetic reversals. The suggestion that these events involve more rapid heat loss from the core (Larson

1991), if correct, implies that a thick nonconductive chemical layer is absent at the base of the mantle. That is, if such a layer existed, the core would sense changes in the temperature of the overlying mantle only by thermal diffusion through the layer. A lag of approximately the layer thickness squared divided by the thermal diffusivity thus would occur between the start of plume ascent and the change in reversal rate.

Boundary Layer Viscosity Contrast

The presence of plumes, particularly the presence of plume heads, indicates that the hot material in the basal boundary layer is two or three orders of magnitude lower than the overlying normal mantle. This situation does not occur in laboratory experiments, rather an equilibrium temperature and viscosity gradient is established with about one order of magnitude contrast (Nataf 1991). Two reasons why a large viscosity contrast and hence plumes may exist in the Earth are qualitatively evident but not well understood (Nataf 1991).

First, the viscosity increases a few orders of magnitude from the low viscosity asthenosphere to the normal mantle just above the basal boundary layer. (Depth dependent viscosity is impractical in the laboratory because large pressures are needed to change viscosity.) This tends to retard flow in the lower mantle and provide a high-viscosity lid over the basal boundary layer. Otherwise, plate motions including sinking slabs would stir the basal mantle to plate velocities. This would preclude relatively stationary hotspots (Duncan & Richards 1991, Richards 1991). In addition, slab material tends to pond in the lower mantle and increase the temperature and viscosity contrast of the lid.

The relationship between stirring and viscosity is easily quantified dimensionally. Stirring from flow in low-viscosity upper layers scales to the inverse of viscosity because the shear stress,

$$\tau = \eta \frac{v}{X} \tag{1}$$

(where η is viscosity, v is velocity, and X is the length scale over which velocity varies), remains approximately constant with depth. Slabs are more effective at stirring because they pond (or thicken) as they enter higher viscosity layers. The velocity times the thickness of the slab (vS) remains constant ($v_0 S_0$) with depth. The shear stress scales to the thickness and the density contrast of the slab as

$$\tau = S \Delta\rho g \approx \frac{v_0 S_0}{v} \Delta\rho g. \tag{2}$$

Combining (1) and (2) gives a velocity that varies inversely with the square root of viscosity. Thus, if hotspot relative velocities are several millimeters per year, a viscosity increase by a factor of about 50 is required between the upper mantle beneath the asthenosphere and the deep lower mantle.

Second, nonsteady convection may increase the viscosity contrast near the base of the mantle (Sleep et al 1988, Nataf 1991). Both the core and mantle of the Earth are cooling. The higher viscosity of the lowermost mantle may cause convection associated with the basal boundary layer to be relatively inefficient, such that plate tectonics would cool the mantle faster than plumes could cool the core. Thus, a nonequilibrium temperature contrast may build up above the core-mantle boundary during the evolution of the Earth.

INTERACTION OF PLUMES AND RIDGES

A key element in the early reasoning in favor of mantle plumes is that hotspot tracks cross ridge axes (Morgan 1971, 1972, 1981; Duncan 1984). This implies that an off-axis hotspot evolves into an on-axis hotspot, and then back into an off-axis one. Therefore, off-axis hotspots, like Hawaii, and on-axis ones, like Iceland, are different aspects of the same phenomenon: plumes. Conversely, a crack in a plate—an alternative mechanism for off-axis hotspots—would not be expected to cross the free edge of a ridge axis. In addition, on-ridge hotspots are relevant here for demonstrating that hotspots really have exceptionally hot material beneath them and for understanding the dynamics of ridge axes.

Excess Source Temperature of Plumes

The ridge axis at on-axis hotspots is subaerial or rather shallow compared with the ~2.5 km depth for normal ridges. Paired submarine plateaus are formed when the thickened crust spreads off axis in both directions. Examples include the Iceland-Faeroe and Iceland-Greenland plateaus, the Rio Grande and Walvis rises from the Tristan hotspot, and the Cocos and Nazca rises from the Galapagos hotspot. That is, part of the uplift is permanent and associated with crust that is thicker than the normal 5–7 km thick oceanic crust. For example, the crustal thickness beneath Iceland is poorly resolved in part because it is not clear whether a 7.2 km s^{-1} layer beginning at 10–15 km depth is crust or anomalous mantle. The crustal thickness may reach 30–35 km beneath the Iceland Faeroe plateau (see Flóvenz & Gunnarsson 1991).

The origin of the excess melting that forms the thickened crust is explained in terms of adiabatic upwelling of hot plume material to the ridge axis (Schilling 1983; Bickle 1986; Klein & Langmuir 1987, 1989;

McKenzie & Bickle 1988). For simplicity, I illustrate the thermodynamics using a eutectic system with a single melting curve before discussing the real Earth. The adiabat of upwelling material intersects the melting curve at some depth because the melting point increases much more rapidly with depth (3°C km^{-1}) than the solid adiabatic gradient (0.3°C km^{-1}). The temperature and pressure of partial molten material lie on the melting curve. The fraction of melt is obtained from energy conservation by assuming for purposes of calculation that the material remains a superheated solid until it partially melts at some depth. The specific heat times the temperature difference between the extrapolated solid adiabat and the melting curve, $C_p\Delta T$, equals the latent heat for the fraction of melt, FL. The fraction of melt is thus

$$F = \frac{C_p \Delta T}{L} \tag{3}$$

and the fraction of melt increases with decreasing depth by

$$\frac{\partial F}{\partial z} = \frac{C_p}{L}\frac{\partial \Delta T}{\partial z}. \tag{4}$$

Beneath ridges, the fraction of melt increases by 1% km^{-1} assuming $L = 0.3$ MJ kg^{-1} and $C_p = 1.15$ kJ kg^{-1} °C^{-1}.

The total thickness of melt generated is approximated by integrating the fraction of melt generated from the depth z_m, where melting begins, to the surface:

$$C = \int_0^{z_m} F\,dz. \tag{5}$$

(The thickness of oceanic crust is more precisely estimated by integrating from the base of the crust, including the kinematic effects of compaction which segregates melt, and assuming that a small fraction of the melt does not segregate.) Melting beginning at 35 km depth generates 6 km of crust in the simplified model presented above. It is also evident that the crustal thickness increases with increasing source temperature of the upwelling material because z_m increases by $(\partial\Delta T/\partial z)^{-1}$. The actual mantle is not eutectic but rather melts over a temperature range. The thermal implications are easily (but messily) included in the model by making ΔT decrease with increasing fraction of melting F. Thus a greater depth of initial melting is needed to generate 6 km of oceanic crust than in the eutectic model.

More importantly the chemical effects of noneutectic melting are observable and give information on the temperature and depth of the source region for basaltic magmas (Klein & Langmuir 1987, 1989; McKenzie &

Bickle 1988). The mantle can be considered to be a mixture of an aluminous phase, clinopyroxene $Ca(Mg, Fe)(SiO_3)_2$, orthopyroxene $(Mg, Fe)SiO_3$, and olivine $(Mg, Fe)_2SiO_4$. At the pressures relevant to ridge axes, the phases melt preferentially in that order and the FeO components melt preferentially with respect to the MgO components. Melts thus become more magnesian with increasing fraction of melting. High MgO glasses are thus clear evidence of high source temperatures, as in Hawaii (Clague & Weber 1991). Unfortunately, high MgO lavas are quite rare and hard to observe directly because olivine also crystalizes first in shallow magma chambers lowering MgO.

The Na_2O component is more useful because it preferentially enters the melt phase relative to the solid phases (Klein & Langmuir 1987, 1989). Thus, the Na_2O abundance decreases with increasing fraction of melting as more Na_2O-depleted solid is added to the melt. This component does not enter olivine and thus is not strongly affected by early fractional crystallization. In practice, Na_2O abundances are adjusted for fractional crystallization to the abundance where MgO is 8%. The more complicated behavior of CaO, FeO, SiO_2, and Al_2O_3 allows the depth of melting to be inferred independently of the fraction of melting determined from Na_2O. Melts found at lower lithospheric depths with gamet remaining in the source region are enriched in light rare earth elements.

Klein & Langmuir (1987, 1989) found that chemical differences among mid-oceanic ridge magmas are explained by adiabatic melting of ascending material. That is, hottest source regions start melting at a greater depth and melt more to produce hotspot ridges. Solid adiabat temperatures vary 300°C on a global basis with plumes about 250°C hotter than typical ridges. In basic agreement with this estimate, Watson & McKenzie (1991) used a convection model of melting beneath Hawaii and chemical composition to find a maximum excess temperature relative to typical ridges of 278°C. The expected correlations exist between Na_2O and axial depth and crustal thickness. The correlation between original axial depth and Na_2O also holds for older oceanic crust (Keen et al 1990). Anomalies to the general trend do exist. Obviously material ascending near cool lithosphere—such as to the tip of a ridge segment at a transform fault—does not remain adiabatic. Chemical and isotopic differences between plume material and normal mantle probably also exist. These differences are useful for the entrainment of tracing plume material independent of temperature anomalies (Rideout & Schilling 1985; Schilling 1985, 1991; Schilling et al 1985). Possible origins of heterogeneities are easily envisioned in terms of plate tectonics—for example, subduction of sediments and altered oceanic crust—but quantification is still lacking.

It is likely that the temperature of normal mantle has decreased with time. Thus the process described above for hotspot ridges may be applicable to normal ridges in the Archean (Sleep & Windley 1982, Bickle 1986). The role of hotspots in generating modern Mg-rich lavas (komatiites) and the possible applicability to Archean komatiites has been pointed out (Campbell et al 1989, Storey et al 1991). However, it is possible that the mantle has cooled faster than the core making the temperature contrast above the core-mantle boundary in the Archean too small to produce plumes (Sleep et al 1988). It is difficult to tell whether an isolated exposure of Archean volcanic rocks was produced by a localized plume or from globally hot mantle. Identification of plume heads by widespread synchronous volcanism appears to be the best way to recognize plumes in the Archean.

Entrainment of Plume Material into Ridge Axes and Hotspot Jumps

The subaerial segments of ridge axis crossing Iceland are several hundred kilometers long. In addition several hundred more kilometers of ridge axis north and south of Iceland are moderately affected by plume material. Thus a long segment of the ridge is supplied by a much narrower plume. Two processes are involved. First the plume material is entrained to a place on the ridge axis. Then the plume material flows along the strike of the ridge axis mixing with normal mantle away from the hotspot.

The physics of entrainment is qualitatively explained by noting that the upwelling at normal ridge axes is passive in the sense that the adiabatic interior supplies the necessary material to form the new oceanic lithosphere (Rideout & Schilling 1985; Schilling 1985, 1991; Schilling et al 1985). The plume supplies material to a segment of the ridge making the normal passive upwelling unnecessary. The normal flow in the asthenosphere away from the ridge axis is thus suppressed and plume material is deflected to the ridge axis (Figure 1).

A kinematic consequence of entrainment is that the observed (track) position of an on-ridge hotspot stays on the ridge axis while a fixed plume position would be expected to cross the axis. In the case of Iceland, absolute velocities imply that movement to the east of a few hundred kilometers from under Greenland to the eastern part of Iceland (Vink 1984). The ridge jump at 36 Ma was west toward the expected hotspot position while the current ridge jump is east toward the expected hotspot.

In addition to ridge jumps which can be qualitatively explained by the plume material creating a zone of weakness in the plate, entrainment implies that a hotspot track is captured by the ridge axis for a period of time. This tends to create a jump in the track of an off-ridge hotspot approaching the ridge (Figure 2). When the hotspot is far away from the ridge it moves with the expected track velocity and does not significantly

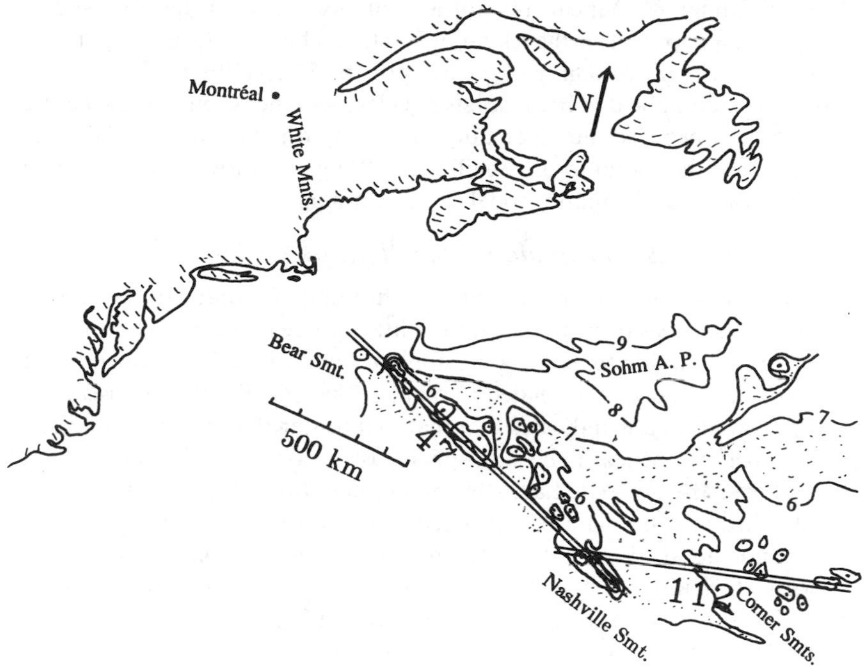

Figure 2 The track of the New England seamount plume (*double lines*) is shown on a depth to basement map modified from Tucholke (1986, Plate 5) by Sleep (1990b). The 5 km contour indicates the base of seamounts. The top of seamounts is indicated by dots. Between Bear and Nashville seamounts the track moved at 47 mm yr^{-1}; between Nashville seamount and the Corner Seamounts the track jumped toward the ridge axis at 112 mm yr^{-1} (Tucholke & Smoot 1990).

supply material to the ridge axis. At some time, the hotspot comes close enough to the ridge axis for entrainment to occur. The observed position of the hotspot then jumps to the ridge axis and remains on the ridge axis until the plume is too far on the other side. A jump off the ridge axis then occurs. The Icelandic hotspot has remained trapped on the ridge axis and has not jumped throughout its history.

The New England-Great Meteor hotspot jumped as it approached the ridge axis from the west (Figure 2). Between 85 and 76 Ma, the hotspot moved at a rate of 112 mm yr^{-1} (compared with the previous rate of 47 mm yr^{-1}) from the off-ridge Nashville seamount to the on-ridge Comer seamounts (Tucholke & Smoot 1990). Only small seamounts occur in the gap (Epp & Smoot 1989). Tristan has jumped in a more complicated way between segments of the ridge axis offset by transform faults [see Figure

6 of O'Conner & Duncan (1990)] and finally off the ridge to the east producing a gap west of the island of Tristan. The New England gap was mistaken for real weakening of the hotspots by Sleep (1990b). The Tristan gap is more complicated because it is not clear whether Gough is a separate hotspot and because the kinematics of a jump off the axis may be more complicated than a jump on. Schilling (1991) gives a low flux for Gough on the basis of a kinematic model of entrainment.

Entrainment, Small-Scale Convection, or Cracks

Plumes as envisioned here are sources of hot material that can flow rather than point sources of heat beneath the lithosphere. This model both provides an explanation for some weak hotspots while making it more difficult to recognize processes unrelated to plumes. Specifically, the discussion of hotspot swells below indicates that material hotter than normal adiabatic mantle persists beneath the lithosphere for a considerable time after it passed over the hotspot. Later this moderately hot material is tapped by cracks through the plate or entrained into the ridge axis. The island of Cocos northeast of the Galapagos hotspot is an example of the former process (Castillo et al 1988), while Rodriguez results from Réunion material entrained to the ridge axis (Morgan 1978).

These mechanisms are contrasted with tapping of normal adiabatic mantle by rifts and cracks in the lithosphere. As noted in the introduction, this mechanism has been applied in the past to even strong hotspots like Hawaii. Petrological indications of source temperature (Klein & Langmuir 1987, 1989; McKenzie & Bickle 1988; Clague & Weber 1991; Watson & McKenzie 1991) are the most definitive way to recognize hot material from plumes. Conversely, a source of stress is often evident, such as the Alpine collision near recent volcanism in the Massif Central in France and the Eifel hotspot near the Rheingraben. Many features including continental breakups may owe their existence to both mechanisms as hotter plume material creates regions of weakness which start cracks for passive upwelling.

A third class of mechanism involves small-scale convection in the asthenosphere driven by cooling at the base of the plate. Stress in the lithosphere caused by drag from the diverging flow at upwellings of this convection as well as the higher temperature of the upwellings themselves was proposed in the past for Hawaii (Solomon & Sleep 1974) and more recently for weak hotspots (Vogt 1991). It first should be noted that this mechanism is hard to distinguish petrologically from ordinary cracking because the upwelling temperature is that of normal adiabatic mantle. Secondly, analysis of topographic and geoidal anomalies in the ocean basins indicates that this mode of convection transports little heat (Davies

1988b). The only region where there is positive evidence of its occurrence is west of the East Pacific Rise where the asthenosphere probably consists of hot entrained plume material rather than normal adiabatic mantle.

More understanding of these issues, particularly with regard to weak hotspots on the continents, are relevant to petroleum exploration. Hotspots, small-scale convection, and cracks all may heat the sedimentary column and the lithosphere causing uplift unconformities and subsequent thermal basin subsidence (cf Crough 1981, Vogt 1991).

HOTSPOT SWELLS

Widespread vertical tectonics are associated with the passage of hotspot track across plates (Crough 1983). The resulting uplift as the hotspot approaches and subsidence after it passes are more easily observed in the ocean where the depth-age relationship of normal oceanic crust and sea level surfaces on atolls are useful references. In particular, a several hundred kilometer wide swell of elevated seafloor flanks the hotspot track; atolls along the axis of the track mark the general subsidence of the swell. Hypotheses regarding the origin of swells were reviewed by Crough (1983) who concluded that thermal expansion was the primary cause of the uplift. I start with this basic mechanism and then discuss extensions and difficulties of this hypothesis that have arisen since 1983.

Basically material from the plume is supposed to heat the lithosphere near the hotspot. The uplift from isostatically compensated thermal expansion is conveniently expressed in terms of an apparent thermal rejuvenated age of the oceanic lithosphere using the depth-age relationship of normal oceanic crust:

$$D = D_0 + 350\,\mathrm{m}\sqrt{\mathrm{age(c), m.y.}}, \tag{6}$$

where D_0 is the depth at the ridge axis. The rejuvenated depth of the swell defines the rejuvenation age,

$$D_r \equiv D_0 + 350\,\mathrm{m}\sqrt{\mathrm{age(r), m.y.}}, \tag{7}$$

and the swell uplift is $E = D - D_r$. To first order, swells and atolls on them subside at the rate predicted for oceanic crust of age(r) at the time of hotspot passage.

Heat Flux from Swell Topography

The hypothesis that the uplift of swells is caused by thermal expansion allows the heat flux to hotspots to be estimated (Davies 1988a, Richards et al 1988, Sleep 1987, 1990a). Assuming isostasy implies that the mass deficiency from hot material at depth is equal to the mass excess of the topography of the swell. The rate at which mass deficiency at depth is produced by the swell is then equal to the track velocity v_p times the mass excess of topography per length of track:

$$B = (\rho_m - \rho_w)\overline{WE}\,v_p. \tag{8}$$

Here ρ_m is the density of the mantle, ρ_w is the density of water, and $\overline{WE}$ is the cross-sectional area of the swell. In practice, the uplift of the swell is obscured by volcanic edifices along the swell axis and flexural downwarps flanking them (Figure 3). (Note: the terminology for this equation is not standardized, the quantity B is called buoyancy flux in papers where Sleep is an author, Davies does not multiply by density and expresses plume magnitude in $m^3\,s^{-1}$; Campbell & Griffiths (1990) multiply by g to express buoyancy flux in $N\,s^{-1}$.) The buoyancy flux is converted to heat flux by multiplying by the specific heat per mass and dividing by the thermal expansion coefficient:

$$H = BC_p/\alpha. \tag{9}$$

Davies (1988a) and Sleep (1990a) have used Equation (8) to determine buoyancy fluxes on a global basis and obtained the conclusion cited above that hotspots are about 6% of the mantle heat flux. The most vigorous hotspot is Hawaii with a buoyancy of around 8 Mg s^{-1}. Réunion has a flux of about 2 Mg s^{-1}. Several weak hotspots including Juan de Fuca have fluxes around 0.5 Mg s^{-1}. Equation (8) can also be applied to swells along old hotspot tracks if the initial uplift can be inferred from the subsidence of atolls through Equation (5) or from the current uplift of the swell relative to its flanks (Sleep 1990b).

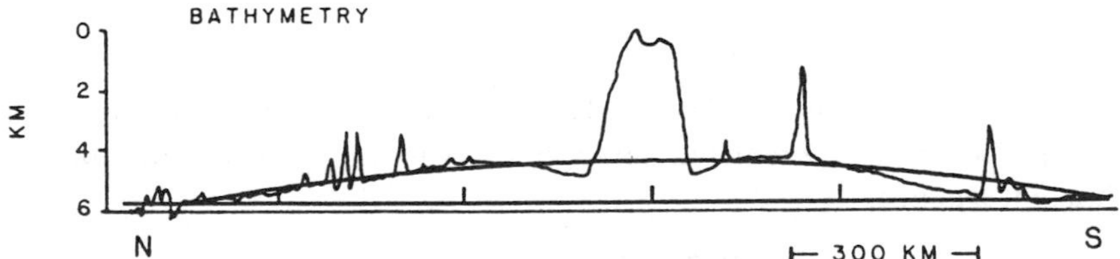

Figure 3 The bathymetry of the Hawaiian swell near Oahu is shown with a parabola to indicate the excess elevation above a constant depth line. The parabola has a halfwidth of 750 km, a height of 1.4 km, and a cross sectional area of 1400 square kilometers. Modified after Watts (1976) and Crough (1983).

Kinematics of Swell Formation

The nose of the Hawaiian swell extending upstream of the active hotspot is crudely parabolic (Figure 4). This shape is explained by the kinematics of flow from a relatively narrow plume through the asthenosphere to the flanks of the swell. A very simple geometric model of this process was presented by Sleep (1987, 1990a) and Richards et al (1988). Flow is represented as radial motion away from the plume,

$$v_{\text{plume}} = \frac{v_f r_f}{r}, \tag{10}$$

where r is the distance from the plume and the velocity is v_f at the reference distance r_f. The volume flux from the hotspot is

$$Q = 2\pi v_f r_f A, \tag{11}$$

where A is the asthenospheric channel thickness. The buoyancy flux is the volume flux times the density difference. The radial flow interacts with flow in the asthenosphere driven by the drag of the overlying plate at

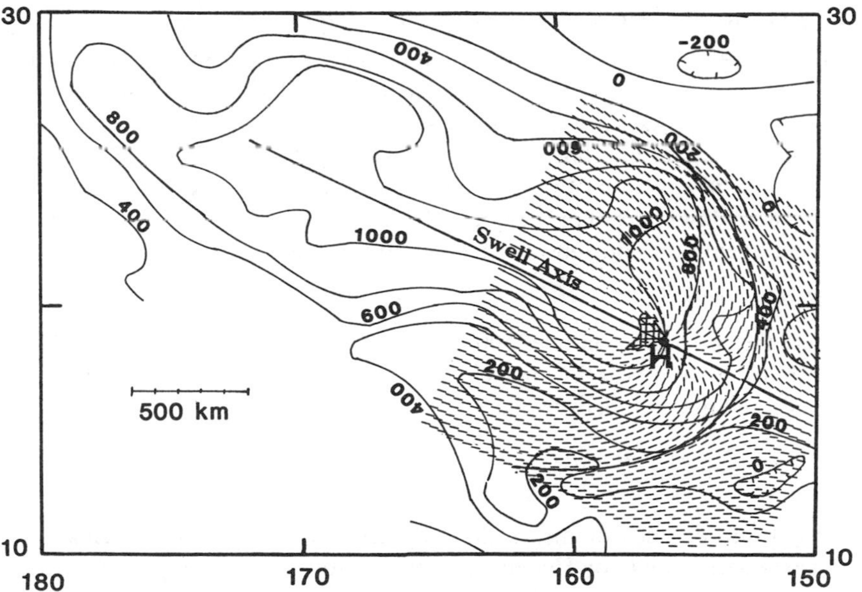

Figure 4 The excess elevation of the Hawaiian swell with the volcanic edifices removed [modified after Schroeder (1984)] is shown with flow directions in the asthenosphere near the nose of the swell [modified after Sleep (1990a)]. The thick parabola-like curve on the nose of the swell separates plume material and normal asthenosphere.

approximately half the plate velocity. Directly upstream of the hotspot, the radial flow from the plume is in the opposite direction of the drag-driven flow and a stagnation point occurs at r_s where

$$\frac{v_f r_f}{r_s} = \frac{v_L}{2}. \tag{12}$$

More generally the streamline through the stagnation point separates asthenosphere from the plume from normal asthenosphere (Figure 4). Watson & McKenzie (1991) obtain a similarly shaped streamline assuming that velocity varies away from the plume as $r^{-0.779}$ rather than r^{-1}. Thus, the shape of the nose of the swell is not a good indication of the detailed dependence of velocity on distance from the plume.

It is evident from this formalism and also Equations (8) and (9) that relatively weak plumes on fast plates should produce narrow swells. This creates one difficulty in that a narrow thermal swell is much more obscured by volcanic edifices and flexure than a wide one. A second difficulty is that very weak hotspots do not produce any tracks at all on fast plates. For example, this would preclude finding the weaker Atlantic hotspots on the Pacific plate (McNutt 1990).

Difficulties with the Thermal Uplift Theory

Crough (1978) and Detrick & Crough (1978) proposed the thermal rejuvenation and subsidence theory because individual atolls and also the axis of the Hawaiian swell appeared to subside at the rate predicted by the theory. An additional prediction of the theory is that the deeper part of the lithosphere is hotter than normal beneath the swell and that heat flow is elevated downstream of the hotspot. I illustrate the general relationship between heat flow and subsidence by giving the age-heat flow relationship,

$$q = q_0(\text{age})^{-1/2}, \tag{13}$$

where the constant q_0 implies 500 mW m^{-2} heat flow on 1 m.y. crust. The depth-age (6) and heat flow-age (13) relationships combine to yield a more general relationship between heat flow and subsidence:

$$q = -\text{constant}\frac{\partial E}{\partial t}. \tag{14}$$

Early heat flow measurements along the axis of the Hawaiian swell show the expected result (Von Herzen et al 1982). However, a later profile across the axis of the swell where the heat flow anomaly is expected to be largest showed much scatter but not the expected excess heat flow (Von Herzen et al 1989). It is hard to envision a time-dependent subsidence mechanism that moves with the lithospheric plate and does not involve thermal contraction by cooling to the surface. Equation (14) indicates that any such mechanism is inappropriate if the latter heat flow data are representative.

Another difficulty of the thermal uplift hypothesis comes from surface wave velocities along the Hawaiian swell (Woods et al 1991). The data show the expected velocity structure of unrejuvenated 80 m.y. lithosphere, rather than the expected rejuvenation age of 25 m.y. However, converted S to P waves indicate that hot material exists at 75 km depth beneath Oahu (Bock 1991).

Thermal subsidence is most easily observed on the rapidly moving Pacific plate where hotspots move quickly away from any dynamic uplift and subsidence mechanisms related to flow from the plume. Thus I return to the locality of Enewetak atoll which Detrick & Crough (1978) used to show that the cause of uplift (and subsidence) moves basically with the lithospheric plate as expected for thermal rejuvenation. As noted by Sleep (1990a), the asthenosphere moves more slowly relative to the hotspot than to the plate. The geometry of hotspot tracks in the Pacific allows this effect to be observed. Track directions before the Hawaii-Emperor bend at 43 Ma are from the NNE at a high angle to the current track direction from the ENE. The lithosphere beneath track segments older than 43 Ma (as well as those somewhat younger) has had time to move away from the plume asthenosphere originally below it. Thus a primarily asthenosphere uplift and subsidence mechanism would imply rapid subsidence after 43 Ma when the plate cleared hot asthenosphere and little subsidence thereafter. Enewetak erupted at 56 Ma (Lincoln & Schlanger 1991), close enough to the bend age to provide a useful test. The later history of Enewetak is compatible with thermal subsidence of rejuvenated lithosphere and if anything more rapid in the last 15 Ma than predicted by the theory (Quinn 1991, Lincoln & Schlanger 1991).

The seismic data from the Hawaiian swell are probably compatible with the subsidence data from Enewetak. The surface wave data are sensitive to intermediate lithospheric depths and thus imply that rejuvenation is not simply using the heat of plume material to change the geotherm to that of young oceanic crust over normal asthenosphere. However, hotter material moving with the lithosphere at greater depths would produce uplift and converted P-waves, but not high temperatures at the depths sensed by the surface wave experiments. More seismic data sensitive to thermal anomalies at greater depths and subsidence data from the same area would help resolve these issues.

CONCLUSIONS

The objective of this review is to discuss the current level of understanding of the causes of hotspots with concentration on physical processes in the interior of the Earth. At one end, certain conclusions can be made with reasonable confidence, most importantly that mantle plumes exist beneath vigorous hotspots with well developed tracks, like Iceland and Hawaii. The properties envisioned for plumes in Figure 1 have an intermediate level of confidence as the dynamics are still only qualitatively understood. Currently unresolved issues include the origin of hotspot swells and the fate of the hot material supplied by plumes. Incomplete understanding of the surficial aspects of hotspots detracts from understanding the deep interior. For example, the number of plumes and hence the structure of flow within the basal boundary layer are poorly understood because it is not evident which weak hotspots are supplied by their own plumes.

I expect the rapid progress toward consensus and understanding to continue in the immediate future. A wide variety of geological and geophysical methods have proved useful. Integration of massive amounts of petrological data into the physical framework, untangling the complicated history of closely spaced Pacific and Atlantic hotspots, and improved subsidence and uplift histories of hotspot swells are important tasks for surface observers. Direct mapping of the basal boundary layer by seismologists seems possible. The physical aspects of the hypothesis are now well enough defined that theoretical fluid dynamicists can include the salient features in their models. Conversely, understanding of hotspots will lead to progress on other topics of Earth science, particularly with regard to the effects of hotspot tracks on field-scale geology and tectonics and the dynamics of the interaction of plumes with flow associated with plate tectonics.

ACKNOWLEDGMENTS

The plume conference in May 1991 at CalTech was quite helpful. I benefited especially from discussions with Geoff Davies, Robert Hill, J.-G. Schilling, David Clague, Mark Richards, David Loper, and Marcia McNutt. The author was supported in part by the National Science Foundation grant EAR-8915570.

Literature Cited

Bickle, M. J. 1986. Implications of melting for stabilization of the lithosphere and heat loss in the Archean. *Earth Planet. Sci. Lett.* 80: 314–24

Bina, C. R., Silver, P. G. 1990. Constraints on lower mantle composition and temperature from density and bulk sound velocity profiles. *Geophys. Res. Lett.* 17: 1153–56

Bock, G. 1991. Long-period S to P converted waves and the onset of partial melting beneath Oahu, Hawaii. *Geophys. Res. Lett.* 18: 869–72

Boehler, R., von Bargen, N., Chopelas, A. 1990. Melting, thermal expansion, and phase transitions of iron at high pressures. *J. Geophys. Res.* 95: 21,731–36

Bonatti, E., Harrison, C. G. A., Fisher, D. E., Honnorez, J., Schilling, J. G., Stipp, J. J., Zentelli, M. 1977. Easter volcanic chain (southeast Pacific): A mantle hot line. *J. Geophys. Res.* 82: 2457–78

Boness, D. A., Brown, J. M. 1990. The electronic band structures of iron, sulfur, and oxygen at high pressures and the Earth's core. *J. Geophys. Res.* 95: 21,721–30

Bukowinski, M. S. T., Wolf, G. H. 1990. Thermodynamically consistent decompression: Implications for lower mantle composition. *J. Geophys. Res.* 95: 12,583–94

Campbell, I. H., Griffiths, R. W. 1990. Implications of mantle plume structure for the evolution of flood basalts. *Earth Planet. Sci. Lett.* 99: 79–93

Campbell, I. H., Griffiths, R. W., Hill, R. I. 1989. Melting in an Archean mantle plume: Heads its basalts, tails its komatiites. *Nature* 339: 697–99

Castillo, P., Batiza, R., Vanko, D., Malavassi, E., Barquero, J., Fernandez, E. 1988. Anomalously young volcanoes on old hot-spot traces, I, Geology and petrology of Cocos Island. *Geol. Soc. Am. Bull.* 100: 1400–14

Chase, C. G. 1979. Subduction, the geoid and lower mantle convection. *Nature* 282: 464–68

Chase, C. G., Sprowl, D. R. 1983. The modern geoid and ancient plate boundaries. *Earth Planet. Sci. Lett.* 62: 314–20

Christensen, U. 1985. Thermal models for the Earth. *J. Geophys. Res.* 90: 2995–3008

Clague, D. A., Dalryple, G. B. 1987. The Hawaiian-Emperor volcanic chain, part 1, Geologic evolution. Volcanism in Hawaii, *U.S. Geol. Surv. Prof. Pap.* 1350: 5–54

Clague, D. A., Dalryple, G. B. 1989. Tectonics, geochronology, and origin of the Hawaii-Emperor chain. In *The Geology of North America*, Vol. N, *The Eastern Pacific Ocean and Hawaii*, ed. E. L. Winterer, D. L. Hussong, R. W. Decker, pp. 188–217. Boulder: Geol. Soc. Am.

Clague, D. A., Weber, W. S. 1991. High MgO glasses from Kilauea volcano, Hawaii. *Nature*. In press

Courtillot, V., Besse, J., Vandamme, D., Montigny, R., Jaeger, J. J., Cappetta, H. 1986. Deccan flood basalts at the Cretaceous/Tertiary boundary? *Earth Planet. Sci. Lett.* 80: 361–74

Crough, S. T. 1978. Thermal origin of hotspot swells. *Geophys. J. R. Astron. Soc.* 55: 451–69

Crough, S. T. 1981. Mesozoic hotspot epeirogeny in eastern North America. *Geology* 9: 2–6

Crough, S. T. 1983. Hotspot swells. *Annu. Rev. Earth Planet. Sci.* 11: 163–93

Davies, G. F. 1988a. Ocean bathymetry and mantle convection, 1, Large-scale flow and hotspots. *J. Geophys. Res.* 93: 10,467–80

Davies, G. F. 1988b. Ocean bathymetry and mantle convection, 2, Small-scale flow. *J. Geophys. Res.* 93: 10,481–88

Davies, G. F. 1990. Mantle plumes, mantle stirring and hotspot geochemistry. *Earth Planet. Sci. Lett.* 99: 94–109

Detrick, R. S., Crough, S. T. 1978. Island subsidence, hotspots and lithospheric thinning. *J. Geophys. Res.* 83: 1236–44

Duncan, R. A. 1984. Age progressive volcanism in the New England seamounts and the opening of the central Atlantic Ocean. *J. Geophys. Res.* 89: 9980–90

Duncan, R. A., Richards, M. A. 1991. Hotspots, mantle plumes, and true polar wander. *Rev. Geophys.* 29: 31–50

Epp, D., Smoot, C. 1989. Distribution of seamounts in the North Atlantic. *Nature* 337: 254–57

Flóvenz, O. G., Gunnarsson, K. 1991. Seismic structure in Iceland and surrounding area. *Tectonophysics* 189: 1–18

Frohlich, C. 1989. The nature of deep-focus earthquakes. *Annu. Rev. Earth Planet. Sci.* 17: 227–54

Griffiths, R. W., Campbell, I. H. 1990. Stirring and structure in mantle starting plumes. *Earth Planet. Sci. Lett.* 99: 66–78

Griffiths, R. W., Campbell, I. H. 1991. On the dynamics of long-lived plume conduits in the convecting mantle. *Earth Planet. Sci. Lett.* 103: 214–27

Hill, R. L. 1991. Starting plumes and continental break-up. *Earth Planet. Sci. Lett.* 104: 381–97

Hooper, P. R. 1990. The timing of crustal extension and the eruption of continental flood basalts. *Nature* 345: 246–49

Houseman, G. A. 1990. The thermal structure of mantle plumes: Axisymmetric or triple junction? *Geophys. J. Int.* 102: 15–24

Jeanloz, R. 1990. The nature of the Earth's core. *Annu. Rev. Earth Planet. Sci.* 18: 357–86

Jealnoz, R., Morris, S. 1986. Temperature distribution in the crust and mantle. *Annu. Rev. Earth Planet. Sci.* 14: 377–415

Jeanloz, R., Richter, F. M. 1979. Convection, composition, and the thermal state of the lower mantle. *J. Geophys. Res.* 84: 5497–5503

Keen, M. J., Klein, E. M., Melson, W. G. 1990. Ocean-ridge basalt compositions correlated with palaeobathymetry. *Nature* 345: 423–26

Klein, E. M., Langmuir, C. H. 1987. Global correlations of ocean ridge chemistry with axial depth and crust thickness. *J. Geophys. Res.* 92: 8089–8115

Klein, E. M., Langmuir, C. H. 1989. Local versus global variations in ocean ridge basalt composition: A reply. *J. Geophys. Res.* 94: 4241–52

Larson, R. L. 1991. Latest pulse of the Earth: Evidence for a mid-Cretaceous superplume. *Geology* 19: 547–50

LeCheminant, A. N., Heaman, L. M. 1989. MacKenzie igneous events, Canada: Middle Proterzoic hotspot magmatism associated with ocean opening. *Earth Planet. Sci. Lett.* 96: 38–48

Lincoln, J. M., Schlanger, S. O. 1991. Atoll stratigraphy as a record of sealevel change: Problems and prospects. *J. Geophys. Res.* 96: 6727–52

Loper, D. E. 1984. The dynamical structures of D″ and deep plumes in a non-Newtonian mantle. *Phys. Earth Planet. Inter.* 34: 57–67

Loper, D. E. 1991. Mantle plumes. *Tectonophysics* 187: 373–84

Loper, D. E., Eltayeb, I. A. 1986. On the stability of the D″ layer. *Geophys. Astrophys. Fluid Dyn.* 36: 229–55

McKenzie, D., Bickle, M. J. 1988. The volume and composition of melt generated by extension of the lithosphere. *J. Petrol.* 29: 625–79

McNutt, M. 1990. Deep causes of hotspots. *Nature* 346: 701–2

McNutt, M., Fischer, K. 1987. The South Pacific superswell. In *Seamounts, Islands, and Atolls*, Geophys. Monogr. Ser., vol. 43, ed. B. Keating, R. Batiza, pp. 25–34. Washington, DC: Am. Geophys. Union

Morgan, W. J. 1971. Convection plumes in the lower mantle. *Nature* 230: 42–43

Morgan, W. J. 1972. Plate motions and deep mantle convection plumes. *Geol. Soc. Am. Mem.* 132: 7–22

Morgan, W. J. 1978. Rodriguez, Darwin, Amsterdam, . . . , A second type of hotspot island. *J. Geophys. Res.* 83: 5355–60

Morgan, W. J. 1981. Hotspot tracks and the opening of the Atlantic and Indian oceans. In *The Sea*, vol. 7, *The Oceanic Lithosphere*, ed. C. Emiliani, pp. 443–87. New York: Wiley

Nataf, H. C. 1991. Mantle convection, plates, and hotspots. *Tectonophysics* 187: 361–72

O'Conner, J. M., Duncan, R. A. 1990. Evolution of the Walvis Ridge-Rio Grande Rise hot spot system: Implications for African and South American plate motions of plumes. *J. Geophys. Res.* 95: 17,475–502

Olson, P., Kincaid, C. 1991. Experiments on the interaction of thermal convection and compositional layering at the base of the mantle. *J. Geophys. Res.* 96: 4347–54

Olson, P., Singer, H. 1985. Creeping plumes. *J. Fluid Mech.* 158: 511–31

Quinn, T. M. 1991. The history of Post-Miocene sea level change: Inferences from stratigraphic modeling of Enewetak atoll. *J. Geophys. Res.* 96: 6713–26

Richards, M. A. 1991. Hotspots and the case for a high viscosity lower mantle. In *Glacial Isostasy, Sea-Leval and Mantle Rheology*, ed. R. Sabadini, K. Lambeck, E. Boschi, pp. 571–87. Dordrecht: Kluwer Academic

Richards, M. A., Duncan, R. A., Courtillot, V. E. 1989. Flood basalts and hotspot tracks: Plume heads and tails. *Science* 246: 103–7

Richards, M. A., Hager, B. H., Sleep, N. H. 1988. Dynamically supported geoid highs over hotspots: Observation and theory. *J. Geophys. Res.* 93: 7690–7708

Richards, M. A., Jones, D. L., Duncan, R. A., DePaolo, D. J. 1991. A mantle plume initiation model for the formation of Wrangellia and other oceanic flood basalt plateaus. *Science*. In press

Rideout, M. L., Schilling, J.-G. 1985. Rare-earth elements, $^{87}Sr/^{86}Sr$, and $^{143}Nd/^{144}Nd$, mantle source variations. *Initial Rep. Deep Sea Drill. Proj.* 82: 483–96

Schilling, J.-G. 1973. Iceland mantle plume: Geochemical evidence along Reykjanes ridge. *Nature* 242: 565–71

Schilling, J.-G. 1985. Upper mantle heterogeneities and dynamics. *Nature* 314: 62–67

Schilling, J.-G. 1991. Fluxes and excess temperatures of mantle plumes inferred from their interaction with migrating mid-ocean ridges. *Nature* 352: 397–403

Schilling, J.-G., Thompson, G., Kingsley, R., Humphris, S. 1985. Hotspot-migrating ridge interaction in the South Atlantic. *Nature* 313: 187–91

Schroeder, W. 1984. The empirical age-depth relation and depth anomalies in the Pacific Ocean basin. *J. Geophys. Res.* 89: 9873–84

Silver, P. G., Carlson, R. W., Olson, P. 1988. Deep slabs, geochemical heterogeneity, and the large-scale structure of mantle convection: Investigation of an enduring paradox. *Annu. Rev. Earth Planet. Sci.* 16: 477–542

Sleep, N. H. 1984. Tapping of magmas from ubiquitous mantle heterogeneities: An alternative to mantle plumes? *J. Geophys. Res.* 89: 9980–90

Sleep, N. H. 1987. Lithospheric heating by mantle plumes. *Geophys. J. R. Astron. Soc.* 91: 1–12

Sleep, N. H. 1988. Gradual entrainment of a chemical layer at the base of the mantle by overlying convection. *Geophys. J.* 95: 437–47

Sleep, N. H. 1990a. Hotspots and mantle plumes: Some phenomenology. *J. Geophys. Res.* 95: 6715–36

Sleep, N. H. 1990b. Monteregian hotspot

track: A long-lived mantle plume. *J. Geophys. Res.* 95: 21,983–90

Sleep, N. H., Richards, M. A., Hager, B. H. 1988. Onset of mantle plumes in the presence of preexisting convection. *J. Geophys. Res.* 93: 7672–89

Sleep, N. H., Windley, B. F. 1982. Archean plate tectonics: Constraints and inferences. *J. Geol.* 90: 363–79

Solomon, S. C., Sleep, N. H. 1974. Some models for absolute plate motions. *J. Geophys. Res.* 79: 2557–67

Stacey, F. D. 1980. The cooling Earth: A reappraisal. *Phys. Earth Planet. Inter.* 22: 89–96

Stacey, F. D., Loper, D. E. 1984. Thermal histories of the core and mantle. *Phys. Earth Planet. Inter.* 36: 99–115

Storey, M., Mahoney, J. J., Kroenke, L. W., Saunders, A. D. 1991. Are oceanic plateaus sites of komatiite formation? *Geology* 19: 376–79

Tucholke, B. E. 1986. Structure of basement and distribution of sediments in the western North Atlantic Ocean. In *The Geology of North America*, Vol. M, *The Western North Atlantic Region*, ed. P. R. Vogt, B. E. Tucholke, pp. 331–40. Boulder: Geol. Soc. Am.

Tucholke, B. E., Smoot, N. C. 1990. Evidence for age and evolution of Corner Seamounts and Great Meteor Seamount chain from multibeam bathymetry. *J. Geophys. Res.* 95: 17,555–70

Vink, G. E. 1984. A hotspot model for Iceland and the Voring Plateau. *J. Geophys. Res.* 89: 9949–59

Vogt, P. R. 1991. Bermuda and Appalachian-Labrador rises: Common non-hotspot processes. *Geology* 19: 41–44

Von Herzen, R. P., Detrick, R. S., Crough, S. T., Epp, D., Fehn, U. 1982. Thermal origin of the Hawaiian swell: Heat flow evidence and thermal models. *J. Geophys. Res.* 87: 6711–23

Von Herzen, R. P., Cordery, M. J., Fang, C., Detrick, R. S., Fang, C. 1989. Heat flow and the thermal origin of hot spot swells: The Hawaiian swell revisited. *J. Geophys. Res.* 94: 13,783–800

Watson, S., McKenzie, D. 1991. Melt generation by plumes: A study of Hawaiian volcanism. *J. Petrol.* 32: 501–37

Watts, A. B. 1976. Gravity and bathymetry in the central Pacific Ocean. *J. Geophys. Res.* 81: 1533–53

White, R., McKenzie, D. 1989. Magmatism at rift zones: The generation of volcanic continental margins and flood basalts. *J. Geophys. Res.* 94: 7685–7729

Whitehead, J. A., Luther, D. S. 1975. Dynamics of laboratory diapir and plume models. *J. Geophys. Res.* 80: 705–17

Williams, Q., Knittle, E., Jeanloz, R. 1991. The high-pressure melting curve of iron: A technical discussion. *J. Geophys. Res.* 96: 2171–84

Wilson, J. T. 1963. A possible origin of the Hawaiian Islands. *Can. J. Phys.* 41: 863–70

Wilson, J. T. 1965. Evidence from ocean islands suggesting movement in the Earth. *Philos. Trans. R. Soc. London Ser. A* 258: 145–65

Woods, M. T., Lévêque, Okal, E. A., Cara, M. 1991. Two-station measurements of Rayleigh wave group velocity along the Hawaiian swell. *Geophys. Res. Lett.* 18: 105–8

Young, C. J., Lay, T. 1987. The core-mantle boundary. *Annu. Rev. Earth Planet. Sci.* 15: 25–36

Annu. Rev. Earth Planet. Sci. 1992. 20:45–84

THE PRIMARY RADIATION OF TERRESTRIAL VERTEBRATES

Robert L. Carroll

Redpath Museum and Department of Biology, McGill University, 859 Sherbrooke Street West, Montreal, H3A 2K6, Canada

KEY WORDS: evolution, amphibians, reptiles, Carboniferous, systematics

INTRODUCTION

The ancestry of all terrestrial vertebrates may be traced to a complex radiation that occurred during the Late Devonian and Early Carboniferous. Unfortunately, knowledge of the fossil record during this period remains very incomplete and we have only brief glimpses of a large number of diverging lineages, for which few relationships are adequately established.

The lineage leading to terrestrial vertebrates can first be recognized in the Middle Devonian, with the emergence of a particular group of bony fish, the osteolepiform rhipidistians. Their basic body form is typical of fish—fusiform trunk, paired and median fins—but the internal skeleton of the paired fins has the same basic structure as in terrestrial vertebrates, with one large proximal and two distal bones in both the front and back fins, comparable with the humerus, ulna, and radius of the arm, and the femur, tibia, and fibula of the leg. The arrangement of the bones of the skull and elements of the vertebrae in osteolepiforms and early land vertebrates are also similar in many features (Panchen & Smithson 1987). There is a rich record of osteolepiform fish in the Late Devonian and throughout the Carboniferous. The Upper Devonian genus *Eusthenopteron* (Figure 1*A*) has been studied extensively as a model for the group of fish that gave rise to land vertebrates (Jarvik 1980). On the other hand, no fossils are known that can be considered intermediate between these clearly aquatic fish and genera that are unequivocally classified as terrestrial vertebrates.

0084–6597/92/0515–0045$02.00

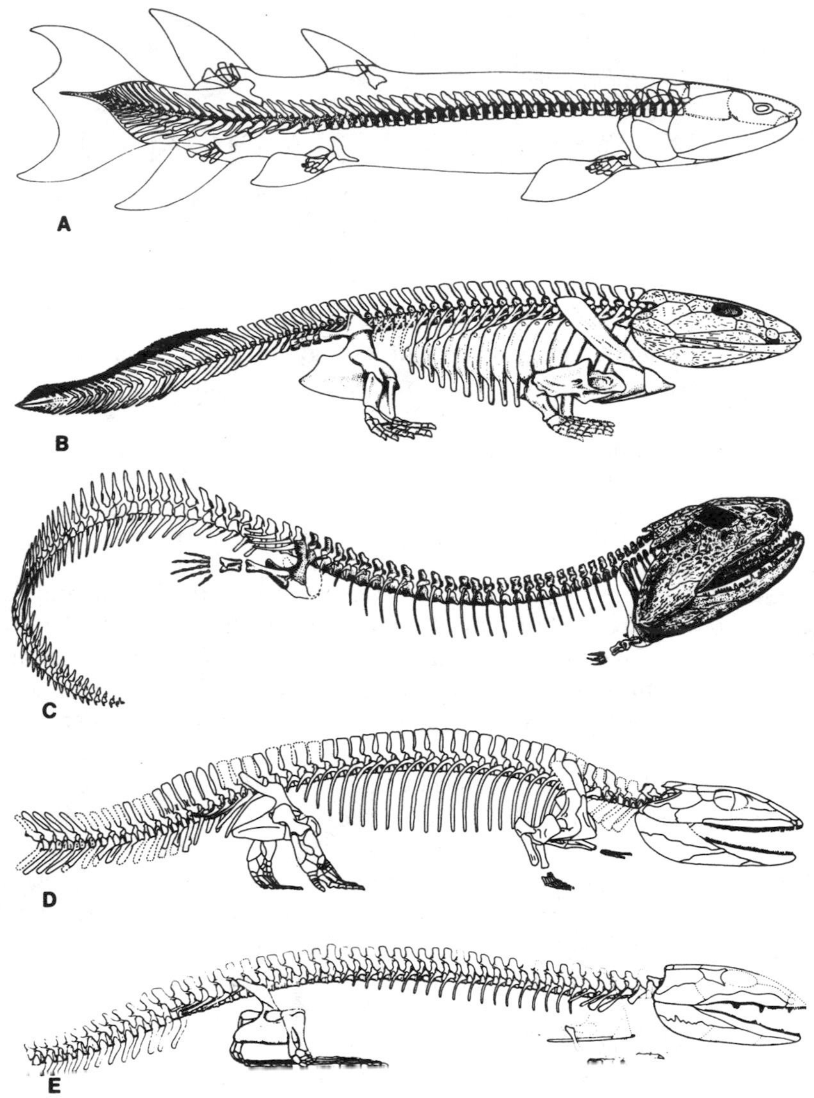

Figure 1 Skeletons of rhipidistian fish and representative "labyrinthodonts." A. The Upper Devonian osteolepiform fish *Eusthenopteron* (modified from Andrews & Westoll 1970). B. The Upper Devonian amphibian *Ichthyostega* (modified from Jarvik 1955). The rear limb has since been shown to have 7 toes. C. The Carboniferous amphibian *Crassigyrinus* (modified from Panchen & Smithson 1990). D. The embolomere *Protorogyrinus* (modified from Holmes 1984). E. The temnospondyl *Caerorhachis* (modified from Holmes & Carroll 1977).

The earliest land vertebrates are known from the Upper Devonian (Famennian) of East Greenland (Jarvik 1980, Clack 1988, Coates & Clack 1990) and Russia (Lebedev 1984) (Figure 1*B*). These animals had short but massive limbs of the basic pattern of subsequent tetrapods—although with a larger number of digits, which may reflect persistently aquatic locomotion. The East Greenland genera are more primitive than all subsequent amphibians in many features, but they don't appear closely related to any of the lineages that were common in the Carboniferous.

Knowledge of the fossil record of terrestrial vertebrates is extremely incomplete in the Lower Carboniferous (Figure 2). Numerous attempts have been made to establish relationships among the many later lineages [see Panchen & Smithson (1988) and references cited therein; and Cannatella & de Queiroz (1989)], but recent descriptions of the earliest known members of the major groups emphasize their distinctive natures. Much more needs to be learned of the latest Devonian and earliest Carboniferous tetrapods before a reliable phylogeny can be established. This review concentrates on the diversity of early tetrapods and some of the major evolutionary trends.

Within the modern fauna, all terrestrial vertebrates can be divided into two major groups: the amphibians and the amniotes. The amniotes—including mammals, birds, and reptiles—are certainly a monophyletic assemblage, distinguished by their ability to lay eggs on land (or in the case of placental mammals and ichthyosaurs, give birth to live young); the young are typically hatched or born at an advanced state of development. The eggs, or internally developing embryos, are surrounded by extra-embryonic membranes—the chorion, allantois, and amnion—that serve for exchange of respiratory gases, retention of water, and protection. Amphibians typically lay their eggs in the water, as did their fish ancestors, and many have a distinct aquatic larval stage.

The first terrestrial vertebrates are assumed to have reproduced via eggs laid in the water. Many Carboniferous tetrapods are known to have had aquatic larval stages (Boy 1986). The majority of Paleozoic tetrapods are thus considered to be amphibians, although they differ greatly in their skeletal anatomy from modern frogs, salamanders, and caecilians (long bodied, limbless tropical amphibians).

There must have been a major divergence early in the evolution of Paleozoic tetrapods, from which one or more lineages led to the modern amphibian orders, and another led to amniotes. On the other hand, a large number of Paleozoic tetrapods may not be closely related to either of the modern tetrapod groups. One of the major unresolved problems in vertebrate evolution is to establish the specific relationships of the ancestral amniotes and the ancestors of the three modern amphibian orders. This

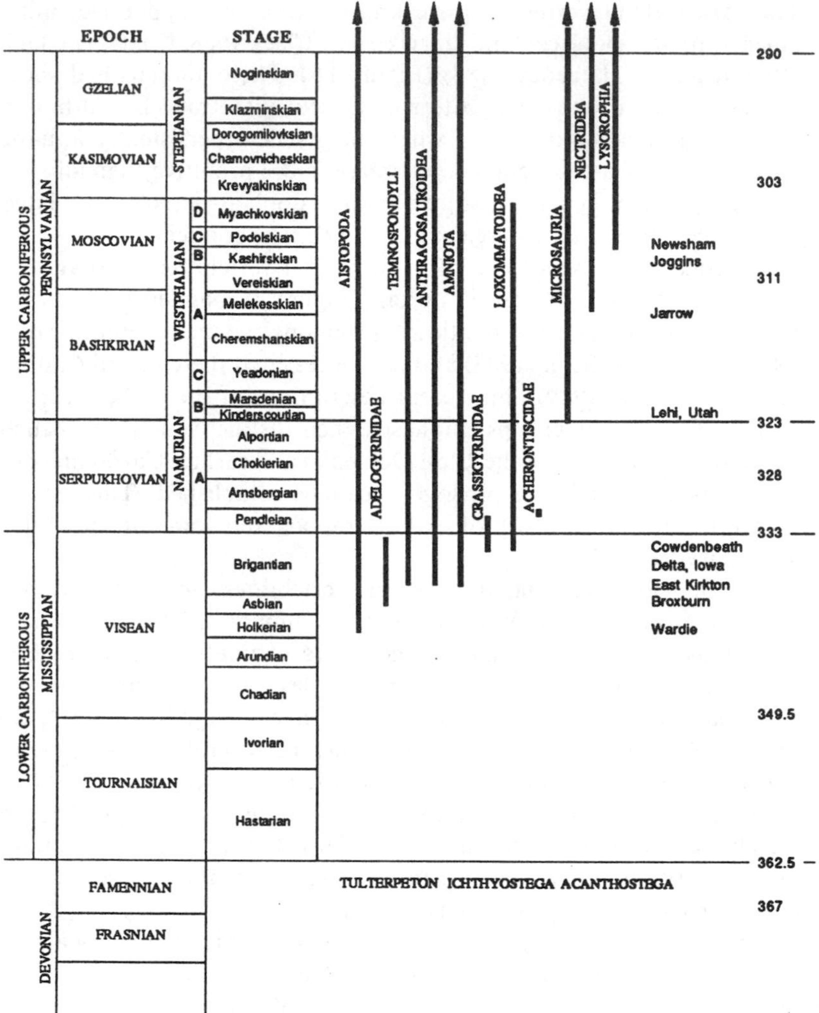

Figure 2 Stratigraphic ranges of major groups of Devonian and Carboniferous tetrapods. Geological time scale based on Harland et al (1989). Selected localities indicated on the right.

problem continues to be clouded by the large number of Paleozoic groups whose relationships with one another have not yet been established.

Discussion of relationships in this paper will follow the principles of phylogenetic analysis espoused by Hennig (1966), Wiley (1981), and Ax (1987), in which emphasis is placed on derived, rather than primitive characters in establishing relationships. One important exception is made to the procedures commonly followed in this methodology. Analysis will begin with the most primitive known genera, rather than starting from the modern groups. Establishing relationships on the basis of living taxa may be the only possible methodology in groups that have a poor fossil record, and it may be practical in taxa in which most of the known lineages have living representatives, but in the case of Paleozoic tetrapods, most groups are without modern descendants. In addition, the anatomy of Paleozoic amphibians can be much more readily understood in relationship to their fish ancestry than by comparison with the very different modern amphibians.

Panchen & Smithson (1987) have discussed the evidence that tetrapods are a monophyletic assemblage, which is the sister-group of the osteolepiform rhipistians. Use of the osteolepiforms as an out-group demonstrates which of the characters of early tetrapods are primitive for the group, and which genera most closely exemplify the primitive condition for terrestrial vertebrates. It has long been clear that none of the early tetrapod groups are ideal ancestors for any of the rest. All exhibit a mosaic of primitive and derived characters that indicate significant periods of independent evolution since the initial divergence of each group.

Although none of the known fossils demonstrate an absolutely intermediate structure between rhipidistian fish and land animals with fully developed limbs, Devonian and Carboniferous amphibians do show progressive modifications of many features of the skeleton from structures that resemble those of rhipidistian fish to those of more advanced, late Paleozoic tetrapods. Such progressive changes are particularly evident in the skull and vertebral column.

ANATOMICAL FEATURES OF SPECIAL IMPORTANCE IN DISTINGUISHING GROUPS OF PALEOZOIC TETRAPODS

Patterns of the Skull

The pattern of the skull bones of osteolepiform rhipidistian fish is basically similar to that of all primitive tetrapods, but is more primitive since the skulls possess a greater number of individual bones and have (like other primitive bony fish) a number of functional units that are movable relative to one another (Figure 3*A–C*). There are four major elements of the skull roof—the paired cheek regions (continuous with the opercula), the snout, and the back of the skull table—that were mobile relative to one another. The braincase was appressed against the skull roof but not suturally attached, and the anterior (ethmoid) and posterior (otic-occipital) portions of the braincase were mobile relative to one another. The hyomandibular

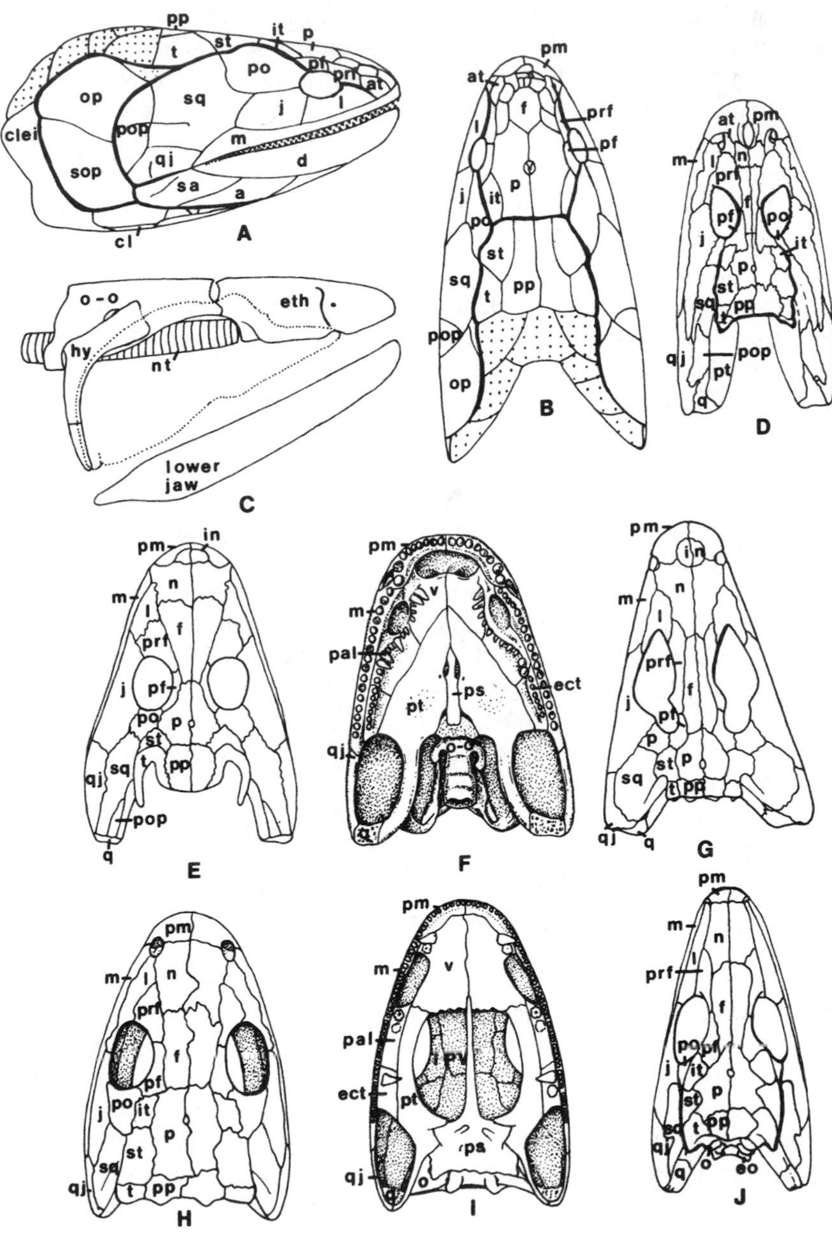
pp
it
p
st
t
po
pf
prf
at
op
sq
j
l
clei
pop
m
d
qj
sop
sa
a
cl
A
pm
at
f
prf
pf
l
j
it
p
po
st
sq
t
pp
pop
op
B
at
pm
m
n
prf
f
pf
po
it
j
p
st
sq
pp
qj
pop
pt
q
D
o-o
eth
hy
nt
lower
jaw
C
pm
in
m
n
l
f
prf
j
pf
po
p
st
qj
sq
t
pp
pop
q
E
pm
v
m
pal
ect
ps
pt
qj
o-o
F
pm
in
m
n
l
prf
f
j
pf
p
st
p
sq
t
pp
qj
q
G
pm
m
l
n
prf
f
pf
j
po
it
p
st
sq
qj
t
pp
H
pm
m
v
pal
pv
ect
pt
ps
qj
q
o
I
pm
m
n
prf
l
f
po
pf
j
it
p
st
sq
qj
t
pp
q
o
eo
J

formed a link between the braincase, the operculum, and the back of the jaws. The palate was mobile relative to the braincase and the cheeks. In the fish, this mobility was related to feeding and respiration in the water. With the loss of gills in terrestrial vertebrates, the mobility of the cheek and operculum no longer functioned in respiration. The increase in the relative length of the anterior portion of the skull in early amphibians rendered the rhipidistian mobility of the braincase inoperative for feeding (Thomson 1967). Mobility between the front and back of the skull table is absent in all land vertebrates, and in fact is absent in one family of rhipidistian fish, the Panderichthyidae (Schultze & Arsenault 1985). Other areas of mobility are progressively lost in land vertebrates, but different patterns of loss occurred in each major group. Loss and/or fusion of bones of the skull roof also occurred in most groups.

The large operculum of the fish is lost in all tetrapods. The anterior margin of the operculum occupied a position between the back of the cheek and the skull table. This area is retained as the squamosal notch in early tetrapods (Godfrey et al 1987). The dorsal portion of this notch retains the position of the spiracular cleft in rhipidistian fish. In some early tetrapods it may have remained open to serve for the passage of air to the lungs when the rest of the skull was submerged (Panchen 1985, p. 548). The squamosal notch is closed in some groups, but in others it is modified to serve for attachment of a tympanic membrane. The preopercular bone

Figure 3 Skulls of a rhipidistian fish and representative "labyrinthodonts." A, B. Lateral and dorsal views of *Eusthenopteron* (modified from Jarvik 1980). Heavy lines indicate divisions between major kinetic units. Coarsely stippled bones linking dermal shoulder girdle and skull table are lost in tetrapods. C. Braincase, hyomandibular, and lower jaw of *Eusthenopteron* (modified from Thomson 1967). Dotted line indicates outline of palatoquadrate. D. Skull roof of *Crassigyrinus* (modified from Panchen 1985). E. Skull roof of *Acanthostega* (modified from Clack 1989). F. Palate of *Ichthyostega* (modified from Jarvik 1980). G. Skull roof of *Loxomma* (modified from Beaumont 1977). H,I. Skull roof and palate of *Dendrerpeton* (modified from Carroll 1987). J. Skull roof of the embolomere *Protorogyrinus* (modified from Holmes 1984). Abbreviations used in figures: *a*, angular; *at*, anterior tectal; *bo*, basioccipital; *cf*, carotid foramen; *cl*, clavicle; *clei*, cleithrum; *d*, dentary; *ect*, ectopterygoid; *eo*, exoccipital; *eth*, ethmoid portion of braincase; *f*, frontal; *FM*, foramen magnum; *hy*, hyomandibular; *in*, internarial bone; *it*, intertemporal; *IPV*, interpterygoid vacuity; *j*, jugal; *l*, lacrimal; *m*, maxilla; *mf*, mandibular foramen; *n*, nasal; *nt*, notochord; *o-o*, otic-occipital portion of braincase; *o* or *opis*, opisthotic; *op*, opercular; *p*, parietal; *pal*, palatine; *pf*, postfrontal; *pl*, pleurosphenoid; *pm*, premaxilla; *po*, postorbital; *pop*, preopercular; *pp*, postparietal; *prf*, prefrontal; *ps*, parasphenoid; *pt*, pterygoid; *PTF*, post-temporal fenestra; *ps*, parasphenoid; *q*, quadrate; *qj*, quadratojugal; *sa*, surangular; *se*, sphenethmoid; *so*, supraoccipital; *sop*, subopercular; *sq*, squamosal; *st*, supratemporal; *sta*, stapes; *t*, tabular; *t-sq*, tabular-squamosal; *TFP*, transverse flange of pterygoid; *v*, vomer; *X*, *XII*, cranial nerves.

(despite its name, an element of the cheek rather than the operculum in rhipidistians) is retained in some primitive tetrapods.

The back of the braincase was not suturally attached to the bones of the skull roof or cheek in rhipidistians, but such a connection has evolved in most tetrapods (Carroll 1980). A variety of different bony connections are established in the various groups. Some emphasize connection between the otic capsule and either the skull table or the cheek (or both), some elaborate connections between the postparietals and the exoccipitals, and others develop a new center of ossification, the supraoccipital, linking the braincase with the back of the skull table.

The hyomandibular, which in fish forms a mechanical link between the back of the braincase and the cheek, operculum, and the lower jaw, is much reduced in all early terrestrial vertebrates. It was once thought that it served the role of an ear ossicle in all primitive tetrapods. Structurally, the hyomandibular is unquestionably homologous with the stapes of land vertebrates, but the large size of this element and its close connection with other skeletal elements in early Carboniferous tetrapods would have made it an inappropriate element in an impedance matching system to detect airborne vibrations. It appears to have continued to have served a supporting role (Carroll 1980) or one associated with respiration (Clack 1989) in several groups of early amphibians. It is reduced in size independently in many groups of later tetrapods.

The two sides of the palate were separated at the midline by the width of the braincase in rhipidistians and articulated with the posterior end of the ethmoid element. The paired vomers, palatines, and pterygoid bones bore large fangs and associated replacement pits. These fangs, like the marginal teeth, show conspicuous infolding of the dentine, for which they are termed labyrinthodont. The labyrinthine infolding of the teeth is absent in many descendant groups and the palatal fangs are lost. Large openings, the interpterygoid vacuities, evolve in some groups of Paleozoic amphibians, and the two sides of the palate may fuse to the base of the braincase.

Vertebrae

The structure of the vertebrae has been one of the most important characteristics used to distinguish the major groups of tetrapods and evaluate their probable relationships. Rhipidistian fish show several patterns of vertebral structure (Andrews & Westoll 1970), but that of the genus *Osteolepis* appears most appropriate as a basis for the derivation of the pattern in early terrestrial vertebrates (Figure 4*A*). In the trunk there are three basic units, all of which develop as paired elements. Dorsally, the neural arches surround the nerve cord. The centra are composed of large anteroventral elements, the intercentra, and posterodorsal pleurocentra. Together they form a thin sheath around the notochord. None of these elements are firmly attached to one another. In the tail, haemal arches extend beneath the intercentra.

In the origin of land vertebrates, articulating processes (zygapophyses) evolved to link the neural arches with one another along the column. In most groups the arches fuse to one another medially. The intercentra typically fuse with one another ventrally to form a large crescentic structure. The subsequent history of the central elements differs from group to group. In some groups the intercentra expand and come to dominate the entire segment; in other groups the pleurocentra extend ventrally at the expense of the intercentra and fuse to form either a crescentic or ringshaped structure. The neural arch evolves a close attachment with either the pleurocentra or the intercentra or both. In some groups of Carboniferous vertebrates, the entire vertebra is ossified as a single element, without any direct evidence of its formation from separate units.

In rhipidistian fish, there was no well-defined bony articulation between the skull and the vertebral column. The notochord extended through the occiput to the level of the pituitary, forming a strong but flexible attachment between the head and trunk. This was further strengthened by the presence of the operculum linking the skull roof and the shoulder girdle. With the loss of the operculum and the separation of the skull and the shoulder girdle, a stronger joint had to evolve between the back of the braincase and the vertebral column. One of the most primitive land vertebrates, *Ichtyostega*, retained the fish condition, but all others reduced the forward extent of the notochord, and evolved a specialized bony articulation between the occiput and the first cervical vertebrae. This articulation differed significantly in divergent groups in relationship to the different patterns of the vertebral centra. In groups that retain multipartite trunk centra, several elements that contribute to the first two cervical vertebrae are involved in forming a complex articulating surface with the occiput. In groups that have a single central element in the trunk region, a single unit forms the primary articulation with the occiput, although accessory elements are sometimes present (Figure 4).

Girdles and Limbs

All of the elements of the amphibian shoulder girdle are evident in the fish, as well as some additional bones linking the skull and girdle that were lost in the transition. Only a single bone is evident in the pelvic region of the fish and it is not attached to the column by a sacral rib, as is the case in all tetrapods with normal limbs. There is a profound difference in the distal ends of the limbs between rhipidistian fish and all land vertebrates. Even the most primitive Devonian tetrapods have ankle and wrist bones and

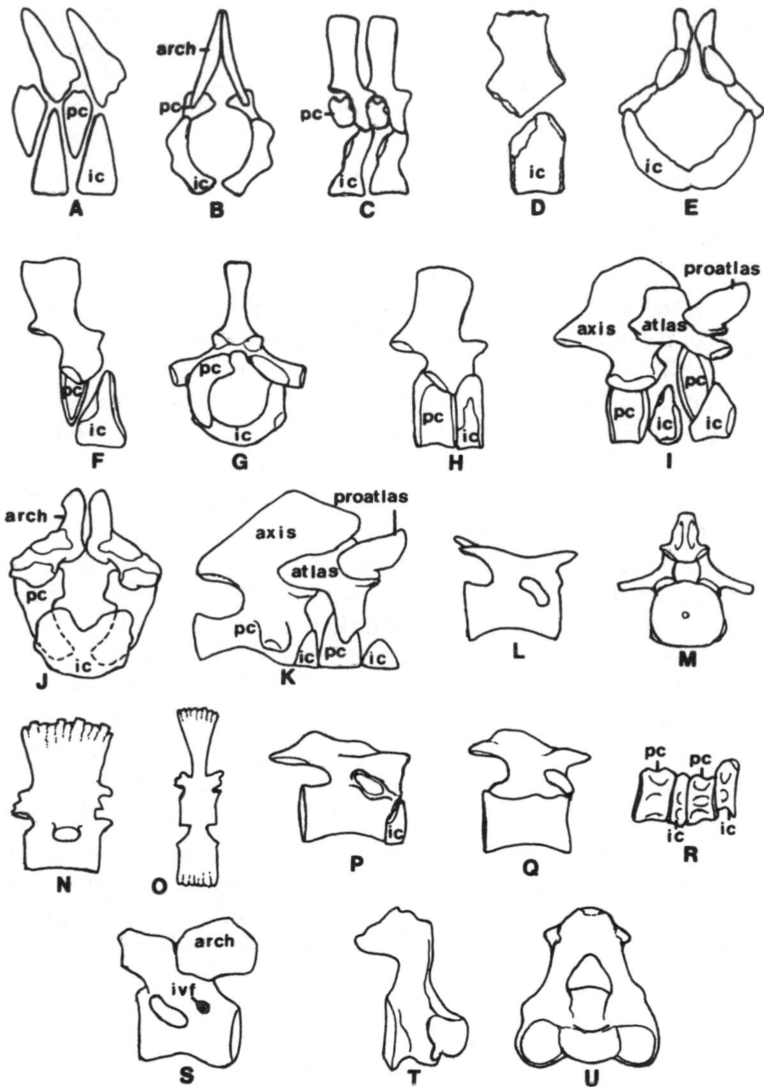

Figure 4 Vertebrae. A, B. Lateral and anterior views of the trunk vertebrae of the Middle Devonian osteolepiform *Osteolepis* (modified from Andrews & Westoll 1970). C. Trunk vertebrae of *Ichthyostega* (modified from Jarvik 1980). D, E. Trunk vertebrae of *Crassigyrinus* in lateral and posterior views (modified from Panchen 1985). F, G. Lateral and anterior views of trunk vertebrae of the temnospondyl *Greererpeton* (modified from Godfrey 1989). H. Lateral view of trunk vertebra of an embolomere. I, J. Lateral and anterior views of the atlas-axis complex of an embolomere (modified from Holmes 1984). K. Atlas-axis complex of the oldest reptile, *Hylonomus*. L. Trunk vertebra of the oldest aïstopod *Lethiscus* (modified from

distinct toes. All rhipidistians have a fin supported externally with dermal rays and the internal skeleton reveals no trace of the precursors of the toes. On the other hand, homologues of proximal wrist and ankle bones are evident, as are unquestionable equivalents of the major limb bones.

Among the early land vertebrates, the pattern of the wrist and ankle are relatively stable, although there is a long term process of reduction in the number of individual bones by loss and fusion. It was long thought that no terrestrial vertebrates had more than five principal digits in the hand and foot. Recently discoveries show that six to eight digits were present in three Devonian genera (Lebedev 1984, Coates & Clack 1990). Digits have been variably lost in later tetrapods.

The femur and particularly the humerus of early land vertebrates were more complicated than those of their descendants, with a number of conspicuous processes for muscle attachment and (in the humerus) foramina for nerves and blood vessels. These are progressively lost in later forms.

In addition to many changes associated with more effective accommodation to a terrestrial way of life, three other major trends are seen among particular groups of Paleozoic tetrapods. One is to return to the water; the second trend is toward significantly smaller body size. The third is to elongate the trunk region and reduce or lose the limbs.

"LABYRINTHODONTS" AND "LEPOSPONDYLS"

Two major assemblages of Paleozoic amphibians have long been recognized: the labyrinthodonts and the lepospondyls. Use of these names in this paper is not intended to support their taxonomic validity, but it is useful to retain them in an informal sense to designate groups of Paleozoic tetrapods that exhibit readily distinguishable anatomical patterns. The labyrinthodonts (Figure 1) retain many primitive features in common with

---

Wellstead 1982). M. Posterior view of trunk vertebra of *Adelospondylus* (modified from Andrews & Carroll 1991). N, O. Trunk and caudal vertebrae of a nectridean (modified from Bossy 1976). P. Trunk vertebra of the long-bodied microsaur *Cardiocephalus*, in which intercentra are well developed (modified from Carroll & Gaskill 1978). Q. Trunk vertebra of a lysorophid (modified from Boyd 1980). R. Trunk centra of *Acherontiscus* (modified from Carroll 1969b). S. First cervical vertebra and accessory arch of the aïstopod *Ophiderpeton* (modified from Carroll 1989). T, U. Lateral and anterior views of the first cervical of the microsaur *Cardiocephalus* (modified from Carroll & Gaskill 1978). Abbreviations: *ic*, intercentrum; *ivf*, intravertebral foramen for spinal nerve; *pc*, pleurocentum.

their fish ancestors: the number and configuration of the bones of the skull roof and palate, labyrinthodont infolding of the teeth, large palatal fangs, and especially vertebrae composed of several centers of ossification. Labyrinthodonts were of moderate to large size (20 cm to 5 m) and the skull was large relative to the length of the trunk. Primitive members of this assemblage were presumably ancestral to all later tetrapods. The labyrinthodonts are probably monophyletic in the sense of having a common ancestry, but would be considered paraphyletic if they gave rise to any subsequent groups (such as the modern amphibians and amniotes) that are recognized as taxonomically distinct.

The lepospondyls are a much more heterogenous assemblage (Figures 5–7). In contrast with the labyrinthodonts, they show extreme variability in the number of presacral vertebrae—from as few as 14 to more than two hundred. In some the limbs were well developed, but in others they were greatly reduced or totally absent. The number of bones in the skull roof is commonly reduced, and their pattern may be radically modified. Despite these conspicuous differences, the lepospondyls have been assumed to belong to a single group because of the common expression of the following features that are derived relative to the condition in labyrinthodonts: absence of labyrinthine infolding of the teeth, loss of palatal fangs, and absence of a squamosal notch. The most important derived character is the nature of the vertebrae. In all lepospondyls, the vertebral centra are in the form of complete cylinders. In all but one genus, a single centrum dominates each segment, and in most genera there is no more than a single center of ossification in the trunk vertebrae. A single vertebral element forms the primary articulation between the neck and the skull. No species has been recognized that bridges the gap between any of the genera recognized as labyrinthodonts and any of the lepospondyl groups.

Despite the presence of numerous derived characters in common, the various lepospondyl orders may have each evolved separately from among more primitive tetrapods. If this were the case, then neither labyrinthodonts nor lepospondyls could be accepted as natural groups.

DEVONIAN AMPHIBIANS

A single specimen of the genus *Tulerpeton* was described by Lebedev (1984) from the Upper Devonian of Russia. It consists of the pectoral girdle, forelimb, and hind limb. The restoration shows close similarities to the Carboniferous embolomeres, except for the presence of six toes in the hand. Until skull and vertebral material is illustrated, it is not possible to establish its phylogenetic relationships with other early tetrapods. This material is now being described in greater detail by Lebedev and Clack.

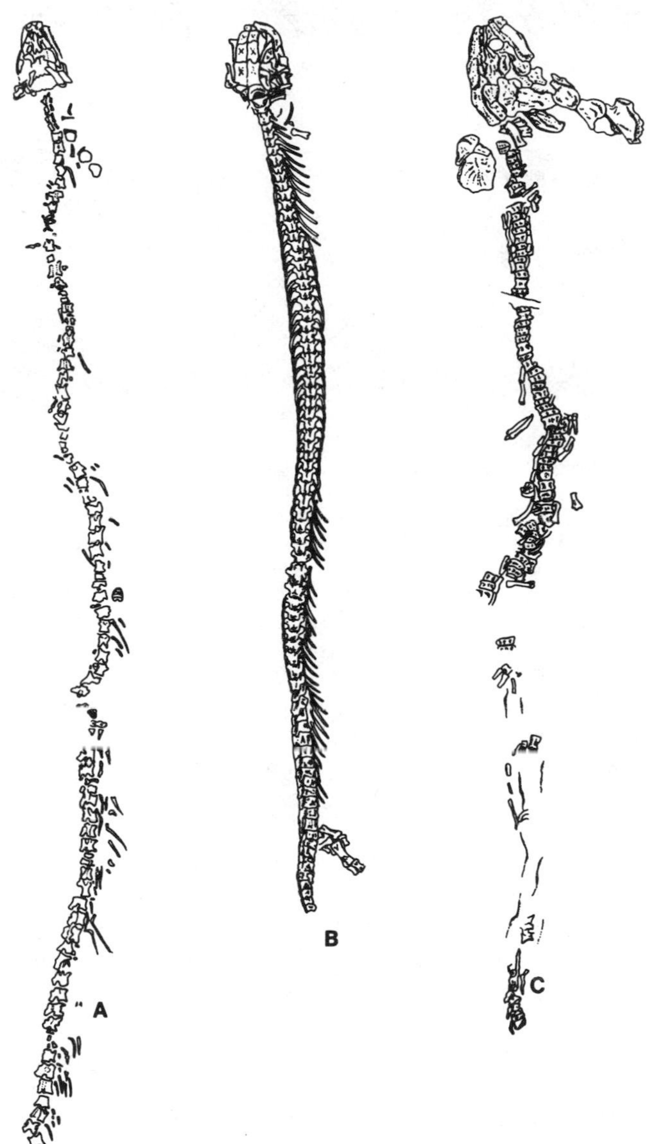

Figure 5 Skeleton of three long-bodied "lepospondyls." A. The oldest aïstopod *Lethiscus* (modified from Wellstead 1982). B. The Pennsylvanian lysorophid *Cocytinus* (modified from Carroll & Gaskill 1978). C. *Acherontiscus* (modified from Carroll 1969b).

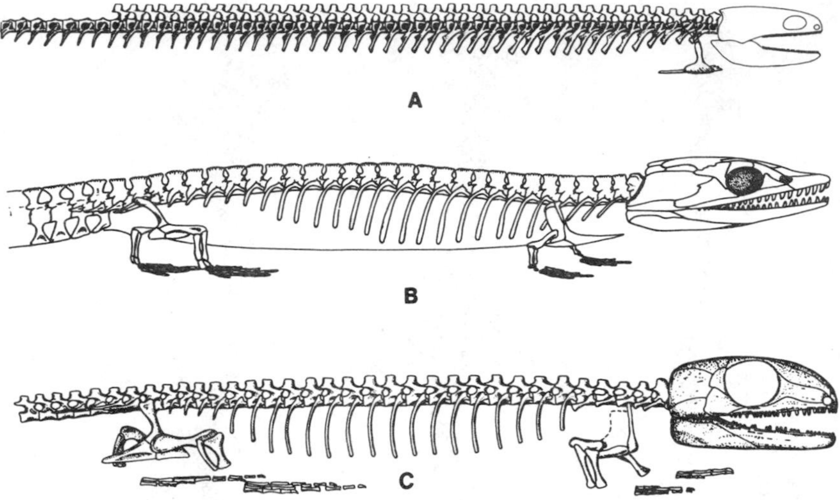

Figure 6 Skeletons of "lepospondyls." A. The oldest adelogyrinid *Palaeomolgophis* (modified from Andrews & Carroll 1991). B. The nectridean *Sauropleura* (modified from Bossy 1976). C. The oldest microsaur, *Utaherpeton* (modified from Carroll et al 1991).

Many more specimens are known from the Upper Devonian of East Greenland. This fauna includes two genera: *Ichthyostega* and *Acanthostega*. They were long grouped together in the order or suborder Ichthyostegalia, but, while sharing some primitive features, they differ from one another in many important characteristics (Clack 1988, Coates & Clack 1991).

Both genera are of moderate size, reaching a length of about a meter, including the tail. Both genera are primitive in having the lateral line canals in the skull in closed tubes, in contrast with the open grooves of later amphibians. The limbs are short but well developed. As in the Russian genus, they have more toes than any later land vertebrates. Seven are present in the foot of *Ichthyostega* and eight in the hand of *Acanthostega* (Coates & Clack 1990). Both have very limited expression of the scapular blade. The vertebrae are multipartite, with pleurocentra and intercentra as in rhipidistians; the neural arches have well developed zygapophyses in *Ichthyostega* but not in *Acanthostega*.

Both genera are primitive in the persistent separation between the ethmoid and otic-occipital portions of the brain case. As in rhipidistians, the anterior (ethmoid) portion in *Ichthyostega* retains surfaces for articulation between the base of the braincase and the palate. The parasphenoid does not extend behind the ethmoid portion of the braincase. *Ichthyostega* also

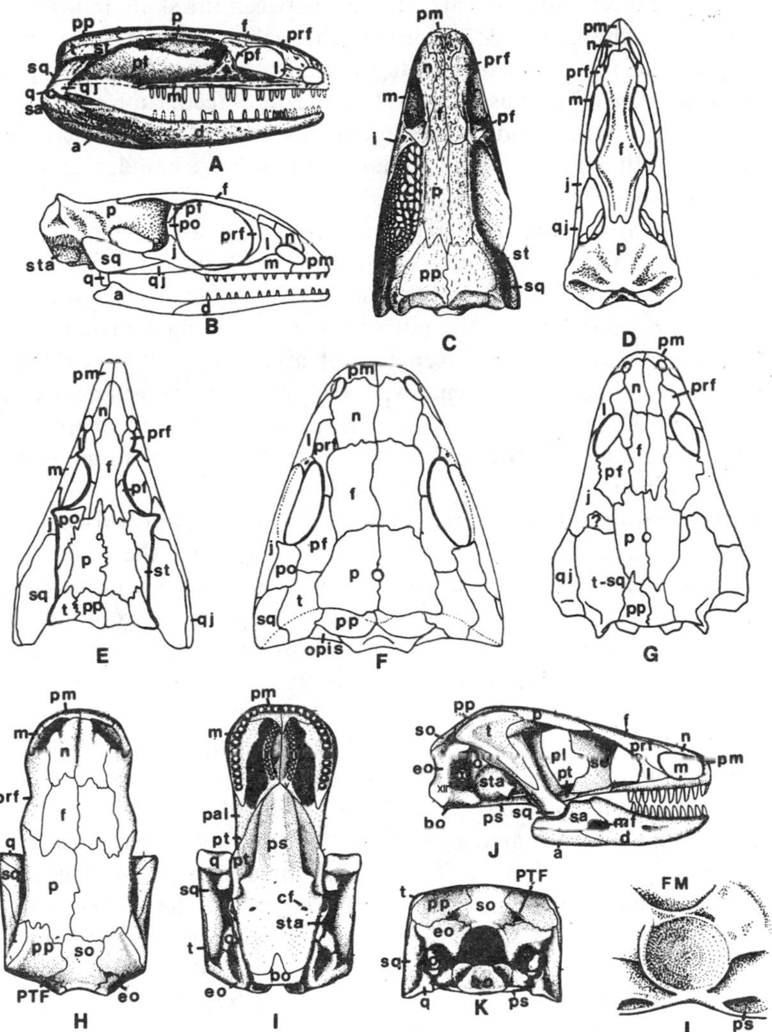

Figure 7 Skulls of "lepospondyls." A, C. Lateral and dorsal views of the aïstopod *Ophiderpeton*. Osteoderms covering temporal fenestra have been omitted on the right side. B, D. Dorsal and lateral views of the skull of *Phlegethontia* (modified from Gregory 1948). E. The nectridean *Sauropleura* (modified from Bossy 1976). F. The primitive microsaur *Tuditanus* (modified from Carroll & Gaskill 1978). G. *Adelospondylous* (modified from Carroll & Gaskill 1978). The postorbital may be absent. H.–K. Dorsal, palatal, lateral, and occipital views of the Permian lysorophid typically referred to as *Lysorophus* (modified from Wellstead 1991). L. Occiput of *Ophiderpeton* showing circular depression which received the notochord. Abbreviations as in Figure 3.

retains a primitive pattern of articulation between the skull and trunk, in which the notochord extends far forward through the base of the braincase. The preopercular is retained in both genera. Both genera retain the primitive closed palate of rhipidistians and have a row of teeth running along the ectopteryoid, palatine, and vomer. In contrast with previous descriptions, individual teeth within this row are larger than the remainder, as in later labyrinthodonts.

Acanthostega is more derived than *Ichthyostega* in having the parasphenoid extend some distance posteriorly beneath the otic-occipital portion of the braincase, but not as far as in later tetrapods. The basioccipital is not fully ossified, so that the notochord may still have penetrated the occiput for a short distance. Remains of atlas and axis arch elements, broadly resembling those of temnospondyls and anthracosaurs, are illustrated by Clack (1988). As in later tetrapods, strongly developed basicranial processes are formed by the basisphenoid and surrounding parasphenoid, which articulate with the palate.

In these features, *Acanthostega* is more advanced than *Ichthyostega*, but this is not the case for the attachment of the back of the braincase. It loosely underlies the postparietals and tabulars in *Acanthostega*, essentially as in *Eusthenopteron*, but in *Ichthyostega* it is supported by complex facets from the underside of the postparietals that are not comparable with any structures in later tetrapods.

Acanthostega is associated with major groups of later tetrapods by the structure of the stapes (Clack 1989). In *Acanthostega*, the anthracosaur *Pholiderpeton* (Clack 1987), the temnospondyl *Greererpeton* (Carroll 1980), and probably *Dendrerpeton* (Godfrey et al 1987), the stapes are short and massive, with an expanded blade that extends across the quadrate ramus of the pterygoid. The pattern in *Greererpeton* suggests an important role in supporting the braincase relative to the dermal bones of the skull. In *Acanthostega*, Clack suggests that it controlled palatal and spiracular movements in breathing. The structure of the stapes in these genera is certainly derived relative to that of the hyomandibular in rhipidistian fish, and may be considered a synapomorphy uniting early labyrinthodonts. The stapes has not been described in *Ichthyostega*.

On the other hand, the cheek and skull table, which are loosely articulated with one another in their fish ancestors and in the earliest members of other groups of early tetrapods, are solidly attached in both *Acanthostega* and *Ichthyostega*, and the intertemporal bone is lost. The postparietals are fused at the midline in *Ichthyostega*, but not in *Acanthostega*. These specializations appear to preclude these genera from being directly ancestral to any of the other early tetrapod groups.

Despite these specific features, the Greenland genera may represent a primitive body form for the labyrinthodonts and perhaps for all early tetrapods. Coates & Clack (1990) suggest that the extra digits in both the front and hind limbs may reflect a persistently aquatic mode of locomotion. This pattern may be primitive for all tetrapods, but might have been separately lost in later lineages.

CARBONIFEROUS "LABYRINTHODONTS"

Crassigyrinidae

Several other, very distinct lineages of labyrinthodonts appear in the Carboniferous, with little evidence of specific affinities. One of the most striking (but consisting of a single species) is the Crassigyrinidae (Panchen & Smithson 1990). (See Figures 1*C*; 2*D*; 4*D*,*E*.) This group has a relatively long trunk and extremely small anterior limbs. The skull is very deep, with the orbits close together, as in the osteolepiform fish *Panderichthys*, and *Elpistostege* (Schultze & Arsenault 1985).

Crassigyrinus is more primitive than *Ichthyostega* and *Acanthostega* in having a line of weakness between the skull table and the cheek and in the retention of an intertemporal bone. As in *Ichthyostega*, the preopercular is retained as is an anterior tectal, a small bone in the snout region common to rhipidistians. As in *Acanthostega*, the basioccipital is incompletely ossified so that the notochord may have penetrated some distance into the occiput, but the parasphenoid extends further back beneath the posterior portion of the braincase.

The braincase is attached to the lateral margin of the skull table through contract between the optic capsule and the tabular. A posttemporal fossa, which penetrates the occiput above the otic capsule in rhipidistians and several groups of primitive tetrapods, appears to be lost. The tabular extends posteriorly beyond the postparietals to form a tabular "horn."

The vertebral pattern is striking. Instead of ossifying both the intercentra and pleurocentra, as occurs in most other Paleozoic labyrinthodonts, there is only a single crescentic element per segment, apparently equivalent to the intercentra of other early tetrapods. It is only loosely attached to the neural arch. The arches themselves are paired, as in rhipidistians, in contrast with the fused condition in most tetrapods. They completely lack postzygapophyses, and the prezygapophyses have nearly vertical articulating surfaces. Although it is possible that the vertebrae retain a condition intermediate between that of rhipidistians and other early tetrapods, the manner of articulation between the arches corresponds broadly with that of some secondarily aquatic tetrapods. In view of the many tetrapod features that *Crassigyrinus* exhibits, it is more parsimonious to assume that its vertebral structure was achieved by reduction of ossification, plausibly associated with a return to a more completely aquatic way of life.

The humerus is primitive in retaining several accessory canals in common with *Ichthyostega* and *Acanthostega* that are lost in all other tetrapods.

Loxommatoidea

The Loxommatoidea are represented by three or four genera, ranging in age from Visean to Moscovian, but most are known only from the skull, which is distinctive in having keyhole shaped orbits (Figure 3*G*). Primitive species retain an intertemporal bone, but all have a solid attachment between the cheek and the skull table. The palate is primitive in the absence of interpterygoid vacuities and the large size of the fangs on the ectoptergyoid, palatine, and vomer. As in *Crassigyrinus*, the otic capsule articulates with the tabular, but the detailed structure of the opisthotic [as illustrated by Beaumont (1977)] differs significantly. In addition to having a single lateral area of contact with the tabular, there is also a specialized surface of articulation with the postparietal. Between these areas of articulation is a large passage, not present in *Crassigyrinus*, that appears to be equivalent to the posttemporal fossa of other primitive tetrapods. Panchen & Smithson (1988) refer to the presence of a tabular horn in loxommatoids, but it is a very short process, clearly distinct from that in *Crassigyrinus*.

Recently discovered material shows that loxommatoids possessed vertebral centra in the shape of wide ventral crescents, similar to those of *Crassigyrinus* (A. C. Milner & Lindsay 1989). Loxommatoids retain lateral line canals, indicating that they were customarily aquatic, but too little is known of the postcranial skeleton to consider their way of life. A possible loxommatoid skull has recently been discovered (but not yet described) from the Brigantian of East Kirkton (A. R. Milner, personal communication).

Temnospondyli

The groups just discussed are known from very short time intervals within the Paleozoic and show little diversity. The two major groups of Paleozoic tetrapods are the temnospondyls and the anthracosauroids. The earliest appearance of both groups is in the recently discovered locality of East Kirkton (Rolfe et al 1990) prior to the occurrence of either *Crassigyrinus* or typical loxommatoids, but these early species have not yet been described. [The entire vertebrate fauna from East Kirkton (near Edinburgh) is the subject of a symposium "Volcanism and early terrestrial biotas" to be held in September 1992. The symposium is sponsored jointly by the Royal Society of Edinburgh and the National Museums of Scotland.

The proceedings will be published in 1993 in the *Transactions of the Royal Society of Edinburgh*.]

Early temnospondyls are characterized by the following derived features: strong attachment of skull roof and cheek; back of braincase supported by contact between ventral lappets of the postparietals and the exoccipitals (initially the opisthotic is not in solid contact with the skull table or cheek); widely open palate [small interpterygoid vacuities are known in some anthracosaurs (Clack 1987) but they are not a common or primitive character of this group]; manus with four toes; and a phalangeal count less than that of anthracosauroids or amniotes.

Primitive temnospondyls retain an intertemporal bone (although it is lost early in the group's evolution), and the pleurocentra and intercentra are both crescentic. Most temnospondyls retain a squamosal notch. The stapes is large in the early genus *Dendrerpeton*, but it is reduced within the group and the squamosal notch assumes the role of an otic notch, supporting a tympanum. The evolution of an impedance matching middle ear early in its history may have been one of the major reasons for the wide radiation of this group.

Most Carboniferous and Lower Permian temnospondyls retain a relatively short vertebral column, with 25 to 32 presacral vertebrae. Most were probably semi-terrestrial. However some families diverged toward either a secondarily fully aquatic or a primarily terrestrial way of life to judge by their limb proportions, presence or absence of lateral line canals, and the nature of the deposits in which they are found. Several families are known primarily from small, perhaps neotenic, species (Boy 1986).

Two lineages diverged early in the evolution of the temnospondyls—the colosteids and *Caerorhachis*. The colosteids, typified by *Greererpeton* (Smithson 1982, Godfrey 1989), were large, long-bodied, fully aquatic forms with very short limbs. They retain a large stapes similar to that of *Acanthostega*.

Caerorhachis (Holmes & Carroll 1977) exhibits the typical temnospondyl open palate and solid attachment of the table and cheek, but has elaborated the pleurocentra as large ventral crescents, as in primitive anthracosauroids. Both colosteids and *Caerorhachis* have closed the squamosal notch.

The evolution of the temnospondyls has recently been reviewed by A. R. Milner (1990), who recognized 40 families and 160 genera, extending into the Early Cretaceous. He identified two major periods of radiation, one in the mid to late Carboniferous and the second in the Late Permian and Early Triassic. The latter resulted primarily in large, secondarily aquatic forms.

Small, late Paleozoic temnospondyls have been suggested as the probable ancestors of one or more of the modern amphibian groups, but a very wide morphological gap separates the known lineages (Milner 1988).

Anthracosauroids

The anthracosauroids include several groups that although distinct from one another in body proportions and presumed habitats are all united by shared derived features of the skull and vertebrae.

In contrast with previously mentioned groups, the anthracosauroids elaborated the pleurocentra at the expense of the intercentra and in most genera it is the dominant central element. In most genera, the skull table is advanced over that of other early tetrapods in the anterior extension of the tabular so that it reaches the parietal rather than being separated from that bone by the supratemporal. The skull resembles that of *Crassigyrinus* in the contact of the tabular to the otic capsule by two facets, the elongation of the tabular horn, and the absence of the posttemporal fossae. In common with amniotes, there is a retractor pit for the origin of the eye muscles within the basisphenoid, beneath the sella turcica. Primitive features include the retention of a mobile articulation between the skull table and cheek, the retention of the intertemporal, a fully mobile basicranial articulation, and the small size of the interpterygoid vacuities.

The best known anthracosaurs are the embolomeres, recognized from the Late Mississippian (lowermost Namurian) into the Early Permian (Figure 1*D*) (Panchen 1980; Holmes 1984, 1989). Within the history of the group, there is progressive change in the configuration of the ossified portions of both the pleurocentra and intercentra from large ventral crescents to complete rings, and elongation of the trunk from 32 to 38 presacrals. The anatomy of the limbs remains similar, but their size (relative to the total trunk length) is reduced suggesting a shift from a semiaquatic to a fully aquatic way of life. The phalangeal count for the group is consistently 2, 3, 4, 5, 3 in the manus, and 2, 3, 4, 5, 5 in the pes. The genera *Jugosuchus* and *Chroniosuchus* from the Upper Permian of Russia may represent late survivors of this group (Clack, personal communication).

Specimens have recently been described from the Brigantian (upper Visean) of Iowa (Bolt et al 1988) that resemble embolomeres in most cranial features, including the weak attachment of the skull roof and cheek, the presence of a tabular horn, and the muted nature of the dermal sculpturing of the skull, but retain the primitive pattern of the skull table in which the tabular is separated from the parietal by the supratemporal. The pleurocentra are primitive in remaining paired, but they extend ventrally toward the midline to form a larger portion of the vertebra than in most temnospondyls. This pattern (termed schizomerous) was postulated by Romer (1964) to represent an intermediate condition between that of early temnospondyls and typical anthracosauroids.

The gephyrostegids (Boy & Bandel 1973) are small anthracosauroids with well-developed limbs and a short vertebral column. They are known primarily from the Upper Carboniferous, but are possibly represented at East Kirkton in the Visean. Their vertebral centra retain the pattern of primitive embolomeres.

The limnoscelids, from the Upper Pennsylvanian and Lower Permian, retain cranial features of the gephyrostegids and embolomeres, but lose the intertemporal bone and partially consolidate the cheek and skull table. The pleurocentra are cylindrical but the intercentra are small crescents. The limnoscelids share several derived features with early amniotes: closure of squamosal notch, presence of a large supraoccipital bone separating the exoccipitals from the skull roof, and a toothed transverse flange of the pterygoid; palatal fangs are lost, but the marginal teeth retain labyrinthine infolding (Fracasso 1987).

Members of the family Seymouriidae are moderately large terrestrial forms from the Lower Permian of North America and Russia. Their pleurocentra are complete cylinders, and the skull roof is firmly attached to the cheek. The stapes is reduced, as in temnospondyls, suggesting the independent evolution of an impedance matching middle ear. The discosauriscids are structurally similar to the seymourids, but they are known primarily from larval or immature stages from aquatic habitats in Europe and China (Spinar 1952, Zhang et al 1984). Panchen & Smithson (1988) have questioned the relationship of seymouriamorphs to other anthracosauroids.

"LEPOSPONDYLS"

Aïstopods

Each of the lepospondyl groups exhibits some primitive features and others that are highly derived, making it impossible to identify any as generally more primitive or closer to labyrinthodonts than any others. The most striking are the aïstopods (Figures 5, 7). All completely lack limbs, although elements of the shoulder girdle may be retained. One genus *Lethiscus*, is known from the mid-Visean (Wellstead 1982). Two families, the Ophiderpetontidae and the Phlegethontiidae, are common and widespread in the Upper Carboniferous (McGinnis 1967). The skulls are very different but highly derived in both. In the Ophiderpetontidae, the orbits are far forward and much of the cheek is covered with a mosaic of small plates rather than normal bones. The adductor chamber is enormous, indicating extremely massive jaw muscles. The palatoquadrate is huge, recalling that of rhipidistian fish. *Phlegethontia* retains more normal skull proportions, but the bones of the cheek form a slender latticework surrounding a solidly ossified braincase. The palate and braincase are loosely articulated, but there are no clearly defined basicranial processes of the basisphenoid.

The basioccipital, exoccipitals, and otic capsule, which appear as distinct areas of ossification in other early tetrapods, are fused to form a single ossification. The otic capsule extends to the cheek. The occiput is recessed beneath the foramen magnum, indicating that the notochord extended at least a short distance foward. The vertebrae consist of elongate centra solidly fused to the neural arches. The first cervical resembles the more posterior trunk vertebrae. The centrum is deeply recessed at both ends. Contact is made between the margins of the centrum and the edge of the conical recess in the occiput. An accessory arch fits between the skull and the first cervical. The neural arches of the anterior cervicals surround openings for the spinal nerves, which pass between the arches in all labyrinthodonts.

Members of the Ophiderpetontidae and Phlegethontiidae both have extra processes from the heads of the ribs giving them a distinctive **k** shape. In contrast, those of the earlier genus *Lethicus* resemble those of other primitive tetrapods, with relatively narrow heads that attach primarily to the anteriorly placed transverse processes.

Aïstopods are snakelike in their body proportions, with up to 230 vertebrae. The Visean genus *Lethiscus* has 58 vertebrae preserved in articulation, although the body may have been considerably longer. No aïstopods retain limbs. *Lethiscus* has small circular elements behind the skull which might be recognized as remnants of the scapula and/or coracoid, but were identified by Wellstead as rhipidistian scales. Boyd (1982) recognized an interclavicle in *Ophiderpeton*. Several specimens of *Ophiderpeton* retain a cleithrum.

Aïstopods are usually considered to be aquatic since they occur most commonly in the coal-swamp localities of Linton, Ohio and Nyrany, Czechoslovakia, but a few remains have been found at East Kirkton, which lacks fish or obviously aquatic amphibians.

Adelogyrinids

Adelogyrinids are restricted to a short span of time in the Visean and Namurian and are known only from a small area near Edinburgh, Scotland (Andrews & Carroll 1991). Like the aïstopods, they have lost their limbs, but the dermal shoulder girdle is retained in its entirety. Like the ophiderpetontids, the orbits are anterior in position; the cheek is fully covered, but with a smaller compliment of individual bones (Figures 6*A*, 7*G*). Only a single bone (termed the tabular-squamosal) occupies the area that in other early tetrapods is formed by the intertemporal, supratemporal, squamosal, and tabular. In some specimens the postorbital is lost as well. Lappets of the postparietals descend to reach the exoccipitals. The braincase may also have been supported by contact between the tabular-squamosal and the otic capsule. The basioccipital is recessed to receive the convex anterior surface of the first cervical centrum. A few Carboniferous labyrinthodonts show traces of the hyoid apparatus, but it is extremely well ossified in all mature adelogyrinids.

The trunk centra are heavily ossified cylinders. There is no trace of intercentra but one isolated caudal centrum that may be assigned to this group has articulation surfaces for a haemal arch. In contrast with aïstopods and nectrideans, the arches are only weakly articulated with the centra. In contrast with all other "lepospondyls," the transverse processes are long and extend from the neural arches as in temnospondyls and anthracosaurs.

Although no specimen is preserved in its entirety, there exists one with at least 70 trunk vertebrae in sequence, without a trace of pelvic girdle or rear limb.

Nectrideans

Nectrideans are first recognized in the early Westphalian (Bossy 1976, Milner 1980, Thayer 1985). The earliest adequately known skeleton belongs to the species *Urochordylus wandesfordii* from the Jarrow locality in Ireland. The highly distinctive caudal vertebrae common to this group have been recognized but not yet described from East Kirkton (A. R. Milner, personal communication). This group is clearly aquatic in basic adaptation, with a very long, laterally compressed tail. The trunk is short, and the limbs relatively long, but slender (Figure 6*B*). Panchen & Smithson (1988) classified the nectrideans close to microsaurs and temnospondyls because of the presence of only four toes in the manus, but the most primitive known species is restored as having five (Bossy 1976).

Nectrideans are immediately distinguished from all other early tetrapods by the fusion of the caudal and haemal arches to the caudal centra to form a symmetrical structure (Figure 4*O*). Accessory articulations above the zygapophyses further link the vertebrae to one another, as in aïstopods. The arches and centra are fused without a trace of suture and openings for the spinal nerves are seen in some arches.

The nature of attachment of the skull table and braincase is only known in a few genera. In *Ptyonius*, the opisthotic extends to the skull roof, but in *Diploceraspis*, the exoccipitals are in articulation with the postparietals. Only a single bone is lost from the skull table in the most primitive genera, the intertemporal. This is the most primitive condition seen among lepospondyls. On the other hand, the skull table resembles that of advanced anthracosauroids in the contact between the tabular and the parietal. The line of mobility between the cheek and the table extends anteriorly to the orbit, with a peg and socket joint between the postorbital and the jugal. Mobility is also evident between the prefrontal and the lacrimal, as in *Eusthenopteron.* It is not possible to establish whether this represents a retention of a primitive pattern or a re-elaboration, based on retention of similar developmental processes.

The palate in early nectrideans has small interpterygoid vacuities and a clearly defined movable articulation with the base of the braincase. In later nectrideans, the interpterygoid vacuities become larger, and in the Keraterpetonidae the ptergyoid forms a sutural articulation with the parasphenoid.

Articulation between the skull and the first cervical vertebra is made via paired condyles formed by the exoccipitals, without contribution from the basioccipital.

Microsaurs

Microsaurs are an extremely heterogeneous assemblage, with more than 25 species in 12 families and 2 suborders (Carroll & Gaskill 1978). The earliest known genus, *Utaherpeton* (Figure 6*C*), is small, with a short vertebral column and limbs with humeri and femora much longer than the distal elements. The trunk centra are cylindrical, but the tail retains separate haemal arches and pleurocentra. Within this group are a variety of body forms—some resembling lizards, others salamanders, and several that are long bodied with short limbs that may have been burrowers.

Microsaurs are distinguished from all other Paleozoic amphibians by the particular pattern of the bones of the temporal region in which a single bone, called the tabular, occupies the position of the intertemporal, supratemporal, and tabular of other early tetrapods. The back of the braincase is supported by contact between the exoccipitals and the postparietals as well as between the otic capsule and the tabular. An extra bone, the supraoccipital, is frequently exposed on the occipital surface. Some microsaurs retain a posttemporal fossa. The basicranial articulation is less mobile than that of anthracosauroids. The occipital condyle is a broad strap-shaped structure formed jointly by the exoccipitals and basioccipital.

Lysorophia

Lyrorophids have long been allied with microsaurs but differ in having much longer bodies—with up to 100 presacral vertebrae, very small limbs, and a much different pattern of the skull (Figures 5*B*; 7*H–K*) (Wellstead 1991). The orbital region is widely open, although with a different pattern of surrounding bones than in aïstopods, and the jaw articulation is at the end of a narrow, anteriorly sloping suspensorium formed by the squamosal and tabular. Anteriorly, the tabular meets the parietal. The maxilla and premaxilla are only weakly attached to the other bones of the skull. The occipital condyle is narrow, but made up of the exoccipitals and basioccipital as in microsaurs. The exoccipitals reach the postparietal, but a large supraoccipital forms the dorsal portion of the occiput.

As in adelogyrinids, the hyoid apparatus is well ossified, but the configuration of the individual elements is much different. The basihyal is fully ossified.

There are no trunk intercentra, but there are haemal arches in the tail. Throughout the column, the neural arches are paired and suturally attached to the centra.

Acherontiscus

Finally, a single specimen has been assigned to its own family, Acherontiscidae (Carroll 1969b) (Figure 5*C*). The skull is poorly known, but it is small relative to the length of the trunk and has a small number of large teeth as in some microsaurs. The orbits are far forward. The nature of attachment of the braincase to the dermal skull, the pattern of articulation between the skull and the trunk, and the structure of the palate are unknown. Elements of the shoulder girdle are preserved, but neither front nor hind limbs are identified with certainty. There are at least 27 trunk vertebrae; more posteriorly, haemal arches are attached to cylindrical caudal centra. *Acherontiscus* is distinguished from other "lepospondyls" by the presence of two cylindrical centra per segment. Neither is firmly attached to the neural arch.

RELATIONSHIPS

Approximately 23 million years separate the fairly well known genera from the Famennian of East Greenland and these of the late Visean of Scotland. Eight lineages appear in the latter half of the Visean, within a period of less than 10 million years, and another three within the first half of the Upper Carboniferous. The relative time of appearance certainly does not correctly reflect the time of differentiation of these groups. The aïstopods are among the most derived of all tetrapods in the loss of girdles and limbs and the modification of the skull, and yet they are the first to appear in the Carboniferous. What the record does indicate is that all the groups are very clearly distinct from one another when they first appear and they may have diverged well down in the Lower Carboniferous or even the Upper Devonian.

Many papers have been written recently on possible interrelationships among Paleozoic tetrapods (Gardiner 1983, Milner et al 1986, Panchen & Smithson 1988, Canatella & de Quieroz 1989, Trueb & Cloutier 1991). Most have expressed dissatisfaction with the previously accepted division into two major groups, the labyrinthodonts and lepospondyls, but no alternative classification is generally accepted.

Carroll & Currie (1991) have suggested a series of procedures for determining relationships in groups with a significant fossil record:

1. Establish the monophyly of the group in question.
2. Establish the polarity of all characters that vary within the group.
3. Recognize subgroups on the basis of unique apomorphies.
4. Determine derived character-states present in the most primitive members of each monophyletic subgroup.
5. Determine possible interrelationships on the basis of derived characters shared by the most primitive known members of each subgroup.

(The term monophyly is used here in the sense of having a common ancestor, rather than in the sense of including all their descendants.)

The features used by Panchen & Smithson (1987) to establish the monophyly of all tetrapods serve to establish the common origin of early land vertebrates. The polarity of most character transformations can be established on the basis of out-group comparison with osteolepiform rhipidistians. The major subgroups are readily distinguished by autapormophies. Unfortunately, it is much more difficult to establish which derived character-states were present in the most primitive members of each monophyletic subgroup because most are so distinct when they first appear in the fossil record. There is a minimum of 25 million years between the divergence of the Devonian genera and the appearance of the major Carboniferous lineages. There is no certain way of knowing which of the derived characters that are present in more than one lineage may have evolved in common with a recognizable sister-group, and which evolved convergently. Where the distribution of derived characters shows extensive congruence, as *within* the particular subgroups, we can be confident of their common ancestry, but this is not the case for relationships *between* the major groups. No matter what relationships are hypothesized, there are a large number of character incongruencies. The same problems were encountered in attempting to establish a phylogeny of the early diapsids (Carroll & Currie 1991).

What appears as the incongruency of characters results from the difficulty of establishing their strict homology in different lineages. The magnitude of this problem differs greatly between the groups termed labyrinthodonts and those designated as lepospondyls. Among the labyrinthodonts, most skeletal characters can be directly compared with those of rhipidistians. All early labyrinthodonts are relatively consistent in their body size, the pattern of the skull bones, and the proportions of the skull, trunk, and limbs, as a result of which the homology of most characters can be readily accepted.

In contrast, the orders included within the lepospondyls differ greatly from all early labyrinthodonts in their small size, and differ greatly among themselves in the pattern of the skull and the proportions of the skull, trunk, and limbs. For this reason, it is difficult to establish the homology of the similar characteristics that are expressed in different groups. It is uncertain whether even the derived traits shared by all lepospondyls are truly homologous.

"Labyrinthodonts"

One way to approach the problem of the relationships among the early tetrapods is to deal separately with labyrinthodonts and lepospondyls. No matter how the lineages grouped as lepospondyls might have diverged, the nature of their relationships to the labyrinthodonts would not effect the relationships of the labyrinthodonts among themselves.

Many modern systematists argue that phylogenetic analysis should be centered on groups with living descendants. Panchen & Smithson (1988) attempted to establish a phylogeny of Paleozoic tetrapods based on the assumption that most lineages may be assignable either to a monophyletic assemblage including the amniotes, or a monophyophyletic assemblage including the modern amphibian orders. On the other hand, it is equally possible that there existed a common lineage of tetrapods prior to this dichotomy. In fact, a major radiation of Paleozoic tetrapods may have occurred before the dichotomy leading to the modern groups.

Current knowledge of the Devonian labyrinthodonts strongly suggests that at least these genera belong to monophyletic lineages that are distinct from those including either the anthracosaurs or temnospondyls, which are prime candidates for being the sister-groups of the living tetrapods. Either the loxommatids and/or the crassigyrinids may also have diverged from the primitive tetrapod lineage prior to the dichotomy between anthracosaurs and temnospondyls.

One of the most difficult problems in establishing the relationships of the crassigyrinids and loxommatids is the nature of their vertebral centra. The presence of only the intercentra might have resulted from the retention of a primitive, rhipidistian condition. In which case, these groups would have been separate from all other tetrapods since the Devonian. Alternatively, this condition might have resulted from the loss of pleurocentra from the vertebral pattern that is present in all other early labyrinthodonts.

If *Crassigyrinius* diverged from a common ancestry with other recognized labyrinthodonts, the most plausible relationship would be with anthracosauroids with which they share the derived configuration of the tabular and its articulation with the otic capsule. This relationship is, however, contradicted by a number of other derived features that are shared by loxommatoids, anthracosaurs, and temnospondyls to the exclusion of *Crassigyrinus*. These include loss of the anterior tectal, preopercular, and all openings in the humerus other than the entepicondylar foramen, and greater ossification of the basioccipital.

Loxommatoids share the solid attachment of the skull table and cheek with temnospondyls, and the strong articulation between the otic capsule and tabular with the anthracosauroids. On the other hand, they must have diverged prior to the appearance of any *known* temnospondyls since they have not evolved large interpterygoid vacuities or reduced the mobility of the basicranial articulation. They must also have diverged from the lineage leading to known anthracosauroids prior to the closure of the posttemporal fossae in that group.

The point of common ancestry between the temnospondyls and anthracosauroids must have occurred prior to the origin of the autapomorphies that characterize either group. Their common ancestors must have retained the primitive mobility between the skull table and cheek, a closed palate, full mobility between the palate and braincase, and the five-toed manus that characterize the anthracosaurs. In addition they must have lacked a solid articulation of the tabular and otic capsule and long tabular horns, since these characters are not present in early temnospondyls. In aodition, the posttemporal fossae must have been retained.

Whatever relationships there were among the Carboniferous labyrinthodonts, their common ancestry must have been at a level when most of their character-states were plesiomorphic. At present, there are no traits that can be securely identified as being uniquely derived which can be used to establish a series of nested dichotomies demonstrating the phylogeny of the Carboniferous labyrinthodont groups. A number of phylogenies could be established using particular character transformations, but all can be contradicted by other traits. This problem cannot be solved until information is available for members of each of the major lineage from the Tournasian that could demonstrate the derived characters of the most primitive members of these groups.

Despite the common anatomical pattern running throughout the labyrinthodonts that indicates an ultimate common ancestry, the specific pattern of divergence of the various groups remains difficult to establish.

"Lepospondyls"

The lepospondyls are a much more diverse and varied assemblage than the labyrinthodonts, making establishment of strict homologies of derived characters between the various groups even more difficult.

There are at least three distinct patterns of relationship that the lepospondyls might exhibit relative to the labyrinthodonts. 1. All lepospondyls might have evolved from a common ancestor which diverged from a single lineage of more primitive early tetrapods. 2. The various lepospondyl lineages might have evolved separately from among the recognized lineages of labyrinthodonts. 3. The various lepospondyl lineages might have evolved separately from among a number of early tetrapod lineages of which we currently have no direct knowledge. The latter case would only appear plausible if the other alternatives were shown to be unlikely.

The hypothesis that the various lepospondyl orders share a unique common ancestry can be tested by evaluating the probable homology of the derived characters which they appear to share in common. These include the following:

- skull lacking squamosal notch;
- absence of large palatal fangs with adjacent replacement pits;
- absence of labyrinthine infolding of dentine;
- loss of one or more of the dermal bones of the skull roof that characterize primitive tetrapods;
- spool-shaped vertebral centra that occupy most if not all of each segment;
- a single unit forming the bony articulation between the trunk and skull.

If all these features could be demonstrated as being homologous, in the sense of having arisen from a common ancestor having this pattern, there would be strong support for the monophyly of this assemblage.

One of the most important factors to consider in establishing the relationships of the lepospondyl groups is their small size relative to that of early labyrinthodonts. From the Upper Devonian through the Lower Carboniferous, there are few, if any, adult labyrinthodonts with a skull length of less than 10 cm. The skulls of Carboniferous lepospondyls are rarely more than one third this size. For a variety of physiological reasons, the brain and major sense organs have a minimum size, below which they cannot function (Hanken 1984). For this reason, the braincase and the sensory capsules dominate the skull in small vertebrates. The effect on skulls less than about 2 cm in length is dramatic (Carroll 1990), but this factor may have influenced skull development and ossification in all lepospondyls.

If all lepospondyl groups had evolved from a single common lineage that had reached small size prior to the divergence of the recognized orders, it would seem probable that the general cranial anatomy and the pattern of the bones of the skull roof would show similar patterns of modification from that of primitive labyrinthodonts. This does not appear to be the case.

Early nectrideans and ophiderpetonid aïstopods, in which the jaw suspension extends progressively behind the level of the occipital condyle during growth, retain the common labyrinthodont pattern. Microsaurs and adelogyrinids primitively have the jaw suspension at the same level as the occiput. In lysorophids, the jaw suspension is far forward. Nectrideans, microsaurs, and adelogyrinids primitively retain a complete covering over the cheek. Large fenestrae are present in all aïstopods and lysorophids, but the pattern of bone loss and the proportions of the skull are much different in these groups.

Comparing relationships between the braincase and surrounding bones of the skull table, cheek, and palate between labyrinthodonts and lepospondyls is difficult because of the relatively much larger brain and sense organs of the lepospondyls. In primitive tetrapods such as the ichthyostegids, temnospondyls, and anthracosaurs, the brain and otic capsule occupy a relatively small proportion of the volume of the skull, and must be attached to the dermal shell of the skull by processes from either the braincase or the surrounding bones. In small lepospondyls, the braincase occupied most of the height of the skull between the palate and the skull roof, and the otic capsules extend nearly to the cheeks. The tabular, postparietal and sometimes a supraoccipital contribute to connecting the dermal skull with the braincase. Specific comparison cannot be made with any of the particular patterns of attachment seen among the labyrinthodonts. No lepospondyl retains the stapes as a large supporting unit of the pattern seen in *Acanthostega* and early anthracosaurs and temnospondyls. The retention of posttemporal fossae in several lepospondyl groups implies derivation from primitive labyrinthodonts or temnospondyls.

Loss of labyrinthine infolding of the teeth, palatal fangs, and the squamosal notch may all be attributed to small size. All of these characters appear late in development in the larger labyrinthodonts for which long growth series are known. For this reason, these features may be the result of convergence, rather than common ancestry, and cannot, of themselves, be taken as evidence of the monophyly of the assemblage. The small size of the skull may also be related to the progressive loss of skull bones, which also occurs among especially small lepospondyls (Carroll 1990).

Many aspects of the skull of lepospondyls can be attributed to small size, but few features support a single common origin or relationships to specific labyrinthodont groups.

The nature of the vertebrae in lepospondyls is more difficult to explain. There may also be some connection between small size and the evolution of spool-shaped centra, but this is not absolute since some very small labyrinthodonts retain multipartite centra and the caudal vertebrae (which are smaller than those in the trunk) retain multipartite centra in some lepospondyls, including the oldest known microsaur, in which the trunk centra are cylindrical (Carroll et al 1991).

Carroll (1989) has argued that vertebral centra among the lepospondyls, like those of living amphibians, amniotes, and teleosts, develop through ossification of tissue formed in the perichordal sheath of the notochord, in contrast with the central elements of labyrinthodonts that may have developed by ossification of chondrifications within the notochordal sheath, as in the living actinopterygian fish *Amia* and *Lepisosteus*. While the specific selection pressure for the formation of cylindrical centra from the perichordal sheath has not been established, the fact that this process has occurred separately in the divergent ancestors of teleosts, living amphibians, and amniotes, raises the possibility that this process may have occurred more than once in primitive members of different lepospondyl groups.

The homology of cylindrical centra among the lepospondyls would be supported by the discovery of close similarities in vertebral structure in all members of this assemblage. Conversely, if the vertebrae in each group were significantly different from each other, despite the broadly similar cylindrical shape of the centra, it opens the possibility that each evolved separately.

The vertebrae of *Acherontiscus* (Figure 4*R*) are fundamentally distinct from those of all other lepospondyls by the presence of two cylindrical centra of nearly equal size. They superficially resemble those of embolomeres, although the skull is very different in its small size and presence of a few large teeth. The presence of two large centra per segment suggests that they were not formed within the perichordal sheath, in contrast with all other genera included within the lepospondyls. Not enough is known of the skull or other elements of the skeleton for close comparison, but there is no strong evidence to ally this genus with any known group of Paleozoic tetrapods.

Leaving *Acherontiscus* aside, there are two patterns of the centra among other lepospondyls. Aïstopods and nectrideans never have more than a single central element per segment in either the trunk or tail. Lysorophids have haemal arches in the tail, but lack intercentra in the trunk. Several microsaurs are also known to have haemal arches, and some have intercentra in the trunk as well. The oldest known microsaur, *Utaherpeton* (Carroll et al 1991), definitely lacks trunk intercentra but haemal arches are present in the tail, together with cylindrical pleurocentra. The articulated specimens of adelogyrinids have no trunk intercentra, but none show the tail.

It might be hypothesized that there existed an evolutionary continuum, with microsaurs illustrating a primitive condition, in which the trunk centra were the first to achieve a fully lepospondylous pattern, followed by the tail. This is conceivable, but it is not confirmed by other aspects of the vertebrae. For example, the nature of articulation between the head and the first cervical is strikingly different between aïstopods and other adequately known lepospondyls. The first cervical centra of aïstopods is very similar to those more posterior in position. It is a simple cylinder, deeply concave at both ends. The occipital condyle is recessed in an equivalent manner, so that the notochord inserted into it in the same way it inserted between the centra in the trunk. The anterior margin of the first cervical in adelogyrinids is not expanded, but appears to form a convex surface, fitting into a shallow recess in the basioccipital. In microsaurs, lysorophids, and nectrideans, the anterior surface of the first cervical centrum is laterally expanded so that it makes contact with the two exoccipitals. This reflects an alternative functional pattern, in which the great width of the articulating surface of the first cervical limits the skull to movement in the vertical plane. The articulation in aïstopods would permit movement in all directions.

There are also significant differences in the nature of the neural arches among the lepospondyl groups. In aïstopods and nectrideans, the arches are fused to the centra throughout the column. In some regions of the column the spinal nerves exit though foramina in the arches, rather than between them, as is the case in most other vertebrates (the arches of some salamanders also have foramina for the spinal nerves). Both nectrideans and aïstopods also have excessory articulating surfaces on the neural spines. They differ radically in the caudal region, for nectrideans fuse the haemal arches to the centra, and aïstopods have no haemal arches.

The neural arches of lysorophids are paired rather than single, in contrast with all early tetrapods except *Crassigyrinus*. On the other hand, the two sides are very securely attached to the centra by complex sutures. There would clearly not have been any mobility between the two sides, as was presumably the case in rhipidistians and *Crassigyrinus*. The neural arches in microsaurs are always fused at the midline, but they may be either fused or suturally attached to the centra. Adelogyrinid arches are only weakly articulated with the centra, but they are fused at the midline, even in very small individuals. Lack of a common pattern of the neural arches in lepospondyls does not preclude their having evolved from a common ancestor, but it makes it difficult to reconstruct the nature of such an ancestor.

The extreme variation in other aspects of vertebral structure makes it necessary to question the common presence of cylindrical centra in lepospondyls as a strong feature supporting the monophyletic origin of this assemblage.

The evidence now available does not support the assumption that the lepospondyl orders shared a unique common ancestry. Can it be demonstrated that any of these groups share a unique ancestry with any known labyrinthodont lineage? Panchen & Smithson (1988) identified one or more derived characters that might unite particular lepospondyl groups to particular labyrinthodonts. Nectrideans were united with microsaurs and temnospondyls on the basis of the common reduction to only four front toes. In fact, the oldest nectridean has five—a primitive condition for Carboniferous tetrapods (Bossy 1976). They united microsaurs with temnospondyls on the basis of the common presence of a link between the postparietals and the exoccipitals. Since some microsaurs also have a connection between the otic capsule and the tabular (an anthracosauroid character) this feature does not carry much weight. They also cite similarity in the shape of the occipital condyle, but this is refuted by the entirely different configuration of the atlas in the two groups.

Some derived characters can be recognized in each of the lepospondyl orders that are also observed in one or other of the labyrinthodont groups. On the other hand, if we add up all the other character changes that would be necessary to transform any known labyrinthodont to any known lepospondyl, that number would far outweigh the number of characters that are suggested as uniting the groups. Phylogenetic analysis could yield a series of variously parsimonious cladograms depicting the relationships of the known labyrinthodonts and lepospondyls. However there is no way to establish whether or not any such relationships are more parsimonious than would be relationships with groups from the Lower Carboniferous for which we do not yet have fossil evidence. With a 23 million year gap, evidence of relationship is much more likely to lie within this period than with the fossils that we now know from later in the Carboniferous. The evidence now available suggests that all the lepospondyl groups diverged late in the Devonian or early in the Carboniferous. None show any obvious affinities with each other or with any of the known labyrinthodont groups.

THE ANCESTRY OF MODERN TETRAPODS

Modern Amphibians

Although the majority of Paleozoic tetrapods are considered to be amphibians, their specific affinities with frogs, salamanders, and caecilians are still not well established. There is a clear distinction between the radiation of tetrapods that began in the Upper Devonian and that which gave rise to the modern orders. The groups termed lepospondyls are not known beyond the Lower Permian. The labyrinthodonts underwent a further radiation in the Late Permian and Triassic, with the last known members in the Cretaceous. Most of the late labyrinthodonts were large and aquatic, and show no significant similarities with modern amphibians.

Anurans (Estes & Reig 1973), urodeles (Ivachnenko 1978), and gymnophonians (Jenkins & Walsh 1990) all appear in the fossil record in the Jurassic. The only other fossil that might pertain specifically to one of the modern orders is *Triadobatrachus* from the Lower Triassic of Madagascar which Rage & Rocek (1989) suggest as the sister-group of frogs.

Most recent reviews of the living orders assume that they have a common ancestry distinct from that of all other tetrapod groups, and most authors assume that this ancestry lies among the small labyrinthodonts of the Late Carboniferous or Early Permian—specifically the dissorophoid temnospondyls. Whatever their relationships, a large morphological gap separates the living orders from all Paleozoic groups. A. R. Milner (1988) recognizes numerous osteological characters that unite the modern amphibian orders: pedicellate teeth; bicuspid teeth; large interpterygoid vacuities partly bordered by palatines (when not absent or fused with pterygoids); absence of supratemporal, jugal, tabular, postparietal and no more than one coronoid; paired occipital condyles; and very abbreviated and straight ribs, not extending into the flank muscle. Of these, only the presence of bicuspid teeth and large interpterygoid vacuities unite them with the dissorophoids. Nearly all the osteological characters of the modern amphibian orders evolved subsequent to their divergence from Paleozoic amphibians.

Alternative schemes of relationship are those that associate caecilians (Carroll & Currie 1975, Walsh 1987) and salamanders (Carroll & Holmes 1980) with microsaurs. These alternatives cannot be tested without additional fossils uniting fully differentiated frogs, caecilians, and salamanders with one or the other groups of Paleozoic amphibians.

A. R. Milner (1988) emphasized the importance of small size in the origin of many of the cranial features of modern amphibians, in which the braincase and sensory organs dominate the skull. Relatively large orbits and palatal vacuities are features in common between frogs and salamanders and both small temnospondyls such as dissorophids, and lepospondyls, especially microsaurs. Dissorophids apparently share with frogs an impedance matching middle ear, to judge by the large otic notch and narrow stapes. Salamanders lack an impedance matching mechanism, but it may have been lost from a condition like that of dissorophids or have been primitively absent, as in the lepospondyl groups. The different patterns of the jaw musculature in frogs and salamanders suggest that the fenestration of the temporal region characteristic of those groups evolved separately in each.

The caecilian or gymnophonians—long bodied, limbless amphibians of the tropics—have long been assumed to be closely related to frogs and salamanders. Their relationship has been particularly difficult to establish because of the absence of a significant fossil record. Estes (1981) and Rage (1986) described isolated vertebrae that demonstrate the presence of caecilians in the Paleocene and Upper Cretaceous. More recently, Jenkins & Walsh (1990) described numerous, nearly complete skeletons of caecilians from the Lower Jurassic. The skull resembles that of modern genera in the presence of an opening for a tentacle and the small size of the orbits, but—in contrast with any living amphibians—all the circumorbital bones common to Paleozoic amphibians are retained. The lower jaw resembles that of modern caecilians in being composed of only two bones and having very long retroarticular processes. Small, but well formed limbs are retained. These specimens give no suggestion of affinities with frogs or salamanders (except for the presence of pedicellate teeth), but the general pattern of the skull roof and palate are very similar to those of the goniorhynchid microsaurs.

Amniotes

All other modern tetrapods belong to a single monophyletic assemblage, the Amniota, represented by reptiles, mammals, and birds. In contrast with amphibians, the ancestors of the major amniote groups are clearly recognizable in the late Paleozoic. The ancestors of mammals are represented by the pelycosaurs, which were the most common and diverse amniotes of the uppermost Pennsylvanian and Permian (Reisz 1986). The ancestors of lizards, *Sphenodon*, crocodiles, dinosaurs, and birds are represented as early as the Upper Pennsylvanian by primitive diapsids such as *Petrolacosaurus* (Reisz 1981). Two possible sister-groups for the turtles have been proposed in recent years: the Captorhinidae (Gaffney & McKenna 1979) and the Procolophonoids (Reisz & Laurin 1991). Both groups are known in the Permian.

All of these groups may trace their ancestry to the protorothyrids of the Carboniferous and Lower Permian (Carroll 1964, 1969a, 1982, 1991). This group is best known from the genera *Hylonomus* (Figure 8) and *Paleothyris* from the Upper Carboniferous, but a more primitive genus, *Westlothiana*, has recently been described from the Lower Carboniferous of Scotland (Smithson 1989, Smithson & Rolfe 1990). The Upper Carboniferous genera are clearly distinct from all other groups of Paleozoic tetrapods in the

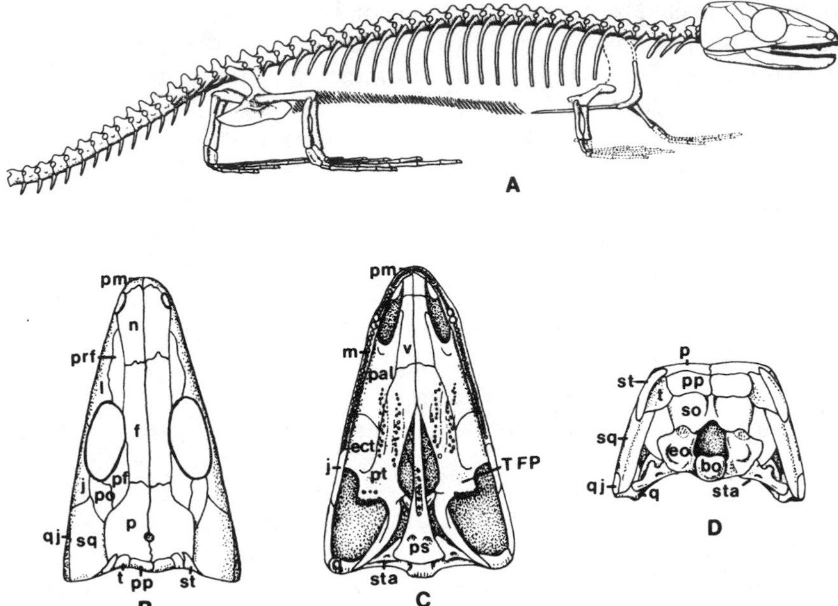

Figure 8 Carboniferous reptiles. A. Skeleton of the protorothyrid *Hyonomus*. B–D. Skull of *Paleothyris*, in dorsal, palatal and occipital views (modified from Carroll 1982). Abbreviations as in Figure 3.

possession of derived features common to later amniotes, including *Sphenodon* and primitive living lizards. These include:

1. loss of the intertemporal, while small tabular, supratemporal, and postparietal are retained;
2. large supraoccipital uniting the back of the skull table and the dorsal margin of the braincase;
3. small convex occipital condyle;
4. toothed transverse flange of pterygoid;
5. fusion of atlas arch and centrum;
6. cylindrical pleurocentra, small crescentic intercentra;
7. fusion of tibiale, proximal centrale, and intermedium of tarsus to form an astragalus.

The general configuration of the atlas-axis complex, the presence of a fully mobile basicranial articulation, and a retractor pit are features shared with anthracosauroids. A transverse flange of the pterygoid and loss of palatal fangs are shared specifically with limnoscelids. Early amniotes were long compared with late Pennsylvanian and early Permian tetrapods such as *Seymouria* and *Limnoscelis.* The discovery of amniotes from the middle Pennsylvanian made close comparison with these later forms of dubious significance, and attention shifted to Middle Pennsylvanian gephyrostegids. The recent discovery of a primitive amniote from the Visean, contemporary with the oldest known anthracosauroids and temnospondyls, leaves no Carboniferous amphibians, other than the oldest aïstopod and some adelogrinids, earlier than the first amniotes. A sister-group relationship with earlier anthracosauroids still seems more plausible than with any other known group of early tetrapod, but it is now impossible to talk about any actual amniote ancestors within the known fossil record. Clearly, the ancestors of this group diverged very early in the Carboniferous, or conceivably even in the Devonian.

As in the case of lepospondyls and the lissamphibia, many of the cranial features of early amniotes may be derived from labyrinthodonts as a result of size reduction.

CONCLUSIONS

Amphibians are known as early as the Upper Devonian, but the gap of nearly 25 million years in the fossil record of tetrapods at the beginning of the Carboniferous makes it very difficult to establish the nature of their initial radiation. Several morphological trends can be seen in most early tetrapod groups: consolidation of the skull, evolution of a bony articulation between the skull and cervical vertebrae, and consolidation of vertebral elements. Secondary aquatic adaptation occurred in many lineages. Other groups become long-bodied and reduced their limbs. Conservative groups that retain many features of their fish ancestors are termed labyrinthodonts. Most belong to two major groups, the temnospondyls and anthracosauroids. Their specific relationships to the poorly known loxommatids and rare crassigyrinids remain speculative. Several more derived groups, collectively termed lepospondyls, apparently diverged from near the base of the tetrapod radiation. Their specific affinities remain unknown. The earliest amniote fossils are known in the Lower Carboniferous, contemporary with the earliest anthracosauroids and temnospondyls. They show more derived features in common with anthracosauroids than with any other Carboniferous group. In contrast, the modern amphibian orders are not known as fossils until the Jurassic. Frogs may be linked to Permo-Carboniferous temnospondyl amphibians via the Lower Triassic genus *Triadobatrachus.* Caecilians may have evolved from long-bodied microsaurs. Salamanders have been hypothesized as having arisen from either microsaurs or temnopsondyls. No single group of Paleozoic amphibians can be shown to include the common ancestors of all of the modern amphibian orders.

ACKNOWLEDGMENTS

I thank L. Trueb and R. Cloutier for allowing me to read their manuscript in press, and R. Holmes, T. Smithson, A. Panchen, and J. Clack for comments on an earlier draft of this manuscript. Dr. Clack was very generous in providing much unpublished information on the anatomy of *Acanthostega* and *Ichthyostega*. The illustrations have been drafted by P. Gaskill. The geological time scale was drafted by R. Lamarche. This research has been supported by grants from the Natural Sciences and Engineering Research Council of Canada.

Literature Cited

Andrews, S. M., Carroll, R. L. 1991. The Order Adelospondyli: Carboniferous lepospondyl amphibians. *Trans. R. Soc. Edinburgh.* In press

Andrews, S. M., Westoll, T. S. 1970. The postcranial skeleton of *Eusthenopteron foordi* Whiteaves. *Trans. R. Soc. Edinburgh* 68: 207–329

Ax, P. 1987. *The Phylogenetic System.* New York: Wiley. 340 pp.

Beaumont, E. H. 1977. Cranial morphology of the Loxommatidae (Amphibia: Labyrinthodontia). *Philos. Trans. R. Soc. London Ser. B* 280: 29–101

Bolt, J. R., McKay, R. M., Witzke, B. J., McAdams, M. P. 1988. *Nature* 333: 768–70

Bossy, K. V. 1976. *Morphology, Paleoecology, and Evolutionary Relationships of the Pennsylvanian Urocordylid Nectrideans (Subclass Lepospondyli, Class Amphibia).* PhD thesis. Yale University. 370 pp.

Boy, J. 1986. Studien über die Branchiosauridae (Amphibia: Temnospondyli) 1. Neue und wenig bekannte Arten aus dem mitteleuropäischen Rotliegenden (?oberstes Karbon bis unteres Perm). *Paläontol. Z.* 60: 131–66

Boy, J. A., Bandel, K. 1973. *Bruktererpeton fiebigi* n. gen. n. sp. (Amphibia: Gephyrostegida) der erste Tetrapode aus dem rheinisch-westfaelischen Karbon (Namur B; W. Deutschland). *Palaeontographica, A* 145: 39–77

Boyd, M. J. 1980. A lysorophid amphibian from the coal measures of northern England. *Palaeontology* 23: 925–29

Boyd, M. J. 1982. Morphology and relationships of the Upper Carboniferous aïstopod amphibian *Ophiderpeton nanum. Palaeontology* 25: 209–14

Cannatella, D. C., de Queiroz, K. 1989. Tetrapod phylogeny and the origin of modern amphibians. *First World Congress of Herpetology*, Univ. Kent, Canterbury, UK (Abstr.)

Carroll, R. L. 1964. The earliest reptiles. *Zool. J. Linn. Soc. London* 45: 61–83

Carroll, R. L. 1969a. A Middle Pennsylvanian captorhinomorph and the interrelationships of primitive reptiles. *J. Paleontol.* 43: 151–70

Carroll, R. L. 1969b. A new family of Carboniferous amphibians. *Palaeontology* 12: 537–48

Carroll, R. L. 1980. The hyomandibular as a supporting element in the skull of primitive tetrapods. In *The Terrestrial Environment and the Origin of Land Vertebrates*, ed. A. L. Panchen, pp. 293–317. Systematics Association Special Volume No. 15. London/New York: Academic. 633 pp.

Carroll, R. L. 1982. Early evolution of reptiles. *Annu. Rev. Ecol. Syst.* 13: 87–109

Carroll, R. L. 1987. *Vertebrate Paleontology and Evolution.* New York. W. H. Freeman. 698 pp.

Carroll, R. L. 1989. Developmental aspects of lepospondyl vertebrae in Paleozoic tetrapods. *Hist. Biol.* 3: 1–25

Carroll, R. L. 1990. A tiny microsaur from the Lower Permian of Texas: size constraints in Palaeozoic tetrapods. *Palaeontology* 33: 893–909

Carroll, R. L. 1991. The origin of reptiles. In *Origins of the Higher Groups of Tetrapods: Controversy and Consensus*, ed. L. Trueb,

H.-P. Schultze. Ithaca: Cornell Univ. Press. In press

Carroll, R. L., Bybee, P., Tidwell, W. 1991. The Oldest Microsaur (Amphibia). *J. Paleontol.* 65: 314–22

Carroll, R. L., Currie, P. J. 1975. Microsaurs as possible apodan ancestors. *Zool. J. Linn. Soc.* 68: 1–40

Carroll, R. L., Currie, P. J. 1991. The early radiation of diapsid reptiles. In *Origins of the Higher Groups of Tetrapods: Controversy and Consensus*, ed. L. Trueb, H.-P. Schultze, pp. 354–425. Cornell: Cornell Univ. Press. In press

Carroll, R. L., Gaskill, P. 1978. The Order Microsauria. *Mem. Am. Philos. Soc.* 126: 1–211

Carroll, R. L., Holmes, R. 1980. The skull and jaw musculature as guides to the ancestry of salamanders. *Zool. Soc. Linn. Soc.* 68: 1–40

Clack, J. A. 1987. *Pholiderpeton scutigerum* Huxley, an amphibian from the Yorkshire coal measures. *Philos. Trans. R. Soc. London Ser. B* 318: 1–107

Clack, J. A. 1988. New material of the early tetrapod *Acanthostega* from the Upper Devonian of East Greenland. *Palaeontology* 31: 699–724

Clack, J. A. 1989. Discovery of the earliest-known tetrapod stapes. *Nature* 342: 425–30

Coates, M. I., Clack, J. A. 1990. Polydactyly in the earliest known tetrapod limbs. *Nature* 347: 66–69

Coates, M. I., Clack, J. A. 1991. Fish-like gills and breathing in the earliest known tetrapod. *Nature* 352: 234–36

Estes, R. 1981. Gymnophiona, Caudata. In *Handbuch des Palaeoherpetologie*, ed. O. Kuhn, 2: 1–115. Stuttgart/New York: Gustav Fischer

Estes, R., Reig, O. 1973. The early fossil record of frogs: A review of the evidence. In *Evolutionary Biology of the Anurans*, ed. J. Vial, pp. 11–63. Columbia: Univ. Missouri Press. 470 pp.

Fracasso, M. 1987. Braincase of *Limnoscelis paludis* Williston. *Postilla* 201: 1–22

Gaffney, E. S., McKenna, M. C. 1979. A Late Permian captorhinid from Rhodesia. *Am. Mus. Nat. Hist. Novit.* 2688: 1–15

Gardiner, B. G. 1983. Gnathostome vertebrae and the classification of the Amphibia. *Zool. J. Linn. Soc.* 79: 1–59

Godfrey, S. J. 1989. The postcranial skeletal anatomy of the Carboniferous tetrapod *Greererpeton burkemorani* Romer, 1969. *Philos. Trans. R. Soc. London Ser. B* 323: 75–133

Godfrey, S. J., Fiorill, A. R., Carroll, R. L. 1987. A newly discovered skull of the temnospondyl amphibian *Dendrerpeton acadianum* Owen. *Can. J. Earth Sci.* 24: 796–805

Gregory, J. T. 1948. A new limbless vertebrate from the Pennsylvanian of Mazon Creek, Illinois. *Am. J. Sci.* 246: 636–63

Hanken, J. 1984. Miniaturization and its effect on cranial morphology in plethodontid salamanders, genus *Thorius* (Amphibia: Plethodontidae) I. Osteological variation. *Biol. J. Linn. Soc.* 23: 55–75

Harland, W. B., Armstrong, R. L., Cox, A. V., Craig, L. E., Smith, A. G., Smith, D. G. 1989. *A Geologic Time Scale*. Cambridge: Cambridge Univ. Press. 263 pp.

Hennig, W. 1966. *Phylogenetic Systematics*. Urbana: Univ. Illinois Press. 263 pp.

Holmes, R. 1984. The Carboniferous amphibian *Proterogyrinus scheelei* Romer, and the early evolution of tetrapods. *Philos. Trans. R. Soc. London Ser. B* 306: 431—527

Holmes, R. 1989. The skull and axial skeleton of the Lower Permian anthracosauroid amphibian *Archer crassidisca* Cope. *Palaeontographica Abt. A* 207: 161–206

Holmes, R., Carroll, R. L. 1977. A temnospondyl amphibian from the Mississippian of Scotland. *Bull. Mus. Comp. Zool.* 147: 489–511

Ivachnenko, K. F. 1978. Urodelans from the Triassic and Jurassic of Soviet Central Asia. *Paleontol. J.* 12: 362–68

Jarvik, E. 1955. The oldest tetrapods and their forerunners. *Sci. Mon.* 80: 141–54

Jarvik, E. 1980. *Basic Structure and Evolution of Vertebrates*, Vol. 1, 2. London: Academic Press. 575 pp., 337 pp.

Jenkins, F. A., Walsh, D. M. 1990. During the Jurassic caecilians had limbs. *J. Vertebr. Paleontol.* 10: 29A (Abstr.)

Lebedev, A. O. 1984. The first find of a Devonian tetrapod vertebrate in the U.S.S.R. *Dokl. Akad. Nauk SSSR, Palaeontol.* 278: 1470–73

McGinnis, H. J. 1967. The osteology of *Phlegethontia*, a Carboniferous and Permian aïstopod amphibian. *Univ. Calif. Publ. Geol. Sci.* 71: 1–47

Milner, A. C. 1980. A review of the Nectridea (Amphibia). In *The Terrestrial Environment and the Origin of Land Vertebrates*, ed. A. L. Panchen, pp. 377–405, Systematics Assoc. Spec. Vol. No. 15. London/New York: Academic. 633 pp.

Milner, A. C., Lindsay, W. 1989. The relationships of the loxommatoidea, evidence from postcranial remains. *First World Congress of Herpetology*, Univ. Kent, Canterbury, UK (Abstr.)

Milner, A. R. 1988. The relationships and origin of living amphibians. In *The Phylogeny and Classification of the Tetrapods*,

Volume 1: *Amphibia, Reptilia, Birds*, ed. M. J. Benton, pp. 59–102, Systematics Assoc. Spec. Vol. No. 35A. Oxford: Clarendon. 373 pp.

Milner, A. R. 1990. The radiation of temnospondyl amphibians. In *Major Evolutionary Radiations*, ed. P. D. Taylor, G. P. Larwood, pp. 321–49, Systematics Assoc. Spec. Vol. No. 42. Oxford: Clarendon

Milner, A. R., Smithson, T. R., Milner, A. C., Coates, M. I., Rolfe, W. D. I. 1986. The search for early tetrapods. *Mod. Geol.* 10: 1–28

Panchen, A. L. 1980. The origin and relationships of the Anthracosaur Amphibia from the late Palaeozoic. In *The Terrestrial Environment and the origin of Land Vertebrates*, ed. A. L. Panchen, pp. 319–50, Systematics Assoc. Spec. Vol. No. 15. London/New York: Academic. 633 pp.

Panchen, A. L. 1985. On the amphibian *Crassigyrinus scoticus* Watson from the Carboniferous of Scotland. *Philos. Trans. R. Soc. London Ser. B* 309: 505–68

Panchen, A. L., Smithson, T. R. 1987. Character diagnosis, fossils and the origin of tetrapods. *Biol. Rev.* 62: 341–438

Panchen, A. L., Smithson, T. R. 1988. The relationships of the earliest tetrapods. In *The Phylogeny and Classification of the Tetrapods, Volume* 1: *Amphibians, Reptiles, Birds*, ed. M. J. Benton, pp. 1–32, Systematics Assoc. Spec. Vol. No. 35A. Oxford: Clarendon. 373 pp.

Panchen, A. L., Smithson, T. R. 1990. The pelvic girdle and hind limb of *Crassigyrinus scoticus* (Lydekker) from the Scottish Carboniferous and the origin of the tetrapod pelvic skeleton. *Trans. R. Soc. Edinburgh* 81: 31–44

Rage, J.-C. 1986. Le plus ancien Amphibien apode (Gymnophiona) fossile. *C. R. Acad. Sci., Paris, t.* 302, *Serie II* 16: 1033–36

Rage, J.-C., Rocek, Z. 1989. Redescription of *Triadobatrachus massionti* (Piveteau, 1936), an anuran amphibian from the early Triassic. *Paleontographica, Abt. A* 206: 1–16

Reisz, R. R. 1981. A diapsid reptile from the Pennsylvanian of Kansas. *Spec. Pap. Univ. Kansas Mus. Nat. Hist.* 7: 1–74

Reisz, R. R. 1986. Pelycosauria. In *Handbuch der Palaeoherpetologie*, ed, O. Kuhn, P. Wellnhofer, 17A: 1–102. Stuttgart/New York: Gustav Fischer

Reisz, R. R., Laurin, M. 1991. *Owenetta* and the origin of turtles. *Nature* 349: 324–26

Rolfe, W. D. I., Durant, G. P., Fallick, A. E., Hall, A. J., Large, D. J., Scott, A. C., Smithson, T. R., Walkden, G. M. 1990. An early terrestrial biota preserved by Visean volcanicity in Scotland. *Geol. Soc. Am. Spec. Pap.* 244: 13–24

Romer, A. S. 1964. The skeleton of the Lower Carboniferous labyrinthodont *Pholidogaster pisciformis*. *Bull. Mus. Comp. Zool.* 131: 129–59

Schultze, H.-P., Arsenault, M. 1985. The panderichthyid fish *Elpistostege*: A close relative of tetrapods? *Palaeontology* 28: 293–309

Smithson, T. R. 1982. The cranial morphology of *Greererpton burkemorani* Romer (Amphibia: Temnospondyli). *Zool. J. Linn. Soc. London* 76: 29–90

Smithson, T. R. 1989. The earliest known reptile. *Nature* 342: 676–78

Smithson, T. R., Rolfe, W. D. Ian 1990. *Westlothiana* gen. nov.: naming the earliest known reptile. *Scott. J. Geol.* 26: 137–38

Spinar, Z. V. 1952. Revision of some Moravian Discosauriscidae. *Rozpr. Ustred. Ustrava. Geol.* 15: 1–159 (In Czech.)

Thayer, D. W. 1985. New Pennsylvanian lepospondyl amphibians from the Swisshelm Mountains, Arizona. *J. Palaeontol.* 59: 684–700

Thomson, K. S. 1967. Mechanisms of intracranial kinesis in fossil rhipidistian fishes (Crossopterygii) and their relatives. *J. Linn. Soc. London, Zool.* 178: 223–53

Trueb, L., Cloutier, R. 1991. A phylogenetic investigation of the inter- and intrarelationships of the Lissamphibia (Amphibia: Temnospondyli). In *Origins of Tetrapods: Controversy and Consensus*, ed. L. Trueb, H.-P. Schultze, pp. 223–313. Ithaca: Cornell Univ. Press. In press

Walsh, D. M. 1987. *Systematics of the caecilians (Amphibia: Gymnophiona)*. PhD thesis. McGill Univ. 228 pp.

Wellstead, C. F. 1982. A Lower Carboniferous aïstopod amphibian from Scotland. *Palaeontology* 25: 193–208

Wellstead, C. F. 1991. Taxonomic revision of the Lysorophia, Permo-Carboniferous lepospondyl amphibians. *Bull. Am. Mus. Nat. Hist.* In press

Wiley, E. O. 1981. *Phylogenetics. The Theory and Practice of Phylogenetic Systematics*. New York: Wiley. 439 pp.

Zhang, F., Li, Y., Wang, X. 1984. The new occurrence of Permian seymouriamorphs in Xinjiiang, China. *Vertebr. Palasiatica* 22: 294–304 (In Chinese)

Annu. Rev. Earth Planet. Sci. 1992. 20:85–112

EFFECT OF TROPICAL TOPOGRAPHY ON GLOBAL CLIMATE

Gerald A. Meehl

National Center for Atmospheric Research, Boulder, Colorado 80307

KEY WORDS: monsoon, Tibetan Plateau, tropical precipitation

INTRODUCTION

The global climate system is a complex interacting set of components that includes the atmosphere, oceans, sea ice and snow, land-surface processes, and chemistry. The source of energy that drives the entire system—solar radiation—is most intense in latitudes where the sun is directly overhead as in the tropics. Since there is an excess of solar energy input at tropical latitudes, the function of the global climate system is to transport this excess heat from the tropics to higher latitudes where it is radiated to space. The agents of this transport are the winds in the atmosphere and the currents in the ocean. Atmospheric circulation systems, the storms common to all mankind, are manifestations of energy transport in the atmosphere. These systems, individually or collectively, affect the activities and welfare of societies all over the planet on time scales of days (individual storms), to seasons (dry or wet "spells"), to years (the Dust Bowl or drought years in central North America during the 1930s), to decades (possible global warming), and to centuries and beyond (ice ages).

Because the driving force of these manifestations of climate variation is energy input to the tropics, considerable effort has been spent in understanding the mechanisms of heating in the tropics. Heating of the land and ocean surfaces is the starting point for the energy cycle. As the surfaces become warmer than the overlying air, convection begins and warm air rises. As it ascends, it cools and the moisture it carries condenses to form clouds. In the tropics, convection can become quite strong, forming

0084–6597/92/0515–0085$02.00

towering cumulus clouds (thunderstorms) with associated heavy rainfall. As an integral part of this process, latent heat is released as moisture condenses in the clouds. This latent heat is a powerful source of energy for the global climate system, and variations of this energy input are important to weather and climate elsewhere on the planet. That is, if the source of energy is varied, the driving force for storm systems is altered and the manifestation of climate in the tropics and elsewhere can be changed significantly.

Tropical topography can affect global climate in at least three important ways. First, it can intensify the solar heating over land by providing an elevated heat source. This magnifies the regional-scale land-sea temperature contrast and facilitates the onset and maintenance of monsoon regimes that produce convective rainfall and latent heating. Variations of this heating can affect weather and climate in other tropical areas, as well as in the extratropics via atmospheric teleconnections. Second, topography can contribute to orographic rainfall and additional latent heating. That is, as moist air is forced up a slope, the air cools, moisture condenses, and rainfall and latent heating occur with consequences similar to those associated with convective heating. Third, topographic barriers in the tropics obstruct the large-scale air flow and force some air to go around rather than over topography. This mechanical effect disturbs the flow downstream and sets up large-scale circulation patterns that extend into the extratropics via planetary long waves.

The large-scale ascent of air in regional monsoon regimes associated with low-level convergence and deep convection results in upper-level divergence and outflow in all directions. Upper-level convergence and subsidence then occurs in the subtropics and some tropical areas, particularly over the ocean, as well as over land areas in the winter hemisphere. This overturning in the north-south plane is termed the Hadley Circulation, and large-scale east-west overturning in the Pacific sector is called the Walker Circulation (Trenberth 1991), with such circulations occurring in other sectors of the tropics simply referred to as large-scale east-west circulations (Krishnamurti 1971).

Figure 1 shows long-term mean observed precipitation from the extreme seasons. Most of the precipitation maxima occur over land regions in the tropics during local summer. During June-July-August (JJA) rainfall is most intense over Central America, Mexico, southwest North America, West Africa, and southern Asia (India-Burma-Thailand area). During December-January-February (DJF), rainfall maxima occur over South America, southern Africa, and Indonesia-Australia. These various precipitation regimes are reviewed in this paper, and the relative role of topography is examined for each. Rainfall regimes over the ocean (the

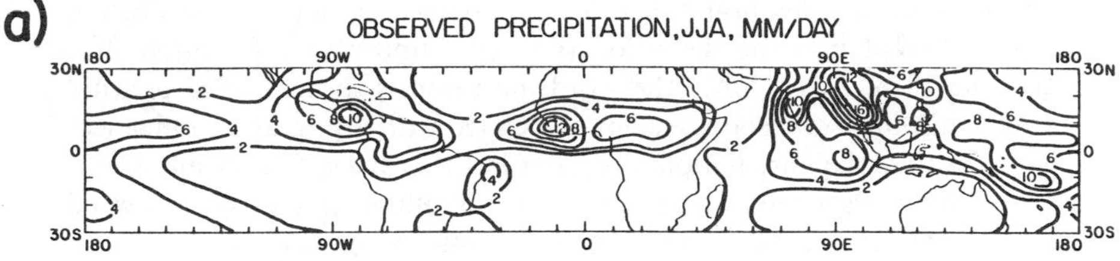

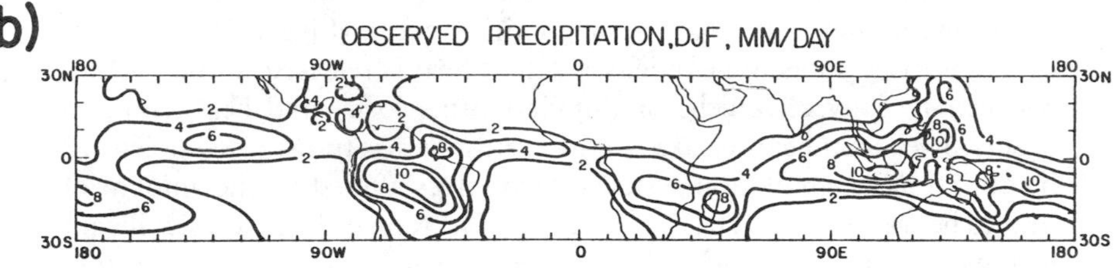

Figure 1 Long-term mean tropical precipitation (mm/day). Data from Jaeger (1976).

Intertropical Convergence Zone—ITCZ—a band of precipitation just north of the equator, and the South Pacific Convergence Zone—SPCZ—a band of precipitation lying diagonally from Papua New Guinea southeastward in the southwest Pacific) will also be mentioned in terms of the role of interannual variability and air-sea interaction in modulating the large topographically anchored land rainfall regimes.

To understand global climate variability, we must study the regional rainfall-producing regimes in the tropics, since variations in tropical rainfall not only affect societies and economies in the tropics, but also because tropical rainfall is a good indicator of the size and location of the energy inputs that drive the global climate system. The major rainfall-producing systems in the tropics are often referred to as "monsoons," or local rainy seasons. The concept of differential heating between land and ocean resulting in monsoon circulations has been traced back to an article by Halley in 1686 in the *Memoirs of the Philosophical Society of Great Britain* (Riehl 1979). Many authors have loosely associated the word monsoon with a seasonal change of wind direction. One of Ramage's (1971) conditions for a monsoon regime is to have at least a 120° wind shift between summer and winter. Such a seasonal reversal of winds from differential heating between land and sea is probably a generally accepted notion of a monsoon. But the perception of a "monsoon season" associated with an annual maximum of rainfall, often life-giving and economically crucial, has also crept into the definition (Global Atmospheric Research Programme 1976, 1978). Since the word monsoon is formally tied to wind direction, winter monsoon seasons associated with topographic influences with ties to global circulation also exist in the tropics. Winter monsoons are not often associated with intense tropical rainfall and are not reviewed in this paper even though they can be dynamically significant.

The mechanism of regional monsoon systems involves the property of heat content, for land surfaces heat faster than ocean surfaces. Thus, air overlying land begins to rise sooner with the onset of spring and summer and draws in moist air from the neighboring ocean (Figure 2). As the air over land rises, moisture condenses, latent heat is released to fuel further convection, convective cells or thunderstorms form, and heavy rainfall occurs. The latent heat released in the atmosphere is transported poleward by atmospheric circulation systems, themselves altered by the amount of heat to be transported. The diurnal cycle of heating over land is significant to how convection is initiated. The mechanism of land-sea temperature contrast is similar to a local sea breeze, but occurs on much larger scales and is more dynamically significant.

The purpose of this review is to better understand the rainfall-producing systems that affect global climate by examining the various monsoon

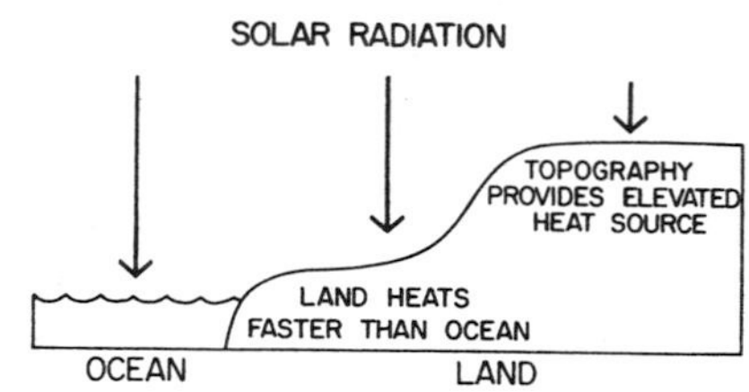

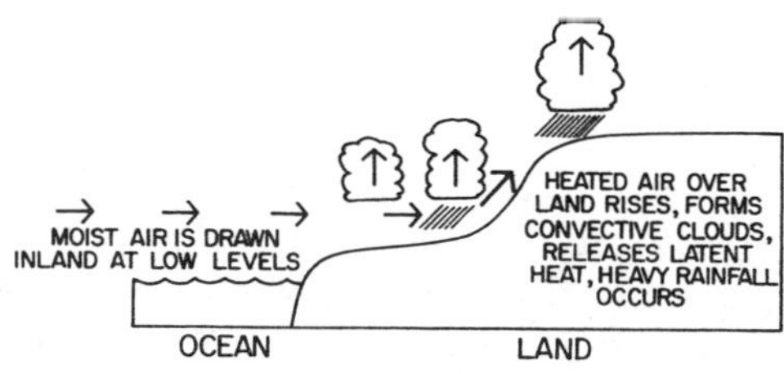

Figure 2 Schematic of land-ocean contrast setting up monsoon circulations.

circulation regimes in the global tropics in terms of the topography and its response to solar input.

INDIAN SUMMER MONSOON AND TIBETAN PLATEAU

In a distance of less than 200 km in southern Asia, the Himalayas rise abruptly from the plains of India to heights of approximately 7 km. The Tibetan Plateau continues to the north extending over roughly 9° of latitude and 35° of longitude at elevations approaching 6 km (see Figure 3). Clearly, an obstacle of such tremendous height and extent must exert a great deal of influence on the atmospheric circulation in that region. It is thought to act not only as a mechanical barrier but as an elevated heat source as well (Murakami 1987a). Both functions will be explored in this section in relation to Indian monsoon rainfall.

Reiter & Gao (1982) show that the high plateau acts as a heat source during summer with its greatest seasonal heating in May prior to monsoon onset (Figure 4). During winter, the cold plateau and relatively warmer air to the south cause a strong temperature gradient associated with a westerly subtropical jet stream which flows to the south (Riehl 1979). In late spring, as the westerlies in the Northern Hemisphere seasonally move northward, the subtropical jet shifts north of the plateau and the monsoon circulation establishes itself over India (Staff Members 1957, Yeh et al 1959). It is this shift, due mainly to the mechanical effects of the plateau barrier, that some researchers believe is the primary triggering mechanism of the monsoon onset over India (Riehl 1979). The importance of the mechanical effects of the elevated Asian topography has been suggested from results of a modeling study where the major upper-level circulations, regionally and globally, were reproduced in a simple model with initial zonal flow and insertion of a large barrier representing the Tibetan Plateau (Krishnamurti et al 1973).

During summer, the strong warming that occurs over the plateau contributes significantly to the monsoon circulation (Chen et al 1985). Both latent heating and sensible heating are associated with low pressure at the surface (He et al 1987), a high pressure cell aloft over the plateau due to hydrostatic considerations (Figure 5), and rising motion which draws relatively cooler, moisture-laden air inland from the ocean expanse to the south (Koteswaram 1958; Luo & Yanai 1983, 1984).

Intense solar radiation incident on the elevated surface heats the plateau (Flohn 1965). It also has been postulated that subsidence of springtime westerly flow over the rim of the Karakoram, Tien Shan, and Himalayan ranges contributes to the sensible heating (Reiter & Gao 1982). However,

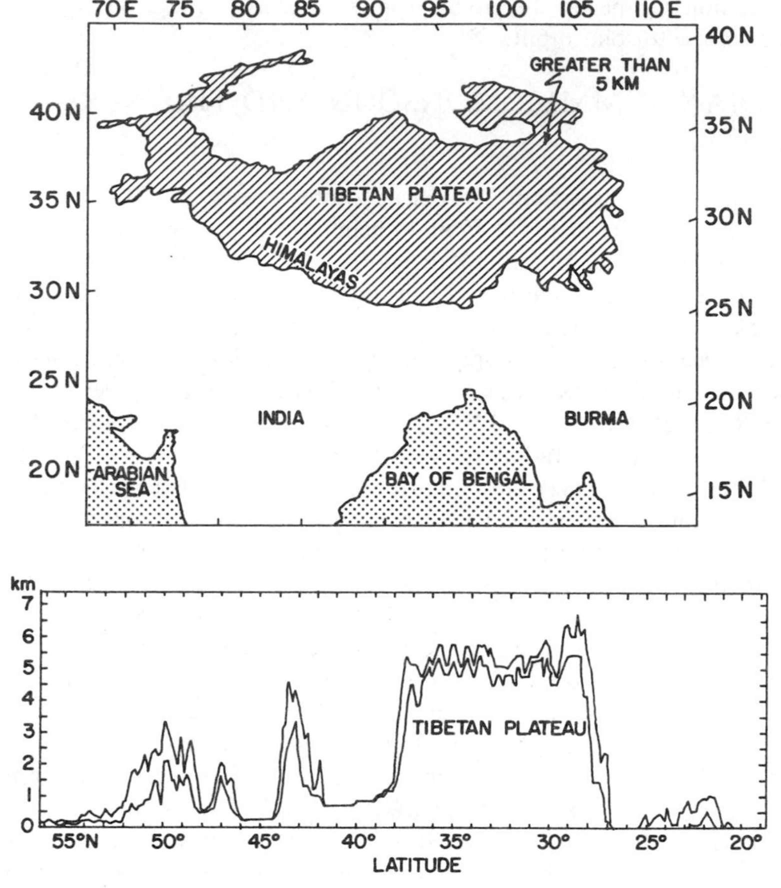

Figure 3 Extent of Himalaya-Tibetan Plateau complex. Cross-hatched areas are elevations higher than 5 km (*top*) (after Atlas of the World 1979); vertical N-S cross-section along 86.25°E longitude (*bottom*). The two lines indicate range of terrain heights reproduced by Wallace et al (1983) from U.S. Navy high-resolution data set.

much evidence seems to indicate that latent heating from large convective cells that develop over and around the plateau may, at times, contribute as much as 80% to the total heating (Global Atmospheric Research Programme 1976). For whatever reason, there can be little doubt that the unique topography of southern Asia acts as a powerful upper-level heat source that contributes to strong land-sea temperature contrast, vigorous monsoon circulation, and intense rainfall during summer (Murakami 1987b).

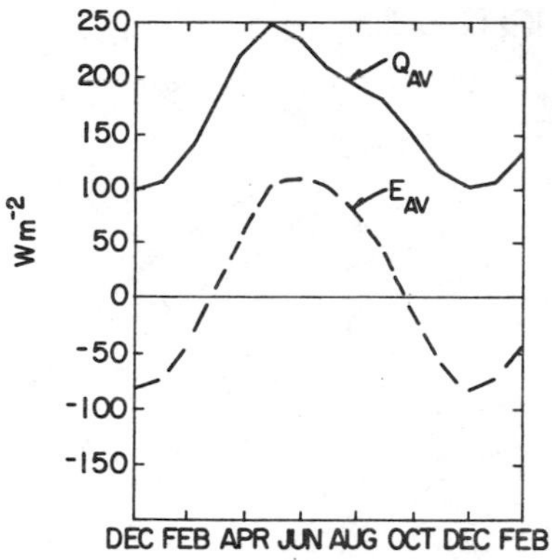

Figure 4 Annual variation of terms of the atmospheric heat budget (W m^{-2}). Note that Q_{AV} (the total heat transfer from plateau to atmosphere, *top curve*) is greatest in May just prior to monsoon onset in June, and the E_{AV} curve (atmospheric heat source) shows the plateau becomes a heat source in late February or early March, and a heat sink around October. Data from a Chinese source reproduced by Reiter & Gao (1982).

To underline the association between the heating of the plateau and the strength of the monsoon, researchers have shown that increased snow cover over the Asian highlands increases albedo, decreases sensible heating, and is associated with decreased rainfall in India and a weaker monsoon (Hahn & Shukla 1976, Barnett et al 1989). In another study, Pant (1983) showed that eddy transports to the north cool the Tibetan Plateau. This weakens and sometimes breaks down the monsoon circulation with associated decreases of precipitation.

In their modeling study, Das & Bedi (1981) have shown that heating of the plateau is necessary to simulate the monsoon circulation, and that increased heating caused by lowering the albedo over the plateau intensifies the monsoon. Yeh (1981) obtained comparable results in a laboratory experiment. As a result of their modeling studies, Kuo & Qian (1981, 1982) conclude that heating (sensible and latent) maintains the monsoon circulation, and that the mechanical effects of the plateau barrier play a modifying role. Additional modeling studies have also pointed to the unique intensity of the upper-level heat source and its importance to the Indian monsoon circulation (Chen et al 1985, Shen et al 1986).

An interesting effect of the east-west mountain barrier is that it insulates the low-latitude air masses from the higher latitudes and helps to maintain an intense trough of low pressure at the surface associated with latent and sensible heating (Ramage 1971). Intrusions of cooler midlatitude air associated with blocking events have been shown to result in "break monsoon" conditions and a disruption of monsoon rains (Raman & Rao 1981). If the mountain barrier is removed—as was done by Hahn & Manabe (1975) in a general circulation model (GCM) experiment—heating still occurs, but the effects are spread over a greater area and not concentrated in the plateau region as low- and high-latitude air masses

200 MB HEIGHT JULY

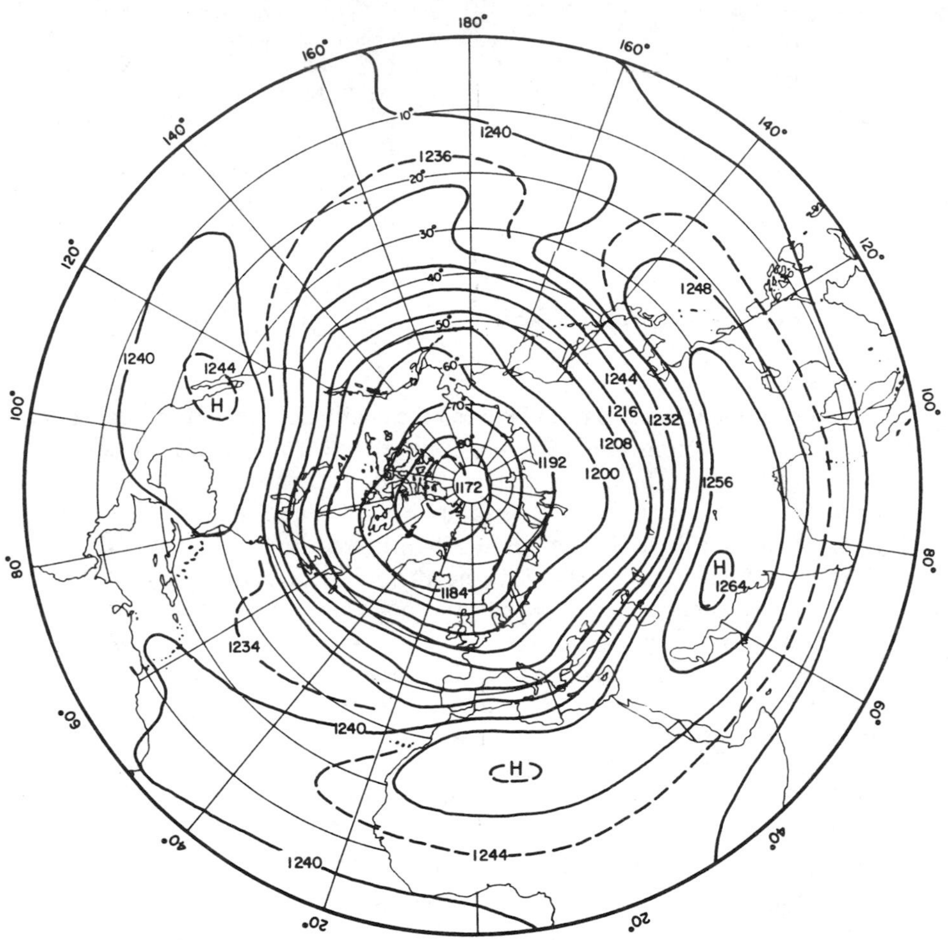

intermix. The monsoon trough forms far to the north and east of its observed position and there is no abrupt monsoonal transition of the subtropical jet. Hahn & Manabe conclude that some of the large-scale monsoon features still occur in the absence of the Himalaya-Tibetan complex, but the monsoon circulation is weakened and, therefore, the mountain barrier must determine the position and intensity of the various monsoon components in the real world.

Two consequences of the elevated heat source of the Tibetan Plateau

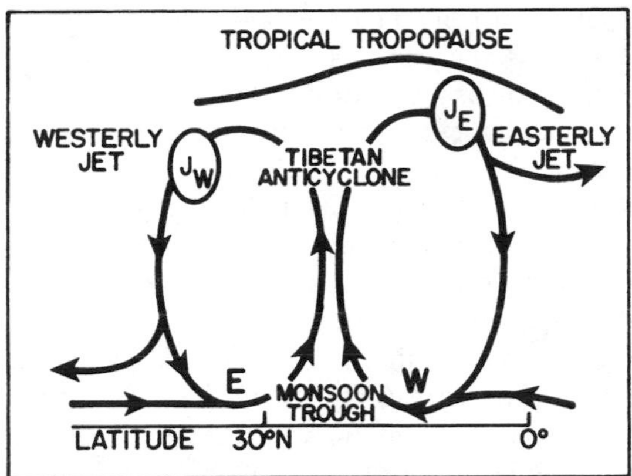

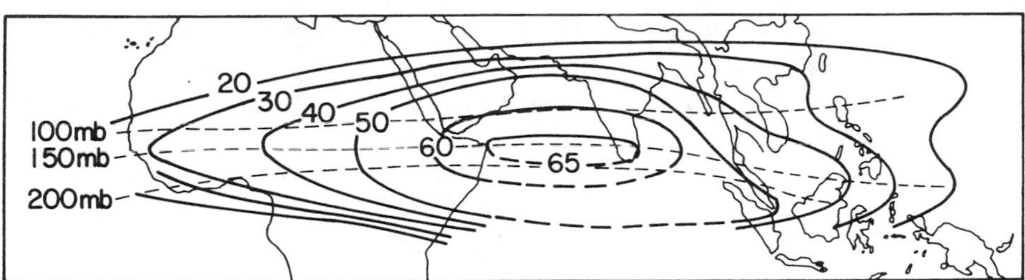

Figure 5 (*a*) Long-term mean 200-mb heights for July from Crutcher & Meserve (1970) (*left*). Note closed contour values of anticyclones (high centers labeled *H*)—over Tibet 1264 m, over Africa 1254 m, and over Mexico 1244 m; air flow can be thought of as roughly paralleling height contours, with clockwise flow around the high centers; (*b*) schematic of vertical circulation in the Indian summer monsoon (*top right*) (after Koteswaram 1960); (*c*) tropical easterly jet during northern summer at various heights (*dashed lines*) in knots (*bottom right*). From Flohn (1964).

mentioned previously are the formation of an upper-level anticyclone and establishment of strong easterly winds to the south over India (Koteswaram 1958, Mason & Anderson 1963). In describing the seasonal development of the upper-level anticyclone, the tropical easterly jet (TEJ), and associated dynamic processes, Flohn (1964) points out that the strong heating of the elevated plateau sets up a temperature gradient between the plateau and the cooler air to the south. The TEJ then forms near 200 mb and is associated with upper-level divergence in the accelerating entrance region over Burma and India and a direct circulation cell in the vertical in those regions with warm air to the north of the jet axis rising and cooler air to the south sinking (Figure 5*b*). Flohn postulates that these vertical motions associated with the TEJ are evidenced by a large precipitation maximum. This seems likely since Tanaka (1982) has shown that a strong TEJ is an indicator of heavy monsoon rainfall. The downstream implications of the TEJ and upper-level anticyclones in general will be discussed in subsequent sections.

Observational and modeling studies indicate that the Himalaya-Tibetan Plateau complex is essential for both the character and intensity of the Indian summer monsoon. Topography serves not only as a mechanical barrier but as an elevated heat source. A combination of these effects seems to be important in maintaining a complicated set of interactions of a number of elements which together form the Indian summer monsoon circulation and associated rainfall. A fluctuation in any one element has ramifications for the behavior of many other components (Krishnamurti & Bhalme 1976, Shukla & Misra 1977, Kanamitsu & Krishnamurti 1978). The fact that GCMs have successfully simulated many elements of the monsoon system when most processes are included (Washington 1981) indicates that the real monsoon depends on many factors, the Himalaya-Tibetan Plateau being one vital element.

As suggested earlier, the Indian summer monsoon rainfall (convective and orographic) contributes to the large-scale vertical motion and heating that affect global climate. Topography contributes to the intensity of these circulations and affects the planetary wave structure due to the mechanical effects of the large topographic barrier (Trenberth & Chen 1988). The topography of eastern tropical Africa also affects the strength of the inflow to India as manifested by the Somali Jet off the coast of eastern Africa (Krishnamurti 1987). Topographic influences on rainfall over Africa will be discussed shortly.

AUSTRALIA

The northern parts of Australia experience what Ramage (1971) calls a "limited summer monsoon." The rainfall maximum during DJF is part of the more extensive Southeast Asian or winter monsoon regime that affects most of Malaysia, Indonesia, Papua New Guinea, the Solomon Islands, and northern Australia (Global Atmospheric Research Programme 1976). Even though Australia undergoes a seasonal monsoon regime, it is somewhat different from the summer monsoon in southern Asia. The most obvious difference is topography. The Great Dividing Range runs roughly north-south in Queensland and extends onto the Cape York Peninsula, and the Barkly Tableland lies south of the Gulf of Carpenteria (Figure 6). No large elevated plateau region exists as in southern Asia, however, and there is no comparable mechanism to form the intense mid- and upper-

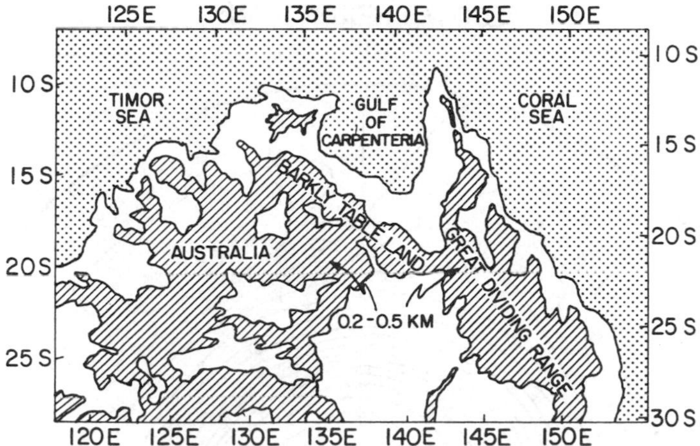

Figure 6 Topography of northern Australia. Cross-hatching indicates elevations greater than 0.2 km and mostly less than 0.5 km. After Atlas of the World (1979).

level heating evident during the Indian monsoon. Yet, a monsoon circulation does exist, the large continental mass does experience heating, and a low-level trough forms as a result (Gunn et al 1989). Sea-level pressure at Darwin in northern Australia has been an indicator of formation of the trough and consequent monsoon onset (Nicholls et al 1982). Allan (1983) showed that heavy convective precipitation in northern Australia is associated with a strong trough at low levels. Sensible heating of the land mass and latent heating from convective activity result in an upper-level anticyclone much as in India but less intense (compare Figure 7 to Figure 5*a*). Nevertheless, the formation of the upper-level anticyclone results in tropical easterlies to the north and an abrupt shift of the subtropical jet poleward (Troup 1961, Hendon & Liebmann 1990). The seasonal shifts of the subtropical jet stream over Australia (Weinert 1968, Brooke 1982) show a southward movement of the jet axis corresponding to the onset of the monsoon in December. This would imply that, for the Indian monsoon, the mechanical effect of the Tibetan Plateau in the northward shift of the subtropical jet may not be as critical as Riehl (1979) and others have postulated. The data from Australia suggest that the heating mechanisms, both sensible and latent, form the upper-level anticyclone which could contribute to forcing the westerlies poleward.

As in the Indian monsoon, the tropical easterlies are an indicator of the strength of the Australian monsoon. Tanaka (1981) and Tanaka & Yoshino (1982) have shown that strong easterly winds at 150 mb over Singapore are associated with heavy rainfall over northern Australia. Similar associations

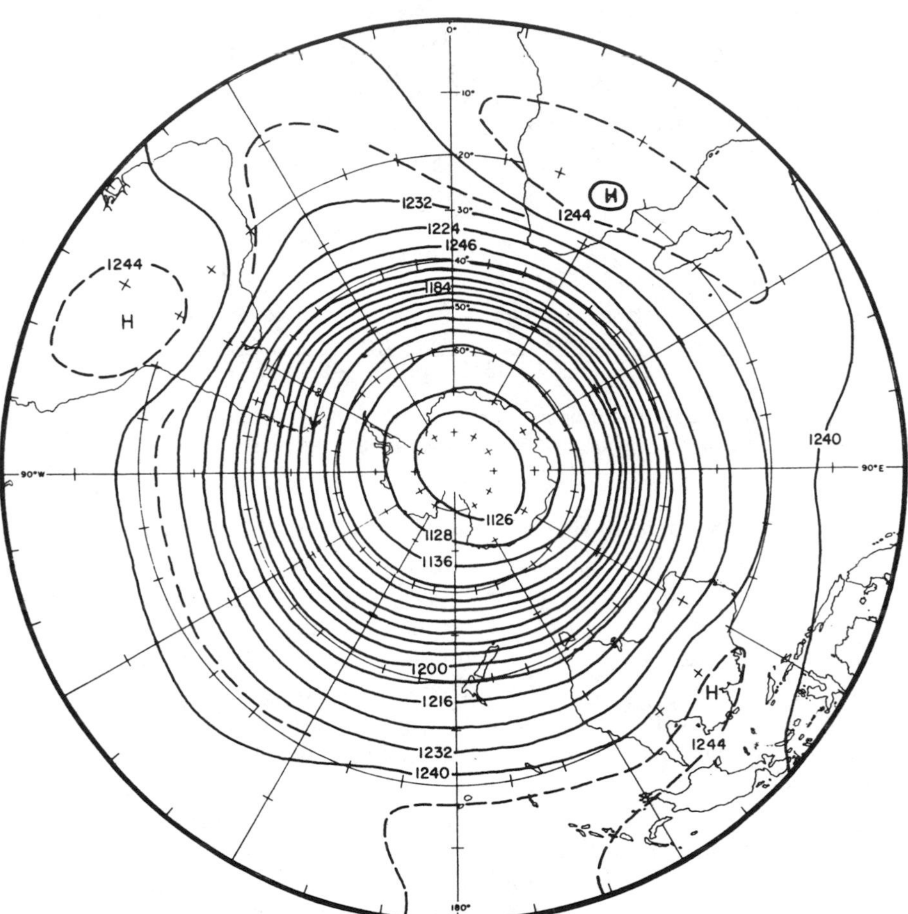

Figure 7 January 200-mb heights from Taljaard et al (1969). Note anticyclone (labeled *H*) over northern Australia with last closed contour of 1244 m, anticyclone over South America with last closed contour of 1244 m, and anticyclone over southern Africa with last closed contour of 1248 m; air flow can be thought of as roughly paralleling height contours, with counterclockwise flow around high centers.

were made earlier between monsoon rainfall in India and the TEJ (Tanaka 1982), but the strength of the tropical upper-level easterlies associated with the Australian monsoon never reaches the intensity or concentration of the TEJ. This fact suggests that Flohn (1964) is correct in his argument that the role of the elevated heat source of Tibet in maintaining the strong TEJ is unique, as inferred by comparing the 200-mb heights in Figure 5*a* over Tibet with those in Figure 7 over Australia. Although heating of Australia occurs and many of the mechanisms are similar, the intensity is much less. The lack of a large elevated heat source and less convective and orographic rain must at least be partially responsible. Another factor that may contribute to the relatively weak Australian monsoon regime compared to India is the region's geography. The vast complex of large mountainous islands immediately north of Australia acts almost like a continent (termed the "island continent" by Ramage 1968). Results from the Winter Monsoon Experiment have suggested that the intensive land-sea breezes set up by the massive jungle islands in Malaysia, Indonesia, and New Guinea, in conjunction with the prevailing large-scale northeasterly low-level flow, force convection virtually around the clock (e.g. Houze et al 1981). Such intense convective activity immediately north of Australia would disrupt the strong cross-equatorial low-level inflow of moisture. This type of flow plays a vital role in contributing to the strength of the Indian monsoon since winds from the south travel across the Indian Ocean virtually unobstructed by islands. Cross-equatorial flow from the north that does reach Australia is associated with a weak monsoon low and deficient rainfall in northern Australia (Allan 1983). In terms of mechanisms for inducing rainfall, there also seems to be little relation between low-level northerly cold surges of Asian origin and monsoon rainfall over Australia (Davidson et al 1983). The main moisture source appears to be associated with low-level westerlies and moisture that originates to the west within the Southern Hemisphere (Allan 1983, Davidson et al 1983). This air has a relatively short fetch over warm tropical water compared to the almost 3000-mile trajectory of moisture over the ocean during the Indian monsoon. However, important rainfall-producing mechanisms that interact with conditions set up by the land-sea heating contrast are passing 30- to 60-day oscillations (Hendon & Liebmann 1990). Sometimes called "Madden-Julian Oscillations" (Madden & Julian 1972), these are eastward-traveling, synoptic-scale disturbances that produce cloudiness and rainfall in the equatorial tropics with a recurrence time of between 30 and 60 days. Yet, geography in general and lack of an elevated heat source in particular appear to keep the Australian summer monsoon from being as intense as the Indian summer monsoon.

WEST AFRICA

Similar to northern Australia, West Africa undergoes a seasonal summer monsoon circulation with a JJA rainfall maximum, and, like Australia,

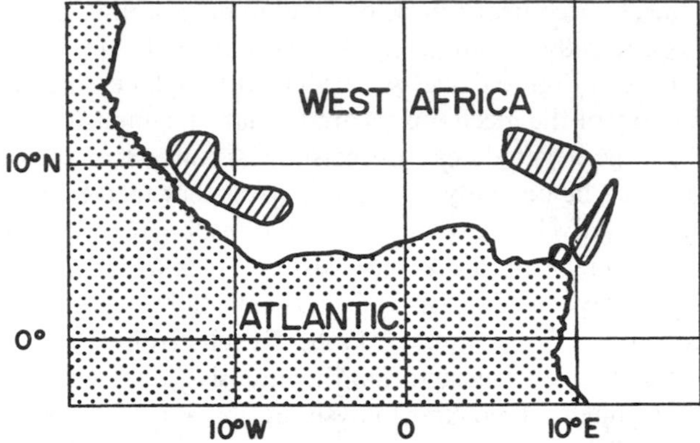

Figure 8 Major topographic features of West Africa. Cross-hatching: 0.5–1.0 km. After Global Atmospheric Research Programme (1978).

West Africa has no significant elevated topographic features (Figure 8). Any orographic effects are mostly local (Global Atmospheric Research Programme 1978).

As in other tropical continental regions, a low-pressure trough forms at the surface over West Africa in response to continental heating from the sun (Desai & Rao 1977), and the northward extent of the trough is associated with rainfall over the land areas (Tanaka et al 1975, Lamb 1978). Sensible heating of the land and latent heating from convection result in the formation of an upper-level anticyclone as in the other monsoon regimes (Figure 5). The subtropical westerlies shift northward (Figure 9), and the TEJ establishes itself over West Africa (Figure 5) and fluctuates in intensity corresponding to the amount of monsoon rainfall (Kidson 1977, Global Atmospheric Research Programme 1981). As with the Australian and Indian monsoons, similar components are present—continental heating, latent heating from convection, low-level trough, upper-level anticyclone, a poleward shift of the subtropical jet, and establishment of strong upper-level equatorial easterlies. But as in the case of Australia, lack of an elevated heat source, such as the Tibetan Plateau, reduces the intensity of the monsoon circulation (e.g. compare 200-mb heights over Africa and Tibet in Figure 5*a*). Also, like Australia, the moist low-level southwesterly inflow from the Atlantic, which is the major moisture source (Flohn et al 1965, Balogun 1981), has a much shorter fetch over water compared to the Indian monsoon (Desai & Rao 1977). In addition, Flohn (1964) postulates that vertical circulation associated with convergence and an indirect circulation near the exit region of the TEJ cause subsidence

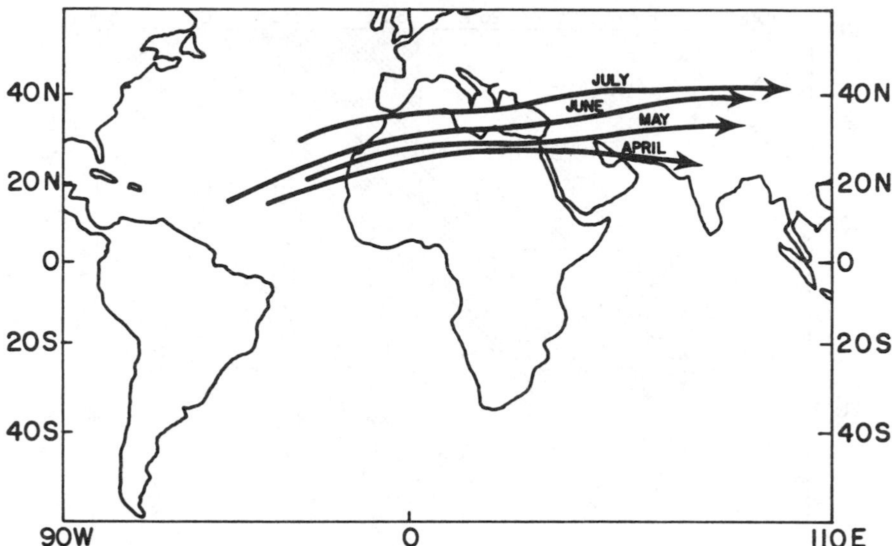

Figure 9 Position of axis of *u*-component wind maximum over northern Africa for April, May, June, and July. Wind data are four-year means. From Gray et al (1976).

over the Sahara which limits the northward extent of the monsoon regime. A significant effect that contributes to the African monsoon's being weaker in relation to the Indian monsoon is the lack of an extensive elevated heat source.

SOUTHERN AFRICA

The rainy season south of the equator in Africa occurs during local summer (DJF) and also involves elements of the other monsoon or rainy season regimes. Topography provides an elevated heat source and also possibly contributes to the formation of a surface trough as a consequence of airflow over the orography (Semazzi 1980). Topography also determines source regions of moisture that penetrate from surrounding ocean areas. The extensive central plateau with elevations higher than 1000 m extends from Tanzania in the east southwestward toward Angola and Namibia (Figure 10). Somewhat farther south is the Highveld in southeastern Africa. These topographic features provide upper-level heat sources that enhance the inflow of moist air from surrounding oceans (Garstang et al 1987). The main source of moisture for central African rainfall is carried in from the Atlantic to the west (Taljaard 1972). A second moisture source is the Indian Ocean east of Africa but, as Taljaard (1955) points out, most

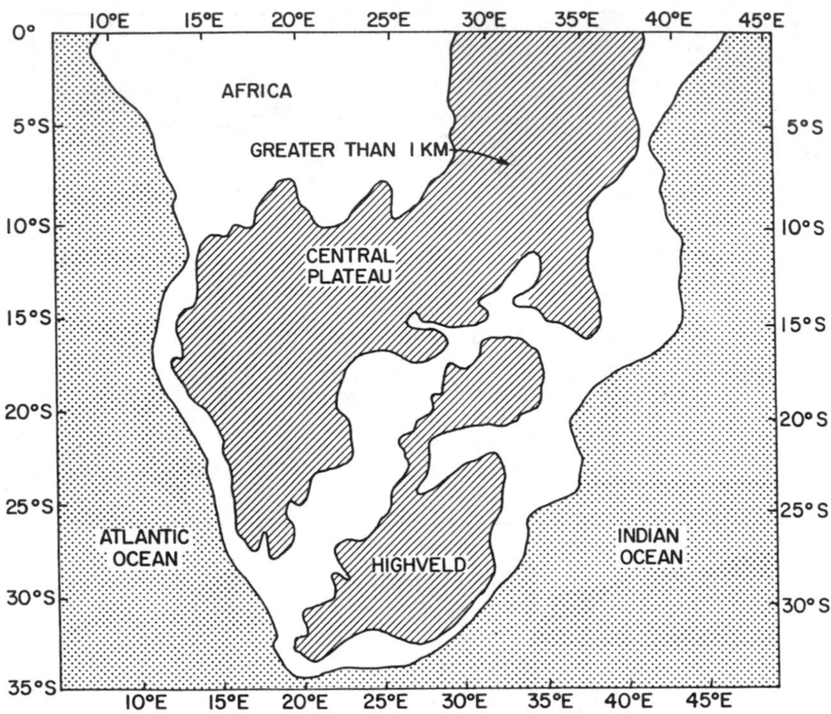

Figure 10 Topography over southern Africa. Hatching denotes areas higher than 1 km. After Atlas of the World (1979).

of the moisture from the Indian Ocean is rained out when the air is lifted near the east coast enroute to the central plateau. Interannual variability of rainfall in southern Africa is considerable (Nicholson 1983, Tyson 1984, 1986, Janowiak 1988) and partly related to the Southern Oscillation (Lindesay 1988). Intraseasonal variability is associated with disturbances of both tropical and midlatitudinal origins (Lyons 1991). As a relative measure of the intensity of the heating, a comparison of the upper-level highs (Figure 7) over southern Africa and northern Australia shows that the elevated topography of southern Africa is associated with a stronger upper-level high compared to the weaker high over the almost flat topography of northern Australia.

SOUTH AMERICA

Like southern Asia, but unlike West Africa or northern Australia, South America has a high plateau called the Altiplano. It extends from roughly

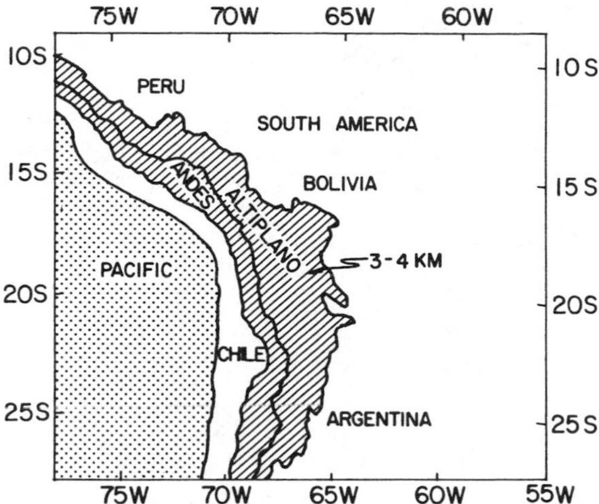

Figure 11 Topography of the Altiplano region of South America. Cross-hatched regions are around 3–4 km of elevation over the Altiplano, and higher in some parts of the Andes at the western rim. After Atlas of the World (1979).

15°S to 21°S and from about 64°W to 68°W (Figure 11) and is much less extensive than the Himalaya-Tibetan Plateau complex. It is not as high as the southern Asia massif—around 4 km compared to nearly 7 km. South America is not usually thought of as having a monsoon regime per se (Ramage 1971), but many of the features associated with monsoons in other regions previously described still occur during local summer (DJF). Like southern Asia, part of the heating of the Altiplano is caused by sensible heat, but the majority is from latent heat release associated with convective activity over and around the plateau (Gutman & Schwerdtfeger 1965). The low-level trough is associated with an upper-level anticyclone (Dean 1971, Kreuels et al 1975), the subtropical jet intensifies to the south (Schwerdtfeger 1961), and upper-level tropical easterlies form to the north (Virji 1981). The heating of the Altiplano causes vertical motion, and moist low-level inflow is drawn in from the north (Virji 1981). Schwerdtfeger (1975) and Rao & Erdogan (1989) give a good summary of these conditions.

The strength of the low-level inflow is weakened since it must travel hundreds of miles over the Amazon River basin where active convection and precipitation take place (Trewartha 1966). The southward advance of the low-level inflow toward the heated Altiplano is also hindered in the eastern part of the continent by the strong Atlantic anticyclone (Hastenrath & Heller 1977), as shown in Figure 12. An unusually strong Atlantic high

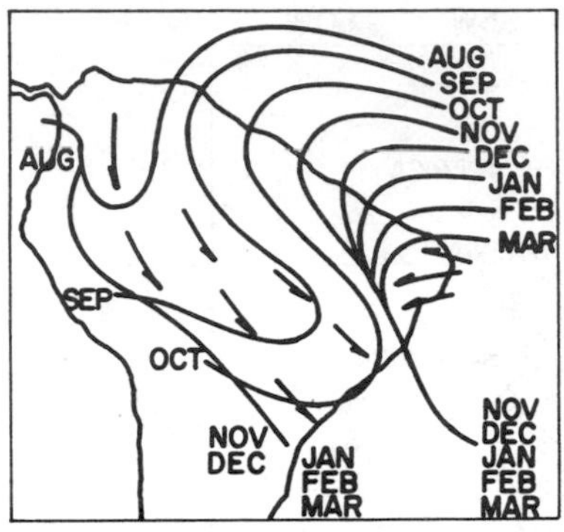

Figure 12 The southward advance over Brazil in spring and summer of the humid unstable equatorial westerlies and intertropical convergence. After Trewartha (1966).

has been associated with droughts in northeastern Brazil called "Sêcas" (Hastenrather & Heller 1977, Chu 1983). In a modeling study, Moura & Shukla (1981) have shown that warm sea-surface temperature anomalies in the tropical Atlantic may set up intense convection over the ocean and subsidence in northeastern Brazil which contributes to the Sêca.

In any case, a number of conditions previously associated with summer monsoon regimes elsewhere occur in South America. The reduced intensity of the resulting circulation and weaker upper-level anticyclone (compare Figure 7 over South America with Figure 5 over Tibet) must first be due, in part, to the limited extent and lower height of the Altiplano compared to the Tibetan Plateau. In addition, the geography of South America, like Australia, forces moist low-level inflow to be drawn across a large continental (island continent in southeast Asia) land mass where moisture is depleted by heavy convective activity. The narrow dimensions of southern South America also allow the strong Atlantic high to inhibit penetration of moist low-level flow in the eastern portions. Virji (1981) used satellite data to reveal a number of features, some listed above, in greater detail than before. He states that the seasonal circulations associated with the Altiplano may be more intense than in the limited station data of previous studies. Rao & Erdogan (1989) analyzed heating over the Altiplano and concluded that latent heating from convection is the biggest contributor to the atmospheric heat source. In their year of analyses, this heating was stronger locally than over Tibet.

MEXICAN OR ARIZONA MONSOON

In northern Mexico, northeastern Arizona, and northwestern New Mexico, a complex of mountain ranges and plateaus serves as an elevated heat source associated with what has been called (and perhaps not quite accurately according to many definitions) the Arizona monsoon (Bryson & Lowry 1955) or the Mexican summer monsoon (Douglas 1983). There is no continuous high plateau in this region, but various massifs, including the Sierra Madre Mountains and Colorado Plateau, form a somewhat dispersed elevated heat source running mostly north-south, covering over 15° latitude and about 10° of longitude with most elevations above 2 km (see Figure 13). As in the other monsoon-type regimes, seasonal heating of the land, in this case somewhat elevated, is associated with a heat low at the surface, and sensible and latent heating cause a mid- and upper-level anticyclone to form over northern Mexico (Bryson & Lowry 1955, Hales 1974) which shifts the subtropical westerlies poleward, as seen in Figure 5*a*. The trough at the surface over northern Mexico assists in establishing low-level cross-equatorial surface flow from the eastern tropical Pacific and western Mexico (Reyes & Cadet 1988). After some views to the contrary in the 1950s (Bryson & Lowry 1955), there now is general agreement that the moisture source near the surface for the Arizona monsoon is from the Pacific Ocean to the southwest (Rasmusson 1967, Hales 1974) where warm surface currents in late summer flow northward (Wyrtki 1965) and southwesterly winds set up by the heating contrast between land and ocean flow onshore in July and August (Horel 1982). The moisture evaporating from the warm ocean is carried by these winds

Figure 13 Topography over northern Mexico and parts of Arizona and New Mexico. Hatching denotes regions higher than 7000 ft (roughly 2.1 km). After Hales (1974).

in the lower troposphere in association with the Mexican anticyclone. This usually causes a rapid onset of rains as far north as Arizona in July (see Figure 14). Moisture for the Arizona monsoon rainfall is carried over land at higher levels in the troposphere from the Gulf of Mexico (Carleton 1986, Adang & Gall 1989, Smith & Gall 1989).

Even though this regime has some of the same characteristics as the monsoon-type regimes reviewed previously, it is less like the others. The geography is such that the heat source consists of a relatively narrow and discontinuous strip of highlands and plateaus (Carleton et al 1990) which is not able to heat to the intensity of the Indian monsoon. Its north-south orientation with a large continental mass lying mostly unobstructed to the north does not insulate the heated areas from intrusions of cool northwesterly flow as in the Indian monsoon. The Arizona or Mexican monsoon circulation is disrupted and rains fail when northwesterly flow prevails over the region (Douglas et al 1982, Douglas 1983). Most of the low-level inflow from the southwest falls orographically in a relatively narrow coastal strip in northern Mexico (Figure 13), surges northeastward in association with the passage of tropical storms off Baja (Hales 1972), or is carried northeastward by mid-level flow associated with the Mexican anticyclone and associated trough off the west coast (Hales 1974). Moist low-level winds have relatively short fetch over the warm tropical Pacific as can be seen, for example, in Horel (1982). A combination of these factors results in a very weak monsoon circulation compared to the Indian

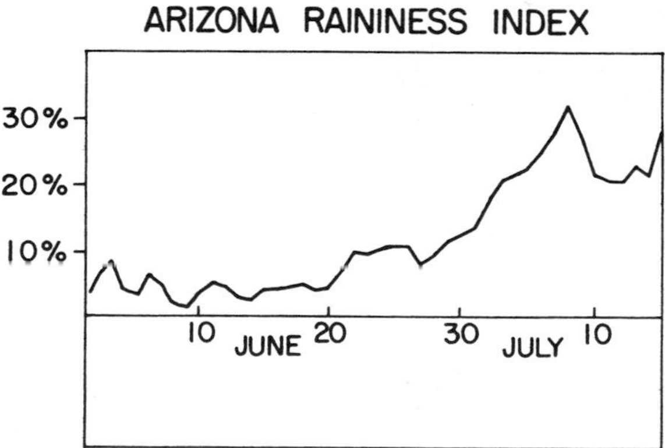

Figure 14 Average daily value of Arizona Raininess Index for 25 years (1930–54). Index is defined as percent of stations reporting at least a trace of precipitation for that day. From Bryson & Lowry (1955).

monsoon. This is also evident in a comparison of the 200-mb heights over Mexico and Tibet in Figure 5*a*.

PRECIPITATION OVER OCEAN

Examination of Figure 1 shows that, even though land areas associated with the monsoon regimes dominate long-term mean tropical precipitation and topography plays a key role in those processes, a great deal of precipitation falls over ocean areas in association with the ITCZ just north of the equator and the SPCZ in the southwest Pacific. During periods of warmer-than-normal sea-surface temperatures in the tropical Pacific (called El Niño-Southern Oscillation Events—ENSO, or simply El Niño) (Philander 1990), precipitation over the tropical Pacific increases markedly (Rasmusson & Carpenter 1982) with effects on the great tropical land precipitation regimes. In particular, the ITCZ merges with a northeastward-shifted SPCZ to form a large precipitation maximum in the central tropical Pacific. Consequently, during ENSO events, rainfall in the Indian monsoon tends to be relatively less (Rasmusson & Carpenter 1983, Ropelewski & Halpert 1987, Meehl 1987a), with decreases over tropical South America (Aceituno 1988) and southern Africa (Kiladis & Diaz 1989). Even precipitation in the Arizona monsoon (Douglas 1983) and West Africa (Lamb & Peppler 1991) is affected by ENSO events. The agents of these long-range effects on tropical precipitation are large-scale east-west circulations in the atmosphere (Meehl 1987b). Tropical precipitation, to a first order, follows a simple axiom: What goes up must come down. That is, if there is intense uplift associated with convection in one region of the tropics, there must be associated large-scale sinking and suppression of convection and precipitation elsewhere. In the long-term mean, the tropical land areas are heated more intensely and experience the greatest regional precipitation with attendant sinking over much of the ocean areas in the tropics. However, if the sea surface temperatures (SSTs) in a certain region (such as the tropical Pacific associated with ENSO) are unusually warm, there is greater-than-normal convection, uplift, and rainfall in that region with attendant subsidence elsewhere in the tropics and subtropics. This reduces rainfall in the monsoon regimes, but it does not mean that monsoons cease. The distribution of land and sea in the tropics with the attendant topography dictates that the seasonal rainfall maxima associated with the monsoon regimes will occur every year. However, in any given year, the rainfall will be either greater or less than normal depending on several factors, one being the occurrence of ENSO events.

Variations of these tropical heat sources associated with changes in intensity and position of regional convection have been shown to affect the extratropical circulation via atmospheric teleconnections (Tribbia 1991). That is, a variation in a tropical heat source can affect the planetary wave structure and cause weather anomalies far from the anomalous heat source in the tropics. Topography and geography in the tropics anchor the main regional convective centers, but interannual variability can be introduced via air-sea interaction with effects on ocean precipitation. This, in turn, interacts with and affects the monsoon regimes via the large-scale east-west atmospheric circulations (Meehl 1987b). A review of extratropical teleconnections is beyond the scope of this paper, but a good recent overview is given by Glantz et al (1991). For the present discussion, it is sufficient to note that variations of heating associated with tropical rainfall regimes partially driven by topography can critically affect global weather and climate.

PALEOCLIMATE AND FUTURE CLIMATE CHANGE

From the preceding comparison of the various monsoon regimes, the unique influence of the Tibetan Plateau seems to be in its incomparable height and vast extent. Because of the great latent and sensible heating associated with this elevated source, there is no other region where such an intense upper-level anticyclone forms. The east-west orientation of the plateau mostly inhibits mixing of low- and high-latitude air masses, thus allowing the low-latitude heating over land to persist and intensify. Therefore, it would have to be assumed that a much lower Tibetan Plateau, which may have existed roughly seven million years ago when the continents were in their present positions (Ruddiman et al 1989), would be associated with less intense upper-level heating, a weakened upper-level anticyclone, a weaker TEJ, and less rainfall over India. In fact, these exact conditions exist when a weaker monsoon occurs over India, intraseasonally or interannually, due to various other external factors that reduce the intensity of the monsoon system (e.g. Krishnamurti & Bhalme 1976, Kanamitsu & Krishnamurti 1978). It should be reasonable to assume that lessening the influence of a major external forcing mechanism, such as the elevated heat source of the Tibetan Plateau, would reduce the intensity of the monsoon. A fairly strong monsoon regime would still exist owing to the unique geographical arrangements of the large Asian continent to the north and unrestricted ocean to the south that provides considerable land-sea temperature contrast and a long fetch of low-level inflow across the Indian Ocean (Trenberth & Chen 1988). This is a rich source of moisture for the monsoon rains. In a model simulation with no Tibetan Plateau, an active monsoon regime still existed over southern Asia (Hahn & Manabe 1975). Raising and lowering the Tibetan Plateau in the context of past climates in a GCM study, however, has shown that the strength of the Indian monsoon depends critically on the elevation of the topography. That is, the lower the Tibetan Plateau the weaker the monsoon circulation (Kutzbach et al 1989).

The effect of tropical topography on future climate may well depend on its interaction with ocean precipitation regimes, as discussed above in relation to ENSO. The current generation of coupled ocean-atmosphere GCMs is capable of internally generating some of the features observed during ENSO (Meehl 1990a). Since these models can be run with increased carbon dioxide and trace gases (Washington & Meehl 1989, Stouffer et al 1989), the effects of possible future climate change can be studied. Because the oceans in such models warm when carbon dioxide is increased and ENSO continues to occur in at least one of these models, the indication is that similar patterns of tropical rainfall anomalies would still be present but with increased magnitude. Dry areas would become drier and wet areas wetter in association with future ENSO events in the tropics (Meehl 1990b). The effects of altering the patterns of tropical heating associated with different regional precipitation amounts and the impacts on the global atmospheric circulation are active areas of study with such coupled GCMs.

CONCLUSION

Tropical topography plays an important role in producing tropical monsoon rainfall regimes whose heat inputs drive the global climate system. The Himalaya-Tibetan Plateau complex in south Asia, unique in elevation and extent, provides an elevated sensible and latent heat source that contributes to the formation of an intense surface trough and upper-level anticyclone, a strong upper-level tropical easterly jet, large land-sea temperature contrasts resulting in strong low-level inflow laden with moisture from the long unrestricted journey from the Indian Ocean to the south, and heavy summer rainfall over southern Asia. The heating and mechanical effects associated with the elevated topography have significant effects on global climate. Other tropical continental areas without elevated topographical heat sources to provide such dramatic effects can still maintain monsoon regimes, but not of comparable intensity (e.g. Australia and West Africa). Regimes with upper-level heat sources of lower elevation and limited extent also cannot generate such strong monsoon circulations (e.g. South America, Mexico, southwestern United States, and southern Africa) even though they are able to maintain many of the corresponding features but with reduced magnitude. The unique height of the extensive Tibetan Plateau seems to be a major forcing feature which, if lowered to elevations of seven million years ago, is associated with a correspondingly weaker Indian summer monsoon. Effects from fluctuations of heating in the regional monsoon regimes in the tropics are communicated globally via atmospheric teleconnections. Alterations of these effects are often associated with air-sea interaction processes (e.g. ENSO), as well as land-surface processes (e.g. Barnett et al 1989). Future climate change as a result of increased carbon dioxide and trace gases is being studied with coupled ocean-atmosphere GCMs. Preliminary results from one of these models (Meehl et al 1991) indicate that patterns of tropical land precipitation will be roughly the same, but that the effects of ocean precipitation anomalies associated with ENSO events could cause larger-amplitude monsoon land precipitation anomalies.

Acknowledgments

The author acknowledges helpful comments from Kevin Trenberth, Harry van Loon, Roger Barry, David Greenland, William Hay, and Warren Washington. The National Center for Atmospheric Research is sponsored by the National Science Foundation. A portion of this study is supported by the Office of Health and Environmental Research of the U.S. Department of Energy as part of its Carbon Dioxide Research Program.

Literature Cited

Aceituno, P. 1988. On the functioning of the Southern Oscillation in the South American sector. *Mon. Weather Rev.* 116: 505–24

Adang, T. C., Gall, R. L. 1989. Structure and dynamics of the Arizona monsoon boundary. *Mon. Weather Rev.* 117: 1423–38

Allan, R. J. 1983. Monsoon and teleconnection variability over Australia during the Southern Hemisphere summers of 1973–77. *Mon. Weather Rev.* 111: 113–42

Atlas of the World 1979. *The Times Concise Atlas of the World, Times Books.* London: Times Books Ltd. 232 pp.

Balogun, E. E. 1981. *Int. Conf. on Early Results of FGGE and Aspects of Its Monsoon Experiments.* Geneva: World Meteorol. Organ. 8.22–8.27

Barnett, T. P., Dümenil, L., Schlese, U., Roecker, E., Latif, M. 1989. The effect of Eurasian snow cover on regional and global climate variations. *J. Atmos. Sci.* 46: 661–85

Brooke, R. R. 1982. A study of the subtropical jetstream in the Australian region. *Aust. Meteorol. Mag.* 30: 223–39

Bryson, R. A., Lowry, W. P. 1955. The synoptic climatology of the Arizona summer precipitation singularity. *Bull. Am. Meteorol. Soc.* 36: 329–39

Carleton, A. M. 1986. Synoptic-dynamic character of "bursts" and "breaks" in the Southwest U.S. summer precipitation singularity. *J. Climatol.* 6: 605–23

Carleton, A. M., Carpenter, D. A., Weser, P. J. 1990. Mechanisms of interannual variability of the southwest United States summer rainfall maximum. *J. Climate* 3: 999–1015

Chen, L., Reiter, E. R., Feng, Z. 1985. The atmospheric heat source over the Tibetan Plateau May–August 1979. *Mon. Weather Rev.* 113: 1771–90

Chu, P. S. 1983. Diagnostic studies of rainfall anomalies in Northeast Brazil. *Mon. Weather Rev.* 111: 1655–64

Crutcher, H. L., Meserve, J. M. 1970. *Selected Level Heights, Temperatures, and Dew Points for the Northern Hemisphere.* NAVAIR 50-1C-52, Revised. Washington, DC: Naval Weather Serv. Command

Das, P. K., Bedi, H. S. 1981. A numerical model of the monsoon trough. In *Monsoon Dynamics*, ed. J. Lighthill, R. P. Pearce, pp. 351–63. Cambridge: Cambridge Univ. Press

Davidson, N. E., McBride, J. L., McAvaney,

B. J. 1983. The onset of the Australian monsoon during winter MONEX: Synoptic aspects. *Mon. Weather Rev.* 111: 496–516

Dean, G. A. 1971. The three-dimensional wind structure over South America and associated rainfall over Brazil. *Report No. 71-4*. Tallahassee: Florida State Univ. 114 pp.

Desai, B. N., Rao, Y. P. 1977. A comparative study of the summer monsoons over Africa and India. *Indian J. Meteorol. Hydrol. Geophys.* 28: 21–30

Douglas, A. V. 1983. *Proc. 7th Climate Diagnostics Workshop*. Washington, DC: U.S. Dept. Commerce. pp. 79–80

Douglas, A. V., Cayan, D. R., Namias, J. 1982. Large-scale changes in North Pacific and North American weather patterns in recent decades. *Mon. Weather Rev.* 110: 1851–62

Flohn, H. 1964. Investigations of the tropical easterly jet. *Bonn. Meteorol. Abhandl.* 4: 1–69

Flohn, H. 1965. Thermal effects of the Tibet Plateau during the Asian monsoon season. *Aust. Meteorol. Mag.* 49: 55–57

Flohn, H., Henning, D., Korff, H. C. 1965. Studies of the water vapor transport over Northern Africa. *Bonn. Meteorol. Abhandl.* 6, 36 pp.

Garstang, M., Kelbe, B., Emmitt, G. D., London, W. B. 1987. Generation of convective storms over the escarpment of northeastern South Africa. *Mon. Weather Rev.* 115: 429–43

Glantz, M. H., Katz, R. W., Nicholls, N., eds. 1991. *Teleconnections Linking Worldwide Climate Anomalies*. Cambridge: Cambridge Univ. Press. 535 pp.

Global Atmospheric Research Programme 1976. *GARP Publ. Series No. 18*. Geneva: World Meteorol. Organ. 126 pp.

Global Atmospheric Research Programme 1978. *GARP Publ. Series No. 21*. Geneva: World Meteorol. Organ. 88 pp.

Global Atmospheric Research Programme 1981. *GARP Special Rept. No. 37*. Geneva: World Meteorol. Organ. 104 pp.

Gray, T. I. Jr., Irwin, J. R., Krueger, A. F., Vanadore, M. S. 1976. *Average Circulation in the Troposphere over the Tropics*. Washington, DC: U.S. Dept. Commerce. 238 pp.

Gunn, B. W., McBride, J. L., Holland, G. J., Keenan, T. D., Davidson, N. E., Hendon, H. H. 1989. The Australian summer monsoon circulation during AMEX Phase II. *Mon. Weather Rev.* 117: 2554–74

Gutman, G. J., Schwerdtfeger, W. 1965. The role of latent and sensible heat for the development of a high pressure system over the subtropical Andes, in the summer. *Meteorol. Rundsch.* 18: 69–75

Hahn, D. G., Manabe, S. 1975. The role of mountains in the South Asian monsoon circulation. *J. Atmos. Sci.* 32: 1515–41

Hahn, D. G., Shukla, J. 1976. An apparent relationship between Eurasian snow cover and Indian monsoon rainfall. *J. Atmos. Sci.* 33: 2461–62

Hales, J. E. Jr. 1972. Surges of maritime tropical air northward over the Gulf of California. *Mon. Weather Rev.* 100: 298–306

Hales, J. E. Jr. 1974. Southwestern United States summer monsoon source—Gulf of Mexico or Pacific Ocean? *J. Appl. Meteorol.* 13: 331–42

Hastenrath, S., Heller, L. 1977. Dynamics of climatic hazards in northeast Brazil. *Q. J. R. Meteorol. Soc.* 103: 77–92

He, H., McGinnis, J. W., Song, Z., Yanai, M. 1987. Onset of the Asian summer monsoon in 1979 and the effect of the Tibetan Plateau. *Mon. Weather Rev.* 115: 1966–95

Hendon, H. H., Liebmann, B. 1990. A composite study of onset of the Australian summer monsoon. *J. Atmos. Sci.* 47: 2227–40

Horel, J. D. 1982. On the annual cycle of the tropical Pacific atmosphere and ocean. *Mon. Weather Rev.* 110: 1863–78

Houze, R. A., Geotis, S. G., Marks, F. G., West, A. K. 1981. Winter monsoon convection in the vicinity of north Borneo. Part I: Structure and time variation of the clouds and precipitation. *Mon. Weather Rev.* 109: 1595–1614

Jaeger, L. 1976. Monatskarten des Nieherschlags für die Ganze Erde. *Ber. Deutch. Wetterd.* 139(18): 38

Janowiak, J. E. 1988. An investigation of interannual rainfall variability in Africa. *J. Climate* 1: 240–55

Kanamitsu, M., Krishnamurti, T. N. 1978. Northern summer tropical circulation during drought and normal rainfall months. *Mon. Weather Rev.* 106: 331–47

Kidson, J. W. 1977. African rainfall and its relation to the upper air circulation. *Q. J. R. Meteorol. Soc.* 103: 441–56

Kiladis, G. N., Diaz, H. F. 1989. Global climatic anomalies associated with extremes in the Southern Oscillation. *J. Climate* 2: 1069–90

Koteswaram, P. 1958. The easterly jet stream in the tropics. *Tellus* 10: 43–87

Koteswaram, P. 1960. The Asian summer monsoon and the general circulation over the tropics. In *Monsoons of the World*, pp. 105–10. New Delhi: India Meteorol. Dept.

Kreuels, R., Fraedrich, K., Ruprecht, E. 1975. An aerological climatology of South America. *Meteorol. Rundsch.* 28: 17–24

Krishnamurti, T. N. 1971. Tropical east-west circulations during the northern summer. *J. Atmos. Sci.* 28: 1342–47

Krishnamurti, T. N. 1987. Monsoon models. In *Monsoons*, ed. J. S. Fein, P. L. Stephens, pp. 467–522. New York: Wiley

Krishnamurti, T. N., Bhalme, H. N. 1976. Oscillations of a monsoon system. Part I. Observational aspects. *J. Atmos. Sci.* 33: 1937–54

Krishnamurti, T. N., Daggupaty, S. M., Fein, J., Kanamitsu, M., Lee, J. D. 1973. Tibetan high and upper tropospheric tropical circulations during northern summer. *Bull. Am. Meteorol. Soc.* 54: 1234–39

Kuo, H. L., Qian, Y.-F. 1981. Influence of Tibetan Plateau on cumulative and diurnal changes of weather and climate in summer. *Mon. Weather Rev.* 109: 2337–56

Kuo, H. L., Qian, Y.-F. 1982. Numerical simulation of the development of mean monsoon circulation in July. *Mon. Weather Rev.* 110: 1879–97

Kutzbach, J. E., Guetter, P. J., Ruddiman, W. F., Prell, W. L. 1989. Sensitivity of climate to late Cenozoic uplift in Southern Asia and the American west: Numerical experiments. *J. Geophys. Res.* 94: 18,393–407

Lamb, P. J. 1978. Case studies of tropical Atlantic surface circulation patterns during recent sub-Saharan weather anomalies: 1967 and 1968. *Mon. Weather Rev.* 106: 482–91

Lamb, P. J., Peppler, R. A. 1991. West Africa. In *Teleconnections Linking Worldwide Climate Anomalies*, ed. M. H. Glantz, R. W. Katz, N. Nicholls, pp. 121–90. Cambridge: Cambridge Univ. Press

Lindesay, J. A. 1988. South African rainfall, the Southern Oscillation and a Southern Hemisphere semi-annual cycle. *J. Climatol.* 8: 17–30

Luo, H., Yani, M. 1983. The large-scale circulation and heat sources over the Tibetan Plateau and surrounding areas during early summer of 1979. Part I. Precipitation and kinematic analyses. *Mon. Weather Rev.* 111: 922–44

Luo, H., Yanai, M. 1984. The large-scale circulation and heat sources over the Tibetan Plateau and surrounding areas during the early summer of 1979. Part II. Heat and moisture budgets. *Mon. Weather Rev.* 112: 966–89

Lyons, S. W. 1991. Origins of convective variability over equatorial Southern Africa during Austral summer. *J. Climate* 4: 23–39

Madden, R. A., Julian, P. R. 1972. Description of global scale circulation cells in the tropics with 40–50 day period. *J. Atmos. Sci.* 29: 1109–23

Mason, R. B., Anderson, C. E. 1963. The development and decay of the 100 mb summertime anticyclone over Southern Asia. *Mon. Weather Rev.* 91: 3–12

Meehl, G. A. 1987a. The tropics and their role in the global climate system. *Geogr. J.* 153: 21–36

Meehl, G. A. 1987b. The annual cycle and interannual variability in the tropical Pacific and Indian Ocean regions. *Mon. Weather Rev.* 115: 27–50

Meehl, G. A. 1990a. Seasonal cycle forcing of El Niño in a global coupled ocean-atmosphere climate model. *J. Climate* 3: 72–98

Meehl, G. A. 1990b. *Proc. Fourteenth Annual Climate Diagnostics Workshop*. Washington, DC: U.S. Dept. Commerce. pp. 40–46

Meehl, G. A., Branstator, G. W., Washington, W. M. 1991. El Niño-Southern Oscillation and CO_2 climate change. *J. Climate*. Submitted

Moura, A. D., Shukla, J. 1981. On the dynamics of droughts in northeast Brazil: Observation, theory, and numerical experiments with a general circulation model. *J. Atmos. Sci.* 38: 2653–75

Murakami, T. 1987a. Orography and monsoons. In *Monsoons*, ed. J. S. Fein, P. L. Stephens, pp. 331–64. New York: Wiley

Murakami, T. 1987b. Effects of the Tibetan Plateau. In *Monsoon Meteorology*, ed. C.-P. Chang, T. N. Krishnamurti, pp. 235–70. New York: Oxford Univ. Press

Nicholls, N., McBride, J. L., Ormerod, R. J. 1982. On predicting the onset of the Australian wet season at Darwin. *Mon. Weather Rev.* 110: 14–17

Nicholson, S. E. 1983. Sub-Saharan rainfall in the years 1976–80: evidence of continued drought. *Mon. Weather Rev.* 111: 1646–54

Pant, P. S. 1983. A physical basis for changes in the phases of the summer monsoon over India. *Mon. Weather Rev.* 111: 487–95

Philander, S. G. 1990. *El Niño, La Niña, and the Southern Oscillation*. San Diego: Academic. 293 pp.

Ramage, C. S. 1968. Role of a tropical "maritime continent" in the atmospheric circulation. *Mon. Weather Rev.* 96: 365–70

Ramage, C. S. 1971. *Monsoon Meteorology*. New York: Academic. 296 pp.

Raman, C. R. N., Rao, Y. P. 1981. Blocking high over Asia and monsoon droughts over India. *Nature* 289: 271–73

Rao, R. V., Erdogan, S. 1989. The atmospheric heat source over the Bolivian Plateau for a mean January. *Boundary-Layer Meteorol.* 46: 13–33

Rasmusson, E. M. 1967. Atmospheric water vapor transport and the water balance of North America: Part I. Characteristics of the water vapor flux field. *Mon. Weather Rev.* 95: 403–26

Rasmusson, E. M., Carpenter, T. H. 1982. Variations in tropical sea surface temperature and surface wind fields associated with the Southern Oscillation/El Niño. *Mon. Weather Rev.* 110: 354–84

Rasmusson, E. M., Carpenter, T. H. 1983. The relationship between eastern equatorial Pacific sea-surface temperatures and rainfall over India and Sri Lanka. *Mon. Weather Rev.* 111: 517–28

Reiter, E. R., Gao, D.-Y. 1982. Heating of the Tibet Plateau and movements of the South Asian high during spring. *Mon. Weather Rev.* 110: 1694–1711

Reyes, S., Cadet, D. L. 1988. The southwest branch of the North American monsoon during summer 1979. *Mon. Weather Rev.* 116: 1175–87

Riehl, H. 1979. *Climate and Weather in the Tropics*. New York: Academic. 611 pp.

Ropelewski, C. F., Halpert, M. S. 1987. Global and regional scale precipitation patterns associated with the El Niño/Southern Oscillation. *Mon. Weather Rev.* 115: 1606–26

Ruddiman, W. F., Prell, W. L., Raymo, M. E. 1989. Late Cenozoic uplift in southern Asia and the American West: Rationale for general circulation modeling experiments. *J. Geophys. Res.* 18,379–91

Schwerdtfeger, W. 1961. Strömmungs und Temperaturfeld der Freien Atmosphäre über den Anden. *Meteorol. Rundsch.* 14: 1–6

Schwerdtfeger, W. S., ed. 1975. Climates of Central and South America. *World Survey of Climatology. Vol. 12.* Amsterdam: Elsevier. 532 pp.

Semazzi, F. H. M. 1980. Stationary barotropic flow induced by a mountain over a tropical belt. *Mon. Weather Rev.* 108: 922–30

Shen, R., Reiter, E. R., Bresch, J. F. 1986. Some aspects of the effects of sensible heating on the development of summer weather systems over the Tibetan Plateau. *J. Atmos. Sci.* 43: 2241–60

Shukla, J., Misra, B. M. 1977. Relationship between sea surface temperature and cloud speed over the Central Arabian Sea and monsoon rainfall over India. *Mon. Weather Rev.* 105: 998–1002

Smith, W. P., Gall, R. L. 1989. Tropical squall lines of the Arizona monsoon. *Mon. Weather Rev.* 117: 1553–69

Staff Members of the Sector of Synoptic and Dynamic Meteorology, Institute of Geophysics and Meteorology, Academica Sinica 1957. On the general circulation over eastern Asia. *Tellus* 9: 432–46

Stouffer, R. J., Manabe, S., Bryan, K. 1989. Interhemispheric asymmetry in climate response to a gradual increase of atmospheric CO_2. *Nature* 342: 660–62

Taljaard, J. J. 1955. Stable stratification in the atmosphere over Southern Africa. *Notos* 4: 217–30

Taljaard, J. J. 1972. Synoptic meteorology of the Southern Hemisphere. In *Meteorology of the Southern Hemisphere*, ed. C. W. Newton, pp. 139–213. Boston: Am. Meteorol. Soc.

Taljaard, J. J., van Loon, H., Crutcher, H. L., Jenne, R. L. 1969. *Climate of the Upper Air: Vol. I, Southern Hemisphere Temperatures, Dew Points, and Heights at Selected Pressures.* NAVAIR 50-1C-55. Washington, DC: Supt. Doc.

Tanaka, M. 1981. Interannual fluctuations of the tropical monsoon circulation over the greater WMONEX area. *J. Meteorol. Soc. Jpn.* 59: 825–31

Tanaka, M. 1982. Interannual fluctuations of the tropical easterly jet and the summer monsoon in the Asian region. *J. Meteorol. Soc. Jpn.* 60: 865–75

Tanaka, M. B., Weare, C., Navato, A. R., Newell, R. E. 1975. Recent African rainfall patterns. *Nature* 255: 201–3

Tanaka, M., Yoshino, M. M. 1982. *Proc. International Conference on the Scientific Results of the Monsoon Experiment.* Geneva: World Meteorol. Organ. 6.9–6.12

Trenberth, K. E. 1991. General characteristics of El Niño-Southern Oscillation. In *Teleconnections Linking Worldwide Climate Anomalies*, ed. M. H. Glantz, R. W. Katz, N. Nicholls, pp. 13–42. Cambridge: Cambridge Univ. Press

Trenberth, K. E., Chen, S.-C. 1988. Planetary waves kinematically forced by Himalayan orography. *J. Atmos. Sci.* 45: 2934–48

Trewartha, G. T. 1966. *The Earth's Problem Climates*. Madison: Univ. Wis. Press. 334 pp.

Tribbia, J. J. 1991. The rudimentary theory of atmospheric teleconnections associated with ENSO. In *Teleconnections Linking Worldwide Climate Anomalies*, ed. M. H. Glantz, R. W. Katz, N. Nichols, pp. 285–308. Cambridge: Cambridge Univ. Press

Troup, A. J. 1961. Variations in upper tropospheric flow associated with the onset of the Australian summer monsoon. *Indian J. Meteorol. Geophys.* 12: 217–30

Tyson, P. D. 1984. The atmospheric modulation of extended wet and dry spells over

South Africa, 1958–1978. *J. Climatol.* 4: 621–35

Tyson, P. D. 1986. *Climate Change and Variability in Southern Africa*. Capetown: Oxford Univ. Press. 220 pp.

Virji, H. 1981. A preliminary study of summertime tropospheric circulation patterns over South America from cloud winds. *Mon. Weather Rev.* 109: 599–610

Wallace, J. M., Tibaldi, S., Simmons, A. J. 1983. Reduction of systematic forecast errors in the ECMWF model through the introduction of an envelope orography. *Q. J. R. Meteorol. Soc.* 103: 683–717

Washington, W. M. 1981. A review of general circulation model experiments on the Indian monsoon. In *Monsoon Dynamics*, ed. J. Lighthill, R. P. Pearce, pp. 111–30. Cambridge: Cambridge Univ. Press

Washington, W. M., Meehl, G. A. 1989. Climate sensitivity due to increased CO_2: Experiments with a coupled atmosphere and ocean general circulation model. *Climate Dynamics* 4: 1–38

Weinert, R. A. 1968. Statistics of the subtropical jet stream over the Australian region. *Aust. Meteorol. Mag.* 16: 137–48

Wyrtki, K. 1965. The annual and semiannual variation of SST in the North Pacific Ocean. *Limnol. Oceanogr.* 10: 307–13

Yeh, T.-C. 1981. Some characteristics of the summer circulation over the Qinghai-Xizang (Tibet) Plateau and its neighborhood. *Bull. Am. Meteorol. Soc.* 62: 14–19

Yeh, T.-C., Dao, S. Y., Li, M.-T. 1959. The abrupt change of circulation over the Northern Hemisphere during June and October. In *The Atmosphere and the Sea in Motion*, pp. 249–67. Rossby Memorial Vol. New York: Rockefeller Inst. Press

Annu. Rev. Earth Planet. Sci. 1992. 20:113–42

IN SITU INVESTIGATION OF NUCLEATION, GROWTH, AND DISSOLUTION OF SILICATE CRYSTALS AT HIGH TEMPERATURES

Ichiro Sunagawa[1]

Kashiwa-cho 3-54-2, Tachikawa, Tokyo 190, Japan

KEY WORDS: crystal growth mechanisms, metastable nucleation, crystal growth rate

1. INTRODUCTION

Imperfections and inhomogeneities in single crystals of minerals and the textures of rocks and ores appear primarily as a result of nucleation and growth of crystals, and secondarily through post-growth histories such as dissolution, exsolution, and phase transition. How and under what conditions mineral crystals nucleated, grew, dissolved, or transformed, how rock textures appeared, and what sort of post-growth histories they experienced have been deduced by analyzing these characteristics. Investigations of earth and planetary materials along this line are as important as equilibrium thermodynamic analyses of mineral parageneses, since they provide information on kinetic processes. Various sophisticated methods have been used to detect and analyze these characteristics, including optical and electron microscopies, electron probe micro-analysis and other surface analysis methods, cathode luminescence topography, X-ray topography, laser-beam tomography, etc. A deductive analysis of this type, however, should be supplemented by a theoretical and experimental understanding

[1] Former affiliation: Institute of Mineralogy, Petrology, and Economic Geology, Tohoku University, Aramaki, Aoba-ku, Sendai, 980 Japan.

0084–6597/92/0515–0113$02.00

of the mechanisms of nucleation, growth, dissolution, and phase transition of crystals.

Experimental verification of crystal growth mechanisms has been one of the main subjects in the science of crystal growth. Principal methods in this field have been: 1. characterization of lattice defects and morphological features of single crystals in relation to growth parameters; 2. investigation of the relations between the rates of nucleation, growth, or dissolution and the driving force (supercooling or supersaturation) and analysis of the obtained data compared to theoretical expectations (since in this type of experiment, these rates are measured from outside on crystals whose growth process is not observed, the method is in principle an indirect ex situ method); and 3. in situ observation of growth or dissolution processes. Concepts important to crystal growth mechanisms—e.g. diffusion boundary layers, the Berg effect, spiral growth, rough and smooth interfaces, the roughening transition of an interface, etc—have emerged or have been verified through these investigations. [For the implications of these terms, the reader may refer to Hartman (1973) or Chernov (1984).] Experimental verification of crystal growth mechanisms by methods 2 and 3 have been mainly made on crystals of simple compounds that grow from vapor and aqueous solution phases at ordinary temperature and one atmosphere of pressure. This is because in such cases precise control of growth conditions is easy to obtain. For crystal growth at high temperatures—such as Czochralski growth of silicon crystals, flux growth, hydrothermal growth, and epitaxial thin film growth—the growth mechanism and defect formation have been investigated mainly by methods 1 and 2 on simple and pure systems.

In the earth and planetary sciences where multi-component systems are of primary interest and conditions fluctuate greatly, equilibrium phase analysis of mineral parageneses has been the principal method used to solve the problem. Dynamic crystallization experiments started in the 1970s. The goals were to understand how morphologies of crystals varied depending on the driving force and composition and to determine how various rock textures were formed (see e.g. Lofgren 1980). In these studies, growth conditions were changed during a run, followed by quenching and texture observations. Experiments of this type, however, gradually faded out because although they were useful demonstrations the data obtained were not quantitative or reproducible enough to be suitable for theoretical analysis.

Following the dynamic crystallization experiments, experiments on silicate systems using method 2 were started. The conditions were well controlled, but the growth rates were measured on quenched samples, and the crystallization process itself was completely concealed. There was a trial to directly observe the crystallization process at high temperatures (Kirkpatrick et al 1976), but the visibility was low and the observations were limited to a narrow temperature range.

In the experimental verification of crystal growth mechanisms of a simple compound from an aqueous solution by means of method 2, growth conditions can be controlled more accurately than for the case of high temperature silicate systems, and the rates of growth or dissolution are measured directly, either by the weighing method or by measuring the advancement of an interface. Yet, it has gradually been realized that the data obtained had a high degree of scatter and the growth rates fluctuated considerably during a run, for reasons not entirely clear (see e.g. Tsukamoto 1989a,b). The uncertainty came from the fact that the experimental methods used were in principle indirect ones. The crystallization process was neither directly observed nor recorded, and growth rates were measured by the weight change of a crystal or advancement rate of a face as a whole, but not on individual growth hillocks on a face. Perfection of the crystal under investigation—particularly the nature of individual lattice defects which generate spiral growth layers—was also not well characterized.

If the whole process of crystallization could be observed in situ at a visibility as high as in the characterization methods of category 1, and the growth conditions could be controlled as precisely as in the method of category 2, then we would have the most powerful experimental method to verify the mechanisms of crystallization. The methods of category 3, applied in the past to verify crystal growth mechanisms, were successful to visualize the spiral growth process or concentration gradient around a growing crystal. However, although good for demonstration purposes, they do not fulfill the above requirements.

In the author's former laboratory at Tohoku University, a series of experimental methods which may serve the above purpose have been developed in the past ten years. These have been applied to direct and in situ investigations of nucleation, growth, and dissolution of crystals in aqueous solution at ordinary temperature and pressure. With these methods, a number of problems have so far been investigated at nanometer order visibility and temperature control of less than 0.01°C (see e.g. Tsukamoto 1989a, Sunagawa 1991). The outline, ability, and advantages of these methods are summarized in Table 1, and compared with those of high temperature in situ investigation methods—the main topic to be discussed in this paper. The problems studied include:

1. To detect cluster formation of nanometer size in a solution and measure cluster size and density, and determine the diffusion constant, laser light scattering and Raman spectroscopy have been used (Azuma et al 1989).

Table 1 Comparison of in situ investigation methods of crystallization processes in aqueous solution at ordinary temperatures, in silicate liquids at high temperatures, and by ex situ investigation methods[a]

	In situ				Ex situ
	Aqueous solution Ordinary temperature	Methods[b]	Silicate system High temperature	Remarks[b]	Remarks
Condition control					
T, range	100°C	microcomputer, thermocouple, thermostated	1600°C	see text (4–7)	homogeneous T distribution
T, fluctuation	<1/100°C	or clean room (1–3)	0.3°C	sharp T gradient	indirect measurement of T, ΔT, R
P	1 atm.		1 ~ 20 atm.		
σ_b	0.0x%	from solubility curve	ΔT (1°C)		
σ_s	measured	Mach-Zehnder (8)	not yet		
u	<0 cm/s	two-bath flow system (9, 10)	not yet	not controllable	
Observable phenomena in bulk solution					
Clustering (r, D)	clusters of 5 nm (r)	light scattering, Raman (11)	observable		nothing observable
Heterogeneous nucleation	observable		observable		
Convection	observed	Schlieren (12, 13)	observed		
Diffusion boundary layer	observed	Mach-Zehnder (8, 9)	observed		

Observable phenomena at the interface and in a bulk crystal					
Growth step pattern	steps of nm height are discernible	PCM, DICM (3, 14, 20)	less visibility		nothing observable
Dissolution step and etch pit	Ibid.	Ibid.	observed		
Morphological change	observed	Ibid., PC, video	observed		
Growth sector	observed	Ibid.	observed		
Growth bands	observed	Ibid.	observed		
Inclusion trapment	observed	Ibid.	observed		
Reaction	observed	Ibid.	observed		
Dislocations	identified	PC, etch pit (15, 16)	not yet observed		
Measurable kinetic rates					
Nucleation					
Incubation time	measured	light scattering	measured (7, 17, 18)	direct observation on video	indirect
Probability			measured		
Growth and dissolution					
R and $-R$, interface	measured	Michelson, video (10, 18)	measured	equal thickness fringes, video	indirect
v and $-v$, spiral steps	measured	Michelson, video (10, 19)	not yet		
R, v, p of individual growth hillock	measured	Michelson (9, 19)	not yet		

[a] Abbreviations: T, temperature; ΔT, temperature difference between equilibrium and growth temperatures; P, pressure; σ_b, supersaturation of a solution apart from a growing crystal; σ_s, supersaturation of a solution adjacent to a growing interface; u, flow velocity of a solution; r, size of cluster; D, diffusion constant of clusters; R and $-R$, normal growth and dissolution rates of an interface or a growth hillock, or an etch pit; v and $-v$, advancing and retarding rates of a spiral layer or a dissolution step; p and $-p$, slope of a growth hillock and an etch pit; PC, PCM, DICM, various optical microscopes (see text); Mach-Zehnder, Michelson are interferometric techniques (see text).

[b] References: (1) Tsukamoto 1989a, (2) Tsukamoto 1989b, (3) Sunagawa 1991, (4) Abe 1988, (5) Tsukamoto et al 1983, (6) Abe et al 1987, (7) Abe et al 1991, (8) Onuma et al 1989a, (9) Onuma 1988, (10) Onuma et al 1990, (11) Azuma et al 1989, (12) Onuma et al 1988, (13) Onuma et al 1989b, (14) Tsukamoto 1983, (15) Maiwa 1987, (16) Maiwa et al 1989, (17) Nakamura 1988, (18) Nakamura et al 1990, (19) Maiwa et al 1990, (20) Tsukamoto & Sunagawa 1985.

2. To visualize the diffusion boundary layer and buoyancy-driven convection around and from a growing or dissolving crystal, a Schlieren technique was used. The concentration gradient was measured by Mach-Zehnder interferometry (Onuma 1988, Onuma et al 1988, 1989b).
3. To investigate the effect of solution flow upon growth or dissolution related phenomena, a two-bath flow system was used (Onuma et al 1989a).
4. Mach-Zehnder interferometry was applied to measure supersaturation adjacent to a growing interface (surface supersaturation) (Onuma 1988, Onuma et al 1989a).
5. The advancement of spiral growth layers with heights of nanometer order have been observed and rates measured under phase contrast and interference contrast microscopes and by Michelson interferometry (Tsukamoto 1983, 1989b, Onuma et al 1990).
6. The in situ identification of the dislocation characteristics in a growing crystal has been accomplished by polarizing microscopy and other methods (Maiwa 1987, Maiwa et al 1989, Onuma et al 1990).
7. To measure the normal growth rates of a face and of an individual growth hillock on the face, its slopes, and the advancing rate of the spiral layers, Michelson interferometry has been used (Maiwa et al 1990, Onuma et al 1990).

In these methods, the visibility is enhanced because various sensitive optical and interferometric methods are extensively used. A particular mention should be made of a newly designed objective lens which minimizes optical aberration due to the presence of the solution and the cover glass which is placed between the crystal surface and the objective lens (Tsukamoto & Sunagawa 1985). Well controlled conditions are obtained by using micro- (or personal) computers and performing experiments in a clean or a thermostated room. The clean room is especially useful in the experiments using interferometry, since this greatly reduces the specks on the interferograms due to dust in the air.

Based on these experiences, we have further developed in situ investigation methods workable at temperatures up to 1600°C. We have directly observed the crystallization process and measured the rates of nucleation, growth, or dissolution in liquids of silicate, oxide, and other components at temperatures of 1600°C, with temperature fluctuation less than 0.5°C. This was done by preparing a thin liquid film in a small electric furnace installed on the stage of an optical microscope. Although the visibility and condition control have not yet reached the level of the in situ observation methods of aqueous solution growth, they offer vast improvement over any previous methods used to investigate kinetic problems at high temperatures.

In this paper, the newly developed in situ investigation methods of kinetic problems at high temperatures and selected results obtained by these methods are presented. The work has been done in close collaboration with K. Tsukamoto, T. Abe, H. Nakamura, M. Kawasaki, M. Abe, and Y. Kono, at the Institute of Mineralogy, Petrology and Economic Geology, Tohoku University, Sendai, Japan.

2. THE IN SITU INVESTIGATION METHOD AT HIGH TEMPERATURES

2.1 *Basic Idea and Outline of the Experimental Setup*

To enable direct in situ observation of crystallization processes at high visibility, a thin film of high temperature silicate liquid is prepared so that observations can be made by transmitted light. A loop heater in an electric furnace installed on the stage of an optical microscope is used for this purpose. The liquid film is held by its surface tension on the loop, and crystals are allowed to nucleate and grow inside of this film at a desired supercooling. The film thickness is determined by the thickness of the loop—generally less than 1 mm. Supercooling is attained either by changing the current in the loop heater or through a thin thermocouple (less than 0.1 mm thick) inserted into the film. The thermocouple has dual purposes: 1. to measure and control temperature and 2. to hold crystals steady for observation. Since the operation is equivalent to performing an operation in a bubble film in a loop wire, care and skill are required.

Owing to the surface tension of a film and the thermal gradient between the heater and the central portion of the film, both Marangonian and thermal convection are present in the film. This is advantageous in a sense, because it ensures homogeneous mixing of the liquid and suppresses heterogeneous nucleation on the heater.

Optical microscopes of various types—including polarization microscopes (PM), phase contrast (PCM) and differential interference contrast microscopes (DICM)—are used for observation to secure high visibility. Schlieren techniques are applied to visualize concentration differences, liquid immiscibility, or convection. Light scattering methods can be used to observe clustering phenomena. Crystallization processes are observed either under the microscope or on a TV screen, using a CCD video camera. The whole process can be recorded on videotape, for repeated investigations and kinetic rate measurements. On a TV screen, temperature and time are superimposed. A microcomputer is used for this purpose; it also is used to regulate the temperature of the liquid film.

A block diagram of the system is illustrated in Figure 1. It consists of the following three parts:

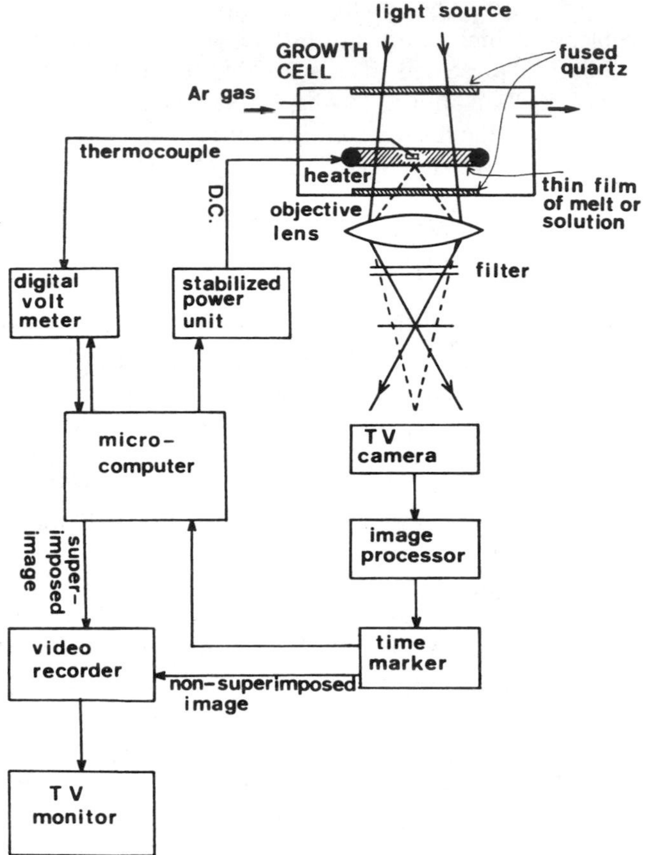

Figure 1 A block diagram of the experimental setup for in situ investigation of crystallization processes at high temperatures. (After Abe et al 1987.)

1. Growth cell: a small electric furnace installed on the microscope stage.
2. Observation and recording system: a transmission optical microscope, a CCD video camera, an image processor, an image marker, a video recorder, and a TV monitor.
3. Temperature control system: a microcomputer, a stabilized power unit connected with a loop heater, and a digital voltmeter connected with a thermocouple.

In the following, details of the respective parts, sample preparation, and some of the experimental techniques are explained.

2.2 *Growth Cell*

In the prototype system, a heating stage of a commercial high temperature microscope was utilized as a growth cell, by replacing the heating element with a loop of $Pt_{80}Rh_{20}$ wire, which gave promising results (Tsukamoto et al 1983). Based on this success, a variety of growth cells were designed to meet different purposes (Abe et al 1987). Figure 2 shows a block diagram of one model. In this model, the tip of a thermocouple inserted into the liquid can be moved with the help of a microscrew. In another model, a small gold-coated spherical mirror is installed around the loop to homogenize the temperature distribution. A conical hollow is prepared in the mirror through which a laser beam is guided to the liquid film for light scattering experiments. Also a growth cell that can withstand vapor pressure up to 20 atmospheres was designed for experiments on systems containing volatile components, like plagioclase. The growth cells of all models are water-cooled from outside.

The atmosphere in the growth cell is controlled with inert gas, supplied through a gas flow meter from a reservoir.

Figure 3 shows the total setup of such an experimental system.

2.3 *Temperature Distribution in the Loop*

Depending on the composition of the liquid film, the temperature distribution between the periphery near the loop and the central portion of a liquid film may be different. The temperature difference between the periphery and the central parts of the film is about 50–100°C at about 1400°C. Within the field of microscopic vision (600×800 μm) the temperature difference is much smaller, in the range of 5–10°C. Owing to the temperature difference and associated thermal convection, heterogeneous nucleation on the loop heater—which is always a serious problem in ex situ experiments—is greatly suppressed and homogeneous liquid mixing is guaranteed.

2.4 *Temperature Controlling System*

Commonly used control systems for temperature regulation—for instance Proportional Integration and Differentiation systems—are made for furnaces with relatively large thermal capacities. In the present system, owing to the small thermal capacity of both the liquid and heater, this is not applicable because a perfect heat balance has to be maintained. For this purpose, a highly stabilized power unit is employed, and the regulation of the current is applied only when the heat balance varies gradually or when abrupt nucleation of crystals takes place releasing latent heat. A DC power supply was chosen so as to minimize the disturbance of the TV image caused by alternating magnetic fields.

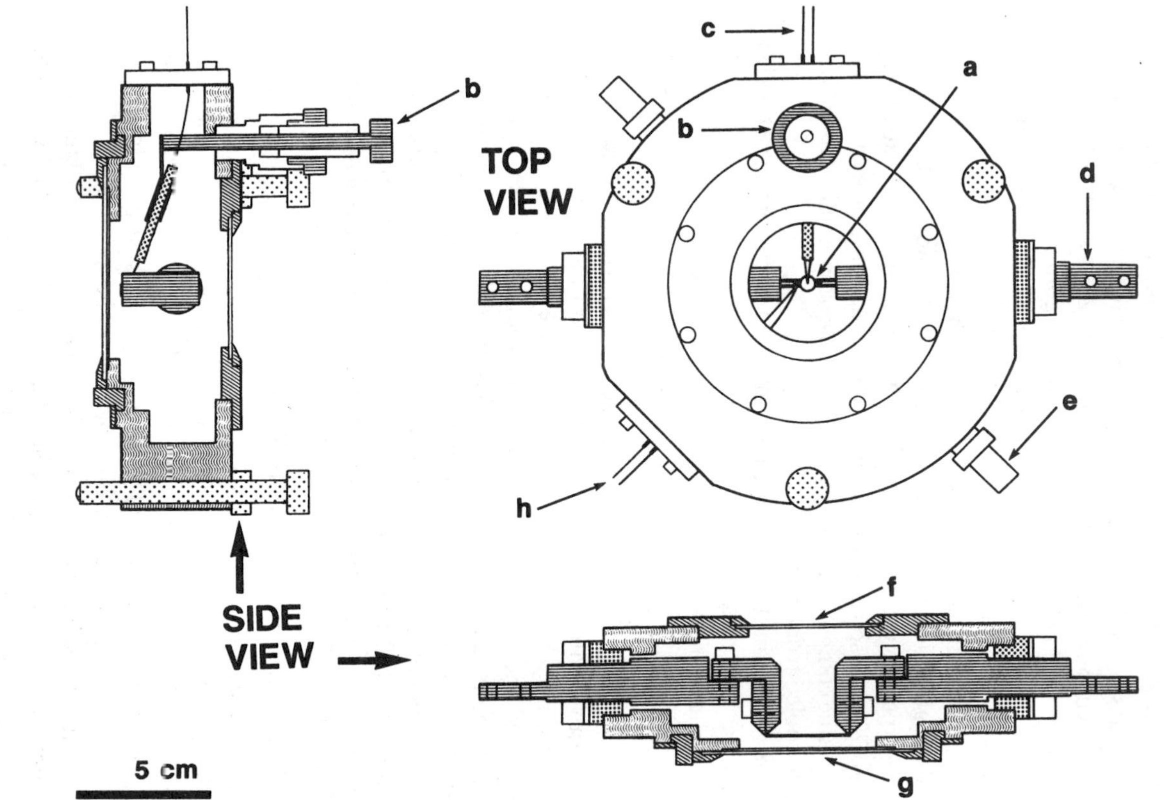

Figure 2 A draft of one model of the growth cell. (*a*) loop heater, (*b*) manipulator, (*c*) thermocouple, (*d*) fixed element for electrical supply, (*e*) element for gas supply, (*f*) window of fused quartz, (*g*) radiation shield and window, (*h*) thermocouple for measuring heater temperature. (After Abe et al 1987.)

Figure 3 Growth cell (*upper*) and total experimental setup (*lower*) of the high temperature in situ investigation method. M: reversed type optical microscope with a growth cell and a CCD video camera, VM: video timer, V: digital multimeter, C: microcomputer, U: video recorder, P: printer, T, E: image processor.

Temperature is measured by a digital voltmeter employing a $Pt_{80}Rh_{20}$ thermocouple (0.075 mm thick). The data are then sent to a microcomputer, which calibrates the measurement and sends a signal to a DC stabilized power unit for controlling electric current in the heater. The stability of the voltmeter and the power supply are 0.003% and 0.005%, respectively. The resolving power of the voltmeter is 0.1 μV. Thus a stability of 0.3°C at 1600°C is guaranteed.

2.5 *Observation and Recording Systems*

To improve visibility in this type of observation, two problems have to be overcome: (*a*) the effect of infrared rays radiating from the high temperature body and (*b*) the aberration associated with observation through the liquid layer and window glass. Problem (*a*) was solved by the use of a filter which cuts off infrared rays. By a combination of appropriate filters, it should be possible to investigate thermal behaviors associated with the crystallization process—although this has not yet been done. Problem (*b*) was overcome by a newly designed objective lens of magnification 20 with a 7 mm-long working distance (Olympus TS-20) (Tsukamoto & Sunagawa 1985). This lens was designed to fit with a differential interference contrast microscope (DICM) of the Nomarski type, but can also be used with other types of microscopes. It remarkably improves the visibility.

Depending on the purpose, any optical microscope can be used. Under the polarization microscope (PM), crystals show equal thickness fringes (see Figure 4). Three-dimensional imagery of crystals and precise measurements of growth or dissolution rates are possible. Phase contrast microscopes (PCMs) and DICMs are useful for observing growth layers with small step heights developing on a face. Figure 5 shows an example of growth layers developing from the corner of a face on a pseudo-hexagonal $CaAl_2Si_2O_8$ crystal. Both bunched thick layers and thinner layers are discernible. Although the visibility is still not as high as in the case of aqueous solution growth, it is significant that layer-by-layer growth was directly detected on a silicate crystal growing at high temperatures. PCMs and DICMs can also be used to detect concentration differences in a liquid film, but such visualizations are easier with the Schlieren technique. This technique can be applied simply by placing a knife-edge and a small pinhole in the optical path. Figure 6 shows three examples: a convection plume (*middle*), a diffusion boundary layer around a growing crystal (*bottom*), and convection and diffusion combined around a dissolving crystal (*top*).

On the TV monitor screen, time and temperature data are superimposed by a combination of an image enhancer, a timer, and a microcomputer. Since the images of the whole crystallization process are recorded together with these data on videotape, repeated investigations of the crystallization

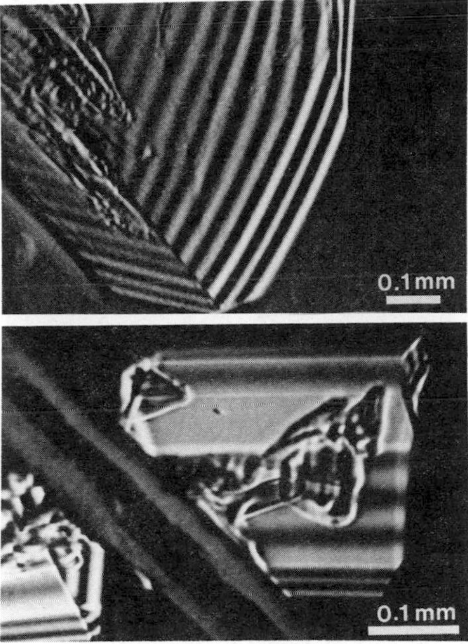

Figure 4 Equal thickness fringes observable under a polarization microscope. Diopside crystal growing in an An_{100} liquid at 1380°C (*upper*), forsterite crystal growing in a Di_{100} liquid at 1305°C. (Courtesy of T. Abe.)

process and precise measurements of nucleation, growth, or dissolution rates are possible.

Only the crystals growing or dissolving within the liquid film are used as subjects of investigations. Identifying those crystals that grow submerged in the liquid film is easy because the surface of a crystal exhibits drastic change as soon as it emerges.

2.6 *Experimental Procedure*

Depending on the purpose of the experiment and the composition of the system, different sample preparation techniques may be adopted. In general, a slice of the appropriate size and thickness to fit the loop is prepared either from a sintered rod of a desired composition or from a glass sample. The slice is placed on the loop heater and is melted for an appropriate duration at a superheat temperature in order to ensure homogeneous melting. The temperature is then lowered to a desired supercooling temperature. For demonstration purposes or just to observe the changes and movement in a film, a thermocouple is unnecessary. For accurate measurements of kinetic rates and conditions, a thermocouple is inserted into the liquid, on which a desired crystal is nucleated or picked

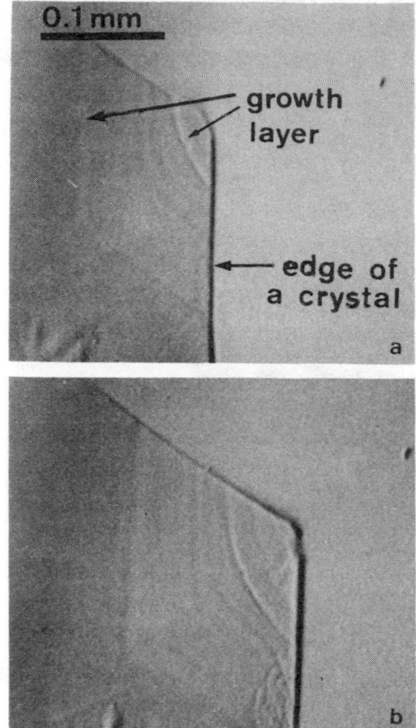

Figure 5 Growth steps observed on the face of a pseudo-hexagonal $CaAl_2Si_2O_8$ crystal (Tsukamoto et al 1973). Note that both thin and bunched thick growth layers originate from the corner of the face and advance inward, forming a hopper face. Frame *b* was taken 10 seconds after frame *a*.

up. Both isothermal and dynamic crystallization experiments are possible while the crystallization process is observed in situ.

The seeding technique is useful and often applied when nucleation of a desired phase is difficult for a given condition. First, the liquid is substantially supercooled so that crystals can easily nucleate. After picking up a desired crystal on the thermocouple, the temperature is then raised to dissolve unnecessary crystals. Crystals that survive can be used as a seed for further growth or to measure growth rates at a given temperature. By this technique, growth or dissolution rates of any phase can be measured over a wide temperature range.

Once ex situ identification is done on quenched samples, in situ identification of phases appearing in a liquid of known composition can be accomplished by determining the morphological characteristics and refractive indexes of the growing crystals. Optical observation, X-ray microdiffractometry, and electron microprobe analysis are applicable to identify the phases on quenched samples. Because of the small volume of a liquid film, quenching is easier and provides more faithful samples than in the case of ex situ experiments.

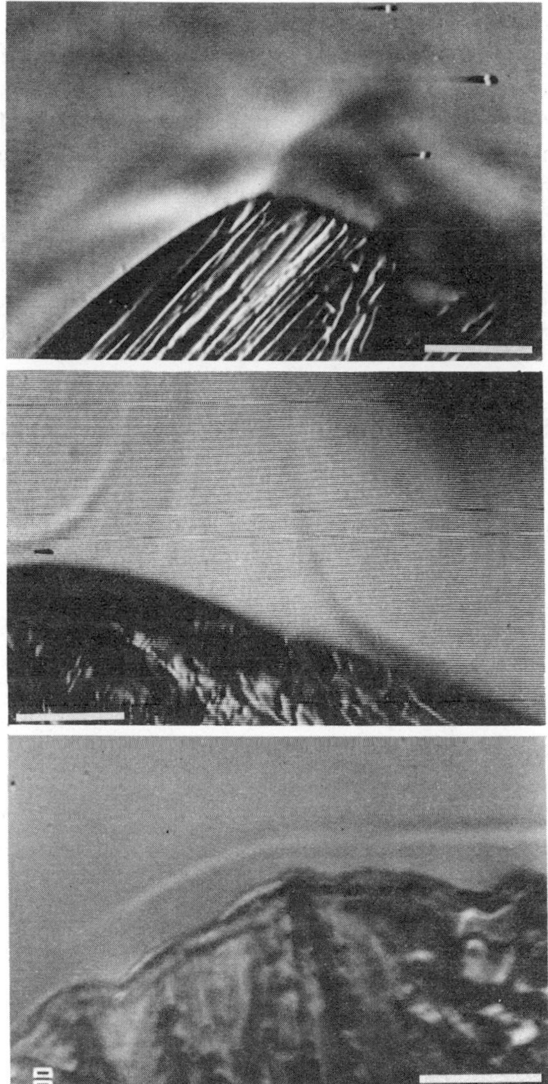

Figure 6 Three Schlieren photographs showing both diffusion and convection around rapidly dissolving forsterite crystals in a $Di_{60}En_{40}$ liquid at 1460°C (*upper*), a convection plume arising from a forsterite spherulite growing in a $Di_{70}En_{30}$ liquid at 1430°C (*middle*), and a diffusion boundary layer around tridymite crystals growing in a Fo_{61}-$Silica_{39}$ liquid. Bar indicates 0.1 mm. (Courtesy of T. Abe.)

2.7 *Advantages of the Method*

The primary advantage of our method is that the whole process of crystallization taking place under a known and well-controlled condition can be directly observed at high visibility and recorded on videotape for repeated investigations and precise rate measurements. It follows from these that a number of problems can be better investigated using direct in situ rather than any of the indirect, ex situ methods.

1. Nucleation, growth, or dissolution of crystals in silicate liquids can be analyzed. Concrete knowledge on the changes and fluctuations that took place during the process—such as the role of the diffusion boundary layer, convection plumes, liquid immiscibility, etc—can be gained.
2. The position and time of appearance of crystals in a liquid film can be directly determined. Nucleation phenomena—such as heterogeneous nucleation, metastable nucleation, rate and probability of nucleation, incubation time, etc—can be investigated directly. This has been impossible in ex situ investigations. Also, fewer experimental runs are required to obtain reliable data on nucleation than in ex situ experiments.
3. Morphological changes of silicate crystals, including bulk external morphology (dendritic, hopper, and polyhedral), internal morphology (growth sector), surface microtopography (growth step patterns on a face), interface roughness, and roughening transition of a face, can be directly investigated in relation to various growth parameters, both in growth and dissolution processes. Formation of growth banding, chemical zoning, growth sectors, and their behaviors in dissolution or phase transformation can be directly investigated.
4. The melting point (liquidus) of any crystalline phase can be determined precisely based on the transformation from a polyhedral to rounded morphology. Figure 7 demonstrates how the melting point of metastably nucleated corundum in a pure anorthite melt was determined. When a metastable phase occurs as one of the stable phases in the relevant equilibrium phase diagram, the metastable liquidus may be evaluated by extrapolating the liquidus surface to the metastable region, assuming that the liquidus changes regularly with composition, at least over a short range. But for metastable phases other than this type, the present method is the only way to determine the liquidus temperature.
5. The rates of nucleation, growth, or dissolution can be measured repeatedly on videotape. Figure 8 shows examples of growth rates versus supercooling relations of stable and metastable polymorphs of $CaAl_2Si_2O_8$ appearing in a liquid of An_{100} composition. The super-

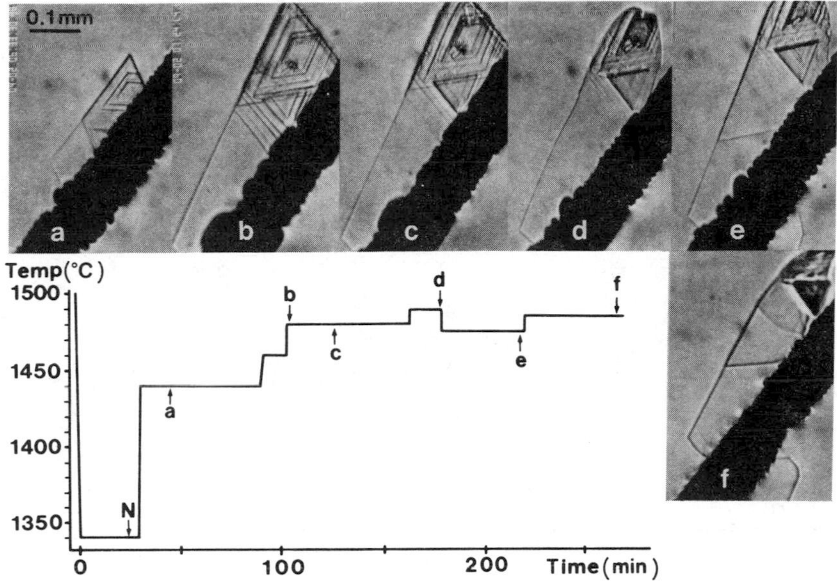

Figure 7 Determination of the liquidus temperature of a metastably nucleated phase based on its morphological transformation from polyhedral to rounded forms due to temperature change for metastably nucleated corundum in an An_{100} liquid. (From Abe et al 1991.)

cooling of metastable phases was obtained from the melting point measured by method 4.

These advantages indicate that this in situ investigation method at high temperature offers a powerful tool to investigate various kinetic problems encountered in earth and planetary sciences, as well as in other scientific and technological disciplines. The advantages of the method are compared in Table 1 with the previous ex situ methods of high temperature crystallization, and also with the in situ observation methods of crystal growth from aqueous solution at ordinary temperature and pressure. To demonstrate the applicability of the method, a few topics are selected and described in the next section.

3. SELECTED TOPICS FROM THE RESULTS

3.1 *Dynamic Movements*

When a liquid is supercooled by simply decreasing the temperature of a loop heater, with no help of a thermocouple, nucleation occurs sporadically in the liquid. Due to the presence of thermal and Marangonian convection, the crystals move and rotate gently in the liquid film and meet different

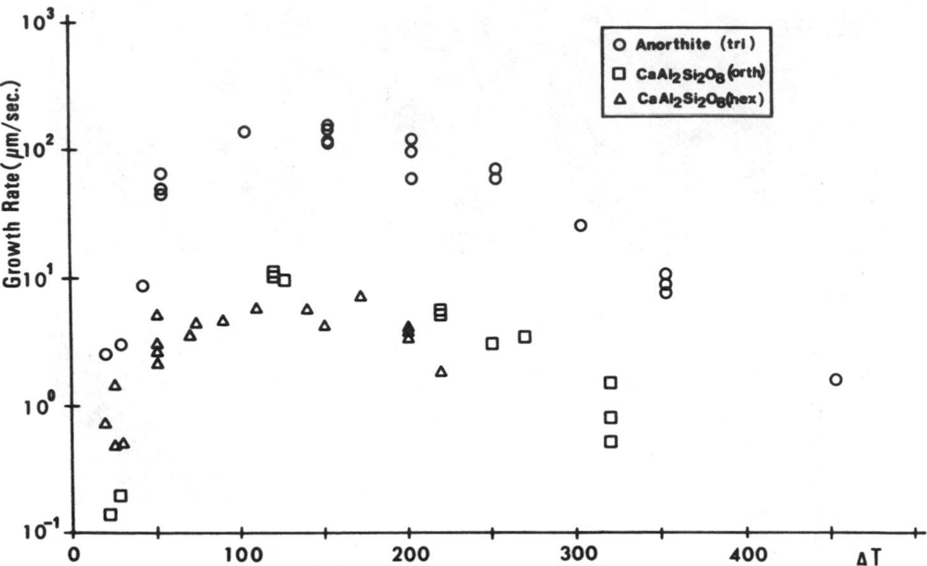

Figure 8 Growth rates versus supercooling relation of three $CaAl_2Si_2O_8$ polymorphs in an An_{100} liquid. The supercoolings of metastable polymorphs were obtained using the respective liquidus temperatures, measured by the in situ investigation method. (From Abe et al 1991.)

temperatures. Thus various dynamic movements and changes are observable, as if the observer were seeing the movements and changes inside a magma chamber. One example is illustrated by the sequence of photographs shown in Figure 9. At a large supercooling ($\Delta T > 250°C$), diopside takes a spherulitic morphology, consisting of a radiating polycrystalline aggregate. If the temperature is increased after their formation, the spherulites dissociate into a large number of tiny individual crystallites, which move away from the mother spherulites, owing to the presence of convection. Decreasing the temperature during the movement, the crystallites then start to grow again, forming larger single crystals in the liquid away from the mother spherulite.

In the gehlenite $Ca_2Al_2SiO_7$–åkermanite $Ca_2MgSi_2O_7$ system, sector-zoned crystals with chemical zoning and growth banding are formed. The process of their formation can be traced clearly. On dissolving these crystals, growth sectors rich in Åk start to dissolve at first, while those rich in Geh remain undissolved. As a result, the crystal dissociates in a sectorial manner.

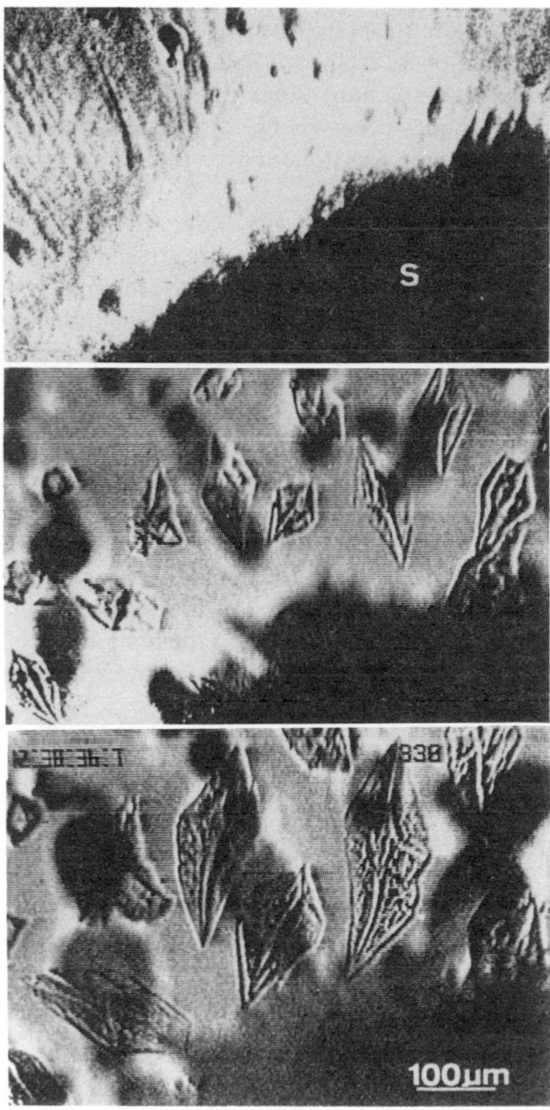

Figure 9 An example used to demonstrate dynamic movements in a liquid film. The diopside spherulite (*s*, in the *top*) dissociates into crystallites, which grow larger single crystals (*middle* and *bottom*) as they move away from the mother spherulite. (From Abe et al 1987.)

3.2 *Metastable Nucleation*

Prior nucleation of a phase (or phases) other than the phase that is expected from the equilibrium phase relation is called *metastable nucleation* in this paper. The phenomenon is more commonly encountered than might be assumed. Some examples observed by the present method are given in Table 2. The phenomenon was noticed even in indirect quenching experiments (Kirkpatrick et al 1983).

In Figure 10, the phases in an An_{100} liquid are indicated by their sequence of appearance on a *T*(temperature)-*t*(time) diagram (Abe et al 1991). The experiments were performed at isothermal conditions. One stable (triclinic anorthite) and two metastable (pseudo-orthorhombic and pseudo-hexagonal) polymorphs of $CaAl_2Si_2O_8$, corundum, and mullite appear in the pure An_{100} liquid, where the triclinic anorthite is the only stable phase. The stable triclinic anorthite never appears as the first phase at temperatures between 1400° and 1100°C. It appears only as the second phase below 1200°C. Even in that case, it rarely nucleates on the earlier formed metastable polymorphs. Only at temperatures below 1100°C, can anorthite nucleate as the first phase. In contrast, the two metastable $CaAl_2Si_2O_8$ polymorphs nucleate easily as the first phase or as the secondary phase after mullite in the temperature range between 1250°C and 1100°C. Mullite and corundum nucleate as the first phase between 1350°C and 1200°C, but are not observed at temperatures lower than 1200°C,

Table 2 Examples of metastable nucleation phenomena observed by the in situ investigation method

Liquid composition (wt%)	Stable phase	Phase appearing prior to the stable phase
An_{100}	Triclinic anorthite (An)	pseudo-hexagonal and pseudo-orthorhombic $CaAl_2Si_2O_8$ polymorphs, Mullite, Corundum
$An_{80}Wo_{20}$	An	two $CaAl_2Si_2O_8$ polymorphs
$Di_{40}An_{60}$	An	Diopside (Di)
$Di_{50}An_{50}$	An	Di
$Di_{60}An_{40}$	An	Di
Di_{100}	Di	Forsterite (Fo)
$Di_{80}Tp_{10}Ak_{10}$	Di	Fo
$Di_{65}Tp_{25}Ak_{10}$	Fo	Perovskite (Pv)
$Di_{55}Tp_{35}Ak_{10}$	Pv	Fo
$Di_{85}En_{15}$	Di	Fo

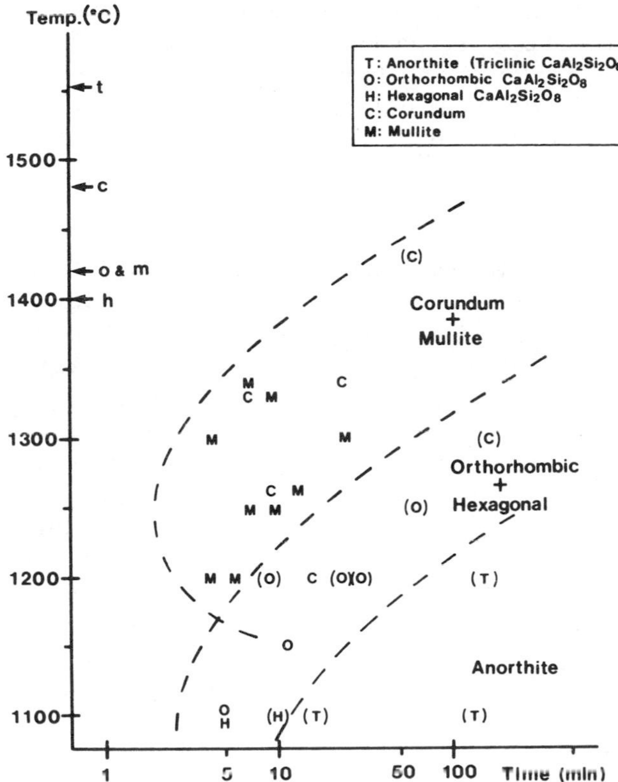

Figure 10 Temperature-time plot of phases appearing in an An_{100} liquid, demonstrating metastable nucleation phenomenon. The liquidus temperatures of respective phases measured by the in situ investigation method are indicated on the vertical axis. (From Abe et al 1991.)

where two metastable $CaAl_2Si_2O_8$ polymorphs dominate. As long as the stable triclinic anorthite is absent, the crystals of metastably nucleated phases grow steadily below the respective liquidus temperatures which were determined by the present in situ investigation method (see Figure 7). In a liquid of $An_{80}Wo_{20}$ composition, essentially similar phenomenon and the same sequence of appearance was observed for $CaAl_2Si_2O_8$ polymorphs, but much longer incubation time was needed. Corundum and mullite were not observed in this composition.

3.3 *Analysis of the Cause of Metastable Nucleation*

As probable causes of metastable nucleation, the following four candidates may be nominated: 1. heterogeneous nucleation of metastable phases on the heater, 2. the transient nature of nucleation of the stable phase, 3. the slower growth rate of the stable phase compared with the metastable phases, and 4. the smaller interfacial energy of metastable phases owing to a competition between the interfacial energy γ and the driving force $\Delta\mu$, both of which are related to the activation energy for nucleation,

$$A_N = \gamma^3/\Delta\mu^2.$$

From in situ observations of both An_{100} and $An_{80}Wo_{20}$ liquids, reason 1. can be excluded. It was also demonstrated that since the growth rate of stable triclinic anorthite is one order of magnitude faster than that of any metastably nucleated phase at a given temperature (see Figure 8), the second candidate can also be eliminated. So the remaining possibility is either 3. the transient nature of nucleation of the stable phase, or 4. the smaller solid-liquid interfacial energy of metastable phases, since the degree of supercooling that corresponds to the driving force is always much larger for the stable phase than for metastable phases (see Figure 10).

Nakamura (1988) and Nakamura et al (1990) analyzed the nature of nucleation in a composition of 65% diopside ($CaMgSi_2O_6$), 25% Ti-pyroxene ($CaTiAl_2O_6$), and 10% åkermanite ($Ca_2MgSi_2O_7$), based on Toschev's theory [Toschev et al 1972, Toschev 1973; see also Tsuchiyama (1983), who applied the theory for the first time to geological materials]. From the equilibrium phase relations for this composition, the first phase to appear in the liquid should be forsterite. However, in the temperature range 965°C–1114°C, the first phase to appear was always perovskite ($CaTiO_3$, Pv). The second phase was in most cases diopside ($CaMgSi_2O_6$, Di), and the stable forsterite (Fo) appeared only as the last phase.

Nakamura (1988) and Nakamura et al (1990) checked the effect of atmosphere, duration of superheat, positions of the thermocouple, and other factors upon nucleation, and confirmed that neither candidate 1. nor 2. could be the cause of the observed metastable nucleation. The growth rate of Fo was about one or more orders of magnitude faster than those of Pv and Di at any temperature investigated.

In Figure 11, the measured probability of nucleation P versus time t relation (a and b), and $-\ln(1-P)$ versus t relations (c and d) obtained for Pv and Di, respectively, at 1065°C are indicated. It is concluded from the two sets of figures that Pv exhibits a transient nature of nucleation, whereas Di shows a steady state nature. Fo, on the other hand, should have a steady state character nucleation—judging from its wide random region. If Fo nucleation had a transient nature, this would result in the appearance of metastable Di and Pv prior to Fo—the opposite of what is observed. Thus candidate 3. (the transient nature of nucleation of the

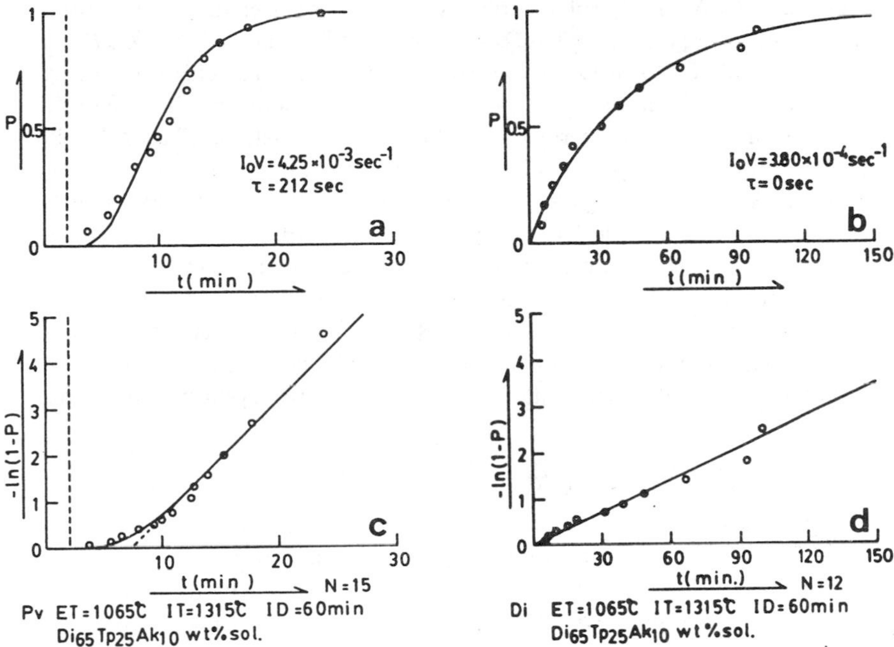

Figure 11 The measured P versus t diagram of Pv (*a*) and Di (*b*), and $-\ln(1-P)$ versus t diagram of Pv (*c*) and Di (*d*) at 1065°C. Pv shows a transient nature, whereas Di shows a steady state nature of nucleation. (After Nakamura et al 1990).

stable phase as an explanation for the observed metastable nucleation) was also excluded.

If Fo has a much larger interfacial energy with the liquid than Pv or Di, metastable nucleation of the latter two may occur, because the activation energy for nucleation is determined by a competition between the interfacial energy and the driving force. The tangent of the steady state part of Figure 11*b* for Pv is larger than that of Figure 11*d* for Di. This implies that the activation energy for nucleation is smaller for Pv than for Di, and therefore the interfacial energy of Pv is smaller than that of Di. From a much wider random region of Fo than Pv and Di, the relative interfacial energy of Fo can be evaluated and is found to be much larger than Pv or Di. It follows from this that the observed metastable nucleation phenomenon is caused by the presence of a much smaller interfacial energy of Pv and Di in the liquid when compared to that of stable Fo.

Abe et al (1991) arrived at the same conclusion for the case of An_{100} and $An_{80}Wo_{20}$ liquids, through the analysis of growth rates of stable and metastable $CaAl_2Si_2O_8$ polymorphs. As will be discussed in a later section, the volume energy term of phase transition from liquid to solid, $-L\Delta T/T_m$, for metastable $CaAl_2Si_2O_8$ polymorphs in the classical nucleation theory, can never exceed that of stable triclinic anorthite. Therefore, the difference between the interfacial energy of the stable and metastable phases should be the controlling factor of the observed metastable nucleation phenomenon in An_{100} and $An_{80}Wo_{20}$ liquids. Namely, the interfacial energies between the metastable polymorphs and the liquid should be much smaller than those between the stable triclinic anorthite and the liquid. Such a situation is realized when the melt structure is closer to the structure of the metastable phases than that of the stable phase. The observed metastable nucleation phenomenon may indicate characteristics of liquid structure of silicates at temperatures where crystallization takes place.

3.4 *Stability of Metastably Nucleated Phases*

When the stable phase is absent in the liquid, metastably nucleated phases grow steadily below the respective liquidus temperatures, as if they were stable phases. However, they behave very differently once the stable phase appears in the liquid. Abe (1988) and Abe et al (1991) investigated the behavior of metastable phases in the presence of the stable phase in An_{100} and $An_{80}Wo_{20}$ liquids. Crystallization in An_{100} liquid represents a case of crystal growth in a melt phase in which heat transfer plays an essential role, whereas an $An_{80}Wo_{20}$ liquid represents a case of solution growth in which mass (solute) transfer plays a more important role than heat transfer. [As to the difference between melt growth and solution growth, the reader may refer to Sunagawa (1984).]

Abe et al (1991) performed the following two types of experiments. 1. By quenching a liquid in which crystals of a metastable phase appeared, a glass sample is prepared. This is reheated at a given heating rate to a desired temperature and kept for a desired duration, followed by quenching and X-ray diffraction analyses. Stable triclinic anorthite nucleates in the glass portion, and the proportion of metastable phase diminishes. 2. Changes of crystals of metastable phases in the presence of stable triclinic anorthite were observed in situ. In this experiment, a seed crystal of triclinic anorthite is stuck onto the thermocouple and immersed into a liquid, in which crystals of a metastable phase are growing.

In both experiments, it was confirmed that crystals of metastable phases started to dissolve, transform, or lose their transparency as soon as crystals of the stable phase appeared and started to grow. The rate of this change varies depending on the conditions. For instance, pseudo-hexagonal

$CaAl_2Si_2O_8$ polymorph survived in experiment 1 even when the sample was kept at 1200°C for 30 minutes. But at 1350–1380°C, the change occurred in a few seconds in experiment 2.

By in situ observation, the following two types of changes were distinguished: (*a*) growth of stable triclinic anorthite concurrently with the melting of the metastable polymorphs, and (*b*) direct transformation without melting. Figure 12 shows examples of the two cases. As seen in the upper row of photographs, a hexagonal crystal of pseudo-hexagonal $CaAl_2Si_2O_8$ polymorph was at first growing at 1392°C, $\Delta T = 8$°C, in an An_{100} liquid (*a*), but as soon as crystals of stable anorthite started to grow, the bulk temperature increased to 1394°C (*b*, *c*), and the former started melting. At lower temperature, however, for example 1380°C, where ΔT for the metastable phase exceeds 20°C, this phenomenon was not observed. In an An_{100} liquid, the melting of the metastable phase occurs only over a narrow temperature range near the melting point of the metastable phase. In an An_{100} liquid, the melting is clearly due to the increase in temperature, for which only the latent heat released by crystallization of stable triclinic anorthite is responsible. On the other hand, in a solution system, like an $An_{80}Wo_{20}$ liquid, both temperature increase and mass (solute) transfer are responsible for melting, since a closed system can be more easily maintained for chemical components than for heat. Thus, melting will take place in a wider temperature range than in an An_{100} liquid.

The three photographs (*d*–*f*) in the lower row of Figure 12 show an example of a case where mechanism (*b*) operates. The transparency of a metastably nucleated crystal is lowered as soon as it contacts a stable anorthite crystal. It appears that direct transformation without melting operates in this case, since the outline of the former crystal is maintained. This process is observed even at lower temperatures where mechanism (*a*) is no longer operative. A similar process was also observed in a liquid of An_{70}-Fo_{10}-$Silica_{20}$ composition.

Two other metastable phases—corundum and mullite—did not show such a distinct change as in the case of $CaAl_2Si_2O_8$ polymorphs. They showed almost negligible change at 1200–1300°C, even if crystals of the stable triclinic anorthite were present in the liquid, even though careful examinations were made under the microscope on the thin sections of the quenched samples. The slow rate of transformation is probably due to the necessity of solute removal and less sensitivity to the temperature change. Dissolution of metastable corundum and mullite will occur when the composition of the residual liquid changes to become depleted in the growth of metastable phases due to the crystallization of the stable phase.

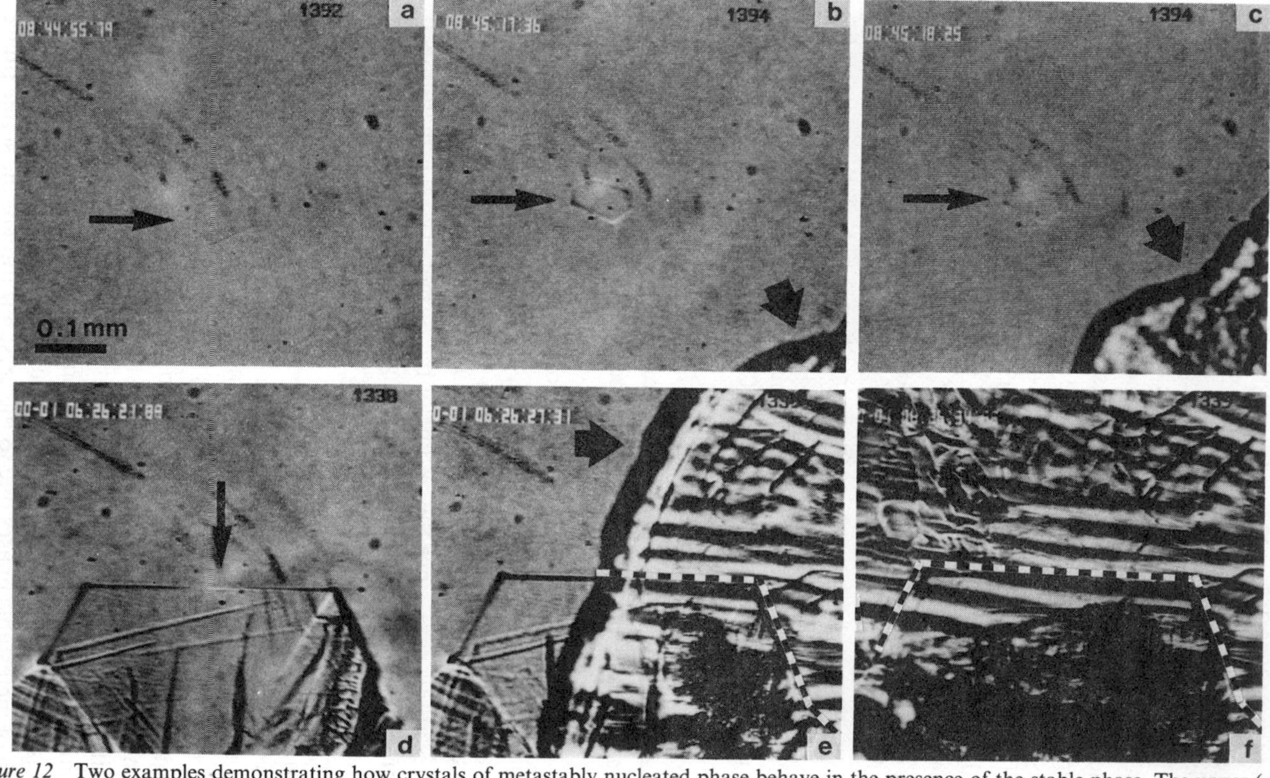

Figure 12 Two examples demonstrating how crystals of metastably nucleated phase behave in the presence of the stable phase. The upper (*a–c*) set shows concurrent melting of the metastable phase and growth of the stable phase. The lower set (*d–f*) shows the loss of transparency of the metastable phase when it contacts the stable phase. Crystals of the metastably nucleated phase (in this case the pseudo-hexagonal $CaAl_2Si_2O_8$ polymorph) are indicated by thin arrows, and those of the stable phase (triclinic anorthite) by thick arrows. (Modified from Abe et al 1991.)

3.5 *Analysis of Growth Rates Versus Driving Force Relations*

Since the growth rates of crystals in a melt phase are strongly dependent on the viscosity—which varies inversely with temperature—the growth rates of the three $CaAl_2Si_2O_8$ polymorphs in Figure 8 were normalized by the viscosities, using the values measured by Cukierman & Uhlmann (1973) and evaluated by Bottinga & Weil (1972). Figure 13 shows the normalized growth rates versus supercooling relations. The relations appear to be almost equal for the three $CaAl_2Si_2O_8$ polymorphs, except for slightly higher values of the stable phase at supercooling greater than 300°C. This is probably due to uncertainty in determining the growth direction.

The normalized growth rate is principally controlled by the latent heat and the supercooling degree and expressed as follows:

$$R_\eta \propto \left[1-\exp\left(-\frac{L\Delta T}{kT_\mathrm{m}\cdot T}\right)\right],$$

where L is the latent heat, k Boltzmann's constant, T the absolute temperature, and T_m the melting point (see e.g. Klein & Uhlmann 1974). Even

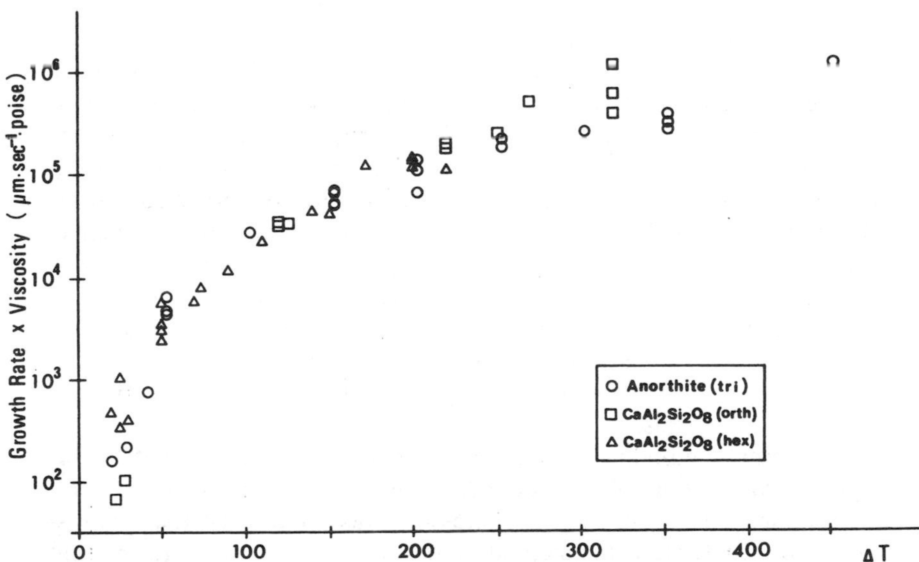

Figure 13 Normalized growth rates R_η versus the calibrated supercooling relation of three $CaAl_2Si_2O_8$ polymorphs in an An_{100} liquid. (From Abe et al 1991.)

when the normalized growth rates are correlated to $\Delta T/kT_mT$, instead of simply ΔT, we do not see any significant differences between the two relations. This implies that the latent heats of the two metastable $CaAl_2Si_2O_8$ polymorphs are not greatly different from that of the stable triclinic anorthite.

The growth rates in the above analysis were measured in the direction of elongation of crystals, where the interface roughness is assumed to be greatest and the continuous or adhesive type growth mechanism is expected to be operative. Analysis of growth rates versus driving force relations in other directions with smoother interfaces and where layer-to-layer growth mechanism is assumed is still underway.

4. CONCLUSIONS

A new experimental method which enables direct and in situ investigations of the processes and kinetics of nucleation, growth, and dissolution of crystals at high temperatures up to 1600°C has been devised. The whole process of crystallization in a liquid film of silicate and other components can be observed at high visibility under optical microscopes of various types, and by applying interferometric, light scattering, and Schlieren techniques. Repeated investigations of the process and precise measurements of the kinetic rates are possible with the use of videotape. High visibility is achieved because the process is observed through a thin liquid film by transmitted light, and various measures to enhance visibility have been designed. Temperature control with fluctuations less than 0.5°C is attained by a combination of a microcomputer, a voltmeter, and a stabilizer. The method can, therefore, provide kinetic data precise enough to confront the theoretical analyses of the mechanisms of nucleation, growth, and dissolution of crystals in magmatic crystallization. Selected results obtained by this method (Section 3) include: observations on the dynamic movements and changes of crystals in a liquid film associated with temperature changes and various types of convection, metastable nucleation phenomenon and the analysis of its cause, stability of a metastably nucleated phase in the presence and absence of a stable phase, and an analysis of growth rates versus driving force relations in the direction of elongation of crystals.

The present method, based on the experience of in situ observations of crystal growth of simple compounds in aqueous solutions at ordinary temperature and one atmosphere of pressure, was developed in the author's laboratory. In Table 1, the methods and advantages of, and information obtainable through in situ investigations of nucleation, growth, and dissolution of crystals are compared for the cases of crystallization in aqueous solutions at ordinary temperature and pressure, in silicate liquids at high temperatures, and for ex situ investigations of silicates.

Owing to difficulties involved in controlling high temperatures, visibility and condition control of high temperature in situ investigation methods have yet to reach the levels of the ordinary temperature in situ investigation method. It is still not possible to directly observe spiral growth, to measure advancing rate of growth steps in relation to surface supersaturation, or to characterize in situ the nature of dislocations in growing crystals. In other words, the technique resides at the level of microscopic investigation, not yet reaching the level needed to investigate the molecular process of crystallization, which is achieved in the in situ investigation of crystallization in aqueous solutions. However, it is clear that the method has various advantages superior to previous ex situ investigation methods in analyzing magmatic crystallization. By further improving the method, we may eventually reach the level of the in situ investigation method of crystallization in aqueous solutions.

ACKNOWLEDGMENTS

The method described in this paper was developed with the collaboration of K. Tsukamoto, T. Abe, H. Nakamura, M. Kawasaki, M. Abe, and Y. Kono, at the Institute of Mineralogy, Petrology, and Economic Geology, Tohoku University, to whom the author expresses his deepest thanks. The help of T. Abe of Yamaguchi University in preparing the figures is greatly acknowledged. Financial support from the Grant-in-Aid for fundamental research given by the Japan Ministry of Education, Culture, and Science is greatly appreciated.

Literature Cited

Abe, T. 1988. *Nucleation, growth and stability of metastable phases in some silicate systems*. PhD Thesis. Tohoku Univ., Sendai. 125 pp.

Abe, T., Tsukamoto, K., Sunagawa, I. 1987. Instrumentation for in-situ observation of growth and dissolution processes of crystals in high temperatures at 1600°C. *Mineral. J.* 13: 479–89

Abe, T., Tsukamoto, K., Sunagawa, I. 1991. Nucleation, growth and stability of $CaAl_2Si_2O_8$ polymorphs. *Phys. Chem. Minerals* 17: 473–84

Azuma, T., Tsukamoto, K., Sunagawa, I. 1989. Clustering phenomenon and growth units in lysozyme aqueous solution as revealed by laser light scattering method. *J. Cryst. Growth* 98: 371–76

Bottinga, Y., Weill, D. F. 1972. The viscosity of magmatic silicate liquids: A model for calculation. *Am. J. Sci.* 272: 438–75

Chernov, A. A. 1984. *Modern Crystallography, III, Crystal Growth*. Heidelberg: Springer-Verlag. 517 pp.

Cukierman, M., Uhlmann, D. R. 1973. Viscosity of liquid anorthite. *J. Geophys. Res.* 78: 4920–23

Hartman, P., ed. 1973. *Crystal Growth: An Introduction*. Amsterdam: North-Holland. 531 pp.

Kirkpatrick, R. J., Robinson, G. R., Hays, J. F. 1976. Kinetics of crystal growth from silicate melts: Anorthite and diopside. *J. Geophys. Res.* 81: 5715–20

Kirkpatrick, R. J., Reck, B. H., Pelly, I. Z., Kuo, L. C. 1983. Programmed cooling experiments in the system $MgO\text{-}SiO_2$: Kinetics of a peritectic reaction. *Am. Mineral.* 68: 1095–1101

Klein, L., Uhlmann, D. R. 1974. Crystallization behavior of anorthite. *J. Geophys. Res.* 79: 4869–74

Lofgren, G. 1980. Experimental studies on the dynamic crystallization of silicate melts. In *Physics of Magmatic Processes*, ed. R. B. Hargraves, pp. 487–551. Princeton NJ: Princeton Univ. Press. 585 pp.

Maiwa, K. 1987. *Role of lattice defects in aqueous solution growth of* $Ba(NO_3)_2$ *crystals*. PhD Thesis, Tohoku Univ., Sendai. 130 pp.

Maiwa, K., Tsukamoto, K., Sunagawa, I. 1989. Observation of screw and mixed dislocations in barium nitrate crystals by means of birefringence and x-ray topography. *J. Cryst. Growth* 98: 590–94

Maiwa, K., Tsukamoto, K., Sunagawa, I. 1990. Activities of spiral growth hillocks on the (111) faces of barium nitrate crystals growing in an aqueous solution. *J. Cryst. Growth* 102: 43–53

Nakamura, H. 1988. *Experimental and theoretical analyses of nucleation and growth behaviors in some silicate systems*. PhD Thesis, Tohoku Univ., Sendai. 268 pp.

Nakamura, H., Tsukamoto, K., Sunagawa, I. 1990. In-situ observation of high temperature silicate solutions. *J. Cryst. Growth* 99: 1227–31

Onuma, K. 1988. *A combined investigation of interface kinetics and environmental phase in aqueous solution growth*. PhD Thesis, Tohoku Univ., Sendai. 195 pp.

Onuma, K., Tsukamoto, K., Sunagawa, I. 1988. Role of buoyancy driven convection in aqueous solution growth: A case study of $Ba(NO_3)_2$ crystal. *J. Cryst. Growth* 89: 177–88

Onuma, K., Tsukamoto, K., Sunagawa, I. 1989a. Measurements of surface supersaturations around a growing K-alum crystal in aqueous solution. *J. Cryst. Growth* 98: 377–83

Onuma, K., Tsukamoto, K., Sunagawa, I. 1989b. Effect of buoyancy driven convection upon the surface microtopographs of $Ba(NO_3)_2$ and CdI_2 crystals. *J. Cryst. Growth* 98: 384–90

Onuma, K., Tsukamoto, K., Sunagawa, I. 1990. Growth kinetics of K-alum crystals in relation to the surface supersaturations. *J. Cryst. Growth* 100: 125–32

Sunagawa, I. 1984. Growth of crystals in nature. In *Materials Science of the Earth's Interior*, ed. I. Sunagawa, pp. 63–105. Tokyo/Dordrecht: Terra Sci./Reidel. 653 pp.

Sunagawa, I. 1991. In situ observation of nucleation, growth and dissolution of crystals in ordinary temperature aqueous solutions and high temperature silicate solutions. In *Dynamic Processes of Materials Transport and Transformation in the Earth's Interior*, ed. F. Marumo, pp. 139–68. Tokyo: Terra Sci. 521 pp.

Toschev, S. 1973. Homogeneous nucleation. In *Crystal Growth: An Introduction*, ed. P. Hartman, pp. 1–49. Amsterdam: North-Holland. 531 pp.

Toschev, S., Milchev, A., Stoyanov, S. 1972. On some probabilistic aspects of the nucleation process. *J. Cryst. Growth* 13/14: 123–27

Tsuchiyama, A. 1983. Crystallization kinetics in the system $CaMgSi_2O_6\text{-}CaAl_2Si_2O_8$: The delay in nucleation of diopside and anorthite. *Am. Mineral.* 68: 687–98

Tsukamoto, K. 1983. In situ observation of mono-molecular growth steps on crystals growing in aqueous solution, I. *J. Cryst. Growth* 61: 199–209

Tsukamoto, K. 1989a. *In situ observation methods developed for the verification of crystal growth mechanisms*. PhD Thesis. Tohoku Univ., Sendai. 189 pp.

Tsukamoto, K. 1989b. In situ direct observation of a crystal surface and its surroundings. In *Morphology and Growth Unit of Crystals*, ed. I. Sunagawa, pp. 451–78. Tokyo: Terra Sci. 690 pp.

Tsukamoto, K., Abe, T., Sunagawa, I. 1983. In situ observation of crystals growing in high temperature melts or solutions. *J. Cryst. Growth* 63: 215–18

Tsukamoto, K., Sunagawa, I. 1985. In situ observation of mono-molecular growth steps on crystals growing in aqueous solution. II. Specially designed objective lens and Nomarski prism for in situ observation by reflected light. *J. Cryst. Growth* 71: 183–90

Annu. Rev. Earth Planet. Sci. 1992. 20:143–58

TECTONIC EVOLUTION OF THE PYRENEES

Pierre Choukroune

Centre Armoricain d'Etude Structurale des Socles, Université de Rennes 1, Campus de Beaulieu, 35042 Rennes Cedex, France

KEY WORDS: deep structure, plate tectonic models, orogenesis

1. INTRODUCTION

The Pyrenees are a part of the Alpine chain of Western Europe which runs from the northern Iberian margin in the West to the Alps in the East (Figure 1). This E-W belt is considered to be a shortened crustal domain that occupies the site of the boundary between separate Iberian and European plates during Cretaceous times. In their central part, the Pyrenees display a fanlike cross-section: most of the faults are northward directed thrusts in the North, while large southward translations of Mesozoic and Cenozoic series can be observed in the South.

The range is divided into several zones, bounded by major faults. From North to South one can distinguish:

- the Aquitaine foreland, affected only by concentric folds and overthrust by
- the North Pyrenean Zone (NPZ) in which Cretaceous flysch deposits are locally highly strained and metamorphosed. Hercynian basement core massifs crop out in the NPZ.
- The Axial Zone is made up of Hercynian rocks overlain by some remnants of Mesozoic cover.
- The South Pyrenean Units (Mesozoic and Cenozoic series) are southward translated and overthrust
- the Ebro foreland where molassic Tertiary strata rest directly on the Hercynian basement.

The crustal architecture of the Pyrenees results from structural evolution

0084–6597/92/0515–0143$02.00

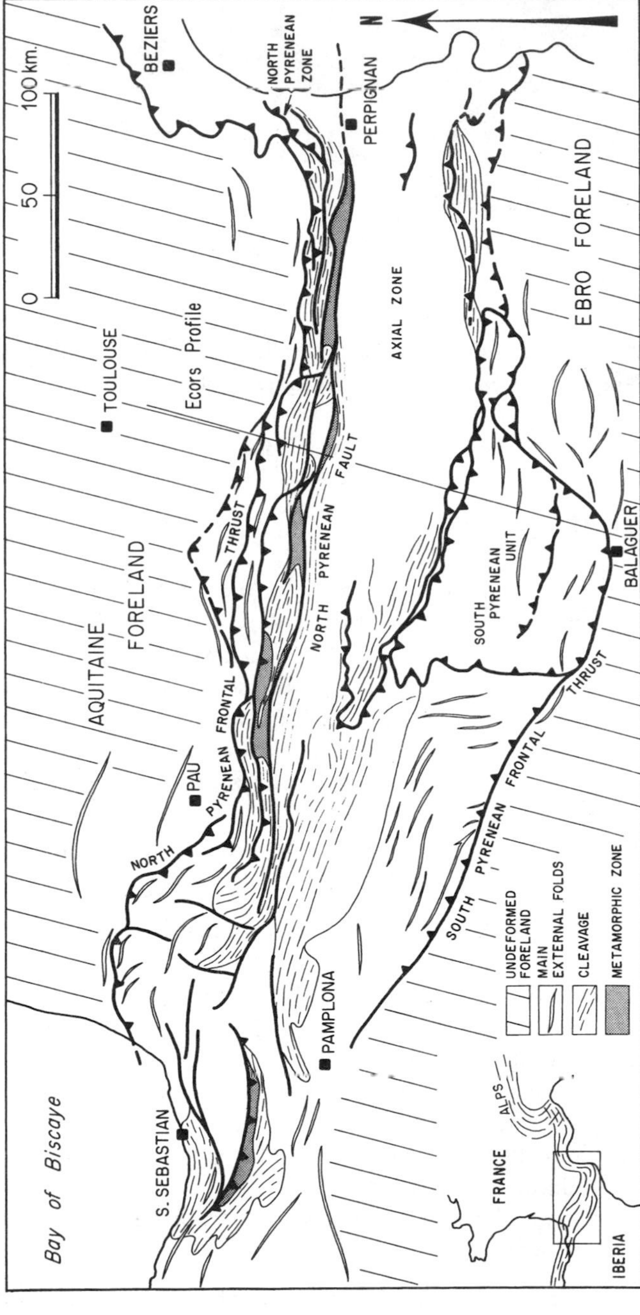

Figure 1 Structural sketch map of the Pyrenees. Shown are the main features of the range. Note the spatial relationship between the North Pyrenean Fault (NPZ) and the metamorphic domain of the North Pyrenean Zone (NPZ).

starting during late Paleozoic (Hercynian) times: The Paleozoic rocks cropping out in the Axial Zone and in the North Pyrenean Zone are affected by South-verging Hercynian folds and thrusts and are locally metamorphosed under granulitic conditions. They contrast with those, outcropping in the Aquitaine foreland, which are of lower grade. This discrepancy is explained by a large disruption of the original Hercynian structural pattern which occurred after the Hercynian orogeny.

Late Triassic to Liassic extensional events followed the Hercynian compression. During Jurassic times, the site of the Pyrenees was part of the SE European margin and experienced shelf sedimentation.

The opening of the North Atlantic Ocean led progressively to the development of narrow and deep E-W basins in the NPZ filled with turbiditic deposits and slope facies breccias (Souquet et al 1980).

Finally, N-S convergence of the Iberian and European plates induced compressional deformation during the Late Cretaceous and Eocene Pyrenean orogeny. In response to these shortening events, flexural foreland basins have developed during Late Cretaceous to Eocene times in the Aquitaine, and during Eocene to Early Miocene in the southern slope of the belt.

There exists many earlier studies on the Pyrenees (Roussel 1903, Bertrand 1907, Carez 1910, Jacob 1928, Casteras 1933), but the first modern structural approach to the overall structure of the belt and its evolution was undertaken by Mattauer (1968).

2. FEATURES OF THE PYRENEAN CHAIN

The Pyrenees are characterized by several unusual structural features.

First, Mesozoic and Cenozoic rocks have not suffered strong internal deformation except in a narrow steeply dipping E-W zone corresponding to the internal metamorphic zone of the belt. This zone is spatially linked to a major vertical discontinuity which bounds the Axial Zone and the North Pyrenean Zone: the North Pyrenean Fault (NPF) (see Figure 1). The extreme narrowness (between 1 and 5 km) of the area where low pressure/high temperature metamorphosed (LP/HT) and highly strained rocks occur, enhanced the role and influence of the NPF during the Pyrenean orogeny. In this narrow LP/HT belt which runs parallel to the NPF, some pieces of lherzolite are also seen to occur. The relationship between these mantle rocks, their emplacement history, and metamorphism has been recently discussed (Minnigh et al 1980, Choukroune 1980, Hall & Benet 1983, Vielzeuf & Kornprobst 1983, Goldberg et al 1986, Bodinier et al 1987, Dauteuil et al 1987).

Within the NPZ strain characteristics change abruptly: The strain gradient is abnormally high and the stretching direction—mainly vertical in the NPZ—is horizontal in the metamorphic belt (Choukroune 1976b). Syntectonic metamorphism is dated at around 90 Ma by K/Ar methods (Albarede & Michard-Vitrac 1978, Montigny et al 1986).

Furthermore, the Pyrenees are one of the few modern mountain chains where no granitic plutonism or migmatization has occurred.

Lastly, no remnants of oceanic crust can be observed although many alkaline magmatic bodies intruded the Cretaceous basins of the North Pyrenean Zone during their filling (Azambre & Rossy 1976).

To summarize, the NPF, which can be traced for 300 km, appears to be one of the major features of the Pyrenean belt: It represents the southern limit of the localized Cretaceous basins, it has concentrated heat and high strain, and its behavior during orogeny has controlled the orientation of the strain axis. Furthermore, seismic refraction and teleseismic surveys indicate that the Moho discontinuity is offset vertically by about 15 km beneath the NPF (Daignieres 1978, Daignieres et al 1982). Van der Voo & Boessenkol (1973) noted a discrepancy of paleomagnetic results on each side of the NPF, which has been used to define the boundary between the Iberian and European plates at the NPF.

Most Pyrenean authors have taken these salient characteristics into account in modeling the geodynamic evolution and reconstruction of the deep structure of the chain on the crustal scale.

3. STRUCTURE AND EVOLUTION OF THE NORTH PYRENEAN ZONE: RECENT DATA

During the last few years, major new results in North Pyrenean geology concern the geometry of the Cretaceous basins, tecto-sedimentary cycles, and the Late Cretaceous basin-modifying tectonics (Souquet et al 1985; Puigdefabregas & Souquet 1986; Debroas 1987, 1990; Johnson & Hall 1989; Deramond et al 1990).

From middle Albian to Cenomanian, grabens, no more than several kilometers wide, were opened. Rhomboidal in shape, they correspond to en echelon pull-apart troughs compatible with an *E W sinistral strike-slip movement* along a fault system limited to the South by the NPF (Debroas 1990) (Figure 2). This configuration was proposed earlier by Choukroune & Mattauer (1978) and compared with the San Andreas fault system near the Salton Sea.

Structural inversion of these troughs started during the late Senonian (Figure 3) and corresponds to the development of folding phases, cleavages, and metamorphism (Choukroune 1976a).

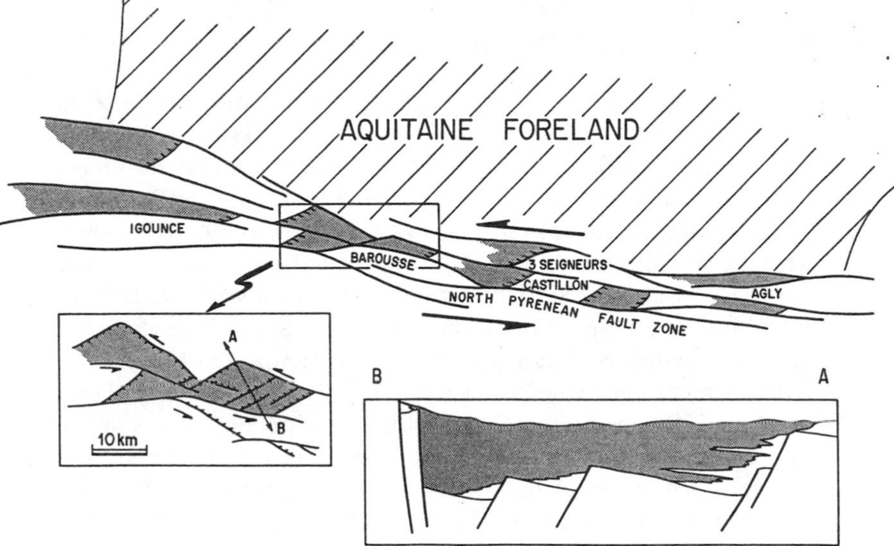

Figure 2 Albian-Cenomanian pull-apart basins in the NPZ: Their geometry is compatible with a transtensional sinistral strike-slip movement. The reconstructed section (Debroas 1990) is taken in the Baronnies graben (central NPZ) during Cenomanian-Turonian times.

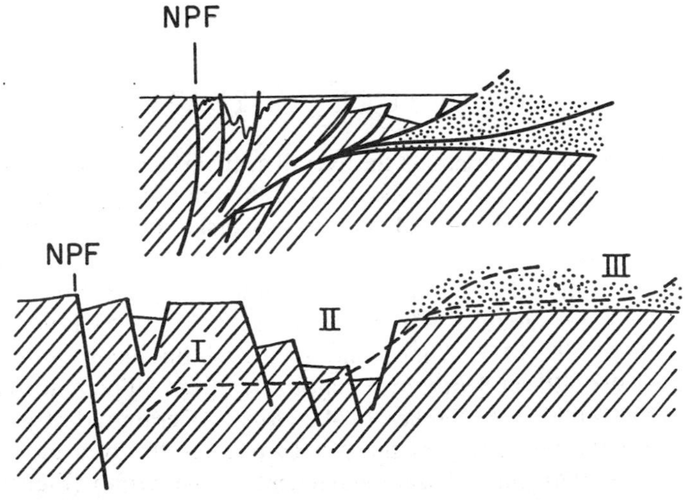

Figure 3 Schematic balanced section across the North Pyrenean Zone near the ECORS profile (modified after Deramond et al 1990).

4. STRUCTURE AND EVOLUTION OF THE SOUTH PYRENEAN ZONE: RECENT DATA

Southward vergence dominates on the southern border of the Axial Zone. The Mesozoic and Cenozoic cover of thrust units (Gavarnie nappe or Nogueras nappes) is affected by a general decollement above the Triassic series and large southward translation (Garrido-Megias & Rios 1972, Seguret 1972, Sole-Sugranes 1978, Camara & Klimowitz 1985) (Figure 4).

If we disregard the root zones of the translated units, no penetrative deformation can be observed. Imbricated thrust systems with related hanging wall folds define a typical thin-skin style for the South Pyrenean Units and an approximately N-S transport direction (Munoz et al 1986).

The décollement zone emerges at the South Pyrenean Frontal Thrust (Figures 1 and 4), on the mainly undeformed Ebro molassic southern foreland basin where Tertiary strata rest directly on the Hercynian basement. As recorded by syntectonic deposits, the emplacement of the South Pyrenean Unit began in the Early Eocene and has continued throughout Tertiary times (Labaume et al 1985).

During the last ten years, most studies have been dedicated to detailed geometrical descriptions of thrust sheets (Seguret 1972, Parrish 1984, Munoz 1988, Teixell 1990), thrust sequences (Munoz et al 1986, Verges & Munoz 1990), and to the sedimentary consequences of this superficial tectonic evolution (Puigdefabregas 1975; Labaume et al 1983, 1985, 1987; Seguret et al 1984; Mutti et al 1984a,b; Puigdefabregas & Souquet 1986; Puigdefabregas et al 1986; Berastegui et al 1990).

5. THE DEEP STRUCTURE: A DEBATE

In spite of a general agreement on the overall near surface structure (shown in the top of Figure 5) and the geometry of the main discontinuities, the deep structure of the Pyrenees is highly controversial. One can distinguish five main models.

The first assumes that Pyrenean deformation results from a limited southward directed subduction (not represented on Figure 5) (Boillot 1984, Boillot & Capdevilla 1977). The second is a model that incorporates a 60-km offset of the NPF at depth, by a major sole thrust dipping north at 6° (Figure 5*A*) (Williams & Fischer 1984). Others (Deramond et al 1985) consider that the thickening of the Axial Zone results from motion on thrust surfaces that join a basal detachment at the Moho (Figure 5*B*). Using an inhomogenous strain model, Seguret & Daignieres (1986) consider that the thrusts die out at depth, in an area of strong ductile defor-

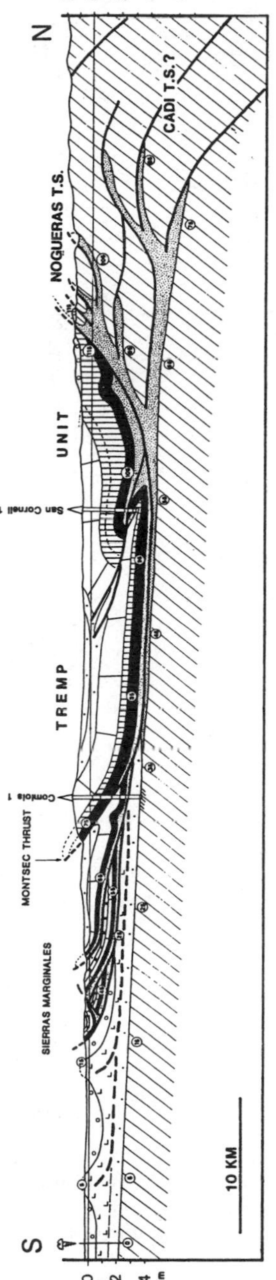

Figure 4 Structural section across the South Pyrenean units near the ECORS profile. (After Seguret 1972.)

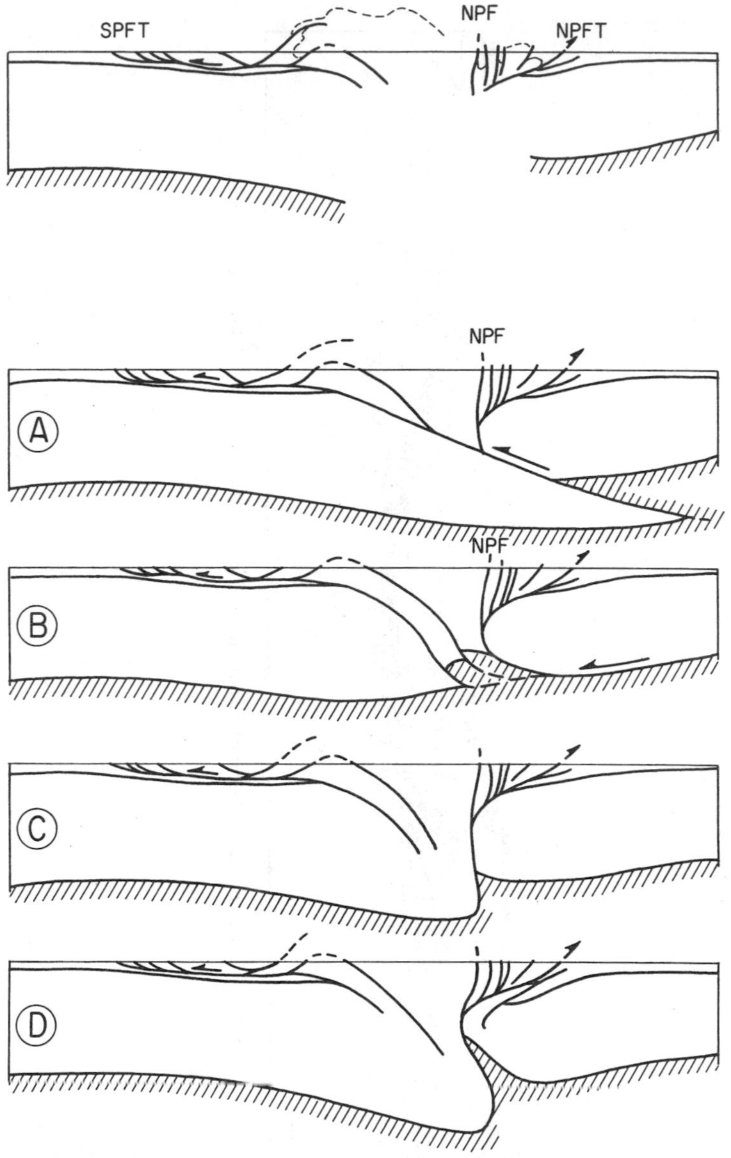

Figure 5 Crustal scale sections of the Pyrenean belt presenting some of the different hypotheses for its deep structure. *A*: according to Williams & Fischer (1984), *B*: according to Deramond et al (1985), *C*: according to Seguret & Daignières (1986), *D*: according to Choukroune et al (1990b, modified). All models use the same superficial geometry shown on the upper part of the figure.

mation (Figure 5*C*). Using the results of the ECORS Pyrenees program, a final model implies a slight deformation of the NPF due to the collision of the Iberian and European plates (Choukroune et al 1989, 1990a,b; Roure et al 1989) (Figure 5*D*).

The ECORS Pyrenees profile is the only deep seismic survey that crosses an entire orogenic belt and constitutes a major advance in Pyrenean geological knowledge. The main results can be summarized as follows (Figure 6):

1. The entire crust shows well-defined reflectors whose geometry confirms the general fan shape of the belt.
2. The Iberian crust is thicker than the European one and both are limited by deep layered material above the Moho discontinuity.
3. Only the Iberian crust is significantly thickened by a southward overstacking of slabs.
4. In the external domains, reflectors accurately define the geometry of major thrusts and structures affecting the Mesozoic and Cenozoic cover of the belt.
5. Both Iberian and European middle-crusts display north dipping reflectors which can be interpreted as Hercynian structures and representing parts of the Ibero-Armorican arc (ECORS Pyrenees team 1988; Choukroune et al 1989, 1990a,b).

Using these results and additional geological and geophysical data. Roure et al (1989) tried to balance a crustal geological section, restoring the geometry to the period after the Albian-Cenomanian and before the Late Cretaceous compressional events: At least 100 km of shortening are required (Figure 7).

More recently, the seismic interpretation has been compared with experimental analog models. It has been shown that the observed asymmetry could be the consequence of deformation of deep brittle layers in the lower crust and upper mantle (Davy et al 1990).

6. BAY OF BISCAY OPENING AND THE EVOLUTION OF THE PYRENEAN DOMAIN: THE MODELS

From the time of Argand (1924), numerous studies have been dedicated to the question of the opening of the Bay of Biscay and its relationships with the structural evolution of the contiguous continental domain (see Debyser et al 1971). In order to present the different hypotheses, it is important to recall the basic data which constrain this major question.

First of all, the formation of oceanic crust in the Bay of Biscay occurred

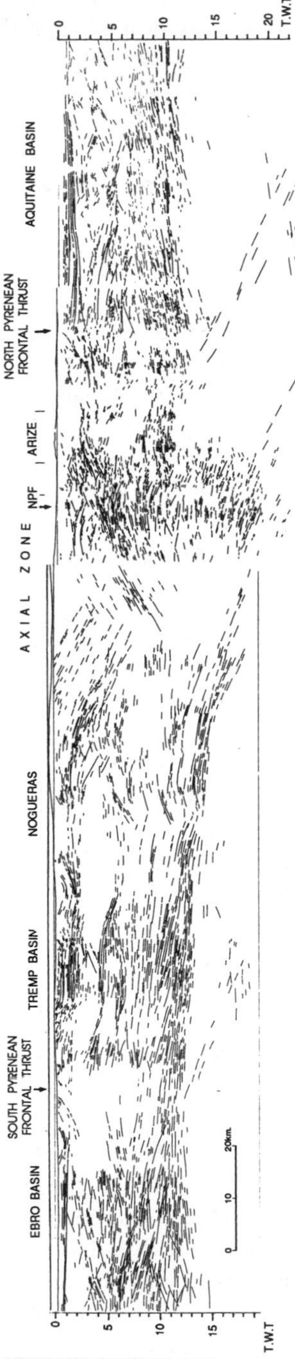

Figure 6 Line drawing of the ECORS Pyrenees profile extracted from unmigrated seismic data (from Choukroune et al 1989).

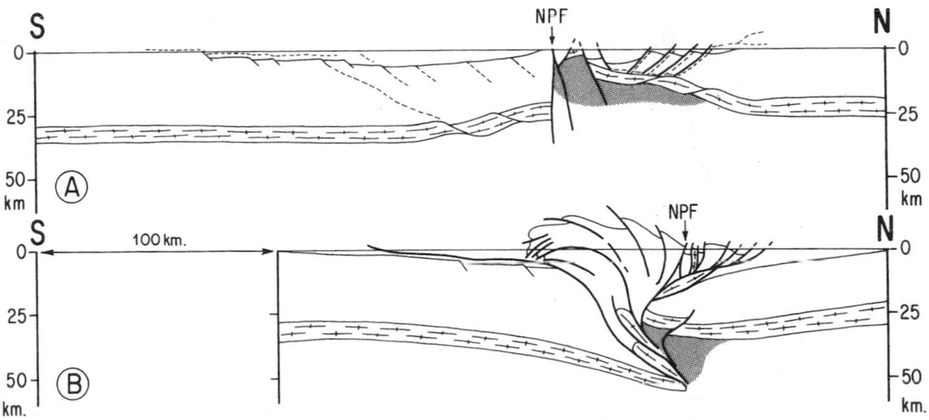

Figure 7 Balanced crustal cross-section along the ECORS profile: (*A*) the preorogenic stage during the probable sinistral strike-slip motion of the Iberian plate along the NPF. (*B*) This plate boundary is then slightly deformed involving 100 km of shortening (modified after Roure et al 1989).

between 120 and 80 My during the early oceanic stages of the North Atlantic (Williams & McKenzie 1971, Williams 1973). This period corresponds to the preorogenic stages of the Pyrenean domain, with a great instability of the North Pyrenean Zone which was occupied by narrow subsiding basins running parallel to, and limited by, the NPF. Secondly, it has been well established (Van der Voo & Zijderveld 1971, Van der Voo & Boessenkol 1973) that the boundary between Iberia and Europe must be situated in the North Pyrenean Zone, in the vicinity of the NPF, and that the total rotation of Iberia relative to Europe is 35°. Thirdly, the opening of the Bay of Biscay and the formation of the Pyrenees are successive events clearly linked to relative displacements between two stable plates: Iberia and Europe.

Two major types of hypotheses have been proposed for the opening of the Bay of Biscay (Figure 8): A first group implies a scissors-opening with a pole of rotation situated either in the Pyrenees (Argand 1924, Choubert 1935, Carey 1958, Schoeffler 1965, Bullard et al 1965, Matthews & Williams 1968, Masson & Miles 1984, Bacon & Gray 1970) or in the eastern prolongation of the belt (Montadert & Winnock 1971, Williams 1973). Recently, Sibuet & Collette (1991) proposed a similar kind of kinematic model using a new survey of magnetic anomalies. This kind of model implies a highly variable amount of extension along the belt during the preorogenic stages, and for some of them, a true oceanization of the

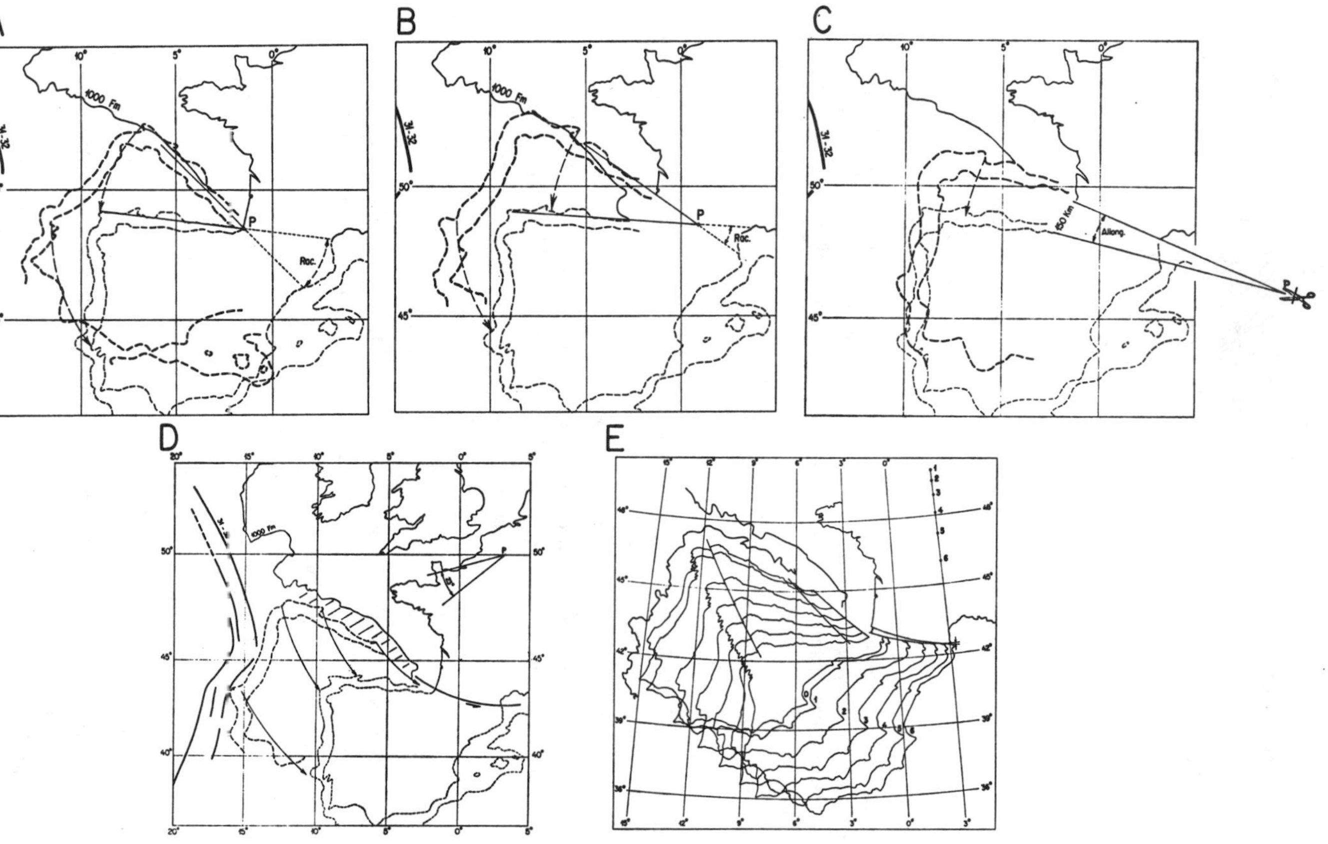

Figure 8 Some of the kinematic models for the opening of the Bay of Biscay. *A*: according to Carey (1958); this model implies a strong and inhomogeneous shortening (rac) during the oceanization. *B*: according to Schoeffler (1965). *C*: according to Montadert & Winnock (1971). *D*: according to Le Pichon et al (1971) who considered the NPF as a transform fault. This model has been modified by Choukroune et al (1973) in order to take into account the structural evolution of the continental domain (*E*).

Pyrenean domain (Figure 8). None of these models take into account the pull-apart geometry of the Cretaceous basins.

A second group of hypotheses describe the opening as a rotation around a pole situated in northern France and implies a left-lateral strike slip motion along the NPF considered as a transform fault between the Iberian and European plates (Le Pichon et al 1971, Choukroune et al 1973, Olivet et al 1984, Sibuet 1989).

Considering the important constraints provided by geological data and structural characteristics of the Pyrenean area, the second type of kinematic models is most likely.

7. CONCLUSIONS

In reviewing the literature on the Pyrenees, several points of agreement are evident while strong controversies still remain. Most authors agree with the geometry of the different units: The South Pyrenean unit is translated southward, the North Pyrenean domain displays a narrow vertical belt where rocks have undergone strong deformation and metamorphism near the NPF, and its northern limit is a major northward directed thrust. The superficial bulk geometry of the chain is therefore well accepted. The timing of the tectonic events is also well established: After the Lower Cretaceous episode of an important localized extension, compressional regimes started in the NPZ during Senonian times (80–85 Ma) and occupied progressively more and more external domains during the Tertiary.

The mode of emplacement, the age, the role, and the significance of the lherzolitic bodies and Hercynian granulitic rocks in the NPZ is still a debated question. Are they a simple result of the Lower Cretaceous intense thinning? Is the emplacement of lherzolites due to diapiric processes? Did they constitute the heat source and were they responsible for the high temperature metamorphic transformations which affected the Mesozoic material? And, if so, why do we observe lherzolitic bodies in unmetamorphosed areas?

The deep structure of the belt is also controversial even though the ECORS profile has provided new constraints. Some support the dominant deep southward vergency hypothesis while others put the importance on a flower-type structure centered on the NPF.

Finally, because of equivocal magnetic anomaly patterns in the Bay of Biscay, the kinematics of the Iberian plate are not unanimously accepted. Nevertheless, hypotheses involving transform motion along the NPF, during preorogenic times, is favored by those who consider compatibility with geological data.

In any case, one may conclude that the Pyrenean belt cannot be easily compared to other belts and is probably unique in its unusual structural style and evolution.

Literature Cited

Albarede, F., Michard-Vitrac, A. 1978. Datation de métamorphisme des terrains secondaires des Pyrénées par les méthodes ^{39}Ar-^{40}Ar et ^{37}Rb-^{87}Sr; ses relations avec les péridotites associées. *Bull. Soc. Géol. Fr.* (7)20: 681–87

Argand, E. 1924. *Congrès Géol. Intern., 13th (1922), Liège*, pp. 169–371

Azambre, B., Rossy, M. 1976. Le magmatisme alcalin d'âge crétacé dans les Pyrénées occidentales et l'Arc basque; ses relations avec le métamorphisme et la tectonique. *Bull. Soc. Géol. Fr.* (7)18: 1725–28

Bacon, M., Gray, F. 1970. A Gravity Survey in the Eastern part of the Bay of Biscay. *Earth Planet. Sci. Lett.* 10: 101–12

Bertrand, L. 1907. Contribution à l'histoire stratigraphique et tectonique des Pyrénées orientales et centrale. *Bull. Serv. Carte Géol. Fr.* 17, 118 pp.

Berastegui, X., Garcia-Senz, J. M., Losantos, M. 1990. Evolution tecto-sédimentaire du bassin en extension d'Organya (Unité Sud Pyrénéenne centrale, Espagne). *Bull. Soc. Géol. Fr.* (8)2: 251–64

Bodinier, J. L., Dupuy, C., Dostal, J. 1987. Geochemistry and petrogenesis of eastern Pyrenean peridotites. *Geochim. Cosmochim. Acta* 51: 279–90

Boillot, G. 1984. Some remarks on the continental margins in the Aquitaine and French Pyrenees. *Geol. Mag.* 121(5): 407–12

Boillot, G., Capdevila, R. 1977. The Pyrenees: subduction and collision? *Earth Planet. Sci. Lett.* 35: 151–60

Bullard, E., Everett, J. E., Smith, A. G. 1965. The fit of continents around the Atlantic. *Philos. Trans. R. Soc. London Ser. A* 258: 41 51

Camara, P., Klimowitz, J. 1985. Interpretacion geodinamica de la vertiente centro-occidental surpirenaica (cuencas de Jaca-Tremp). *Estud. Geol.* 41: 391–404

Carey, W. S. 1958. The orocline concept in geotectonics. *Roy. Soc. Tasmania Proc.* 89: 255–88

Carez, L. 1910. Résumé de la Géologie des Pyrénées françaises. *Bull. Soc. Géol. Fr.* (4)10: 670–81

Casteras, M. 1933. Recherches sur la structure du versant nord des Pyrénées centrales et orientales. *Bull. Serv. Carte Géol. Fr.* 37(389)

Choubert, B. 1935. Recherches sur la genèse des chaînes paléozoîques et antécambriennes. *Rev. Géogr. Phys. et Géol. Dyn.* 8: 5–50

Choukroune, P. 1976a. Structure et évolution tectonique de la zone nord-pyrénéenne (analyse de la déformation dans une portionde chaîne à schistosité subverticale). *Mém. Soc. Géol. Fr.* 127, 116 pp.

Choukroune, P. 1976b. Strain patterns in the Pyrenean Chain. *Philos. Trans. R. Soc. London Ser. A* 283: 271–80

Choukroune, P. 1980. Comment on "Quenching: An additional model for the emplacement of the lherzolite at Lherz." *Geology* 8: 514–15

Choukroune, P., Seguret, M., Galdeano, A. 1973. Caractéristiques et évolution structurale des Pyrénées. *Bull. Soc. Géol. Fr.* 7: 601–11

Choukroune, P., Mattauer, M. 1978. Tectonique des plaques et Pyrénées: sur le fonctionnement de la faille transformante nord-pyrénéenne; comparaisons avec les modèles actuels. *Bull. Soc. Géol. Fr.* (7)20: 689–700

Choukroune, P., ECORS Team 1989. The ECORS Pyrenean deep seismic profile reflection data and the overall structure of an orogenic belt. *Tectonics* 8: 23–39

Choukroune, P., Pinet, B., Roure, F., Cazes, M. 1990a. Major hercynian structures along the ECORS Pyrenees and Biscaye lines. *Bull. Soc. Géol. Fr.* (8)6: 313–20

Choukroune, P., Roure, F., Pinet, B., ECORS Team 1990b. Main results of the ECORS Pyrenees Profile. *Tectonophysics* 173: 411 23

Daignières, M. 1978. Géophysique et faille Nord-Pyrénéenne. *Bull. Soc. Géol. Fr.* 20: 677–80

Daignières, M., Gallart, J., Banda, E., Hirn, A. 1982. Implications of the seismic structure for the orogenic evolution of the Pyrenean range. *Earth Planet. Sci. Lett.* 57: 88–100

Dauteuil, O., Raymond, D., Ricou, L. E. 1987. Brèches de fracturation hydraulique dans la zone métamorphique des Pyrénées, exemples à l'Est du Saint-Barthélémy. *C. R. Acad. Sci. Paris* 304: 1025–28

Davy, P., Choukroune, P., Suzanne, P. 1990. Hypothèses mécaniques de déformation de la lithosphère appliquées à la Formation des Pyrénées. *Bull. Soc. Géol. Fr.* (8)6: 219–28

Debroas, E. J. 1987. Modèle de bassin triangulaire à l'intersection de décrochement divergents pour le fossé albo-cenomanien de la Ballongue (zone nord pyrénéenne, France). *Bull. Soc. Géol. Fr.* 5: 887–98

Debroas, E. J. 1990. Le Flysch noir albo-cénomanien témoin de la structuration albienne à sénonienne de la zone Nord-Pyrénéenne en Bigorre (Hte Pyrénées, France). *Bull. Soc. Géol. Fr.* (8)6: 273–85

Debyser, J., Le Pichon, X., Montadert, L. 1971. *Histoire Structurale du Golfe de Gascogne*. Paris: Technip

Déramond, J., Baby, P., Specht, M., Crouzet, G. 1990. Géométrie des chevauchements dans la Zone nord-pyrénéenne ariégeoise précisée par le profil ECORS. *Bull. Soc. Géol. Fr.* (8)6: 287–94

Déramond, J., Graham, R. H., Hossack, J. R., Baby, P., Crouzet, G. 1985. Nouveau modèle de la chaîne des Pyrénées. *C. R. Acad. Sci. Paris* 301(16): 1213–16

ECORS Pyrenees Team 1988. The ECORS deep reflection seismic survey across the Pyrenees. *Nature* 331: 508–10

Garrido-Megias, A., Rios, L. M. 1972. Sintesis geologica del Secondario y Terciario entre los rios Cinca y Segre (Pirineo central de la vertiente surpirenaica, provincias de Huesca y Lerida). *Bol. Geol. Min.* 83(1): 1–47

Goldberg, J. M., Maluski, H., Leyreloup, A. F. 1986. Petrological and age relationship between emplacement of magmatic breccia, alkaline magmatism, and static metamorphism in the north pyrenean zone. *Tectonophysics* 129: 275–90

Hall, C. A., Bennet, V. C. 1983. Significance of lherzolite at the Etang de Lhers, Central Pyrenees, Southern France. *Earth Planet. Sci. Lett.* 45: 349–54

Jacob, Ch. 1928. Structure de versant septentrional des Pyrénées centrales. *C. R. Somm. Soc. Géol. Fr.* 137–38

Johnson, J. A., Hall, C. A. 1989. The structural and sedimentary evolution of the Cretaceous North Pyrenean Basin, southern France. *Geol. Soc. Am. Bull.* 101: 231–47

Labaume, P., Mutti, E., Seguret, M., Rosell, J. 1983. Megaturbidites carbonatées du bassin turbiditique de l'Eocène inférieur et moyen sud-pyrénéen. *Bull. Soc. Géol. Fr.* (7)25: 927–41

Labaume, P., Seguret, M., Seyve, C. 1985. Evolution of a turbiditic foreland basin and analogy with an accretionary prism: Example of the Eocene South-Pyrenean Basin. *Tectonics* 4: 661–85

Labaume, P., Mutti, E., Seguret, M. 1987. Megaturbidite: A depositional model from the Eocene of the SW-Pyrenean Foreland Basin, Spain. *Geomarine Lett.* 7: 91–101

Le Pichon, X., Bonnin, J., Francheteau, J., Sibuet, J. C. 1971. Une hypothèse d'évolution tectonique du golfe de Gascogne. See Debyser et al 1971, 6(11): 1–44

Masson, D. G., Miles, P. R. 1984. Mesozoic seafloor spreading between Iberia, Europe and North America. *Mar. Geol.* 56: 279–87

Mattauer, M. 1968. Les traits structuraux essentiels de la chaîne pyrénéenne. *Rev. Géogr. Phys. et Géol. Dyn.* 10(1): 3–11

Matthews, D. H., Williams, C. A. 1968. Linear magnetic anomalies in the Bay of Biscay: a qualitative interpretation. *Earth Planet. Sci. Lett.* 4: 315–20

Minnigh, L. D., Van Calsteren, P. W. C., Den Tex E. 1980. Quenching: An additional model for the emplacement of the lherzolite at Lherz. *Geology* 8: 18–21

Montadert, L., Winnock, E. 1971. L'histoire structurale du Golfe de Gascogne. See Debyser et al 1971, p. VI.16.1–18

Montigny, R., Azambre, B., Rossy, M., Thuizat, R. 1986. K-Ar study of Cretaceous magmatism and metamorphism from the Pyrenees: age and length of rotation of the Iberian peninsula. *Tectonophysics* 129: 257–74

Munoz, J. A. 1988. *Reunion Extraordinaria SGF-SGE Ecors-Pirineos*. Boussens-Balaguer. 35 pp.

Munoz, J. A., Martinez, A., Verges, J. 1986. Thrust sequences in the eastern Spanish Pyrenees. *J. Struct. Geology* 8: 399–405

Mutti, E., Ricci Lucchi, F., Seguret, M., Zanzucchi, G. 1984a. Seismoturbidites: a new group of resedimented deposits. *Mar. Geol.* 55: 103–16

Mutti, E., Sgavetti, M., Remacha, E. 1984b. La relazione tra piattaforme deltizie e sistemi torbiditici nel Bacino Eocenico Sud-Pirenaico di Tremp-Pamplona. *G. Geol.* 46(2): 3–32

Olivet, J. L., Bonnin, J., Beuzart, P., Auzende, J. M. 1984. *Rapport Scientifique et Technique C.N.E.X.O.*, Paris, 54. 108 pp.

Parish, M. 1984. A structural interpretation of a section of the Gavarnie Nappe and its implications for Pyrenean geology. *J. Struct. Geol.* 6: 247–55

Puigdefabregas, C. 1975. La sedimentacion molasica en la Cuenca de Jaca. *Pirineos* 104: 1–188

Puigdefabregas, C., Munoz, J. A., Marzo,

M. 1986. Thrust belt development in the eastern Pyrenees and related depositional sequences in the southern foreland. In *Foreland Basins*, ed. P. A. Allen, P. Homewood, I.A.S. Special Publ. 8: 229–46

Puigdefabregas, C., Souquet, P. 1986. Tecto-sedimentary cycles and depositional sequences of the Mesozoic and Tertiary from the Pyrenees. *Tectonophysics* 129: 173–203

Roure, F., Choukroune, P., Berastegui, X., Munoz, J. A., Villien, A., et al. 1989. ECORS Deep seismic data and balanced cross-sections; Geometric constraints on the evolution of the Pyrenees. *Tectonics* 8(1): 41–50

Roussel, J. 1903. Tableau stratigraphique des Pyrénées. *Bull. Serv. Carte Géol. Fr.* XV(97)

Schoeffler, J. 1965. Une hypothèse sur la tectogenèse de la chaîne pyrénéenne et de ses abords. *Bull. Soc. Géol. Fr.* 7: 917–20

Seguret, M. 1972. *Etude Tectonique des Nappes et Séries Décollées de la Partie Centrale du Versant Sud des Pyrénées: Caractère Synsédimentaire, Role de la Compression et de la Gravité*. Ser. Geol. Str., vol. 2, USTL Montpellier. 155 pp.

Seguret, P., Daignières, M. 1986. Crustal scale balanced cross-sections of the Pyrenees: discussion. *Tectonophysics* 129: 303–18

Seguret, M., Labaume, P., Madariaga, R. 1984. Eocene seismicity in the Pyrenees from megaturbidites of the South Pyrenean basin. *Mar. Geol.* 55: 117–31

Sibuet, J. C. 1989. Paleoconstraints during rifting of the Northeast Atlantic passive margins. *J. Geophys. Res.* 94: 7265–77

Sibuet, J. C., Collette, B. J. 1991. Triple junctions of Bay of Biscay and North Atlantic: new constraints on the kinematic evolution. *Geology* (19)5: 522–25

Sole-Sugranes, L. S. 1978. Gravity and compressive nappes in the central southern Pyrenees. *Am. J. Sci.* 278: 609–37

Souquet, P., Debroas, E. J., Boirie, J. M., Pons, Ph., Fixari, G., et al 1985. Le groupe du Flysch noir (Albo-Cénomanien) dans les Pyrénées. *Bull. Cent. Rech. Explor.-Prod. Elf-Aquitaine* 9(1): 183

Teixell, A. 1990. Chevauchements alpins dans la terminaison occicentale de la Zone Axiale Pyrénéenne. *Bull. Soc. Géol. Fr.* (8)6: 241–50

Van der Voo, R., Boessenkool, A. 1973. Permian paleomagnetic result from the western Pyrenees delineating the plate boundary between the Iberian peninsula and stable Europe. *J. Geophys. Res.* 78: 518–27

Van der Voo, R., Zijderveld, J. D. A. 1971. Renewed paleomagnetic study of the Lisbon volcanics and implications for the rotation of Iberian peninsula. *J. Geophys. Res.* 76: 3913–21

Verges, J., Munoz, J. A. 1990. Séquences de chevauchement dans les Pyrénées sud-centrales. *Bull. Soc. Géol. Fr.* (8)6: 265–72

Vielzeuf, D., Kornprobst, J. 1984. Crustal splitting and the emplacement of Pyrenean lherzolites and granulites. *Earth Planet. Sci. Lett.* 67: 87–96

Williams, C. A. 1973. A possible triple junction in the N-E Atlantic West of Biscay. *Nature* 244: 86–88

Williams, C. A., McKenzie, D. 1971. The evolution of the North-East Atlantic. *Nature* 232: 168–73

Williams, G. D., Fischer, M. W. 1984. A balanced section across the Pyrenean orogenic belt. *Tectonics* 3(7): 773–80

Annu. Rev. Earth Planet. Sci. 1992. 20:159–79

METAMORPHIC BELTS OF JAPANESE ISLANDS

Shohei Banno

Department of Geology, Kyoto University, Kyoto 606, Japan

Takashi Nakajima

Geochemistry Department, Geological Survey of Japan, Tsukuba 305, Japan

KEY WORDS: high-pressure metamorphism, low-pressure metamorphism, paired metamorphism, island arc

INTRODUCTION

In 1961 Miyashiro presented the concept of paired metamorphic belts based on the mode of occurrences of high-pressure metamorphic belts along the Circum-Pacific orogenic belts, particularly in Japan. According to him the Japanese islands are composed essentially of three paired metamorphic belts: the late Paleozoic Hida–Sangun and late Mesozoic Ryoke–Sanbagawa pairs in Honshu, and the late Mesozoic Kamuikotan–Hidaka in Hokkaido. At that time he could use only 11 K-Ar mica ages.

Detailed petrographic work has continued since that time, and many radiometric ages are now available in the literature. From the late 1970s to the early 1980s, two major breakthroughs were made in Japanese basement geology; the first was in conodont-biostratigraphy followed by radiolaria-biostratigraphy, and the second was due to the increased productivity of K-Ar dating. Finally, the geosyncline concept was abandoned, and has been replaced by plate-tectonics approaches for explaining why particular metamorphic belts are located where they are today. Uyeda & Miyashiro (1974), Saito & Hashimoto (1982), and Maruyama & Seno (1986) pioneered this research. These and later works revealed that many metamorphic areas actually consist of several units.

0084–6597/92/0515–0159$02.00

We will use the terms "belt" for an elongated metamorphic area, and "unit" for a metamorphic area of any shape and size. Both a belt and a unit may be terrane rooted in the lithosphere or a thrust sheet. A unit may also be a tectonic block having a distinct *P-T-t* path. "Zone" is used to denote a geologically distinct area, including regions of unmetamorphosed rocks. As Japan is an island arc, many geological units occur as a belt or zone rather than as a circular area.

After a general review, we will introduce recent petrological works of three belts from well studied areas: the Sanbagawa, Ryoke, and Hidaka metamorphic belts. The first two may form a pair of late Mesozoic age, and the third represents a cross section through island arc crust. Brief reviews of the metamorphism of the Abukuma, Sangun, and Shimanto belts are available elsewhere (Banno & Nakajima 1991).

METAMORPHIC BELTS AND UNITS

In the Japanese Islands, metamorphic rocks occur in two modes; one forms a belt consisting of a single metamorphic unit, and the other is a zone where various metamorphic and unmetamorphosed units are mixed by tectonic processes in one way or another. Herein, the latter type of zone is denoted as a zone containing multiple metamorphic units. Figure 1 schematically shows the metamorphic belts of the Japanese Islands. The ages of protolith and metamorphism, and the baric type of each metamorphic belt are listed in Table 1. Those of zones with multiple metamorphic units are listed in Table 2. These belts and zones are briefly discussed below. The metamorphic geology and petrology of the Sanbagawa, Ryoke, and Hidaka belts will be reviewed in more detail in later sections. The details of the ages of metamorphic rocks in Japan will be published by T. Nakajima by mid–1992.

Zones Containing Multiple Metamorphic Units

As shown in Table 2, we designate the Hida Marginal, Kurosegawa, Nagasaki, and Motai-Yakuki zones as belonging to this category. Some other zones may be added to this category or eliminated. In these zones, various types of rock occur as fault-bounded blocks or inclusions in serpentinite. A typical example is the Hida Marginal zone, where high-pressure schists of ca. 320 Ma, eclogite, granitoid of 400 Ma, low- to medium-pressure garnet amphibolite and metagabbro of 400–440 Ma, and serpentinite occur along with the fragments of the late Paleozoic Akiyoshi complex (Komatsu 1990).

The Kurosegawa zone includes unmetamorphosed Silurian and Devonian sediments and volcanics and various igneous and metamorphic rocks:

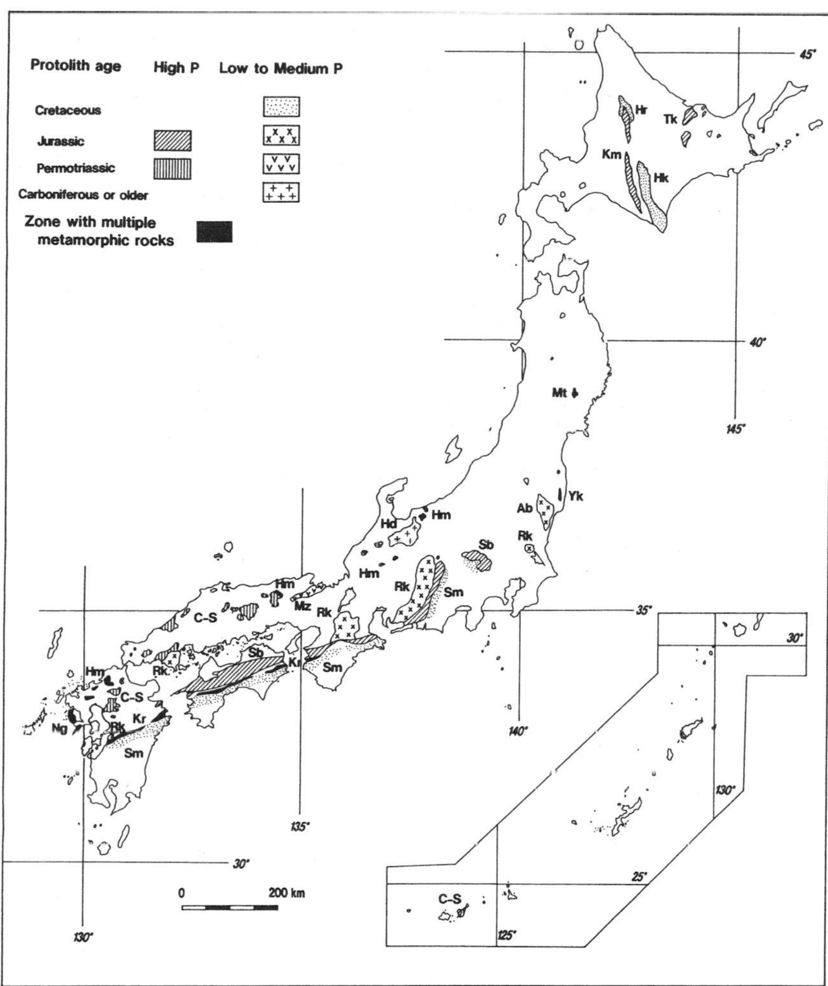

Figure 1 Metamorphic belts in the Japanese Islands. Abbreviations are: Tk, Tokoro; Km, Kamuikotan; Hk, Hidaka; Mt, Motai; Yk, Yakuki; Ab, Abukuma; Hd, Hida; Hm, Hida Marginal; Rk, Ryoke; Sb, Sanbagawa; Sm, Shimanto; Kr, Kurosegawa; Mz, Maizuru; C-S, Chizu-Suo; Ng, Nagasaki.

granitoid and gneiss of ca. 400 Ma, high-pressure schists with ca. 370, 310, and 210 K-Ar mica ages, staurolite-bearing and other medium *P-T* schists, garnet + cpx-bearing granulite, serpentinite, and low-grade schists of 185–230 Ma K-Ar muscovite ages. The 185–230 Ma schists are not associated with serpentinite and are developed in the northern part of the Kurosegawa

Table 1 Metamorphic belts of Japanese Islands—belts with single metamorphic units

Name	Baric type	Facies series[†]	Protolith	Metamorphic age (Ma)	References[d]
Tokoro	high P *a	PA/EG	Cretaceous	90-95 (K-Ar)	1
Kamuikotan tectonic block	high P *a ,c	LG/EG EA	Late Jurassic/Early Cretaceous	70-110 (K-Ar) 130-140 (K-Ar)	2, 3 3
Hidaka Main	low P	GS/AM/GR	Late Cretaceous/Paleogene	17-50 (K-Ar, Rb-Sr)	4
Hidaka West	low P	GS/AM/GR		17-24 (K-Ar)	4
Horokanai	low P	ZL/PP/GS/AM/GR	Jurassic	100 (^{39}Ar-^{40}Ar)	2
Abukuma	low P	GS/AM/GR	Jurassic and ?	90-116 (K-Ar, Rb-Sr, Sm-Nd) both metamorphics and granitoids	5
Hida-Oki	low P	AM/GR	?	170-240 (K-Ar, Rb-Sr) 410 (Sm-Nd)	6 7
Unazuki	medium P	GS/AM	Carboniferous	210-240 (Rb-Sr)	6
Yakuno (Maizuru)	low-medium P	PA/GS/AM	Permian	220-270 (K-Ar)	8
Suo	high P *b	PA/EG	Permian	200-240 (K-Ar)	9, 10 11
Chizu	high P *b	PA/EG	Jurassic ?	160-190 (K-Ar)	9
Ryoke	low P	GS/AM(/GR)	Jurassic	60-110 see text both metamorphics and granotoids	12, 13
Sanbagawa eroded unit	high P *b	PA/EG/EA/EC EA	Jurassic - ? ?	60-90 (K-Ar, ^{39}Ar-^{40}Ar) 84-157 (K-Ar, ^{39}Ar-^{40}Ar) see text	14 - 18 19, 20
Chichibu	medium or high P	PP?/PA	Jurassic	140 (K-Ar)	21
Shimanto	medium P ?	PP/PA/GS	Cretaceous/Paleogene	63-75 (K-Ar)	22

*[a] Aragonite stable but no jadeite+quartz.
*[b] Aragonite unstable but lawsonite and crossite.
*[c] Include jadeite+quartz in tectonic block.
[†] Abbreviation of facies name: ZL, zeolite; PP, prehnite-pumpellyite; PA, pumpellyite-actinolite; GS, greenschist; EG, epidote-glaucophane; EA, epidote-amphibolite; AM, amphibolite; GR, granulite; EC, eclogite.
For references, see Table 2.

Table 2 Zones containing multiple metamorphic units

Name	Age (Ma)	Rock types	References[d]
Kurosegawa	400-440 (K-Ar, Rb-Sr)	Granitoid, Garnet amphibolite, Gneiss	23, 24
	400 (K-Ar, Rb-Sr)	Garnet two mica schist, Staurolite schist	23, 24
	340-380 (K-Ar)	High pressure schist	25
	315-330 (K-Ar)	High pressure schist	25
	210-240 (K-Ar)	Jadeite-glaucophane schist	23
	above units occur associated with serpentinite		
	185-225 (K-Ar)	PP and PA schist, formerly regarded as Chichibu Group	15
Hida Marginal	240-426 (K-Ar, Rb-Sr)	Metagabbro, Amphibolite	7, 8, 26
	300-350 (K-Ar, Rb-Sr)	High pressure schist	9
	?	Eclogite, Garnet Amphibolite	
	370-420 (K-Ar, Rb-Sr)	Granitoid	
		Serpentinite	
Motai-Yakuki	280 (K-Ar)	High pressure schist	
	100 (K-Ar)	High pressure schist	
		Serpentinite-Peridotite	
Nagasaki	60-80 (K-Ar)	High pressure schist	
	440-480 (K-Ar)	Metagabbro, Serpentinite	8
		Serpentinite	

[d] List of major data source for Tables 1 and 2.

1. Sakakibara (1991); 2. Ishizuka (1987); 3 Imaizumi & Ueda (1981); 4. Osanai et al (1991); 5. Hiroi & Kishi (1989); 6. Shibata et al (1970); 7. Asano et al (1990); 8. Shibata et al (1977); 9. Shibata & Nishimura (1989); 10. Nishimura et al (1990); 11. Faure et al (1986); 12. Shiba et al (1979); 13. Nakajima et al (1990); 14. Itaya & Takasugi (1988); 15. Suzuki et al (1990); 16. Takasu & Dallmeyer (1990); 17. Isozaki & Itaya (1990); 18. Isozaki & Itaya (1991); 19. Takasu & Dallmeyer (1991); 20. Yokoyama & Itaya (1990); 21. Isozaki et al (1990); 22. Shibata et al (1988); 23. Maruyama et al (1984); 24. Hayase & Nohda (1969); 25. Ueda et al (1980); 26. Nishina et al (1990).

zone for several tens to one hundred kilometers in Shikoku as revealed recently (Isozaki & Itaya 1991). The Kurosegawa zone runs subparallel to Southwest Japan, but obliquely cuts the Sanbagawa and Chichibu belts in western Shikoku and Kyushu. Its tectonic interpretation has changed from time to time (Maruyama et al 1984, Taira et al 1983, Banno 1991). Recently, Maruyama et al (1990) and Isozaki & Itaya (1991) proposed that it is a tectonic outlier of the Hida Marginal zone, and the Suo and Chizu zones located on the top of the nappe sequence of Southwest Japan.

Other zones listed here are not as variegated as above, but still accompany several metamorphic units. The Nagasaki zone has a 60–80 Ma K-Ar age and an epidote-glaucophane type facies series along with metagabbro of 400–480 Ma hornblende K-Ar age as fault-bounded blocks.

Belts With a Single Metamorphic Unit

Among the metamorphic belts listed in Table 1, the Tokoro, Kamuikotan, Suo + Chizu, Ryoke, Sanbagawa, Chichibu, and Shimanto belts were derived from accretionary complexes. The Tokoro belt is predominantly a basic complex of possibly seamount origin. The high-pressure part lies in the albite-aragonite field with epidote-glaucophane at higher grade, but most of the area is occupied by zeolite to prehnite-pumpellyite facies (Sakakibara 1991). The Kamuikotan belt was derived from an accretionary complex, and is accompanied by 140 Ma tectonic blocks of garnet amphibolite. The coherent sequence is in the albite-aragonite field, but quartz + jadeite occurs in a few tectonic blocks.

Nishimura (1990) has demonstrated that the metamorphic rocks formerly called the Sangun belt actually consists of at least two distinct units, the ca. 300 Ma schist in northern Kyushu and in the Hida Marginal belt, and 160–230 Ma units with two apparent maxima, the Chizu unit of 180 Ma and the Suo unit of 220 Ma. They are considered to be a metamorphosed Permian Akiyoshi complex, but could include a Jurassic protolith in the Chizu unit. The 160–230 Ma unit also occurs in Ishigaki Island, Southwest Ryukyu (Nishimura et al 1989).

The Chichibu belt, which has been regarded as an unmetamorphosed equivalent of the Sanbagawa belt, is composed of three distinct units of ca. 120 Ma, 140 Ma, and 185–222 Ma (Isozaki & Itaya 1991). The ca. 120 Ma unit which occupies the northernmost part next to the Sanbagawa belt may be a part of the latter, but the 185–220 Ma unit is added to the Kurosegawa zone in Table 2. The significance of the 140 Ma unit is not clear. The redefinition of the Chichibu zone will be proposed in the near future, but in Figures 1 and 2, we include it in the Sanbagawa belt.

The Shimanto accretionary complex consists of Cretaceous and

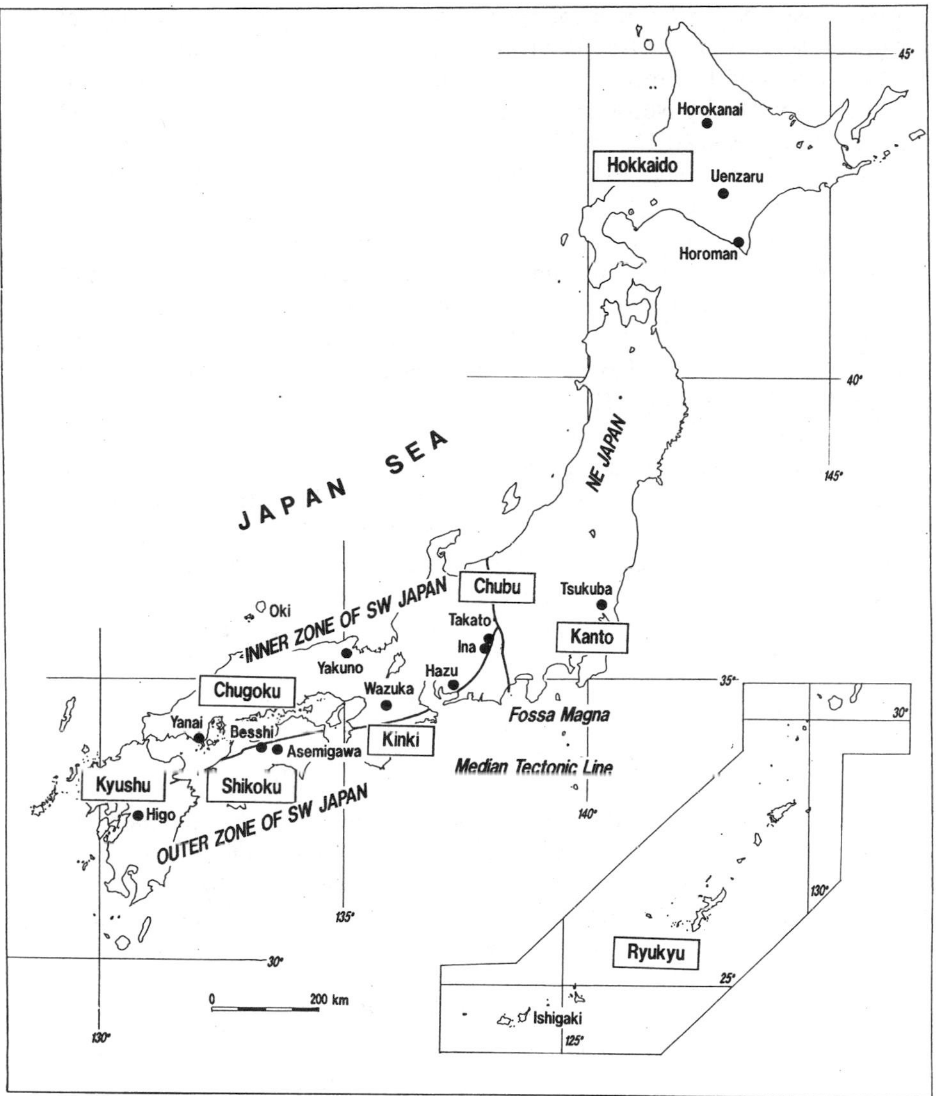

Figure 2 Locality names referred to in the text.

Miocene complex. Pumpellyite-actinolite facies phyllite develop locally in the former.

The Hida-Oki complex is made of gneiss of the amphibolite facies and locally of the granulite facies, and is intruded by granitoid of ca. 180 Ma.

Andalusite occurs but it is not certain whether its origin is due to regional or contact metamorphism. The Hida-Oki complex is expected and believed to be a part of the Precambrian Eurasian craton. This may well be correct, but a great many isotopic ages that have been collected in the past 30 years have revealed 180 and 220 Ma maxima in K-Ar ages, and 410 Ma Sm-Nd (Asano et al 1990) without providing any dated Precambrian metamorphism.

The Unazuki schists at the eastern margin of the Hida belt is the only kyanite-sillimanite metamorphic belt, and includes chloritoid and staurolite-bearing pelitic schists. It contains Carboniferous bryozoa and has a 220–240 isotopic age. The Unazuki belt is petrologically similar to the Ogcheon zone of Korea, and provides a clue for reconstructing the pre–Japan Sea geology (Hiroi 1983).

The central Abukuma metamorphic belt consists of the higher-grade Takanuki series—consisting predominantly of pelitic and psammitic gneiss, and the Gosaisyo series—predominantly with basalt which accompanies Jurassic radiolaria. The predominant facies series is the famous low-pressure type, ranging from greenschist to amphibolite and low granulite. Kyanite occurs locally in the Takanuki series, but has been ascribed to temporal high pressure caused by the obduction of the oceanic Gosaisyo series on the Takanuki series of an accretionary complex (Hiroi & Kishi 1989).

The Yakuno, Horokanai, and Hidaka West ophiolite have the common feature that the metamorphic grade increases with ophiolite stratigraphy until the granulite facies is reached at the peridotite layer. The Yakuno ophiolite is a higher pressure facies series than the others—as judged from the absence of prehnite-pumpellyite facies in it. The Yakuno ophiolite and overlying Permo-Triassic sedimentary rocks constitute the Maizuru zone, which thrust over the late Paleozoic ultra-Tamba and Mesozoic Mino-Tamba accretionary complex (Ishiwatari 1990).

The Birthplace of the Protolith and Metamorphic Belts

The geological and lithological features of the metamorphic belts mentioned above and certain paleomagnetic and paleontological features constrain the birthplaces of protolith and metamorphism.

1. The Japanese Islands were located at the Eurasian continental margin before 20–15 Ma, the date of the opening of the Japan Sea.
2. In Southwest Japan, the pile of nappe with younger accretionary complex lower in the sequence is a general feature (Hayasaka & Hara 1982, Charvet et al 1987).
3. High-pressure metamorphism took place in subduction zones.

4. The paleolatitude of oceanic components of the accretionary complex, such as the Mino-Tamba and Shimanto, is equatorial (Hirooka 1990).

Some metamorphic units, such as the Hida-Oki belt and the Takanuki series of the central Abukuma belt, may be fragments of the Asian continent, but the majority of Japanese regional metamorphic belts developed from accretionary complexes, composed of trench fill sediments and oceanic components transported from the south by the plates converging on the eastern border of Asia. During the middle Miocene, they split apart from the Asian continent and the Japan Sea opened.

This is the most simplified story and is tolerant to various modifications. For example, the present position could also be due to syn- or post-tectonic strike slip movement as exemplified by the Median Tectonic Line (MTL), which was left-lateral in the Cretaceous. Such movement may explain the chaotic mixing of assorted metamorphic units occurring in the zone with multiple metamorphic units. We don't know yet how much displacement of the MTL and other transcurrent faults occurred. This makes it difficult to answer the crucial question of whether or not the Ryoke and Sanbagawa belt were paired.

METAMORPHIC GEOLOGY AND PETROLOGY OF THREE PARTICULAR AREAS

Past petrological studies of Japanese metamorphic belts have already produced data characterizing their baric type and have briefly discussed the tectonics of the Japanese Islands. More elaborate *P-T-D* (deformation)-*t* paths of individual belts are necessary, however, to decipher what took place in the depths of orogenic belts. In this paper we focus on three particular Japanese metamorphic belts: the Sanbagawa and Ryoke belts which run on the south and north sides of the MTL for 700 and 1000 km, respectively, and the Hidaka belt in Hokkaido.

Sanbagawa (Sambagawa) Metamorphic Belt

The Sanbagawa belt underwent high-pressure intermediate type metamorphism, in which jadeite + quartz and probably aragonite were unstable, but epidote + crossite was stable. The petrology of this belt has been studied in most detail in central Shikoku, especially in the Bessi (Besshi) and Asemigawa areas. The metamorphic geology of the other parts of the Sanbagawa belt is more or less similar to that of central Shikoku, and the same mineral zones are used for their mapping.

PROTOLITH The main rock types of the Sanbagawa belt are metamorphosed basaltic rocks, shale, greywacke sandstone, chert, rare conglomerate, and limestone. These form a coherent sequence throughout the Sanbagawa belt. Bedded cupriferous-pyrite ore of the Bessi (Besshi)-type is also a member of this coherent sequence. In the higher grade area in central Shikoku, tectonic blocks of higher grade than the coherent schists occur. These are eclogitized gabbro, granulitized gabbro, dunite with garnet clinopyroxenite, and kyanite-bearing eclogitic mica schists (Takasu 1989). Only metabasite and meta-ultramafics were regarded as tectonic blocks.

Two types of datable fossils are known. One is Triassic conodonts in low-grade limestone, and the other is late Jurassic radiolaria in metavolcanic sediments of the Mikabu Greenstone complex of Shikoku. As both are oceanic components, the depositional age in the trench should be younger than these ages (Isozaki & Itaya 1990).

METAMORPHIC ZONATION The areal distribution of metamorphic grades can be mapped in terms of the chlorite (Chl-z.), garnet (Grt z.), albite-biotite (Ab-Bt z.), and oligoclase-biotite zones (Ol-Bt z.) using pelitic assemblages. The chemistry of amphibole in basic schist is sensitive to metamorphic temperature and is used to define the grade of metamorphism. From low to high grade, the mineral zones and associated amphiboles species are: pumpellyite+actinolite (Chl z.), epidote+winchite+hematite (Chl z.), epidote+crossite+hematite (Chl z. Grt z.), albite+epidote+barroisite (Ab-Bt z.), and oligoclase+epidote+hornblende (Ol-Bt z.). X_{an} is ca. 0.15–0.28 in pelitic schists of the Ol-Bt zone. The tectonic blocks are now mostly equilibrated in the epidote amphibolite facies under the same physical conditions as the coherent sequence they were emplaced into. However, they also retain relics of the facies they once belonged to, namely the eclogite and granulite facies. A contact aureole is developed around the tectonic block of eclogitized metagabbro, Sebadani body (Takasu 1989). The *P-T-t* trajectories of two groups of Sanbagawa schists are shown in Figure 3. The retrograde trajectory in this figure is less steep than those previously given, because the retrograde trajectory locally passes the crossite field in central Shikoku (Hara et al 1990).

An inverted thermal structure has been recognized in Shikoku, where the biotite zone schists occur structurally within the garnet zone. Hara et al (1977, 1990) proposed that this is a part of a pile of nappes that formed in central Shikoku. However, so far, no distinct gap of metamorphic grade has been recognized in the whole of central Shikoku, even where a discontinuity is suggested from the structural and geochronological points of view (Banno & Sakai 1989).

GEOCHRONOLOGY The impact of isotope chronology has also been large with regard to the higher grade area. Itaya & Takasugi (1988) have shown

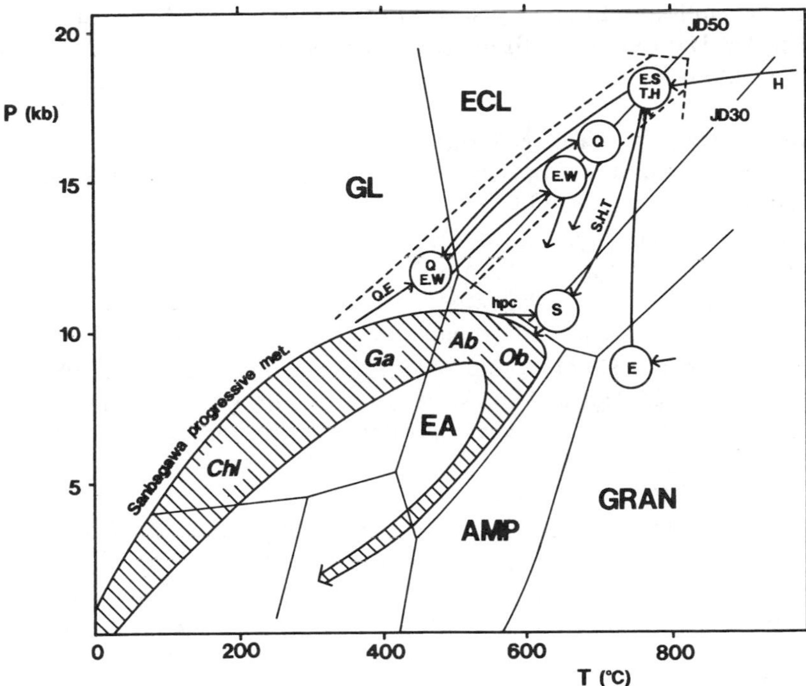

Figure 3 The *P-T* trajectory of the Sanbagawa progressive metamorphism and the tectonic blocks in central Shikoku (from Takasu 1989, Figure 2). ECL, GL, EA, AMP, and GRAN = eclogite facies, glaucophane schist facies, epidote amphibolite, amphibolite, and granulite facies, respectively. Sanbagawa progressive met. = metamorphic field gradient of the Sanbagawa progressive metamorphism. Chl, Ga, Ab, Ob = chlorite, garnet, albite-biotite, and oligoclase-biotite zones, respectively. S, T, W, E, H, N, and Q represent individual tectonic blocks. hpc = high-pressure contact metamorphism by the Sebadani (S) mass.

that the K-Ar muscovite age of the central Shikoku ranges from 67–90 Ma with a greater age for higher grade material. Takasu & Dallmeyer (1990) distinguish three groups of ^{40}Ar-^{39}Ar ages within the schists. The hornblende ^{40}Ar-^{39}Ar age fall in the 80 to 95 Ma range in most of central Shikoku, which they called the Besshi nappe. In contrast, the psammitic formation of the lower-chlorite zone (their Oboke nappe) has a muscovite ^{40}Ar-^{39}Ar age of around 70 Ma. They have further found ^{40}Ar-^{39}Ar isotope correlation ages of 133 and 160 Ma for hornblende in epidote amphibolite clasts in Eocene conglomerate overlying the Sanbagawa schists, and proposed the former existence of the Kuma nappe now eroded away. The Oboke nappe has similar radiometric ages to the Cretaceous through early Paleogene Shimanto accretionary complex, which developed to the south and in some places directly underlies the Sanbagawa belt. This led Hara et al (1990) to regard the Oboke nappe as a part of the Shimanto belt.

KINEMATIC STUDY Takagi & Hara (1979) described albite porphyroblasts that showed rotation about a N-S axis. Faure (1983) described a curvilinear fold including sheath folds in eastern Shikoku, and considered that the major movement of the schists was top to the east, which carried the Besshi nappe [nomenclature of Takasu & Dallmeyer (1990)] over the Oboke nappe. He considered that after the east-west flow, north-south compression gave rise to the exhumation of the Sanbagawa belt. On the contrary, Wallis (1990) considers that the major movement of the Sanbagawa belt in central Shikoku was top to the west, and that ductile deformation that post-dated the EW flow was only locally associated with high strain.

SCENARIO OF EVOLUTION Possibly in the late Jurassic to early Cretaceous, the protolith of the Sanbagawa belt was deposited in a trench along the eastern border of Asia. The subduction went as deep as 50 km to produce kyanite-bearing eclogitic mica schist, and associated eclogitized metagabbro. In the major part of the Sanbagawa schists, the coherent sequence reached around 10 kbars recorded in the zoisite + kyanite + quartz assemblage of metagabbro. At the time of peak metamorphism, the higher grade Ol-Bt and Ab-Bt zone schists were at lower pressure than the lower grade Grt zone (Banno et al 1986). This may sound unusual, but is possible if the isotherm in the subduction zone is inclined with respect to the isobar. Then followed a large scale top-to-west shear of the metamorphic belt, which gave rise to the formation of large sheath folds. The Sanbagawa schists have EW mineral lineation in both prograde and retrograde stages, suggesting their exhumation was accompanied by EW shear.

This scenario does not tell us where metamorphism took place. We know that the MTL was left-lateral in Cretaceous, but have no reliable estimate of the associated displacement. Thus, we are not certain whether the Ryoke and Sanbagawa belts were paired. Many models accounting for the exhumation of the Sanbagawa belt, which was largely completed by the Eocene, have been proposed: oblique subduction, collision of the alleged Kurosegawa continent and Eurasia, change of direction of plate motion, underplating by the Shimanto complex that is now developed south of the Sanbagawa belt, decoupling of the subducted ridge, etc (Maruyama 1990). Such events potentially may all have affected the tectonics of the Sanbagawa belt. However, none can account for the exhumation by itself. Much thought as to how and why it occurred is needed.

The Ryoke Belt

The Ryoke belt is a low-pressure, i.e. andalusite-sillimanite type, regional metamorphic belt accompanied by abundant granitoid. The metamorphic rocks of the central Abukuma Plateau were once considered to be a part of the Ryoke belt, but we will reserve this view since the former belt is much older (90–115 Ma) than the neighboring strict sense Ryoke belt (60 Ma). In the Ryoke belt the granitoids occupy a far larger area than the regional metamorphic rocks. The metamorphic rocks are confined to several areas, such as Kyushu (Higo), Yanai, Kinki (Wazuka etc), Chubu (Hazu, Ina, Takato), and Tsukuba.

PROTOLITH The Ryoke metamorphic rocks gradually change to the Jurassic Mino-Tamba complex, which is composed of Jurassic trench fill matrix and exotic chert, limestone, and greenstone of Carboniferous to Triassic. The greenstones and limestones are rare in the Ryoke belt except in western Kyushu. Epidote-glaucophane schists, probably of the Chizu and Suo zones were also affected by the Ryoke metamorphism in Kyushu.

GRANITOIDS The granitic rocks associated with gneiss and migmatite are commonly gneissose and intruded concordantly to the general trend of gneisses, while massive granitoids crosscut the trend of metamorphosed and unmetamorphosed rocks. Most granitoid in the Ryoke metamorphic belt is of ilmenite series and I-type, although some gneissose granitoid contain garnet and muscovite. The high-grade part of the Ryoke belt is a migmatite zone.

METAMORPHIC ZONATION AND THERMAL STRUCTURE Most of the Ryoke metamorphic belt can be mapped in terms of the chlorite, chlorite-biotite, and biotite zone for the lower- to medium-grade areas. In the Wazuka area, Wang et al (1986) has shown that biotite was first formed in psammitic rock by the reaction, K-feldspar + chlorite + quartz = biotite + muscovite + H_2O. Cordierite zone and sillimanite zone were used for medium to higher grade. In the Yanai area, the sillimanite stability field is further divided into two, lower-grade zones where sillimanite + biotite + Ksp is stable and a higher-grade cordierite–garnet zone where sillimanite + biotite is no longer stable. In the Ryoke belt, the first appearance of andalusite and sillimanite is not clearly differentiated. Muscovite breakdown and andalusite and sillimanite transition are often considered to have taken place at the same grade, but muscovite is rather common in much of the sillimanite zone. Probably prograde and retrograde muscovite have not been distinguished properly. In the Hazu area, where staurolite occurs in pelitic schist, the muscovite breakdown took place well within

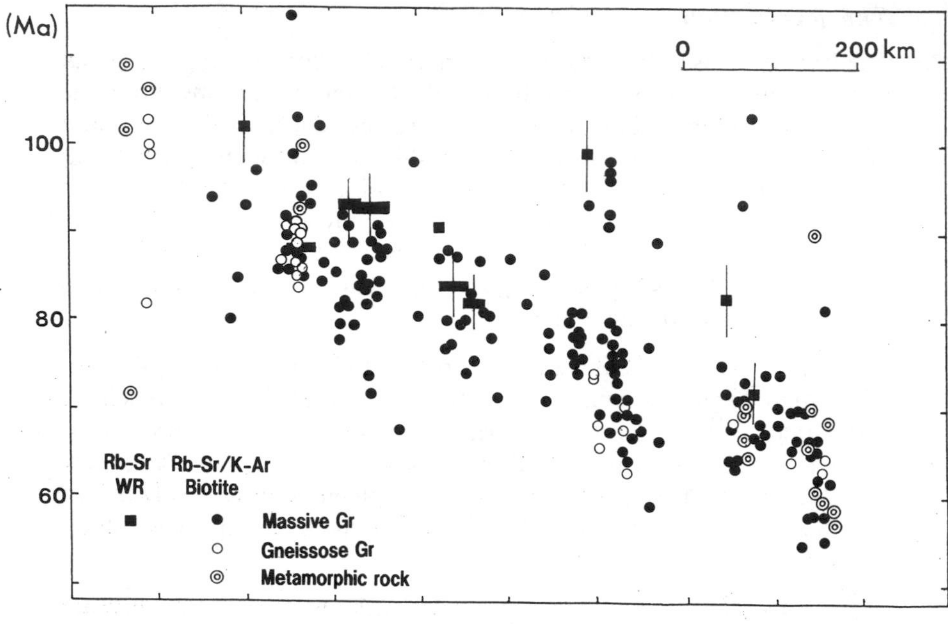

Figure 4 Along-arc lateral variation of radiometric ages of the Ryoke granitic and metamorphic rocks.

the sillimanite zone (Asami et al 1982). The amphibolite facies is the dominant highest grade but orthopyroxene-bearing rocks occur in western Kyushu (Yamamoto 1962).

Contact metamorphism around granitoid intrusion is common even in high-grade areas, but the regional isograd is independent of the shape of granitoid body.

The thermal axis of the metamorphic belt lies north of the MTL in the Yanai and Kinki regions, and coincides with the MTL in the Chubu region. In the Yanai region, the high-grade zone lies 50 km north of the MTL. The geobarometry in the Yanai area by Nakajima suggests 7–10 km burial of the sillimanite zone, implying a 50–70°C km^{-1} geotherm.

GEOCHRONOLOGY It is noteworthy that the isotopic age of Ryoke granitic and metamorphic rocks show a systematic along-arc decrease from 100 Ma in westernmost Chugoku to 60 Ma in Chubu (Nakajima et al 1990). Rb-Sr whole rock isochron ages were obtained for massive granitoid but not for gneissose granitoid and metamorphics. However, since the cooling history of these rocks—as indicated by K-Ar hornblende, biotite, and K-feldspar—are similar, we may assume that the age of massive granitoid is similar to the age of gneissose granitoid.

The Rb-Sr and K-Ar biotite ages of the granitic and metamorphic rocks follow the same eastward younging trend, with a 5 to 10 Ma time lag from the Rb-Sr isochron age for the entire Ryoke belt. Therefore, the tectonic setting responsible for the Ryoke magmatism and metamorphism cannot be explained by the steady subduction model and needs an episodic event such as the encounter of the Kula-Pacific ridge with the Eurasia plate and subsequent ridge-trench-trench migration along the eastern margin of the Asian continent.

Hidaka Belt in Hokkaido

The southwestern region of Hokkaido is geologically the extension of Northeast Japan, and its basement is composed of a Jurassic accretionary complex intruded by Cretaceous granitoid. However, the central and eastern parts of Hokkaido comprise metamorphic belts that are absent in Northeast Japan: the Kamuikotan high-pressure schists, the Horokanai metamorphosed ophiolite, the Hidaka belt, and the Tokoro belt. As is seen in Figures 1 and 2, the so-called Hidaka belt is divided into the Western belt ("zone" in Komatsu et al 1989) and Main belt (zone), respectively. The Western belt is a metamorphosed ophiolite, and the Main belt consists of island arc crust which thrust from east to west over the Western belt (Komatsu et al 1989). The metamorphosed ophiolite consists of gabbro, layered gabbro and ultramafic rocks, and tectonized peridotite, and is partly overturned. Its facies series includes greenschist, epidote amphibolite, amphibolite and lower granulite facies.

HIDAKA BELT, ISLAND ARC CRUST In Hokkaido, the tectonic debate has focused on whether the Hidaka belt is an island arc that collided with an island arc or was formed in situ in an accretionary prism that developed on the east of Northeast Japan, but further to the north. The Main belt is composed of two sequences, an upper metasediment sequence of two mica schist and two mica gneiss, and a lower sequence of basic and pelitic granulite, amphibolite intercalated with pelitic gneiss, and the hornblende gneiss which is the alternation of pelitic biotite gneiss and biotite amphibolite. These two sequences are thought to have been tectonically stacked within the accretionary complex. The lower sequence has abundant oceanic components, i.e. metabasite. The upper sequence is transitional to the unmetamorphosed Nakanogawa Formation of the Cretaceous (although Paleogene has been recently reported; see Kiminami et al 1990). Komatsu et al (1989) designated the major regional metamorphism as M0 stage.

According to Komatsu et al (1989), shear zones, which are developed at various levels and are layer-parallel and dextral, were formed by a D1 event—i.e. post-dating the major metamorphism. This was accompanied by retrograde metamorphism (M1) at slightly lower temperature than the M0 metamorphism. Thus D1 is a horizontal flow which could have detached the present Hidaka belt from the lowest crustal layer. The D2 phase deformation is represented by mylonite that develops at the base, i.e. at the contact with the Western belt. The mylonitization (M2) proceeded in the greenschist facies. Post-dating the mylonite, pseudotachylite was formed, filling veins that crosscut the mylonite (Toyoshima 1990).

Igneous intrusions are common. In the lower to middle structural level, S-type tonalite which locally contains abundant metamorphic enclaves, is common. I-type granitoid were intruded discordantly to both the upper sequence and concordant S-type granitoid. The granulite facies area includes high-temperature peridotite bodies, which are solid intrusions of mantle material. The Uenzaru and Horoman bodies, especially the latter, have been studied in detail petrologically.

METAMORPHIC ZONATION The mineral zones are not standardized. The zones shown in Figure 5 follow Komatsu et al (1989), who distinguished zones I, II, III, and IV in terms of the paragenesis of pelitic rocks. The characteristic assemblages are from the lower to higher grade as follows: chlorite+phengite, biotite+muscovite (+ And, Grt, Crd), biotite+ garnet+sillimanite (+ Crd or Kfs), and garnet+cordierite+Kfs (+ Sil or Bt). However, Shiba (1988) who has worked in the southern part of the belt, distinguished five mineral zones for the area comprising zones II, III, and IV. The *P-T-D-t* trajectory by Osanai et al (1991) shown in Figure

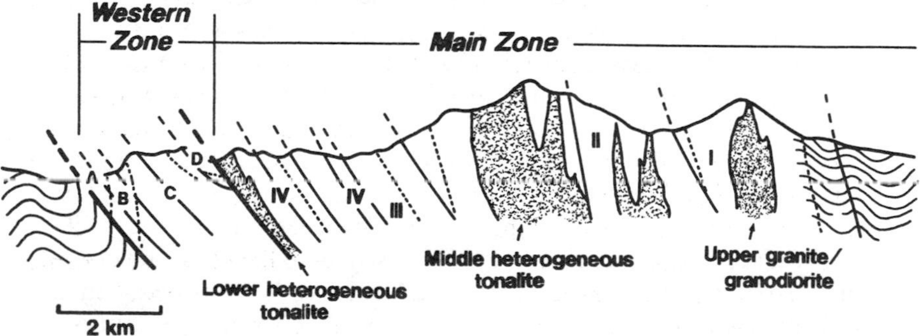

Figure 5 Cross section of the Hidaka belt (Komatsu et al 1989, Figure 1). I to IV and A to D represent metamorphic mineral zones in the order of increasing metamorphic grade, for the Main Zone and West Zone, respectively.

6 is similar to that of Shiba (1988). Both authors consider that the peak *P-T* path leaves the muscovite+quartz field in the andalusite field and goes to biotite+almandine ($X_{Mn} = 0.1$)+sillimanite field. The geothermobarometry suggests that granulite formed at 700°C and 7 kbar. Both of these authors postulate isobaric heating. Osanai et al's path for an indi-

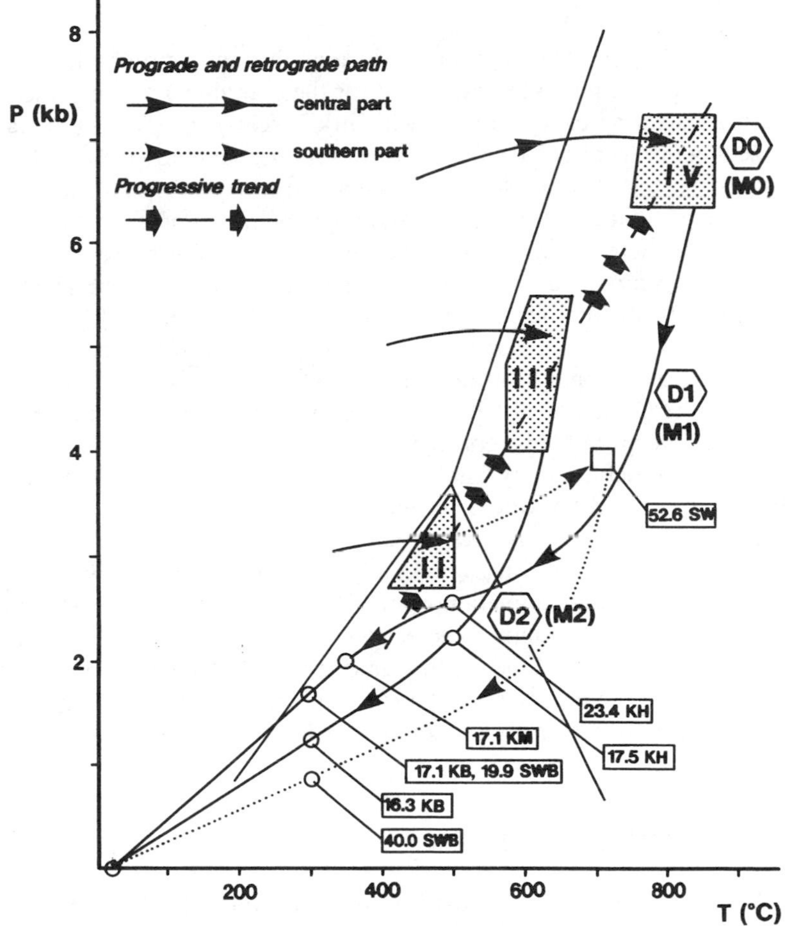

Figure 6 The *P-T-t* trajectory of the Hidaka Main belt (modified from Figure 7 of Osanai et al 1991). Metamorphic events of M0, M1, M2 and deformation events of D0, D1, D2 are shown. Arrowed dotted line shows the reheating process by granitic intrusion in the southern area. Numbers in the boxes are radiometric ages. K = K-Ar, S = Rb-Sr, W = whole rock isochron, H = hornblende, B = biotite, M = muscovite, and WB = whole rock-biotite isochron ages.

vidual zone goes from the kyanite to sillimanite field, whereas Shiba's proceeds from the andalusite to sillimanite field. This may be favored by a tectonic model with the accretionary prism being heated by a source such as the subducting ridge, but this has not been substantiated petrologically. The analysis of Osanai et al's staurolite breakdown fits a wide range of dP/dT. The sphalerite geobarometry on which Shiba (1988) bases his model is popular but faces serious doubt in its thermodynamic ground (Banno 1988).

MIGMATITE Migmatite is common in the granulite and amphibolite facies areas. The 700°C and 7 kbar estimated for the granulite and the presence of garnet + orthopyroxene + quartz with little K-feldspar suggests that partial melting occurred at such depth. However, Osanai et al (1991) did not consider that this anatexis formed magma that intruded into a shallower depth of the Hidaka belt. In addition to rock composed of leucosome and melanosome, there also occurs a quartz-rich pod of millimeter size objects, which are accompanied by idiomorphic and normally zoned plagioclase ($X_{an} = 0.80$–0.50), biotite, and rare K-feldspar. Tagiri et al (1989) considered this patch to be the melt produced in gneiss in situ. This could be due to the heat of tonalitic granitoid intrusion in the middle of the structural level (cf Osanai et al 1991). A general decrease of large lithophile elements in the granulite is reported (Ba, K, Sr depletion by Osanai et al; and Cs, Rb, K depletion by Tagiri et al). For us, however, it seems that only Cs, Rb, K, and Ba were depleted.

Much detailed work has been accumulated in the past 15 years. The Hidaka belt has granulite facies rocks at the structural bottom (west), and greenschist facies and unmetamorphosed rocks at the top (east), and presents an essentially continuous section of island arc crust. Much more research remains to be done on this unique example of island arc crust, which may help the modeling of the regional metamorphism and magma genesis in accretionary prisms, such as the Ryoke belt and possibly the Shimanto belt.

CONCLUDING REMARKS

The Japanese islands are composed of many metamorphic belts. Most of them were derived from late Paleozoic to Cenozoic accretionary complexes which developed along the eastern border of the Asian continent. They are either andalusite-sillimanite type, low-pressure metamorphic belts, or intermediate high-pressure belts where epidote + glaucophane and locally aragonite are stable. No regional jadeite-quartz area exists, though such an assemblage occurs in tectonic blocks in high-pressure areas.

Besides metamorphic belts consisting of a single metamorphic unit, there are regions where small areas and blocks of various metamorphic rocks are abundant within serpentinite and sedimentary sequences.

In these past 30 years, the study of the metamorphic belts of Japan has succeeded in allowing us to identify the geological and petrological characteristics of these belts. It is helping to establish a more quantitative understanding of regional metamorphism, which is the process going on in the depth of the orogenic zone underneath the continental margin.

Of course many problems await more extensive studies. The genesis of zones with multiple metamorphic units and the validity of the concept of paired metamorphism in Japan are two of the most urgent problems which in particular lie ahead of us.

ACKNOWLEDGMENTS

We are deeply obligated to the following individuals: Simon R. Wallis, T. Hirajima, T. Ikeda, M. Shiba, A. Takasu, and A. Ishiwatari for their interpretation of regional data; M. Komatsu, Y. Osanai, and A. Takasu for allowing the use of their figures; and Y. Isozaki, T. Itaya, A. Takasu, R. D. Dallmeyer, and M. Sakakibara for access to unpublished information.

Literature Cited

Asami, M., Hoshino, M., Miyakawa, K., Suwa, K. 1982. Metamorphic conditions of staurolite schists of the Ryoke metamorphic belt in the Hazu-Hongusan area, central Japan. *J. Geol. Soc. Jpn.* 88: 437–50

Asano, M., Tanaka, T., Suwa, K. 1990. Sm-Nd and Rb-Sr ages of the Hida metamorphic rocks in the Wada-gawa area, Toyoma Prefecture. *J. Geol. Soc. Jpn.* 96: 957–66[1]

Banno, S. 1991. Metamorphism in island arcs—the Japanese islands. In *Dynamic Process of Material Transport in the Earth's Interior*, ed. F. Marumo, pp. 433–52. Tokyo: Terapub

Banno, S., Nakajima, T. 1991. Metamorphic belts of Japan. *Episodes*. In press

Banno, S., Sakai, C. 1989. Geology and metamorphic evolution of the Sambagawa belt, Japan. See Daly et al 1989, pp. 519–32

Banno, S., Sakai, C., Higashino, T. 1986. Pressure temperature trajectory of the Sanbagawa metamorphism deduced from garnet zoning. *Lithos* 19: 51–63

Charvet, J., Faure, M., Caridroit, M., Guidi, A. 1987. Some tectonic and tectonogenetic aspects of southwest Japan: an alpine-type orogen in an island arc position. In *Formation of Active Ocean Margin*, ed. N. Nasu, pp. 791–817. Tokyo: Terrapub

Daly, J. S., Cliff, R. A., Yardley, B. W. D., eds. 1989. *Evolution of Metamorphic Belts*. Geol. Soc. London Sp. Pub. 43. 566 pp.

Faure, M. 1983. The Mesozoic orogeny in the outer zone of SW Japan in eastern Shikoku. *J. Gakugei, Tokushima Univ.* 34: 1–35

Faure, M., Monie, J., Fabbri, O. 1986. Microtectonics and ^{40}Ar-^{39}Ar dating of high pressure metamorphic rocks of the south Ryoko Arc and their bearing on the pre-Eocene geodynamic evolution of Eastern Asia. *Tectonophysics* 156: 143–66

Hayasaka, Y., Hara, I. 1982. *89th Ann. Meeting, Geol. Soc. Jpn.* Niigata. 556 pp.[2]

Hara, I., Hide, K., Takeda, K., Tsukuda, E., Shiota, T. 1977. Tectonic movement in the Sambagawa Belt. In *The Sambagawa Belt*,

[1] In Japanese with English abstract.

[2] In Japanese.

ed. K. Hide, pp. 307–90. Hiroshima: Hiroshima Univ. Press. 440 pp.[1]

Hara, I., Shiota, T., Hide, K., Okamoto, K., Takeda, K., et al 1990. Nappe structure of the Sanbagawa belt. *J. Metamorph. Geol.* 8: 441–56

Hara, I., Shiota, T., Takeda, K., Okamoto, K., Hide, K. 1990. Sambagawa belt. See Ichikawa et al 1990, pp. 137–64

Hayase, I., Nohda, S. 1969. Geochronology on the "Oldest Rocks" of Japan. *Geochem. J.* 3: 45–53

Hiroi, Y. 1983. Subdivision of the Hida metamorphic complex, central Japan, and its bearing on the geology of far east in Pre-Sea of Japan time. *Tectonophysics* 78: 317–33

Hiroi, Y., Kishi, S. 1989. P-T evolution of the Abukuma metamorphic rocks in northeast Japan: metamorphic evidence for oceanic crust obduction. See Daly et al 1989, pp. 481–86

Hirooka, K. 1990. Paleomagnetic studies of Pre-Cretaceous rocks in Japan. See Ichikawa et al 1990, pp. 401–6

Ichikawa, K., Mizutani, S., Hara, I., Hada, S., Yao, A., eds. 1990. *Pre-Cretaceous terranes of Japan*. Osaka: Osaka City Univ. 413 pp.

Imaizumi, M., Ueda, Y. 1981. On the K-Ar ages of the rocks of two kinds existing in the Kamuikotan metamorphic rocks located in the Horokanai district, Hokkaido. *J. Jpn. Assoc. Miner. Petrol. Econ. Geol.* 76: 88–92[1]

Ishiwatari, A. 1990. Yakuno ophiolite and related rocks in the Maizuru terrane. See Ichikawa et al 1990, pp. 109–20

Ishizuka, H. 1987. Igneous and metamorphic geology of the Horokanai ophiolite in the Kamuikotan zone, Hokkaido, Japan: a synthetic study. *Mem. Fac. Sci. Kochi Univ.* 8: 1–70

Isozaki, Y., Itaya, T. 1990. Chronology of Sanbagawa metamorphism. *J. Metamorph. Geol.* 8: 401–11

Isozaki, Y., Itaya, T., Kawato, K. 1990. Metamorphic age of Jurassic accretionary complex in the Northern Chichibu belt, Southwest Japan. *J. Geol. Soc. Jpn.* 96: 557–60[1]

Isozaki, Y., Itaya, T. 1991. Pre-Jurassic klippe in northern Chichibu belt in west-central Shikoku, Southwest Japan—Kurosegawa terrane as a tectonic outlier of the pre-Jurassic rocks of the Inner zone. *J. Geol. Soc. Jpn.* 97: 431–50[1]

Itaya, T., Takasugi, H. 1988. Muscovite K-Ar ages of the Sanbagawa schists, Japan and argon depletion during cooling and deformation. *Contrib. Mineral. Petrol.* 100: 281–90

Kiminami, K., Kawabata, S. Miyashita, S. 1990. Discovery of Paleogene radiolarians from the Hidaka Supergroup and its significance with special reference to the ridge subduction. *J. Geol. Soc. Jpn.* 96: 323–26[1]

Komatsu, M. 1990. Hida "Gaien" belt and Joetsu belt. See Ichikawa et al 1990, pp. 25–40

Komatsu, M., Osanai, Y., Toyoshima, T. M., Miyashita, S. 1989. Evolution of the Hidaka metamorphic belt, northern Japan. See Daly et al 1989, pp. 487–93

Komatsu, M., Sakakibara, M., Fukuzawa, H., Toyoshima, T. 1990. Opening of Chishima Basin and the tectonics of the Hidaka metamorphic belt. *Chikyu.* 12: 501–6[2]

Maruyama, S. 1990. *97th Annu. Meeting, Geol. Soc. Jpn.* Toyama. p. 484[2]

Maruyama, S., Banno, S., Matsuda, T., Nakajima, T. 1984. Kurosegawa zone and its bearing on the development of the Japanese Islands. *Tectonophysics* 110: 47–60

Maruyama, S., Isozaki, Y., Soma, T. 1990. *97th Annu. Meeting, Geol. Soc. Jpn.* Toyama. pp. 32–33[2]

Maruyama, S., Seno, T. 1986. Orogeny and relative plate motions: example of the Japanese Islands. *Tectonophysics* 127: 305–29

Miyashiro, A. 1961. Evolution of metamorphic belt. *J. Petrol.* 2: 277–311

Nakajima, T., Shirahase, T., Shibata, K. 1990. Along-arc variation of Rb-Sr and K-Ar ages of Cretaceous granitic rocks in Southwest Japan. *Contrib. Mineral. Petrol.* 104: 381–89

Nishimura, Y. 1990. Sangun metamorphic rocks: terrane problem. See Ichikawa et al 1990, pp. 63–80

Nishimura, Y., Itaya, T., Isozaki, Y., Kameya, A. 1989. Depositional age and metamorphic history of 220 Ma high P/T metamorphic rocks: an example of the Nishikicho area, Yamaguch Prefecture, Southwest Japan. *Mem. Geol. Soc. Jpn.* 33: 143–66[1]

Nishina, K., Itaya, T., Ishiwatari, A. 1990. *97th Annu. Meeting, Geol. Soc. Jpn.* Toyama

Osanai, Y., Komatsu, M., Owada, M. 1991. Metamorphism and granite genesis in the Hidaka metamorphic belt, Hokkaido, Japan. *J. Metamorph. Geol.* 9: 111–24

Saito, Y., Hashimoto, M. 1982. South Kitakami Region: An allochthonous terrane in Japan. *J. Geophys. Res.* 87: 3691–96

Sakakibara, M. 1991. Metamorphic petrology of the northern Tokoro metabasites, eastern Hokkaido, Japan. *J. Petrol.* 32: 333–64

Shiba, M. 1988. Metamorphic evolution of

southern part of the Hidaka belt, Hokkaido, Japan. *J. Metamorph. Geol.* 6: 273–96

Shiba, M., Ueda, Y., Onuki, H. 1979. K-Ar ages of metamorphic rocks from the Tsukuba district, Ibaraki Prefecture. *J. Jpn. Assoc. Mineral. Petrol. Econ. Geol.* 74: 122–25[1]

Shibata, K., Igi, S., Uchiumi, S. 1977. K-Ar ages of hornblendes from gabbroic rocks in Southwest Japan. *Geochem. J.* 11: 57–64

Shibata, K., Nishimura, Y. 1989. Isotopic ages of the Sangun crystalline schists, Southwest Japan. *Mem. Geol. Soc. Jpn.* 33: 317–41[1]

Shibata, K., Nozawa, T., Wanless, R. K. 1970. Rb-Sr geochronology of the Hida metamorphic belt, Japan. *Can. J. Earth Sci.* 7: 1381–1401

Shibata, K., Sugiyama, Y., Takagi, H., Uchiumi, S. 1988. Isotopic ages of rocks along the Median Tectonic Line in the Yoshino area, Nara Prefecture. *Bull. Geol. Surv. Jpn.* 39: 759–81[1]

Suzuki, H., Isozaki, Y., Itaya, T. 1990. Tectonic superposition of the Kurosegawa terrane upon the Sanbagawa metamorphic belt in eastern Shikoku, Southwest Japan—K-Ar ages of weakly metamorphosed rocks in the northeastern Kamikatsu Town, Tokushima Prefecture. *J. Geol. Soc. Jpn.* 96: 145–53[1]

Tagiri, M., Shiba, M., Onuki, H. 1989. Anatexis and chemical evolution of pelitic rocks during metamorphism and migmatization in the Hidaka belt, Hokkaido. *Geochem. J.* 23: 321–38

Taira, A., Saito, Y., Hashimoto, M. 1983. The role of oblique subduction and strike-slip tectonics in the evolution of Japan. *Geodynamic Ser. AGU* 11: 303–16

Takagi, K., Hara, I. 1979. Relationship between growth of albite porphyroblasts and deformation in a Sambagawa schist, central Shikoku, Japan. *Tectonophysics* 58: 113–25

Takasu, A. 1989. P-T histories of peridotite and amphibolite tectonic blocks in the Sambagawa metamorphic belt, Japan. See Daly et al 1989, pp. 533–38

Takasu, A., Dallmeyer, R. D. 1990. ^{40}Ar-^{39}Ar age constraints for the tectonothermal evolution of the Sambagawa metamorphic belt, central Shikoku, Japan: a Cretaceous accretionary prism. *Tectonophysics* 185: 303–24

Takasu, A., Dallmeyer, R. D. 1991. ^{40}Ar-^{39}Ar mineral ages of metamorphic clasts from the Kuma Group (Eocene), central Shikoku, Japan: implications for tectonic development of the Sambagawa accretionary complex. *Lithos*. In press

Toyoshima, T. 1990. Pseudotachylite from the Main zone of the Hidaka metamorphic belt, Hokkaido, Northern Japan. *J. Metamorph. Geol.* 8: 507–23

Ueda, Y., Nakajima, T., Maruyama, S., Matsuoka, K. 1980. K-Ar ages of muscovite from greenstone in the Ino Formation and schists blocks associated with the Kurosegawa tectonic zone near Kochi city, central Shikoku. *J. Jpn. Assoc. Mineral. Petrol. Econ. Geol.* 75: 230–33

Uyeda, S., Miyashiro, A. 1974. Plate tectonics and Japanese islands; a synthesis. *Bull. Geol. Soc. Am.* 85: 1159–70

Wallis, S. R. 1990. Folding and stretching in the Sambagawa metamorphic belt: the Asemigawa region, central Shikoku. *J. Geol. Soc. Jpn.* 96: 345–52

Wang, G.-F., Banno, S., Takeuchi, K. 1986. Reactions to define the biotite isograd in the Ryoke metamorphic belt, Kii Peninsula, Japan. *Contrib. Mineral. Petrol.* 93: 9–17

Yamamoto, H. 1962. Plutonic and metamorphic rocks along the Usuki-Yatsushiro tectonic line in the western part of central Kyushu. *Bull. Fukuoka Gakugei Univ.* 12: 93–172

Yokoyama, K., Itaya, T. 1990. Clasts of high-grade Sanbagawa schist in Middle Eocene conglomerates from the Kuma Group, central Shikoku, Southwest Japan. *J. Metamorph. Geol.* 8: 467–74

Annu. Rev. Earth Planet. Sci. 1992. 20:181–219

THE CHARACTER OF THE FIELD DURING GEOMAGNETIC REVERSALS

Scott W. Bogue

Department of Geology, Occidental College, Los Angeles, California 90041

Ronald T. Merrill

Geophysics Program, University of Washington, Seattle, Washington 98195

KEY WORDS: geomagnetism, paleomagnetism, polarity transition, transition zone

INTRODUCTION

It is well established that the geomagnetic field, which is predominantly dipolar at the earth's surface, has alternated between normal and reverse polarity through geologic time (Figure 1). Evidence of this behavior is preserved in the ancient magnetism of rocks. The remanence of radiometrically-dated lava flows, for example, shows that the field has reversed at least 23 times over the past 5 million years (Figure 2). Reversals are not periodic; it is apparent in Figure 2 that lengths of recent polarity intervals range from 10^4 to 10^6 years and that the intervals follow a very irregular pattern. This behavior contrasts sharply with that of the sun's field, which switches polarity every 11 years.

Paleomagnetic observations discussed in more detail below show that a geomagnetic reversal takes thousands of years to complete. What happens to the geomagnetic field during the transitional period? Does the entire field just rotate from one orientation to the other? Does the field disappear

0084–6597/92/0515–0181$02.00

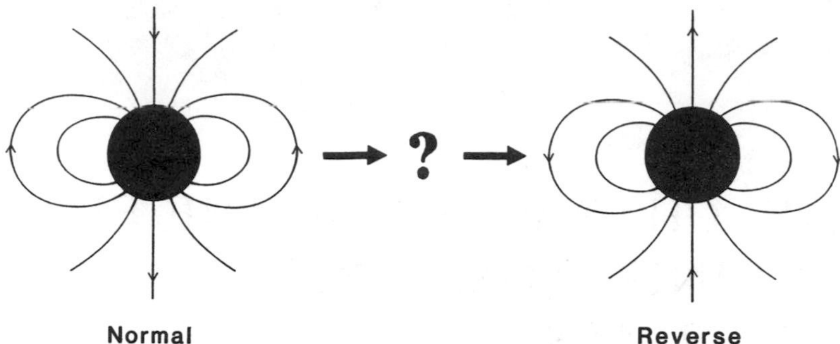

Figure 1 Field lines for normal and reverse dipole field. The geomagnetic field has alternated between these two polarity states thousands of times in the geologic past. Reversals proceed rapidly, and so the field has only spent a few percent of geologic time in transition.

and then reemerge with opposite polarity? If there is a transitional field present, is it simple or complex? How are changes in direction and intensity related? Does the field vary smoothly or in a stop-and-go fashion? Does geomagnetic behavior during one reversal resemble that during the next? These questions are particularly challenging ones for paleomagnetists precisely because reversals occur so quickly. The field spends little time in the transitional stage, and so the chance of paleomagnetic preservation in the incomplete geologic record is small. Furthermore, paleomagnetic fidelity suffers owing to the low field intensity and rapid directional change that seem to characterize the transitional field.

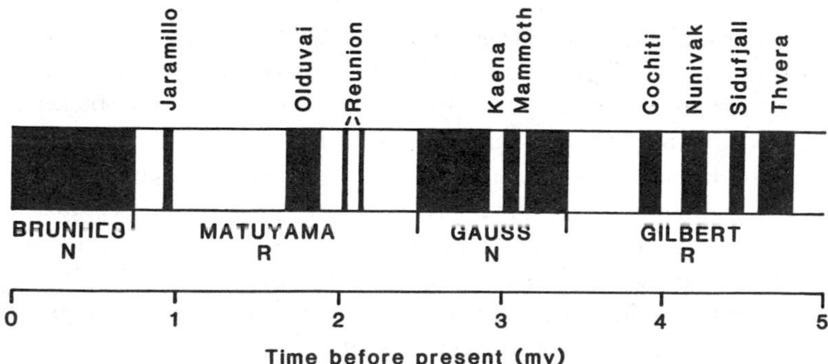

Figure 2 The last 5 million years of geomagnetic polarity history (time scale of Harland et al 1982). Black represents normal polarity; white represents reverse polarity. Polarity chrons are labeled horizontally; subcrons are labeled vertically.

In spite of these difficulties, interest in the transitional geomagnetic field has remained high for the last two decades. One enduring hope has been that a better understanding of the reversal process will shed light on the workings of the field-generating process in the earth's outer core. Observations of the field during those times when it is varying most rapidly will provide constraints on the conductivity of the lower mantle and velocities of flow at the top of the fluid core. The recognition that some animals and plants navigate using magnetic fields has heightened interest in the character of the magnetic environment during transitional intervals. There is even a purely geological motivation: It has long been hoped that distinctive features of particular reversals might prove useful in resolving the ambiguities inherent to magnetostratigraphic correlations. We suspect that for most paleomagnetists, however, the primary motivation for studying reversals is an innate curiosity about what is one of the most dramatic, global-scale changes the earth undergoes on a frequent basis.

The scientific effort to understand reversals combines the study of paleomagnetic transition zones (rock sequences magnetized during reversals), observations of the behavior of the stable (i.e. nontransitional) field, and knowledge of physical conditions and processes of the deep earth. In this review, we begin with some general background on the geodynamo and rock magnetism for the reader unfamiliar with these topics. We then summarize current knowledge of the transitional field. This discussion will center on those studies that have been especially influential in shaping ideas. We will touch on problems related to the fidelity of the various paleomagnetic recorders, an issue that is at the forefront of much current research. Finally, we will show how knowledge of the transitional geomagnetic field might lead to a better understanding of the processes occurring in the earth's outer core.

There are several excellent reviews of this subject. Recent encyclopedia chapters by Laj (1989) and Hoffman (1989) deal directly with polarity transitions. A very readable summary in Scientific American (Hoffman 1988), provides an overview of current thinking and attests to the widespread interest in reversals. Comprehensive reviews of the literature are produced every four years for the International Union of Geodesy and Geophysics; Bogue & Hoffman (1987) and Clement & Constable (1991) are the most recent two. General background on the fundamentals of paleomagnetism and rock magnetism can be found in the early chapters of McElhinny (1973), while O'Reilly (1984) provides a more detailed and recent review of rock magnetism in particular. Finally, the reader wishing an introduction to geomagnetism and dynamo theory may consult Merrill & McElhinny (1983).

GENERAL BACKGROUND

The Earth's Core and Main Magnetic Field

The magnetic field measured at the earth's surface, which has contributions from external sources (e.g. ionospheric currents) and the crust and upper mantle (e.g. remanent magnetizations), arises primarily from fluid motions in the outer core. This source region begins at a depth of 2881 km and extends to 5150 km, below which the solid inner core extends to the earth's center at 6371 km. Both the outer and inner parts of the core are believed to be good electrical conductors (3×10^5 S/m) and to consist primarily of iron. Temperatures in the core remain somewhat controversial due to experimental inconsistencies (e.g. Boehler et al 1989, Williams et al 1991) and to uncertainty about the amount of light material (e.g. oxygen, hydrogen, sulfur, or other elements) alloying with iron. Nevertheless, it is widely accepted that the temperature in the core is at least several thousand degrees C, too high for any known material to retain a permanent magnetization. The magnetic field in the core, therefore, is almost certainly produced by a dynamo, an engine that converts mechanical energy (i.e. fluid motion) into electrical currents and magnetic field. The heat that drives the geodynamo can ultimately be traced to cooling of the earth, and the convective flow in the outer core is likely initiated by compositional buoyancy of less-dense material released by freezing of the iron-rich inner core (e.g. Braginsky 1964, Gubbins 1977).

Primarily because of its mathematical complexity, no complete dynamo model exists for the earth. Nevertheless, it is useful to briefly consider the magnetic induction equation, a relation that must be satisfied in any viable model. The equation

$$\frac{\partial \mathbf{H}}{\partial t} = k\nabla^2\mathbf{H} + \nabla \times (\mathbf{v} \times \mathbf{H}) \tag{1}$$

where $\mathbf{H}$ is the magnetic field, t is time, k is the magnetic diffusivity, and $\mathbf{v}$ is the fluid velocity, describes the change in the magnetic field resulting from diffusion [the first term on the right side of (1)] and induction [the second term on the right side of (1)]. If the fluid velocity is zero, $\mathbf{H}$ can only change by free decay (i.e. diffusion). That the geomagnetic field has existed for at least 3 billion years (Merrill & McElhinny 1983), reversing polarity many times, attests to the importance of the field regeneration (i.e. induction) process, which involves motion of the electrically-conducting core fluid through the magnetic field of its own making. The resulting electrical current produces new magnetic field that can enhance or partially cancel the preexisting field. The equation is blind to the sign of **H**: A flow that amplifies normal field will also amplify reverse field (Merrill et al 1979, 1990), so that polarity reversal is probably a very common phenomenon among planetary and stellar dynamos.

In addition to the magnetic induction equation, a viable dynamo model must also satisfy several other equations (such as the Navier-Stokes equation) which are coupled to each other and involve nonlinear feedback. The complete problem is mathematically very complex, and realistic solutions have not been obtained. Indeed, theory cannot yet answer a question as simple as: If the earth's core were twice as large, or if the mean temperature of the core were twice as high, would the intensity of the mean geomagnetic field be larger, smaller, or the same? There is certainly no dynamo theory that makes reliable, detailed predictions of the field during reversals. Nevertheless, the integration of (incomplete) dynamo theory and (imperfect) paleomagnetic observations has been increasing for the last 15 years, and the result has been a better understanding of the processes acting in the earth's deep interior.

Principles of Rock Magnetism

The historical record of earth's magnetic field is limited to the last few hundred years (e.g. Bloxham & Gubbins 1985), and so a geomagnetic polarity reversal has never been observed. One must therefore turn to the record of ancient geomagnetic field directions and intensities preserved in rocks. For the paleomagnetist, the most important magnetic property of a rock is its remanent magnetization, i.e. the magnetization present in zero external field. This magnetization (M) decays with time (t) in zero field according to

$$M = M_0 \exp(-t/\tau) \tag{2}$$

where M_0 is the initial magnetization and τ is the relaxation time. A reasonable theory for τ exists only for rocks containing uniformly magnetized grains. This theory (Néel 1949) predicts that τ for an ensemble of identical, uniformly magnetized grains each with volume V is given by

$$\tau = f^{-1} \exp(VM_s h_c/2kT), \tag{3}$$

where T is temperature, k is the Boltzmann constant, M_s the saturation magnetization, h_c the microscopic coercivity, and f is a constant of the order 10^{-8} sec. The h_c measures how hard it is to rotate the magnetization within a grain, and typically increases with decreasing temperature. Small changes in T, V, and other parameters produce very large changes in τ; a temperature decrease of only a few tens of degrees can take τ from several seconds to millions of years. Because τ can change so dramatically with temperature, a magnetization is often said to be locked in during cooling at a "blocking temperature." In practice, because V and h_c vary from grain to grain, there will be a spectrum of blocking temperatures in a rock.

Rocks commonly contain several components of magnetization that can be analyzed by stepwise demagnetization. The two most common techniques are thermal demagnetization and alternating field (AF) demagnetization. A step in thermal demagnetization involves cooling a sample in zero field from T_1, thus randomizing the magnetization in grains with blocking temperatures less than T_1. This process is repeated at increasingly higher temperatures until the sample is completely demagnetized. Geometrical analysis of the change in magnetization after each step can yield the remanence directions acquired in different temperature intervals. Alternating field demagnetization is a similar process, except that grain magnetizations are sorted by their h_c rather than blocking temperature. The choice between the two techniques depends on whether the components of interest differ most in blocking temperature or h_c.

What is the origin of the various components of magnetization in a rock? Remanence acquired when a rock is formed is called a primary magnetization; those acquired later are called secondary magnetizations. The primary magnetization in a lava flow forms during cooling as grains reach their blocking temperatures. This kind of magnetization, called a thermoremanent magnetization (TRM), is usually aligned very closely to the ambient geomagnetic field. Reheating by an overlying lava flow after the field has reversed could produce a nearly antiparallel secondary magnetization that is also a TRM. The statistical preference of small magnetic grains to align with the field as they settle out of quiet water is called a depositional remanent magnetization (DRM). In many sediments, this alignment can be reset after deposition by compaction, bioturbation, and other disturbances. The resulting magnetization, called a post-DRM (pDRM), often forms within a few meters of the sediment-water interface and constitutes the oldest remanence in most sediments.

Chemical remanent magnetization (CRM), which can be either primary or secondary, is the remanence produced by any chemical change. An example is the magnetization acquired during the growth of magnetic minerals in a magnetic field (e.g. Stokking & Tauxe 1990). It can be seen from Equation 3 that a decrease in T with V constant (leading to a TRM) produces the same change in stability (i.e. τ) as does an increase in V with T constant (leading to a CRM). It follows that CRM and TRM can have very similar stabilities, making them difficult to separate by partial demagnetization. It is a mistake, however, to ignore the remanence of a sample just because chemical change is evident. For example, the iron-titanium oxides carrying the primary magnetization of a rock can oxidize without changing the remanence direction even though the intensity is drastically altered (e.g. Johnson & Merrill 1972, Ödzemir & Dunlop 1985).

Rocks that have acquired at least part of their magnetization during a reversal of the geomagnetic field provide direct information on the geometry and variation of the intermediate field. The kind of information however, depends on the nature of the recording medium. As discussed below, the transitional interval only lasts for several thousands of years. A thin lava flow can erupt and acquire a primary TRM in less than a year, providing a virtually instantaneous reading of the ancient field. Sequences of flows are not erupted uniformly; a typical pattern is for several flows to be erupted in rapid sequence and then be followed by a hiatus. The result is a digital record of field change in which the time interval between data is poorly known. In contrast, the remanence of a marine sediment typically locks in over an interval ranging from hundreds to thousands of years, depending on the sedimentation rate and the extent of post-depositional remagnetization. In effect, the record of field variation is seen through a low band-pass filter that acts to smooth the signal (e.g. Hyodo 1984, Hoffman & Slade 1986, Weeks et al 1988a, Rochette 1990). Chemical changes also affect the character of paleomagnetic information. In lava flows, oxidation is minimized because the cooling time is short. In slowly-cooled intrusive rocks, such as granite, chemical alteration of magnetic minerals is more common simply because the minerals spend a much longer time at high temperature. The chemical environment in sediments is complex and highly variable; changes in Eh, acidity, and other factors can profoundly alter the size and composition of the magnetic fraction (Henshaw & Merrill 1980, Karlin & Levi 1985). In short, paleomagnetists generally find that rock magnetic problems make the remanence of intrusive and sedimentary rocks more difficult to analyze than that of extrusive rocks.

In addition to the direction of the ancient geomagnetic field, the remanence of some rocks can provide good estimates of the ancient field intensity. In general, reliable paleointensities are difficult to obtain because of the problem of chemical alteration, which always affects the intensity of the remanence even when the direction is unchanged. All viable absolute paleointensity techniques involve reproducing the natural remanence acquisition process in the laboratory. As a practical matter, this restricts absolute paleointensity work to rocks that carry a TRM.

The most reliable paleointensity method is the double-heating experiment of the Thelliers (Thellier & Thellier 1959). Provided that the external field is much smaller than h_c, then it is easy to show (theoretically and experimentally) that the TRM of a rock is linearly proportional to H (i.e. $M_{trm} = CH$). Remarkably, this relation is often satisfied for any discrete interval of the blocking temperature distribution, although each interval has its own constant of proportionality C_i. One version of the Thelliers' experiment (Coe 1967) determines C_i for each blocking temperature interval by giving the sample an artificial TRM in a known laboratory field. The natural remanence in the same interval is found by heating and cooling the sample in zero field and measuring the component lost ($M_{n,i}$). The ancient field intensity (H_n) is then given by $H_n = M_{n,i}/C_i$. By determining H_n for many blocking temperature intervals, one can check that the C_i's have not altered (in a nonlinear fashion) during the course of the heatings and also isolate intervals (usually at the low temperature end) where secondary magnetization has contaminated the TRM. [Merrill & McElhinny (1983) provide a more complete description of absolute paleointensity techniques; Chauvin et al (1990) serves as a recent example of careful paleointensity work.]

Relative paleointensities can sometimes be obtained from sediments. The basic idea of relative techniques is to use the variation of natural remanent magnetization (NRM), say down a long marine core, as a measure of the variation of the ancient field intensity. Usually, one normalizes the NRM by a factor that corrects for changes in the volume of magnetic material present in the sediment. The acquisition of anhysteretic remanent magnetization is a popular choice for this purpose because it can be measured quickly and at room temperature (e.g. O'Reilly 1984). This procedure can give reliable results provided that chemical changes have not altered the magnetic properties of the sediment. Many studies have shown that this condition is often not met; for example, the remanence intensity of many sediments undergoes a large decrease during diagenesis (Karlin & Levi 1985, Karlin 1990). This process is neither reversible nor reproducible in the laboratory, and so it is not possible to correct for. Another problem is that sediments smooth the geomagnetic signal, averaging over time intervals (hundreds to thousands of years) that are significant with respect to the length of the transitional interval. Low remanence intensities can result when samples contain opposing components acquired at only slightly different times but when the field had nearly opposite directions. Sediments that provide reliable intensity information during times of normal field variation may not be useful during reversals when the field is much more variable. Recent work (King et al 1983, Tauxe & Valet 1989) has made it possible for paleomagnetists to identify those sedimentary units that are most likely to record accurately the variation of the field intensity, and so there is some hope for the future.

The vast majority of paleomagnetists believe that some lava flows provide reliable absolute paleointensities, and that some sedimentary cores provide reliable relative paleointensities. As is to be expected, however, there is considerable disagreement over the validity of particular estimates simply because none of the methods is foolproof. The pro-con exchange of Aitken et al (1988) and Walton (1988), for example, reflects one aspect of this controversy.

EVIDENCE FROM NONTRANSITIONAL FIELDS

Spatial Variation of the Stable Geomagnetic Field

In 1838, Gauss showed how to separate the geomagnetic potential into its spherical harmonic components. From this analysis one can obtain the geocentric, inclined dipole (the degree 1 terms) that best approximates the geomagnetic field in a least squares sense. The residual, or the nondipole field (NDF), is described by terms of degree 2 (quadrupoles), terms of degree 3 (octupoles), and so on. (We will refer to these higher degree terms collectively as "multipoles," even though the term normally applies to dipoles as well.) As the spherical harmonic degree increases, the corresponding field becomes increasingly complex (see Figures 1 and 3); terms up through degree 8 are adequate to describe most of the present day NDF. In general, it takes $(n+1)^2-1$ measurements of field direction to resolve the spherical harmonic terms out to degree n. For example, there are three dipole (one axial, aligned with earth's spin axis, and two equatorial) and five quadrupole terms. Therefore, eight measurements would

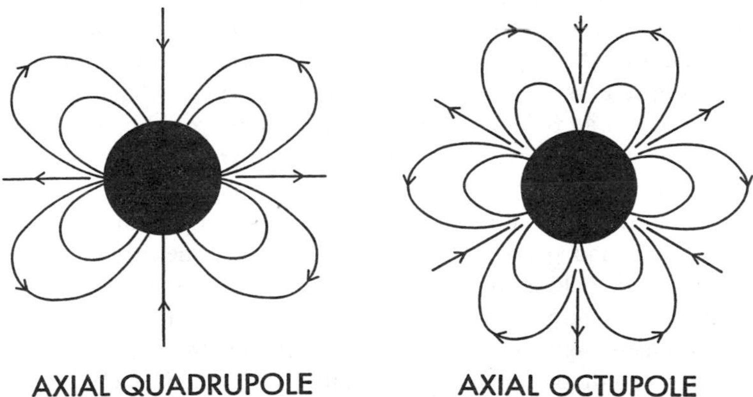

Figure 3 Field lines for axial magnetic fields of degree 2 and 3. The nondipole field can be thought of as the sum of an axial quadrupole, an axial octupole, plus fields of higher order and degree. Axial quadrupole and octupole fields predominate in some models of the transitional field.

be required to determine the dipole and nondipole terms for a spherical harmonic analysis truncated at degree 2.

One way to picture the NDF is to contour a particular magnetic element, such as the vertical component. Viewed in this manner, the present nondipole field consists of six large maxima (upward or downward) in the vertical nondipole field intensity. The intensity of a typical feature varies rapidly enough that its lifetime is probably in the 10^3–10^4 year range. It is now widely recognized that, statistically, the historical nondipole field has drifted westward at 0.18°/yr (Bullard et al 1950), although more detailed analyses suggest that most of this drift occurs in the Atlantic hemisphere (Bloxham & Gubbins 1985). Paleomagnetic data are not yet capable of proving that westward drift is a long-term feature of the geomagnetic field (e.g. Lund 1989).

The separation of the field into dipole and nondipole parts is purely mathematical, but it has guided the intuition of those speculating about the nature of the transitional field by offering a picture of what the most variable part of the geomagnetic field is like. Furthermore, the separation has motivated a class of conceptual models for reversals (e.g. Cox 1968, 1969; Hillhouse & Cox 1976). These models derive from the notion that the dipole and nondipole fractions of the field are to some extent independent. If this notion were true, and if the dipole part of the field were to disappear or become insignificant during a reversal, then the typical transitional field might resemble the modern NDF. The average nondipole field vector today has a strength of about 8 microteslas, a value similar to the lowest transitional field intensities determined from paleomagnetism. Furthermore, the westward drift and other variations of NDF features (were they to continue through the reversal) would cause the field direction at a typical site to follow a complicated, indirect path between its two stable orientations (e.g. Weeks et al 1988b, Constable 1990), a characteristic seen in some but not all paleomagnetic records. There would be no coherence to the directional change at distant sites as there would be if the field remained dipolar.

Although attractive in its simplicity, the idea that the dipole and nondipole fields act independently is unrealistic. As described above, the source of the magnetic field is thought to be electric currents in the outer core, and all current loops give rise to both dipole and nondipole fields. In addition, paleomagnetic results now suggest that the magnitudes of the dipole and nondipole fields are correlated through time (McFadden & McElhinny 1982). Finally, there appear to be persistent nondipole components of the field that change sign with the dipole, producing time-averaged normal and reverse field directions that are roughly antisymmetric (Schneider & Kent 1990; Merrill et al 1979, 1990). Thus, neither theory nor observation are consistent with any reversal model [such as that of Cox (1968, 1969)] that requires the dipole and nondipole fields to vary independently. Instead, there is the implication that dipole and nondipole components of the geomagnetic field result from the same dynamo system. On the other hand, paleomagnetic studies also show a slight asymmetry between the mean normal and reverse fields, as if part of the axial quadrupole did not reverse with the main dipole. Schneider & Kent (1988) find some evidence that this quadrupolar component is prominent during reversals, although the paleomagnetic data as a whole do not support the hypothesis that the transitional field always has a high degree of axial symmetry.

Secular Variation

STABLE VERSUS TRANSITIONAL FIELD As described above, features of the nondipole field drift westward and change size. The dipole field also exhibits time variation, as evidenced by the observation that the time-averaged dipole is aligned along the rotation axis even though the present best-fit dipole is inclined 11°. Both effects contribute to the overall secular variation (SV) of the geomagnetic field. A simple question to ask is whether the time variation of the geomagnetic field during reversals is similar in character to that observed during times of stable field. One clue comes from observations of SV recorded in the sediments of Mono Lake, California. These sediments record an "excursion" of the geomagnetic field (Denham 1974, Liddicoat & Coe 1979); that is, a large departure of the geomagnetic field from its stable orientation that is not part of a polarity reversal. If caused by rotation of the dipole field, excursions would be seen to occur simultaneously all over the earth. At the other extreme, field components of high degree and large magnitude can produce geographically-restricted phenomena, even local field reversal (e.g. Harrison & Ramirez 1975). Paleomagnetic data are inadequately distributed in space and time to convincingly test these conceptual models in most cases. Nevertheless, many paleomagnetists (e.g. Hoffman 1981, Coe et al 1984) have argued that at least some excursions represent "aborted reversals" resulting from a change in the dynamo process very similar to that occurring during a full polarity reversal.

Lund et al (1988), examining the record of SV in sediments deposited immediately following the Mono Lake excursion, believe they can identify a distinctive pattern of field variation associated with the excursion that has recurred at least four times with a period of about 3000 years. They suggest a link between the unusual field behavior of the excursion and the usual variation observed during times of stable field. To the extent that the Mono Lake excursion represents a reversal-like change in the geodynamo,

Lund et al's (1988) results imply that variation of the transitional field might resemble SV.

Valet et al (1986) argue that some aspects of SV continue right across polarity transitions. Their evidence comes from an analysis of field variation before, during, and after a late Miocene reversal recorded in marine clays on Crete. The authors found that directional variations (measured with respect to the average pre-reversal direction) with periods between 2000 yr and 4500 yr characterized both the stable and transitional fields. The amplitude of the oscillations increased during the transitional interval as would be expected if the NDF remained intact while the main dipole field was diminished. Unfortunately, because the polarity transition itself is only several thousand years in duration, the analysis is not statistically robust. Nevertheless, there remains the interesting suggestion that some of the field variation characteristic of stable field persists through the transitional interval, a property that is complementary to that discussed by Lund et al (1988).

DYNAMO FAMILIES AND REVERSALS It is reasonable to assume that the present geomagnetic field is representative of those that have existed in the recent geological past. Since the field has long-term symmetry about the rotation axis, it is also reasonable to take the variation of the field along a line of latitude as representative of the SV of the field as it would be observed from a particular site at that latitude. These assumptions are supported by McFadden et al (1988) who showed that the latitudinal variation of SV determined by this method agrees very well with paleomagnetic results from lava flows less than 5 my old. Furthermore, McFadden et al (1988) show that the SV can be split into two components, one that is zero at the equator and increases nearly linearly with latitude and one that is latitudinally-invariant. The former is due to field components described by spherical harmonic terms for which the degree minus the order is odd—a group of terms that is referred to as the "primary dynamo family" (Figure 4). Field components belonging to this family include the primary axial dipole, some nonaxial quadrupoles, the axial octupole, and others. The latitudinally-invariant SV is due to field components described by spherical harmonic terms for which the degree minus the order is even, the "secondary dynamo family" (Figure 4). This group includes the axial quadrupole, equatorial dipoles, and others. These two families have been of interest to dynamo theoreticians (e.g. Roberts & Stix 1972) for many years because they provide information on symmetries or lack of symmetries in the core velocity field.

Several lines of evidence (Merrill & McFadden 1988) suggest that the primary and secondary families are only weakly coupled (that is, vary

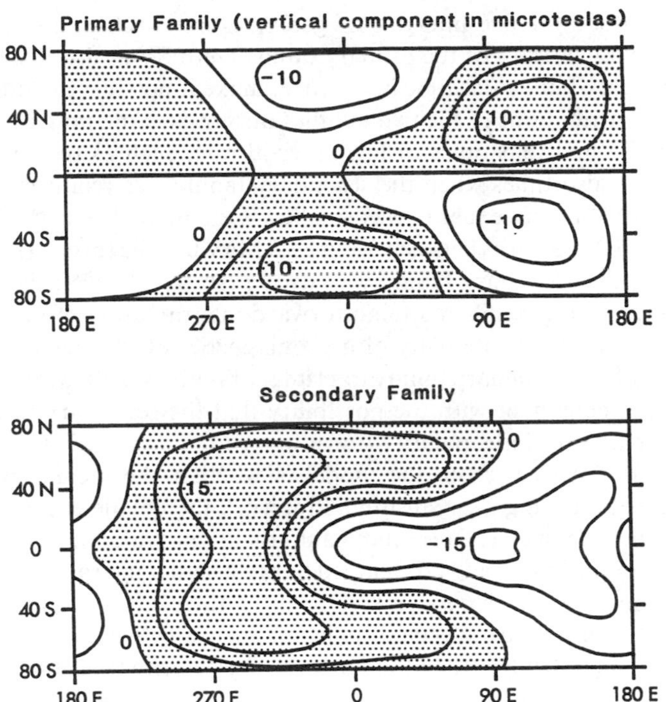

Figure 4 Contour maps showing the intensity of the vertical component of the primary and secondary field families for the 1975 International Geomagnetic Reference Field. The axial dipole is not included in the primary field map. Shaded areas indicate inward directed field; the contour interval is 5 microteslas.

almost independently) during times when the main geomagnetic field is not reversing. One observation consistent with this independence is that the axial quadrupolar part of the geomagnetic field (secondary family) has changed sign (i.e. reversed polarity) since 1838, whereas the axial dipole (primary family) has not. On the other hand, paleomagnetic analyses (Merrill et al 1979, 1990; Schneider & Kent 1990) of mean normal and reverse geomagnetic fields show that the axial dipole and much of the axial quadrupole have changed sign together for the past 5 million years. Merrill & McFadden (1988) suggest therefore that the two families are coupled during polarity transitions and that the coupling may be related to the reversal process.

Strong support for a link between the coupling of magnetic field families and reversals is provided by analysis of the paleosecular variation (PSV) recorded by lavas. McFadden et al (1991) have compiled estimates of PSV versus latitude for five intervals of geologic time back to 195 Ma. The special characteristics of the primary family (zero SV at the equator) and the secondary family (latitude invariance) make it possible to estimate the relative contributions of the two families during each interval. McFadden et al (1991) find that times of low reversal rate, such as the Cretaceous period, are also times when the secondary family was relatively reduced (so that PSV at the equator was smaller; see Figure 5). In more recent geological time, such as between 5 and 22.5 Ma when the reversal frequency was high, the secondary family was strong (so that PSV at the equator was large; see Figure 5). A straightforward interpretation of these observations is that the probability of a geomagnetic reversal increases when the ratio of the secondary family to primary family is high. Moreover, the evidence is consistent with the possibility that the secondary to primary ratio increases during a polarity transition.

What is not clear from the model is whether the secondary family predominates during the transitional interval, thereby determining intermediate field geometry. Current dynamo theory provides no help. Indeed, the theory is not sufficiently developed to say whether interactions between the dynamo families have *anything* to do with reversals. It has become crucial, therefore, to determine paleomagnetically whether one family becomes much stronger than the other during polarity transitions. Recent

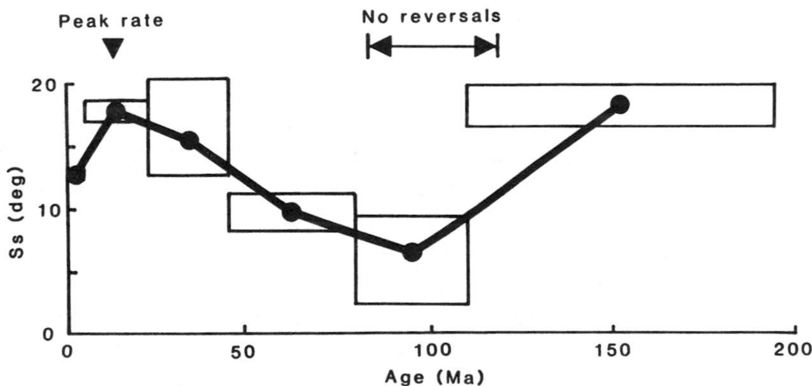

Figure 5 Angular standard deviation of VGPs (S_s) due to fields of the secondary family versus time. The secondary family was weak during the long Cretaceous Normal Superchron (118 Ma to 83 Ma), an interval of 35 my during which the field did not reverse. The secondary family is strong at 12 Ma, when reversal frequency peaked at nearly 6 reversals/my. The primary family follows the opposite pattern; high at 100 Ma and low at 12 Ma. [Figure adapted from McFadden et al (1991).]

suggestions that the transitional field has a strong equatorially-symmetric octupole component (Clement 1991) or a strong equatorial dipole component (Tric et al 1991, Laj et al 1991) are consistent with the hypothesis that the secondary family controls the field during the intermediate stage.

EVIDENCE FROM POLARITY TRANSITIONS

Duration of Reversals

It has long been clear that the geomagnetic field spends a very small amount of time in transition between its two stable states. Cox & Dalrymple (1967) demonstrated this property of the field on statistical grounds by analyzing the paleomagnetism of 88 dated lava flows from the last 3.6 million years. Only one of these flows had both a remanence direction far from that expected for normal or reverse field and a radiometric age that placed its origin near a polarity boundary. A straightforward statistical inference based on this rarity yielded 4600 years as an estimate of the average length of the transitional interval or, more precisely, that portion of the transitional interval associated with unusual field directions. When compared with the magnetic polarity history of the field for the last several million years, this result implies that the field is in transition only about 2% of the time.

Studies of paleomagnetic transition zones have yielded direct estimates of reversal lengths. Sediments of the deep sea, which are deposited more-or-less continuously at rates easily determined from the polarity stratigraphy, are especially useful for this purpose. Slowly-deposited sediments, however, are problematical. Ninkovich et al (1966), for example, found that the interval of transitional directions in cores from the north Pacific averaged about 1000 years long. In contrast, the interval of reduced field intensity (as judged from the strength of the remanence of the sediments) was much longer, about 20,000 years. A very similar pattern was observed for the transition at the onset of the Olduvai subchron by Herrero-Bervera et al (1987). In both these cases, the sedimentation rates were near 1 cm/kyr; thus even the smallest samples average over a significant amount of time. Nearly antiparallel components of magnetization characteristic of normal and reverse field recorded in the same sample could produce the directional and intensity pattern described by Ninkovich et al (1966). Because of this problem, Rochette (1990) has suggested a minimum sedimentation rate of 5 cm/kyr for reasonable fidelity in deep-sea sediments.

Estimates of reversal length from complete sedimentary transition zones where the sedimentation rate is greater than 5 cm/kyr generally range between 1300 years (lower Sidufjall, Linssen 1988) to 10,000 years (upper Olduvai, Tric et al 1991). It is not clear whether this range is due to how the transitional interval is defined, geographical variation, differences from reversal to reversal, or rock magnetic effects. Clement & Kent (1984) have presented results from seven deep-sea cores obtained near longitude 180E at latitudes ranging between 33S and 45N. As observed from within 35 degrees of the equator, the Matuyama-Brunhes reversal appears to have lasted about 4.9 kyr. Observed from higher northern latitudes, it appears to have lasted more than 4 kyr longer. Geographical variations are to be expected if multipolar fields are significant during reversals and if estimates of the duration are based on directional data only. This can be seen by considering how an axial dipole (Figure 1) and an axial quadrupole (Figure 3) would superimpose as the dipole shrinks to zero and then grows in the opposite orientation. At the poles, where the field directions for the two sources are parallel or antiparallel, the directional reversal would be instantaneous. At the equator, where the two field directions are perpendicular, the directional change would be gradual and last longer.

In any case, the data of Clement & Kent (1984) are from slowly-deposited sediments; the highest accumulation rate among the cores is 1.2 cm/kyr. The effect of time-averaging of the geomagnetic signal is certainly a concern. Valet et al (1989) conducted a similar study on the same reversal, using four sites from the northern hemisphere with sedimentation rates averaging near 5 cm/kyr. The correlation between latitude and reversal length is not apparent in these data, suggesting that the result of Clement & Kent (1984) reflects a rock magnetic rather than geomagnetic phenomenon. A firm resolution of this issue is important because a lack of geographical variation would provide weak evidence that a dipolar field predominated during the Matuyama-Brunhes transition.

It is much more difficult to estimate reversal lengths in volcanic transition zones. Currently available radiometric dating techniques are not capable of resolving age differences of several thousand years in rocks old enough to have recorded geomagnetic reversals. Even if an average eruption rate can be determined magnetostratigraphically for a thick section of lavas, the episodic nature of volcanic activity makes it hazardous to interpret the time spanned by the relatively few eruptive units in the transition zone. The best estimate of reversal length from lava flows comes from the Steens Mountain transition zone, which has an unusually large number of flows recording the transitional field. In their analysis of this section, Mankinen et al (1985) estimated an average eruption rate using the presumably normal Miocene SV recorded in flows above and below the transition zone. Mankinen et al (1985) find that this reversal took 3600 years (based on the directional change) to 4400 years (based on field intensity) to occur, a figure that is certainly within the range of estimates derived on statistical grounds and from sediments.

Overall, three lines of evidence suggest that the typical reversal lasts roughly 4000 years. The significance of this result can be appreciated by comparing it with the time the geomagnetic field would take to decay if flow in the outer core were to cease. Using the method of McFadden et al (1985) and conductivities of 3×10^5 s/m to 5×10^5 s/m, we find that free decay of the lowest radial mode of the geomagnetic dipole to 15% its usual value would take 24,500 to 47,500 years. That polarity transitions are much shorter than this implies that the reversal process in the core is dynamic. In particular, reversals do not appear to reflect a slowing or cessation of fluid convection in the outer core. Some studies (e.g. Dunn et al 1971) have found an interval of reduced intensity that is many times longer than the interval of transitional field directions. This pattern, if truly representative of field behavior, suggests that reversals start with a proportional decrease in both the dipole and nondipole fields. As will be discussed in more detail below, most transition zones in rapidly-deposited sediments and volcanic rocks show a different pattern, with changes in direction and intensity being more nearly synchronous. Additional data from high-resolution paleomagnetic recorders are needed to document the existence of systematic variations in reversal length with latitude or age.

Rates of Directional Change

In addition to providing information on the length of reversals, paleomagnetic records show that the rate of direction change is highly variable during the course of a transition. Because they record the ancient field continuously, rapidly-deposited sediments would seem to provide the best measure of rates of field change. Some sets of evenly spaced samples from transition zones show similar, serially-correlated directions while others show wildly divergent directions. Clement & Kent (1986), Laj et al (1988), and Tric et al (1991) are examples of recent papers that have commented on this aspect of sedimentary records, and interpreted the contrasting behavior as evidence of variable field change. As will be discussed below, however, evidence from volcanic rocks suggests rates of change that are much too fast to be recorded by typical sediments, even when very small samples are taken. At 5 cm/kyr, a relatively rapid accumulation rate, even millimeter-thick samples will average 20 years of field history, a period longer than some episodes of the claimed rapid field change.

The evidence for variable field change from most volcanic records is equivocal because eruptions are episodic. Nevertheless, Bogue & Coe (1982) interpreted large gaps in sequences of paleofield directions obtained from lava flows on Kauai to be evidence of rapid field change in the transitional interval. The lack of baked soil horizons, which would be present if flows were exposed to weathering for long periods of time—is a hint that the gaps do not simply represent eruption hiatuses.

Support for this interpretation comes from the much more complete transition zone preserved in the basaltic flows of Steens Mountain, Oregon (Mankinen et al 1985). As at Kauai, the Steens Mountain record includes sequences of flows containing serially-correlated directions that are followed by flows with very different directions (see Figure 6). Furthermore, it is clear at Steens Mountain that the largest directional changes occur when the field intensity is low. Using rates based on the SV recorded before and after the transition, Mankinen et al (1985) concluded that episodes of rapid field movement ("geomagnetic impulses") could produce over 90° of angular change in only a few years. Even more astonishing is the claim (Mankinen et al 1985, Coe & Prévot 1989) that the transitional field may

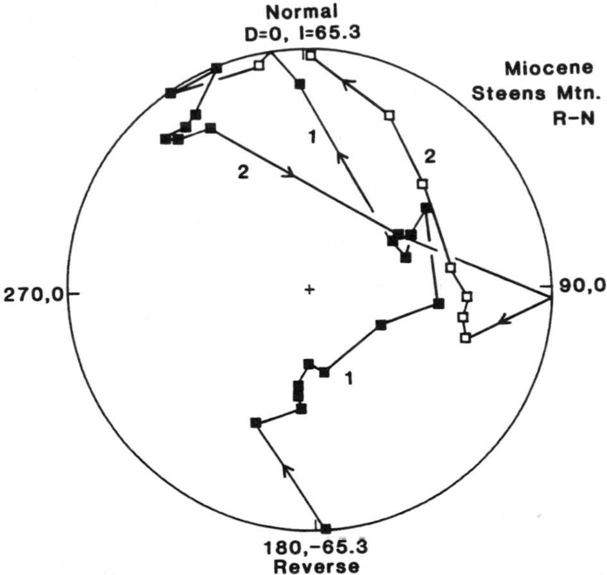

Figure 6 Equal-area plot of transitional directions for the Steens Mountain R-N reversal. The directions have been rotated about a horizontal E-W axis so that the normal and reverse axial dipole directions plot at the top and bottom of the figure [D′,I′ space of Hoffman (1984)]. Solid (*open*) symbols are on the lower (*upper*) hemisphere. The directions at the beginning and end of the sequence are averages of flows bracketing the transition zone. During this reversal, the field vector rotated from a reverse to a normal-like direction (phase 1), and then "rebounded" to intermediate directions (phase 2).

change substantially within the short time (days or weeks) that it takes a typical lava flow to cool below the blocking temperature of titano-magnetite. The evidence consists of two lava flows that record ancient field directions in their quickly-chilled exteriors that differ from those in their more slowly-cooled interiors. If the nonuniform magnetization of these flows is a pure TRM that accurately records the paleofield directions, then straightforward cooling models suggest that the transitional field can change as fast as 0.2 to 3 degrees per day.

Field change this rapid is clearly problematical. One can better appreciate the difficulty by considering it in the context of "fast dynamos" (Soward 1989). Fast dynamos are probably not directly applicable to the earth, but their theory provides an upper bound on the rate at which geomagnetic change can occur. In fast dynamos, the diffusion of the magnetic field is taken to be zero. A simple dimensional analysis of the terms remaining in Equation 1 yields an estimate of T, the shortest time for a significant change in magnetic field:

$$T = L/v, \tag{4}$$

where v is the magnitude of the convection velocity and L is the appropriate length scale. For an upper limit on the rates, it is reasonable to let $L = 1000$ km; there is surely little power at the earth's surface in the high-degree harmonics associated with core features smaller than this. A field change of 21 degrees per week (as claimed for the Steens Mountain record) requires fluid velocities that are 10^4 times greater than those that are estimated (from SV) to be present in the outer core today. Although flow this rapid during a polarity transition cannot be ruled out, it seems very improbable. Moreover, conventional estimates of lower mantle conductivity (e.g. Merrill & McElhinny 1983) imply that the lower mantle effectively screens out core signals with periods of less than about a year. Because of these concerns, Merrill & McFadden (1990) raise the possibility that the non-uniform magnetization seen in a few lavas at Steens Mountain could reflect an episode of chemical remagnetization rather than rapid field change. This kind of explanation was examined in a preliminary way by Coe & Prévot (1989), and is the subject of ongoing research by them.

Another aspect of the Steens Mountain transition zone relevant to this discussion is the prominent directional "rebound" first described by Watkins (1969). As documented in more detail by Mankinen et al (1985), the rebound followed an episode (phase 1 on Figure 6) during which the field vector rotated from its reversed (up and south) to normal (down and north) orientation. Paleointensity data (discussed in detail below) suggest that at this point the field may have been dipolar but not fully stable. The next episode of directional change, the rebound (phase 2 on Figure 6), consisted of a long loop that took the field more than 90° away from the normal-like direction. The rebound can also be described in terms of virtual geomagnetic poles (VGPs), the poles of hypothetical geocentric dipoles that correspond to the observed field directions. The rebound took the VGP from high northerly latitudes to the central Pacific, to positions offshore Peru and western Africa, and back to high northerly latitudes. This direction swing took place while field intensity was low and may have included two of the "impulses" of very rapid field change described above.

How many swings or rebounds does the transitional field typically make? Transition zones in very slowly-deposited deep-sea sediments (~ 1 cm/kyr) often show repeated swings between high-latitude VGP positions. In a record of the lower Olduvai transition from the north-central Pacific (Herrero-Bervera et al 1987), the VGP is seen to cross the equator as many as 10 times during the course of the transition. A well-studied transition zone in slowly-cooled intrusive rock (Dunn et al 1971, Dodson et al 1978) also displays numerous directional swings. What these two records have in common is that individual samples contain multiple remanence components acquired over a substantial length of time. A plausible explanation of the numerous swings is that they result from variations in the recorded magnitudes of opposing remanence components. In contrast, samples from lava flows or rapidly-deposited sediments record a short interval of geomagnetic field behavior. In general, reversal records from these media display no more than two or three large swings. A detailed volcanic record of the upper Jaramillo normal-to-reverse (N-R) transition from French Polynesia (Chauvin et al 1990), or the sedimentary record of the upper Olduvai transition (R-N) from the north Atlantic (Clement & Kent 1986) are fine examples (Figures 7 and 8).

Intensity of the Transitional Field

Almost all paleomagnetic records of reversals display a decrease in remanence intensity that accompanies transitional field directions (e.g. Wilson et al 1972). This pattern strongly suggests that the field itself does weaken substantially during reversals. It has proven difficult, however, to be more specific about intensity changes because reliable data are so few. As discussed above, only relative paleointensities can be inferred from sedimentary data, and even those are difficult to interpret because the field varies so rapidly during reversals. Consequently, one must turn to volcanic rocks to find the most trustworthy record of field intensity during polarity transitions.

Certainly the most detailed study of transitional field intensity appears to be that from Steens Mountain, Oregon (Prévot et al 1985). This section of basaltic lavas, first studied by Watson (1969), has recorded a Miocene

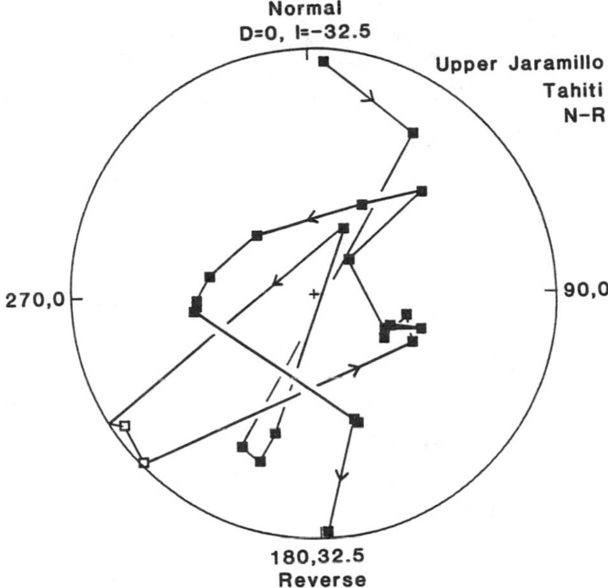

Figure 7 Equal-area plot of transitional directions for the Upper Jaramillo N-R reversal, as recorded by lava flows in Tahiti (see Figure 6 for explanation of the plot). During this reversal, the transitional field made two large swings back toward the normal field direction. Most intermediate directions lie near the N-S vertical plane (they plot near the line connecting the points labeled *Reverse* and *Normal*), as would be expected if the transitional field retained a degree of axial symmetry.

R-N reversal in great detail. Prévot et al (1985) determined paleointensities for over 70 lava flows using the double-heating technique of Thellier & Thellier (1959). There are 35 values from the transitional interval itself, with 20 or more from the intervals just preceding and following the main reversal (see Figure 9). For the approximately 5000 years preceding the onset of the transition, the average paleointensity (31.5 microteslas) was one-third lower and varied less than is expected for the Miocene. During this same interval, the dispersion of field directions was higher than usual. This correlation suggested to Prévot et al (1985) that the main dipole part of the field was diminished relative to the more rapidly-varying nondipole part of the field. It is possible that this behavior is related to the reversal process itself and therefore represents a precursor to the main transition.

For the 3500 years following the reversal, the field intensity was much closer to expected Miocene values. Large fluctuations, however, occurred over very short intervals. In two instances, successive lava flows show the field nearly tripling in strength while changing direction by only a few

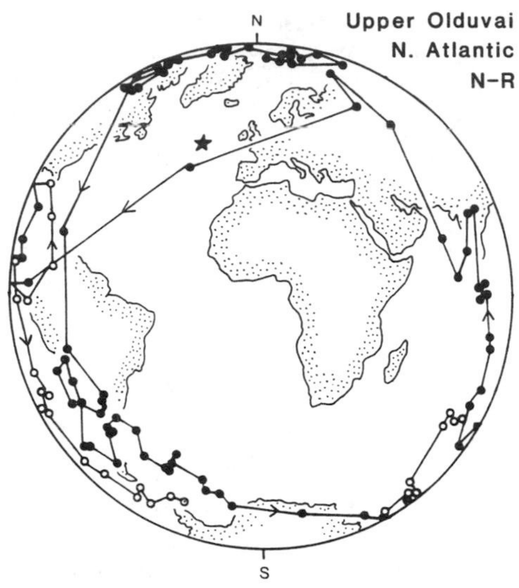

Figure 8 A VGP path for the upper Olduvai transition (N-R) from deep-sea sediments of the north Atlantic. Closed (*open*) symbols plot on the near (*far*) hemisphere; the sampling site is marked by a star. After moving from the north polar to south polar region, the VGP makes a large return loop (rebound) to high northern latitudes. Most of the VGPs lie far from the longitude of the sampling site (or its antipode), indicating that the transitional field was not axisymmetric. Figure adapted from Clement & Kent (1986).

degrees. The apparent period of this variation (hundreds of years) may be compared with the much longer apparent periods (at least thousands of years) typical of the geomagnetic field over the last 10,000 years. This curious behavior gives one the impression that the newly-established normal dipole was unstable in magnitude but not in orientation. Another interpretation is that the large intensity fluctuations are due to the rapid growth and decay of nondipole field components that happened to produce field at Steens Mountain directed very near the dipole direction. Finally, it is possible that these rapid changes are a product of some rock magnetic effects as discussed in the preceding section.

During the transitional phase itself, the field direction proceeded directly from its starting reverse orientation to a normal-like direction (phase 1), and then followed a large loop out to transitional orientations and back (phase 2). Phase 1 begins with the field intensity down by 30% to 50% of its pretransitional value and the field moving 80° from its starting orientation. The intensity gets as low as 5.3 microteslas (about 17% the average pretransitional value). At the end of phase 1, when a normal-like direction is achieved, the field intensity is near the pretransitional value. During

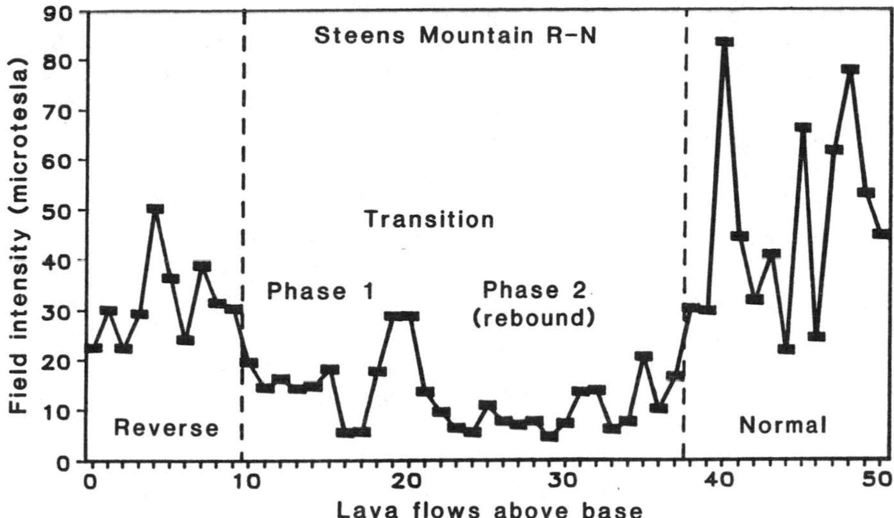

Figure 9 Variation of the ancient field intensity during the Steens Mountain reversal. See Figure 6 for the directional behavior during phases 1 and 2.

phase 2, the field intensity again drops, getting as low as 4.5 microteslas (14% the pretransitional value) during an interval when the direction may have been changing very rapidly (as discussed above). The average intensity during the entire transitional phase is 10.9 microteslas, or about 15% the average Miocene value.

How typical is the transitional field behavior recorded at Steens Mountain? Unfortunately, there are very few absolute paleointensity data available for direct comparison. During an R-N transition recorded in basaltic lavas on the Hawaiian island of Kauai, the ancient field intensity dropped from 43.1 microteslas to 10.1 microteslas and then rose to 29.3 microteslas while remaining within 30° of the reverse axial dipole direction (Bogue & Coe 1984). In contrast, the first drop in intensity at Steens Mountain was accompanied by steady movement of the field toward opposite polarity. Chauvin et al (1990) report minimum paleointensities ranging from 3 to 8 microteslas for reversals recorded in Tahiti at 18S. These values are lower than the average from Steens Mountain (at 43N), perhaps a hint that transitional field intensity varies with latitude. Comparison of the normalized magnetization strength of lavas from Iceland and Tahiti provides additional support for this idea (Chauvin et al 1990).

Shaw (1975), studying the "R3-N3" reversal in Iceland, found a large number of successive lava flows that yielded very similar, equatorial VGP positions. While the field direction was stationary at this intermediate position, the ancient field intensity rose from 5.4 microteslas, a low transitional value, to 38.6 microteslas and then returned to 9.5 microteslas. The paleointensity method (Shaw 1974) used in this study, in which all samples are remagnetized in a single heating to above 580°C, is generally considered less reliable (e.g. Coe & Grommé 1973) than the stepwise double-heating method used for the other paleointensity results discussed above. Nevertheless, Shaw (1975) interprets the R3-N3 data as evidence of a briefly-stable intermediate state which the geomagnetic field assumed in between its stable reverse and normal states. In general terms, the pattern of changing field intensity and steady field direction is similar to that seen at Steens Mountain. The hypothetical third state of the field, however, is not in evidence; high paleointensities are not associated with equatorial VGPs in the Steens Mountain transition zone.

Many of the features observed at Steens Mountain have also been observed in sedimentary records. For example, Okada & Niitsuma (1989), found that normalized remanence intensity decreased by 80% in the Matuyama-Brunhes transition at the Boso Peninsula, Japan. Laj et al (1988) found evidence in sediments from Greece that episodes of rapid directional change occur when the field intensity is low. There are also differences, however. In many but not all sedimentary records, the interval of low field intensity is more than twice as long as the interval of transitional directions. This relation is also seen in the Miocene reversal recorded in the Tatoosh intrusion (Dunn et al 1971). Although the data are few, there is no evidence in volcanic records of long intensity lows, leading most observers to conclude that the difference likely results from the time-averaging inherent to sedimentary and intrusive records. Better knowledge of the transitional field intensity is clearly desirable. It is unfortunate indeed that young volcanic transition zones (the best source of paleointensity data) are very rare, and that this scarcity may severely limit what can be known.

Structure of the Transitional Field

INTRODUCTION A primary goal of paleomagnetic research on reversals has been to document the geometry of the transitional field. Motivating this effort has been the hope that distinctive field structure might differentiate alternative models of the reversal process. One example of such a model is that of Gubbins (1987), which has also been discussed by Fuller (1987). In this scheme, reversals are caused by the poleward migration of patches of inverse magnetic flux at the surface of the outer core. Analysis of the historic geomagnetic field suggests that two such patches with inward-directed flux in the southern hemisphere of the core (which has predominantly outward-directed flux) are responsible for the 5% decline in the dipole moment since 1838. If reversals begin in a similar way, with the positions of inverse patches controlled by slowly-varying features of the lower mantle, then the structure of the transitional field might be complex but similar from one reversal to the next.

Other models make more specific predictions. The kinematic dynamo model of Olson & Hagee (1990), for example, produces a transitional field that has a strongly octupolar character. The recent model of Merrill & McFadden (1988) involves an interaction between the primary and secondary dynamo families which have opposite symmetry about the equator (Figure 4). Paleomagnetic data may eventually be capable of demonstrating whether one family predominates during reversals, thus lending support to the model. One thing is certain: Knowledge of the geodynamo is far from perfect and, as for the past few decades, the questions motivating paleomagnetic work will continue to evolve. In sketching a history of these ideas below, we recount a debate that has come full circle, with a venerable hypothesis finding new support in the most recent paleomagnetic results.

EARLY IDEAS A seemingly straightforward approach to determining the geometry of the transitional field would be to use paleomagnetic measurements to find the coefficients for the lowest degree spherical harmonic terms, say those that describe the dipolar, quadrupolar, and octupolar components of the intermediate field. It is sobering to recall, however, that one needs 15 spot measurements just to characterize the field out through degree 3 (octupoles). Many more measurements than this would be needed to assess the errors. Finally, one would face the nearly impossible task of ensuring that the observed paleofield directions were precisely synchronous. The key to avoiding these formidable problems came in the recognition that if the field were dipolar, then field directions measured anywhere on the globe would yield coincident VGPs. Geographically coincident sequences or "paths" of VGPs would characterize a purely dipolar transition, whereas dissimilar VGP paths would characterize a nondipolar transition. This simple test is possible without knowledge of the individual spherical harmonic terms.

Some of the earliest work seemed to indicate that the transitional field was dipolar. Creer & Ispir (1970) showed five examples of reversals with similar, longitudinally-confined VGP paths. This pattern of field behavior suggests that the field somehow rotates directly from one stable orientation to another. Steinhauser & Vincenz (1973) examined the available data in a statistical way and came to a similar conclusion, speculating further that the intermediate (equatorial) dipoles have displayed two preferred orientations over the past hundred million years. Because they presented simple models amenable to further testing, these two studies played an important role in shaping the early stages of debate on the geometry of the transitional field.

DEMISE OF THE TRANSITIONAL DIPOLE HYPOTHESIS Although conclusions about global magnetic fields are best drawn from large data sets, a comparison of just two records of the Matuyama-Brunhes reversal (Hillhouse & Cox 1976) convinced most workers that the transitional field was not dipolar. This study, published only three years after the Steinhauser & Vincenz paper, compared the remanent magnetism of two sedimentary sequences: one from the Boso Peninsula in Japan; and the other from Lake Tecopa in southeastern California. The key strata in both sections were deposited and magnetized 730,000 years ago during the Matuyama-Brunhes transition. The transitional VGPs for the pair of sites formed two bands separated by some 180° of longitude (see Figure 10). The clear implication of this result was that the transitional field had not remained dipolar during the reversal and instead had assumed a more complicated geometry.

Hillhouse & Cox (1976) speculated that during reversals, the westward drifting part of the nondipole field decreases along with the main dipole, leaving behind a nondipole field that is relatively stationary. This model, which came to be known as the "standing field" model, was motivated by an analysis of the historic nondipole field (Yukutake & Tachinaka 1969)

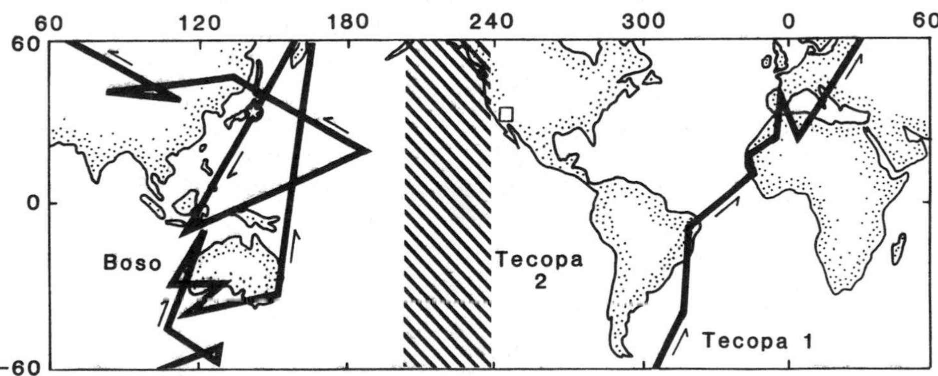

Figure 10 Transitional VGP paths for the Matuyama-Brunhes reversal as recorded in sediments at Lake Tecopa, California (*open square*) and the Boso Peninsula, Japan (*star*). Tecopa 1 is the path of Hillhouse & Cox (1976); the hatched area (Tecopa 2) marks the region traversed by the complicated VGP path of Valet et al (1988a). The gross dissimilarity between the VGP path from Japan and either path from Lake Tecopa suggests that the transitional field was not dipolar.

that divided it into drifting and nondrifting parts. The standing field model made a simple and testable prediction: that reversals of opposite sense should produce sequences of transitional field directions that are identical except for their reversed order. Transition zones recording successive reversals thus provided a direct test of the model. Some data (e.g. Bogue & Coe 1982) were consistent with the model, but others (e.g. Valet & Laj 1984) suggested that a standing field did not predominate during most reversals.

AXISYMMETRIC TRANSITIONAL FIELD A major conceptual advance was made by Hoffman (1977), Hoffman & Fuller (1978), and Fuller et al (1978) who pointed out that large groups of nondipole terms could be distinguished paleomagnetically without the need to determine the terms individually. In particular, they investigated zonal transitional fields, i.e. those that retain symmetry about the rotation axis during a reversal. Zonal fields can be divided into two groups: fields that are symmetric about the equator versus fields that are antisymmetric. Reversals were modeled as a smooth conversion of dipole field to intermediate fields of one group and then to dipole field of opposite polarity. One motivation for this model came from the dynamo theory of Parker (1969) and Levy (1971), in which reversals began in a localized region of the core and then propagated or "flooded" outward to eventually affect the entire source region. In one variation of the model, the northern and southern hemispheres of the core would be sources of opposing dipole flux; the resulting field would be zonal and symmetric about the equator; an axial quadrupolar component would predominate. Another simple variation produced an antisymmetric zonal transitional field with a strongly octupolar character.

For the paleomagnetist, this model had appeal because it was easily testable. The requirement of zonal symmetry meant that field directions everywhere would remain close to the north-south vertical plane during the reversal. Expressed in terms of VGPs, transitions would be characterized by longitudinally-confined VGPs that always passed near the sampling site (a "near-sided" path) or its antipode (a "far-sided" path). As illustrated in Figure 11, a systematic correlation of VGP paths with site location and sense of reversal was sufficient to determine which group of zonal fields was predominant during reversals and, with additional assumptions, where in the core the reversals initiated.

A high point in this approach to analysis of transitional fields occurred when Hoffman (1979) showed that a generalized zonal flooding model could simulate the longitudinal position of five VGP paths for the Matuyama-Brunhes transition. In this generalized model, some nonaxisymmetric field components are present because the reversal propagates outward from a particular point in the source region. The paleomagnetic data

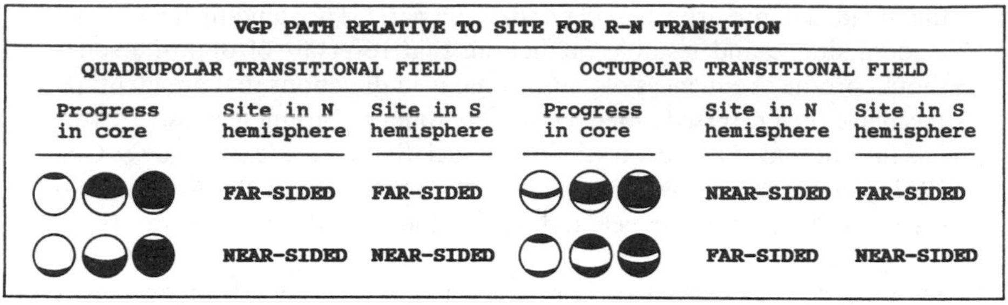

VGP PATH RELATIVE TO SITE FOR R-N TRANSITION					
QUADRUPOLAR TRANSITIONAL FIELD			OCTUPOLAR TRANSITIONAL FIELD		
Progress in core	Site in N hemisphere	Site in S hemisphere	Progress in core	Site in N hemisphere	Site in S hemisphere
	FAR-SIDED	FAR-SIDED		NEAR-SIDED	FAR-SIDED
	NEAR-SIDED	NEAR-SIDED		FAR-SIDED	NEAR-SIDED

Figure 11 Diagram for R-N transitions showing how the position of the transitional VGP path can be used to distinguish quadrupolar from octupolar transitional fields, and to determine where in the core the reversal process began [white (shaded) areas produce reverse (normal) dipole moment]. For example, antipodal paths from northern and southern hemisphere sites are consistent with an octupolar geometry. Furthermore, near-sided paths from the northern sites would suggest that the reversal initiated in the equatorial region of the outer core. For N-R transitions, near-sided paths become far-sided and vice versa. Figure adapted from Hoffman & Fuller (1978).

seemed to require a predominantly octupolar transitional field with an initiation point in the equatorial Atlantic. Although longitudinally-confined as expected from the model, the middle section of the VGP path from Lake Tecopa (Hillhouse & Cox 1976) lay about 40° east of that predicted by the model, the largest discrepancy among the data. Valet et al (1988a) reanalyzed the section at Lake Tecopa using modern paleomagnetic techniques, including detailed thermal demagnetization of all samples. This new work suggests that the AF demagnetization employed by Hillhouse & Cox (1976) did not fully isolate the primary magnetization of the sediments from younger components. The revised VGP path is longitudinally-confined and is centered 120° west of its former position, close to the sampling site at Lake Tecopa (and still far from the VGP path from Boso Peninsula) as would be expected if the transitional field were predominantly axisymmetric. Ironically, the fit to Hoffman's (1979) generalized model with nonzonal elements was slightly poorer than before.

NONZONAL TRANSITIONAL FIELD Although the zonal transitional field hypothesis could successfully account for certain subsets of the data, a growing body of observational evidence shows that nonzonal field components are significant during most reversals. Valet et al (1988b) and Laj et al (1988) have discussed paleomagnetic results for reversals of Miocene age recorded in marine clays from the Mediterranean region. Nonaxisymmetric intermediate fields appear to be important in all seven reversals analyzed; the VGP paths lie 90° east or west of the sampling sites. Herrero-Bervera & Theyer (1986) found evidence of nonzonal components in four younger reversals (bounding the Olduvai and Jaramillo subchrons) recorded by slowly-deposited sediments of the Pacific basin. As can be seen in Figure 8, nonzonal components are very apparent in the upper Olduvai transition as recorded in rapidly-deposited sediments from the north Atlantic.

Even the Matuyama-Brunhes transition, which figured so prominently in the development of the zonal field hypothesis, is now known to have a nonaxisymmetric character. Valet et al (1989), Clement & Kent (1991), and Clement (1991) have discussed a record of this reversal from deep-sea sediments deposited at the equator near longitude 337E. The transitional field directions transform to a longitudinally-confined VGP path passing through India. The VGP paths for other sites, located at mid-latitudes in the northern and southern hemispheres, are located nearly 180° away. Clement (1991) has shown that the observed pattern of field directions is consistent with a nonzonal model in which an equatorial octupole term predominates. These observations certainly raise the possibility that site-dependent VGP paths were characteristic of the transitional field only at mid-latitudes. In a more general sense, these developments have served as a reminder that geographically well-distributed data are needed to infer global field structure.

Finally, Hoffman (1986), Roperch & Duncan (1990), and Chauvin et al (1990) have all raised the possibility that zonal field components predominate early in transitional intervals with nonzonal components becoming increasingly significant later. The very detailed transition zone from Steens Mountain (Mankinen et al 1985) shows a sequence of intermediate field directions that is consistent with this idea. (See Figure 6: Purely axisymmetric field directions would plot on the line connecting the points labeled *Reverse* and *Normal.*)

DIPOLAR HYPOTHESIS REVIVED Recently, Tric et al (1991) have assembled a large subset of the data available for reversals of the past 10 my. Transitional VGP paths in this data set tend to lie near the broad longitudinal sector that includes North and South America (see Figure 12). Taken at face value, this observation appears to support the conclusion that Creer & Ispir (1970) and Steinhauser & Vincenz (1973) made two decades ago: that the transitional field is predominantly dipolar. Furthermore, Tric et al (1991) and Laj et al (1991) suggest that the propensity of the transitional field to point toward the Americas may somehow be related to unusual conditions in the deep earth below this region. Seismic evidence suggests that the lower mantle there is colder than nearby areas (e.g. Dziewonski & Woodhouse 1987), and some models of fluid flow in the outer core show the underlying region to be one where north-south streams predominate (e.g. Bloxham & Jackson 1991).

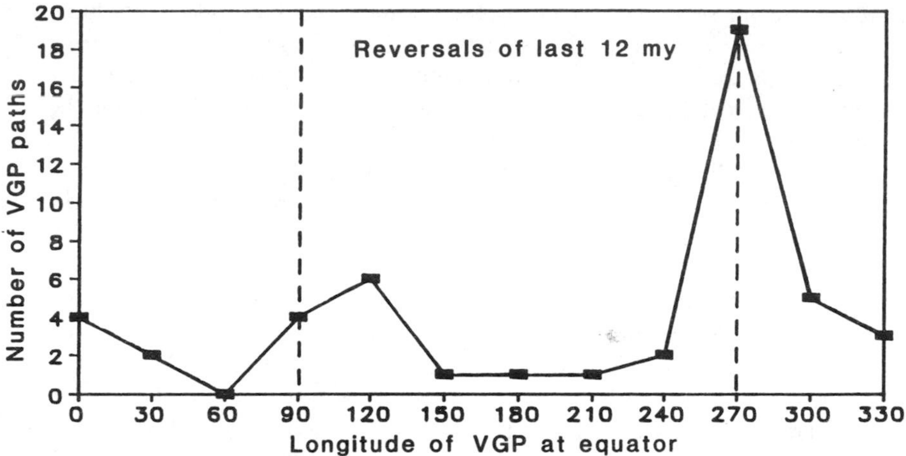

Figure 12 The evidence for dipolar transitional field. The plot shows where VGP paths from the last 12 million years cross the equator. There are maxima at 270E (centered on the Americas) and near 90E (opposite the Americas). The data are from the compilation of Tric et al (1991).

The large data set gathered by Tric et al (1991) provides a heretofore unseen statistical view of the transitional field. Nevertheless, several problems are apparent. In addition to the tendency to cluster over the Americas, the VGP paths also show a preference for the longitudinal band on the opposite side of the globe, centered on the region between India and Australia. The presence of antipodal VGPs suggests that the transition field can at least locally be dominated by nondipole terms (although a strong dipolar component is not ruled out). Also, the geographical distribution of sites is poor. Only 7 of 48 sites are within 20° of the equator, and only 5 come from the southern hemisphere. As demonstrated for the Matuyama-Brunhes transition by Clement (1991), data from equatorial sites can be critical for demonstrating the nondipolar or nonzonal character of the transitional field. In addition, some data are of questionable quality because of rock magnetic complications. Finally, the connection to deep-earth features is not firm. In particular, the inferred velocity structure at the top of the fluid core varies significantly from model to model depending on the assumptions used to constrain the model (e.g. Bloxham & Jackson 1991).

At this point it is useful to be more precise in our usage of the terms "dipolar" and "nondipolar." Remember that the geomagnetic field is predominantly dipolar at the earth's surface, and nondipolar at the core-mantle boundary. This difference exists because the various spherical harmonic terms vary with distance from the center of the earth. The magnitude of the dipole term at the surface (radius *Re*) is reduced by the factor $(Rc/Re)^3$ compared with its magnitude at the top of the core (radius *Rc*). The corresponding factors for quadrupole and octupole terms are $(Rc/Re)^4$ and $(Rc/Re)^5$. Suppose that, during a reversal, the spherical harmonic spectrum at the core-mantle boundary were flat out to degree 10, with higher degree terms being negligible. In this case the dipole field would be 1/10 the total field at the core-mantle boundary and roughly 1/2 the total field at the earth's surface. Clearly, this dipole must point somewhere at every instant during the reversal. Despite the problems mentioned above, it is perhaps not so surprising that Tric et al (1991) have detected a dipole component in the transitional field at the earth's surface.

Now consider what would be observed if a few spherical harmonic terms of very high degree were added to our hypothetical core spectrum. At the core-mantle boundary, this transition field would certainly be less dipolar than the typical stable field. Nevertheless, the VGP paths from many sites would be geographically similar and thus appear to indicate a dipolar transition. The field at a few sites, however, would produce very different VGP paths, thereby providing the key evidence of the additional nondipolar elements. This pattern is exactly what is observed for the Matuyama-Brunhes transition; the equatorial site discussed by Clement (1991) yielded a VGP path 180° away from the rest. In short, transitional field that has a substantial dipolar component at the earth's surface may in fact be considerably more complex than typical stable field. We address the persistence of particular transitional dipole orientations in the following section.

Time Evolution of Transitional Fields

Much of the discussion so far has been predicated on the assumption that some features of the transitional field repeat from reversal to reversal. Given the nonlinearities intrinsic to the geodynamo process, however, it is entirely possible that reversals are relatively disorderly phenomena. Therefore, a key question is whether there are systematic features of the transitional field—characteristics that repeat or evolve in a coherent way through time. Such features, if found, are of additional interest because their rate of change may be related to dynamic properties of the deep earth. Features that vary rapidly are most likely controlled directly by flow in the outer core, whereas those that vary much more slowly are likely modulated by changing conditions in the lower mantle.

Early studies of transition zones in Iceland (reviewed by Dagley & Lawley 1974), for example, showed the transitional field to be highly variable from reversal to reversal, with VGP paths exhibiting large longitudinal swings and loops. In contrast, volcanic records of successive Pliocene reversals from Kauai (Bogue & Coe 1982), yielded very similar intermediate field directions. The difference between this result and those from Iceland may be evidence of secular change in the reversal process, or may be related to latitudinal variation in transitional field behavior (as suggested by Hoffman 1979).

Although volcanic rocks have provided some useful data, most studies of successive reversals have focused on sedimentary units. For example, a very interesting set of results is now available for sites in the Mediterranean region. Valet & Laj (1984), Valet et al (1988b), Linssen (1988), Laj et al (1988), and Tric et al (1991) have provided records of late Tertiary reversals spanning approximately 10 my, all viewed from a limited geographic area. Detailed VGP paths exist for 12 reversals out of the approximately 50 that have occurred during the interval. Most of the transition zones studied have yielded VGP paths that are longitudinally confined, and so it is reasonable to look for trends in the longitudinal positions of the paths. This property of the reversals is shown in Figure 13, where the longitude at which each VGP path crosses the equator is plotted against time. In contrast to Figure 12 or the analysis of Tric et al (1991), Figure 13 weights each reversal equally. This approach provides an especially clear picture of how the transitional field evolves from reversal to reversal.

For the 12 my old reversals from Zackinthos (Laj et al 1988), the VGP makes multiple crossings of the equator. For the normal-to-reverse transitions, the crossing point appears to drift westward at a rate of the same order (0.1 deg/yr) as the westward drift of the nondipole field. The drift appears to continue across two successive N-R transitions. Equally rapid drift in the opposite direction occurs during the intervening reverse-to-normal transition. The next youngest data come from Italy (Linssen 1988) and Crete (Valet et al 1988b). Over a 3 my interval, the equatorial crossing point appears to drift slowly (10^{-4} deg/yr) eastward for 4 N-R and 3 R-N transitions. At 7 ma, the VGP paths for N-R and R-N transitions appear "out of phase," lying 180° apart longitudinally. During the next 3 my, the N-R paths drift eastward faster than the R-N, so that by 4.6 Ma the VGP paths appear "in phase," occupying the same longitudinal band regardless of the sense of reversal. After 4.6 Ma, it appears that VGP paths for both N-R and R-N reversals are again drifting westward.

If the pattern suggested by Figure 13 is indicative of how transitional behavior evolves at a site, then three important conclusions follow. First, there would seem to be no special significance to successive reversals of opposite sense having similar or antipodal VGP paths, a property that is

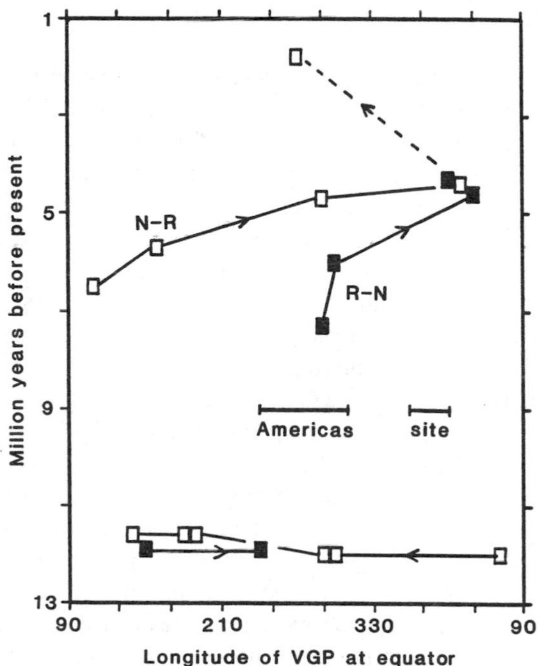

Figure 13 Time evolution of transitional field behavior. The plot shows where transitional VGP paths from sites in the Mediterranean region cross the equator. Solid symbols are for R-N transitions; open symbols are for N-R transitions. The crossing points for reversals of both senses appear to drift in one direction for millions of years, and then change direction. There is no strong preference for equatorial crossings near the Americas or at the site longitude.

the focus of discussions by Valet & Laj (1984), Linssen (1988), and Valet et al (1988b). The drift apparent in Figure 13 would occasionally produce such effects, but not in general. Second, there is no suggestion in the figure that VGP paths prefer particular longitudinal bands as there is for the global data set compiled by Tric et al (1991). The site longitude and the longitude of the Americas are both marked on Figure 13, and the VGP paths show no tendency to drift more slowly or stop within these particular sectors (or their opposites). Third, if the apparent correlation of transitional behavior from reversal to reversal is verified by future work, it would seem to indicate that the geodynamo has a "memory" which persists over a several million year time scale. One possibility is that this "memory" is related to the lateral variation of temperature or topography at the core-mantle boundary, although these properties are not expected to change as fast as the transition data would seem to imply (Loper & Stacey 1983).

SUMMARY

In spite of decades of hard work, there is still much to be learned about the nature of the field during geomagnetic reversals. The problem is a difficult one; geodynamo theory is not far enough advanced to make detailed predictions, high-resolution paleomagnetic records are scarce, and rock magnetic problems often make the data difficult to interpret. Nevertheless, the study of reversals remains an active field of research. A key motivation is the growing awareness of interactions between the earth's mantle and core, an insight that has stimulated interest in many geomagnetic phenomena. We anticipate, therefore, that knowledge of the transitional field will continue to grow.

Paleomagnetic evidence shows that polarity transitions are characterized by an interval of directional change and reduced field strength. The best estimates for the duration of the transitional interval average near 4000 years, a figure that has not changed much in spite of 25 years of research. The range of estimates, however, is large; values as low as 1000 years and longer than 10,000 years have been reported. Of the many factors that may contribute to this variation, geographical differences are particularly interesting because they relate to the geometry of the transitional field.

Both absolute and relative paleointensity determinations suggest that the field intensity drops to 10–20% of its usual value for a site during a reversal. Free decay of the geomagnetic dipole to values this low would take ten times longer than the average transition appears to last, an observation that strongly suggests that the reversal process is not associated with a cessation of convection in the outer core. In this regard, suggestions that field variation during transitions is similar in character to that during stable intervals are interesting and deserve further attention. Also relevant is the evidence from volcanic rocks that reversals are somehow linked to interactions between the two dynamo families.

The transitional field almost certainly involves multipolar sources in the core. The geometry of this field at the earth's surface, however, is the focus of ongoing controversy. One issue to be resolved, for example, is whether terms of a single dynamo family become prominent during reversals. There seems to be general agreement that both zonal and nonzonal terms are significant, although some data suggest that early in transition the field may be predominantly zonal. The apparent attraction toward the Americas of transitional VGPs suggests that the transitional field has a substantial dipolar component. This geometry is to be expected simply because the transitional field is observed far from the earth's core so that high-degree nondipole fields are strongly attenuated. Other evidence, however, suggests that high-degree field components are sometimes important during reversals. For example, two or three directional rebounds, perhaps reflecting the passing of small-scale (i.e. high degree) magnetic features under the observing site, are seen in many reliable transitional records. In brief, we speculate that the field during a typical transition is 50% dipolar or more, but that nondipole fields also contribute significantly.

A related issue is whether the dipolar part of the transitional field is systematically oriented toward a certain part of the globe, perhaps reflecting lateral heterogeneity in the boundary conditions at the core-mantle interface. Tric et al (1991) claim that the field has exhibited such a geographical bias for the past 12 my. As viewed from the Mediterranean region, however, VGP paths for the past 12 my appear to drift eastward and westward with time, showing no preference for particular longitudinal bands. It is clear that substantially more work will be needed to reconcile these observations.

Perhaps the most surprising suggestion to emerge from analyses of polarity transition data is that the field may sometimes change sharply over a period as short as a few weeks. We find that field variation of this kind would require an increase in core convection velocity of three to four orders of magnitude relative to times of stable field. An alternative explanation of the key paleomagnetic data involves plausible, but as yet undocumented, rock magnetic effects.

ACKNOWLEDGMENTS

The National Science Foundation has provided financial support for our work on geomagnetic reversals. We thank Phil McFadden, Ken Hoffman, and Brad Clement for their help in preparing this review.

Literature Cited

Aitken, M. J., Allsop, A. L., Bussell, G. D., Winter, M. B. 1988. Determination of the intensity of the Earth's magnetic field during archaeological times: reliability of the Thellier technique. *Rev. Geophys.* 26: 3–12

Bloxham, J., Gubbins, D. 1985. The secular variation of the Earth's magnetic field. *Nature* 317: 777–81

Bloxham, J., Jackson, A. 1991. Fluid flow near the surface of Earth's outer core. *Rev. Geophys.* 29: 97–120

Boehler, R., Besson, J. M., Nicol, M., Nielsen, M., Itie, J. P., et al. 1989. X-ray diffraction of γ-Fe at high temperatures and pressures. *J. Appl. Phys.* 65: 1795–97

Bogue, S. W., Coe, R. S. 1982. Successive palaeomagnetic reversal records from Kauai. *Nature* 295: 399–401

Bogue, S. W., Coe, R. S. 1984. Transitional paleointensities from Kauai, Hawaii, and geomagnetic reversal models. *J. Geophys. Res.* 89: 10,341–54

Bogue, S. W., Hoffman, K. A. 1987. Morphology of geomagnetic reversals. *Rev. Geophys.* 25: 910–16

Braginsky, S. 1964. Magnetohydrodynamics of the earth's core. *Geomag. Aeron.* 4: 698–712

Bullard, E. C., Freedman, C., Gellman, H., Nixon, J. 1950. The westward drift of the Earth's magnetic field. *Philos. Trans. R. Soc. London Ser. A* 243: 67–92

Chauvin, A., Roperch, P., Duncan, R. A. 1990. Records of geomagnetic reversals from volcanic islands of French Polynesia 2. Paleomagnetic study of a flow sequence (1.2–0.6 Ma) from the island of Tahiti and discussion of reversal models. *J. Geophys. Res.* 95: 2727–52

Clement, B. M., Constable, C. G. 1991. Polarity transitions, excursions and paleosecular variation of the earth's magnetic field. *Rev. Geophys.* Suppl.: US Natl. Rep. to Int. Union Geodesy Geophys. 1987–1990: 433–42

Clement, B. M., Kent, D. V. 1984. Latitudinal dependency of geomagnetic polarity transition durations. *Nature* 310: 488–91

Clement, B. M., Kent, D. V. 1986. Geomagnetic polarity transition records from five hydraulic piston core sites in the North Atlantic. In *Initial Reports of the Deep Sea Drilling Project*, ed. W. F. Ruddiman, R. B. Kidd, E. Thomas, et al, 94: 831–52. Washington: US Gov. Print. Off.

Clement, B. M. 1991. Geographical distribution of transitional VGPs: Evidence for non-zonal equatorial symmetry during the Matuyama-Brunhes geomagnetic reversal. *Earth Planet. Sci. Lett.* 104: 48–58

Clement, B. M., Kent, D. V. 1991. A southern hemisphere record of the Matuyama-Brunhes polarity reversal. *Geophys. Res. Lett.* 18: 81–84

Coe, R. S. 1967. Paleo-intensities of the earth's magnetic field determined from Tertiary and Quaternary rocks. *J. Geophys. Res.* 72: 3247–62

Coe, R. S., Grommé, C. S. 1973. A comparison of three methods of determining geomagnetic paleointensities. *J. Geomag. Geoelectr.* 25: 415–35

Coe, R. S., Grommé, S., Mankinen, E. A. 1984. Geomagnetic paleointensities from excursion sequences in lavas on Oahu, Hawaii. *J. Geophys. Res.* 89: 1059–69

Coe, R. S., Prévot, M. 1989. Evidence suggesting extremely rapid field variation during a geomagnetic reversal. *Earth Planet. Sci. Lett.* 92: 292–98

Constable, C. G. 1990. A simple statistical model for geomagnetic reversals. *J. Geophys. Res.* 95: 4587–96

Cox, A. 1968. Lengths of geomagnetic polarity intervals. *J. Geophys. Res.* 73: 3247–60

Cox, A. 1969. Geomagnetic reversals. *Science* 163: 237–45

Cox, A., Dalrymple, G. B. 1967. Statistical analysis of geomagnetic reversal data and the precision of potassium-argon dating. *J. Geophys. Res.* 72: 2603–14

Creer, K. M., Ispir, Y. 1970. An interpretation of the behavior of the geomagnetic field during polarity transitions. *Phys. Earth Planet. Inter.* 2: 283–93

Dagley, P., Lawley, E. 1974. Paleomagnetic evidence for the transitional behavior of the geomagnetic field. *Geophys. J. R. Astron. Soc.* 36: 577–98

Denham, C. R. 1974. Counter-clockwise motion of paleomagnetic directions 24,000 years ago at Mono Lake, California. *J. Geomag. Geoelectr.* 26: 487–98

Dodson, R., Dunn, J. R., Fuller, M., Williams, I., Ito, H., et al. 1978. Paleomagnetic record of a late Tertiary field reversal. *Geophys. J. R. Astron. Soc.* 53: 373–412

Dunn, J. R., Fuller, M., Ito, H., Schmidt, V. A. 1971. Paleomagnetic study of a reversal of the earth's magnetic field. *Science* 172: 840–45

Dziewonski, A. M., Woodhouse, J. H. 1987. Global images of the earth's interior. *Science* 236: 37–48

Fuller, M. 1987. A mechanism for reversals? *Nature* 326: 132–33

Fuller, M., Williams, I., Hoffman, K. A. 1979. Paleomagnetic records of geomagnetic field reversals and the morphology of the transitional fields. *Rev. Geophys. Space Phys.* 17: 179–203

Gubbins, D. 1977. Energetics of the earth's core. *J. Geophys.* 43: 453–64

Gubbins, D. 1987. Mechanism for geomagnetic polarity reversals. *Nature* 326: 167–69

Harland, W. B., Cox, A. V., Llewellyn, P. G., Pickton, C. A. G., Smith, A. G., Walters, R. 1982. *A Geologic Time Scale*. Cambridge: Cambridge Univ. Press. 131 pp.

Harrison, C. G. A., Ramirez, E. 1975. Areal coverage of spurious reversals of the earth's magnetic field. *J. Geomag. Geoelectr.* 27: 139–51

Henshaw, P. C., Merrill, R. T. 1980. Magnetic and chemical changes in marine sediments. *Rev. Geophys. Space Phys.* 18: 483–504

Herrero-Bervera, E., Theyer, F. 1986. Non-axisymmetric behaviour of Olduvai and Jaramillo polarity transitions recorded in north-central Pacific deep-sea sediments. *Nature* 322: 159–62

Herrero-Bervera, E., Theyer, F., Helsley, C. E. 1987. Olduvai onset polarity transition: two detailed paleomagnetic records from North Central Pacific sediments. *Phys. Earth Planet. Inter.* 49: 325–42

Hillhouse, J., Cox, A. 1976. Brunhes-Matuyama polarity transition. *Earth Planet. Sci. Lett.* 29: 51–64

Hoffman, K. A. 1977. Polarity transition records and the geomagnetic dynamo. *Science* 196: 1329–32

Hoffman, K. A. 1979. Behavior of the geodynamo during reversal: a phenomenological model. *Earth Planet. Sci. Lett.* 44: 7–17

Hoffman, K. A. 1981. Paleomagnetic excursions, aborted reversals and transitional fields. *Nature* 294: 67–68

Hoffman, K. A. 1984. A method for the display and analysis of transitional paleomagnetic data. *J. Geophys. Res.* 89: 6285–92

Hoffman, K. A. 1986. Transitional field behaviour from Southern Hemisphere lavas: evidence for two-stage reversals of the geodynamo. *Nature* 320: 228–32

Hoffman, K. A. 1988. Ancient Magnetic Reversals: Clues to the Geodynamo. *Sci. Am.* 256: 76–83

Hoffman, K. A. 1989. Geomagnetic polarity reversals: theory and models. In *The Encyclopedia of Solid Earth Geophysics*, ed. D. E. James, pp. 547–55. New York: Van Nostrand Rheinhold

Hoffman, K. A., Fuller, M. 1978. Transitional field configurations and geomagnetic reversal. *Nature* 273: 715–18

Hoffman, K. A., Slade, S. B. 1986. Polarity transition records and the acquisition of remanence: a cautionary note. *Geophys. Res. Lett.* 13: 483–86

Hyodo, M. 1984. Possibility of reconstruction of the past geomagnetic field from homogeneous sediments. *J. Geomag. Geoelectr.* 36: 45–62

Johnson, H. P., Merrill, R. T. 1972. Magnetic and mineralogical changes associated with low-temperature oxidation of magnetite. *J. Geophys. Res.* 77: 334–41

Karlin, R. 1990. Magnetite diagenesis in marine sediments from the Oregon continental margin. *J. Geophys. Res.* 95: 4405–19

Karlin, R., Levi, S. 1985. Geochemical and sedimentological control of the magnetic properties of hemipelagic sediments. *J. Geophys. Res.* 90: 10,373–92

King, J. W., Banerjee, S. K., Marvin, J. 1983. A new rock-magnetic approach to selecting sediments for geomagnetic paleointensity studies: application to paleointensity for the last 4000 years. *J. Geophys. Res.* 88: 5911–21

Laj, C. 1989. Geomagnetic polarity reversals: observations. In *The Encyclopedia of Solid Earth Geophysics*, ed. D. E. James, pp. 535–47. New York: Van Nostrand Rheinhold

Laj, C., Guitton, S., Kissel, C., Mazaud, A. 1988. Complex behavior of the geomagnetic field during three successive polarity reversals, 11–12 m.y.b.p. *J. Geophys. Res.* 93: 11,655–66

Laj, C., Mazaud, A., Weeks, R., Fuller, M., Herrero-Bervera, E. 1991. Geomagnetic reversal paths. *Nature* 351: 447

Levy, E. H. 1971. Kinematic reversal schemes for the geomagnetic dipole. *Astrophys. J.* 171: 635–42

Liddicoat, J., Coe, R. S. 1979. Mono Lake geomagnetic excursion. *J. Geophys. Res.* 84: 261–71

Linssen, J. H. 1988. Preliminary results of a study of four successive sedimentary geomagnetic reversal records from the Mediterranean (Upper Thvera, Lower and Upper Sidufjall, and Lower Nunivak). *Phys. Earth Planet. Inter.* 52: 207–31

Loper, D. E., Stacey, F. D. 1983. The dynamical and thermal structure of deep-mantle plumes. *Phys. Earth Planet. Inter.* 33: 304–17

Lund, S. P. 1989. Paleomagnetic secular variation. In *Encyclopedia of Solid Earth Geophysics*, ed. D. E. James, pp. 867–88. New York: Van Nostrand Rheinhold

Lund, S. P., Liddicoat, J. C., Lajoie, K. R., Henyey, T. L., Robinson, S. W. 1988. Paleomagnetic evidence for long-term (10^4 year) memory and periodic behavior in the earth's core dynamo process. *Geophys. Res. Lett.* 15: 1101–4

Mankinen, E. A., Prévot, M., Grommé, C. S., Coe, R. S. 1985. The Steens Mountain (Oregon) geomagnetic polarity transition 1. Directional history, duration of episodes, and rock magnetism. *J. Geophys. Res.* 90: 10,393–416

McElhinny, M. W. 1973. *Palaeomagnetism and Plate Tectonics*. London: Cambridge Univ. Press. 358 pp.

McFadden, P. L., McElhinny, M. W. 1982. Variations in the geomagnetic dipole 2: Statistical analysis of VDMs for the past 5 million years. *J. Geomag. Geoelectr.* 34: 163–89

McFadden, P. L., Merrill, R. T., McElhinny, M. W. 1985. Nonlinear processes in the geodynamo: palaeomagnetic evidence. *Geophys. J. R. Astron. Soc.* 83: 111–26

McFadden, P. L., Merrill, R. T., McElhinny, M. W. 1988. Dipole/quadrupole family modeling of paleosecular variation. *J. Geophys. Res.* 93: 11,583–88

McFadden, P. L., Merrill, R. T., McElhinny, M. W., Lee, S. 1991. Reversals of the Earth's magnetic field and temporal variations of the dynamo families. *J. Geophys. Res.* 96: 3923–33

Merrill, R. T., McElhinny, M. W. 1983. *The Earth's Magnetic Field: Its History, Origin and Planetary Perspective*. London: Academic. 401 pp. 1st ed.

Merrill, R. T., McElhinny, M. W., Stevenson, D. J. 1979. Evidence for long-term asymmetries in the earth's magnetic field and possible implications for dynamo theories. *Phys. Earth Planet. Inter.* 20: 75–82

Merrill, R. T., McFadden, P. L. 1988. Secular variation and the origin of geomagnetic field reversals. *J. Geophys. Res.* 93: 11,589–97

Merrill, R. T., McFadden, P. L. 1990. Paleomagnetism and the nature of the geodynamo. *Science* 248: 345–50

Merrill, R. T., McFadden, P. L., McElhinny, M. W. 1990. Paleomagnetic tomography of the core-mantle boundary. *Phys. Earth Planet. Inter.* 64: 87–101

Néel, L. 1949. Theorie du traînage magnétique des ferromagnétiques en grains fins avec applications aux terres cuites. *Ann. Geophys.* 5: 99–136

Ninkovich, N., Opdyke, N. D., Heezen, B. C., Foster, J. H. 1966. Paleomagnetic stratigraphy, rates of deposition and tephrachronology in North Pacific deep-sea sediments. *Earth Planet. Sci. Lett.* 1: 476–92

Okada, M., Niitsuma, N. 1989. Detailed paleomagnetic records during the Brunhes-Matuyama geomagnetic reversal, and a direct determination of depth lag for magnetization in marine sediments. *Phys. Earth Planet. Inter.* 56: 133–50

Olson, P., Hagee, V. L. 1990. Geomagnetic polarity reversals, transition field structure, and convection in the outer core. *J. Geophys. Res.* 95: 4609–20

O'Reilly, W. 1984. *Rock and Mineral Magnetism.* New York: Chapman & Hall. 200 pp.

Özdemir, Ö., Dunlop, D. J. 1985. An experimental study of chemical remanent magnetizations of synthetic monodomain titanomagnetites with initial thermoremanent magnetizations. *J. Geophys. Res.* 90: 11,513–23

Parker, E. N. 1969. The occasional reversal of the geomagnetic field. *Astrophys. J.* 158: 815–27

Prévot, M., Mankinen, E. A., Coe, R. S., Grommé, C. S. 1985. The Steens Mountain (Oregon) geomagnetic polarity transition 2. Field intensity variations and discussion of reversal models. *J. Geophys. Res.* 90: 10,417–88

Roberts, P. H., Stix, M. 1972. α-effect dynamos, by the Bullard-Gellman formalism. *Astron. Astrophys.* 18: 453–66

Rochette, P. 1990. Rationale of geomagnetic reversals versus remanence recording processes in rocks: a critical review. *Earth Planet. Sci. Lett.* 98: 33–39

Roperch, P., Duncan, R. A. 1990. Records of geomagnetic reversals from volcanic islands of French Polynesia 1. Paleomagnetic study of a polarity transition in a lava sequence from the island of Huahine. *J. Geophys. Res.* 95: 2713–26

Schneider, D. A., Kent, D. V. 1988. Inclination anomalies from Indian Ocean sediments and the possibility of a standing nondipole field. *J. Geophys. Res.* 93: 11,621–30

Schneider, D. A., Kent, D. V. 1990. The time-averaged paleomagnetic field. *Rev. Geophys.* 28: 71–96

Shaw, J. 1974. A new method for determining the magnitude of the paleomagnetic field: application to five historic flows and five archaeological samples. *Geophys. J. R. Astron. Soc.* 39: 133–41

Shaw, J. 1975. Strong geomagnetic fields during a single Icelandic polarity transition. *Geophys. J. R. Astron. Soc.* 40: 345–50

Soward, A. M. 1989. On dynamo action in a steady flow at large magnetic Reynolds number. *Geophys. Astrophys. Fluid Dynam.* 49: 3–22

Steinhauser, P., Vincenz, S. A. 1973. Equatorial paleopoles and behavior of the dipole field during polarity transitions. *Earth Planet. Sci. Lett.* 19: 113–19

Stokking, L. B., Tauxe, L. 1990. Properties of chemical remanence in synthetic hematite: testing theoretical predictions. *J. Geophys. Res.* 95: 12,639–52

Tauxe, L., Valet, J. P. 1989. Relative paleointensity of the Earth's magnetic field from marine sedimentary records: a global perspective. *Phys. Earth Planet. Inter.* 56: 59–68

Thellier, E., Thellier, O. 1959. Sur l'intensité du champ magnétique terrestre dans le passé historique et géologique. *Ann. Geophys.* 15: 285–376

Tric, E., Laj, C., Jehanno, C., Valet, J.-P., Kissel, C., et al 1991. High resolution record of the Upper Olduvai transition from Po valley (Italy) sediments: support for dipolar transition geometry? *Phys. Earth Planet. Inter.* 65: 319–36

Valet, J. P., Laj, C., Langereis, C. G. 1988b. Sequential geomagnetic reversals recorded in Upper Tortonian marine clays in western Crete (Greece). *J. Geophys. Res.* 93: 1131–51

Valet, J. P., Tauxe, L., Clark, D. R. 1988a. The Matuyama-Brunhes transition recorded from Lake Tecopa sediments (California). *Earth Planet. Sci. Lett.* 87: 463–72

Valet, J. P., Tauxe, L., Clement, B. M. 1989. Equatorial and mid-latitude records of the last geomagnetic reversal from the Atlantic Ocean. *Earth Planet. Sci. Lett.* 94: 371–84

Valet, J.-P., Laj, C. 1984. Invariant and changing transitional field configurations in a sequence of geomagnetic reversals. *Nature* 311: 552–55

Valet, J.-P., Laj, C., Tucholka, P. 1986. High-resolution sedimentary record of a geomagnetic reversal. *Nature* 322: 27–32

Walton, D. 1988. The lack of reproducibility in experimentally determined intensities of the Earth's magnetic field. *Rev. Geophys.* 26: 15–22

Watkins, N. D. 1969. Non-dipole behavior during an upper Miocene geomagnetic polarity transition in Oregon. *Geophys. J. R. Astron. Soc.* 17: 121–49

Weeks, R. J., Fuller, M., Williams, I. 1988a. The effect of recording medium upon reversal records. *Geophys. Res. Lett.* 15: 1255–58

Weeks, R. J., Fuller, M., Williams, I. 1988b. A model for transitional field geometries involving low-order zonals and drifting nondipole harmonics. *J. Geophys. Res.* 93: 11,613–20

Williams, Q., Knittle, E., Jeanloz, R. 1991. The high-pressure melting curve of iron: A technical discussion. *J. Geophys. Res.* 96: 2171–84

Wilson, R. L., Dagley, P., McCormack, A. G. 1972. Paleomagnetic evidence about the source of the geomagnetic field. *Geophys. J. R. Astron. Soc.* 28: 213–24

Yukutake, T., Tachinaka, H. 1969. Separation of the earth's magnetic field into the drifting and the standing parts. *Bull. Earthquake Res. Inst.* 47: 65–97

Annu. Rev. Earth Planet. Sci. 1992. 20:221–43

COSMIC-RAY EXPOSURE HISTORY OF ORDINARY CHONDRITES

K. Marti and T. Graf

Department of Chemistry, University of California, San Diego, La Jolla, California 92093-0317

KEY WORDS: collisional history, noble gases, parent bodies

INTRODUCTION

The cosmic-ray exposure age (T_e) of a meteorite measures the time interval in space between formation of a meteoroid as a meter-sized object (by removal from a shielded location within the parent body) and its capture by the Earth. Cosmic-ray produced noble gases have figured prominently for more than thirty years in studies of meteorite origin, orbital and parent-body histories, in the dating of lunar craters, and in the study of regolith dynamics. Exposure ages are in general determined from the concentrations of cosmic-ray produced noble gases and appropriate production rates, adjusted for variable shielding, in cases where this record is available. Over the years, several analyses of T_e systematics were carried out. Conclusions regarding collisional destruction rates (Eberhardt & Hess 1960, Gault 1969), origin of meteorites (Arnold 1965, Wänke 1966, Wetherill 1985), parent-body stratigraphy (Crabb & Schultz 1981), and collisional breakup events (Geiss et al 1960, Anders 1964) were based on the distribution and clustering of exposure ages. Variations in T_e systematics of different petrographic types among the chondrites as well as for gas-rich chondrites were considered by Crabb & Schultz (1981). These authors found that every petrographic type contains solar-type gases and they concluded that the existing evidence supported a model of chondrite parent-body fragmentation and reassembly into megabreccias. It has been

0084–6597/92/0515–0221$02.00

suggested several times that L-chondrites of very low gas retention ages also tend to have low T_e ages (Anders 1964, Wänke 1966, Crabb & Schultz 1981). Since a priori no relationship of these ages is expected, this may point toward distinct thermal and collisional histories of subgroups of chondrites. Events capable of resetting the ^{39}Ar-^{40}Ar clock are well documented (Turner et al 1978; Bogard et al 1976, 1980; Turner 1988).

This review discusses updated exposure ages of ordinary chondrites, the largest group of meteorites. It includes all published data (Schultz & Kruse 1989) on non-Antarctic meteorites and recent unpublished data. Data from Antarctica are excluded to eliminate any bias due to multiple falls. However, Schultz et al (1991) recently compared T_e statistics of Antarctic and non-Antarctic H-chondrites and observed rather similar histograms. Calculations are from Graf & Marti (1989a, 1992). Distributions of exposure ages have evolved over the years as new data and better production rates became available (Geiss et al 1960, Kirsten et al 1963, Anders 1964, Tannenbaum 1967, Crabb & Schultz 1981). Some of the properties that were discovered long ago have stood the test of time, while others remain hypotheses which need to be assessed further with new data and improved statistics.

THE COSMIC-RAY RECORD

The records of cosmic-ray particles were studied in extraterrestrial matter, i.e. meteorites and lunar samples. We consider two types of energetic particles, the galactic cosmic rays (GCRs) which originate outside the solar system and the solar cosmic rays (SCRs), emitted by major flares on the sun. These particles have sufficient energy to penetrate solar system matter, but only particles with higher energies (≥ 10 MeV) can produce nuclear reactions which are of interest in determining cosmic-ray ages. The GCR particles have high energies and are continuously incident on the solar system, but their flux and energy spectrum is modulated by the interplanetary magnetic field and, therefore, by solar activity. On the other hand the SCR particles have much lower energies and higher fluxes; their effects can be observed only in the surface layers of extraterrestrial matter, e.g. the lunar surface and in gas-rich meteorites, which represent regolith breccias.

The initial sources of GCR particles and the mechanisms of acceleration to very high energies are still incompletely understood, but the current view is that they originate in supernovae shock waves throughout the interstellar medium. Their modulation in the heliosphere is due to scattering by the interplanetary magnetic field and this modulation is the dominant source of the observed GCR flux variability. The fluxes of 0.2–

0.5 GeV/nucleon particles are modulated by about one order of magnitude during a solar cycle, while at energies greater than 5 GeV/nucleon no significant modulation is observed (e.g. Reedy et al 1983). The energy spectrum of GCR particles is approximately a power law in energy: $E^{-2.5}$. Since most secondary particles are produced by particles with $E > 1$ GeV/nucleon, the overall modulation effect is not very large.

The temporal variations of GCRs over a variety of time scales have been reviewed by Jokipii & Marti (1986) and Zanda et al (1989), those of SCRs by Reedy & Marti (1991). It is clear that changes in the intensity of GCRs in the local interstellar medium could be masked by solar modulation. If GCRs originate in supernova shock waves, the intensity of GCRs could vary, depending on the frequency of supernova explosions in our vicinity and on the parameters of the local interstellar medium. Although studies in meteorites indicate that the GCR flux was constant over a million year time scale (Reedy et al 1983), the ^{40}K, ^{41}K measurements in iron meteorites (Voshage 1967) indicate that the GCR flux was smaller in the distant past. Lavielle et al (1985) argued that this may reflect a recent increase in the GCR flux, since the K and Ar isotopic records in iron meteorites cannot be explained either by meteorite orbital changes or by space erosion.

GCR Reactions in Meteorites

The interaction processes are determined by the energy, charge, and mass of GCR particles and by the chemical composition of the target meteoroid. Charged particles lose energy by ionizing atoms as they pass through matter and those of large Z and low energy are stopped rapidly. Nuclear reactions between incident particles of high energy and target nuclei involve the formation of secondaries and of residual nuclei with smaller masses than those of the target nuclei. The cascade of nuclear interactions of secondary particles propagates to considerable depths, but below a depth of ~ 1 kg cm^{-2} (corresponding to 2–3 m in solids), very few primary or secondary particles are left and, therefore, the record of cosmic-ray reactions is limited to the surface region of larger objects and to small (meter-sized) objects in space. The exposure of meteoritic matter to GCRs produces a spectrum of reaction products, both stable and radioactive. Stable reaction products can only be identified by their characteristic isotopic signatures, if the target material has a low natural abundance of the element of interest. Important target elements are major elements like O, Mg, Al, Si, Ca, and Fe, but important information is also obtained from minor and trace elements such as Sr, Y, Zr (for Kr isotopes), Ba, REE (for Xe isotopes), or isotopes of elements with large neutron capture cross-sections. A useful approach to the evaluation of the complexity of GCR-induced reactions is the laboratory bombardment of thick targets and of artificial meteoroids by accelerator protons (e.g. Kohman & Bender 1967, Michel et al 1989). The nuclear interactions of primary GCR particles and the buildup of the cascade of secondary particles within a meteoroid are responsible for the observed variations of the rates of interaction. Therefore, the production rates of GCR-produced nuclides vary with the size of and the position within the meteoroid. Size and depth dependencies of production rates and model calculations of reaction rates by GCR and secondary particles have been published by Graf et al (1990a,b); see also the review by Vogt et al (1990).

Production Rates

Ordinary chondritic meteorites are rather uniform in chemical composition (e.g. Fredrickson et al 1990) and production rates of noble gases do not vary by much because of target element abundance variations. Since systematic differences are observed in the metal/silicate ratio we use different rates for H-, L-, and LL-chondrites. The variability in the adopted production rates mainly reflect large variations in shielding conditions. With an assumed constant GCR flux in interplanetary space the concentrations of a radionuclide R and of a stable nuclide S are obtained as

$$[R] = P_R \lambda^{-1}(1-e^{-\lambda t}), \tag{1}$$

$$[S] = P_S t \tag{2}$$

where P_R and P_S are the production rates of the radio- and stable nuclides, respectively, λ the decay constant, and t the exposure time.

Several methods were developed to determine production rates in meteorites (e.g. ^{3}H-^{3}He, ^{22}Na-^{22}Ne, ^{36}Cl-^{36}Ar, ^{81}Kr-^{83}Kr). In all methods it is assumed that the exposure geometry remained unchanged over the entire period of exposure. No other assumptions regarding shielding are necessary for the ^{81}Kr-^{83}Kr method, since the shielding dependencies of $P_R = P_{81}$ and $P_S = P_{83}$ essentially cancel in the P_{81}/P_{83} ratio that enters in the equation for the ^{81}Kr-^{83}Kr exposure age T_e (Marti 1967):

$$T_e = (P_{81}/P_{83})([^{83}\mathrm{Kr}]/[^{81}\mathrm{Kr}])_c \lambda_{81}^{-1} \tag{3}$$

$$P_{81}/P_{83} = 0.475([^{80}\mathrm{Kr}]+[^{82}\mathrm{Kr}])[^{83}\mathrm{Kr}]^{-1} \tag{4}$$

(for $T_e \gg \lambda_{81}^{-1} = 3.07 \times 10^5$ a). ^{81}Kr-^{83}Kr ages (3) can be used to determine production rates P_R and P_S of other nuclides with Equations (1) and (2). This approach was used by Nishiizumi et al (1980) and by Eugster (1988). The latter author based the production rates entirely on ^{81}Kr-^{83}Kr ages. In another approach to determine the production rate of ^{21}Ne (P_{21}), t in Equation (1) is substituted with $t = T_e = [S]P_S^{-1}$ from Equation

(2). This permits a fit to a two-parameter curve (P_R and P_S). Several radionuclides (e.g. ^{26}Al, ^{10}Be, ^{53}Mn) are suitable for this method (Nishiizumi et al 1980, Moniot et al 1983). Production rates P_{21} calibrated with ^{81}Kr-^{83}Kr ages are consistent with those calibrated against ^{22}Na-^{22}Ne ages, ^{10}Be-ages, and ^{53}Mn-ages, but not with P_{21} obtained from ^{26}Al-ages (Nishiizumi et al 1980). This inconsistency has recently been interpreted to reflect pre-irradiation effects in some meteorites with short (<4 Ma) exposure ages (Graf & Marti 1989b). It should be pointed out that the present production rates are substantially smaller than older production rates in the literature and, therefore, exposure ages of chondrites discussed in this paper are systematically longer compared to e.g. Crabb & Schultz (1981).

Returning to the dependence of production rates on size of the meteoroid and shielding depth within a body, variations of 10–30% from average production rates are common and occasionally larger effects are observed. In order to obtain a 10% precision (or better) in exposure ages one must allow for these variations. Direct measurements of the depth within a meteoroid at which a sample was irradiated are not possible, because of unknown ablation losses during the meteorite passage through the atmosphere. As discussed below, in many cases the shielding depth can be estimated from the densities of nuclear tracks (Fleischer et al 1967, Bhattacharya et al 1973, Bhandari et al 1980) or, in cases of heavy shielding, from neutron capture effects (thermal and epithermal). Total depth profiles for GCR-produced nuclides are available only in few cases and such profiles can be used as calibration standards against which one can test calculations based on semiempirical models (Signer & Nier 1960; Graf et al 1990a,b). There are several meteoritic signatures that indicate variations in the shielding depth, but the most widely used indicator of shielding in chondrites is the ratio $(^{22}\text{Ne}/^{21}\text{Ne})_c$, where the subscript c indicates that only the cosmic-ray produced portion of meteoritic neon has to be used for this purpose (Eberhardt et al 1966, Nyquist et al 1973, Nishiizumi et al 1980, Eugster 1988). Production rates are usually evaluated in correlation plots against $(^{22}\text{Ne}/^{21}\text{Ne})_c$ as the shielding parameter. However, it is important to realize that these production rates are not uniquely determined by the neon shielding parameter and that the limitations need to be carefully assessed, especially for low $(^{22}\text{Ne}/^{21}\text{Ne})_c$ ratios. These limits were evaluated in detail by Graf et al (1990b). In practice it is often difficult to set the limits, especially for data published before the isotopic abundances of atmospheric Ne were redetermined (Eberhardt et al 1965).

In this review we adopt the production rates used by Graf & Marti (1992) which are similar to those of Eugster (1988) except for P_{38}. The adopted rates are given for H-, L-, and LL-chondrites and are expressed in units of 10^{-9} cm^3 STP g^{-1} Ma^{-1}:

$$P_3 = F_3[0.209 - 0.043(^{22}\text{Ne}/^{21}\text{Ne})_c]$$
$$F_3(\text{H}) = 0.98,\ F_3(\text{L}) = 1.00,\ F_3(\text{LL}) = 1.00 \quad (4)$$

$$P_{21} = F_{21}\,0.161[21.77(^{22}\text{Ne}/^{21}\text{Ne})_c - 19.32]^{-1}$$
$$F_{21}(\text{H}) = 0.94,\ F_{21}(\text{L}) = 1.00,\ F_{21}(\text{LL}) = 1.00 \quad (5)$$

$$P_{38} = F_{38}[0.0113 - 0.0064(^{22}\text{Ne}/^{21}\text{Ne})_c]$$
$$F_{38}(\text{H}) = 1.09,\ F_{38}(\text{L}) = 1.00,\ F_{38}(\text{LL}) = 0.99. \quad (6)$$

Atmospheric Ablation of Meteorites

Meteoroids undergo strong physical changes during their passage through the atmosphere (e.g. Bronshten 1983). A wide range of meteorite velocities and masses, coupled with dynamical and chemical properties which control the dissipation of kinetic energy, lead to a large variability of ablation losses (e.g. Bhandari et al 1980). Bhandari et al estimated that ablation losses generally exceed by an order of magnitude the total mass loss by space erosion during cosmic-ray exposure in space. These authors observed a relation between the meteoroid mass, the cosmic-ray track production rate, and the cosmic-ray produced $(^{22}\text{Ne}/^{21}\text{Ne})_c$ ratio. This empirical relation and model calculations by Bhattacharya et al (1973) were used to estimate (preatmospheric) meteoroid masses for a large set of meteorites (Bhandari et al 1980).

Complex Exposure Histories

In several meteorites the calculated exposure ages and inferred exposure geometries as obtained by different methods disagree; in such cases complex multistep exposure histories are generally implied. We mentioned earlier that temporal or spatial variations of the GCR flux might be a factor in such variations, but we will not further consider this option for ordinary chondrites (the time frame of the last 50 Ma). Some calculations of the evolution of meteorite orbits imply that multistep (complex) exposure histories should not only be common for irons but also for stony meteorites (occurring in more than 25% of the meteorites) (Wetherill 1980, 1985). For a small number of chondrites complex exposure histories are well established, sometimes possible complexities are indicated, and many times these may have been overlooked because of inadequate assessment capabilities. The fraction of complex cases is not currently established, and some of the exposure ages reported here may eventually turn out to be complex. We adopt one-stage exposures to GCRs for all chondrites, except in cases where complexities are clearly established. The large H4-chondrite Jilin provides a good example of a complex history; it also demonstrates the usefulness of studies of radioisotopes of varying half-lives (Begemann et al 1985, and refs therein). H-chondrites with short exposure ages (<4 Ma), which show discrepancies in their inferred ^{26}Al-ages, are discussed by Graf & Marti (1992).

THERMAL HISTORY OF CHONDRITES

Ordinary chondrites derive from at least three different bodies, each having its own chemical and oxygen isotopic signatures. Isotopic uniformity of H3 to H6 and L3 to L6 chondrites are consistent with closed-system metamorphism within each parent body, except for LL3-chondrites which differ slightly from LL4 to LL6 and may indicate a partially open system (Clayton et al 1991). Distinct parents are also suggested by specific characteristics pertaining to the early thermal evolutions of the H, L, and LL parents and of the petrographic subtypes of each meteorite class (Wood & Pellas 1991, Turner 1988, Lipschutz et al 1989). Significant information on internal structures and sizes of ordinary chondrite parent bodies is obtained by comparing ^{39}Ar-^{40}Ar ages of objects coming from the same parent source. Pellas & Fieni (1988) observed that for 11 out of 12 H-chondrites, ^{39}Ar-^{40}Ar ages overlap (independent of the petrographic type) at $(4.48 \pm 0.40) \times 10^9$ a. This indicates that the Ar closure temperature for large portions of the parent body was achieved within a period of ~ 80 Ma. In the model of a layered parent-body structure (Wood 1967, Pellas & Storzer 1981) it is assumed that the petrographic types of chondrites represent samples from different layers which experienced varying degrees of metamorphism. In this model the information obtained from each petrographic type may provide insight on parent-body evolution. The ^{39}Ar-^{40}Ar method is in general not sensitive enough to establish whether the different petrographic types were located at different depths in the parent body. There appears to be a clear-cut difference between ^{39}Ar-^{40}Ar ages of LL3 through LL5 and LL6-chondrites. For the L-body, the data indicate that early and late events and associated thermal pulses perturbed the ^{39}Ar-^{40}Ar clocks making any chronological interpretation very difficult. Recent work on U-Pb ages of H-chondrites lend support to a differential cooling history for types H4, H5, and H6 (Göpel et al 1991).

Late Thermal Events

The ^{39}Ar-^{40}Ar technique determines both ^{40}Ar and K in the same sample and by means of a stepwise temperature release is capable of distinguishing the time of an episodic ^{40}Ar loss from the entire sample or from only the least retentive ^{40}Ar sites. Turner (1969) found six L-chondrites that indicated degassing events between 440 and 540 Ma ago and two L-chondrites that indicated degassing events 300 Ma ago. None of these meteorites had been completely degassed, and the ^{39}Ar-^{40}Ar ages were obtained from theoretical curves fitted to the data.

Gopalan & Wetherill (1971) performed Rb-Sr isotopic studies on 15 L-chondrites, including 10 black chondrites. Whole rock data for black L "falls" conformed to the 4.56 Ga isochron defined by other chondrites. From measurements on different density separates of the heavily shocked Farmington and Orvinio chondrites, these authors concluded that the Sr isotopes were partially homogenized very recently in their history. More recently, Nakamura et al (1990) studied collisional events on the L-parent body by means of an internal isochron for an impact-melted L-chondrite (Point of Rocks) yielding a precise 460 Ma age for this event. Anders (1964) and Heymann (1967) concluded that the parent body of black L-chondrites was probably involved in a collisional degassing event 520 ± 60 m.y. ago, which also produced the shock effects in these meteorites. In addition, it was noted that a sizable fraction of the nonblack L-chondrites ($\sim 25\%$) show U-He gas retention ages close to 520 m.y., and therefore may have been involved in the same event that was considered distinct from those collisional events that initiated the cosmic-ray exposure of L-chondrites. Bogard et al (1976) investigated 14 lightly to heavily shocked ordinary chondrites using the ^{39}Ar-^{40}Ar dating technique. At least 9 meteorites show degassing ages less than 600 m.y., which are interpreted as the times of major collisional events among meteorite parent objects, probably in the asteroid belt. These 9 meteorites experienced collisional degassing of 89–98% of their radiogenic ^{40}Ar, but in no case was the degassing complete. Therefore individual whole rock K-Ar ages are inaccurate measurements of the times of collisional events. Obvious groupings of degassing ages among chondrites exist. Examples of both H- and L-chondrites show ages of ~ 40 Ma and ~ 30 Ma, several L-chondrites show ages of ~ 500 Ma (Bogard et al 1976, 1980; Turner 1988). It seems that several collisions occurred and that several parent objects were involved.

The systematics of Ar retention was studied in detail in the shocked L6-chondrite Peace River which shows a well defined low-temperature plateau age of 450 Ma (Turner 1988). The Ar extracted by laser probe melting of microgram-sized samples shows a large variability of K on this scale, but correlated variations of $^{40}Ar_r$. The data agreed with the 450 Ma "stepped-heating" age and did not provide evidence of Ar loss later than 450 Ma ago. In addition to the chronological information obtained from Ar studies, Turner (1988) also obtained an estimate of the energy deposition required to nearly completely outgas the shocked L-chondrites. His estimate of ~ 200 kJ/kg chondrite is considerably larger than the typical energy deposition in ejecta of moderate-sized lunar craters and considerably larger than the energy required to fragment bodies in space. A heat pulse of this magnitude such as associated with the degassed L-chondrites is unlikely to be due to impact heating on the parent body. A more likely alternative is a major impact event, capable of breaking up the parent body. As discussed later on, this alternative is supported by the small number of gas-rich meteorites among L-chondrites and it is consistent with the frequency of expected collisional events in the asteroid belt, as for example implied by the systematics of cosmic-ray exposure ages of iron meteorites.

The interpretation of collisional events was not universally accepted because of dynamical problems (see Turner 1988) involved in storing collisional fragments for ~400 Ma in Earth-crossing orbits. This problem can be resolved, if collisional fragments are either large (few km radius) or can be stored in a region of the asteroid belt from which they are easily perturbed into Earth-crossing orbits on a short time scale, such as the chaotic zone discussed by Wisdom (1983). An alternative view is that chondritic parent bodies now exist and that fragments (e.g. L-chondrites) are produced by collisions, injected into the "chaotic zone" and then perturbed into Earth-crossing orbits. In this case the inferred ~400 Ma age may either represent the thermal event on the parent body or may indicate incomplete outgassing of $^{4}He_r$ and $^{40}Ar_r$ at the time of the cratering event that produced the meteorites and injected them into the chaotic zone. Wetherill (1985) evaluated the dynamical probability that ordinary chondrites may directly originate from the asteroid belt and found a value of ~10%, while the remainder are obtained by subsequent fragmentation of larger Earth-crossing asteroid fragments.

In the absence of ^{39}Ar-^{40}Ar age data the radiogenic gases $^{4}He_r$ (from U and Th decay) and $^{40}Ar_r$ (from ^{40}K) can serve as monitors of the thermal evolution of meteoroids and of their parent bodies. Whenever available the concentrations of radiogenic $^{4}He_r$ and $^{40}Ar_r$ were used to assess the thermal history of groups of meteorites. A typical concentration of $^{4}He_r$ accumulated over the age of a chondrite is $\sim 17 \times 10^{-6}$ cm^3 STP g^{-1}, that of $^{40}Ar_r \sim 70 \times 10^{-6}$ cm^3 STP g^{-1}. Lower concentrations reflect incomplete retention of radiogenic gases and thermal events.

EXPOSURE AGE (T_e) HISTOGRAMS

L-Chondrites

Although we consider that the asteroid belt is in the ultimate source of ordinary chondrites, because of apparently multiple fragmentation, the cosmic-ray record in general dates the breakout from a shielded location within the immediate parent object. The exposure-age distribution of L-chondrites was most recently studied by Graf & Marti (1990a, 1992). We discuss the overall T_e histogram and then subdivide the L-chondrites in order to consider the salient features that are recognized only in subsamples of L-chondrites and assess specific features which might relate to parent-body properties and history. In Figure 1 the histogram is plotted on a logarithmic scale of 10% age resolution which corresponds to uncertainties in the highest quality data (quality A). Figure 1 shows that the majority (73%) of L-chondrite exposure ages are >10 Ma. It was long known that this record differs from that of the H-chondrites (Wänke 1966, Crabb & Schultz 1981). This distribution is inconsistent with a steady state L-meteoroid reservoir of Earth-crossers. In the past there was considerable debate whether a continuum represents the T_e histogram or whether there exist clusters in the data (e.g. Tannenbaum 1967, Crabb & Schultz 1981). We note that the upper end of the distribution is characterized by a peak at $T_e \sim 40$ Ma and a sharp drop-off above. A dip is indicated below the 40 Ma peak. This is the record expected for a major collisional event at this time. Additional clusters are indicated at $T_e \sim 5$ Ma and in the 20–30 Ma region, but the resolution in Figure 1 does not permit the identification of additional collisional events. We now turn to the T_e histograms of subsamples, first of petrographic types L3–L6 (Figure 2), and then we inspect the records of meteorites with high and low retention of radiogenic ^4He_r and $^{40}Ar_r$ (Figure 3), which, as discussed earlier, imply distinct thermal histories.

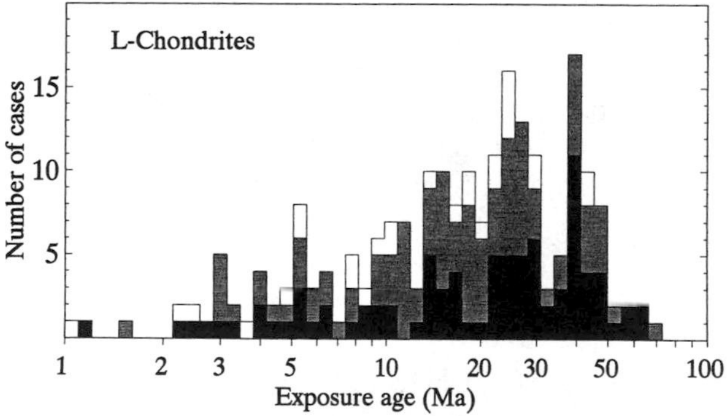

Figure 1 Exposure age (T_e) histogram of L-chondrites. Data are grouped after Graf & Marti (1989a) into quality classes: high quality A (*black*), intermediate quality B (*grey*), low quality C (*open*). These quality assignments are based on the agreement of exposure ages calculated by different methods. The age scale is logarithmic with a resolution of 10% of the age, corresponding to estimated uncertainties of quality A data.

The L6-chondrites—by far the largest subgroup of L-chondrites—are considered first, since they provide a sample with reasonable statistics. Distributions for the less abundant L5, L4, and L3 types will then be compared to the L6 histograms. The T_e histogram of the L6 group is shown in Figure 2 and reveals an overall structure similar to that of the total L group, e.g. a preponderance of ages $T_e > 10$ Ma. Very sharp drop-offs are observed on both sides of the $T_e \sim 40$ Ma peak and establish a major collisional event for the L6 subgroup. The only other conspicuous feature is a rather broad cluster, centered at $T_e \sim 28$ Ma, which was noted also in Figure 1. In comparing the other petrographic types we note that the 40 Ma peak is visible in the histogram of the L5-chondrites but is missing in the L4 histogram, although here the statistics are rather poor. Therefore, the recorded major collisional event ~40 Ma ago produced

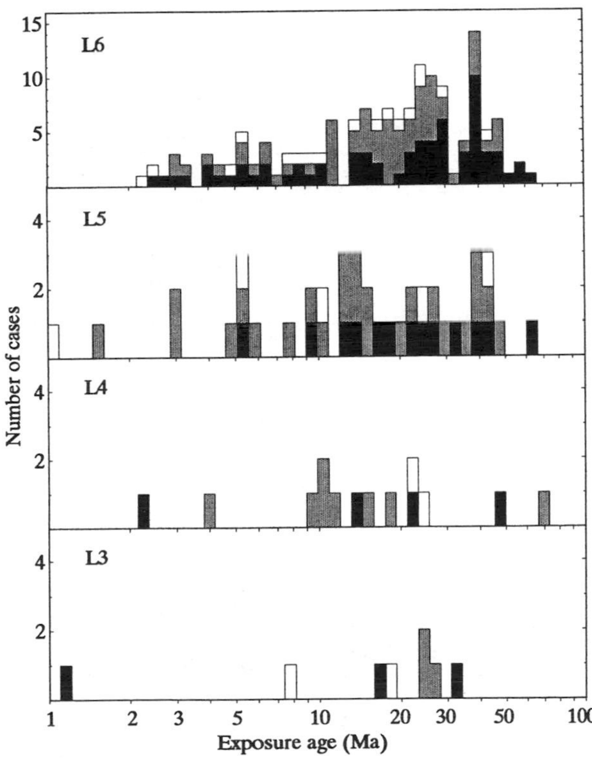

Figure 2 Exposure age (T_e) histograms of different petrographic types of L-chondrites shown in Figure 1.

mainly petrographic types 6 and 5. Indications for other clusters of exposure ages are present in Figure 2, but these are not observed uniformly among the various types. Specifically, the broad cluster at ~28 Ma in the L-chondrite histogram is again only featured in the L6 and L5 subgroups.

We earlier presented the evidence for shock effects and associated complex thermal histories of many L-chondrites based on ^{39}Ar-^{40}Ar ages and the low retention of radiogenic ^4He_r and $^{40}Ar_r$. For the L6-chondrites many authors observed trends revealing shorter exposure ages for chondrites with low ^4He retention ages (Anders 1964; Wänke 1966, 1968; Schultz 1976; Crabb & Schultz 1981). Figure 3 provides a good illustration of this trend as ages for L-chondrites containing >40 (units of 10^{-6}

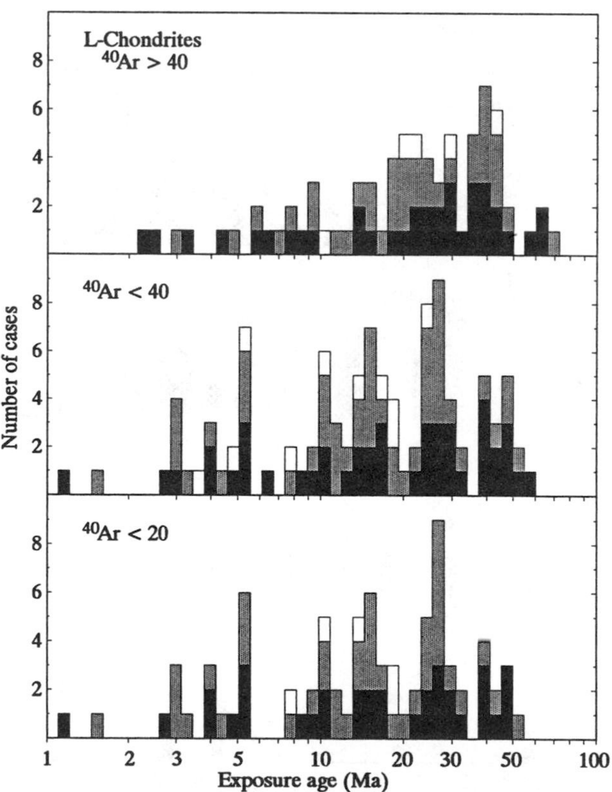

Figure 3 Exposure age (T_e) histograms of subgroups of L-chondrites with different retention of radiogenic ^{40}Ar. Top panel: $^{40}Ar > 40$, middle panel: $^{40}Ar < 40$, bottom panel: $^{40}Ar < 20$; all concentration units are 10^{-6} cm^3 STP g^{-1}.

cm^3 STP g^{-1}) ^{40}Ar (*top panel*) are compared to L6-chondrites with incomplete ^{40}Ar retention (<40; *center panel*), and to those with substantial losses (<20; *bottom panel*). The cluster at ~28 Ma discussed above appears to be dominated by poor ^{40}Ar retainers since relative to the 40 Ma event, the peak "grows in" when we move down the panels. A similar "grow-in" situation is observed for the cluster at 5 Ma, which is missing in the top panel. We note that many members of the 28 and 5 Ma clusters have ^{40}Ar < 10 ($\times 10^{-6}$ cm^3 STP g^{-1}) and it appears that these events indicate collisional fragmentation of a parent body which has earlier suffered catastrophic out-gassing (breakup). Ten L-chondrites show ^{40}Ar < 10 and ^{4}He$_r$ < (10^{-6} cm^3 STP g^{-1}). All these samples reveal T_e < 15 Ma and obviously support the above trend. We discussed earlier that the ^{4}He$_r$ concentrations of these meteorites indicate very recent degassing (<500 Ma) and that the ^{39}Ar-^{40}Ar systematics of some members of this group reveal plateau ages corresponding to ≤300 or ~500 Ma, which indicate catastrophic events (Heymann 1967, Bogard et al 1976). The Malakal meteorite, a member of this group, reveals a ^{39}Ar-^{40}Ar plateau at ~50 Ma, which is within error limits, consistent with the 40 Ma exposure-age peak (Bogard et al 1976).

H-Chrondites

The T_e histogram of H-chondrites (Graf & Marti 1992) is shown on a logarithmic age scale in Figure 4. The H-distribution is quite distinct from that of the L-chondrites with a major well-known peak at ~7 Ma and another peak at ~33 Ma, which is followed by a sharp drop-off. As the distributions for different petrographic types appear to differ, we discuss the H3–H6 histograms separately in Figure 5.

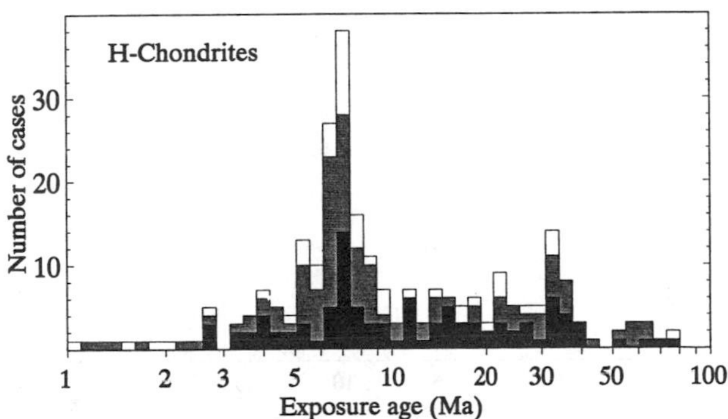

Figure 4 Exposure age (T_e) histogram of H-chondrites. See caption of Figure 1 for quality classes; age scale is logarithmic.

We observe exposure-age peaks at ~7 Ma and ~33 Ma for all petrographic types. However, a cluster at ~4 Ma is observed only in the histograms of H5- and H6-chondrites. Figure 5 also reveals that exposure-age distributions for petrographic types differ in the frequency of ages $T_e < 10$ Ma, which might reflect a differential sampling of the petrographic types in the ~7 Ma collision.

Graf & Marti (1990b) studied various subdivisions of the H-chondrites of varying ^{40}Ar retention, as previously illustrated for L6-chondrites. The 33 Ma peak was observed in all sets and we interpret this as evidence for an H-parent break-up which produced statistically good samples of all petrographic types and fragments of varying radiogenic gas signature. It is of course possible that in some cases, major losses of radiogenic gases occurred in this postulated breakup, since ~20% of the measured H-

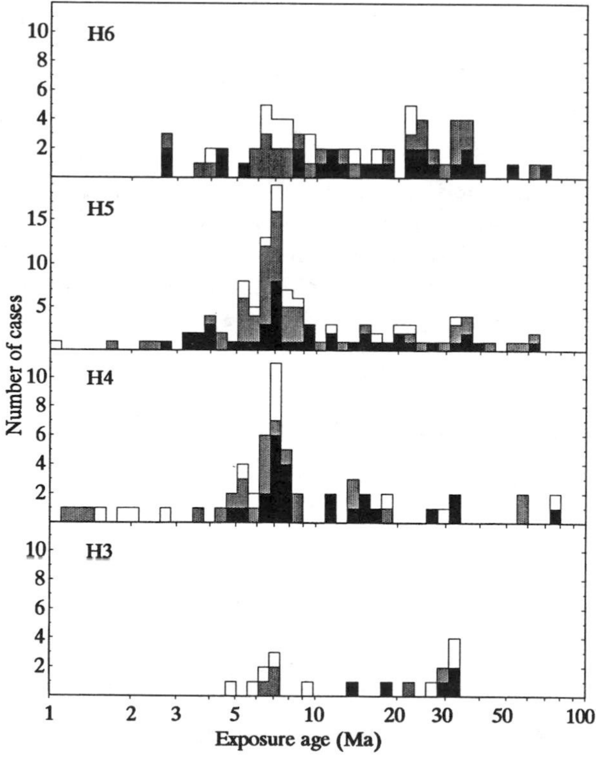

Figure 5 Exposure age (T_e) histograms of different petrographic types of H-chondrites shown in Figure 4.

chondrites have ^{39}Ar-^{40}Ar ages ≤ 500 Ma (Turner 1988). On the other hand, very few of the H-chondrites involved in the ~ 7 Ma breakup have suffered major gas losses. In an analysis of the 7 Ma age peaks, Graf & Marti (1991a) noted a minor offset in the peak-centers of the H4 and H5 groups, which is not resolved in Figure 5. Schultz et al (1990) have extensively studied the noble gas records of H4 falls and found a clear peak in their histogram at ~ 8 Ma. We will return to this offset when we discuss chondrite orbits.

LL-Chondrites

The T_e histogram of LL-chondrites (Graf & Marti 1991b) is shown in Figure 6 and shows that the exposure ages of about 30% of all LL-chondrites cluster in a peak at ~ 15 Ma. The peak shape is consistent with a collisional breakup, but the age distribution is not consistent with that expected for a quasi-continuous supply of meteorites. One (St. Mesmin) out of three solar-gas bearing LL-chondrites is in this peak. The systematics of ^{40}Ar retention ages shows that only minor heating occurred in the ~ 15 Ma collision of the parent of the LL-chondrites, since only two strongly degassed members (Paragould and Oberlin) are observed, and it is not known when these losses occurred. No clear differences are observed for petrographic types 4 to 6. A cluster of exposure ages (mainly of LL6) is observed at ~ 10 Ma, and smaller clusters at ~ 28 Ma and 40–50 Ma. Interestingly, the positions of the latter two clusters agree with clusters or peaks found in the T_e histograms of L-chondrites. We further note that a minor chondrite group, the L/LL chondrites, which was considered by Wasson & Wang (1991) to be possibly associated with the LL-chondrites,

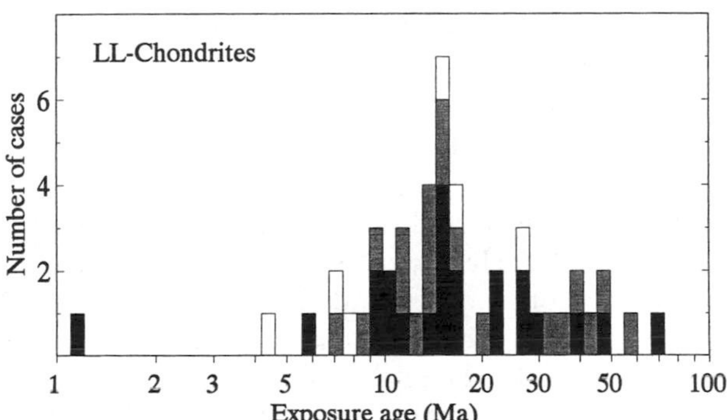

Figure 6 Exposure age (T_e) histogram of LL-chondrites. See caption of Figure 1 for quality classes; age scale is logarithmic.

records a gap at the position of the major ($T_e = 15$ Ma) exposure-age peak of the LL chondrites.

Effects of Pre-Irradiation

Multistage irradiation histories were documented for some meteorites. Nishiizumi et al (1980) pointed out that pre-irradiation effects are more easily revealed in cases of short exposure ages, and that pre-irradiation might account for observed shifts in P_{21} production rates, when calibrated with ^{26}Al activities. Wetherill (1980) carried out an investigation of the statistical expectation of complex cosmic-ray exposure of meteorites as a result of collisional fragmentation of larger objects. His predictions suggest that multiple exposures should be a common phenomenon (>25%) and that this conclusion is insensitive to variations in fragmentation parameters or of size distributions of bodies ≥100 m in diameter. The pre-irradiation records in such cases imply a more shielded early irradiation stage and as a consequence the T_e data used in this study represent only lower limits of the total time of cosmic-ray exposure. The expected effects of unrecognized pre-irradiation histories is an erosion of the T_e-clusters, interpreted to represent major collisional events. If such pre-exposures are common, then the exposure age clusters might actually be more pronounced than reported in this work. Graf & Marti (1989b, 1992) discuss the systematics of short (<4 Ma) exposure ages of H-chondrites and conclude that many of the meteorites with low $(^{22}Ne/^{21}Ne)_c$ ratios (<1.08) are likely to have experienced complex exposure histories.

DYNAMICAL CONSIDERATIONS

Orbits of Ordinary Chondrites

The orbits of three ordinary chondrites were astrometrically determined (Ceplecha 1961, McCroskey et al 1971, Halliday et al 1981). Other methods also were used to derive orbital information. Simonenko (1975) calculated orbital elements from visually determined radiants of observed ordinary chondrite falls. Meteorites (most of them assumed to be ordinary chondrites) were identified by different methods among the Prairie Network fireballs and MORP fireballs and their orbital elements were determined (Wetherill & ReVelle 1981; Halliday et al 1983, 1989). All meteoroid orbital data reveal perihelia strongly concentrated near the Earth's orbit and semi-major axes from less than 1 AU to greater than 3 AU (Wetherill 1985). This observation is confirmed by the relative frequency of ordinary chondrites falling between noon and midnight (p.m. falls/total falls). The time of fall parameter has long been recognized to be a sensitive indicator for proposed orbits (Wood 1961, Wetherill 1969). This is in particular true for the clustering of the perihelia of ordinary chondrites near 1 AU. If the entry velocity of a meteoroid is small (<12 km s^{-1}) it will be strongly deflected by the Earth's gravity. At high velocities (>17 km s^{-1}), the radial and perpendicular velocity components are sufficiently high to result in an approach almost perpendicular to the Earth's motion. In both cases the flux of meteorites will be similarly distributed for the a.m. and p.m. hemispheres. In the intermediate velocity range the perihelion or aphelion must be close to 1 AU. Meteoroids with perihelia ~1 AU will predominantly be p.m. falls and those with aphelia ~1 AU will predominantly be a.m. falls. Graf & Marti (1991, 1992) compiled the following average fall parameters (p.m. falls/total falls) from the *Catalogue of Meteorites* (Graham et al 1985 and updates): H-chondrites, 0.60; L-chondrites, 0.69; LL-chondrites, 0.68. In this compilation, only falls between 6 a.m. and 6 p.m. are considered because the statistics for night-time falls are probably biased. All three groups show a clear predominance of p.m. falls.

Orbital Clues to the H5-Parent Body

We have already discussed the differences in the T_e histograms for the petrographic subtypes of H-chondrites. A calculation of their respective fall parameters should reveal any orbital differences among the subgroups. The following fall parameters are obtained from the *Catalogue of Meteorites* (Graham et al 1985 and updates): H6, 0.673; H5, 0.483; H4, 0.667. Values for types H3, 4, and 6 are consistent with L- and LL-chondrites, with ~2/3 p.m. falls, but the H5 record is distinctly different and actually shows a slight a.m. excess. An analysis of the H5 systematics by Graf & Marti (1991a) revealed that the low value of the fall parameter is almost entirely due to H5-chondrites in the ~7 Ma exposure-age peak. According to these authors, the ~7 Ma exposure-age peak consists of two components: an H5 component dominated by a.m. falls and an H4 component of predominantly p.m. falls, with a peak-to-peak separation of only ~0.5 Ma. Independent of the probability argument for a mutual collisional event, the low p.m. fraction indicates an orbital evolution of the H5 parent before the ~7 Ma collision.

Solar Heating Effects

Perihelia of meteoroid falls with an a.m. predominance are expected to be smaller than those of falls with a p.m. excess. For example, Wetherill (1969) calculated a p.m.-fraction of 0.42 for meteoroids originating on the Apollo asteroid 1959LM. The perihelion distance of 1959LM is 0.7 AU and for such close approaches to the sun, losses of gas due to solar heating are possible. Gas losses during the ~7 Ma interval since the breakup of the H5-parent would result in preferential losses of $^{3}He_c$ relative to $^{21}Ne_c$, both of which are only produced during cosmic-ray exposure. The possibility that $^{4}He_{r}$ loss in some H-chondrites is due to solar heating was suggested earlier (Hintenberger et al 1966, Wänke 1966). Graf & Marti (1992) discuss the correlation of the $(^{3}He/^{21}Ne)_{c}$ ratio versus the $(^{22}Ne/^{21}Ne)_{c}$ shielding parameter (Eberhardt et al 1966) and observe an offset between H4 p.m. falls and H5 a.m. falls with $T_{e} \sim 7$ Ma. Furthermore, $^{4}He_{r}$ concentrations in these H5-chondrites are systematically lower by $\sim 20\%$ if compared to those of the H4 group. Both observations indicate that these H5-chondrites on average lost about 20% of their $^{3}He_{c}$ and $^{4}He_{r}$ due to solar heating effects during their ~ 7 Ma cosmic-ray exposure interval. These solar heating effects support the above model of the evolution of the H5 and H4 parent objects. However, there is no difficulty in reconciling the evidence with an ultimate origin in the same H-asteroid, except that multiple fragmentation and differential orbital evolution is required.

OBSERVATIONS ON THE REGOLITH HISTORY

The distribution of solar gas bearing meteorites was discussed by Crabb & Schultz (1981). Because of the short stopping range of the solar wind plasma (<0.1 μm), these gases must have been acquired on the regolith surface of their parent bodies. Crabb and Schultz note a good correlation between the T_{e} histograms of gas-rich ($\sim 15\%$) and the entire class of H-chondrites. On the other hand, very few L-chondrites (~ 7) contain solar-type gases and none of the shocked or degassed L-chondrites (e.g. those of $T_{e} \sim 28$ Ma or 5 Ma). These observations support an origin of L-chondrites from the interior of a parent body that was broken up in the asteroid belt. Fragments thereof were later injected into Earth-crossing orbits. It is interesting to note that most solar-gas bearing L-chondrites have relatively old K-Ar ages, except one (RKP 80207).

Many chondrites contain materials from more than one petrographic type and from different classes of meteorites. These chondrites provide information on parent-body regolith histories. Binns (1967) studied the collection in the British Museum and found evidence for regolith processes in $\sim 20\%$ of the ordinary chondrites. Lipschutz et al (1989) reviewed the information on foreign materials in ordinary chondrites and noted that the largest abundance of these xenoliths are reported to be carbonaceous chondritic, but that $\sim 36\%$ of these appear to be related to ordinary chondrites. Some of the xenoliths had lost varying amounts of the radiogenic gases ^{4}He and ^{40}Ar, but the evidence of survival of exotic clasts indicates low relative velocities of these projectiles when they impacted the regoliths of chondrite parent bodies (Wilkening 1977, Wasson & Wetherill

1979), consistent with expectations from collisional debris in the asteroid belt. The inclusion of regolith breccias into the T_e histograms raises questions regarding the possibility of pre-irradiation effects during their regolith history. However, the evidence that most gas-rich meteorites are observed in the assigned major collisional events indicates that the pre-irradiation effects are difficult to discern. Nevertheless, we excluded a few meteorites that show evidence for regolith processes and mixtures of host materials (e.g. Djermaja, Fayetteville, Pantar, Weston).

CONCLUSIONS

Observational constraints on the delivery of asteroids and meteoroids include exposure ages, orbital distributions, and data on composition. We have discussed the exposure age histograms separately for H-, L-, and LL-chondrites. None of these histograms is consistent with a continuous delivery of asteroidal material to the Earth, as the observed T_e histograms clearly disagree with expected exponential distributions (shown in Figure 7) for a variety of orbital lifetimes. We conclude that T_e histograms are dominated by stochastic events and that the continuous supply of asteroidal material can account only for a minor background of the T_e histograms. It is difficult, therefore, to infer collisional or dynamical half-lives for chondrite populations from measured T_e histograms. We have attempted to identify major collisional events among the major classes of ordinary chondrites in order to estimate the frequency of stochastic events. The H-chondrites recorded two major events, ~7 Ma and ~33 Ma ago; among the L-chondrites, events at ~28 Ma and ~40 Ma ago were identified, and the LL-chondrites suffered a collision ~15 Ma ago.

Orbital maturity in the inner solar system is documented by the

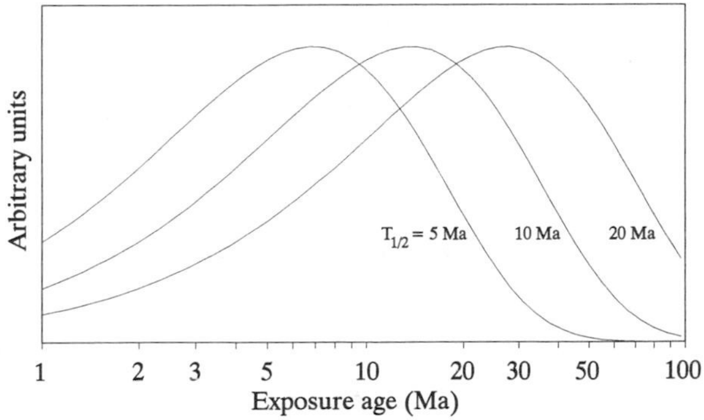

Figure 7 Calculated histograms for different assumed orbital lifetimes ($T_{1/2}$ = 5, 10, 20 Ma) which correspond to distributions expected for a continuous supply of meteoroids.

p.m./total fall ratio among observed meteorite falls. All chondrite classes exhibit a uniform ratio of 2/3, except type H5 chondrites which reveal a ≤ 0.5 p.m./total fall ratio. This shift in the time of fall statistics suggests a strongly evolved orbit for the H5-parent (perihelion distance <1 AU) at the time of collision, ~ 7 Ma ago. The H4- (and other H) chondrites also recorded a collisional event at about the same time and, therefore sequential breakups and differential orbital evolution of the H4 and H5 are indicated, but both may be offsprings of a larger grandparent.

The evidence for stochastic fragmentation and delivery of meteoroids to the Earth may suggest that the present predominance of chondritic samples among the meteorite falls may represent an excursion from a more typical population distribution, such as implied by spectral data among main-belt asteroids (e.g. Gaffey et al 1989). However, the T_e records show evidence for rather recent collisional events on three distinct chondritic parent bodies (H, L, LL), and these statistics hardly represent a coincidence. On the other hand, there is no clear evidence for very recent (<4 Ma) collisional activity and it is unlikely that fall statistics of chondrites evolved significantly on a 1 Ma time scale as proposed by Lipschutz & Samuels (1991). The T_e histograms and the recorded collisional events suggest that chondritic parents are probably members of the population of near-Earth asteroids (NEAs). The H5-parent is expected among those with clearly evolved orbital elements and the evidence for solar heating effects among H5-chondrites lends support for this interpretation.

The records on the thermal history of parent asteroids retained in the form of radiogenic components (^4He_r, $^{40}Ar_r$) and ^{39}Ar-^{40}Ar systematics reveal that in some cases major gas losses did occur during the last 0.1–1 Ga. The evidence for relatively minor gas losses during collisional events ~ 7, ~ 15, and ~ 40 Ma ago, suggest that heat pulses in stochastic events recorded in T_e histograms are minor, but much more frequent than major heating events. In order to understand the irradiation records of gas-rich breccias the development of asteroidal regoliths is required. On the other hand, it is difficult to assess the records in favor or against models of gravitational reassembly of disrupted asteroids (Crabb & Schultz 1981, Taylor et al 1987). Since most collisional events involved more than one petrographic type, we can argue that the parents were either large (few km) or that they were megabreccias with a signature of more than one, but not necessarily all petrographic types.

Acknowledgments

We thank O. Eugster and R. Wieler for their unpublished data. We profited from discussions with K. Nishiizumi, P. Pellas, and L. Schultz who also provided a data disk of the noble gas compilation by Schultz and Kruse. This work was supported by NASA grant NAG-9-41 and by Calspace CS-13-90.

Literature Cited

Anders, E. 1964. Origin, age, and composition of meteorites. *Space Sci. Rev.* 3: 583–714

Arnold, J. R. 1965. The origin of meteorites as small bodies. II. The model. *Astrophys. J.* 141: 1536–47

Begemann, F., Zhaohui, L., Schmitt-Strecker, S., Weber, H. W., Zitu, X. 1985. Noble gases and the history of Jilin meteorite. *Earth Planet. Sci. Lett.* 72: 247–62

Bhandari, N., Lal, D., Rajan, R. S., Arnold, J. R., Marti, K., Moore, C. B. 1980. Atmospheric ablation in meteorites: A study based on cosmic ray tracks and Neon isotopes. *Nucl. Tracks* 4: 213–62

Bhattacharya, S. K., Goswami, J. N., Lal, D. 1973. Semiempirical rates of formation of cosmic ray tracks in spherical objects exposed in space: preatmospheric and post-atmospheric depth profiles. *J. Geophys. Res.* 78: 8356–63

Binns, R. A. 1967. Structure and evolution of non-carbonaceous meteorites. *Earth Planet. Sci. Lett.* 2: 23–28

Bogard, D. D., Hirsch, W. C. 1980. ^{40}Ar/^{39}Ar dating, Ar diffusion properties, and cooling rate determinations of severely shocked chondrites. *Geochim. Cosmochim. Acta* 44: 1667–82

Bogard, D. D., Husain, L., Wright, R. J. 1976. ^{40}Ar-^{39}Ar dating of collisional events in chondrite parent bodies. *J. Geophys. Res.* 81: 5664–78

Bronshten, V. A. 1983. *Physics of Meteoric Phenomena*. Dordrecht: Reidel

Ceplecha, Z. 1961. Multiple fall of Pribram meteorites photographed. *Bull. Astron. Inst. Czech.* 12: 21–47

Clayton, R. N., Mayeda, T. K., Goswami, J. N., Olsen, E. J. 1991. Oxygen isotope studies of ordinary chondrites. *Geochim. Cosmochim. Acta* 55: 2317–37

Crabb, J., Schultz, L. 1981. Cosmic-ray exposure ages of the ordinary chondrites and their significance for parent body stratigraphy. *Geochim. Cosmochim. Acta* 45: 2151–60

Eberhardt, P., Eugster, O., Geiss, J., Marti, K. 1966. Rare gas measurements in 30 stone meteorites. *Z. Naturforsch.* 21a: 414–26

Eberhardt, P., Eugster, O., Marti, K. 1965. A redetermination of the isotopic composition of atmospheric Neon. *Z. Naturforsch.* 20a: 623–24

Eberhardt, P., Hess, D. C. 1960. Helium in stone meteorites. *Astrophys. J.* 131: 38–46

Eugster, O. 1988. Cosmic-ray production rates for ^{3}He, ^{21}Ne, ^{38}Ar, ^{38}Kr, and ^{126}Xe in chondrites based on ^{81}Kr-Kr exposure ages. *Geochim. Cosmochim. Acta* 52: 1649–62

Fleischer, R. L., Price, P. B., Walker, R. M., Maurette, M., Morgan, G. 1967. Tracks of heavy primary cosmic rays in meteorites. *J. Geophys. Res.* 72: 355

Fredricksson, K., Fredriksson, B. J. 1990. Bulk composition of chondrites. *Meteoritics* 25: 365–66 (Abstr.)

Gaffey, M. J., Bell, J. F., Cruikshank, D. P. 1989. Reflectance spectroscopy and asteroid surface mineralogy. In *Asteroids II*, ed. P. B. Binzel, T. Gehrels, M. S. Matthews, pp. 98–127. Tucson: Univ. Arizona Press

Gault, D. E. 1969. On cosmic-ray exposure ages of stone meteorites. *Meteoritics* 4: 177 (Abstr.)

Geiss, J., Oeschger, H., Signer, P. 1960. Radiation ages of chondrites. *Z. Naturforsch.* 15A: 1016–17

Gopalan, K., Wetherill, G. W. 1971. Rb-Sr studies on black hypersthene chondrites: Effects of shock and reheating. *J. Geophys. Res.* 76: 8484–92

Göpel, C., Manhes, G., Allegre, C. J. 1991. Constraints on the time of accretion and thermal evolution of chondrite parent bodies by precise U-Pb dating of phosphates. *Meteoritics*. In press (Abstr.)

Graf, T., Marti, K. 1989a. Exposure ages of H-chondrites and parent body structure. *Lunar Planet. Sci.* XX: 353–54 (Abstr.)

Graf, T., Marti, K. 1989b. H-Chondrites: Exposure ages and thermal events on parent bodies. *Meteoritics* 24: 271–72 (Abstr.)

Graf, T., Marti, K. 1990a. Exposure ages and collisional history of L-chondrites parent bodies. *Lunar Planet. Sci.* XXI: 431–32 (Abstr.)

Graf, T., Marti, K. 1990b. Collisional histories of chondrites. *Meteoritics* 25: 368 (Abstr.)

Graf, T., Marti, K. 1991a. The H5 parent collision 7 Ma ago. *Lunar Planet. Sci.* XXII: 473–74 (Abstr.)

Graf, T., Marti, K. 1991b. Exposure ages of LL- and L/LL-chondrites and implications for parent body histories. *Meteoritics*. In press

Graf, T., Marti, K. 1992. In preparation

Graf, T., Singer, P., Wieler, R., Herpers, U., Sarafin, R. et al. 1990a. Cosmogenic nuclides and nuclear tracks in the chondrite Knyahinya. *Geochim. Cosmochim. Acta* 54: 2511–20

Graf, T., Baur, H., Signer, P. 1990b. A model for the production of cosmogenic nuclides in chondrites. *Geochim. Cosmochim. Acta* 54: 2521–34

Graham, A. L., Bevan, A. W. R., Hutchison, R. 1985. *Catalogue of Meteorites*. British Mus./Univ. Arizona Press. 4th ed.

Halliday, I., Blackwell, A. T., Griffin, A. A. 1989. The typical meteorite event, based on photographic records of 44 fireballs. *Meteoritics* 24: 65–72

Halliday, I., Griffin, A. A., Blackwell, A. T. 1981. The Innisfree meteorite fall: A photographic analysis of fragmentation, dynamics, and luminosity. *Meteoritics* 16: 153–70

Halliday, I., Griffin, A. A., Blackwell, A. T. 1983. Meteorite orbits from observations by camera networks. In *Highlights of Astronomy*, ed. R. M. West, pp. 399–404. Dordrecht: Reidel

Heymann, D. 1967. On the origin of hypersthene chondrites: Ages and shock effects of black chondrites. *Geochim. Cosmochim. Acta* 29: 1203–8

Hintenberger, H., Schultz, L., Wänke, H. 1966. Messung der Diffusionverluste von radiogenen und spallogenen Edelgasen in Steinmeteoriten II. *Z. Naturforsch.* 21A: 1147–59

Jokipii, J. R., Marti, K. 1986. Temporal variations of cosmic rays over a variety of time scales. In *The Galaxy and the Solar System*, ed. R. Smoluchowski, J. N. Bahcall, M. S. Matthews, pp. 116–28. Tucson: Univ. Arizona Press

Kirsten, T., Krankowsky, D., Zähringer, J. 1963. Edelgas- und Kalium-Bestimmungen an einer grösseren Zahl von Steinmeteoriten. *Geochim. Cosmochim. Acta* 27: 13–42

Kohman, T. P., Bender, M. L. 1967. Nuclide production by cosmic rays in meteorites and the moon. In *High Energy Nuclear Reactions in Astrophysics*, ed. B. S. P. Shen, pp. 169–244. New York: Benjamin

Lavielle, B., Marti, K., Regnier, S. 1985. Ages d'exposition des meteorites de fer: histoires multiples et variations d'intensite du rayonnement cosmique. In *Isotopic Ratios in the Solar System*, pp. 15–20. Centre Natl. d'Etudes Spatiales

Lipschutz, M. E., Gaffey, M. J., Pellas, P. 1989. Meteoritic parent bodies: nature, number, size and relation to present-day asteroids. In *Asteroids II*, ed. R. P. Binzel, T. Gehrels, M. S. Matthews, pp. 740–77. Tucson: Univ. Arizona Press

Lipschutz, M. E., Samuels, S. M. 1991. Ordinary chondrites: Multivariate statistical analysis of trace element contents. *Geochim. Cosmochim. Acta* 55: 19–34

Marti, K. 1967. Mass-spectrometric detection of cosmic-ray-produced ^{81}Kr in meteorites and the possibility of Kr-Kr dating. *Phys. Rev. Lett.* 18: 264–66

McCrosky, R. E., Posen, A., Schwartz, G., Shao, C. Y. 1971. Lost City meteorite—its recovery and a comparison with other fireballs. *J. Geophys. Res.* 76: 4090–4108

Michel, R., Peiffer, F., Theis, S., Begemann, F., Weber, H. et al. 1989. Production of stable and radioactive nuclides in thick stony targets ($r = 15$ and 25 cm) isotropically irradiated with 600 MeV protons. Simulation of the production of cosmogenic nuclides in meteorites. *Nucl. Instrum. Meth.* B42: 76–100

Moniot, R. K., Kruse, T. H., Tuniz, C. 1983. The ^{21}Ne production rate in stony meteorites estimated from ^{10}Be and other radionuclides. *Geochim. Cosmochim. Acta* 47: 1887–95

Nakamura, N., Fujiwara, T., Nohda, S. 1990. Young asteroid melting event indicated by Rb-Sr dating of the Point of Rocks meteorite. *Nature* 345: 51–52

Nishiizumi, K., Regnier, S., Marti, K. 1980. Cosmic ray exposure ages of chondrites, pre-irradition and constancy of cosmic ray flux in the past. *Earth Planet. Sci. Lett.* 50: 156–70

Nyquist, L., Funk, H., Schultz, L., Signer, P. 1973. He, Ne and Ar in chondritic Ni-Fe as irradiation hardness sensors. *Geochim. Cosmochim. Acta* 37: 1655–85

Pellas, P., Fieni, C. 1988. Thermal histories of ordinary chondrite parent asteroids. *Lunar Planet. Sci.* XII: 825–27 (Abstr.)

Pellas, P., Storzer, D. 1981. ^{244}Pu fission track thermometry and its application to stony meteorites. *Proc. R. Soc. London Ser. A* 374: 253–70

Reedy, R. C., Arnold, J. R., Lal, D. 1983. Cosmic-ray record in solar system matter. *Science* 219: 14

Reedy, R. C., Marti, K. 1991. Solar cosmic ray fluxes during the last ten million years. In *The Sun in Time*, ed. C. P. Sonett, M. S. Giampapa, M. S. Matthews, pp. 260–87. Tucson: Univ. Arizona Press

Schultz, L. 1976. On the cosmic-ray exposure

age distribution in ordinary chondrites. *Meteoritics* 11: 359–60 (Abstr.)

Schultz, L., Kruse, H. 1989. Helium, Neon, and Argon in meteorites. A data compilation. *Meteoritics* 24: 155–72

Schultz, L., Weber, H. W., Begemann, F. 1990. Cosmic ray exposure ages of H3- and H4-chondrites. *17th Int. Conf. Cosmochronology, Isotope Geol.* 90 (Abstr.)

Schultz, L., Weber, H. W., Begemann, F. 1991. Noble gases in H-chondrites and potential differences between Antarctic and non-Antarctic meteorites. *Geochim. Cosmochim. Acta* 55: 59–66

Signer, P., Nier, A. O. 1960. The distribution of cosmic-ray-produced rare gases in iron meteorites. *J. Geophys. Res.* 65: 2947–64

Simonenko, A. N. 1975. Orbital elements of 455 meteorites. Atlas 67. Moscow: Nauka

Tanenbaum, A. S. 1967. Clustering of the cosmic ray ages of stone meteorites. *Earth Planet. Sci. Lett.* 2: 33–35

Taylor, G. J., Maggionre, P., Scott, E. R. D., Rubin, A. E., Keil, K. 1987. Original structures and fragmentation and reassembly histories of asteroids: Evidence from meteorites. *Icarus* 69: 1–13

Turner, G. 1969. Thermal histories of meteorites by the ^{39}Ar-^{40}Ar method. In *Meteorite Research*, ed. P. M. Millman, pp. 407–17. Dordrecht-Holland: Reidel

Turner, G. 1988. Dating of secondary events. In *Meteorites and the Early Solar System*, ed. J. F. Kerridge, M. S. Matthews, pp. 276–88. Tucson: Univ. Arizona Press

Turner, G., Enright, M. C., Cadogan, P. H. 1978. The early history of chondrite parent bodies inferred from ^{40}Ar-^{39}Ar ages. *Proc. Lunar Planet. Sci. Conf.* 9: 989–1025

Vogt, S., Herzog, G. F., Reedy, R. C. 1990. Cosmogenic nuclides in extraterrestrial materials. *Rev. Geophys.* 28: 253–75

Voshage, H. 1967. Bestrahlungsalter und Herkunft der Eisenmeteorite. *Z. Naturforsch.* 22a: 477–506

Wänke, H. 1966. Der Mond als Mutterkörper der Bronzit-Chondrite. *Z. Naturforsch.* 21a: 93–110

Wänke, H. 1968. Radiogenic and cosmic ray exposure ages of meteorites, their orbits and parent bodies. In *The Origin and Distribution of the Elements*, ed. L. H. Ahrens, pp. 411–21. Oxford, New York: Pergamon

Wasson, J. T., Wang, S. 1991. The histories of ordinary chondrite parent bodies: U, Th-He age distributions. *Meteoritics* 26: 161–67

Wasson, J. T., Wetherill, G. W. 1979. Dynamical, chemical and isotopic evidence regarding the formation locations of asteroids and meteorites. In *Asteroids*, ed. T. Gehrels, pp. 926–74. Tucson: Univ. Arizona Press

Wetherill, G. W. 1969. Relationships between orbits and sources of chondritic meteorites. In *Meteorite Research*, ed. P. M. Millman, pp. 573–89. Dordrecht-Holland: Reidel

Wetherill, G. W. 1980. Multiple Cosmic-Ray Exposure Ages of Meteorites. *Meteoritics* 15: 386–87 (Abstr.)

Wetherill, G. W. 1985. Asteroidal source of ordinary chondrites. *Meteoritics* 20: 1–22

Wetherill, G. W., ReVelle, D. O. 1981. Which fireballs are meteorites? A study of the Prarie Network photographic meteor data. *Icarus* 48: 308–28

Wilkening, L. L. 1977. Meteorites in meteorites: Evidence for mixing among the asteroids. In *Comets, Asteroids, Meteorites*, ed. A. H. Delsemme, pp. 389–95. Toledo, Ohio: Univ. Toledo Press

Wisdom, J. 1983. Chaotic behavior and the origin of the 3/1 Kirkwood gap. *Icarus* 56: 51–74

Wood, J. A. 1961. Stony meteorite orbits. *Mon. Not. R. Astron. Soc.* 122: 79–88

Wood, J. A. 1967. Chondrites: their metallic minerals, thermal histories, and parent planets. *Icarus* 6: 1–49

Wood, J. A., Pellas, P. 1991. What heated the parent meteorite planets? In *The Sun in Time*, ed. C. P. Sonett, M. S. Giampapa, M. S. Matthews, pp. 740–60. Tucson: Univ. Arizona Press

Zanda, B., Malinie, B., Audouze, J. 1989. Propagation of high-energy particles inside solid matter: cosmic-ray-induced spallation in iron meteorites. *Earth Planet. Sci. Lett.* 94: 171–88

Annu. Rev. Earth Planet. Sci. 1992. 20:245–87

CADMIUM AND $\delta^{13}C$ PALEOCHEMICAL OCEAN DISTRIBUTIONS DURING THE STAGE 2 GLACIAL MAXIMUM

Edward A. Boyle

Department of Earth, Atmospheric and Planetary Sciences, Massachusetts Institute of Technology, Cambridge, Massachusetts 02139

KEY WORDS: paleoceanography, ocean circulation, paleochemistry, carbon isotopes

INTRODUCTION

Changes in heat and salt transport by ocean currents have been implicated in ice age climate cycles (Weyl 1968, Newell 1974, Broecker et al 1990b). In addition to the direct physical effects of the ocean on climate, the ocean carbon system controls atmospheric CO_2, and the CO_2 fluctuations documented in ice core bubbles (Barnola et al 1987, Neftel et al 1988) are thought to be a significant factor in climate oscillations (Hansen et al 1984). Dramatic changes in oceanic circulation patterns and carbon system chemistry have occurred during the Late Quaternary glacial/interglacial global climate cycles (Boyle & Keigwin 1982, Curry & Lohmann 1983, Shackleton & Pisias 1984, Boyle & Keigwin 1987, Duplessy et al 1988, Curry et al 1988, Broecker et al 1990a). These changes in the ocean system may be significant driving forces in the progression of ice age CO_2 cycles (Toggweiler & Sarmiento 1985, Ennever & McElroy 1985, Wenk & Siegenthaler 1985, Boyle 1988a,b, Broecker & Peng 1990). The oceanic distributions of $\delta^{13}C$, Cd, and other carbon-system-linked properties are controlled by the interaction between biological uptake and decomposition with the general circulation of the ocean. Because records of past oceanic Cd and $\delta^{13}C$ are preserved in shells of planktonic and benthic foraminifera

0084–6597/92/0515–0245$02.00

from oceanic sediment sequences, a significant body of paleochemical data on these two tracers has been obtained during the past decade. Here, published data for $\delta^{13}C$ are merged with an extensive new data set for deepwater Cd, and a comparison is made of the glacial maximum distributions of these tracers. These data provide a picture of the three-dimensional characteristics of global paleochemical distributions during glaciation.

METHODS

A global chronostratigraphic framework is essential to this task. Although ^{14}C dating is the fundamental chronometer of the late Pleistocene [as recalibrated by coral $^{230}Th/U$ dating (Bard et al 1990)], this slow and expensive technique can be bypassed in low-resolution studies of deep-ocean cores by foraminiferal $\delta^{18}O$ measurements. Given the dominant role of ice volume in controlling $\delta^{18}O$ of seawater, the glacial maximum can be identified by maximum $\delta^{18}O$ values (Shackleton 1967, Shackleton & Opdyke 1973, Chappell & Shackleton 1986). Hence the strategy employed here (and in other studies of this type) is to identify the glacial maximum from $\delta^{18}O$ of benthic or planktonic fossils, and then use benthic foraminiferal $\delta^{13}C$ and Cd analyses from these intervals to characterize bottom water chemistry. It should be emphasized that the concern of this paper is with global-scale changes between Pleistocene glacial maximum conditions and Holocene interglacial conditions. For this reason, high frequency variations within these periods or during deglaciation will not be considered, and chronologies adopted here should be considered as having a coarse resolution, with an accuracy and resolution no better than several thousand years.

Several deepwater $\delta^{13}C$ studies of the glacial maximum ocean have been published. The most geographically diverse data sets were compiled by Duplessy et al (1988) and Curry et al (1988) and supplemented for the Indian Ocean by Kallel et al (1988). This $\delta^{13}C$ data is reported as the permil (‰) deviation from the isotope ratio in the PDB standard. The goal of this study is to obtain a comparably diverse data set for cadmium to compare with the carbon isotope results. This goal required maximization of geographic coverage within a reasonable time, so it was not possible to analyze detailed downcore records in most cases. Sixty-five sediment cores with published $\delta^{18}O$ stratigraphies (locations shown in Figure 1 and listed in Table 4) were sampled and analyzed for benthic foraminiferal Cd as close to the oxygen isotope maximum as possible. In most cases, additional samples were taken a few tens of cm above and below the apparent glacial maximum (all taken within oxygen isotope stage 2); hence samples range from a few thousands of years before the glacial maximum until the earliest

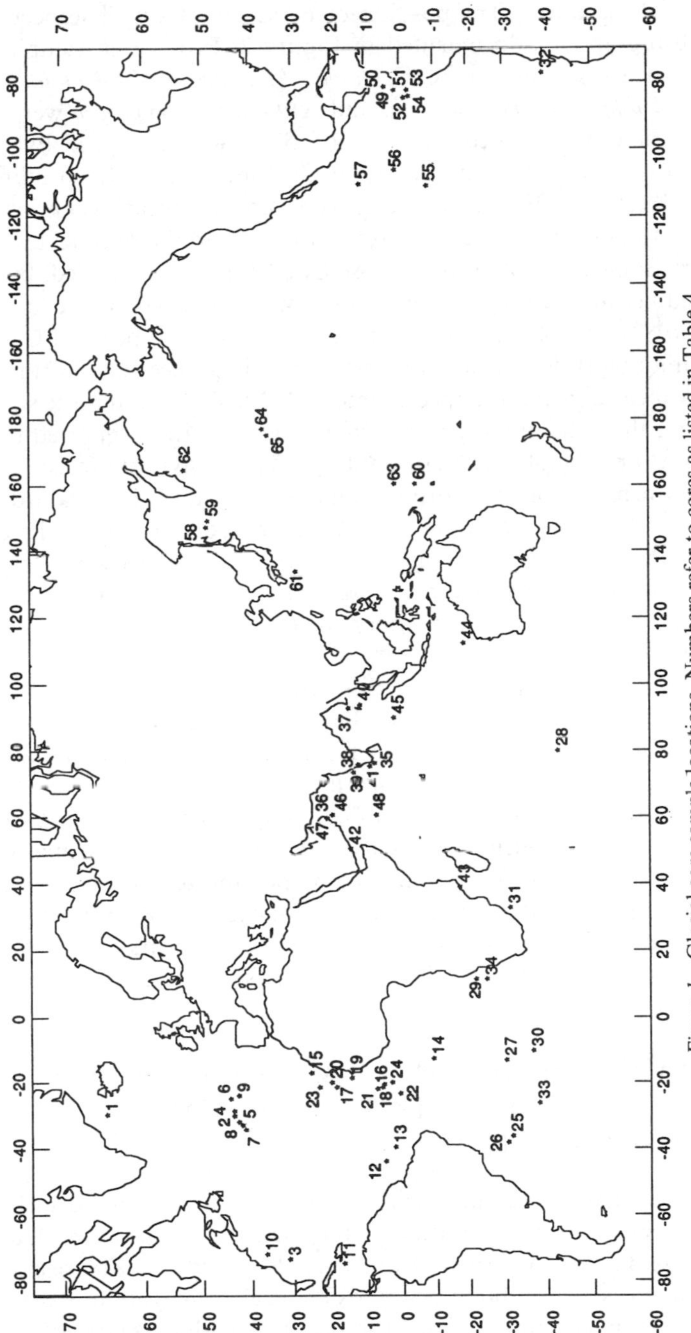

Figure 1 Glacial core sample locations. Numbers refer to cores as listed in Table 4.

stages of deglaciation. The position of these samples with respect to the $\delta^{18}O$ stratigraphy is documented in Figure 2. From each sample, individuals of *Uvigerina* spp., *Cibicidoides kullenbergi, Cibicidoides wuellerstorfi, Cibicidoides pachyderma*, and *Nutallides umbonifera* were handpicked wherever they occurred. All of these were analyzed (whenever sufficient specimens were available) for Cd/Ca by the method outlined by Boyle & Keigwin (1985/6). Replicates were run wherever possible; a typical core has 10 pooled analyses of various species and their replicates within the core. Some cores other than those listed here were sampled, but were excluded because of excessive levels of manganese carbonate overgrowth (Boyle 1983). Measurements are reported as μmol Cd per mol Ca in the shell (μmol/mol). However, as noted below, it is necessary to transform this raw measurement into an estimate of Cd in the seawater the shell was grown in; this estimate will be referred to as CdW and reported in units of nmol Cd per kg of seawater (nmol/kg). The analytical precision of the graphite furnace atomic absorption analyses of Cd/Ca in standard ARS-1 is about 3.5% over a period of a few years (comparable to the period over which most of these samples were run); replicate analyses of separate picks from the same sample rarely agree as well as this, signifying some variability between individual specimens (Boyle 1988a, Boyle & Keigwin 1985/6).

There are potential pitfalls in this strategy. 1. Significant deepwater variability may occur on time scales that are short relative to the resolution of typical deep-ocean cores and to fundamental limits for global ocean $\delta^{18}O$ synchroneity. At very high sedimentation rate sites, where higher-frequency variability may be recorded, only three spot samples within stage 2 may not represent the true average values. 2. Because many cores have been heavily sampled, it is not always possible to make paleochemical analyses at the glacial maximum because of the lack of core sample. For these reasons, one should not over-interpret fine points of this data set; only large-scale signals confirmed by nearby samples and sites should be considered seriously.

FORAMINIFERAL CALIBRATION: EVIDENCE FOR DEPTH-DEPENDENT CADMIUM INCORPORATION

Previous studies have established the utility of $\delta^{13}C$ and Cd in benthic foraminifera as tracers of bottom water chemical composition (Duplessy & Shackleton 1984, Graham et al 1981, Boyle 1988a). The response of foraminiferal shells to bottom-water chemical composition is calibrated by comparison of core-top fossil foraminifera with estimates of bottom-

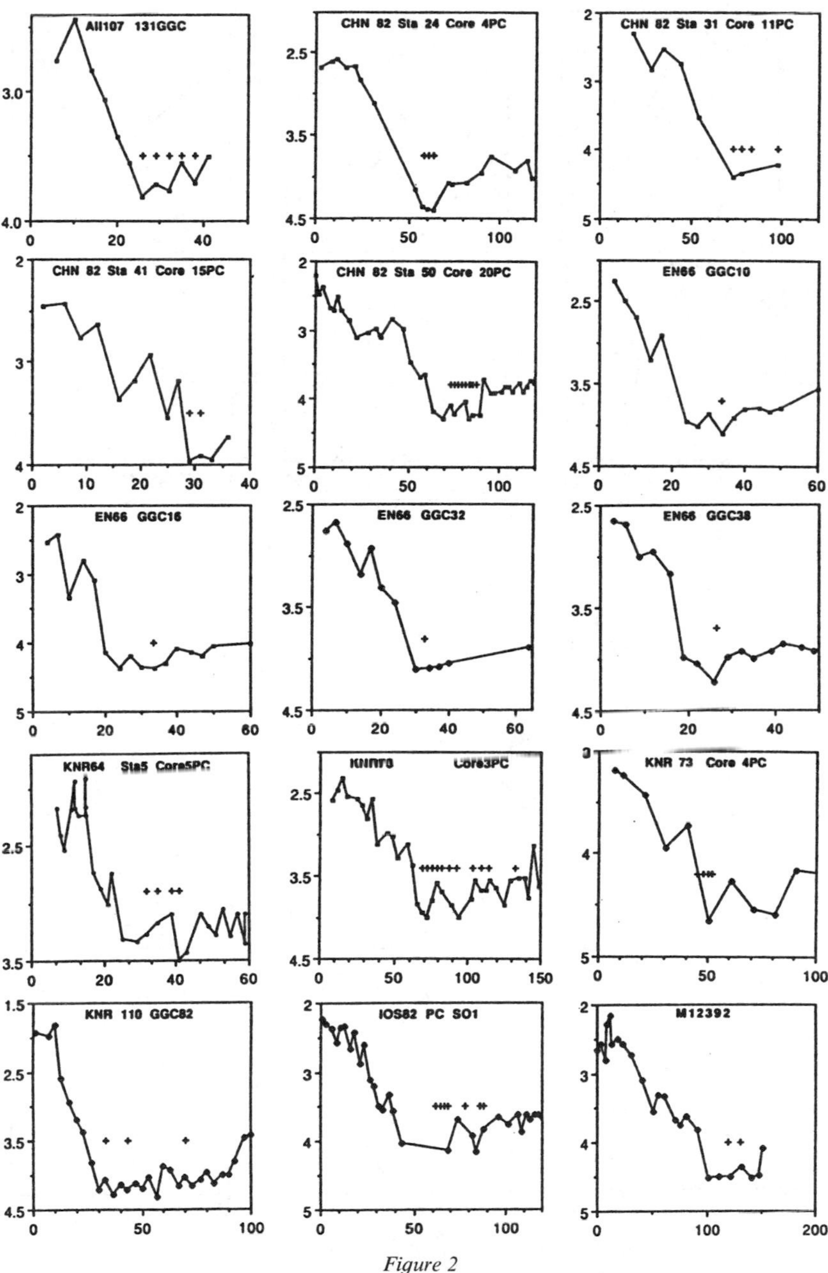
All107 131GGC
CHN 82 Sta 24 Core 4PC
CHN 82 Sta 31 Core 11PC
CHN 82 Sta 41 Core 15PC
CHN 82 Sta 50 Core 20PC
EN66 GGC10
EN66 GGC16
EN66 GGC32
EN66 GGC38
KNR64 Sta5 Core5PC
Core3PC
KNR 73 Core 4PC
KNR 110 GGC82
IOS82 PC SO1
M12392

Figure 2

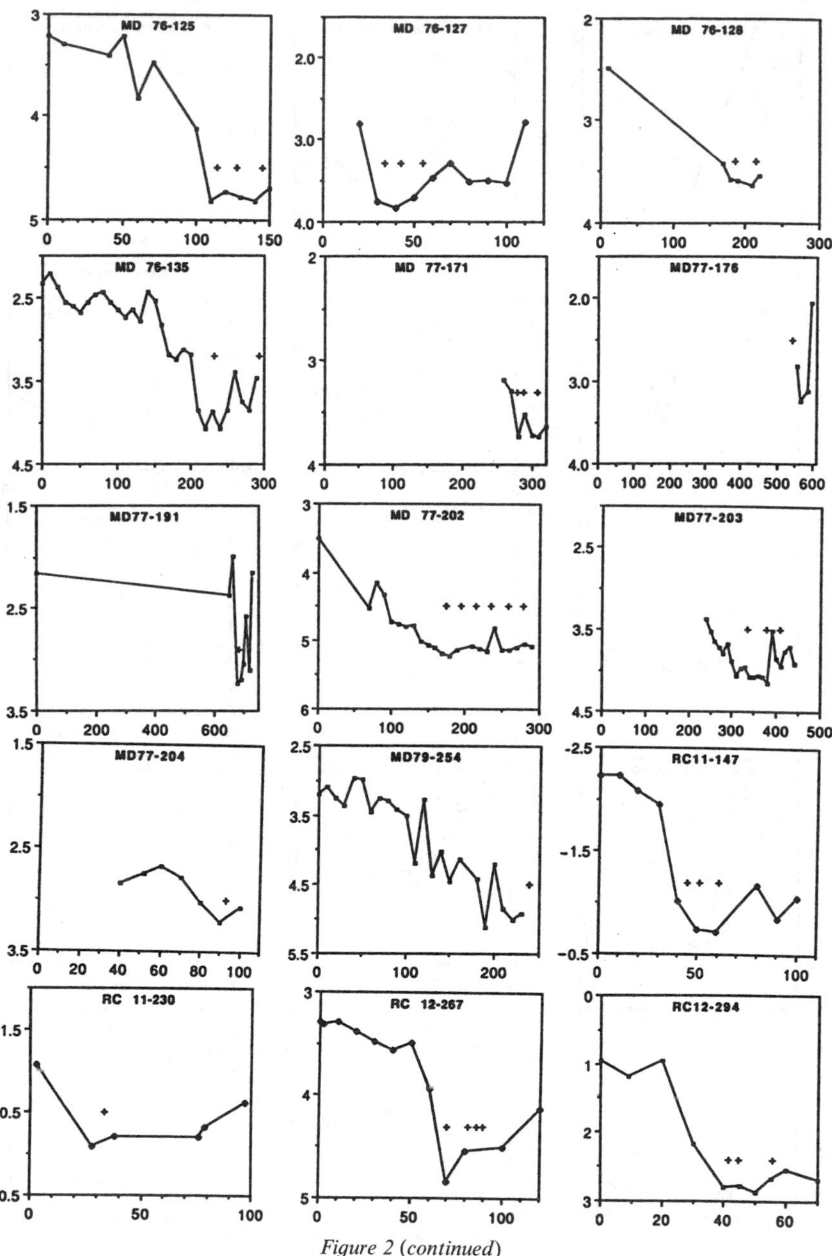

Figure 2 (continued)

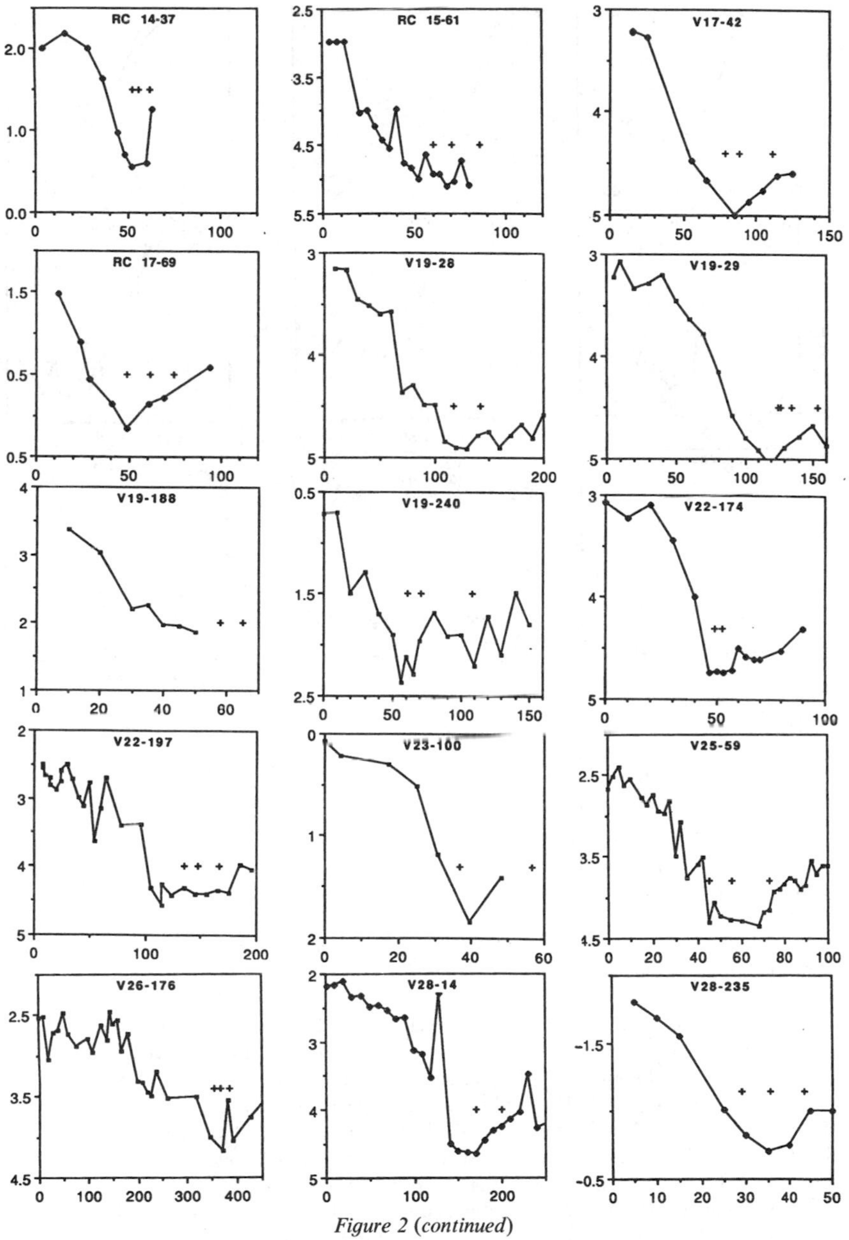

Figure 2 (continued)

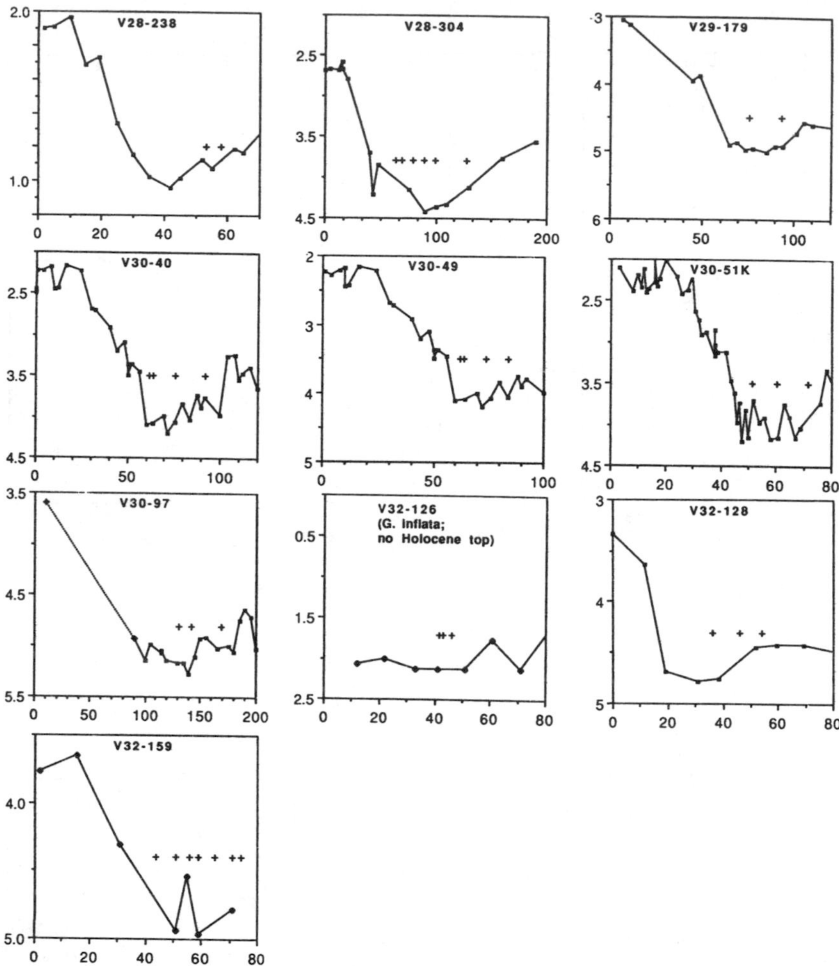

Figure 2 Oxygen isotope documentation of cores. Cores are listed in alphabetical and numerical order. Isotope data from Bloemendal et al (1988), Boyle & Keigwin (1982), Boyle (1984), Boyle & Keigwin (1985/6), Boyle & Keigwin (1987), Curry & Lohmann (1983), Curry et al (1988), Duplessy et al (1988), Kallel et al (1988), Keigwin (1987), Keigwin & Jones (1989), Luz & Shackleton (1975), Labeyrie et al (1987), Mix et al (1986), Mix & Fairbanks (1985), Mix & Ruddiman (1985), Mix et al (1990), Morley & Shackleton (1984), Oppo & Fairbanks (1987), Oppo & Fairbanks (1990), Oppo et al (1990), Parkin & Shackleton (1973), Peng et al (1977), Prell et al (1980), Romine (1982), Shackleton & Opdyke (1973), Shackleton (1977a), Shackleton (1977b), Shackleton et al (1983), Streeter & Shackleton (1979), Thompson et al (1979), Thompson (1981). Core V24-1 age estimate based on bulk C^{14} date given by Balsam (1981). Note that additional documentation of $\delta^{18}O$ and glacial Cd samples is shown in Figures 11 and 12. The $\delta^{18}O$ data is from many benthic and planktonic species which were not always indicated in the original publications; hence the significance should be restricted to the position of Cd samples relative to the $\delta^{18}O$ maximum.

water composition. For $\delta^{13}C$, genera (and perhaps species within genera) have different offsets from bottom-water composition (Duplessy & Shackleton 1984, Graham et al 1981). The offset between *Uvigerina* and other genera appears to depend on the organic carbon content and/or productivity of the overlying waters (Zahn et al 1986). For this reason, current practice is to use data from *Cibicidoides* wherever possible; $\delta^{13}C$ data from this genus are used exclusively here.

Calibration of deepwater calcitic species for Cd is simpler with respect to genus: Major calcitic genera show similar relationships to bottom-water chemistry with no obvious offsets between species [Figure 3; data from Boyle (1988a) supplemented by data in Table 1]. The average value of the empirical distribution coefficient

$$D = \frac{(Cd/Ca)_{foram}}{(Cd/Ca)_{water}}$$

for calcitic species from core tops deeper than 3000 m depth (below the

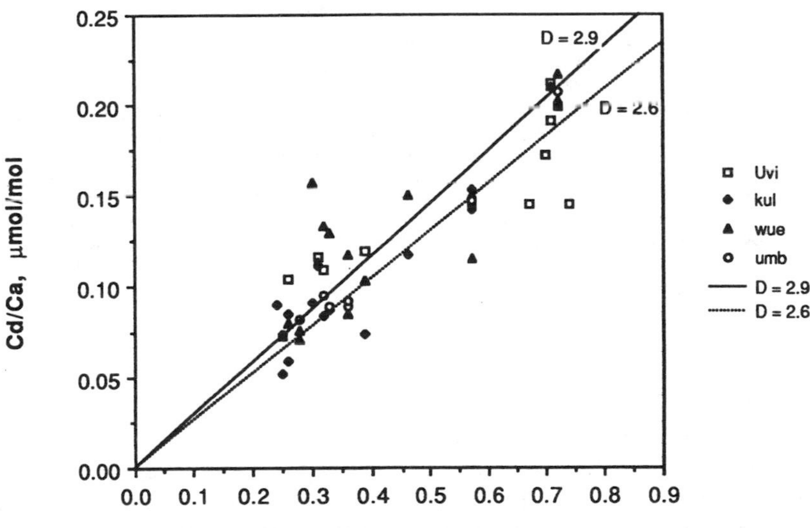

Figure 3 Cd (foram) vs Cd (estimated for bottom water) for samples > 3000 m. The average value $D = 2.9$ for all calcitic species is shown as the solid line; a dotted line is shown for a distribution coefficient that is 10% lower. See text for method of estimation of bottom water Cd.

Table 1 New core-top Cd data[a]

Core	Sample depth (cm)	Lat	Long	Depth (m)	Estimated P (μmol/kg)	Cd/Ca (μmol/mol) Uvi	kul	wue	umb
All107-104GC	1-3	-31°16'	-35°55'	1389	2.22	.053	.084	.060	
MD77-203	+ 9-11	13°05'	73°11'	1712	2.81			.199	
MW88-4	+ 0-1	12°39'	117°56'	3530	2.87		.105	.123	
MW88-7	+ 0-1	11°00'	118°29'	990	2.72	.102	.101		
MW88-8	+ 0-1	11°35'	118°43'	1305	2.90	.135	.122	.114	
MW88-9	+ 0-1	11°38'	118°38'	1465	2.90		.115	.139	
MW88-10	+ 0-1	11°43'	118°31'	1605	2.97		.163	.211	
MW88-11	+ 0-1	11°53'	118°20'	2165	2.87			.272	
MW88-12	+ 0-1	11°56'	118°13'	2495	2.87			.249	
MW88-13	+ 0-1	10°36'	118°17'	990	2.72	.106	.114	.083	
MW88-14	+ 0-1	7°42'	116°25'	825	2.14	.093	.079		
MW88-19	+ 0-1	7°49'	116°10'	1835	2.93		.154	.178	
V19-27	+ 5-7	- 0°28'	-82°04'	1373	2.74	.135	.105	.112	
V19-30	+ 4-7	- 3°23'	-82°21'	3071	2.51	*.140	.157		
V21-29	3-6	0°57'	-89°21'	712	2.69	.151	.177		
V21-30	9-11	- 1°13'	-89°41'	617	2.72	.124	.387?	.137	
V28-304	6-12	28°32'	134°08'	2942	2.60			*.140	
V30-51K	+ 0-3	19°52'	-19°55'	3409	1.44	.109	*.084	.133	.095
V32-159	+ 5-8	48°40'	147°24'	1235	2.91	.135			
V32-161	+ 15-17	48°17'	149°04'	1600	2.84	.169			

[a] See Table 1 of Boyle (1988a) for core storage and depth convention information for All, RC, and V cores. MW cores stored at the University of South Carolina by R. Thunell. Latitude and longitude are in degrees and minutes, with south and west negative. Cd/Ca is given for *Uvigerina* spp. (Uvi), *Cibicidoides kullenbergi* or *Cibicidoides pachyderma* (kul), *Cibicidoides wuellerstorfi* (wue), and *Nutallides umbonifera* (umb). This table includes all data collected between July 21, 1988 and July 18, 1991. See Boyle (1988c) for further core-top data used in this paper.

[+] Cores for which the author is aware of published isotopic stratigraphies indicating more than 15 cm of Holocene sediment.

* Duplicate or triplicate analyses of these foraminifera samples.

ocean surface) is $D = 2.9$. (Note: in order to calculate D from the data in Figure 3, one must normalize Cd in the seawater to the calcium concentration in seawater, 0.01 mol/kg.) However, this value, based on averaging the distribution coefficient for each core top, seems to overestimate foraminifera for some cores at high Cd; the true distribution coefficient might be as much as 10% lower. The value of $D = 2.9$ will be used here for deepwater fossils, but data interpretation requires consideration of the possibility that D may be somewhat lower.

Calibration of core tops from shallower depths [see Table 1 for new data in addition to that of Boyle (1988a)] demonstrates that the distribution coefficient is a function of the depth of the core below the sea surface (Figure 4). Although thermodynamic considerations allow equilibrium distribution coefficients to be a function of temperature and pressure, available thermodynamic data are not adequate to define exactly what equilibrium variability is to be expected. Hence it is difficult to determine

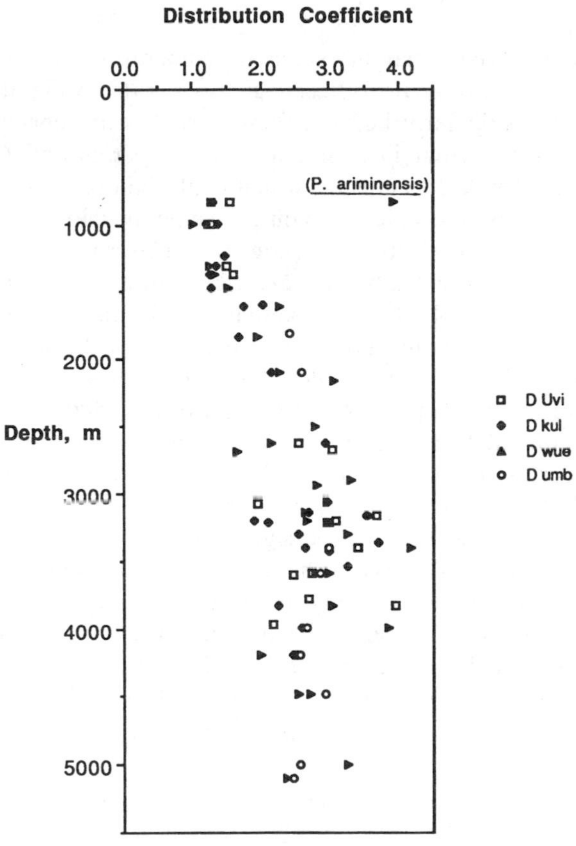

Figure 4 Distribution coefficient vs depth for major calcitic benthic foraminifera. Note that *C. pachyderma*, which occurs mainly in cores <2500 m depth, is plotted with the same symbol as *C. kullenbergi*, which mainly occurs in cores >2500 m depth. Also note that the one high *D* at shallow depths was for the only (unreplicated) analysis of *P. ariminensis*.

Table 2 Comparison of distribution coefficients at different depths and temperatures

Core	Species	Depth (m)	Temperature (deg C)	Distribution coefficient
CHN82 Sta41 Core 15PG	wue	2155	3.65	2.04
KNR64 Sta5 Core 5PG	wue	3047	3.70	3.14
CHN82 Sta21 Core 1PC	Uvi	830	10.8	1.28
MW88-14GGC	Uvi	825	6.8	1.56

whether the observed variability is due to temperature, pressure, or some nonthermodynamic biological effect (e.g. growth rate, availability and type of food, etc), particularly because these variables are correlated in the ocean. It is of particular interest that both *Uvigerina* and *Cibicidioides* show the same depth-dependence, which might be taken as an argument against an organism-specific mechanism. Some anecdotal evidence suggests that depth or other depth-correlated parameters may be more important than temperature (Table 2). Near-bottom temperatures at cores CHN82-41-15 and KNR64-5-5 are similar, but the distribution coefficient in the shallow core is significantly lower than that for the deeper core. The distribution coefficient for the shallow warm core CHN82-1 (core-top data for this core is listed in Boyle 1988a) is close to that for the shallow but colder core MW88-14GGC. However, the scatter inherent in such calibrations casts doubt on such a cursory examination, and we cannot be confident that temperature has been ruled out as a significant parameter. Even if depth proves to be more significantly correlated with D than temperature, it is also unclear whether pressure or some other depth-dependent parameter is of primary importance.

For the purposes of this paper, Cd distribution coefficients in the glacial ocean are estimated from a simple empirical D (water-depth) relation derived from this core-top data set (Table 3). It would be interesting to know if foraminiferal $\delta^{13}C$ also displays some sort of depth-dependence

Table 3 Empirical foraminiferal Cd distribution coefficients used in this work

Depth (m)	Distribution coefficient
<1150	1.3
1150–3000	$1.3 + (\text{depth} - 1150)\ (1.6/1850)$
>3000	2.9

in its response. Core-top benthic foraminiferal $\delta^{13}C$ from shallow cores has not been specifically tested for depth dependency relative to bottom-water chemistry. Examination of the core-top data set of Duplessy & Shackleton (1984) does not reveal an obvious depth-dependence; however, this data set did not have many cores at shallower depths. Slowey (1990) notes that cores from 400–1500 m depths on the Bahama Banks region are about 0.2–0.3‰ heavier than bottom waters, and most Holocene samples from core V19-27 (Mix et al 1990; Figure 12) are about 0.4–0.5‰ heavier than modern bottom waters. However, at the moment it is not possible to confidently establish a depth-dependence correction for $\delta^{13}C$, therefore the ensuing discussion will follow the current practice of carbon isotope workers and assume that *Cibicidoides* $\delta^{13}C$ is directly comparable to bottom-water $\delta^{13}C$ with no depth-dependence.

RESULTS

Glacial maximum data for Cd are listed by species in Tables 6–11 (see Appendix). Individual species are compared to averages for *Uvigerina* spp. and *C. wuellerstorfi* in Figure 5. For most cores, inter-species differences are not systematic, with most samples scattering about the 1:1 line [note this lack of inter-specific differences was seen previously in downcore work reported in Boyle (1988a)]. For the following cores, the between-species differences are substantial, so their average values should be considered cautiously: V19-29, V19-240, V25-59, MD77-176, MD79-254, RC14-37, RC17 69. Table 4 lists average results for all analyses from each core, along with published $\delta^{13}C$ data, estimates for modern bottom-water properties at the site, and core-top $\delta^{13}C$ and Cd data (where available). Modern bottom-water Cd and $\delta^{13}C$ were estimated at the stations by 1. estimating bottom P at the same depth from the nearest GEOSECS station, 2. using a "global" average Cd-P relationship (Boyle 1988a) to estimate bottom-water P, and 3. using a regression of $\delta^{13}C$ vs P at core-top locations where Duplessy et al (1984) estimated bottom-water $\delta^{13}C$ from the data published by Kroopnick (1985). It should be understood that real deviations from the regression lines exist, and that within a given region somewhat better estimates might be obtained by using local rather than global regressions. In the context of the global-scale view in this paper, the differences between the local and global relationships are small and do not justify complicating the procedure. In deriving Cd/Ca averages, analyses which deviated on the high side of the data set were excluded for reasons of potential contamination; these rejected analyses (flagged by question marks in Tables 6–11) comprised only 7% of the total data set, and the ensuing discussion would not be affected significantly were these data retained.

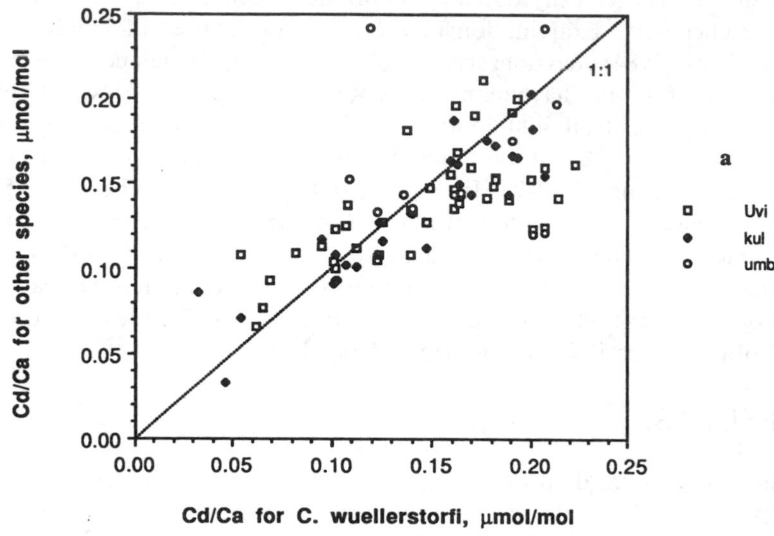

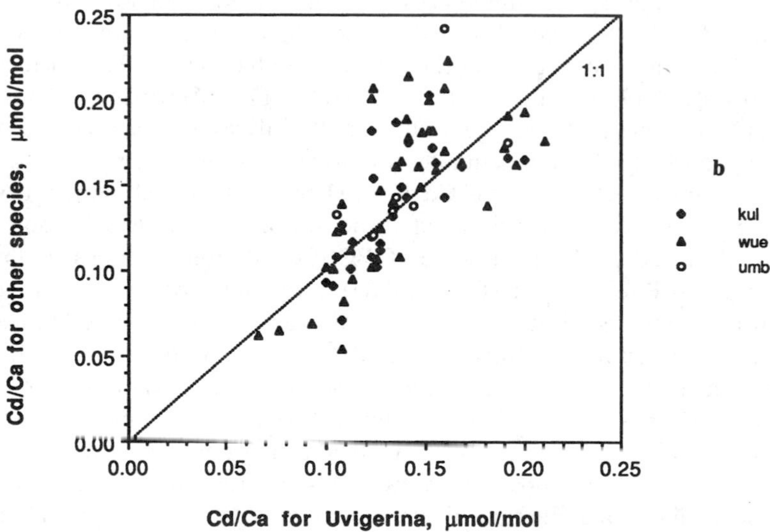

Figure 5 Interspecies comparison plots. Note that *C. pachyderma* data is plotted with the same symbol as *C. kullenbergi*. (*a*) Other species vs *C. wuellerstorfi*. (*b*) Other species vs *Uvigerina* spp.

Table 4 Data summary table

#	Core	Water depth (m)	Lat. (dec. deg.)	Long. (dec. deg.)	Sample depths (cm)	Modern Cd (nmol/kg)	Core top Cd/Ca (μmol/mol)	Core top CdW (nmol/kg)	Glacial Cd/Ca (μmol/mol	Std. dev.	*n*	Glacial CdW (nmol/kg)	Glac.-Int. CdW (nmol/kg)	Modern C^{13}	Core-top C^{13} permil PDB	Glacial C^{13} .....
	Northern Atlantic:															
1	V28-14	*1855	64.78	-29.57	170 - 200	0.22			0.046	0.027	4	0.24	0.02	1.09	1.12	1.13
2	CHN82 Sta41 Core15PC	2153	43.37	-28.23	27 - 31	0.25	0.051	0.24	0.041	0.008	2	0.19	-0.06	1.02	0.63	1.19
3	V24-1	3012	30.50	-73.50	161 - 198	0.25			0.073	0.013	9	0.25	0.00	1.00		
4	CHN82 Sta50 Core20PC	3070	43.50	-29.87	73 - 87	0.25			0.101	0.014	18	0.35	0.10	1.02	1.23	0.66
5	CHN82 Sta31 Core11PC	3209	42.38	-31.80	73 - 98	0.25	0.052	0.18	0.100	0.008	4	0.34	0.09	1.01	1.18	0.76
6	V29-179	3331	44.00	-24.53	76 - 93	0.24	0.090	0.31	0.108	0.016	9	0.37	0.13	1.03	1.10	0.63
7	V30-97	3371	41.00	-33.93	131 - 169	0.25			0.097	0.011	8	0.34	0.09	1.01		
8	CHN82 Sta24 Core4PC	3427	41.72	-32.85	58 - 64	0.25	0.074	0.26	0.111	0.005	7	0.38	0.13	1.01	1.11	0.48
9	IOS82 PC SO1	3540	42.38	-23.52	62 - 88	0.26	0.085	0.29	0.137	0.015	14	0.47	0.21	0.97	0.76	0.44
10	V26-176	3942	36.00	-72.00	355 - 385	0.25			0.094	0.010	3	0.32	0.07	1.00		0.18
	Tropical Atlantic:															
11 w	KNR64 Sta5 Core5PG	*1800	16.52	-74.80	35 - 39	0.25			0.039	0.010	4	0.21	-0.04	0.86	0.89	1.31
12 w	KNR110 82GGC	2816	4.34	-43.49	8 - 71	0.27			0.088	0.010	3	0.32	0.05	0.93	1.11	0.21
13 w	V25-59	3824	1.37	-39.48	46 - 73	0.26	0.075	0.26	0.168	0.042	13	0.58	0.32	0.96	1.03	0.07
14 c	V22-174	2630	-10.07	-12.82	49 - 53	0.35	0.089	0.34	0.093	0.017	5	0.36	0.01	0.77	0.80	0.72
15 e	M12392	2573	25.17	-16.83	120 - 131	0.28			0.077	0.014	3	0.30	0.02	0.90	0.96	0.36
16 e	EN66 38GGC	2931	4.92	-20.50	27 - 27	0.31	0.127	0.45	0.203	0.029	4	0.71	0.40	0.84	0.98	0.69
17 e	V30-49	3093	18.43	-21.08	62 - 84	0.31			0.164	0.027	22	0.57	0.26	0.84		0.21
18 e	EN66 16GGC	3152	5.47	-21.13	34 - 34	0.32	0.107	0.37	0.210	0.025	5	0.72	0.40	0.83	0.94	0.43
19 e	V22-197	3167	14.17	-18.58	135 - 167	0.31	0.113	0.39	0.112	0.014	13	0.39	0.08	0.84	0.49	0.17
20 e	V30-51K	3409	19.87	-19.92	52 - 72	0.32	0.101	0.35	0.136	0.011	14	0.47	0.15	0.83		
21 e	EN66 10GGC	3527	6.65	-21.90	34 - 34	0.32	0.121	0.42	0.198	0.033	7	0.68	0.36	0.82	0.82	0.19
22 e	V30-40	3706	0.20	-23.13	62 - 92	0.32			0.150	0.026	16	0.52	0.20	0.82		-0.11
23 e	V23-100	4579	22.68	-21.30	37 - 57	0.34			0.152	0.027	8	0.52	0.18	0.79		0.00
24 e	EN66 32GGC	5003	2.47	-19.73	33 - 33	0.36	0.100	0.34	0.138	0.026	3	0.48	0.12	0.76	0.86	-0.32
	Southern Ocean:															
25	AII107 65GGC	2795	-32.03	-36.19	34 - 43	0.29			0.120	0.018	6	0.44	0.15	0.88		
26	AII107 131GGC	2925	-30.88	-38.05	26 - 38	0.30			0.101	0.016	11	0.36	0.06	0.87	0.96	0.85
27	V19-240	3103	-30.58	-13.28	61 - 108	0.39			0.164	0.043	11	0.56	0.17	0.70		0.02
28	RC11-120	3135	-43.52	79.87	67 - 102	0.57	0.151	0.52	0.149	0.019	8	0.51	-0.06	0.38	0.46	-0.42
29	RC13-228	3204	-22.33	11.20	102 - 132	0.39	0.099	0.34	0.117	0.027	28	0.40	0.01	0.71	0.50	-0.31
30	RC12-294	3308	-37.27	-10.10	42 - 56	0.46	0.128	0.44	0.163	0.024	11	0.56	0.10	0.58	0.81	-0.23
31	RC17-69	3308	-31.50	32.60	49 - 75	0.45			0.193	0.041	7	0.67	0.22	0.59		
32	RC15-61	3771	-40.62	-77.20	60 - 86	0.64			0.161	0.020	11	0.55	-0.09	0.25		
33	RC12-267	4144	-38.68	-25.78	70 - 89	0.65	0.152	0.52	0.122	0.033	6	0.42	-0.23	0.24		
34	RC13-229	4191	-25.50	11.33	59 - 86	0.57	0.137	0.47	0.155	0.036	18	0.54	-0.03	0.37	0.39	-0.36

Table 4 *(continued)*

#	Core	Water depth (m)	Lat. (dec. deg.)	Long. (dec. deg.)	Sample depths (cm)	Modern Cd (nmol/kg)	Core top Cd/Ca (μmol/mol)	Core top CdW (nmol/kg)	Glacial Cd/Ca (μmol/mol	Std. dev.	n	Glacial CdW (nmol/kg)	Glac.-Int. CdW (nmol/kg)	Modern C^{13}	Core-top C^{13} permil PDB	Glacial C^{13} .....
	Indian Ocean (mainly northern):															
35	MD77-191	1254	7.30	76.43	684 - 684	0.83			0.064	0.028	3	0.46	-0.37	-0.08		0.24
36	MD77-204	1430	19.18	58.26	93 - 93	0.94			0.070	0.011	1	0.45	-0.49	-0.28		0.00
37	MD77-176	1537	14.50	93.10	542 - 542	0.83			0.078	0.021	3	0.48	-0.35	-0.09		
38	MD76-127	1610	12.08	75.90	35 - 55	0.87			0.112	0.013	19	0.66	-0.21	-0.16		-0.12
39	MD76-128	1712	13.08	73.18	186 - 265	0.86	0.199	1.11	0.098	0.012	13	0.55	-0.31	-0.15	0.09	-0.15
40	MD77-171	1760	11.47	94.08	279 - 308	0.80			0.098	0.016	4	0.53	-0.27	-0.04		
41	MD76-125	1878	8.35	75.20	114 - 145	0.78			0.110	0.027	13	0.57	-0.21	0.01	0.29	-0.08
42	MD76-135	1895	14.27	50.32	233 - 294	0.70			0.120	0.011	12	0.61	-0.09	0.15	-0.14	-0.07
43	MD79-254	1934	-18.00	39.00	239 - 239	0.63			0.124	0.022	2	0.62	-0.01	0.27		-0.02
44	RC11-147	1953	-19.03	112.45	45 - 61	0.70			0.131	0.022	16	0.66	-0.04	0.14		
45	RC14-37	2226	1.47	90.17	53 - 62	0.74			0.178	0.042	14	0.80	0.06	0.07		
46	MD77-202	2427	19.13	60.42	175 - 278	0.82			0.152	0.004	18	0.63	-0.19	-0.06	0.17	-0.39
47	MD77-203	2442	20.42	59.33	335 - 408	0.82			0.148	0.015	10	0.61	-0.21	-0.06		-0.38
48	V19-188	3356	6.87	60.67	47 - 65	0.71			0.158	0.014	5	0.54	-0.17	0.12		
	Eastern Tropical Pacific															
49	V19-27	1373	0.47	-82.07	99 - 133	0.84	0.117	0.78	0.123	0.017	15	0.83	-0.01	-0.10	0.40	-0.03
50	V17-42	1814	3.53	-81.18	78 - 112	0.82	0.200	1.07	0.181	0.031	12	0.96	0.14	-0.08		-0.24
51	V19-29	2673	-3.58	-83.93	125 - 154	0.74	0.227	0.87	0.171	0.023	9	0.65	-0.09	0.07	0.02	
52	V19-28	2720	-2.37	-84.65	118 - 142	0.76			0.188	0.027	8	0.71	-0.05	0.04	0.19	
53	V19-30	3071	-3.38	-82.35	127 - 158	0.74	0.145	0.50	0.186	0.032	22	0.64	-0.10	0.07	0.03	-0.43
54	TR163-31B	3210	-3.62	-83.97	80 - 100	0.71	0.211	0.73	0.198	0.025	31	0.68	-0.03	0.13	0.11	-0.23
55	RC11-230	3259	-8.80	-110.80	24 - 34	0.70			0.194	0.025	2	0.67	-0.03	0.15		-0.15
56	KNR73 3PC	3606	0.37	-106.18	69 - 132	0.70	0.172	0.59	0.187	0.043	27	0.64	-0.06	0.15	0.29	-0.27
57	KNR73 4PC	3681	10.85	-110.27	45 - 55	0.71	0.191	0.66	0.213	0.017	5	0.73	0.02	0.13	0.29	-0.21
	Western North Pacific:															
58	V32-159	1235	48.67	147.40	44 - 74	0.91	0.135	0.98	0.143	0.041	18	1.04	0.13	-0.23		
59	V32-161	1600	48.28	149.07	145 - 237	0.84	0.169	1.00	0.136	0.025	21	0.80	-0.04	-0.10		
60	V28-235	1746	-5.45	160.48	29 - 44	0.80			0.162	0.023	13	0.89	0.09	-0.03		0.16
61	V28-304	2942	28.53	134.13	64 - 128	0.78	0.140	0.49	0.137	0.006	8	0.48	-0.30	0.00	0.29	0.07
62	Rama 44	2980	53.01	164.65	111 - 126	0.77			0.145	0.014	4	0.50	-0.27	-0.18	0.17	-0.37
63	V28-238	3120	1.02	160.48	52 - 59	0.76			0.149		1	0.51	-0.25	0.04		-0.49
64	V32-128	3623	36.47	177.17	36 - 46	0.76	0.162	0.56	0.139	0.009	2	0.48	-0.28	0.04	0.31	-0.17
65	V32-126	3370	35.32	174.90	42 - 46	0.73			0.138	0.008	3	0.48	-0.25	0.09		-0.25

* Actual depth of core is 3047 m; nominal depth is sill depth of Caribbean overflow.

w: western basin; c: central; e: eastern basin.

Data tables from this paper will be made available on a 3.5″ floppy disk (Apple II, Macintosh, or MS-DOS format) upon request to the author.

Data from these cores are considered as composite vertical profiles for six groups corresponding to hydrographic regions (Table 4, Figures 7 and 8). Note that these regions are broad; for example, the term "Northwest Pacific" is attached to cores ranging from 5°S to 53°N in the western Pacific. Phosphorus profiles from representative GEOSECS stations within these regions (Bainbridge 1980, Craig et al 1982, Weiss et al 1982) are shown for comparison (Figure 6). By grouping cores from such broad regions, real horizontal variability will be seen as "scatter" within the vertical profiles. Glacial maximum $\delta^{13}C$ and CdW data in Figures 6 and 7 are compared to a reference line based on an estimate for $\delta^{13}C$ and Cd for the modern ocean for each region (the $\delta^{13}C$ line in these figures is

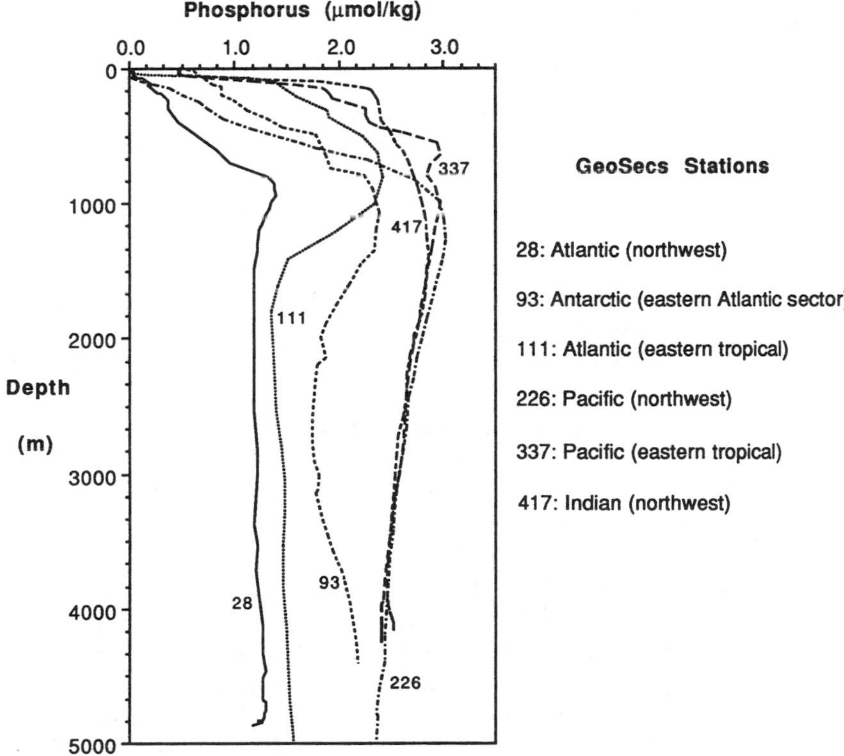

Figure 6 Geosecs station phosphorus profiles from regions considered in this discussion.

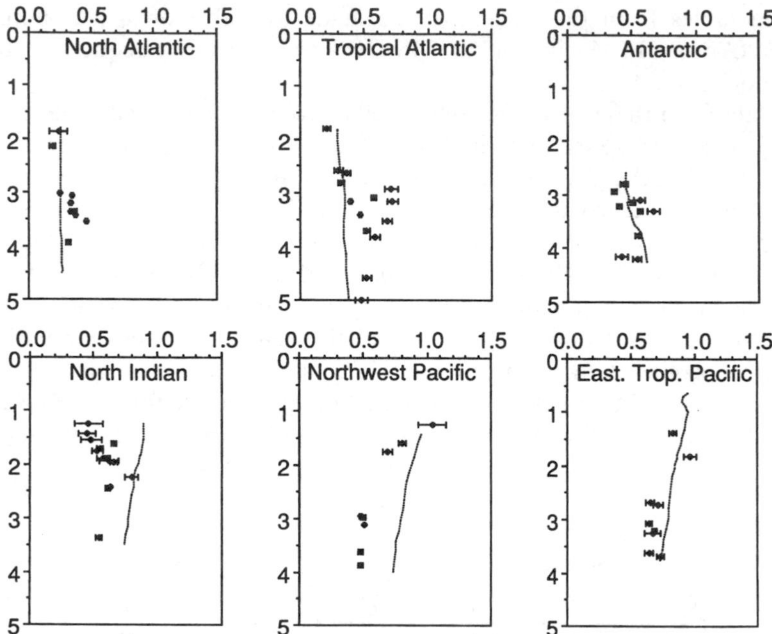

Figure 7 Glacial maximum CdW profiles. Dotted line signifies modern bottom water Cd as estimated by method outlined in text. Error bars are the one-sigma reduced standard deviations ($\sigma/\sqrt{n}$) of the averages reported in Table 4. Units: nmol Cd/kg.

shifted 0.3% below modern values to compensate for the shift in global average $\delta^{13}C$ during the last glacial maximum). Core-top data was not plotted to avoid complicating the figures, but where available, they are listed in Table 4. Note that in this paper, $\delta^{13}C$ data are plotted with a reversed scale so that the graphical sense of change for CdW and $\delta^{13}C$ are in the same direction, rather than being mirror images. In all figures, $\delta^{13}C$ and CdW are scaled so that changes in both tracers are visually comparable based on the modern correlation between tracers, and $\delta^{13}C$ and CdW scale ranges are kept fixed throughout the paper.

Discussion of these data will be divided into four sections: 1. Novel features of the new Cd data set, 2. global ocean averages for $\delta^{13}C$ and CdW, 3. aspects for which $\delta^{13}C$ and CdW evidence are concordant, and 4. major differences between CdW and $\delta^{13}C$ evidence. In the discussion, the phrase "the nutrient content" is used as a shorthand reference to the modern global-scale correlation between high P, Cd, and low $\delta^{13}C$, without intending to ignore significant real differences between these tracers.

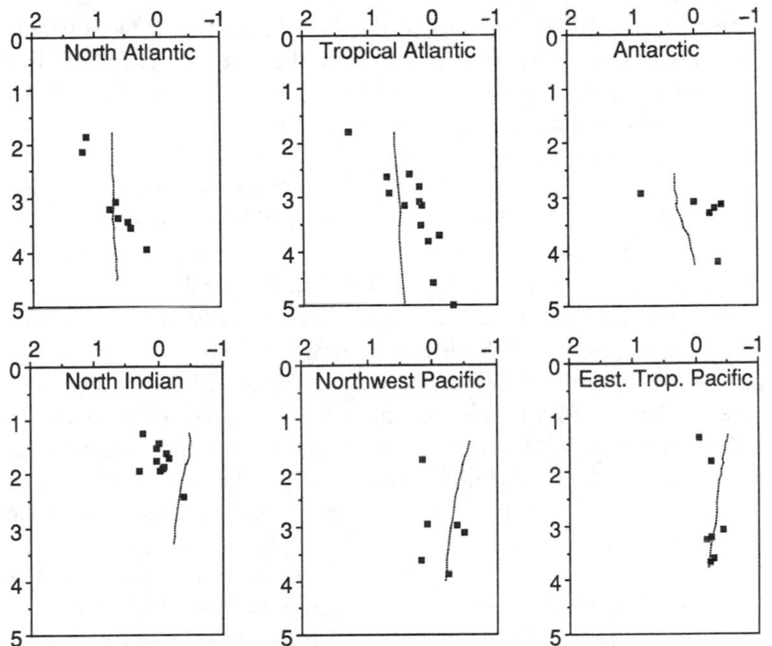

Figure 8 Glacial maximum $\delta^{13}C$ profiles. Dotted line signifies modern bottom water $\delta^{13}C$ as estimated by method outlined in text, shifted by -0.3% to compensate for the global average $\delta^{13}C$ shift. Note scale reversal.

NOVEL FEATURES IN OCEANIC CADMIUM EVIDENT FROM THE NEW DATA

Previously published data on glacial oceanic Cd were confined mainly to the North Atlantic and deep Eastern Tropical Pacific (Boyle & Keigwin 1982, Boyle & Keigwin 1985/6, Boyle & Keigwin 1987, Boyle 1988a). Although this data set contains some new data for these regions, it is largely compatible with results from previously published studies, namely, 1. the Cd content of Eastern Tropical Pacific deepwaters was perhaps slightly lower than in the modern ocean, 2. the Cd content of glacial deep North Atlantic waters was about 30% higher than in the modern ocean, and 3. the upper waters (1500–2500 m) of the glacial North Atlantic were Cd-depleted compared to deeper waters, and may be Cd-depleted even relative to those of the modern ocean.

For other geographic regions, some novel observations are possible from this new data set:

1. Glacial Antarctic deepwaters had about the same Cd levels as did the modern ocean.

2. Upper waters (1000–2000 m) of the Northern Indian Ocean had Cd levels lower by at least 40% compared to the modern ocean, and Indian Ocean deepwaters appear to also be as much as 30% lower in Cd. These data confirm the inference on nutrient status of upper waters of the Northern Indian Ocean derived from $\delta^{13}C$ studies (Kallel et al 1988).
3. Glacial age deepwaters (>1500 m) of the Northwest Pacific appear to have had about 30% lower Cd concentrations than in the modern ocean; in particular it appears that the deep Northwest Pacific had lower Cd concentrations than the Eastern Tropical Pacific (Figure 9). A bias in the estimate for the foraminiferal Cd distribution coefficient could exaggerate the CdW difference between the modern ocean and the glacial maximum fossil evidence seen in Figure 7 [if $D = 2.6$ for deep cores, rather than 2.9 as assumed here, then the glacial-interglacial (G-I) difference would be closer to 20%]. On the other hand, uncertainties in D do not apply to the difference between the two Pacific regions, because the same bias would apply to both regions. The weakest aspect of this observation of low glacial CdW in the Northwest Pacific is the lack of good core-top evidence from this region (see Table 4 and Figure 12); the two available deep core-top numbers from this region do not show such a large glacial reduction in foraminiferal CdW as is implied by comparison to the modern water column. The downcore record

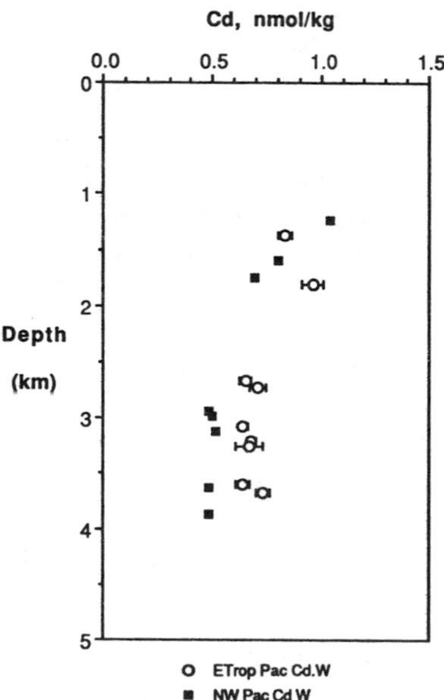

Figure 9 Comparison of Northwest Pacific and Eastern Tropical Pacific Glacial Maximum Cd profiles.

from core Rama 44 (Figure 12) does suggest some increase in Cd during deglaciation, but the core-top data is not available from that core. The lack of good core tops (and perhaps the lack of a shift in the few existing core tops) is largely due to the high degree of dissolution in the modern Pacific, so that core tops are heavily dissolved (except in regions of high carbonate productivity). It is likely that many sediment mixed-layer benthic foraminifera specimens are mixed upward from late glacial times; this may then account for low apparent core-top numbers from this region. Nonetheless, this lack of a good core-top record establishing a large reduction in bottom water CdW is a weak point of the present data set. Establishing a good continuous modern-through-glacial record of Cd in this region is worthy of high priority in future work. If this region were filled by a nutrient-depleted water mass, it would be highly significant and perhaps could be accounted for by the initiation of a new North Pacific deepwater source.

4. It appears that upper waters (<2500 m) of the Pacific were only slightly depleted in CdW relative to the modern ocean, if at all; if there is a depletion, it is larger in the Northwest Pacific than in the Eastern Tropical Pacific.

GLOBAL OCEAN AVERAGES FOR CARBON ISOTOPES AND CADMIUM

The global oceanic inventory and global average vertical distribution of $\delta^{13}C$ and CdW are estimated from these profiles after appropriate volumetric weighting. The ocean was first divided into three major basins (Atlantic, Pacific, and Indian), and the average volume in each basin was estimated for 1–2, 2–3, 3–4, and 4–5 km intervals (Table 5). The upper 1 km was ignored in these averages because there is little glacial age data from this interval. Each of the six profiles was assigned to represent fractions of each of these ocean basins. Average modern and glacial $\delta^{13}C$ and CdW values for each depth interval were estimated (by eye) from the profiles. Where no glacial data were available for a depth interval, the value from the nearest adjacent depth interval was used. This method of assignment of volume fractions and estimation of averages is somewhat arbitrary, but by setting out the calculations explicitly, it is easy to assess the sensitivity of resulting averages to these assumptions.

Modern oceanic average values for $\delta^{13}C$ and Cd as calculated in this fashion are 0.25‰ and 0.63 nmol/kg, respectively. These estimates may be contrasted with those from another method. The oceanic average phosphorus concentration has been estimated to be 2.2 μmoles/kg from detailed assessment of results of the GEOSECS expeditions (Takahashi et al 1981).

Table 5 Global ocean inventory estimate. Numbers in parentheses are for the glacial ocean

	N. Atlantic	Trop. Atlantic	Antarctic	N. Indian	NW Pacific	ET Pacific
	Volume fractions:					
1-2 km	0.0260	0.0260	0.0479	0.0438	0.0796	0.0796
2-3 km	0.0237	0.0237	0.0445	0.0417	0.0756	0.0756
3-4 km	0.0197	0.0197	0.0384	0.0373	0.0689	0.0689
4-5 km	0.0115	0.0115	0.0229	0.0229	0.0452	0.0452
Cadmium:						
	Cd Values for Depth Zones:					
1-2 km	0.20 (0.20)	0.25 (0.20)	0.30 (0.40)	0.80 (0.55)	0.84 (0.80)	0.83 (0.87)
2-3 km	0.25 (0.23)	0.30 (0.30)	0.30 (0.40)	0.75 (0.60)	0.82 (0.55)	0.75 (0.65)
3-4 km	0.26 (0.37)	0.32 (0.55)	0.50 (0.55)	0.70 (0.50)	0.75 (0.45)	0.70 (0.65)
4-5 km	0.28 (0.50)	0.35 (0.50)	0.62 (0.47)	0.70 (0.50)	0.75 (0.45)	0.70 (0.65)
	Mean Oceanic Cd:		0.63 (0.55)			
	Mean Cd Depth Profile:					
1-2 km			0.64 (0.62)			
2-3 km			0.62 (0.51)			
3-4 km			0.62 (0.53)			
4-5 km			0.65 (0.52)			
Carbon Isotopes:						
	C13 Values for Depth Zones:					
1-2 km	1.10 (1.60)	0.85 (1.60)	0.70 (-0.40)	-0.10 (0.20)	-0.15 (0.25)	-0.10 (-0.15)
2-3 km	1.00 (1.00)	0.85 (0.70)	0.70 (-0.40)	0.00 (-0.50)	-0.10 (-0.10)	0.05 (-0.30)
3-4 km	1.00 (0.60)	0.83 (0.20)	0.60 (-0.40)	0.10 (-0.40)	0.05 (-0.30)	0.15 (-0.30)
4-5 km	0.90 (0.20)	0.78 (0.00)	0.40 (-0.40)	0.10 (-0.40)	0.05 (-0.30)	0.15 (-0.30)
	Mean Oceanic 13:		0.25 (-0.05)			
	Mean C13 Depth Profile:					
1-2 km			0.20 (0.27)			
2-3 km			0.25 (-0.10)			
3-4 km			0.30 (-0.22)			
4-5 km			0.25 (-0.27)			

Given the correlations between $\delta^{13}C$, Cd, and P, this phosphorus average corresponds to average $\delta^{13}C$ and Cd values of 0.2‰ and 0.6 nmol/kg, respectively. Agreement of the averages estimated by these different methods is good, so that the weighted global averaging scheme employed in Table 5 is reasonable. The glacial distribution gives mean $\delta^{13}C$ and CdW estimates of 0.04‰ and 0.55 nmol/kg, respectively, for changes of -0.30‰ for $\delta^{13}C$ and -13% for Cd. This estimated $\delta^{13}C$ shift may be compared to the -0.32‰ estimate by Duplessy et al (1988) based on much the same data set. Boyle (1988a) previously estimated that there was little change in mean oceanic Cd, based on a few North Atlantic and Tropical Pacific cores. The apparent decrease in mean oceanic Cd calculated here is small and on the order of the likely accuracy of foraminiferal calibration (e.g. if the distribution coefficient for depths >3000 m is 2.6, rather than 2.9 as assumed here, then glacial oceanic CdW in this depth range would be underestimated by 10%), and hence the conclusion to be drawn from this estimate is that the oceanic Cd inventory does not change much between glacial and interglacial times.

Figure 10 illustrates vertically-resolved ocean average nutrient distributions for $\delta^{13}C$ and CdW. The two tracers imply quite different ocean-average vertical nutrient redistributions between glacial and interglacial times: $\delta^{13}C$ implies a net shift of nutrients from intermediate waters into deeper waters, but CdW implies no significant vertical shift. This difference is largely due to results from the glacial Pacific, where $\delta^{13}C$ data imply that intermediate depth nutrient levels are somewhat lower than deeper

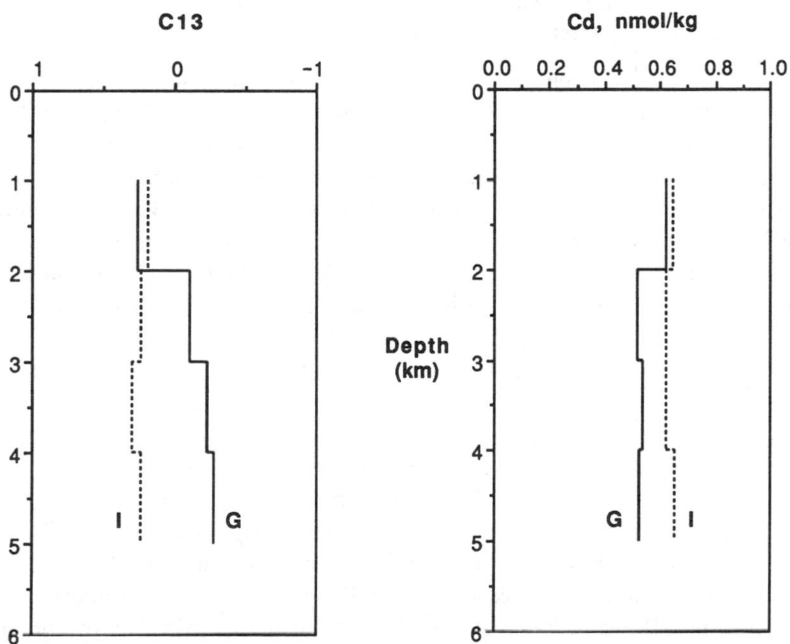

Figure 10 Global average $\delta^{13}C$ and Cd in glacial oceanic deepwaters. Dotted lines (I = Interglacial) refer to the modern ocean; solid lines to the glacial ocean (G = Glacial) as inferred from benthic foraminiferal evidence. $\delta^{13}C$ reported as ‰ relative to PDB standard.

waters, and CdW data imply that Pacific intermediate waters have higher nutrients than deeper waters. This difference has a major impact on the "vertical nutrient redistribution" model for glacial-interglacial atmospheric p_{CO_2} changes; the CdW evidence implies that although certain regions had significant vertical nutrient shifts, the ocean as a whole did not, hence minimizing the atmospheric p_{CO_2} response.

CONCORDANT CARBON ISOTOPE AND CADMIUM EVIDENCE

$\delta^{13}C$ and CdW data agree on the following features of glacial ocean paleochemical distributions:

1. Upper waters of the North Atlantic, Tropical Atlantic, and Northern Indian Oceans were significantly depleted in nutrients compared to the deepest waters in those basins during glacial times.
2. Deep waters of the North Atlantic were enriched in nutrients relative to today.
3. In the shallowest waters of the North and Tropical Atlantic, and in the intermediate depth waters of the Northern Indian Ocean, nutrients are depleted compared to modern values.

These features of oceanic chemical distributions in glacial times are supported by both tracers and can be considered well-established; apart from the new Northern Indian Ocean CdW evidence presented here, data supporting these glacial ocean paleochemical distributions have been noted in Boyle & Keigwin (1982), Curry & Lohmann (1983), Boyle & Keigwin (1987), Oppo & Fairbanks (1987), Zahn et al (1987), Duplessy et al (1988), Curry et al (1988), Kallel et al (1988), Oppo & Fairbanks (1990), and Oppo et al (1990). With the exception of the Northern Indian Ocean, these features have been documented by detailed downcore analyses as well as by glacial maximum data. The North Atlantic data may be interpreted as signifying increased ventilation of Atlantic intermediate waters by nutrient-depleted Northern Source water, and reduced (but not eliminated) ventilation of the deep North Atlantic by nutrient-depleted source waters. The suggestion that the upper Atlantic nutrient depletion might have been due to increased outflow of nutrient-depleted Mediterranean Outflow (Zahn et al 1987, Oppo & Fairbanks 1987) appears ruled out by low Ba content of the upper Atlantic (Lea & Boyle 1990), as well as by volumetric considerations (Boyle & Keigwin 1987) and considerations of hydraulic flow control (Thunell & Williams 1989). Kallel et al (1988) discussed several possibilities for reduced nutrient content of Northern Indian Ocean Intermediate Water: "(1) dense water formed by increased evaporation and possible decreased temperatures of the surface water of the Northern Arabian Sea, (2) expansion to the north of 10°S of the Antarctic Intermediate Water (AAIW) in the absence of a northern local source, and (3) increased flow of intermediate water from the Pacific Ocean through the Indonesian Archipelago." The new CdW evidence cannot distinguish between (1) and (2), but this new data indicating persistently high CdW in glacial Pacific intermediate waters now rules out option (3). It also may be worth noting that the CdW data presented here does not suggest such a sharp hydrographic front as did the $\delta^{13}C$ data presented by Kallel et al (1988).

DISCORDANT CARBON ISOTOPE AND CADMIUM EVIDENCE

$\delta^{13}C$ and CdW data are in substantial disagreement on the following features of glacial ocean paleochemical distributions:

1. $\delta^{13}C$ data (including both that data presented here and in other publications) indicate that Antarctic and Indian Ocean deepwaters were as nutrient-enriched as anywhere in the glacial ocean. CdW data indicate only minor changes in the nutrient content of glacial Antarctic deepwaters compared to the modern ocean, lower nutrient content for glacial Indian Ocean deepwater relative to today, and that deep waters from both basins were lower in nutrients than in the deep Eastern Tropical Pacific during glacial times.
2. Contrasting with the modern ocean, the available glacial $\delta^{13}C$ data suggest that upper waters of the Pacific ocean had nutrient levels comparable or slightly depleted relative to deepwater of this basin. CdW data indicate that although there may have been a slight reduction in nutrient levels in the intermediate Pacific, intermediate-water nutrient levels have been consistently higher than nutrient levels of deeper waters.
3. $\delta^{13}C$ data indicate that the nutrient content of deepwaters of the Northwest Pacific and Eastern Tropical Pacific were similar to those of the modern ocean (Keigwin 1987), with little difference between the two regions ($\delta^{13}C$ data from core V28-304 may be an exception to this generalization). CdW data, on the other hand, clearly indicate somewhat lower nutrient content for the Eastern Tropical Pacific and indicate substantially lower nutrient content for the Northwest Pacific, compared to the modern ocean, and that between the two regions, the Northwest Pacific had lower CdW concentrations than the Eastern Tropical Pacific.

4. As noted above, $\delta^{13}C$ data imply that on a global average, upper waters of the ocean had much lower nutrient levels than deeper waters, in contrast to the modern ocean where the reverse distribution occurs. CdW data imply little net vertical redistribution between glacial and interglacial times.

KEY DISCREPANCIES BETWEEN $\delta^{13}C$ AND Cd: DOWNCORE RECORDS

The regionally discordant features of $\delta^{13}C$ and CdW glacial maximum data sets are confirmed by detailed downcore studies in each region.

$\delta^{13}C$ and CdW in Cores RC13-228, RC13-229, and RC11-120: Antarctic Deepwater Paleochemical Composition Compared to the Eastern Tropical Pacific

Two downcore records from the South Atlantic and one from the South Indian Ocean illustrate this discrepancy (Figure 11; Tables 12–14). RC13-228 (Curry et al 1988) and RC13-229 (Oppo et al 1990) are in the Eastern South Atlantic (south flank of the Walvis Ridge); RC11-120 (Curry et al 1988) is from the subantarctic Indian Ocean. The *Cibicidoides* $\delta^{13}C$ records from these cores are compared to *Cibicidoides* $\delta^{13}C$ records from the Eastern Tropical Pacific [cores V19-30 and TR163-31b (Curry et al 1988)] because $\delta^{13}C$ variations in cores from the Eastern Tropical Pacific are thought to reflect predominantly changes in global mean $\delta^{13}C$. The CdW records from the Southern Ocean sites are also compared to the CdW records from the same Eastern Tropical Pacific cores [note that in the case of Cd, various species are used—generally more than one is used in each core—because there are no consistent inter-specific differences; the actual data for each species are listed in Tables 12–14 and in Boyle (1988a)]. It should be emphasized that each Cd datum is based on replicate analyses, so that subtle differences between average datums are significant, particularly when several values all show the same tendency. All Southern Ocean cores display a large $\delta^{13}C$ shift (nearly 1‰) towards more negative values during glacial times; in contrast, Cd variations are subtle. Despite topographic isolation from the North Atlantic, the 3204-m site of core RC13-228 is presently overlain by a high percentage of Northern Source water ($\sim$65%, according to Broecker et al 1991); as a result, late Holocene $\delta^{13}C$ is about 0.6‰ more positive and CdW is about 0.4 nmol/kg lower than Pacific deepwater. Glacial $\delta^{13}C$ values at this site move toward Pacific values, but remain about 0.3‰ more positive. Glacial CdW data, on the other hand, remain much lower at this site than in Pacific deep waters, which would indicate persistent ventilation by low-nutrient North Atlantic water (or a

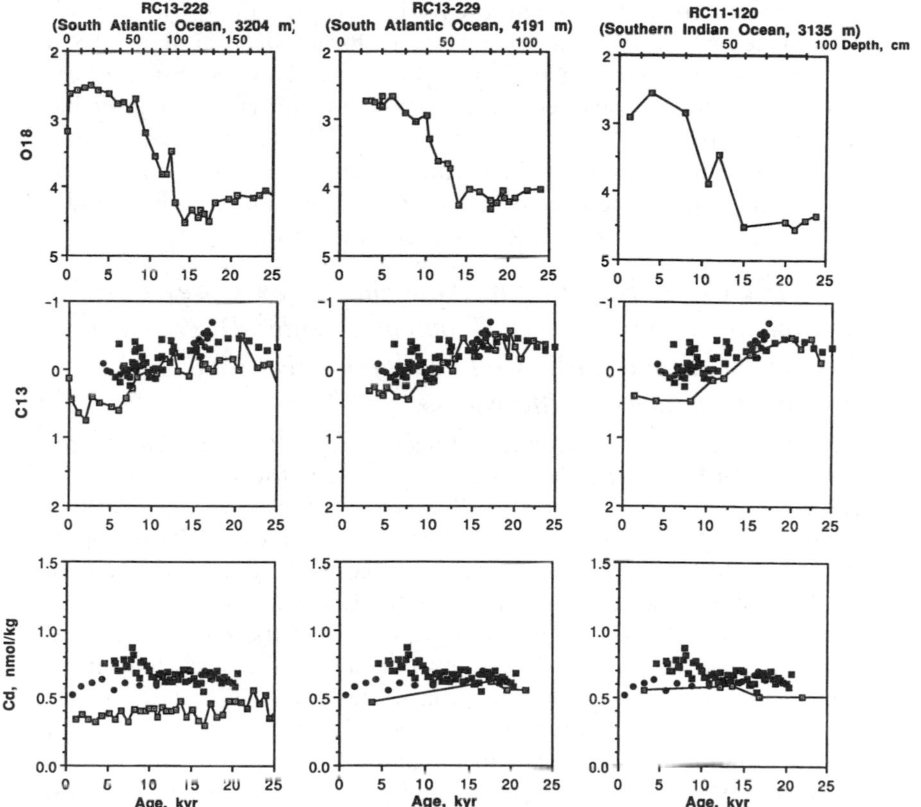

Figure 11 Downcore comparisons of paleochemical data from the Southern Ocean with data from the Eastern Tropical Pacific: cores RC13-228, RC13-229, and RC11-120 (Southern Ocean) shown as open squares with connecting line; Eastern Tropical Pacific cores TR163-31B (*dark squares*) and V19-30 (*dark circles*) shown as symbols only. Benthic foraminiferal $\delta^{18}O$ and $\delta^{13}C$ from *Cibicidoides* spp. reported as ‰ relative to PDB standard (no species offset applied). See Tables 12–14 for Cd species used.

new nutrient-depleted southern water deepwater source). At the site of deeper South Atlantic core RC13-229 (4191 m), which is presently overlain by waters dominated by end-member Antarctic Bottom Water (see references in Oppo et al 1990), late Holocene $\delta^{13}C$ is about 0.4‰ heavier than deep Pacific core tops; during the glacial maximum, there is no significant $\delta^{13}C$ difference between this site and deep Pacific sites. Late Holocene CdW at this site is lower than Pacific deepwater by about 0.25 nmol/kg; the difference is reduced during glacial times, but this site remains about 0.1 nmol/kg lower in CdW than Eastern Tropical Pacific deepwater. At the site of subantarctic Indian Ocean core RC11-120, overlain by Circumpolar Deep Water in the modern ocean, late Holocene $\delta^{13}C$ is about 0.5‰ heavier and CdW is about 0.25 nmol/kg lower than Eastern Tropical Pacific deepwater; during glacial times, $\delta^{13}C$ in this core is indistinguishable from glacial tropical Pacific deepwater while CdW remains significantly lower ($\sim$0.2 nmol/kg) than in the Pacific. Hence in all of these cores there is a clear discrepancy between $\delta^{13}C$ data and CdW data if both are to be interpreted in terms of the nutrient content of the water.

$\delta^{13}C$ and CdW from Intermediate and Deep Cores from the Northwest and Eastern Tropical Pacific: Discrepancies Between $\delta^{13}C$ and CdW Concerning Vertical and Inter-Basin Nutrient Differences

Figure 12 compares intermediate and deep $\delta^{18}O$, $\delta^{13}C$, and CdW records from the Northwest and Eastern Tropical Pacific (no *Cibicidoides* $\delta^{13}C$ record is available from the intermediate waters of the Northwest Pacific). According to $\delta^{13}C$ data, Northwest and Eastern Tropical Pacific deepwaters in both regions have similar nutrient contents, both in the modern day (Figure 6; Craig et al 1982), and late Holocene and during glacial times (Keigwin 1987), and intermediate waters of the Eastern Tropical Pacific are lower in nutrients than deepwaters during the glacial maximum, deglaciation, and early Holocene (Mix et al 1990). CdW data, on the contrary, indicate that the glacial Northwest Pacific had somewhat lower nutrients than in this region (Figure 7) and lower than nutrients in both the Northwest and Eastern Tropical Pacific (Figure 12). In the Northwest Pacific, this comparison is flawed by the lack of good core-top evidence for both $\delta^{13}C$ and Cd and the absence of an intermediate-water *Cibicidoides* $\delta^{13}C$ record (Morley et al 1991). Continuous downcore records from this area might not support the discrepancy based on glacial maximum and early Holocene samples. On the other hand, downcore evidence and glacial maximum evidence is quite thorough in the Eastern Tropical Pacific, and it is more difficult to appeal to limited data to explain the discrepancy between $\delta^{13}C$ and CdW data. It is conceivable that the shallow water distribution coefficient $D(z)$ estimated earlier is in error, but $D(z)$ would have to be in error nearly a factor of two at the site of V19-27 to be reconciled with $\delta^{13}C$ data. In view of the consistency of the calibration (Figure 4), that possibility seems remote.

DISCUSSION OF POSSIBLE CAUSES OF $\delta^{13}C$ AND CdW DISCORDANCY

How are discrepancies between $\delta^{13}C$ and CdW data to be explained? It should be emphasized that many of these features are well-established

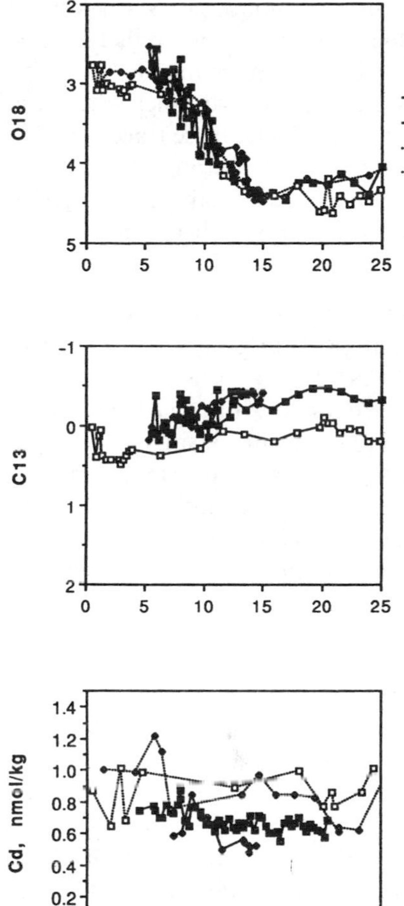

Figure 12 Downcore comparison of paleochemical data from Northwest Pacific and Eastern Tropical Pacific. Shallow cores indicated by open symbols; deep cores shown by filled symbols. Eastern Tropical Pacific cores indicated by squares; Northwest Pacific cores indicated by diamonds. Benthic foraminiferal $\delta^{18}O$ and $\delta^{13}C$ from *Cibicidoides* spp. (TR163-31B, V19-27, V19-30) reported as ‰ relative to PDB standard (no species offset applied); *Uvigerina* spp. $\delta^{18}O$ data from V32-161 plotted with 0.64‰ subtracted to be comparable to *Cibicidoides* spp.

characteristics of glacial maximum sediments; one cannot appeal to inaccuracy of a single data set or insufficient data to explain away differences between the tracers. (*a*) At worst, one (or both) tracers are misleading us, i.e. benthic foraminifera do not always reflect chemical characteristics of bottom water accurately. Some concerns over reliability of *Cibicidioides* $\delta^{13}C$ have been expressed (Keigwin et al 1991; M. Sarnthein, personal communication). Although there are fewer core-top data that cast doubt upon reliability of foraminiferal Cd, the smaller data base available for this element may hide problems lurking in unexamined cores. (*b*) A more positive possibility would be that benthic foraminifera are in fact reliable indicators of bottom water for both properties, but we may be mistaken in our expectation that both tracers must behave similarly. Such a situation might arise if geochemical processes that discriminate these tracers (which are not important today) become more important during glacial times. This possibility is more in the realm of speculation than that of certain knowledge, but some known processes do discriminate between $\delta^{13}C$ and Cd:

1. Gas exchange between ocean and atmosphere. This process can alter $\delta^{13}C$ of surface waters without affecting Cd. In general, the extent to which gas exchange alters $\delta^{13}C$ of surface waters is thought to be limited (Broecker 1982). Charles & Fairbanks (1990) argue that modern subantarctic surface waters are shifted toward more positive $\delta^{13}C$ values by this process, but they also argue that the extent of gas exchange in this region was not significantly different during glacial times. R. S. Keir (personal communication 1991) uses a complex box model to infer that gas exchange might be a significant factor in controlling vertical $\delta^{13}C$ variations within the ocean.
2. Differential isotopic and chemical enrichments of $\delta^{13}C$ relative to Cd. Modern Antarctic plankton are about 8‰ lighter in $\delta^{13}C$ than plankton elsewhere (Sackett et al 1965; Rau et al 1982, 1990, 1991, 1992). If plankton of this isotopic composition were to become quantitatively more important in parts of the glacial ocean than today, it might be possible to create waters that are light in $\delta^{13}C$ without being enriched in Cd. The upper limit for this effect would be reached for a water mass in which the preformed phosphorus content was zero and all of the metabolic carbon added to the water was of this light composition in which case an anomaly of 0.7‰ could be generated. It seems likely that the "real world" magnitude of such an effect would be smaller, although perhaps not entirely negligible.
3. Decomposition of terrestrial carbon. Decomposition of organic carbon from tree-like organic carbon (that has very little Cd) could alter $\delta^{13}C$ without affecting Cd. It seems improbable that the flux of tree-like organic carbon could be sufficiently massive and persistent to cause anomalies over large oceanic regions. However, this possibility cannot be ruled out completely.

Clearly, further progress in interpretation of deepwater paleochemical signals requires that these discrepancies between $\delta^{13}C$ and Cd be understood. Although some thought should be given to potential mechanisms that give rise to real signals of this magnitude, highest priority should be given to examination of the reliability of these tracers under the widest range of modern conditions.

CONCLUSIONS

1. The incorporation of Cd into calcitic benthic foraminifera is depth-dependent, with at least a factor of two less Cd incorporated into shells at 1000 m compared to those below 3000 m. The mechanism of this depth-dependence is uncertain, but appears as if it may be more likely related to pressure or other depth-correlated properties than to temperature.
2. The cadmium content of deep Antarctic waters was about the same during glacial and interglacial times, and remains lower than in deep-waters of the Eastern Tropical Pacific. This result is at variance with $\delta^{13}C$ data showing a shift towards lighter values (in excess of the shift in the global mean) during glacial times, which are as light as in the Eastern Tropical Pacific at that time.
3. The cadmium content of Northern Indian Ocean above 2500 m was about 40% lower than in the modern ocean. This result confirms previous conclusions based on carbon isotope data, and implies a new source of nutrient-depleted intermediate water to this region.
4. The cadmium content of glacial age deepwaters of the Northwest Pacific were 20–30% lower than today, and were lower in Cd than glacial age deepwaters of the Eastern Tropical Pacific. This result implies a new source of nutrient-depleted water to this region and is at variance with $\delta^{13}C$ data which indicate no change in the nutrient content and that the Northwest and Eastern Tropical Pacific had similar $\delta^{13}C$ values.
5. $\delta^{13}C$ and Cd data both indicate that (*a*) intermediate waters of the North Atlantic are somewhat nutrient-depleted compared to today and clearly depleted compared to deeper waters; (*b*) deepwaters of the North and Tropical Atlantic are enriched in nutrients relative to today; and (*c*) intermediate waters of the Northern Indian Ocean are nutrient-depleted compared to today.
6. Resolving discrepancies between $\delta^{13}C$ and Cd data should be a high priority in future paleochemical studies.

ACKNOWLEDGMENTS

The work could not have been done without the many hours at sea of those who collected the cores. The published oxygen isotope stratigraphies were also essential. Najla Adzaboi and Pei Ling Du picked the benthic foraminifera. Lloyd Keigwin provided picked benthics from core Rama

44, and Bob Thunell provided those from the MW core-tops. Laurent Labeyrie arranged for access to the MD core collection. Delia Oppo and Dan McCorkle offered helpful suggestions for revision of the manuscript. This research was supported by the National Science Foundation through grants to the author (OCE 8710328) and to the core repositories at Lamont-Doherty Geological Observatory, Woods Hole Oceanographic Institution, and the University of Rhode Island. Curation of the MD cores is provided by the CNRS, France. The IOS core curation is supported by Great Britain.

APPENDIX

Table 6 *Uvigerina* spp. data

Core	Depth (cm)	Cd/Ca (μmol/mol)			
CHN82-4	58	.103	.110	.118	
	61	.075	.114	.109	
	64	.200?	.111	.115	
CHN82-11	73	.108			
	78	.105			
	83	.094			
	98	.091			
CHN82-20	73	.108			
	75	.093			
	79	.114			
	81	.109			
	83	.097			
	85	.115			
	87	.094			
EN66-10GGC	34	.157	.161		
EN66-38GGC	27	.152			
IOS82-1	62	.128	.145	.147	.132
	64	.148	.151	.142	
	67	.154	.147		
	69	.146	.136		
	78	.126			
	86	.106			
	88	.111			
KNR73-3	69	.182	.217	.213	
	72	.190	.217		
	75	.203	.207		
	79	.178	.202	.170	
	82	.186			
	87	.174	.182		
	92	.208	.279?		
	103	.153			
	114	.194			
	132	.186			
KNR73-4	45	.210			
	48	.221			
	50	.185	.219		
	52	.228			
	55	.112?			
M12392	131	.093			
MD76-125	114	.098	.110		
	127	.115	.120	.854?	
	145	.122	.105		
MD76-127	35	.118	.110		
	43	.106	.094		
	55	.098	.105	.104	
MD76-128	186	.109	.093	.088	
	215	.094	.108		
MD76-135	232	.115	.115		
	294	.108	.093		
MD77-176	542	.108			
MD77-182	63	.068?			
	75	.134?			
MD77-191	684	.069	.062		
MD77-202	175	.158			
	195	.131	.141	.132	
	215	.155	.170		
	235	.151			
	258	.126	.147		
	278	.143	.152		
MD77-203	335	.143	.160		
	378	.151			
	408	.148	.134		
MD77-204	93	.070			
MD79-254	239	.108			
RAMA 44	111	.125	.153		
	126	.152	.151		
RC11-120	67	.120			
	88	.158	.142		
	102	.130			
RC11-147	45	.132			
	51	.128	.127	.111	
	61	.128	.138		
RC11-230	34	.211			
RC12-267	70	.132			
	81	.136			
RC12-294	42	.154	.173		
	45	.150	.134		
	56	.144	.134		
RC12-339	40	.143			
	45	.122			
	55	.098	.133		
RC13-228	102	.183	.561?	.132	.152
	107	.143	.109	.112	.113
	112	.110	.092	.339?	
	117	.092	.112	.144	
	122	.100	.422?	.388?	.112
	127	.148			
	132	.123	.148		
RC13-229	59	.150	.160	.145	
	72	.132	.156	.202	.138
	86	.144	.164		

Table 6 (*continued*)

Core	Depth (cm)	Cd/Ca (μmol/mol)				Core	Depth (cm)	Cd/Ca (μmol/mol)			
RC14-37	53	.134	.154			V22-197	147	.118	.100		
	56	.142	.196				135	.132	.124		
	62	.147	.193				167	.139	.126		
RC15-61	60	.152	.126	.152		V24-1	174	.075	.072		
	71	.152	.178				198	.076	.079		
	86	.131	.179				161	.078	.084		
RC17-69	49	.136				V25-59	46	.113			
	75	.146					73	.117			
TR163-31B	80	.191	.216	.196			56	.130	.134		
	84	.203	.201	.158		V26-176	355	.098			
	87	.222	.208	.143	.145		367	.177?	.100		
	87	.303?	.180	.212			385	.085			
	90	.210	.201			V28-235	29	.146	.151		
	94	.170	.189				36	.150	.183		
	97	.235	.173	.228	.246		44	.167			
	100	.193	.199	.186		V28-238	56	.149			
V17-42	112	.153	.140			V28-304	69	.134			
	88	.157					90	.140			
	78	.168	.140				99	.128			
V19-27	99	.106	.115				128	.134			
	127	.129				V29-179	76	.105	.132		
	112	.130					93	.117	.105	.107	
	133	.154				V30-40	62	.142	.161		
V19-28	118	.192	.200				64	.141			
	127	.224?					76	.140	.122		
	142	.230	.216	.169	.171		92	.124	.113		
V19-29	135	.156	.184			V30-49	62	.164	.206	.177	
	127	.178					64	.133			
	125	.183	.201				74	.154	.140		
	154	.189	.175				84	.172	.198		
V19-30	158	.138	.231	.240	.169	V30-51K	52	.137	.131		
	127	.157	.229	.195	.160		61	.136	.129		
	144	.188	.176	.193	.214	V30-97	131	.094	.089	.089	.120
	137	.219	.163	.191	.199		142	.091	.096		
V19-188	58	.150	.152	.151			169	.109	.091		
	65	.154				V32-159	44	.154	.150		
	47	.357?	.437?				51	.287?	.171		
V19-240	108	.106	.150				56	.126	.111	.122	
	61	.144	.164				59	.123			
	71	.328?	.302?				65	.209	.226	.212	
V21-30	31	.111	.122				71	.107			
	48	.111					74	.105	.088	.110	
	53	.127				V32-161	145	.150	.135		
	36	.136	.133				155	.173	.148	.171	
	43	.167					165	.143	.147	.139	
V22-174	49	.110	.108				175	.127	.153	.149	
	53	.159?					185	.160	.129		
							186	.132			
							196	.085	.096	.131	
							206	.109	.101		
							220	.169			
							237	.905?			

Table 7 *Cibicidoides kullenbergi* data

Core	Depth (cm)	Cd/Ca (μmol/mol)	
CHN82-20	77	.069	
	81	.101	
	84	.104	
KNR110-82	34	.094	
	44	.094	
	71	.077	
KNR64-5PG	32	.034	
	35	.042	
	39	.023	
KNR73-3	72	.147	
	79	.185	
RC11-120	102	.149	
RC12-339	45	.188	
	55	.121	
RC13-228	102	.130	
	107	.087	
	112	.099	
	117	.128	
	122	.066	
RC13-229	59	.183	
	86	.144	
RC14-37	53	.155	
	62	.141	
RC15-61	60	.159	.187
	86	.169	
TR163-31B	97	.165	
V19-188	47	.504?	
V22-197	147	.100	
	135	.115	
V23-100	57	.142	
V25-59	46	.219	
	56	.145	
V27-86	50	.190?	
V28-14	200	.086	
V28-235	29	.156	
	36	.129	
V29-179	76	.117	
V30-40	62	.243?	
	92	.187	
V30-49	62	.190	
	74	.110	.190
	74	.153	
	84	1.624?	

Table 8 *Cibicidoides pachyderma* data

Core	Depth (cm)	Cd/Ca (μmol/mol)	
MD76-125	114	.107	
	127	.165?	
	145	.095	
MD76-127	35	.104	.137
	43	.090	.100
	55	.123	.096
MD76-128	186	.104	.075
	215	.101	.090
MD76-135	232	.148	.129
	294	.114	.117
MD77-171	279	.068	
	308	.118	
MD77-176	542	.071	
MD77-181	117	.181?	
	145	.109?	
MD77-182	75	.175?	
MD77-202	175	.232?	
RC11-147	45	.120	.105
	51	.112	
	61	.113	
V17-42	78	.237	.183
	88	.193	
	112	.198	
V19-27	112	.104	.108
	127	.110	
	133	.141	

Table 9 *Cibicidoides* spp. data

Core	Depth (cm)	Cd/Ca (μmol/mol)	
RC12-267	70	.079	
	89	.119	
V19-240	61	.138	.151
	71	.653?	.827?
	108	.207	.204
V30-51K	61	.152	
	72	.132	

Table 10 *Cibicidoides wuellerstorfi* data

Core	Depth (cm)	Cd/Ca (μmol/mol)	Core	Depth (cm)	Cd/Ca (μmol/mol)
AII107-	34	.096 .120	MD77-181	125	.188?
65GGC	39	.122 .144		145	.222? .156?
	43	.133 .102	MD77-182	75	.194?
AII107-	26	.108 .109	MD77-183	115	.141?
131GGC	29	.088 .113	MD77-191	684	.062
	32	.122 .118	MD77-202	175	.177
	35	.084 .076		195	.170 .162
	38	.112 .101 .079		215	.170 .158
CHN82-15	29	.036		235	.154 .134
	31	.047		278	.421?
CHN82-20	73	.076	MD77-203	335	.161
	75	.098		378	.141 .138
	77	.127		408	.164 .140
	79	.093	MD79-254	239	.139
	81	.121	RC11-120	67	.178
	83	.102		88	.145
	85	.092		102	.168
	87	.102	RC11-147	45	.132 .128
EN66-10GGC	34	.206 .209 .183 .229		51	.150 .266? .142
EN66-16GGC	34	.218 .228 .172 .233		61	.159 .174
EN66-32GGC	33	.109	RC11-230	24	.349?
EN66-38GGC	27	.148 .214 .237		34	.176
KNR64-5PG	32	.026	RC12-267	81	.172
	35	.039		86	.095
	39	.053	RC12-294	42	.203 .162
	41	.038 .040		45	.175 .163 .203
KNR73-3	69	.184	RC13-228	107	.121 .093
	79	.238		112	.151
	82	.150		117	.021?
M12392	120	.067		122	.072
	131	.071		127	.119
MD76-125	114	.096		132	.087
	127	.131 .102	RC13-229	59	.282? .101 .092
	145	.096 .134		72	.184 .140
MD76-127	35	.108 .144	RC13-229	86	.234 .121? .205
	43	.242? .124	RC14-37	53	.183 .218
	55	.112 .127		56	.201 .225
MD76-128	186	.081 .114		62	.243 .269
	215	.103 .110	RC15-61	60	.294? .182
MD76-135	232	.114 .130	RC17-69	49	.219 .244
	294	.111 .140		62	.179
MD77-171	279	.381?		75	.199 .228
	289	.098 .107			
MD77-176	542	.054			

Table 10 (*continued*)

Core	Depth (cm)	Cd/Ca (μmol/mol)	Core	Depth (cm)	Cd/Ca (μmol/mol)
TR163-31B	80	.173	V25-59	46	.206 .203
	84	.177		56	.171 .204
	87	.202 .188		73	.214 .205
	94	.181	V27-60	189	.061
	97	.196	V27-86	50	.116 .073 .087
	100	.231		122	.108
V17-42	78	.182 .222	V28-14	170	.028
	88	.196		200	.033 .036
	112	.283? .487?	V28-235	29	.134 .158
V19-27	99	.108 .129		36	.185 .149
	109	.128		44	.212 .183
	112	.120	V28-304	64	.141
	127	.108		79	.129
	133	.158		90	.141
V19-28	118	.148 .192?		99	.145
	127	.304?	V29-179	76	.187? .105
	142	.278? .176		93	.107 .073
V19-30	127	.152 .220	V30-40	62	.119 .131
	137	.210		64	.181
	144	.213? .134		76	.155 .162
	158	.148 .170		92	.184 .196
V19-188	47	.767?	V30-49	62	.145 .173 .148
	58	.182		64	.327?
V19-240	61	.119		74	.126 .190 .211 .171
	71	.479? .666?		84	.136 .182 .148
	108	.161 .255	V30-51K	52	.160 .145
V22-174	49	.070 .083		61	.139 .140
	53	.093		72	.127 .126
V22-197	135	.108 .108	V32-126	42	.142
	147	.096 .097		43	.129
	167	.099	V32-128	36	.153
V23-100	37	.180		46	.166
	57	.121 .194	V32-161	196	.108
V24-1	174	.040 .081			
	198	.074			

Table 11 *Nutallides umbonifera* data

Core	Depth (cm)	Cd/Ca (μmol/mol)
EN66-10GGC	34	.242
EN66-16GGC	34	.197
EN66-32GGC	33	.160 .144
KNR73-3	79	.138
	82	.203
	109	.166 .187
	132	.182
MD76-127	55	.133
RC11-147	45	.242?
	61	.003?
RC13-229	86	.121
V23-100	37	.172 .142
	57	.139 .125
V25-59	46	.468?
	73	.120
V32-159	74	.101
	56	.122
V30-51K	72	.135 .115
V30-40	92	.143
V32-126	46	.143
V32-159	65	.166 .164

Table 12 Cd data from core RC13-228

Depth (cm)	Cd/Ca (μmol/mol) Uvi[a]				kul		wue		avg[b]	sd[c]	n[d]	r[e]
7	.119				.074		.103		.099	.023	3	
12	.150?				.111				.111		1	1
17	.111				.085				.098	.018	2	
22	.115	.097			.087	.057	.103		.092	.022	5	
27	.093	.267?			.125	.106			.108	.016	3	1
32	.121				.087	.151?	.128		.112	.022	3	1
37	.097				.090		.109		.099	.010	3	
42	.129				.105				.117	.017	2	
47	.156?				.092				.092		1	
52	.121				.113		.121		.118	.005	3	
57	.121				.113		.111		.115	.005	3	
62	.140				.106		.101		.116	.021	3	
67	.128				.120		.118		.122	.005	3	
72	.130	.136			.122	.112	.108		.122	.012	5	
77	.130				.082		.103		.105	.024	3	
82	.124				.126				.125	.001	2	
87	.143				.103		.099		.115	.024	3	
92	.136	.129			.092		.102		.115	.021	4	
97	.127						.113		.120	.010	2	
102	.132	.152	.183?	.561?	.130				.138	.012	3	
107	.102	.112	.143?	.109	.087		.093	.121	.104	.013	6	
112	.110	.092	.339?		.099	.151			.120	.027	3	1
117	.144	.092	.112		.128		.021		.095	.057	5	1
122	.112	.100	.388?	.422?	.066		.072		.087	.022	4	
127	.148						.119		.134	.021	2	
132	.123						.087		.105	.025	2	
137	.130						.087		.108	.030	2	
142	.147						.133		.140	.010	2	
147	.156						.122		.139	.024	2	
152	.136						.137		.137	.001	2	
157	.122						.119		.120	.002	2	
162	.147						.176		.161	.021	2	

[a] Uvi = *Uvigerina* spp.; kul = *Cibicidoides kullenbergi*; wue = *Cibicidoides wuellerstorfi*.
[b] avg = average of analyses at a given depth.
[c] sd = standard deviation.
[d] n = number of analyses included in average.
[e] r = number of analyses excluded from average.

Table 13 Cd data from core RC13-229[a]

Depth (cm)	Cd/Ca (μmol/mol) Uvi				Cib		wue			umb		avg	sd	n	r
5	.137	.150	.144		.141	.143	.122	.102		.144	.149	.137	.015	9	0
59	.150	.160	.145		.183		.282?	.101	.092			.139	.067	7	1
72	.156	.202	.138	.132			.184	.140				.159	.028	6	0
86	.164	.144			.144		.234	.121	.205			.169	.043	6	0

[a] Uvi = *Uvigerina* spp.; Cib = *Cibicidoides* spp.; wue = *Cibicidoides wuellerstorfi*; umb = *Nutallides umbonifera*; avg = average for sample; sd = standard deviation of samples included in mean; n = number of samples included in mean; r = number of samples excluded from mean.

Table 14 Cd data from core RC11-120

Depth (cm)	Cd/Ca (μmol/mol) Uvi	Cib	wue
12		.153	.150 .195
46	.152 .154	.157	.210
53	.196 .154	.173	.161
67	.120		.178
88	.158 .142		.145
102	.130	.149	.168

Table 15 Cd data from core V19-27

Depth (cm)	Cd/Ca (μmol/mol) Uvi	Cpa[a]	wue
6	.135	.105	.112
24			.097
36.5	.139		.128
49			.102
60.5	.130		.163
77.5	.139		.126
93.5	.159		.138
98.5	.106,.115		.108,.129
108.5			.128
111.5	.130	.104,.108	.120
126.5	.139	.110	
127.5			.108
132.5	.154	.141	.158

[a] Cpa = *Cibicidoides pachyderma*.

Table 16 Cd data from core V19-30[a]

Depth (cm)	Cd/Ca (μmol/mol)				avg	sd	n
4.5	.134	.123			.129	.008	2
6	.139	.163			.151	.017	2
14	.167	.173			.170	.004	2
23.5	.168	.169	.197		.178	.016	3
33	.204	.193	.189	.151	.184	.023	4
44.5	.160	.159	.174	.158	.163	.008	4
54.5	.172	.155	.203	.175	.176	.020	4
68	.167	.179	.184		.177	.009	3
77	.182	.231	.184		.199	.028	3
84.5	.181	.164			.172	.012	2
94.5	.180	.181	.173		.178	.004	3
105.5	.194	.189	.167		.183	.014	3
115	.197	.168	.197		.187	.017	3
127	.195	.160	.157	.229	.185	.034	4
137	.191	.199	.219	.163	.193	.023	4
144.5	.193	.214	.188	.176	.193	.016	4
157.5	.138	.231	.240	.169	.195	.049	4

[a] *Uvigerina* spp.

Table 17 Cd data from core Rama 44[a]

Depth (cm)	Cd/Ca (μmol/mol)
31	.171
36	.174
51	.202
56	.186, .180
61	.143
91	.163
96	.157
106	.151
111	.125, .153
126	.152, .151
141	.525?

[a] *Uvigerina* spp.

Table 18 Core age-depth models used in this paper

Core	Depth (cm)	Age (ka)
RC13-228	0.0	0.0
	55.0	8.7
	100.0	13.4
	200.0	28.0
RC13-229	2.9	3.2
	14.5	7.5
	40.0	12.7
	45.0	15.0
	90.0	22.2
RC11-120	5.0	0.9
	30.0	8.5
	45.0	11.4
	95.0	25.1
V19-27	10.0	0.9
	59.0	3.9
	101.0	20.2
V19-30	0.0	0.0
	115.0	14.7
	157.5	17.5
Rama 44	0.0	5.1
	21.0	5.9
	71.0	12.6
	136.0	14.8
V32-161	0.0	0.0
	60.0	5.5
	110.0	8.5
	140.0	12.5
	170.0	16.8
	244.0	30.0
TR163-31b	0.0	5.5
	150.0	19.6

Literature Cited

Bainbridge, A. E. 1980. *GEOSECS Atlantic Ocean Expedition, Vol. 2, Sections and Profiles*. Washington DC: US Gov. Print. Off.

Balsam, W. 1981. Late quaternary sedimentation in the western North Atlantic: stratigraphy and paleoceanography. *Paleogeog. Paleoclimatol. Paleoecol.* 35: 215–40

Bard, E., Hamelin, B., Fairbanks, R. G., Zindler, A. 1990. Calibration of the ^{14}C timescale over the past 30,000 years using mass spectrometric U-Th ages from Barbados corals. *Nature* 345: 405–10

Barnola, J. M., Raynaud, D., Korotkevitch, Y. S., Lorius, C. 1987. Vostok ice core: a 160,000-year record of atmospheric CO_2. *Nature* 329: 408–14

Bloemendal, J., Lamb, B., King, J. 1988. Paleoenvironmental implications of rock-magnetic properties of Late Quaternary sediment cores from the Eastern Equatorial Atlantic. *Paleoceanogr.* 3: 61–87

Boyle, E. A. 1983. Manganese carbonate overgrowths on foraminifera tests. *Geochim. Cosmochim. Acta* 47: 1815–19

Boyle, E. A. 1984. Benthic hydrography and the 41 kyr obliquity cycle. In *Climate Processes and Climate Sensitivity, Am. Geophys. Union Monogr.* 29: 360–68

Boyle, E. A. 1988a. Cadmium: chemical tracer of deep-water paleoceanography. *Paleoceanography* 3: 471–89

Boyle, E. A. 1988b. Vertical oceanic nutrient fractionation and glacial/interglacial CO_2 cycles. *Nature* 331: 55–56

Boyle, E. A. 1988c. The role of vertical chemical fractionation in controlling late Quaternary atmospheric carbon dioxide. *J. Geophys. Res.* 93: 15,701–14

Boyle, E. A., Keigwin, L. D. 1982. Deep circulation of the North Atlantic over the last 200,000 years: geochemical evidence. *Science* 218: 784–87

Boyle, E. A., Keigwin, L. D. 1985/6. Comparison of Atlantic and Pacific paleochemical records for the last 250,000 years: changes in deep ocean circulation and chemical inventories. *Earth Planet. Sci. Lett.* 76: 135–50

Boyle, E., Keigwin, L. D. 1987. North Atlantic thermohaline circulation during the last 20,000 years linked to high latitude surface temperature. *Nature* 330: 35–40

Broecker, W. S. 1982. Glacial to interglacial changes in ocean chemistry. *Prog. Oceanogr.* 11: 151–97

Broecker, W. S., Peng, T.-H. 1990. The cause of glacial to interglacial atmospheric CO_2 change: a polar alkalinity hypothesis. *Glob. Biogeochem. Cycles* 3: 215–40

Broecker, W. S., Peng, T. H., Trumbore, S., Bonani, G., Wolfli, W. 1990a. The distribution of radiocarbon in the glacial ocean. *Global Biogeochem. Cycles* 4: 103–17

Broecker, W. S., Bond, G., Klas, M., Bonani, G., Wolfli, W. 1990b. A salt oscillator in the glacial Atlantic? The concept. *Paleoceanography* 5: 469–78

Broecker, W. S., Blanton, S., Smethie, W., Ostlund, G. 1991. Radiocarbon decay and oxygen utilization in the deep Atlantic ocean. *Glob. Biogeochem. Cycles.* In press

Chappell, J., Shackleton, N. J. 1986. Oxygen isotopes and sea level. *Nature* 324: 137–40

Charles, C. D., Fairbanks, R. G. 1990. Glacial to interglacial changes in the isotopic gradients of southern ocean surface water. In *Geological History of the Polar Oceans: Arctic vs. Antarctic*, pp. 519–37. Dordrecht: Kluwer

Craig, H., Broecker, W. S., Spencer, D. W. 1982. *GEOSECS Pacific Ocean Expedition, Vol. 4, Sections and Profiles.* Washington, DC: US Gov. Print. Off.

Curry, W., Lohmann, G. P. 1983. Reduced advection into Atlantic Ocean deep eastern basins during last glaciation maximum. *Nature* 306: 577–80

Curry, W. B., Duplessy, J. C., Labeyrie, L. D., Shackleton, N. J. 1988. Changes in the distribution of ^{13}C of deep water CO_2 between the last glaciation and the Holocene. *Paleoceanography* 3: 317–42

Duplessy, J. C., Labeyrie, L., Blanc, P. L. 1988. Norwegian Sea Deep Water variations over the last climatic cycle: paleoceanographical implications. In *Long and Short Term Variability of Climate*, pp. 83–116. Berlin: Springer-Verlag

Duplessy, J. C., Matthews, R. K., Prell, W., Ruddiman, W. F., Caralp, M., Hendy, C. H. 1984. C-13 record of benthic forminifera in the last interglacial ocean: implications for the carbon cycle and global deep water circulation. *Quat Res.* 21: 225–43

Duplessy, J. C., Shackleton, N. J. 1984. Carbon-13 in the world ocean during the last interglaciation and the penultimate glacial maximum. *Progr. Biometeorol.* 3: 348–54

Duplessy, J. C., Shackleton, N. J., Fairbanks, R. G., Labeyrie, L., Oppo, D., Kallel, N. 1988. Deepwater source variations during the last climatic cycle and their impact on the global deepwater circulation. *Paleoceanography* 3: 343–60

Ennever, F., McElroy, M. 1985. Changes in atmospheric CO_2: Factors regulating the glacial to interglacial transition. In *The Carbon Cycle and Atmospheric CO_2: Natu-*

ral Variations Archean to Present, ed. E. Sundquiest, W. S. Broecker, pp. 154–62. Washington, DC: Am. Geophys. Union

Graham, D. W., Corliss, B. H., Bender, M. L., Keigwin, L. D. 1981. Carbon and oxygen isotopic equilibria of recent deep-sea benthic foraminifera. *Mar. Micropaleontol.* 6: 483–97

Hansen, J., Lacis, A., Russel, G., Stone, T. P., Fung, I., et al. 1984. Climate Sensitivity: analysis of feedback mechanisms. In *Climate Processes and Climate Sensitivity*, ed. J. Hansen, T. Takahashi, pp. 130–63. Am. Geophys. Union. Monogr.

Kallel, N., Labeyrie, L. D., Juillet-Laclerc, A., Duplessy, J. C. 1988. A deep hydrological front between intermediate and deep water masses in the glacial Indian Ocean. *Nature* 333: 651–55

Keigwin, L. D. 1987. North Pacific Deep Water formation during the latest glaciation. *Nature* 330: 362–64

Keigwin, L. D., Jones, D. A. 1989. Glacial-Holocene stratigraphy, chronology, and paleoceanographic observations on some North Atlantic sediment drifts. *Deep-Sea Res.* 36: 845–67

Keigwin, L. D., Jones, G. A., Lehman, S. J., Boyle, E. A. 1991. Deglacial meltwater discharge, North Atlantic Deep Circulation, and Abrupt Climate Change. *J. Geophys. Res.* 96: 16,811–26

Kroopnick, P. M. 1985. The distribution of ^{13}C in the world oceans. *Deep Sea Res.* 32: 57–84

Labeyrie, L. D., Duplessy, J. C., Blanc, P. L. 1987. Variations in mode of formation and temperature of oceanic deep waters over the past 125,000 years. *Nature* 327: 477–82

Lea, D. W., Boyle, E. A. 1990. Foraminiferal reconstruction of barium distributions in water masses of the glacial oceans. *Paleoceanography* 5: 719–42

Luz, B., Shackleton, N. J. 1975. $CaCO_3$ solution in the tropical east Pacific during the past 130,000 years. In *Dissolution of Deep-Sea Carbonates*, ed. W. V. Sliter, A. W. H. Bé, W. H. Berger, pp. 142–50. Cushman Found. Foraminiferal Res., Spec. Publ.

Mix, A., Fairbanks, R. G. 1985. North Atlantic surface-ocean control of Pleistocene deep-ocean circulation. *Earth Planet. Sci. Lett.* 73: 231–43

Mix, A. C., Ruddiman, W. F. 1985. Structure and timing of the last deglaciation: oxygen isotope evidence. *Quat. Sci. Rev.* 4: 59–108

Mix, A., Ruddiman, W. F., McIntyre, A. 1986. Late Quaternary paleoceanography of the tropical Atlantic, 1: spatial variability of annual mean sea-surface temperatures, 0–20,000 years B.P. *Paleoceanography* 1: 43–46

Mix, A. C., Pisias, N. G., Zahn, R., Rugh, W., Lopez, C. 1990. Carbon-13 in Pacific deep and intermediate waters, 0–370 kyr bp: implications for ocean circulation and Pleistocene CO_2. *Paleoceganography* 6: 205–26

Morley, J. J., Heuser, L. E., Shackleton, N. J. 1991. Late Pleistocene/Holocene radiolarian and pollen records from sediments in the Sea of Okhotsk. *Paleoceanography* 6: 121–31

Morley, J. J., Shackleton, N. J. 1984. The effect of accumulation rate on the spectrum of geological time series: evidence from two South Atlantic cores. In *Milankovitch and Climate*, ed. A. Berger, J. Imbrie, J. Hays, G. Kukla, B. Saltzman, pp. 467–80. Reidel

Neftel, A., Oeschger, H., Staffelbach, T., Stauffer, B. 1988. CO_2 record in the Byrd ice core 50,000–5,000 years BP. *Nature* 331: 609–11

Newell, R. E. 1974. Changes in the poleward energy flux by the atmosphere and ocean as a cause of climate change. *Quat. Res.* 4: 117–27

Oppo, D., Fairbanks, R. G. 1987. Variability in the deep and intermediate water circulation of the Atlantic Ocean during the past 25,000 years: Northern hemisphere modulation of the Southern Ocean. *Earth Planet. Sci. Lett.* 86: 1–15

Oppo, D. W., Fairbanks, R. G. 1990. Atlantic ocean thermohaline circulation of the last 150,000 years: relationship to climate and atmospheric CO_2. *Paleoceanography* 5: 277–88

Oppo, D. W., Fairbanks, R. G., Gordon, A. L., Shackleton, N. J. 1990. Late Pleistocene Southern Ocean ^{13}C variability. *Paleoceanography* 5: 43–54

Parkin, D. W., Shackleton, N. J. 1973. Trade wind and temperature correlations down a deep-sea core off the Saharan coast. *Nature* 245: 455–57

Peng, T. H., Broecker, W. S., Kipphut, G., Shackleton, N. 1977. Benthic mixing in deep-sea cores as determined by C^{14} dating and its implications regarding climate stratigraphy and the fate of fossil fuel CO_2. In *The Fate of Fossil Fuel CO_2*, ed. N. Andersen, N. Mallahof, pp. 355–75. New York: Plenum

Prell, W. L., Hutson, W. H., Williams, D. F., Be, A. W. H., Geitzenauer, K., Molfino, B. 1980. Surface circulation of the Indian Ocean during the last glacial maximum, approximately 18,000 yr. B.P. *Quat. Res.* 14: 309–36

Rau, G. H., Froelich, P. N., Takahashi, T., des Marais, D. J. 1991. Does sedimentary organic $\delta^{13}C$ record variations in Quaternary Ocean $[CO_2(aq)]$? *Paleoceanography* 6: 335–47

Rau, G. H., Sweeney, R. W., Kaplan, I. R. 1982. Plankton $^{13}C/^{12}C$ ratio changes with latitude: differences between northern and southern oceans. *Deep-Sea Res.* 289: 1035–39

Rau, G. H., Takahashi, T., Des Marais, D. J. 1990. Latitudinal variations in plankton C^{13}: implications for CO_2 and productivity in past oceans. *Nature* 341: 516–18

Rau, G. H., Takahashi, T., Des Marias, D. J., Sullivan, C. W. 1991. Particulate organic matter $\delta^{13}C$ variations across the Drake Passage. *J. Geophys. Res.* In press

Romine, K. 1982. Late quaternary history of atmospheric and oceanic circulation in the eastern Equatorial Pacific. *Mar. Micropaleontol.* 7: 163–87

Sackett, W. M., Eckelmann, W. R., Bender, M. L., Be, A. W. H. 1965. Temperature dependence of carbon isotope composition in marine plankton and sediments. *Science* 148: 235–37

Shackleton, N. J. 1967. Oxygen isotope analyses and Pleistocene temperatures reassessed. *Nature* 215: 15–17

Shackleton, N. J. 1977a. Carbon-13 in Uvigerina, tropical rainforest history and the equatorial Pacific carbonate dissolution cycles. In *The Fate of Fossil Fuel CO_2*, ed. N. Andersen, N. Mallahof, pp. 401–27. New York: Plenum

Shackleton, N. J. 1977b. The oxygen isotope stratigraphic record of the Late Pleistocene. *Philos. Trans. R. Soc. London Ser. B* 280: 169–82

Shackleton, N. J., Imbrie, J., Hall, M. A. 1983. Oxygen and carbon isotope record of East Pacific core V19-30: implications for the formation of deep water in the late Pleistocene North Atlantic. *Earth Planet. Sci. Lett.* 65: 233–44

Shackleton, N. J., Opdyke, N. D. 1973. Oxygen isotope and paleomagnetic stratigraphy of equatorial Pacific core V28-238: oxygen isotope temperatures and ice volumes on 10^6 yr scale. *Quat. Res.* 3: 39–55

Shackleton, N. J., Pisias, N. G. 1984. Atmospheric carbon dioxide, orbital forcing, and climate. In *Carbon Dioxide and the Carbon Cycle, Natural Variations Archaen to Present*, ed. E. Sundquist, W. S. Broecker, pp. 313–18. Washington, DC: Am. Geophys. Union

Slowey, N. C. 1990. *The Modern and Glacial Thermoclines along the Bahamas Banks.* PhD thesis. Mass. Inst. Technol./Woods Hole Oceanogr. Inst.

Streeter, S. S., Shackleton, N. J. 1979. Paleocirculation of the deep North Atlantic: 150,000-year record of benthic foraminifera and oxygen-18. *Science* 203: 168–71

Takahashi, T., Broecker, W. S., Bainbridge, A. E. 1981. Supplement to the alkalinity and total carbon dioxide concentration in the world oceans. In *Carbon Cycle Modelling*, ed. B. Bolin, pp. 159–200. New York: Wiley

Thompson, P. R. 1981. Planktonic foraminifera in the western North Pacific during the past 150,000 years: comparison of modern and fossil assemblages. *Paleogeog. Paleoclimatol. Paleoecol.* 35: 241–79

Thompson, P. R., Be, A. W. H., Duplessy, J. C., Shackleton, N. J. 1979. Disappearance of pink-pigmented Globigerioides ruber at 120,000 yrs. b.p. in the Indian and Pacific Oceans. *Nature* 280: 554–58

Thunell, R. C., Williams, D. F. 1989. Glacial-Holocene salinity change in the Mediterranean Sea: hydrographic and depositional effects. *Nature* 338: 493–96

Toggweiler, R., Sarmiento, J. L. 1985. Glacial to interglacial changes in atmospheric carbon dioxide: the critical role of ocean surface water in high latitudes. In *Carbon Dioxide and the Carbon Cycle, Natural Variations Archaen to Present*, ed. E. Sundquist, W. S. Broecker, pp. 163–84. Washington, DC: Am. Geophys. Union

Weiss, R. F., Broecker, W. S., Craig, H., Spencer, D. W. 1982. *GEOSECS Indian Ocean Expedition, Vol. 6, Sections and Profiles.* Washington, DC: US Gov. Print. Off.

Wenk, T., Siegenthaler, U. 1985. The high-latitude ocean as a control of atmospheric CO_2. In *Carbon Dioxide and the Carbon Cycle: Natural Variations Archaen to Present*, ed. E. Sundquist, W. S. Broecker, pp. 185–94. Washington, DC: Am. Geophys. Union

Weyl, P. K. 1968. The role of the oceans in climatic change: a theory of the ice ages. *Meteorol. Monogr.* 8: 37–62

Zahn, R., Sarnthein, M., Erlenkeuser, H. 1987. Benthic isotope evidence for changes of the Mediterranean outflow during the late Quaternary. *Paleoceanography* 2: 543–60

Zahn, R., Winn, K., Sarnthein, M. 1986. Benthic foraminiferal ^{13}C and accumulation rates of organic carbon: Uvigerina peregrina group and Cibicidoides wuellerstorfi. *Paleoceanography* 1: 27–42

Annu. Rev. Earth Planet. Sci. 1992. 20:289–328

GIANT PLANET MAGNETOSPHERES

Fran Bagenal

Department of Astrophysical, Planetary and Atmospheric Sciences, University of Colorado, Boulder, Colorado 80309

KEY WORDS: Jupiter, Saturn, Uranus, Neptune

INTRODUCTION

As its name suggests, a planet's magnetosphere is the region of space influenced by the planet's magnetic field. The Voyager tour of the outer solar system has confirmed that, like Earth, all four giant planets have extensive magnetospheres due to their strong magnetic fields, generated by convective motions in an electrically-conducting region in the planet's interior. The magnetospheres of Jupiter, Saturn, Uranus, and Neptune vary in size, form, and content but share common features: They are all large structures dominated by strong planetary magnetic fields that contain thermal plasma; there are processes that accelerate the thermal plasma to produce populations of energetic particles which are trapped in radiation belts around the planets; there are significant interactions between the plasma and satellites that are embedded in the magnetospheres; and each magnetosphere produces similar types of plasma waves, radio emissions, and aurora. With the Voyager spacecraft having made measurements with the same complement of instruments at the four planets we can now make a comparison of the family of magnetospheres. The topic of planetary magnetospheres has been reviewed previously by Siscoe & Slavin (1979), Stern & Ness (1982), Bagenal (1985a), Connerney (1987), Hill & Dessler (1991), and McNutt (1991).

The orientations of the planet and its magnetic field control the morphology and dynamics of a planet's magnetosphere. We first consider how the giant planet magnetospheres fall into two categories—the large, symmetric magnetospheres of Jupiter and Saturn and the smaller, irregular

0084–6597/92/0515–0289$02.00

magnetospheres of Uranus and Neptune. We then discuss the characteristics of the plasma and our current understanding of the magnetospheric processes for each planet in turn. Finally, we compare the energetic particle populations, radio emissions, and remote sensing of magnetospheric processes in these giant planet magnetospheres.

PLANETARY MAGNETIC FIELDS

Table 1 shows a comparison of planetary magnetic fields. While the net magnetic moment of each of the outer planets is many times greater than that of the Earth, the planets' large radii result in magnetic fields at the surface (or cloud tops) that are all on the order of a Gauss. None of the planetary magnetic fields are purely dipolar but the dipole (first order) approximation gives an indication of the strength (B_0) and orientation of the field. The regularity of the planet's magnetic field can be gauged by the tilt of the dipole axis from the rotation axis and by deviation of the minimum and maximum surface field strengths from B_0 and $2B_0$

Table 1 Comparison of planetary magnetic fields

	Earth	Jupiter[a]	Saturn[a]	Uranus[a]	Neptune[a]
Radius, R_{Planet} (km)	6,373	71,398	60,330	25,559	24,764
Spin Period (Hours)	24	9.9	10.7	17.2	16.1
Magnetic Moment /M_{Earth}	1[b]	20,000	600	50	25
Surface Magnetic Field (Gauss)					
Dipole Equator, B_0	0.31	4.28	0.22	0.23	0.14
Minimum	0.24	3.2	0.18	0.08	0.1
Maximum	0.68	14.3	0.84	0.96	0.9
Dipole Tilt and Sense[c]	+11.3°	-9.6°	-0.0°	-59°	-47°
Distance (A.U.)	1[d]	5.2	9.5	19	30
Solar Wind Density (cm^{-3})	10	0.4	0.1	0.03	0.005
R_{CF}	8 R_E	30 R_J	14 R_S	18 R_U	18 R_N
Size of Magnetosphere	11R_E	50-100 R_J	16-22 R_S	18 R_U	23-26 R_N

[a] Magnetic field characteristics from Acuña & Ness (1976), Connerney et al (1982, 1987, 1991).
[b] $M_{Earth} = 7.906 \times 10^{25}$ Gauss $cm^3 = 7.906 \times 10^{15}$ Tesla m^3.
[c] Note: Earth has a magnetic field of opposite polarity to those of the giant planets.
[d] 1 A.U. $= 1.5 \times 19^8$ km.

respectively. Jupiter and Saturn have magnetic fields like that of the Earth, where the magnetic axis is roughly aligned with the rotation axis and has only moderate deviation from a dipole. Uranus and Neptune on the other hand, have very irregular magnetic fields with magnetic axes at large angles from their rotation axes and large deviations from a dipole field.

The details of a planet's magnetic field are determined by fitting magnetometer data obtained along spacecraft trajectories with a spherical harmonic expansion model of the magnetic field (e.g. Connerney 1981). Low-orbit satellites allow the Earth's magnetic field to be modeled to high order (Langel & Estes 1985). For the outer planets, spacecraft flybys provide only the lower-order terms (Connerney 1981; Connerney et al 1987, 1991). When the Earth's field is scaled to the core-mantle boundary (believed to be the outer boundary of the geodynamo), the spectrum is very flat, i.e. there is the same amount of power at all spatial scales (e.g. Langel & Estes 1982). This implies that the convective motions which drive the dynamo are small in scale. Connerney et al (1991) point out that should there be a similarly-flat spectrum at the boundary of dynamo regions of other planets, then the observed harmonic structure of Jupiter's and Saturn's fields imply that the outer boundaries of their dynamos are at 0.8 R_J and 0.4 R_S respectively. This is consistent with the expected locations of the pressure-induced transition from molecular (low conductivity) to metallic (high conductivity) hydrogen in each planet.

Following this line of argument, Connerney et al (1991) find that the large nondipolar components of the magnetic fields of Uranus and Neptune imply the existence of a flat spectrum very close to the planet's surface—an unlikely location of the dynamo region. Connerney et al (1991) argue that it is perhaps more realistic to attribute the nondipolar and highly inclined fields of Uranus and Neptune to a fundamental difference in their dynamo mechanisms. Discovery of the irregular field of Uranus, tempted some to ascribe the irregularity to our chancing upon Uranus during a reversal of the magnetic field, similar to the reversals found in the Earth's geologic record (Schultz & Paulikas 1990, Rädler & Ness 1990). The discovery of a second irregular field at Neptune, however, makes this idea implausible. Connerney et al (1991) took Parker's (1969) suggestion that a large tilt implies large convective cells and argue that for Uranus and Neptune large cells would be consistent with the lower conductivity of the liquid mantle region of these planets, which is thought to comprise water, ammonia, and methane.

In conclusion, there appear to be two different types of planetary dynamos. The planets with highly-conducting dynamo regions—Earth (iron), Jupiter, and Saturn (both metallic hydrogen)—have dynamos with small length scales which produce largely dipolar magnetic fields with small tilts with respect to the rotation axis. Stevenson (1982) further argues that in the particularly symmetric case of Saturn, the nonaxisymmetric components are attenuated by differential rotation of an outer conductive shell. Dynamos in planets with poorly-conducting mantles, such as Uranus and Neptune, operate over larger scales and generate nondipolar magnetic fields that are highly inclined to the planet's rotation axis. Perhaps planetary magnetic fields (the external manifestation of interior processes) could provide information about physical conditions inside planets, but we are currently hampered by our limited understanding of magnetic dynamos. Furthermore, we are unfortunately running out of planets on which to test dynamo theories.

MAGNETOSPHERIC MORPHOLOGY

The term *magnetosphere* was coined by Gold (1959) to describe the region of space wherein the principal forces on a plasma are electrodynamic in nature and are a result of the planet's magnetic field. Planetary magnetospheres are embedded in the solar wind, which is the outward expansion of the solar corona [see papers in Pizzo et al (1988) for reviews of the solar wind]. At Earth's orbit and beyond, the solar wind has an average speed of about 400 km s^{-1}. The density of particles (mainly electrons and protons) is observed to decrease, from values of about 3–10 cm^{-3} at the Earth, as the inverse square of the distance from the Sun, consistent with a steady radial expansion of the solar gas into a spherical volume. The solar wind speed, while varying between about 300 and 700 km s^{-1}, always greatly exceeds the speed of waves characteristic of a low density, magnetized, and completely ionized gas (Alfvén waves). Thus a shock is formed upstream of an obstacle, such as a planetary magnetosphere that is imposed on the super-Alfvénic solar wind flow. A planetary *bow shock* can be described in fluid terms as a discontinuity in bulk parameters of the solar wind plasma in which mass, momentum, and energy are conserved. Entropy, however, increases as the flow traverses the shock with the solar wind plasma being decelerated and heated so that the flow can be deflected around the magnetosphere. Thus a shock requires dissipative processes and the presence of a magnetic field allows dissipation to occur on a scale much smaller than a collisional scale length. Although planetary bow shocks do not play a significant role in magnetospheric processes, the crossings of spacecraft through planetary bow shocks have provided an opportunity to study the exotic plasma physics of high Mach number collisionless shocks that cannot be produced in a laboratory [for reviews of collisionless shocks see Stone & Tsurutani (1985); for discussion of giant planet bow shocks also see Russell et al (1982), Slavin et al (1985), Bagenal et al (1987), Moses et al (1990)].

To first approximation, the magnetic field of a planet deflects the plasma flow around it, carving out a cavity in the solar wind (Figure 1). The layer of deflected solar wind behind the bow shock is called the *magnetosheath* and the boundary between the magnetosphere and the solar wind plasma is called the *magnetopause*. The solar wind generally pulls out part of the planetary magnetic field into a long cylindrical *magnetotail*, extending far downstream behind the planet.

Well before Biermann (1957) provided cometary evidence of a persistent solar wind, Chapman & Ferraro (1931) considered how a strongly magnetized body would deflect a flow of particles from the Sun and made an estimate of the location of the magnetopause stagnation point—the boundary between the magnetosphere and the solar wind in the direction of the Sun. They proposed that a dipolar magnetic field (of strength B_0 at the planet's equatorial radius R_p) would stand off the flow to a distance R_{CF} where the external ram pressure of the solar wind balances the internal pressure of the planet's magnetic field: $R_{CF}/R_p = (B_0^2/8\pi m_i n_{sw} V_{sw}^2)^{1/6}$ (where m_i, n_{sw}, and V_{sw} are the ion mass, density, and flow speed of the solar wind). This approximation not only assumes that the magnetic pressure of the solar magnetic field that is embedded in the solar wind is negligible but also that the particle pressure inside the magnetosphere is small. Furthermore, the physical processes that may operate at the mag-

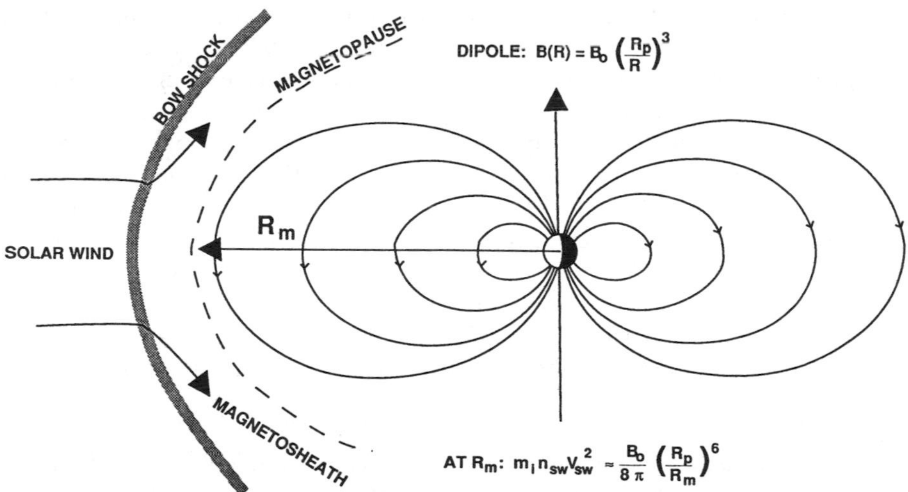

Figure 1 The Chapman-Ferraro solar wind stand-off distance for a dipole magnetic field.

netosphere boundary, such as electric currents resulting from the interconnection of the solar and planetary magnetic fields, are ignored. In reality, the observed magnetopause stand-off distance, R_m, is found to be a factor of 1–2 larger than R_{CF} (see Table 1). Jupiter is the only notable exception, where the plasma pressure inside the magnetosphere is sufficient to further "inflate" the magnetosphere. This makes the magnetosphere of Jupiter a huge object—about 1000 times the volume of the Sun with a tail that extends at least 6 AU in the antisunward direction, beyond the orbit of Saturn. If the Jovian magnetosphere were visible, from Earth its angular size would be twice that of the Sun even though it is at least four times farther away. The magnetospheres of the other giant planets are much more modest (while still dwarfing that of the Earth), having a similar scale of about 20 times the planetary radius—comparable to the size of the Sun.

While the size of a planetary magnetosphere depends on the strength of a planet's magnetic field, the configuration and internal dynamics depend on the field orientation (illustrated in Figure 2) which is described by two angles: the tilt of the magnetic field with respect to the planet's spin axis and the angle between the planet's spin axis and the solar wind direction which is generally within a few degrees of radially outward from the Sun. Since the direction of the spin axis with respect to the solar wind direction only varies over a planetary year (many Earth years for the outer planets) and the planet's magnetic field is assumed to vary only on geological time scales, these two angles are constant for the purposes of describing the magnetospheric configuration at a particular epoch. Earth, Jupiter, and Saturn have both small dipole tilts and small obliquities. This means that the orientation of the magnetic field with respect to the solar wind does not vary appreciably over a planetary rotation period and that seasonal effects are small. Thus Earth, Jupiter, and Saturn have symmetric and quasi-stationary magnetospheres, with Earth and Jupiter each exhibiting only a small wobble due to their $\sim 10^\circ$ dipole tilts. In contrast, the large dipole tilt angles of Uranus and Neptune mean that the orientation of their magnetic fields with respect to the interplanetary medium varies considerably over a planetary rotation period, thus making highly asymmetric and time-variable magnetospheres. Furthermore, Uranus' large obliquity means that the configuration of its magnetosphere will have strong seasonal changes over its 84-year orbit. Below we compare the morphologies of these two topologically-distinct types of magnetospheres.

Symmetric Magnetospheres

Magnetospheric configuration is generally well-described by magnetohydrodynamics (MHD) in which the magnetic field can be considered to be frozen into the plasma flow. Thus we need to consider the processes

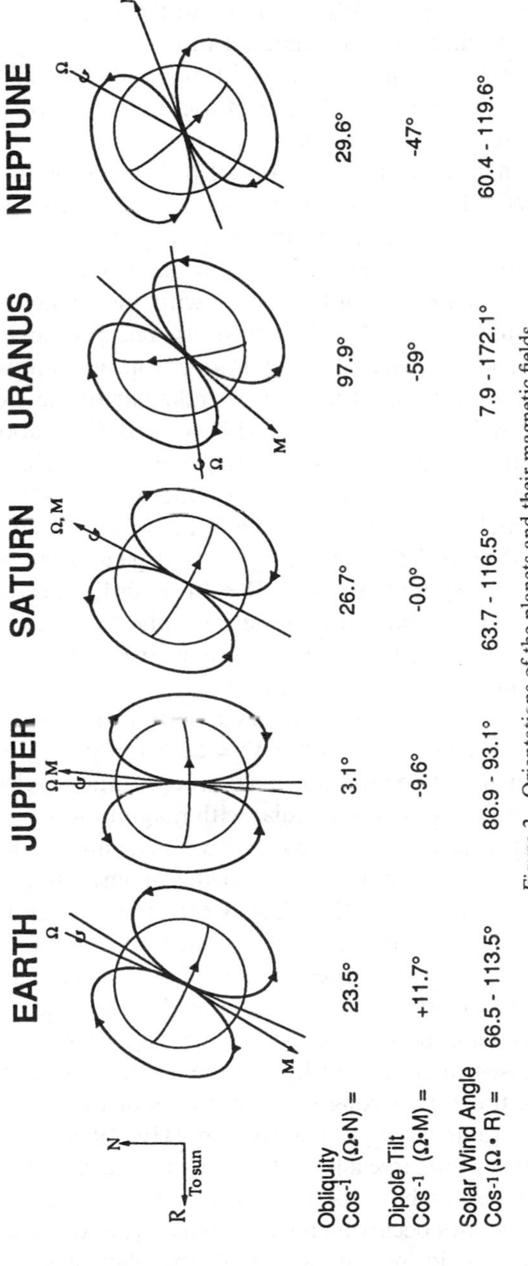

Figure 2 Orientations of the planets and their magnetic fields.

controlling magnetospheric flows [for further discussion of this topic see Vasyliunas (1983)]. The two largest sources of momentum in planetary magnetospheres are the planet's rotation and the solar wind. The nature of any large-scale circulation of material in the magnetosphere depends on which momentum source is tapped. For planetary magnetospheres, corotation of plasma with the planet is a useful first approximation with any departures from strict corotation occurring when certain conditions break down. It may be helpful to think of plasma in the magnetosphere as mass that is coupled by means of magnetic field lines to a giant flywheel (the planet) with the ionosphere acting as the clutch.

For magnetospheric plasma to rotate with the planet, the upper region of the neutral atmosphere must corotate with the planet and must be closely coupled to the ionosphere by collisions. The electrical conductivity of the ionosphere σ^{i} is large so that in a corotating ionosphere (with velocity $\mathbf{V}^{i}$) any horizontal currents (perpendicular to the local magnetic field) are given by Ohm's law, $\mathbf{J}^{i}_{\perp} = \sigma^{i}_{\perp}(\mathbf{E}^{i}+\mathbf{V}^{i} \times \mathbf{B})$. Just above the ionosphere the conductivity perpendicular to the magnetic field in the (collision-free) magnetosphere, $\sigma^{m}_{\perp}$, is essentially zero and $\mathbf{E}^{m} = -\mathbf{V}^{m} \times \mathbf{B}$. Because the plasma particles are far more mobile in the direction of the local magnetic field, the parallel conductivity $\sigma^{m}_{\parallel}$ is large and the field lines can be considered to be equipotentials ($\mathbf{E} \cdot \mathbf{B} = 0$). Thus the electric field in the magnetosphere can be mapped into the ionosphere (Figure 3*a*). Because the ionosphere is relatively thin, the electric field $\mathbf{E}^{m}$ just above the ionosphere is the same as $\mathbf{E}^{i}$ so that we can write $\mathbf{J}^{i}_{\perp} = \sigma^{i}_{\perp}(\mathbf{V}^{i}-\mathbf{V}^{m}) \times \mathbf{B}$. The condition for corotation of the magnetospheric plasma is that the ratio J^{i}/σ^{i} be sufficiently small so that $\mathbf{V}^{m} = \mathbf{V}^{i} = \mathbf{\Omega} \times \mathbf{r}$. For a dipolar magnetic field that is aligned with the rotation axis, the corotational electric field (in the equatorial plane) is therefore radial with magnitude $E_{co} = \Omega B_0/r^2$.

It is clear that large ionospheric conductivities facilitate corotation. The large $\sigma^{m}_{\parallel}$ also means that any currents in the magnetosphere that result from mechanical stresses on the plasma are directly coupled by field-aligned currents to the ionosphere. Thus corotation breaks down when mechanical stresses on the magnetospheric plasma drive ionospheric currents that are sufficiently large for the ratio J^{i}/σ^{i} to become significant. Such conditions might occur in regions of the magnetosphere where there are large increases in mass density due to local ionization of neutral material, where there are strong radial motions of the plasma, or where there are sharp gradients in plasma pressure (Hill 1979). When the magnetosphere imposes too large a load, the ionospheric clutch begins to slip.

Next let us consider how the momentum of the solar wind may be harnessed by processes occurring near the magnetopause where the external solar magnetic field interconnects with the planetary magnetic field.

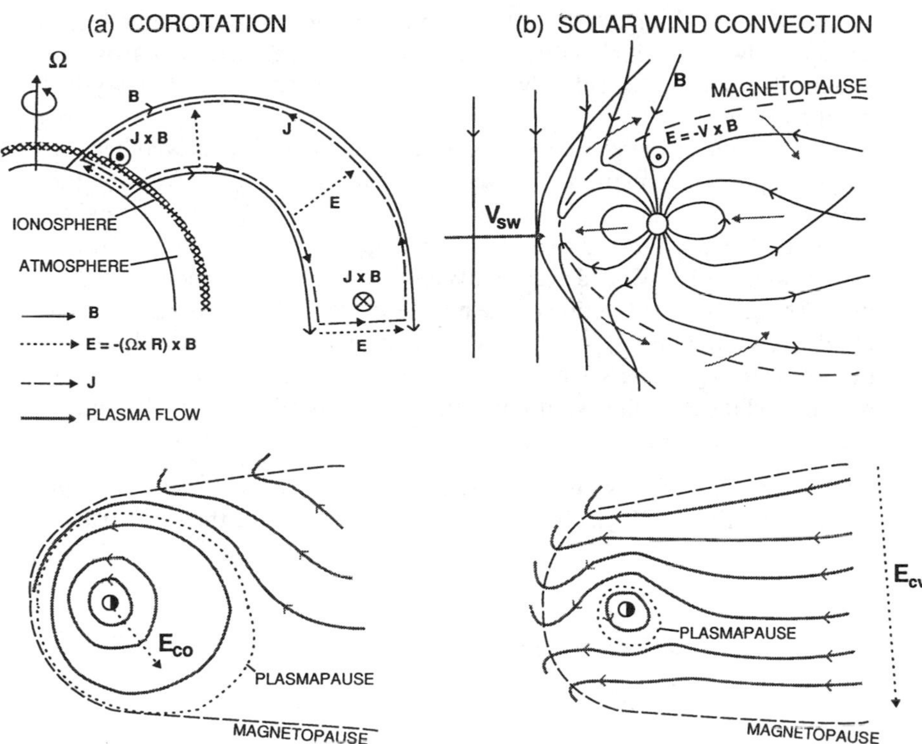

Figure 3 Large scale magnetospheric circulation driven by (*left*) corotation and (*right*) the solar wind. In each case, the upper (*lower*) diagram shows the meridianal view (the view in the equatorial plane). Magnetic field lines are continuous dark arrows; the directions of plasma flow are shown with grey arrows. (Upper left adapted from Belcher 1987; below adapted from Brice & Ioanidis 1970).

Figure 3*b* shows that at the poles the planetary magnetic field lines are open to the solar wind. The solar wind drives a plasma flow across the polar caps and the field lines from the polar region move in the direction of the solar wind flow, being pulled by the solar wind over the poles and back into the extended magnetotail. Conservation of flux requires that field lines are further cut and reconnected in the tail.

The MHD condition of the field being frozen to the flow can be written as $\mathbf{E}+\mathbf{V}\times\mathbf{B}=\mathbf{0}$, which allows the convection electric field to be written $\mathbf{E}_{\rm cv}=-\eta\mathbf{V}_{\rm sw}\times\mathbf{B}_0/R_{\rm m}^3$ (where η is the efficiency of the reconnection process in harnessing the solar wind momentum, $\sim$0.1 for the Earth). In simple magnetospheric models $E_{\rm cv}$ is assumed constant throughout the magnetosphere. The corresponding circulation is given by the $\mathbf{E}\times\mathbf{B}$ drift,

$V_{cv} = \eta V_{sw}(r/R_m)^3$ where R_m is the magnetopause distance. After being carried tailward at high latitudes, the plasma then drifts towards the equatorial plane and eventually returns in a sunward flow to the dayside magnetopause.

Comparison of the corresponding electric fields indicates whether the magnetospheric circulation is driven primarily by the solar wind or the planet's rotation. Since E_{co} is proportional to R^{-2} and E_{cv} proportional to R^3, it seems reasonable to expect that corotation dominates close to the planet while solar wind driven convection dominates outside a critical distance R_c. Thus the fraction of the magnetosphere that corotates is given by $R_c/R_m = (R_m\Omega/\eta V_{sw})^{1/2}$, which simply means that magnetospheres of rapidly rotating planets with strong magnetic fields are dominated by rotation while the solar wind controls the plasma flow in smaller magnetospheres of slowly rotating planets.

THE EARTH The following is a brief description of the Earth's magnetosphere (Figure 4) which will allow comparison with the giant planet

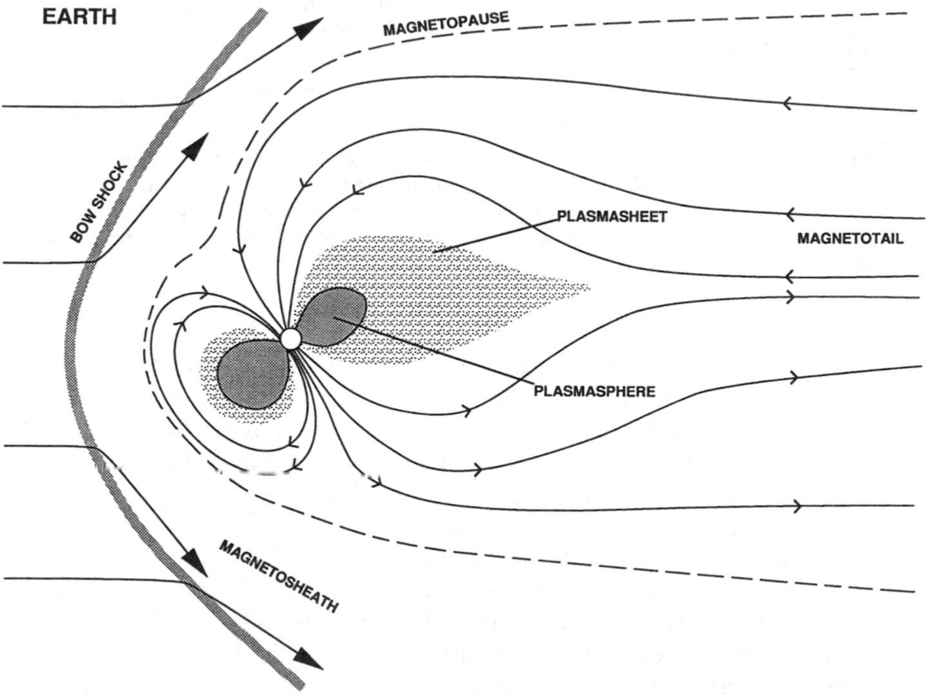

Figure 4 The magnetosphere of the Earth.

magnetospheres. Reviews of the terrestrial magnetosphere can be found in the 1991 IUGG Report and references therein. For the Earth, $R_c/R_m \approx 0.4$ and there exists a region close to the planet where the plasma corotates with the planet, the magnetic field lines remain closed, and large densities of plasma can build up over time. This is the *plasmasphere*. At a distance of about 4 R_E there is a sharp boundary, the *plasmapause*, where the plasma density drops abruptly (from ~ 100 cm^{-3} to ~ 1 cm^{-3}) and outside of which the circulation, driven by the solar wind, is sunward at the equator and antisunward at high latitudes. This means that a large proportion of the Earth's magnetosphere is strongly influenced by the solar wind and will respond to changes in solar wind conditions. In particular, the dayside reconnection rate and hence the convection electric field varies with the orientation of the interplanetary magnetic field, with maximum reconnection occurring when the planetary and solar magnetic fields are oppositely-directed. Under some interplanetary conditions the convection electric field probably results from a purely viscous interaction (which must involve collision-free, micro-scale processes and is poorly understood) between the solar wind and the magnetospheric plasma rather than reconnection. For a given solar wind condition, however, the reconnection on the dayside magnetopause appears to be quasi-steady. By contrast, the kinetic energy of the solar wind that is stored in the tension of stretched magnetic field lines in the tail is violently released episodically in what are known as *magnetospheric substorms*. Oppositely-directed magnetic field lines are believed to reconnect in the center of the magnetotail and the plasma in the reconnected flux tubes is accelerated away from the reconnection point (10–15 R_E). A major focus of studies of the Earth's magnetosphere is understanding the details of how the magnetosphere is coupled to the solar wind and the processes whereby the magnetosphere responds to variations in the interplanetary medium.

JUPITER AND SATURN Long before spacecraft visited the giant planets, theorists estimated R_c/R_m to be much greater than unity for Jupiter and concluded that the magnetosphere of Jupiter would be dominated by rotation throughout and relatively unaffected by the solar wind (Gledhill 1967, Melrose 1967, Brice & Ioannidis 1970). Similarly, Siscoe (1979) predicted that the other giant planet magnetospheres should also be rotation-dominated with $R_c/R_m > 1$. The Voyager Plasma Science instruments confirmed that the bulk motion of the plasma in the magnetospheres of both Jupiter and Saturn is largely azimuthal but measured significant deviations from rigid rotation (McNutt et al 1981, Richardson 1986). The magnetospheres of Jupiter and Saturn are sketched in Figures 5 and 6 respectively. In the case of Jupiter the plasma flow was measured to be

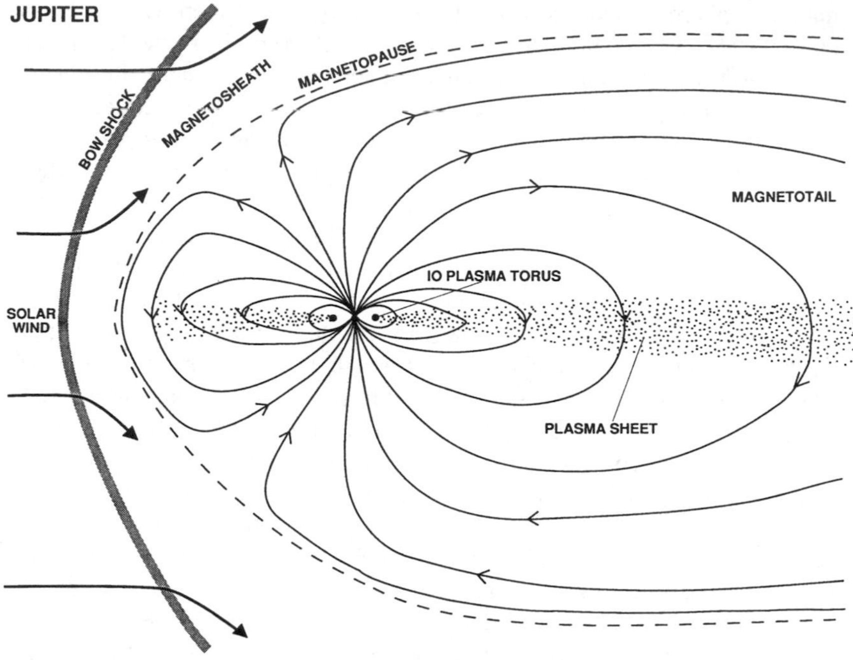

Figure 5 The magnetosphere of Jupiter.

within 1% of rigid corotation in the inner region at 5 R_J (Bagenal 1985b). At larger distances, the Voyager observations confirmed Hill's (1979) prediction that the angular momentum required to accelerate plasma to higher azimuthal velocities in order to maintain corotation with the planet becomes an increasing strain on the frictional coupling between the Jovian ionosphere and neutral atmosphere and the flow lags behind corotation. McNutt et al (1979) reported departure from corotation occurring from about 12 R_J outwards with the azimuthal flow tending towards a constant speed of about 200 km s^{-1} beyond 20 R_J. The mechanical stresses in the magnetosphere that cause this departure from corotation are the large plasma source near Jupiter's satellite Io and the subsequent outward transport of this material (Hill 1980). At Saturn, the observed ~30% deviations of the azimuthal flow from rigid corotation are due to the local production of plasma in the vicinity of Saturn's icy satellites outside ~6 R_S (Richardson 1986). (The characteristics of these magnetospheric plasmas are discussed below.)

The question then arises that if the flow in the bulk of the magnetosphere is azimuthal, what happens in the tail? Spacecraft have passed only through

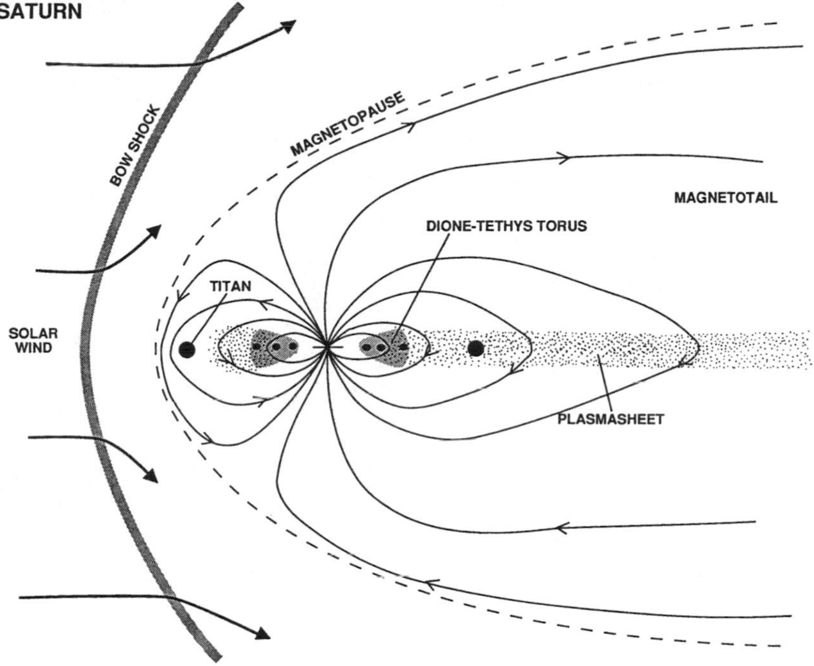

Figure 6 The magnetosphere of Saturn.

the dawn sectors of the Jovian and Saturnian magnetospheres so we can only conjecture about the true structure of their magnetotails. There are two main theories, illustrated in Figure 7. Vasyliunas (1983) proposes a planetary wind model in which at some distance from the planet the kinetic energy of the rotational flow becomes greater than the energy of the magnetic field (i.e. when the corotational speed equals the local Alfvén speed). The flow then "breaks" and reconnects the magnetic field; material is then disconnected from Jupiter and flung down the tail. Alternatively, Cheng & Krimigis (1989) argue that one can just extend the Brice & Ioannidis (1970) corotation-convection model and have solar wind driven convection bringing in solar wind material on the dusk side of the magnetotail and a magnetospheric wind on the dawnside. Whilst Cheng & Krimigis (1989) present the composition of energetic particles that was measured by Voyager as it passed down the dawn magnetotail as evidence supporting their model, we will have to wait until an orbiter such as *Galileo* or *Cassini* passes through the dusk sector to reveal the nature of the magnetotails of these rotation-dominated planets.

There is also the issue of magnetospheric flow near the dayside mag-

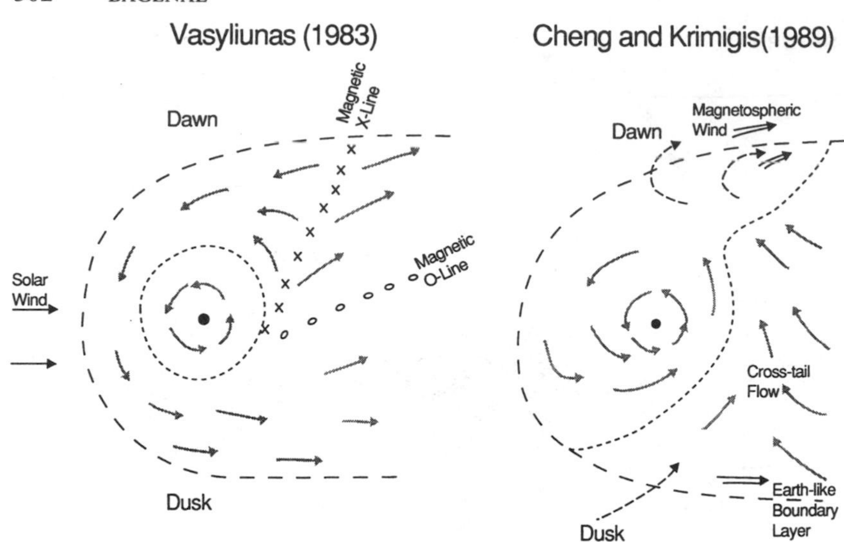

Figure 7 Two models for the configuration of the magnetotails of symmetric, rotation-dominated magnetospheres such as Jupiter and Saturn. (*Left*) From Vasyliunas 1983; (*right*) from Cheng & Krimigis 1990.

netopause. In the afternoon sector one expects the corotational flow to approximately match the magnetosheath flow in both magnitude and direction. One therefore expects little interaction between the interior and exterior flows across the magnetopause (though to date no spacecraft have crossed the afternoon magnetopause and actually measured the flow). In the morning sector the corotation flow is oppositely directed to the solar wind plasma just outside the magnetopause. Strong shears in plasma flow lead to instabilities. Goertz (1983) explains large fluctuations in plasma parameters observed in the outer regions of Saturn's magnetosphere in terms of very turbulent flows with blobs of magnetospheric plasma being detached and carried antisunward by the magnetosheath flow.

Asymmetric Magnetospheres

Early studies were based on experience of Earth, Jupiter, and Saturn and only considered symmetric magnetospheres. Uranus sent magnetospheric theorists back to basics. In the symmetric case, the convection electric field, and hence convective motions, are quasi-steady in the inertial reference frame. Once the magnetic tilt angle becomes appreciable this is no longer true.

URANUS In the case of Uranus where the rotation axis is currently nearly parallel to the solar wind direction, the solar wind driven convection and the direction of planetary rotation are orthogonal and hence decoupled: In the reference frame rotating with Uranus (in which there is no corotation electric field) the solar wind driven convection is quasi-steady and permeates throughout the magnetosphere (Hill 1986, Vasyliunas 1986, Selesnick & Richardson 1986). Thus the plasma in Uranus' magnetosphere corotates with the planet once every 17 hours but on a longer time scale (days) the plasma is circulated through the magnetosphere by the solar wind driven convection. Elements of plasma exhibit helical trajectories, spiraling sunward at the magnetic equator and antisunward at high latitudes. Figure 8 shows two configurations of Uranus' magnetosphere separated by half a planetary rotation. To first approximation, the magnetosphere of Uranus resembles that of the Earth but revolves every 17 hours around the planet-Sun line.

This simple picture was modified by Selesnick & McNutt (1987) to include effects due to drifts of the hot plasma population, which leads to electric currents and partial shielding of the inner region ($<5\ R_U$) from the convection electric field. If the inner region is shielded from the solar wind driven convection then appreciable densities of cold plasma can accumulate. Conversely, magnetospheric plasma that originates in the outer region is deflected around, and hence excluded from, the inner shielded region. Such a quasi-steady shielding model is consistent with the enhanced densities of cold plasma measured in the inner region of Uranus' magnetosphere and the abrupt decrease in hot (keV) plasma inside $\sim 5\ R_U$ (McNutt et al 1987, Selesnick & McNutt 1987). However, Sittler et al

URANUS

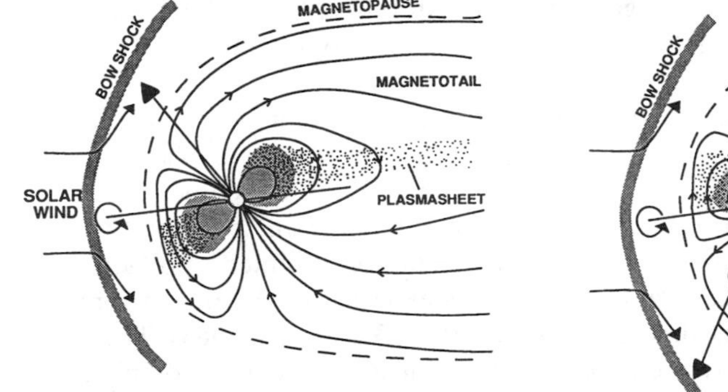

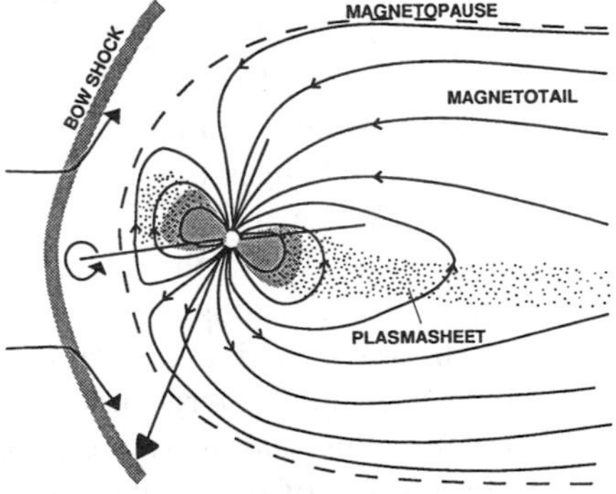

Figure 8 The magnetosphere of Uranus. Left and right are separated by half a planetary rotation.

(1987) and McNutt et al (1987) point out that features of the Voyager plasma data are also reminiscent of the highly time-dependent substorms in the Earth's magnetosphere. Further, Mauk et al (1987) and Cheng et al (1987) note substorm-like signatures in the energetic particle data at Uranus. By developing his model of convection to include the nondipolar magnetic field and time-dependent injection of plasma from the magnetotail, Selesnick (1988) is able to match many of the features observed by Voyager. Nevertheless, the few hours of data obtained on a single passage just give us a glimpse of the Uranian system. If, as we suspect, the magnetosphere of Uranus is as dynamic as the Earth's, then a statistical study of data obtained on multiple passages through the region will be necessary to properly distinguish spatial and temporal variations.

Uranus' magnetotail shows strong similarities to the Earth's magnetotail: The lobes of oppositely-directed magnetic field are separated by a cross-tail current and a sheet of enhanced plasma density (Ness et al 1986, Bridge et al 1986, Voigt et al 1987, Behannon et al 1987). The plasma sheet lies in the magnetic equatorial plane near Uranus but bends parallel to the solar wind flow tailward of distances beyond 10–15 R_U. The fundamental difference between the Uranian magnetotail and that of the Earth is that the whole tail structure rotates in space approximately about the Uranus-Sun line because of the near-alignment of the Uranian spin axis with the solar wind flow. For further discussion of the configuration of Uranus' magnetosphere see the review by Belcher et al (1991) and references therein.

NEPTUNE If a planetary rotation axis is not approximately aligned with either the magnetic dipole axis or the solar wind flow direction then there exists no reference frame in which the plasma flow is steady (Selesnick 1990). At the time of the Voyager 2 encounter in 1989, Neptune's northern hemisphere was close to midwinter with the rotation axis tipped 113° from the Sun, i.e. 67° from the radial, solar wind direction. The 47° tilt of the magnetic dipole means that the angle between the solar wind and the dipole axis changes between 20° and 114° over the 16.1 hour planetary rotation. When the angle is near 90° the configuration is, momentarily, symmetrical like Earth, Jupiter, and Saturn. When the angle is small we have a unique configuration with the magnetic axis pointed "pole-on" into the solar wind, a configuration that was expected for Uranus before Voyager 2 found a large dipole tilt (Siscoe 1975, Voigt et al 1983). These configurations lead to very different magnetic field topologies. Complete reconfiguring of the magnetosphere must occur every planetary rotation (Ness et al 1989, Belcher et al 1989).

Figure 9 shows the two extreme configurations that occur 8 hours

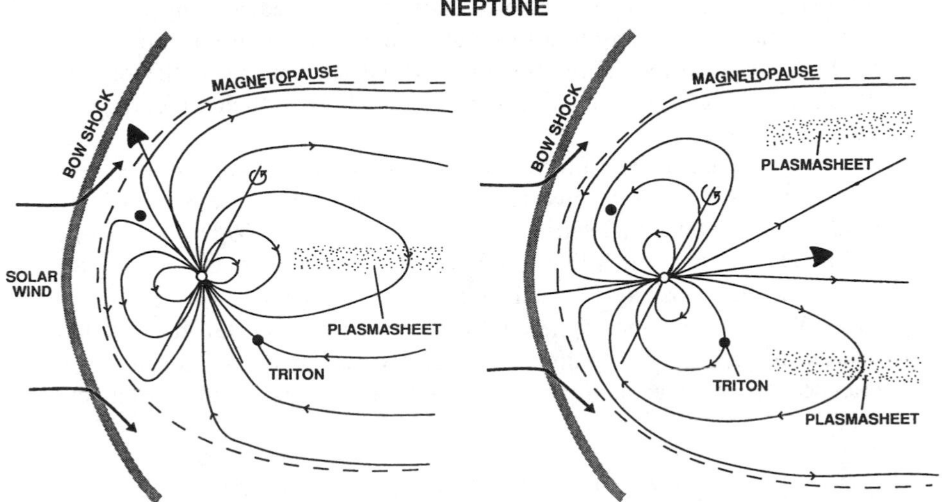

Figure 9 The magnetosphere of Neptune. Left and right are separated by half a planetary rotation.

apart. A theoretical model of plasma convection in Neptune's changing magnetosphere developed by Selesnick (1990) can be summarized as follows. The corotation velocity is everywhere greater than convection but convection has a cumulative effect over several planetary rotations, leading to a net sunward transport of plasma in the magnetic equatorial plane. The maximum reconnection of solar and planetary magnetic fields occurs for the "Earth-like" configuration. During the "pole-on" configuration there is only a small region of the magnetopause where the planetary and solar fields are antiparallel. The occurrence of each configuration corresponds to a specific magnetic longitude passing through local noon. Therefore, in the corotating reference frame convection varies systematically with time, being stronger when the longitudes corresponding to the Earth-like configuration cross local noon. Thus an element of plasma that was at local noon during Earth-like configuration will drift away from Neptune while an element that was at midnight will drift (more slowly) towards the planet. Viewed from the nonrotating reference frame the plasma spirals inward or outward depending on its location at the time of Earth-like configuration. Therefore, in Selesnick's (1990) model, convection and hence variations in density are expected to be strongly longitude dependent. Alternatively, Hill & Dessler (1990) assume a longitudinal asymmetry in the distribution of plasma and derive a four-cell convection pattern that corotates with the planet. Richardson et al (1991) point out, however, that while the above convection models predict transport rates comparable to those required to match the Voyager plasma data ($\tau \approx 1$ day), the observations show no longitudinal asymmetry in the plasma density and indicate only inward transport of plasma which is inconsistent with both models.

The dramatic changes in the configuration of the magnetotail occurring every planetary rotation must further complicate the dynamics of Neptune's magnetosphere (Belcher et al 1989, Ness et al 1989, Voigt & Ness 1990). During the Earth-like configuration the magnetotail is like the Earth's with oppositely-directed magnetic fields separated by a current sheet. When the magnetosphere is pole-on, the magnetotail has a cylindrical configuration with planetward-directed field on the outside and field lines leaving the planet on the inside separated by a cylindrical current sheet.

THERMAL PLASMA CHARACTERISTICS

It is rather misleading to describe a magnetosphere as an empty cavity from which the solar wind is excluded. Magnetospheres contain considerable amounts of plasma which have "leaked in" from various sources (Table 2). Firstly, the magnetopause is not entirely "plasma-tight." Solar wind plasma enters through the polar cusp and, whenever the interplanetary magnetic field has a component antiparallel to the planetary field at the

Table 2 Plasma characteristics of planetary magnetospheres

	Earth	Jupiter	Saturn	Uranus	Neptune
Maximum Density (cm^{-3})	1-4000	>3000	~100	3	2
Primary Sources	O^+, H^+ Ionosphere[a]	O^{n+}, S^{n+} Io	O^+, H_2O^+ H^+ Dione, Tethys	H^+ H cloud	N^+, H^+ Triton
Secondary Sources	H^+ Solar wind[a]	H Ionosphere	N^+, H^+ Titan	H^+ Solar wind	H^+ Solar wind
Source Strength (ions/s)	2×10^{26}	$>10^{28}$	10^{26}	10^{25}	10^{25}
(kg/s)	5	700	2	0.02	0.2
Lifetime	days[b] hours[c]	10-100 days	30 days - years	1-30 days	~1 day

[a] Chappell et al (1987).
[b] Filling time for plasmasphere.
[c] Convective time outside plasmapause.

magnetopause, magnetic reconnection is likely to occur and solar wind plasma will leak into the magnetosphere. Secondly, although ionospheric plasma is generally cold and gravitationally bound to the planet, a small fraction has sufficient energy to escape up magnetic field lines and into the magnetosphere. Thirdly, the interaction of magnetospheric plasma with any natural satellites that are embedded in the magnetosphere can generate significant quantities of plasma.

Before discussing the plasma characteristics of each giant planet in turn, it should be noted that the bulk of magnetospheric plasmas are generally found not far from equilibrium, i.e. their particle distribution functions are observed to be approximately Maxwellian (though the ion and electron populations often have different temperatures). This fact is remarkable considering that the sources are usually expected to be monoenergetic and time scales for equilibration by means of Coulomb collisions are usually much longer than transport time scales. At the same time, planetary magnetospheres support a variety of plasma waves which have various energy sources and cover a wide range of frequencies (see the review by Kurth & Gurnett 1991). Interactions between these waves and particle populations are thought to be responsible for thermalizing the bulk of the plasma as well as accelerating or scattering particles at higher energies. For, in addition to the thermal populations (which make up the bulk of the plasma by number density), all planetary magnetospheres contain populations of energetic (MeV) particles which often dominate the energy density (discussed further in a separate section below).

Jupiter

The study of Jupiter's magnetospheric plasma has an interesting history. When Burke & Franklin (1955) discovered Jupiter to be a source of radio emission, it was soon realized that this radio emission must come from energetic charged particles in a strong magnetic field. This remarkable discovery came before Van Allen's detection of the Earth's radiation belts and the in situ verification of the solar wind (Neugerbauer & Snyder 1962). A more puzzling discovery came a few years later when Bigg (1964) revealed that the decametric component of the Jovian radio emission was influenced by Io, the innermost of the four Galilean satellites. The decimetric component of the radio emission was assumed to be synchrotron radiation from electrons with energies of $\sim$10 MeV that gyrate around dipolar magnetic field lines at a distance of a few Jovian radii. This basic picture of a strong magnetic field trapping a large, energetic particle population was confirmed by Pioneers 10 and 11 which reached Jupiter in 1973 and 1974, respectively. The Pioneers also revealed that farther from the planet the magnetic field is considerably stretched out so that the

Jovian magnetosphere is shaped more like a disk than a sphere. Although the large size and radial distension implied the presence of a substantial amount of plasma at lower energies, the Pioneer plasma detector provided littlc information on the thermal population. Nevertheless, the theorists had already come out strongly in favor of a magnetosphere dominated by the planet's rotation (Gledhill 1967, Melrose 1967, Brice & Ioannidis 1970). A few months before the Pioneer 10 encounter, Brown (1974) detected optical emission from neutral sodium atoms in the vicinity of Io using a ground-based telescope. The first direct evidence of the presence of high densities of ionized material at low energies near Jupiter came with the discovery by Kupo et al (1976) of optical emission from S^+ ions. Mekler et al (1977) and Brown (1976), borrowing techniques from studies of more remote astronomical gaseous nebulae, concluded that the S^+ emission comes from a dense (500–3000 cm^{-3}) ring of cold (few eV) plasma that corotates with Jupiter inside the orbit of Io. In 1979 the Voyager 1 spacecraft confirmed that Io is the major source of plasma in the Jovian magnetosphere—a fact that seemed less surprising when the Voyager cameras revealed the satellite's active volcanoes. Bright ultraviolet emission (Broadfoot et al 1979) and local plasma measurements (Bridge et al 1979) revealed an extensive torus of hotter (~80 eV) plasma outside the inner ring of cold plasma (Figure 10).

THE IO-PLASMA INTERACTION Bigg's observation that Io modulates the intensity of Jovian decametric radio emission initiated many early models of the satellite's interaction with the magnetospheric plasma (Marshall &

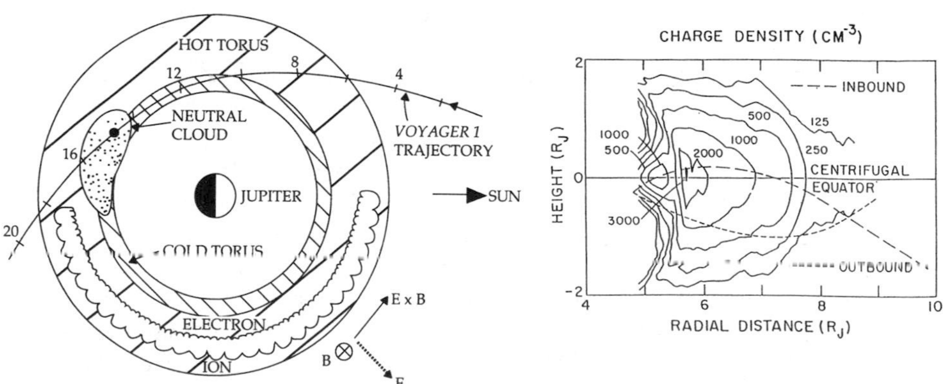

Figure 10 (*Left*) Sketch in Jupiter's equatorial plane of the Voyager 1 trajectory through the Io plasma torus. (From Belcher 1987.) (*Right*) Contours of local charge density in the Io plasma torus derived from data obtained on the inbound trajectory (*long dashes*) of the Voyager 1 spacecraft. (From Bagenal et al 1985.)

Libby 1967, Piddington & Drake 1968, Goldreich & Lynden-Bell 1969, Gurnett 1972, Goertz & Deift 1973). These early studies assumed Io to be a perfect conductor and the ambient plasma density to be very low. They examined how Io's motion in the planetary magnetic field might cause the satellite to act as a unipolar generator and investigated the possibility that large field-aligned currents might directly connect the satellite to the planet. Following Drell's description of a large conducting body generating Alfvén waves as it moves through a magnetic field (Drell et al 1965), Marshall & Libby (1967) were the first to propose that Io might generate large amplitude Alfvén waves that propagate along the magnetic field to the ionosphere of Jupiter where the radio bursts are triggered. However, in applying the theory to Io, the early theorists were hampered by ignorance of the properties of Io and the surrounding plasma.

The perturbations of the magnetic field and plasma flow that were measured in the vicinity of Io when Voyager 1 passed beneath the satellite confirmed the theoretical expectations of a strong interaction between Io and the magnetospheric plasma (Ness et al 1979, Bridge et al 1979). Indeed, further analysis indicated that an Alfvénic disturbance was radiated by Io, carrying a 10^6 amp field-aligned current towards the ionosphere of Jupiter (Neubauer 1980, Belcher et al 1981, Acuna et al 1981, Barnett 1986). Moreover, the observed high plasma densities implied that the propagation speed of Alfvén waves is small in the torus (Bagenal & Sullivan 1981). This means that by the time an Alfvén wave has traveled from Io to Jupiter's ionosphere (where it is reflected) and back, Io has moved along its orbit so that the field aligned currents do not form a closed loop through Io as was first suggested by Goldreich & Lynden-Bell (1969) but rather form open-ended Alfvén wings similar to Drell's model (reviewed by Belcher 1987).

Although it seems that to first approximation Io is a good conductor, in detail the Io-plasma interaction is complicated by the presence of Io's atmosphere (reviewed by Schneider et al 1989 and Cheng & Johnson 1989). Io's volcanoes are believed to be ultimately responsible for a tenuous ($\sim 10^{-9}$ bar) atmosphere of mostly SO_2, either via direct venting or sublimation of volcanic frosts deposited on the surface. The atmosphere is probably patchy and is expected to vary with Io's volcanic activity (e.g. Ingersoll 1989, Moreno et al 1991). Io orbits well inside Jupiter's magnetosphere, embedded in the corotating magnetospheric plasma and high fluxes of energetic particles. The issue is whether these particles reach the surface of Io. Theoretical studies (Sieveka & Johnson 1985, McGrath & Johnson 1987, Moreno et al 1991) and recent observations (Ballester et al 1990) suggest that the atmosphere is collisionally thick (with an exobase at $<0.5\ R_{Io}$), particularly on the dayside and/or above volcanic plumes, so that the impinging charged particles do not reach the surface but collide with atmospheric constituents, heating the upper atmosphere (Johnson 1989) and sputtering energetic atoms and molecules (at a rate of 10^{28}–10^{29} s^{-1}), and forming an extended neutral corona around Io. Presumably, the main constituents of this sputter-corona are products of SO_2 dissociation. Emissions from extended clouds of neutral oxygen (Brown 1981) and sulfur (Durrance et al 1983) have been detected. However, because of their efficient scattering of sunlight, two minor constituents, sodium and potassium, are more readily visible. Since its discovery in 1973 (Brown 1974), the bright sodium cloud has been studied as a tracer of neutral-ion processes in the vicinity of Io (see reviews by Brown et al 1983; Schneider et al 1989, 1991a). Most recently, Schneider et al (1991b) have interpreted jet-like features in images of the sodium cloud as evidence that a substantial amount of sodium escapes in the form of sodium-bearing molecules (rather than sodium atoms). Two important implications of this observation are that the impinging plasma must penetrate deep into Io's atmosphere and that a substantial amount of molecular SO_2 might also be sputtered off rather than just its dissociation products.

The size of Io's neutral sputter-corona is limited by rapid electron-impact ionization and charge-exchange with the corotating ions. This leads to a fundamental problem of stability (Huang & Siscoe 1987, but also see Cheng 1988). The problem is well-stated by Schneider et al (1989): "At the core of the complex reactions between ions and neutrals is the basic fact that the plasma torus generates itself: the corotating ions lift the neutrals off Io, and the electrons ionize them. In this narrow view, we have an unstable positive feedback loop, where ions beget more ions, ad infinitum." Since emissions from the neutral sodium cloud and the torus plasma have not been increasing over the years, there must be a mechanism stabilizing the system. The three main candidate mechanisms for maintaining stability are: (*a*) a means of transporting plasma away from Io, such as the fluxtube interchange instability, that depends nonlinearly on plasma density so that the removal rate grows faster than the source as density increases (Huang & Siscoe 1987), (*b*) the decrease in electron temperature at higher densities which leads to a decreasing ionization rate, limiting the source (Barbosa et al 1983, Smith & Strobel 1985), and (*c*) the atmosphere of Io acts as a buffer by controlling the supply of neutrals either because an increase in flux of corotating ions leads to a lower exobase and smaller effective impact cross-section (Johnson 1989) or because increased ionization in the vicinity of Io leads to further deflection and cooling of the flow around Io (Bagenal 1989, Linker et al 1989). In order to understand the complex processes that couple Io and its atmosphere to the magnetospheric plasma we need detailed models of the Io-plasma interaction [e.g. further development of

3-dimensional MHD numerical simulations similar to that of Linker et al (1988, 1989)] as well as measurements of the response of neutral clouds and the Io torus to changes in volcanic activity (Schneider et al 1989).

THE IO PLASMA TORUS As the Voyager spacecraft approached Jupiter the ultraviolet spectrometer detected powerful emission ($3–6 \times 10^{12}$ Watts) from sulfur and oxygen ions in a toroidal region encompassing the orbit of Io (see reviews by Brown et al 1983 and Strobel 1989). When the Voyager 1 spacecraft flew though the plasma torus, the Plasma Science instrument made local measurements of both the electrons and the various ionic species: $O^{+,2+}$, $S^{+,2+,3+}$, and an ion with a mass/charge ratio of 64, which could be SO_2^+ and/or S_2^+ (see reviews by Belcher 1983 and Bagenal 1989). From the plasma measurements it is clear that the Io plasma torus is divided into two distinct regions with a sharp boundary at 5.7 R_J, inside Io's orbit at 5.9 R_J (see Figure 10). The large outer region of the warmer ($\sim$80 eV) plasma produces the UV and much of the optical emissions while the colder ($\sim$few eV) plasma inside emits only at optical wavelengths.

Observations confirm early predictions that the distribution of plasma along magnetic field lines is limited by the strong centrifugal forces which tend to confine the plasma to the region of the field line farthest from Jupiter's spin axis—the centrifugal equator (Gledhill 1967). To first approximation, the plasma density decreases exponentially with distance from the centrifugal equator $n(z) = n_0 \exp -(z/H)^2$, where the scale height H is given by $H = (2kT_i/3m_i\Omega^2)^{1/2}$ for a spin rate of Ω and ions of mass m_i and temperature T_i (Hill & Michel 1976). Thus the warmer ions in the outer region of the torus have a large scale height and are more spread out along the field than the cold ions inside Io's orbit (see Figure 10).

The plasma is either produced directly in the interaction between Io's atmosphere and the magnetospheric plasma or by ionization of the extended neutral clouds. The lack of enhanced UV emission near Io limits the source strength from the first mechanism. It is estimated that a total of 10^{28}–10^{29} ions must be produced by Io per second to maintain the plasma torus. In either case, when the neutrals are ionized they experience a Lorentz force as a result of their motion relative to the local magnetic field; this force causes the ions to gyrate about the magnetic field at a speed equal to the magnitude of the neutral's initial velocity relative to the surrounding plasma. The ion is accelerated until its guiding center motion matches the plasma rest frame, corotating with Jupiter. Because a particle's gyroradius is mass-dependent, the new ion and its electron are separated after ionization. Hence there is a radial current caused by the ions being "picked-up" by the magnetic field. This radial current across the torus is linked by field-aligned currents to the ionosphere of Jupiter where the

J × **B** force is in the opposite direction to the planet's rotation (see Figure 3*a* and 10 *left*). Thus the planet's angular momentum is tapped electrodynamically by the newly ionized plasma.

Oxygen and sulfur ions picked-up by the magnetic field gain gyroenergies of 260 and 520 eV respectively. The initial velocity distributions are expected to be highly anisotropic (Siscoe 1977) and unstable to the generation of plasma waves (e.g. Barbosa et al 1985, Barbosa & Kruth 1990), and the different ionic species and electrons are not in thermodynamic equilibrium. One expects Coulomb collisions and wave-particle interactions to change the distribution to a more stable one: firstly, to pitch-angle scatter particles into an isotropic distribution; secondly, to produce equipartition of energy for each species (i.e. Maxwellian distributions); and thirdly, to produce equipartition of energy between species of different mass. Inside 5.7 R_J, the ion species are close to equilibrium having Maxwellian distributions with the same temperature (Bagenal 1985b). Although the separate species' distribution functions are not resolved in the Voyager data, in the warm region of the torus complete thermal equilibrium is unlikely because the time scales for equilibration are probably longer or of the same order as the time scale for transport (Bagenal & Sullivan 1981, Bagenal 1989).

While initial calculations of the energy injection rate from the ionization of new material at the pick-up energy were able to balance the radiation output (Barbosa et al 1983, Smith & Strobel 1985), Shemansky (1988) calculates, using more accurate (higher) values for the radiative efficiency of sulfur ions, that an appreciable additional source of energy is required to explain the observed plasma conditions. Proposed solutions to this "energy crisis" currently include enhanced charge-exchange reactions in Io's exosphere (Shemansky 1988, but also see Bagenal 1989), collisional heating by inwardly-diffusing energetic particles (Smith et al 1988), and local magnetic pumping (Ip 1990). Attempts to model the cold, inner torus (Richardson & Siscoe 1983, Barbosa & Moreno 1988) have encountered difficulties explaining (*a*) substantial densities of O^{2+} when the electron temperature is <1 eV; (*b*) the presence of SO_2^+ (or S_2^+) ions at 5.3 R_J when one expects any molecules sputtered to be rapidly dissociated near Io; and (*c*) the presence of a hot component to the ion distribution indicating a local source of pick-up ions. These models are far from reproducing the detailed measurements made by Voyager (Bagenal 1985b, 1989) and from ground-based telescopes (Brown et al 1983, Trauger 1984).

PLASMA SHEET While the high densities (>1000 cm^{-3}) are confined to a toroidal region within ~ 1 R_J of Io's orbit, the iogenic material extends out to at least 40 R_J, forming a thin (<5 R_J thick) sheet of warm (10's of eV) plasma with densities decreasing to a few per cm^{-3} by 20 R_J and dominated by sulfur and oxygen ions (Belcher 1983). This is Gledhill's (1967) magnetodisc, where the centrifugal forces on the corotating plasma stretch out the magnetic field at the equator. In addition to the warm (10's of eV) iogenic plasma there is a hot ($\sim$30 keV) thermal plasma (with Maxwellian distributions) in the middle magnetosphere, beyond 10 R_J (see the review by Krimigis & Roelof 1983). This hot plasma has an energy density greater than the local magnetic field (i.e. $\beta > 1$) and inflates the magnetosphere, making it flatter at the poles (Engle 1991) and more compressible than a vacuum dipole magnetosphere (Caudal 1986). In a self-consistent model of the magnetodisc, Caudal (1986) finds that the radial Maxwell stresses of the stretched magnetic field are balanced by a combination of pressure gradient forces from the hot plasma ($\sim$70%) and to a lesser extent ($\sim$30% overall) centrifugal stresses exerted by the warm plasma (see also McNutt 1983, Mauk & Krimigis 1987, and Khurana & Kivelson 1989a).

The intriguing issue is, What is the source of the hot plasma? The density gradient indicates a source in the outer magnetosphere (Krimigis & Roelof 1983). The ion composition is not measured directly at the keV energies of the hot thermal plasma, but at higher energies the sulfur and oxygen concentrations were found to be strongly enhanced over solar abundances, implicating Io. Furthermore, analyses of plasma waves (Khurana & Kivelson 1989b) and the structure of the plasma sheet (Caudal & Connerney 1989) indicate that protons comprise only 20–50% of the composition which rules out the solar wind as the main source. Alternatively, Barbosa et al (1984) propose that the torus ions are recycled in the outer magnetosphere. When corotating ions in the torus undergo charge-exchange reactions with Io's neutral clouds, the neutralized atom keeps most of its momentum but is no longer confined by the magnetic field and hence is ejected from the Jovian system. Clear evidence of this wind of fast neutrals is provided by recent observations of a faint flaring disk of neutral sodium atoms extending out to at least 400 R_J (Mendillo et al 1990). Eviatar & Barbosa (1984) estimate that about 2% of the neutral sulfur and oxygen wind will be ionized in the outer magnetosphere where they will pick up keV energies corresponding to the local azimuthal speed of 300 km s^{-1}. Adiabatic heating of such ions as they are transported inwards can provide the energies of the hot plasma in the plasma sheet. To provide the observed high fluxes of particles at MeV energies, however, a variety of acceleration processes have been proposed, reviewed in Dessler (1983) and by McNutt (1991).

Currently, we are left with a double puzzle: The warm plasma does not cool adiabatically as it expands out into the magnetosphere (one expects plasma to cool from 80 eV at 6 R_J to 2 eV at 20 R_J), and Caudal & Connerney (1989) find that as the hot (20 keV) plasma is transported inwards the change in pressure with flux-tube volume is not adiabatic ($\gamma < 1$ outside $10 R_J$ (see also Paranicas et al 1990).

RADIAL TRANSPORT The distribution of plasma in the inner and middle magnetosphere, particularly the torus structure and the presence of iogenic plasma in the middle magnetosphere, indicates that plasma is being transported outward, perpendicular to the magnetic field. Furthermore, the sharp boundary close to Io's orbit between the two regions of the torus suggests a sharp change from slow ($\tau \approx$ year) inward diffusion inside 5.7 R_J to rapid ($\tau \approx$ 10–100 days) outward diffusion outside 5.7 R_J, requiring two mechanisms for radial transport (Richardson et al 1980). The situation is summarized by Fazakerley (1990) as follows (abridged):

> The transport mechanism operating inside Io's orbit is consistent with diffusion driven by fluctuations in Jupiter's ionosphere, which was originally proposed to account for the inward transport of energetic particles supplying the Jovian radiation belts. The second (outward) transport mechanism is not understood but it is required to: tap centrifugal potential energy, transport iogenic plasma more rapidly than ionospheric driven diffusion, and produce the observed variations in the rate of transport with distance. Most models are variations of MHD centrifugally-driven interchange motion, ranging from large-scale convection systems to transport by a multitude of small, independent flux tubes (see review by McNutt 1991). However, it is difficult to reconcile the long-lasting azimuthal symmetry of the Io torus with the existence of large scale convection cells. At the same time, small flux tubes are unlikely to remain as coherent structures on the time scales envisaged for interchange motions in MHD models, due to particle drift motion. The most serious difficulty with (pure) MHD models, however, is that Voyager observations indicate that fluctuations in the density of iogenic plasma are too slight, on all length scales, to be compatible with any *strictly* MHD interchange theories (Richardson & McNutt 1987). Moreover, both the ionospheric dynamo model and the family of MHD models assume adiabatic motion and hence fail to account for the rise in temperature with distance, which has been observed in the torus and plasma sheet.

The transport mechanism proposed by Fazakerley (1990) (short wavelength interchange transport driven by electrostatic drift waves) is one in a spate of recent studies that invoke *non*-MHD processes (either on their own or in conjunction with MHD interchange) to circumvent the limitations imposed by Richardson & McNutt (1987). [These studies are too numerous to cite and are summarized by McNutt (1991).] At the same time, Cheng & Johnson (1989) argue that the Richardson & McNutt (1987) observations do not rule out all MHD interchange models but may just put a limit on the diffusion rate. Currently, the mechanism for radial transport and the source of hot plasma in the middle magnetosphere remain the two (probably related) major unsolved issues of the Jovian magnetosphere.

Saturn

In the two years between September 1979 and August 1981 the magnetosphere of Saturn was explored by three spacecraft—Pioneer 11 and Voyagers 1 & 2 (see reviews by Scarf et al 1984 and Schardt et al 1984). Overall, Saturn's magnetosphere was found to be similar to the Jovian magnetosphere: Satellites are the major source of magnetospheric plasma and the plasma dynamics are dominated by the planet's rotation. Nevertheless, the magnetosphere of Saturn is considerably smaller and the multiple sources of plasma much weaker and less distinct than Io. Moreover, Saturn's magnetosphere is less compressible than Jupiter's. The dayside magnetopause is close to the Chapman-Ferraro distance and varies as $p^{-1/6}$ with the solar wind pressure. This behavior indicates that the planetary field stands off the solar wind with little contribution from the internal plasma pressure.

The magnetosphere of Saturn is separated by a boundary at about 15 R_S into two regions: an inner region where the plasma density and temperature vary smoothly with distance and an outer region where the plasma densities and temperatures vary erratically. While there is wide agreement that the sources of plasma in the inner region of the magnetosphere are the icy satellites Dione and Tethys (with lesser contributions from Rhea, the small inner satellites, and the rings), there is strong debate over the source and loss processes. In the outer region, debate also remains about the role of Titan.

DIONE-TETHYS TORUS The icy satellites of Saturn are continuously bombarded by energetic particles, corotating plasma, and solar radiation, which sputter off substantial amounts of water from the surfaces and form a disk-shaped cloud in Saturn's equatorial plane of neutral molecular and atomic products of the dissociation of H_2O (Johnson et al 1989). It is agreed that the ionization of these neutrals provides the observed plasma between 4 and 8 R_S which has a density of ~few cm^{-3}, a temperature of ~10's of eV, and comprises ~20% light ions (mass 1–2) and ~80% heavy ion species with masses between 14 and 18 (Richardson 1986). Theoretical calculations of Richardson et al (1986) predict an ion composition of light ions (75% H^+, 25% H_2^+) and heavy ions (40% O^+, 40% H_2O^+, 18% OH^+, and 2% O_2^+). In their model, which assumes slow radial transport ($\tau \approx$ years), the densities inward of ~8 R_S are limited by collisional processes with the neutral water vapor cloud (dissociative recombination of the molecular ions and charge exchange for the atomic ions) with radial transport as the major loss mechanism outside 8 R_S. Barbosa (1990), on the other hand, argues for fast radial diffusion ($\tau \approx 30$ days) and a predominantly O^+ plasma. While the Voyager plasma instrument could easily distinguish light from heavy ions, the mass resolution is insufficient to distinguish between masses 16 to 18. Clearly, the issue of radial transport is again a critical one at Saturn as it is at Jupiter. While signatures in energetic particle data of satellite absorption provide estimates of radial transport rates for MeV particles, these estimates span orders of magnitude (see Barbosa 1990 as well as review by Van Allen 1984). Moreover, the transport mechanism (as yet unknown) may be energy-dependent.

In the outer region of Saturn's magnetosphere, the variability in plasma properties have several explanations. Schardt et al (1984) suggests a response to solar wind variations. Goertz (1983) proposes that blobs of denser, colder magnetospheric plasma are being detached and swept away in the turbulent (dawn sector) region where there is a strong shear between the corotation inside the magnetosphere and the antisunward magnetosheath flow. An alternative explanation presented by Eviatar et al (1982) is that the dense material comes from Titan.

TITAN When Voyager 1 flew close to Titan it passed through a "wake" downstream of the satellite in the magnetospheric flow. Strong perturbations of the magnetic field and plasma flow were measured and ions with mass 28 (probably N_2^+ or H_2CN^+) were detected indicating that 10^{24} ions s^{-1} are produced in the complex interaction of the magnetosphere with Titan's thick atmosphere (see the review by Neubauer et al 1984). However, there is debate over whether Titan is the major source of plasma for the outer magnetosphere. First of all, Titan's overall role in the magnetosphere of Saturn is limited by the fact that its 20.3 R_S orbit places Titan just beyond the average 18.8 R_S subsolar stand-off distance and hence Titan often spends part of its orbit outside the magnetopause (Slavin et al 1985). Moreover, Richardson (1986) claims a better fit to the ion spectra in the outer magnetosphere with water group ions (masses 16–18) coming from the icy satellites than with N^+ (mass 14) from Titan. The problem that O^+ ions should be rapidly removed by the resonant charge-exchange with neutral hydrogen detected at Titan's orbit (Broadfoot et al 1981) is raised by Shemansky et al (1985) who conclude that the neutral hydrogen cloud is much denser and extends from Saturn's atmosphere outwards. Eviatar & Richardson (1990) may have found a solution to this dilemma by suggesting that the major ion is H_2O^+ or H_3O^+ (rather than O^+) which has a longer lifetime against charge-exchange with neutral hydrogen. Meanwhile, Barbosa (1987, 1990) argues that a neutral cloud of nitrogen extending from Titan's exosphere is a source of N+ ions in the outer magnetosphere.

RINGS Inside $\sim 4\ R_S$ the plasma temperature drops and the density (~ 100 cm^{-3}) is concentrated in a $<0.5\ R_S$ thin sheet in Saturn's equatorial plane (Bridge et al 1982, Richardson & Sittler 1990). The low plasma temperatures have been ascribed to interactions with ring material. Eviatar & Richardson (1990) propose that water ions form a dense "ionosphere" above Saturn's rings. This could explain several other puzzles of the Saturnian system such as the radiative cooling of plasma inside the orbit of Tethys which allows the E-ring to survive in the plasma gap between Tethys and the main rings. Investigations of magnetospheric processes associated with an extended ring system of particulate matter form a new area of research. If electrodynamic processes are responsible for certain phenomena observed in Saturn's rings then they may play a role in the cosmogeny of the ring system (see reviews by Mendis et al 1984 and Esposito 1993).

Uranus

Two factors make Uranus' magnetosphere rather empty: There are only a few, small icy satellites and the solar wind driven convection quickly circulates material through the magnetosphere in a few days (for reviews of Uranus' magnetosphere see Bergstrahl et al 1991). The low-density plasma was observed to be all protons, with an upper limit of 10^{-2} cm^{-3} on heavy ions, which is consistent with low sputtering rates for the icy satellites (Cheng 1987). There are two distinct plasma populations: a cold population (10's of eV) that is observed throughout the magnetosphere reaching maximum densities of about 2 cm^{-3} and a hot (keV) population that appears to be excluded from the region inside 5 R_U (sketched in Figure 8).

There are two possible sources of the low-energy plasma: ionization of the unusually dense neutral hydrogen corona of Uranus (Broadfoot et al 1986) and outflow of cold plasma from Uranus' ionosphere. McNutt et al (1987) estimate that a convection time of 1–10 days is consistent with a source strength of 10^{25} s^{-1} by electron impact ionization of the neutral hydrogen corona and the observed densities of 0.1–1 cm^{-3}. Cheng (1987) estimates that the outflow of ionospheric plasma could be a comparable source for residence times of 30 days and puts an upper limit on the amount of solar wind plasma that reaches the inner magnetosphere (<6 R_U) at 10^{-5} of the solar wind hitting the magnetosphere. Cheng (1987) concludes that most of the solar wind material that enters is transported by the convection and may not reach the inner magnetosphere.

The spatial distribution and energy spectrum of the hot (keV) population are consistent with ionization of more distant regions of the hydrogen cloud (where the ions pick up larger corotational energies) followed by adiabatic heating due to compression while being convected inward from the nightside of the magnetosphere (Selesnick & McNutt 1987). Alternatively, one should note that in the inner regions of the Earth's plasma sheet one finds keV material (including O^+ ions) that has escaped from the ionosphere at high latitudes and been convected towards the equator in the inner magnetotail where it is heated by adiabatic compression in the sunward return flow of the solar wind driven convection (Cowley 1980). Mauk et al (1987) point out several features of the energetic particle data obtained in the plasma sheet and magnetotail at Uranus that are reminiscent of the Earth. At Uranus, unfortunately, the same ion species, H^+, is produced from the solar wind, the ionosphere, and the neutral hydrogen corona so we have to work harder to test if Uranus' magnetosphere is so similar to that of the Earth.

Neptune

Although Neptune's large satellite Triton orbits at 14.6 R_N, well inside the magnetosphere, it appears that rapid transport does not allow large plasma densities to build up. Protons and N^+ ions were detected at typical densities of 0.1 cm^{-3}, reaching a few cm^{-3} very close to Neptune (Belcher et al 1989, Richardson et al 1991). Richardson et al (1991) show that variations in both density and temperature with distance are consistent with Triton being a 10^{25} ions s^{-1} source of plasma with strong inward diffusion on a time scale of a few days. The N^+ ions are thought to be produced directly in the interaction of the magnetospheric plasma with the satellite's atmosphere (Yung & Lyons 1990), while the protons come from the ionization of a large hydrogen cloud ($n \approx 300$ cm^{-3}) which extends inward from Triton's orbit to about 8 R_N (Zhang et al 1991). Inside a distance of ~7 R_N there appear to be significant losses, due to pitch-angle scattering which causes the ions to stream along the magnetic field into the atmosphere of Neptune or due to charge-exchange if Neptune has a dense neutral hydrogen corona (Richardson et al 1991). The difficulty with this simple picture is that several predictions of strong longitudinal asymmetries in plasma density (Broadfoot et al 1989, Hill & Dessler 1990, Richardson et al 1990, Selesnick 1990) are not observed. Moreover, the convection models of Selesnick (1990) and Hill & Dessler (1990) predict outward diffusion on the dayside of the magnetosphere whereas the data indicate inward diffusion (Richardson et al 1991). Clearly, considerable further study is required in order to understand the plasma configuration and dynamics of Neptune's magnetosphere.

ENERGETIC PARTICLE POPULATIONS

All magnetospheres have significant populations of particles with energies well above the thermal population, at keV–MeV energies (see reviews in

Dessler 1983, Gehrels & Mathews 1984, Bergstrahl et al 1991, as well as Mauk et al 1991). These particles are largely trapped by the strong planetary magnetic field in long-lived radiation belts (summarized in Table 3). Where do these energetic particles come from? Since the interplanetary medium includes energetic particles of solar and galactic origins, an obvious possibility is that these energetic particles were "captured" from the external medium. In the cases of the giant planets, the observed high fluxes are hard to explain without additional internal sources. Compositional evidence also implies that some fraction of the thermal plasma is accelerated to high energies, either by tapping the rotational energy of the planet, in the cases of Jupiter and Saturn, or by processes in the distorted magnetic field in the tail, in the cases of Earth, Uranus, and Neptune. Particle drifts in a nonuniform magnetic field lead to ions and electrons drifting in opposite directions around the planet, producing an azimuthal electric current, called a *ring current*. If the energy density of the energetic particle populations is comparable to the magnetic field energy (i.e. $\beta > 1$), then the ring current produces a magnetic field (ΔB) that significantly perturbs the planetary magnetic field. Table 3 shows that this is the case for Jupiter and Saturn, where the high particle pressures inflate and stretch out the magnetic field and generate a strong ring current in the magnetodisc. While Uranus and Neptune have significant radiation belts, the energy density remains small compared with the magnetic field (i.e. $\beta \ll 1$) and the ring current is very weak. Dessler & Parker (1959) related the

Table 3 Energetic particle characteristics

	Earth	Jupiter	Saturn	Uranus	Neptune
Phase Space Density[a] 100 MeV/G ions $c^2(cm^2 s\ sr\ MeV^3)^{-1}$	2×10^4	2×10^5	6×10^4	8×10^2	8×10^2
β[b]	<1	>1	>1	~0.1	~0.2
ΔB (nT)	10-23	200	10	<1	<0.1
$E_{Particle} / E_{Magnetic\ Field}$	$2\text{-}5 \times 10^{-4}$	3×10^{-4}	3×10^{-4}	$<3 \times 10^{-5}$	$<10^{-5}$
Auroral Power (Watts)	10^{10}	10^{14}	10^{11}	10^{11}	$<10^{8}$

[a] From Cheng et al (1987). Neptune value from A. F. Cheng (private communication).
[b] In the body of the magnetosphere. Higher values are often found in the tail plasma sheet and, in the case of the Earth, at times of enhanced ring current.

magnetic field produced by the ring current to the kinetic energy of the trapped particle population, scaled to the dipole magnetic energy external to the planet. Applying the Dessler-Parker relation to planetary magnetospheres we find that while the total energy content of magnetospheres varies by many orders of magnitude and the sources are very different, it appears that the particle energy builds up to only 1/1000 of the magnetic field energy in each magnetosphere. Earth, Jupiter, and Saturn all have energetic particle populations close to this limit (e.g. Connerney et al 1983). The radiation belts of Uranus and Neptune are much less than this limit, perhaps because it is harder to trap particles in nondipolar magnetic fields (Connerney et al 1991).

Where do these energetic particles go? The majority appear to diffuse inwards towards the planet. Loss processes for energetic particles in the inner magnetospheres are satellite absorption, charge exchange with neutral clouds, and scattering by waves so that the particles stream into the upper atmospheres of the planets where they can excite auroral emission and deposit large amounts of energy (e.g. Cheng et al 1987).

REMOTE SENSING OF GIANT PLANET MAGNETOSPHERES

The Voyager ultraviolet spectrometer measurements of molecular and atomic hydrogen emissions from Jupiter, Saturn, and Uranus show strong enhancements near the planet's magnetic poles. The nature of the spectra and the location of the emissions first indicate that they are auroral and, further, reveal the characteristics of the particles that are precipitating from the magnetosphere into the upper atmosphere. While it is not clear which species (electrons, protons, or heavy ions) are responsible for each component of the auroral emissions, Table 3 shows that the planets with higher radiation belt fluxes have brighter aurora. The UV emissions from Neptune are reported by Broadfoot et al (1989) and Sandel et al (1991) to be very weak with no clear auroral signature. The roles of Triton and the nondipolar magnetic field in the origin of Neptune's weak aurora are currently under debate (Sandel et al 1991, Cheng 1991a,b).

The giant planets are strong radio sources as illustrated by Figure 12 (see reviews by Zarka 1991 and Leblanc 1991). The radio sources of the Earth and Saturn are located in the auroral polar regions of both hemispheres and the beams are fixed in local time—on the nightside for the Earth and on the dayside for Saturn. For Jupiter the presence of Io and the plasma torus lead to many radio emissions. Uranus' main radio source corotates with the planet and is located on the night polar region of the planet. Among the proposed generation mechanisms the most often

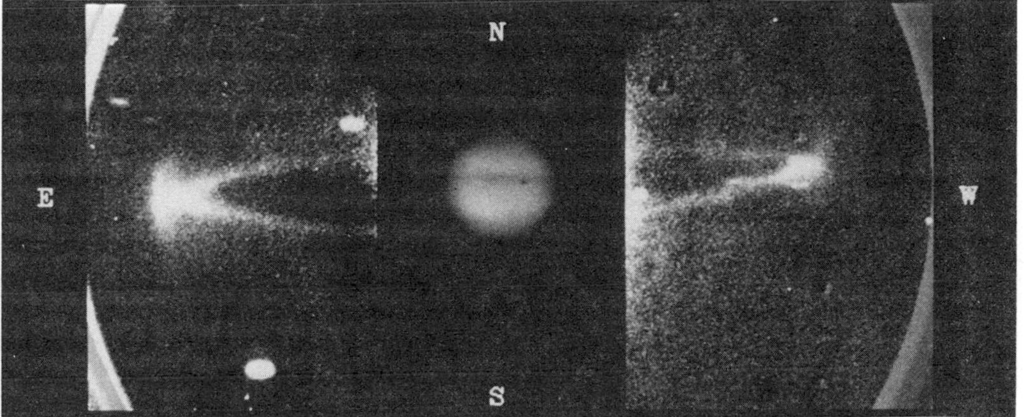

Figure 11 Image of optical emission (6731 Å) from S^+ ions in the Io plasma torus obtained on January 11, 1990 by Schneider and Trauger at the Catalina Observatory. The (circular) instrument field of view is 15 R_J wide. Io's orbital plane is horizontal. The "triptych" is a single image, but the left and right panels are enhanced differently to show detail, and Jupiter's image in the central panel is exposed through a strip of neutral density filter which transmits 10^{-4} of the light.

cited for the radio emission is the cyclotron maser instability. This mechanism predicts beaming of the emission in a narrow range of angles almost perpendicular to the magnetic field in the source region. Solar wind control of the radio emissions is very strong for Earth and Saturn but the degree of correlation of the solar wind fluctuations with Jupiter's radio emissions is much less. Again, the strength of radio emission from a planet seems to be related to the magnetospheric particle fluxes.

Flybys of the giant planets are rare. In 1996 the *Galileo* spacecraft will go into orbit around Jupiter and *Cassini* is currently scheduled to orbit Saturn in 2004. In the meantime, important observations of the magnetospheres of Jupiter and Saturn can be made from Earth (or Earth orbit in the case of X-ray and UV emissions) (Brown et al 1983, Feldman & Bagenal 1990). Figure 11 shows an image of S^+ emission from the Io torus obtained by N. Schneider and J. Trauger with a ground-based telescope. Jupiter's magnetosphere is a dynamic object. To understand its variability it is necessary to study the coupled, highly nonlinear system of the Io torus, the magnetosphere, and Jupiter's ionosphere (sketched in Figure 13). One of the aims of the International Jupiter Watch is to make synoptic measurements of emissions from the Io torus at the same time as measuring the auroral emissions at radio, infrared, ultraviolet, and X-ray wavelengths (Russell et al 1990). Although the large distances to Uranus and Neptune severely limit the emissions that can be detected from Earth, it would be

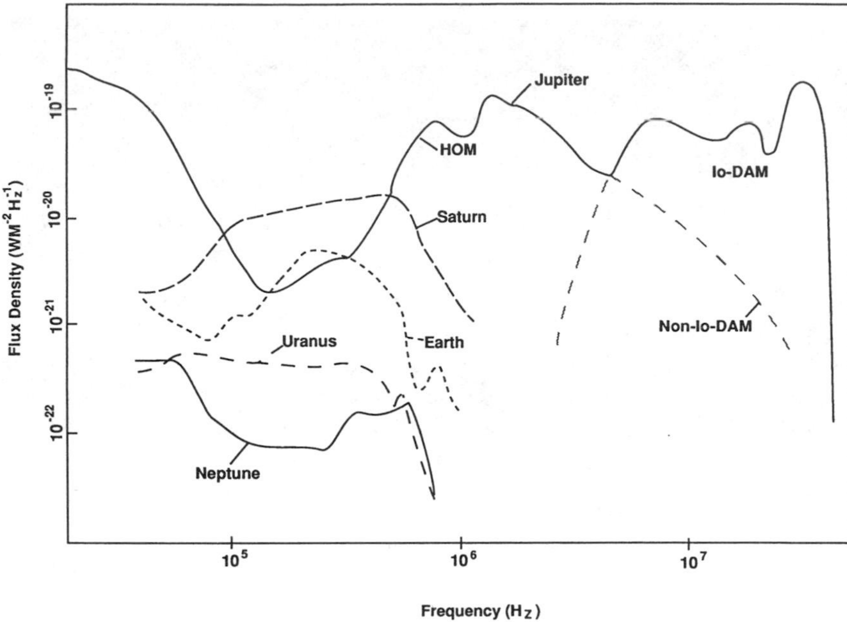

Figure 12 Averaged spectra of the auroral radio emissions computed from 2 days of Voyager Planetary Radio Astronomy data recorded from a range of 100–200 planetary radii. (Adapted from Zarka 1991.)

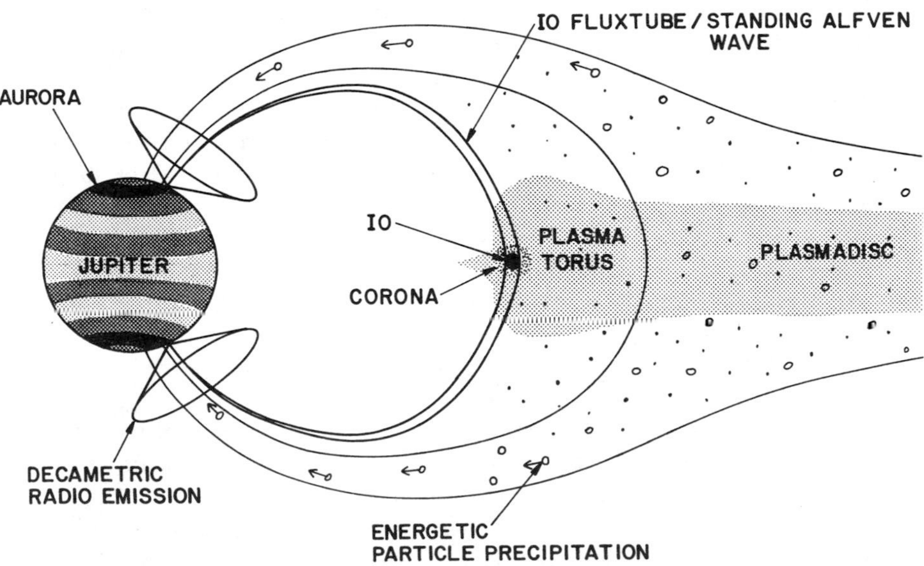

Figure 13 The connections between Io, the magnetosphere, and Jupiter's atmosphere.

valuable to send orbiting spacecraft to these irregular magnetospheres, particularly at a season when Uranus' rotation axis is pointed away from the solar wind direction to produce a configuration very different from that explored by Voyager 2 in 1986.

ACKNOWLEDGMENTS

The author is very grateful for helpful comments from Jack Connerney, Andy Cheng, Alex Dessler, Arkee Eviatar, Melissa McGrath, Ralph McNutt Jr., John Richardson, Nick Schneider, and Richard Selesnick. She acknowledges support by NASA grant NAGW 1622.

Literature Cited

Acuña, M. H., Ness, N. F. 1976. The main magnetic field of Jupiter. *J. Geophys. Res.* 81: 2917–22

Acuña, M. H., Neubauer, F. M., Ness, N. F. 1981. Standing Alfvén wave current system at Io: Voyager 1 observations. *J. Geophys. Res.* 86: 8513–22

Atreya, S. K., Pollack, J. B., Matthews, M. S., eds. 1989. *Origin and Evolution of Planetary and Satellite Atmospheres*. Tuscon: Univ. Arizona Press. 881 pp.

Bagenal, F. 1985a. Planetary Magnetospheres. In *Solar System Magnetic Fields*, ed. E. Priest, pp. 224–56. Dordrecht: Reidel. 290 pp.

Bagenal, F. 1985b. Plasma conditions inside Io's orbit: Voyager measurements. *J. Geophys. Res.* 90: 311–24

Bagenal, F. 1989. Torus-magnetosphere coupling. See Belton et al 1989, pp. 196–210

Bagenal, F., Sullivan, J. D. 1981. Direct plasma measurements in the Io torus and inner magnetosphere of Jupiter. *J. Geophys. Res.* 86: 8447–66

Bagenal, F., McNutt, R. L., Belcher, J. W., Bridge, H. S., Sullivan, J. D. 1985. Revised ion temperatures for *Voyager* plasma measurements in the Io plasma torus. *J. Geophys. Res.* 90: 1755–57

Bagenal, F., Belcher, J. W., Sittler, E. C., Lepping, R. P. 1987. The Uranian bow shock: Voyager 2 inbound observations of a high Mach number shock. *J. Geophys. Res.* 92: 8603–12

Ballester, G. E., Strobel, D. F., Moos, H. W., Feldman, P. D. 1990. The atmospheric abundance of SO_2 on Io. *Icarus* 88: 1–23

Barbosa, D. D. 1987. Titan's atomic nitrogen torus: Inferred properties and consequences for the Saturnian aurora. *Icarus* 72: 53–61

Barbosa, D. D. 1990. Radial diffusion of low-energy ions in Saturn's magnetosphere. *J. Geophys. Res.* 95: 17,167–77

Barbosa, D. D., Coroniti, F. V., Eviatar, A. 1983. Coulomb thermal properties and stability of the Io plasma torus. *Astrophys. J.* 274: 429–42

Barbosa, D. D., Eviatar, A., Siscoe, G. L. 1984. On the acceleration of energetic ions in Jupiter's magnetosphere. *J. Geophys. Res.* 89: 3789–3800

Barbosa, D. D., Coroniti, F. V., Kurth, W. S., Scarf, F. L. 1985. *Astrophys. J.* 289: 392–408

Barbosa, D. D., Moreno, M. A. 1988. A comprehensive model of ion diffusion and charge exchange in the cold Io torus. *J. Geophys. Res.* 93: 823–36

Barbosa, D. D., Kurth, W. S. 1990. Theory and observations of electrostatic ion waves in the cold Io turns. *J. Geophys. Res.* 95: 6443–50

Barnett, A. 1986. In situ measurements of the plasma bulk velocity near the Io flux tube. *J. Geophys. Res.* 91: 3011–19

Behannon, K. W., Lepping, R. P., Sittler, E. C. Jr., Ness, N. F., Mauk, B. H., et al. 1987. The magnetotail of Uranus. *J. Geophys. Res.* 92: 15,354–66

Belcher, J. W. 1983. The low-energy plasma in the Jovian magnetosphere. See Dessler 1983, pp. 68–105

Belcher, J. W. 1987. The Jupiter-Io connection: An Alfvén engine in space. *Science* 238: 170–76

Belcher, J., Goertz, C. K., Sullivan, J. D., Acuña, M. H. 1981. Plasma observations of the Alfvén wave generated by Io. *J. Geophys. Res.* 86: 8508–12

Belcher, J. W., Bridge, H. S., Bagenal, F., Coppi, B., Divers, O., et al. 1989. Plasma observations near Neptune: Initial results from Voyager 2. *Science* 246: 1478–83

Belcher, J. W., McNutt, R. L. Jr., Richardson, J. D., Selesnick, R. S., Sittler, E. C., Bagenal, F. 1991. The plasma environment in Uranus. See Bergstrahl et al 1991

Belton, M. J. S., West, R. A., Rahe, J., ed. 1989. *Time-Variable Phenomena in the Jovian System*. NASA SP-494. 403 pp.

Bergstrahl, J., Miner, E., Mathews, M., eds. 1991. *Uranus*. Tucson: Univ. Arizona Press

Biermann, L. 1957. Solar Corpuscular Radiation and the Interplanetary Gas. *Observatory* 77: 109

Bigg, E. K. 1964. Influence of the satellite Io on Jupiter's decametric emission. *Nature* 203: 1008–10

Brice, N. M., Ioannidis, G. A. 1970. The magnetospheres of Jupiter and Earth. *Icarus* 13: 173–83

Bridge, H. S., Belcher, J. W., Lazarus, A. J., Sullivan, J. D., Bagenal, F., et al. 1979. Plasma observations near Jupiter: Initial results from Voyager 1 encounter with Jupiter. *Science* 204: 972–76

Bridge, H. S., Bagenal, F., Belcher, J. W., Lazarus, A. J., McNutt, R. L., et al. 1982. Plasma observation near Saturn: Initial results from Voyager 2. *Science* 215: 563–70

Bridge, H. S., Belcher, J. W., Coppi, B., Lazarus, A. J., McNutt, R. L. Jr., et al. 1986. Plasma observations near Uranus: initial results from Voyager 2. *Science* 233: 89–93

Broadfoot, A. L., Belton, M. J. S., Takacs, P. Z., Sandel, B. R., Shemansky, D. E., et al. 1979. Extreme ultraviolet observations from Voyager 1 encounter with Jupiter. *Science* 204: 979–82

Broadfoot, A. L., Sandel, B. R., Shemansky, D. E., Holberg, J. B., Smith, G. R., et al. 1981. *Science* 212: 206–11

Broadfoot, A. L., Herbert, F., Holberg, J. B., Hunten, D. M., Kumar, S., et al. 1986. Ultraviolet spectrometer observations of Uranus. *Science* 233: 74–79

Broadfoot, A. L., Atreya, S. K., Bertaux, J. L., Blamont, J. E., Dessler, A. J., et al. 1989. Ultraviolet spectrometer observations of Neptune and Triton. *Science* 246: 1459–66

Brown, R. A. 1974. In *Exploration of the Planetary System*, ed. A. Woszczyk, C. Iwaniszewska, pp. 527–31. Hingham, MA: Reidel

Brown, R. A. 1976. A model of Jupiter's solar nebula. *Astrophys. J.* 206: L179–83

Brown, R. A. 1981. The Jupiter hot plasma torus: observed electron temperature and energy flows. *Astrophys. J.* 244: 1072–80

Brown, R. A., Pilcher, C. B., Strobel, D. F. 1983. Spectrophotometric studies of the Io torus. See Dessler 1983, pp. 197–225

Burke, L. F., Franklin, K. L. 1955. Observations of a variable radio source associated with the planet Jupiter. *J. Geophys. Res.* 60: 213–17

Caudal, G. 1986. A self-consistent model of Jupiter's magnetodisc including the effects of centrifugal force and pressure. *J. Geophys. Res.* 91: 4201–21

Caudal, G., Connerney, J. E. P. 1989. Plasma pressure in the environment of Jupiter inferred from Voyager 1 magnetometer observations. *J. Geophys. Res.* 94: 15,055–61

Chapman, S., Ferraro, V. C. A. 1931. A new theory of magnetic storms. *Terr. Magn. Atmos. Electr.* 36: 77–97

Chappell, C. R., Moore, T. E., Waite, J. H. 1987. The ionosphere as a fully adequate source of plasma for the Earth's magnetosphere. *J. Geophys. Res.* 92: 5896–5909

Cheng, A. F. 1987. Proton and oxygen plasmas at Uranus. *J. Geophys. Res.* 92: 15,309–14

Cheng, A. F. 1988. Two classes of models for temporal variability of the Io torus. *J. Geophys. Res.* 93: 12,751–60

Cheng, A. F. 1991a. Triton torus and Neptune aurora. *Geophys. Res. Lett.* 17: 1669–72

Cheng, A. F. 1991b. Global magnetic anomaly and aurora of Neptune. *Geophys. Res. Lett.* 17: 1697–1700

Cheng, A. F., Krimigis, S. M., Mauk, B. H., Keath, E. P., Maclennan, C. G., et al. 1987. Energetic ion and electron phase space densities in the magnetosphere of Uranus. *J. Geophys. Res.* 92: 15,313–28

Cheng, A. F., Johnson, R. E. 1989. Effects of magnetosphere interactions on origin and evolution of atmospheres. See Atreya et al 1989, pp. 682–722

Cheng, A. F., Krimigis, S. M. 1989. A model of global convection in Jupiter's magnetosphere. *J. Geophys. Res.* 94: 12,003–12

Connerney, J. E. P. 1981. The magnetic field of Jupiter: A generalized inverse approach. *J. Geophys. Res.* 86: 7679–93

Connerney, J. E. P. 1987. The magnetospheres of Jupiter, Saturn, and Uranus. *J. Geophys. Res.* 25: 615–38

Connerney, J. E. P., Ness, N. F., Acuña, M. H. 1982. Zonal harmonic model of Saturn's magnetic field from Voyager 1 and 2 observations. *Nature* 298: 44–46

Connerney, J. E. P., Acuña, M. H., Ness, N. F. 1983. Currents in Saturn's magnetosphere. *J. Geophys. Res.* 88: 8779–89

Connerney, J. E. P., Acuña, M. H., Ness, N. F. 1987. The magnetic field of Uranus. *J. Geophys. Res.* 92: 15,329–36

Connerney, J. E. P., Acuña, M. H., Ness, N.

F. 1991. The magnetic field of Neptune. *J. Geophys. Res.* 96: 19,023–42

Cowley, S. W. H. 1980. Plasma populations in the simple open model magnetosphere. *Space Sci. Rev.* 26: 217–75

Dessler, A. J., ed. 1983. *Physics of the Jovian Magnetosphere*. Cambridge: Cambridge Univ. Press. 544 pp.

Dessler, A. J., Parker, E. N. 1959. Hydromagnetic theory of geomagnetic storms. *J. Geophys. Res.* 64: 2239–52

Drell, S. D., Foley, H. M., Ruderman, M. A. 1965. Drag and propulsion of large satellites in the ionosphere: An Alfvén propulsion engine in space. *J. Geophys. Res.* 70: 3131–45

Durrance, S. T., Feldman, P. D., Weaver, H. A. 1983. Rocket detection of ultraviolet emission from neutral oxygen and sulfur in the Io torus. *Astrophys. J.* 267: L125–29

Engle, I. M. 1991. Idealized Voyager Jovian magnetosphere shape and field. *J. Geophys. Res.* 96: 7793–7802

Esposito, L. 1993. Planetary Rings. *Annu. Rev. Earth Planet. Sci.* In preparation

Eviatar, A., Siscoe, G. L., Scudder, J. D., Sittler, E. C., Sullivan, J. D. 1982. The plumes of Titan. *J. Geophys. Res.* 87: 8091–8103

Eviatar, A., Barbosa, D. D. 1984. Jovian magnetospheric neutral wind and auroral precipitation flux. *J. Geophys. Res.* 89: 7393–98

Eviatar, A., Richardson, J. D. 1990. Water group plasma in the magnetosphere of Saturn. *Ann. Geophys.* 8: 725–32

Fazakerley, A. N. 1990. *Plasma instabilities and transport in the magnetosphere of Jupiter*. PhD thesis. Imperial Coll. Sci., Tech. & Medicine

Feldman, P. D., Bagenal, F. 1991. EUV planetary astronomy. In *Extreme Ultraviolet Astronomy*, ed. R. F. Malina, S. Bowyer, pp. 252–60. New York: Pergamon

Gehrels, T., Mathews, M. S., eds. 1984. *Saturn*. Tucson: Univ. Ariz. Press

Gledhill, J. A. 1967. Magnetosphere of Jupiter. *Nature* 214: 155–56

Goertz, C. K. 1983. Detached plasma blobs in Saturn's turbulence layer. *Geophys. Res. Lett.* 10: 455–58

Goertz, C. K., Deift, P. A. 1973. Io's interaction with the magnetosphere. *Planet Space Sci.* 21: 1399–1415

Gold, T. 1959. *Symposium on the exploration of space* 64: 1665–74

Goldreich, R., Lynden-Bell, D. 1969. Io, a Jovian unipolar conductor. *Astrophys. J.* 156: 59–78

Gurnett, D. A. 1972. Sheath effects and related charge-particle acceleration by Jupiter's satellite Io. *Astrophys. J.* 175: 525–33

Hill, T. W. 1979. Inertial limit on corotation. *J. Geophys. Res.* 84: 6554–58

Hill, T. W. 1980. Corotation lag in Jupiter's magnetosphere: A comparison of observation and theory. *Science* 207: 301–2

Hill, T. W. 1986. The magnetosphere of Uranus: A resonantly driven oscillator? *Eos Trans. Am. Geophys. Union* 67: 341

Hill, T. W., Michel, F. C. 1976. Heavy ions from the Galilean satellites and the centrifugal distortion of the Jovian magnetosphere. *J. Geophys. Res.* 81: 4561–65

Hill, T. W., Dessler, A. J. 1990. Convection in Neptune's magnetosphere. *Geophys. Res. Lett.* 17: 1677–80

Hill, T. W., Dessler, A. J. 1991. Plasma motions in planetary magnetospheres. *Science* 252: 410–15

Huang, T. S., Siscoe, G. L. 1987. Types of planetary tori. *Icarus* 70: 366–78

Ingersoll, A. P. 1989. Io meteorology: How atmospheric pressure is controlled locally by volcanos and surface frost. *Icarus* 81: 289–313

Ip, W.-H. 1990. Neutral gas-plasma interaction: the case of the Io plasma torus. *Adv. Space Res.* 10: 15–23

IUGG Report 1991. U.S. Natl. Rep. IUGG 1987–1990: Solar-Planetary Relationships. *Rev. Geophys.* (Suppl.)

Johnson, R. E. 1989. Plasma ion heating of an SO_2 atmosphere on Io. *Geophys. Res. Lett.* 16: 1117–20

Johnson, R. E., Pospieszalska, M. K., Sittler, E. C., Cheng, A. F., Lanzerotti, L. J., Sieveka, E. M. 1989. The neutral cloud and heavy ion inner torus at Saturn. *Icarus* 77: 311–29

Khurana, K. K., Kivelson, M. G. 1989a. Ultra-low frequency waves in Jupiter's middle magnetosphere. *J. Geophys. Res.* 94: 5241–54

Khurana, K. K., Kivelson, M. G. 1989b. On Jovian plasma sheet structure. *J. Geophys. Res.* 94: 11,791–803

Krimigis, S. M., Roelof, E. C. 1983. Low-energy particle population. See Dessler 1983, pp. 106–56

Kupo, I., Mekler, Y., Eviatar, A. 1976. Detection of ionized sulfur in the Jovian magnetosphere. *Astrophys. J. Lett.* 205: L51–53

Kurth, W. S., Gurnett, D. A. 1991. Plasma waves in planetary magnetospheres. *J. Geophys. Res.* 96: 18,977–91

Langel, R. A., Estes, R. H. 1982. A geomagnetic field spectrum. *Geophys. Res. Lett.* 9: 250–53

Langel, R. A., Estes, R. H. 1985. The near-earth magnetic field at 1980 determined from MAGSAT data. *J. Geophys. Res.* 90: 2495–2509

Leblanc, Y. 1991. The radio sources in the magnetospheres of Jupiter, Saturn and Uranus after Voyager mission. *Adv. Space Phys.* 10: 39–48

Linker, J. A., Kivelson, M. G., Walker, R. J. 1988. An MHD simulation of plasma flow past Io: Alvén and slow mode perturbations. *Geophys. Res. Lett.* 15: 1311–14

Linker, J. A., Kivelson, M. G., Walker, R. J. 1989. The effect of mass loading on the temperature of a flowing plasma. *Geophys. Res. Lett.* 16: 763–66

Marshall, L., Kibby, W. F. 1967. Stimulation of Jupiter's radio emission by Io. *Nature* 214: 126

Mauk, B. H., Krimigis, S. M. 1987. Radial force balance within Jupiter's magnetosphere. *J. Geophys. Res.* 92: 9931–41

Mauk, B. H., Krimigis, S. M., Keath, E. P., Cheng, A. F., Armstrong, T. P., et al. 1987. The hot plasma and radiation environment of the Uranian magnetosphere. *J. Geophys. Res.* 92: 15,283–308

Mauk, B. H., Keath, E. P., Kane, M., Krimigis, S. M., Cheng, A. F., et al. 1991. *J. Geophys. Res.* In press

McGrath, M. A., Johnson, R. E. 1987. Magnetospheric plasma sputtering of Io's atmosphere. *Icarus* 69: 519–31

McNutt, R. L. 1983. Force balance in the magnetospheres of Jupiter and Saturn. *Adv. Space Res.* 3: 55–58

McNutt, R. L. 1991. The magnetospheres of the outer planets. See IUGG Report 1991, pp. 985–97

McNutt, R. L., Belcher, J. W., Sullivan, J. D., Bagenal, F., Bridge, H. S. 1979. Departure from rigid corotation of plasma in Jupiter's dayside magnetosphere. *Nature* 280: 803

McNutt, R. L., Belcher, J. W., Bridge, H. S. 1981. Positive ion observations in the middle magnetosphere of Jupiter. *J. Geophys. Res.* 86: 8319–42

McNutt, R. L., Selesnick, R. S., Richardson, J. D. 1987. Low-energy plasma observations in the magnetosphere of Uranus. *J. Geophys. Res.* 92: 4399–4410

Mekler, Y., Eviatar, A., Kupo, I. 1977. Jovian sulfur nebula. *J. Geophys. Res.* 82: 2809–14

Melrose, D. B. 1967. Rotational effects on the distribution of thermal plasma in the magnetosphere of Jupiter. *Planet. Space Sci.* 15: 381–93

Mendillo, M., Baumgardner, J., Flynn, B., Hughes, J. 1990. The extended sodium nebula of Jupiter. *Nature* 348: 312–14

Mendis, D. A., Hill, J. R., Ip, W.-H., Goertz, C. K., Grün, E. 1984. Electrodynamic processes in the ring system of Saturn. See Gehrels & Mathews 1984, pp. 546–92

Moreno, M., Schubert, G., Baumgartner, J., Kivelson, M. G., Paige, D. A. 1991. Io's volcanic and sublimation atmospheres. *Icarus* 93: 63–81

Moses, S. L., Coroniti, F. V., Kennel, C. F., Kurth, W. S., Gurnett, D. A. 1990. Comparison of plasma wave measurements in the bow shocks at Earth, Jupiter, Saturn, Uranus, and Neptune. *Geophys. Res. Lett.* 17: 1653–56

Ness, N. F., Acuña, M. H., Lepping, R. P., Burlaga, L. F., Behannon, K. W., et al. 1979. Magnetic field studies at Jupiter by Voyager 1: Preliminary results. *Science* 204: 982–86

Ness, N. F., Acuña, M. H., Behannon, K. W., Burlaga, L. F., Connerney, J. E. P., et al. 1986. Magnetic fields at Uranus. *Science* 233: 85–89

Ness, N. F., Acuña, M. H., Burlaga, L. F., Connerney, J. E. P., Lepping, R. P., Neubauer, F. M. 1989. Magnetic fields at Neptune. *Science* 246: 1473–78

Neubauer, F. M. 1980. Nonlinear standing Alfvén wave current system at Io: Theory. *J. Geophys. Res.* 85: 1171–78

Neubauer, F. M., Gurnett, D. A., Scudder, J. D., Hartle, R. D. 1984. Titan's magnetospheric interaction. See Gehrels & Mathews 1984, pp. 760–87

Neugebauer, M., Snyder, C. W. 1962. The mission of Mariner II: Preliminary observations. *Science* 138: 1095–96

Paranicas, C. P., Cheng, A. F., Mauk, B. H., Krimigis, S. M. 1990. Ion space densities in the Jovian magnetosphere. *J. Geophys. Res.* 95: 20,833–38

Parker, E. N. 1969. The occasional reversal of the geomagnetic field. *Astrophys. J.* 158: 815–27

Piddington, J. H., Drake, J. F. 1968. Electrodynamic effects of Jupiter's satellite Io. *Nature* 217: 935–37

Pizzo, V. J., Holzer, T., Sime, D. G., eds. 1988. *Proc. Sixth Int. Solar Wind Conf.* High Altitude Obs. Boulder: NCAR

Rädler, K.-H., Ness, N. F. 1990. The symmetry properties of planetary magnetic fields. *J. Geophys. Res.* 95: 2311–18

Richardson, J. D. 1986. Thermal ions at Saturn: Plasma parameters and implications. *J. Geophys. Res.* 91: 1381–89

Richardson, J. D., Siscoe, G. L., Bagenal, F., Sullivan, J. D. 1980. Time dependent plasma injection by Io. *Geophys. Res. Lett.* 7: 37–40

Richardson, J. D., Siscoe, G. L. 1983. The problem of cooling the cold Io torus. *J. Geophys. Res.* 88: 2001–9

Richardson, J. D., Eviatar, A., Siscoe, G. L.

1986. Satellite tori at Saturn. *J. Geophys. Res.* 91: 8749–55

Richardson, J. D., McNutt, R. L. Jr. 1987. Observational constraints on interchange models at Jupiter. *Geophys. Res. Lett.* 14: 64–67

Richardson, J. D., Sittler, E. C. 1990. A plasma density model for Saturn based on Voyager observations. *J. Geophys. Res.* 95: 12,019–31

Richardson, J. D., Eviatar, A., Delitsky, M. L. 1990. The Triton torus revisited. *Geophys. Res. Lett.* 17: 1673–76

Richardson, J. D., Belcher, J. W., Zhang, M., McNutt, R. L. 1991. Low-energy ions near Neptune. *J. Geophys. Res.* 96: 18,993–19,011

Russell, C. T., Hoppe, M. M., Livesey, W. A. 1982. Overshoots in planetary bow shocks. *Nature* 296: 45–48

Russell, C. T., Caldwell, J. J., de Pater, I., Goguen, J., Klein, M. K., et al. 1990. International Jupiter watch: A program to study the time variability of the Jovian system. *Adv. Space Res.* 10: 239–42

Sandel, B. R., Herbert, F., Dessler, A. J., Hill, T. W. 1991. Aurora and airglow on the night side of Neptune. *Geophys. Res. Lett.* 17: 1693–96

Scarf, F. L., Frank, L. A., Gurnett, D. A., Lanzerotti, L. J., Lazarus, A., Sittler, E. C. Jr. 1984. Measurements of plasma, plasma waves and suprathermal charged particles in Saturn's inner magnetosphere. See Gehrels & Mathews 1984, pp. 318–53

Schardt, A. W., Behannon, K. W., Lepping, R. P., Carbary, J. F., Eviatar, A., Siscoe, G. L. 1984. The outer magnetosphere. See Gehrels & Mathews 1984, pp. 416–62

Schulz, M., Paulikas, G. A. 1990. Planetary magnetic fields: A comparative view. *Adv. Space Res.* 10: 55–64

Schneider, N. M., Smyth, W. H., McGrath, M. A. 1989. Io's atmosphere and neutral clouds. See Belton et al 1989, pp. 75–99

Schneider, N. M., Hunten, D. M., Wells, W. K., Schultz, A. B., Fink, U. 1991a. The structure of Io's corona. *Astrophys. J.* 368: 298–315

Schneider, N. M., Trauger, J. T., Wilson, J. K., Brown, D. I., Evans, R. W., Shemansky, D. E. 1991b. Molecular origin of Io's fast sodium. *Science* 253: 1394–97

Selesnick, R. S. 1988. Magnetospheric convection in the nondipolar magnetic field of Uranus. *J. Geophys. Res.* 93: 9607–20

Selesnick, R. S. 1990. Plasma convection in Neptune's magnetosphere. *Geophys. Res. Lett.* 17: 1681–84

Selesnick, R. S., Richardson, J. D. 1986. Plasmasphere formation in arbitrarily oriented magnetospheres. *Geophys. Res. Lett.* 13: 624–27

Selesnick, R. S., McNutt, R. L. Jr. 1987. Voyager 2 plasma ion observations in the magnetosphere of Uranus. *J. Geophys. Res.* 92: 15,249–62

Shemansky, D. E. 1988. Energy branching in the Io plasma torus: The failure of neutral cloud theory. *J. Geophys. Res.* 93: 1773–84

Shemansky, D. E., Smith, G. R., Hall, D. T. 1985. Extended atomic hydrogen cloud in the Saturn system and a possible analogous distribution at Uranus. *Eos, Trans. Am. Geophys. Union* 66: 1108

Sieveka, E. M., Johnson, R. E. 1985. Nonisotropic coronal atmosphere on Io. *J. Geophys. Res.* 90: 5327–31

Siscoe, G. L. 1975. Particle and field environment of Uranus. *Icarus* 24: 311–24

Siscoe, G. L. 1977. On the equatorial confinement and velocity space distribution of satellite ions in Jupiter's magnetosphere. *J. Geophys. Res.* 82: 1641–45

Siscoe, G. L. 1979. Towards a comparative view of planetary magnetospheres. In *Solar System Plasma Physics Vol. II*, ed. C. F. Kennel, L. J. Lanzerotti, E. N. Parker. Amsterdam: North-Holland. 402 pp.

Siscoe, G. L., Slavin, J. A. 1979. Planetary magnetospheres. *Rev. Geophys. Space Phys.* 17: 1677–93

Sittler, E. C., Ogilvie, K. W., Selesnick, R. S. 1987. Voyager 2 observations. *J. Geophys. Res.* 92: 15,263–81

Slavin, J. A., Smith, E. J., Spreiter, J. R., Stahara, S. S. 1985. Solar wind flow about the outer planets: Gas dynamic modelling of the Jupiter and Saturn bow shocks. *J. Geophys. Res.* 90: 6275–86

Smith, R. A., Strobel, D. F. 1985. Energy partitioning in the Io plasma torus. *J. Geophys. Res.* 90: 9469–93

Smith, R. A., Bagenal, F., Cheng, A. F., Strobel, D. F. 1988. On the energy crisis in the Io plasma torus. *Geophys. Res. Lett.* 15: 545–48

Stevenson, D. J. 1982. Reducing the nonaxisymmetry of a planetary dynamo and an application to Saturn. *Geophys. Astrophys. Fluid Dyn.* 21: 113–27

Stern, D. P., Ness, N. F. 1982. Planetary Magnetospheres. *Annu. Rev. Astron. Astrophys.* 20: 139–61

Stone, R. G., Tsuritani, B. T., eds. 1985. *Collisionless Shocks in the Heliosphere.* Washington, DC: Am. Geophys. Union. 114 pp.

Strobel, D. F. 1989. Energetics, luminosity and spectroscopy of Io's torus. See Belton et al 1989, pp. 183–95

Trauger, J. T. 1984. The Jovian nebula: A post-Voyager perspective. *Science* 226: 337

Van Allen, J. A. 1984. Energetic particles in

the inner magnetosphere of Saturn. See Gehrels & Mathews 1984, pp. 281–317

Vasyliunas, V. M. 1983. Plasma distribution and flow. See Dessler 1983, pp. 395–453

Vasyliunas, V. M. 1986. The convection-dominated magnetosphere of Uranus. *Geophys. Res. Lett.* 13: 621–23

Voigt, G.-H., Hill, T. W., Dessler, A. J. 1983. The magnetosphere of Uranus. *Astrophys. J.* 266: 390–401

Voigt, G.-H., Behannon, K. W., Ness, N. F. 1987. Magnetic field and current structures in the magnetosphere of Uranus. *J. Geophys. Res.* 92: 15,337–46

Voigt, G.-H., Ness, N. F. 1990. The magnetosphere of Neptune: Its response to daily rotation. *Geophys. Res. Lett.* 17: 1705–8

Yung, Y. L., Lyons, J. R. 1990. Triton: Topside ionosphere and nitrogen escape. *Geophys. Res. Lett.* 17: 1717–20

Zarka, P. 1991. The auroral radio emissions from planetary magnetospheres. *Adv. Space Res.* In press

Zhang, M., Richardson, J. D., Sittler, E. C. Jr. 1991. Voyager 2 electron observations in the magnetosphere of Neptune. *J. Geophys. Res.* 96: 19,085–100

Annu. Rev. Earth Planet. Sci. 1992. 20:329–64

THE STRUCTURE OF MID-OCEAN RIDGES

Sean C. Solomon

Department of Earth, Atmospheric, and Planetary Sciences, Massachusetts Institute of Technology, Cambridge, Massachusetts 02139

Douglas R. Toomey

Department of Geological Sciences, University of Oregon, Eugene, Oregon 97403

KEY WORDS: crust, mantle, seismic velocity, seismic attenuation, magma chambers

INTRODUCTION

Mid-ocean ridges provide an important window into the processes of mantle convection and magmatism. The formation, cooling, and eventual subduction of oceanic lithosphere dominate both the large-scale dynamics of the upper mantle and the Earth's global heat loss. The melt generated during pressure-release melting of the upwelling mantle beneath mid-ocean ridges contributes most of the annual magmatic flux of the planet. The emplacement and differentiation of that melt at ridge axes reflects a complex interplay of cooling and deformational processes, which are strongly influenced in turn by the rate of plate separation and the detailed geometry in map view of the zone of crustal accretion.

Seismology offers a powerful tool to elucidate the processes of crustal and lithospheric formation and evolution at mid-ocean ridges. Compressional and shear wave speeds are sensitive to bulk composition, fracture and pore volume, and the presence of partial melt. The rates of attenuation of seismic waves are also strong functions of temperature and fracture porosity. Where there is a preferred orientation to anisotropic

0084–6597/92/0515–0329$02.00

minerals, such as shear-induced ordering of mantle olivine crystals, or to prominent structures, such as near-surface fissures or faults, the velocity and attenuation of seismic waves can display an anisotropy diagnostic of the governing process. The ability of a seismic experiment to resolve seismic properties at depth is limited principally by the seismic wavelength and also by experiment design and depth of interest.

While seismology has long been a primary source of information on the vertical structure of oceanic crust and the dependence of crustal and mantle structure on lithosphere age, a number of seismic experiments and related studies in the last 5 years have considerably altered our view of the detailed structure of mid-ocean ridge axes and of the processes by which new oceanic lithosphere forms. Multichannel reflection seismology is becoming increasingly used to map prominent structural boundaries, such as the tops of magma bodies in the axial crust and the top of the crust-mantle transition. Body wave tomography is being applied to large data sets from active seismic experiments to produce high-resolution images of seismic velocities and attenuation in two and three dimensions. Surface wave phase velocities and arrival times of direct and surface-reflected body waves at teleseismic distances, both in regional studies and as parts of global upper mantle tomographic inversions, are being employed to investigate the ocean ridge mantle in increasingly fine detail.

This paper reviews these recent results on the structure of mid-ocean ridges. After a brief introduction to the now-classical models of oceanic crustal structure and of ocean crustal formation, we turn to the new view of ridge-axis crustal structure from high-resolution seismology, particularly to the variation of that structure with spreading rate and along-axis at a given spreading rate. The details of these structural variations now constitute strong constraints on the nature of axial magma chambers and the relation of magmatism to morphological segmentation of the rise axis. We then review recent results on upper mantle structure beneath ridges, including variations with seafloor age, indications from anisotropy for directions of mantle flow, and long-wavelength along-axis variations in structure and their implications for lateral heterogeneity in mantle temperature and composition. Finally, we close with a summary of several outstanding issues in our understanding of mid-ocean ridge processes and offer suggestions for future seismic experiments to address these issues.

STRUCTURE OF MATURE OCEANIC CRUST AND UPPER MANTLE

It has been known for several decades that mature oceanic crust far from the axes of crustal accretion is vertically layered and is typically about 6–

7 km thick, largely independent of spreading rate (Raitt 1963, Reid & Jackson 1981). With improvements in seismic instrumentation, experiment design, and analysis techniques, the early view of the oceanic crust as a small number of homogeneous layers (Raitt 1963) was replaced by structural models involving smooth variations in velocity with depth and sharply depth-dependent vertical gradients in velocity (Spudich & Orcutt 1980). A typical distribution of *P*-wave velocity with depth in oceanic crust is shown in the upper part of Figure 1.

The interpretation of velocity structure in terms of lithology has been based on analogy with structures in ophiolite complexes and by comparison of velocities measured in laboratory samples from dredged rocks, oceanic drill cores, and ophiolites with the velocities measured in refraction experiments (Fox et al 1973, Christensen & Salisbury 1975, Spudich & Orcutt 1980, Casey et al 1981, Bratt & Purdy 1984). One such interpretation is given in Figure 1. The upper oceanic crust, or layer 2 of Raitt (1963), is thought to grade downward from volcanic flows and pillow basalts into a complex of sheeted diabase dikes. This upper portion of the classical ophiolite lithostratigraphy has been observed in a 1-km-deep drill hole in oceanic crust (Anderson et al 1982). The mid to lower crust, or layer 3 of Raitt (1963), has properties appropriate to the massive to cumulate gabbro layer seen in ophiolite complexes, and the uppermost oceanic mantle has a seismic velocity appropriate for depleted peridotite. On the basis of seismic refraction data, particularly amplitude-range information, and field studies in ophiolites, the boundaries between the principal layers likely display both considerable vertical complexity and significant lateral variability (Casey et al 1981, Bratt & Purdy 1984).

Seismic refraction studies in the oceans have shown that the uppermost oceanic mantle is azimuthally anisotropic, with the fast direction for *P* waves generally coincident with the spreading direction at the time of formation of the overlying crust (Hess 1964, Raitt et al 1969, Shimamura 1984, Shearer & Orcutt 1985). This pattern of anisotropy has also been observed for Rayleigh waves (Forsyth 1975, Nishimura & Forsyth 1988) and lithospheric guided waves (Butler 1985). The anisotropy is believed to result from the preferential alignment of the high-velocity *a*-axis of mantle olivine crystals during the shear strain accompanying the formation and growth of oceanic lithosphere (Hess 1964, Francis 1969). The range of wave types, periods, and distance intervals over which azimuthal anisotropy has been documented suggests that this pattern of anisotropy extends throughout most, if not all, of the mantle portion of oceanic lithosphere.

The vertical structure of oceanic crust is the result of injection, cooling, eruption, and solidification of basaltic magma generated by adiabatic decompression of the column of upwelling mantle beneath the mid-ocean

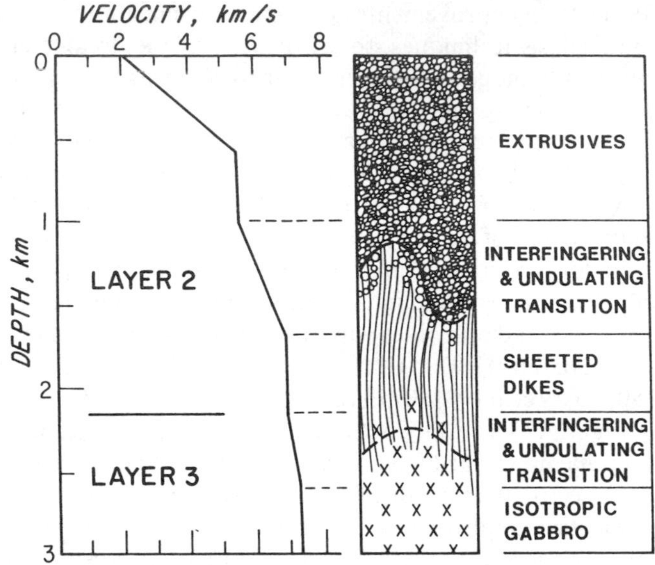

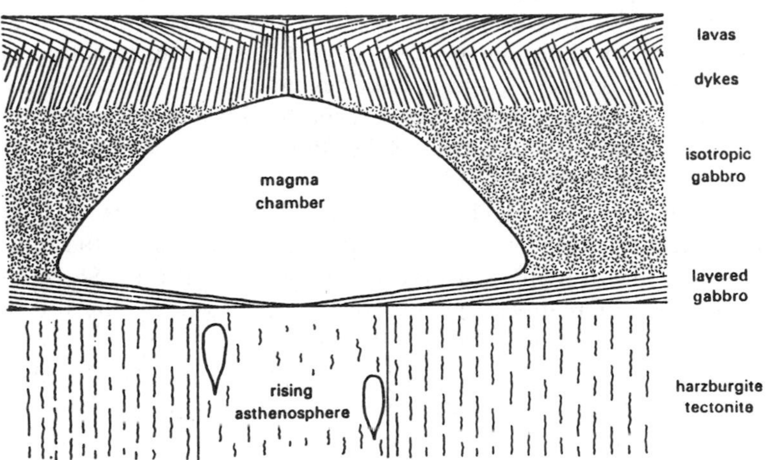

Figure 1 (*Top*) Generalized *P*-wave velocity structure of the upper oceanic crust and a geological interpretation (after Bratt & Purdy 1984) based on deep-sea drilling results and studies of ophiolite complexes. (*Bottom*) A possible schematic structure of a large ridge-axis magma chamber and its relation to oceanic crustal layering (from Cann 1974).

ridge axis. Models for crustal formation developed 10–20 years ago were influenced strongly by field geological observations in ophiolite complexes. They featured large magma chambers (Figure 1) which at least episodically filled the mid to lower oceanic crust and had cross-axis dimensions of a few to several tens of kilometers, depending on ridge spreading rate (Cann 1974, Kidd 1977, Casey & Karson 1981). According to these models, eruptions from these magma chambers form the effusive upper crustal layer and the underlying system of vertical dikes, and slow cooling and solidification of the chamber along its walls and at its base form the massive gabbro layer of the middle oceanic crust and the underlying mafic and ultramafic cumulate layers, respectively, with the downward transition to principally ultramafic material marking the top of the seismologically defined mantle. Such magma chambers should have strong signatures in the seismic structure of ridge-axis regions, and the search for and delineation of axial magma chambers (AMCs) has been the focus of a number of recent seismic experiments. The region of upwelling, melt generation, and vertical magma transport in the ridge-axis upper mantle poses a greater challenge to geophysical observation and has to date been comparatively unstudied.

RECENT RESULTS ON RIDGE-AXIS CRUSTAL STRUCTURE

In the last 10 years it has been increasingly recognized that the mid-ocean ridge is not a simple two-dimensional structure but rather varies systematically along its axis. The application of high-resolution bathymetric swath mapping and deep-towed and submersible observations of fine-scale volcanic, tectonic, and hydrothermal features has revealed that the ridge axis may be divided into distinct segments some 10–100 km in length bounded by structural features or morphological boundaries such as an offset in the location or trend in the axis of most recent crustal accretion. Several scenarios linking this segmentation to the production and transport of melt in the upper mantle have been advanced. According to these scenarios, ascending melt is focused, either by enhanced rates of melt production (Francheteau & Ballard 1983) or enhanced rates of upwelling (Whitehead et al 1984, Crane 1985, Schouten et al 1985), into magmatic centers spaced 10–100 km along the axis. Each magmatic center supplies the majority of heat and melt to the crust beneath a single ridge segment. Within each segment, the characteristics of faulting, hydrothermal circulation, and magmatic accretion vary systematically with distance from the magmatic center (Francheteau & Ballard 1983, Schouten et al 1985).

Recent work on the seismic structure of the crust along the axis of active spreading centers has illuminated and sharpened the concept of segmentation and its relationship to axial magmatism. Both the forms of segmentation and the seismic structure of the crust differ between fast-spreading (full divergence rates of about 70 mm/yr and greater) and slow-spreading (divergence rates less than about 70 mm/yr) ridges. We therefore divide the discussion to follow by spreading rate. Typical examples of the two classes of ridge systems are, respectively, the East Pacific Rise and the Mid-Atlantic Ridge.

Fast-Spreading Rises

Thermal models for mid-ocean ridges predict that shallow crustal temperatures are generally higher along faster spreading rises and that axial magma chambers may be steady-state features (Sleep 1975). The East Pacific Rise has thus been the focus of a number of different types of seismic experiments designed to investigate the cross-sectional shape of axial magma chambers, the along-axis variability in structure, and the relationship of structural variability to seafloor bathymetry, morphology, and tectonics and to the chemistry of seafloor basalts. Most of these studies have concentrated on a limited section of the northern East Pacific Rise between about 9° and 13°N (Figure 2). Important recent experiments include the multichannel seismic reflection imaging of the cross-axis and along-axis continuity of the top of the axial magma chamber (Detrick et al 1987, Mutter et al 1988), the modeling of *P*-wave travel times and amplitudes in two sets of expanding spread profiles (ESPs) shot along and parallel to the rise axis (Harding et al 1989, Vera et al 1990), and the two-dimensional (Burnett et al 1989, Caress et al 1992, Wilcock et al 1991) and three-dimensional (Toomey et al 1990) tomographic imaging of seismic velocity and attenuation in the rise axis region.

SHALLOW CRUST The uppermost kilometer of young oceanic crust, comprising the extrusive flows and pillow basalts and at least a portion of the sheeted dike complex, is the result of volcanic construction. This depth interval undergoes a systematic and geologically rapid evolution in seismic structure near the rise axis. The top 200–400 m of the igneous crust of the rise axis region is characterized by low seismic velocities ($V_p \sim$ 2–3 km/s), low Q, and a high bulk porosity of 10–20% (Herron 1982, Harding et al 1989, Vera et al 1990). Within a narrow zone centered approximately along the rise axis, however, the velocity and Q averaged over the upper 1 km are both high ($V_P \sim$ 4–5 km/s, $Q \sim$ 40–50) relative to the velocity and attenuation averaged over this depth interval only 1–2 km from the neovolcanic zone (McClain et al 1985, Burnett et al 1989, Toomey et al 1990,

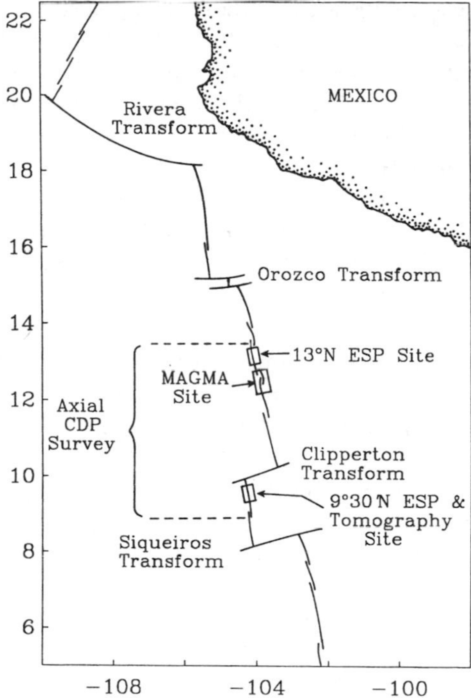

Figure 2 Sites of important seismic experiments along the axis of the northern East Pacific Rise (from Caress et al 1991). Two-dimensional (Burnett et al 1989, Caress et al 1992) and three-dimensional (Toomey et al 1990) tomography experiments were conducted at the MAGMA site (McClain et al 1985) and at 9°30′N, respectively.

Wilcock et al 1991); see Figure 3. Where along-axis coverage is available, the axial high-velocity (Toomey et al 1990), high-Q (Wilcock et al 1991), shallow-crustal anomaly is a persistent feature of the rise crest. The preferred interpretation is that the sheeted dike complex at zero age lies at shallower depths than beneath normal seafloor (Cann 1974, Kidd 1977, Toomey et al 1990). The rapid change within 1–2 km of the axis to lesser upper-crustal seismic velocities and Q is, by this view, the result of repeated eruptions of porous basalts which progressively thicken the extrusive carapace. Resolving the spatial and temporal nature of these axial variations in seismic characteristics and relating the seismic properties to bulk porosity and permeability should provide important constraints on models for hydrothermal circulation through the shallow crust of the rise axis region.

In addition to volcanic eruptions, both tectonic and hydrothermal processes may induce age-dependent variations in porosity or composition of the upper oceanic crust that are detectable by seismic methods (McClain et al 1985, Burnett et al 1989, Wilcock et al 1991). In support of this view are several observations indicating that shallow crustal seismic velocities continue to evolve to distances well beyond a few kilometers from the axis (Houtz & Ewing 1976, Herron 1982, Purdy 1982). Several hypotheses have been put forward to explain aging of the shallow crust, including increases in bulk porosity due to the formation of cracks and, perhaps at least partly on a longer time scale, metamorphic alteration and mineral precipitation during hydrothermal circulation.

AMC REFLECTOR Despite the predictions from thermal models and observations in ophiolites for a melt-filled axial magma chamber as much as 10 km wide and several kilometers in thickness (Cann 1974, Sleep 1975, Bryan & Moore 1977, Casey & Karson 1981, Pallister & Hopson 1981), the various seismic experiments designed to test these predictions along the northern East Pacific Rise have to date imaged only a comparatively narrow, axially discontinuous melt lens (Hale et al 1982, Herron et al 1978, Detrick et al 1987, Kent et al 1990). The best evidence for the existence of even a small axial magma body comes from the high-amplitude, subhorizontal upper crustal reflector seen on multichannel seismic profiles across the rise axis; the amplitude of the reflector versus range and its phase reversal relative to the seafloor reflection at some locations indicate that it marks the roof of a partially to fully molten layer. An example of a multichannel seismic reflection image of the top of the East Pacific Rise AMC in cross section is shown in Figure 4. Between about 9° and 13°N, the depth of the axial reflector ranges between 1.6 and 2.4 km beneath the seafloor (Detrick et al 1987), with increasing depth to the reflector correlating with increased bathymetric depth. Carefully migrated multichannel seismic data constrain the maximum width of the AMC reflector to be less than 3–4 km across the axis (Hale et al 1982, Herron et al 1978, Detrick et al 1987); more recent modeling of diffraction hyperbolas generated by the reflector's edge indicate an upper limit to the cross-axis width of as little as 800–1200 m (Kent et al 1990). Under the assumption that the base of the melt-filled sill marks a sharp contrast in seismic impedance, the reflected waveforms along the axis are consistent with a melt lens only 10–50 m thick (Kent et al 1990); however, if the bottom of the lens changes gradually from melt to a crystalline mush over a depth interval of several hundred meters or more, then the nature of this transition is poorly constrained by reflected wave amplitudes.

Reflected wave amplitude and phase data also constrain the physical characteristics of the AMC, including the nature of the solid-melt interface

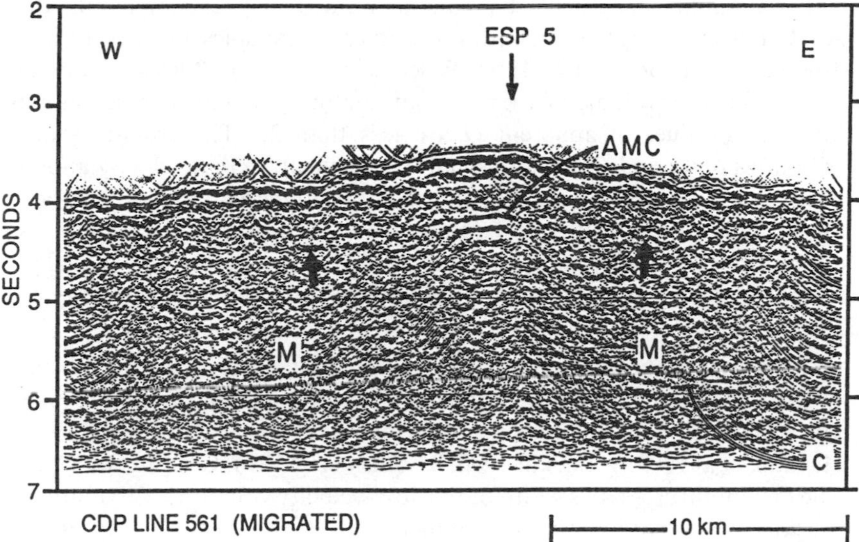

Figure 4 Multichannel seismic reflection profile across the East Pacific Rise near 9°30′N (from Detrick et al 1987). The high-amplitude, sub-horizontal reflector present beneath the rise axis is interpreted as the top of the axial magma chamber (AMC). Reflections from the Moho (*M*) can be traced to within 2–3 km of the rise axis. The arrows indicate a weak reflector interpreted as the top of a frozen magma chamber. This profile crosses the rise axis very near the cross-sections shown in Figure 3.

at the reflector boundary and the material properties of melt within the lens. At normal incidence the *P*-wave reflection coefficient is controlled largely by the contrast in compressional velocity across the reflector. At nonzero offset or range, however, the reflection coefficient is sensitive to discontinuities in the shear modulus, a circumstance that allows inferences regarding the state of the magma lens. At 9°30′N, where a bright AMC reflector is observed on normal-incidence reflection profiles, the nonzero-offset data indicate that the melt is fully molten, i.e. the shear modulus is effectively zero (Vera et al 1990). In contrast, near 13°N, the normal-incidence AMC reflector is considerably weaker and the inferred shear velocities are most consistent with a partially molten lens (Harding et al 1989). Along the northern East Pacific Rise the reflection data are consistent with a vertical-incidence reflection coefficient that varies between 0.2 and 0.4 (Harding et al 1989, Vera et al 1990), with smaller and larger values being consistent with a partially and fully molten lens, respectively. These differences in the AMC reflector are likely expressions of a temporal as well as a spatial variability to crustal magmatism.

Tomographic images of the volume immediately encompassing the AMC reflector (Figure 3) show pronounced anomalies in velocity and attenuation (Toomey et al 1990, Wilcock et al 1991). *P*-wave velocities averaged over a volume of 1 km^3 are anomalously low by as much as 1.5 km/s, and values of apparent Q are less than 20. The cross-axis distributions of V_P and Q are somewhat incongruent, in that velocity anomalies occupy a larger volume than the most pronounced attenuation anomalies. Such differences may result from a dissimilar response of compressional velocity and Q to temperature and melt fraction, as well as from a contribution to apparent attenuation from scattering of energy by the melt lens.

AXIAL LOW-VELOCITY VOLUME The upper crustal melt lens beneath the East Pacific Rise axis caps a significantly larger volume of anomalously low compressional wave velocity (McClain et al 1985, Burnett et al 1989, Harding et al 1989, Vera et al 1990, Toomey et al 1990, Caress et al 1992). The maximum cross-axis width of this low-velocity volume (LVV) is about 5–7 km, well exceeding that of the melt lens; the depth to the top of the LVV gradually increases with distance from the axis (Figure 3). The average compressional velocity within the LVV is about 5.5 km/s. The LVV, however, is not a homogeneous structure. Rise-parallel ESP data (Harding et al 1989, Vera et al 1990) and tomographic imaging (Toomey et al 1990) indicate that velocity within the LVV increases with axial distance. Tomographic data indicate that vertical velocity gradients are positive at mid- to lower-crustal depths (Toomey et al 1990). The structure of the base of the LVV is not known, but analysis of along-axis ESP data (Harding et al 1989, Vera et al 1990) suggests that the lower boundary may be within the lower crust. On the basis of laboratory measurements of compressional wave velocity in basaltic rock and melt (Murase & McBirney 1973), the seismic properties of the axial LVV are consistent with elevated temperatures but do not require the presence of significant partial melt.

Less is known about the *P*-wave attenuation and shear wave velocity structure of the mid to lower crust of the East Pacific Rise axis, but existing information contrasts somewhat with the general characteristics of the *P*-wave LVV. The axial attenuation anomaly appears more subdued in the lower crust than in the volume immediately surrounding the AMC reflector (Wilcock et al 1991), but this may at least partly reflect variations in scattering. The single measurement of shear wave structure at the East Pacific Rise axis (Bratt & Solomon 1984), at the site of a bright AMC reflector on multichannel seismic reflection images, suggests that a large volume of anomalously low *S*-wave velocity is not present at mid-crustal depths.

MOHO AND UPPER MANTLE At least parts of the crustal LVV are underlain by a distinct crust-mantle boundary, or Moho. Cross-axis multichannel reflection data frequently show a Moho signal beneath the rise-axis region (Detrick et al 1987), but this signal is generally absent in the immediate area of the AMC reflector (Figure 4), either because of the absence of a reflecting Moho or because of the attenuation of energy beneath the AMC reflector. Further, observations of Moho arrivals (*PmP*) are reported for ESP lines along and near the rise axis (Harding et al 1989, Vera et al 1990). Amplitude modeling suggests that the Moho is not a single discontinuity but is rather a gradual transition zone about 1 km thick (Vera et al 1990). A summary of the cross-axis structure of the East Pacific Rise near 9°30′N (Figure 5) includes a 1-km-wide AMC reflector, a 6–7-km-wide LVV, and a normal crustal thickness (~6 km) and distinct Moho attained within 1–2 km of the rise axis.

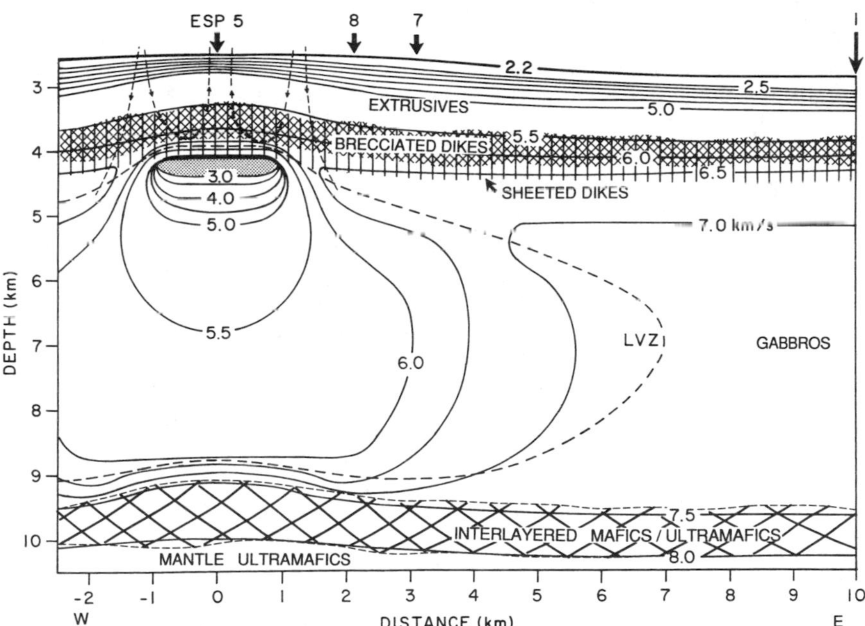

Figure 5 Interpretative summary of the *P*-wave velocity structure (contours in km/s) across the East Pacific Rise axis near 9°30′N (from Vera et al 1990). The summary incorporates the results of cross-axis multichannel seismic reflection profiles (Figure 4) and along-axis ESP data (locations indicated by numbered arrows). The shaded area beneath the axis indicates the inferred magma lens. Conjectural patterns of hydrothermal circulation are indicated by dashed lines and arrows.

There are indications from P-wave delay time (Toomey et al 1990) and attenuation (Wilcock et al 1991) tomography for anomalous structure near the Moho beneath the rise axis at 9°30′N. The depth and extent of this anomaly are poorly defined at present. It is apparent, however, that the mid- to lower-crustal LVV is underlain by a second velocity inversion located near the Moho, perhaps within the uppermost mantle. Similarly, the upper crustal low-Q zone overlies a zone of relatively higher Q in the lower crust but apparently lower Q at near-Moho depths.

ALONG-AXIS VARIATIONS The East Pacific Rise is known to vary along its axis in its overall depth, its cross-sectional shape, and its distribution of tectonic and hydrothermal features (Macdonald et al 1988, Haymon et al 1991). Piecewise linear segments of rise are offset by transform faults and overlapping spreading centers of a variety of dimensions (Macdonald et al 1988) or are delimited by modest changes in the trend of the morphological axis without a distinct offset (Langmuir et al 1986). Such tectonic and morphologic segmentation of the rise is also accompanied by segmentation of seismic structure (Figure 6). The AMC reflector terminates near fracture zones (Detrick et al 1987) and is discontinuous at major overlapping spreading centers (Mutter et al 1988). Near morphologically-defined deviations from axial linearity (devals), which can be

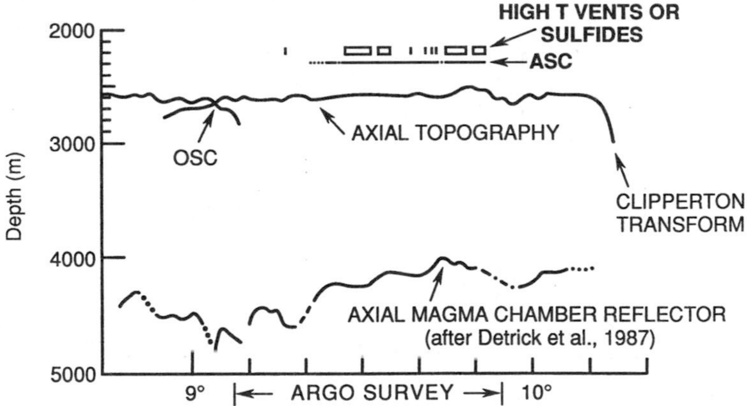

Figure 6 Summary (Haymon et al 1991) of variations in bathymetry, depth to the AMC reflector (from Detrick et al 1987), development of an axial summit caldera (ASC) or graben, and locations of high-temperature hydrothermal vents and sulfide deposits along the East Pacific Rise axis between about 9° and 10°N. The along-axis extent of the survey by the deep-towed ARGO imaging system (the results of which were used to locate the ASC, vent fields, and sulphide deposits) is shown by arrows. The position of a prominent overlapping spreading center (OSC) is also indicated.

accompanied by changes in the average major element chemistry of seafloor basalts (Langmuir et al 1986), the AMC reflector can be discontinuous, abruptly dipping, or continuous (Detrick et al 1987).

Overall the AMC reflector is present along slightly more than 60% of the rise axis between 9° and 13°N (Detrick et al 1987). The reflector appears to be discontinuous for the 90-km section of rise between the Clipperton transform and an overlapping spreading center at 9°03′N—a section that includes several morphologic discontinuities. Detrick et al (1987) proposed that a continuous magma chamber exists along this section of the East Pacific Rise and that this tectonically bounded segment constitutes a spreading center cell that is centrally fed by mantle-derived melt. In contrast, the AMC reflector is absent over the 70-km section of rise to the immediate north of the Clipperton transform (Detrick et al 1987). Other postulated indicators of crustal magma, such as axial depth, axial summit morphology, and width of the axial summit graben (Macdonald et al 1988), also point to a greater melt volume to the south of the Clipperton than to the north. Such variations have not been fully explained but suggest that melt delivery from the mantle is time dependent on scales of 10–100 km.

A different view of axial thermal and magmatic segmentation is emerging from the results of seismic tomography experiments (Toomey et al 1990), which are sensitive to volumetric variations in velocity rather than the sharp contrasts in seismic impedance that are the targets of seismic reflection profiles. The area of the tomographic experiment at 9°30′N was a 15-km-long portion of the rise between the Clipperton and the 9°03′N overlapping spreading center and included two morphologically-defined devals that also coincide with offsets of the axial summit graben (Haymon et al 1991). Three-dimensional images show an axially segmented LVV with the distal ends of the seismic anomaly coinciding with the positions of discontinuities in the trend of the rise axis and the summit graben. These results imply thermal segmentation of the rise axis at the 10-km scale and suggest that this scale of segmentation is imposed by the dynamics of melt generation and transport rather than by tectonic structures in the lithosphere. In this view, the 90-km-long tectonic segment between the Clipperton transform and the 9°03′N overlapping spreading center would encompass several volcanically-defined segments, each centered on a distinct site of injection of melt to upper crustal levels.

Slow-Spreading Ridges

At slower spreading rates the axial crust is significantly cooler than along faster spreading rises (Sleep 1975). As a result, there is a strong mechanical lithosphere of significant thickness. Extension of this strong layer gives rise to earthquakes, large-scale normal faulting, and the development of a median rift valley. Unlike faster spreading rises where plate divergence is accommodated almost entirely by magmatic accretion, along slower spreading ridges the cumulative throw of seafloor normal faults (Macdonald & Luyendyk 1977) and the rate of seismic moment release (Solomon et al 1988) indicate that about 20% of plate divergence is accommodated by lithospheric stretching. Scenarios for segmentation of slow-spreading ridges (Crane 1985, Whitehead et al 1984, Schouten et al 1985) and recent bathymetric and gravity surveys (Sempere et al 1990, Lin et al 1990) suggest systematic variations in the thermal and mechanical properties of the lithosphere along a single ridge segment. Lithosphere near the magmatic center should be relatively hotter and mechanically weaker and should display evidence for more recent magmatism, whereas lithosphere more distant from the magmatic center should be cooler and mechanically stronger.

Several recent experiments have addressed the seismic velocity, density, or mechanical structure of the Mid-Atlantic Ridge and its variation within and between major ridge segments (Figure 7). These include conventional seismic refraction (Purdy & Detrick 1986), marine gravity (Kuo & Forsyth 1988, Lin et al 1990, Morris & Detrick 1991), multichannel seismic reflection (Detrick et al 1990), microearthquake location and delay time tomography (Toomey et al 1985, 1988; Kong et al 1992), and earthquake centroid depth determination from body waveform inversion (Huang et al 1986, Huang & Solomon 1988). An increasingly multidisciplinary focus on selected sites has greatly accelerated our understanding of the interrelations among axial magmatic, tectonic, and hydrothermal processes and of the systematics of axial segmentation.

MEDIAN VALLEY STRUCTURE: BATHYMETRY The median valley of the Mid-Atlantic Ridge is centered on an inner floor of variable width (5–15 km) that contains the neovolcanic zone and is bounded by a nested series of generally rotated, seismogenic, normal-fault blocks that rise outward to the crestal mountains. The width and relief of the median valley are typically 20–30 km and 2000–3000 m, respectively; although the relief may be as little as hundreds of meters and as great as 4000–5000 m. The range in the shape and relief of cross-axis profiles provides an indication of the variable thermal and mechanical structure along slow-spreading ridges.

High-resolution multibeam bathymetric maps of the Mid-Atlantic Ridge (Sempere et al 1990, Lin et al 1990) document the form of morphologic segmentation of slow-spreading ridges (Figure 8). Maxima in the along-axis depth profile coincide with spreading center discontinuities, such as transform faults or non-transform offsets of the neovolcanic zone, while minima are sited approximately midway between discontinuities.

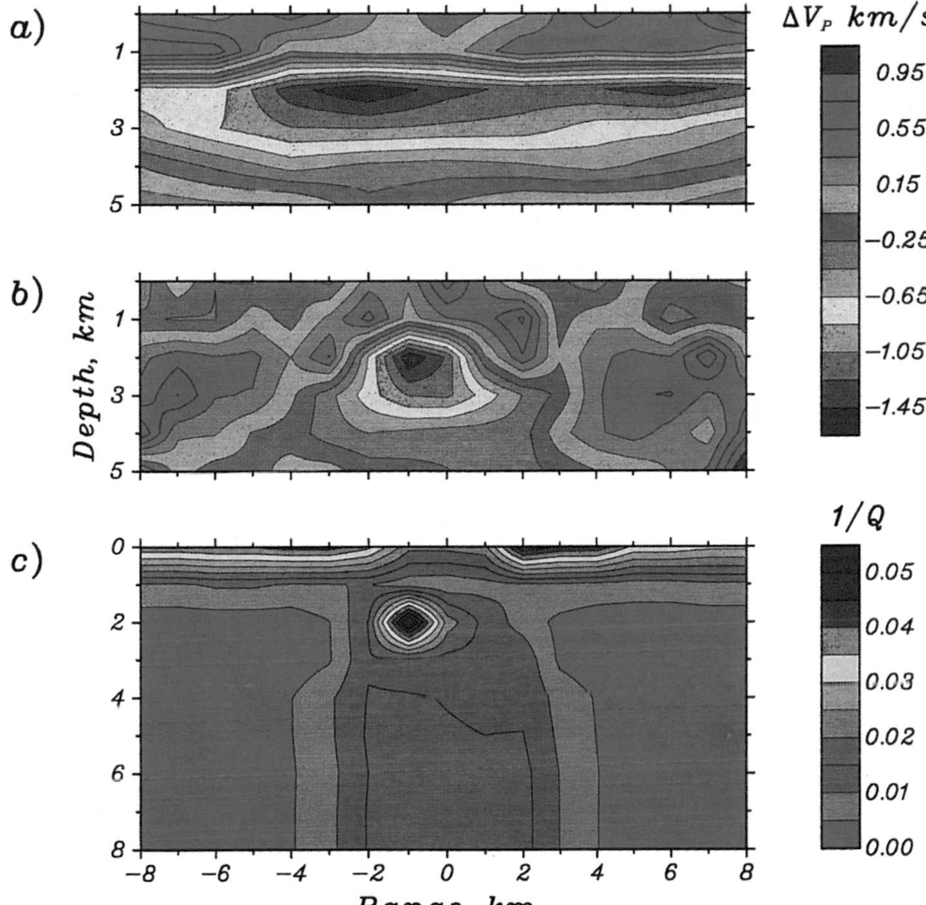

Figure 3 Vertical cross sections through the *P*-wave velocity (V_P) and attenuation (Q^{-1}) structures obtained by seismic tomographic imaging of the East Pacific Rise axis at 9°30′N (Toomey et al 1990, Wilcock et al 1991). (*a*) Along-axis section of the V_P structure 1 km to the west of the rise axis and passing through the center of the axial low-velocity volume. The colors represent departures of the three-dimensional model from the average one-dimensional, depth-dependent velocity structure; positive differences are faster than average, negative slower than average; 0.2 km/s contour interval. Two deviations from axial linearity (devals) occur at about −6 and 4 km range. (*b*) Cross-axis section of the V_P structure passing 2 km south of the experiment center. (*c*) Cross-axis section of apparent Q^{-1} structure. The two-dimensional structure was obtained by constrained, nonlinear tomographic inversion of *P*-wave spectral slope data along paths crossing the rise-axis 0–5 km south of the experiment center.

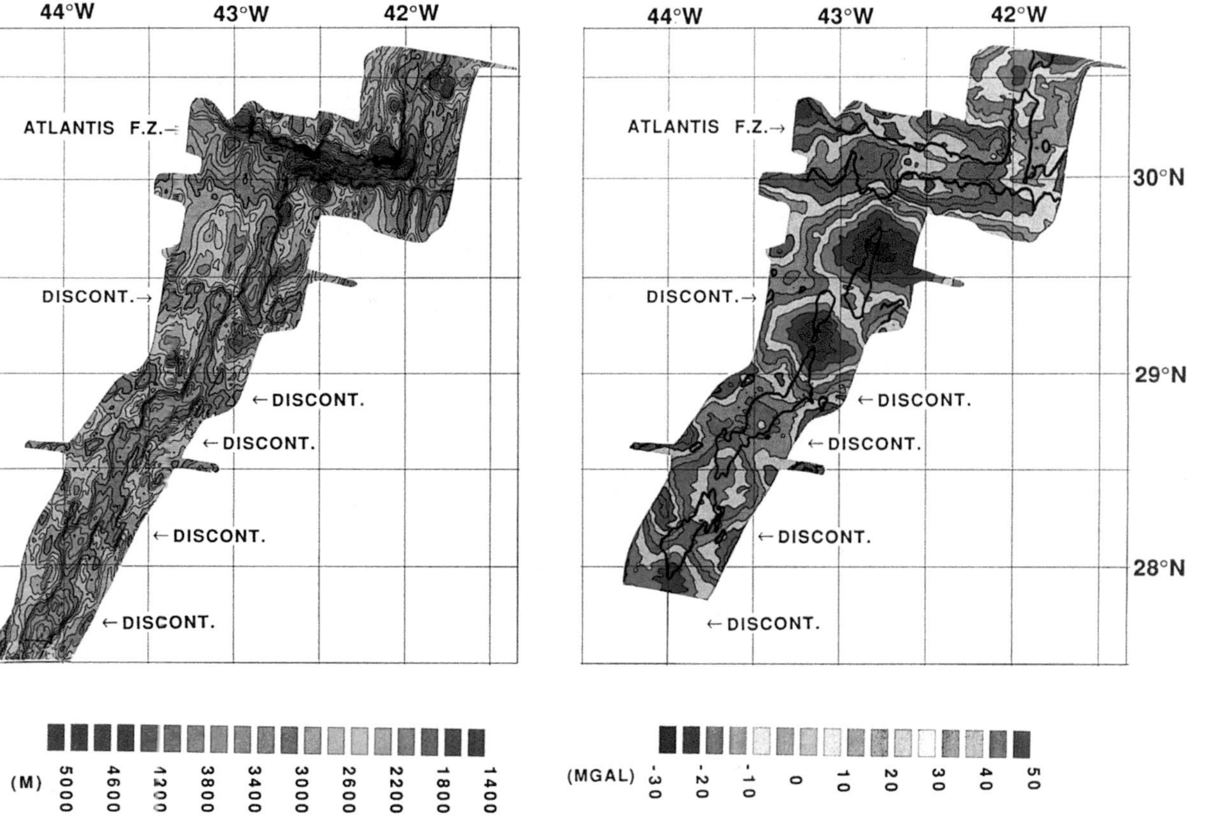

Figure 8 Sea Beam bathymetric map (*left*) and mantle Bouguer gravity anomaly map (*right*) of the Mid-Atlantic Ridge between about 28° and 31°N (from Lin et al 1990). The ridge axis is segmented into sections 20–80 km in length; gravity anomaly lows are centered over each segment. The contour interval is 200 m for bathymetry and 5 mgal for mantle Bouguer anomaly; the zero level for mantle Bouguer anomaly is arbitrary.

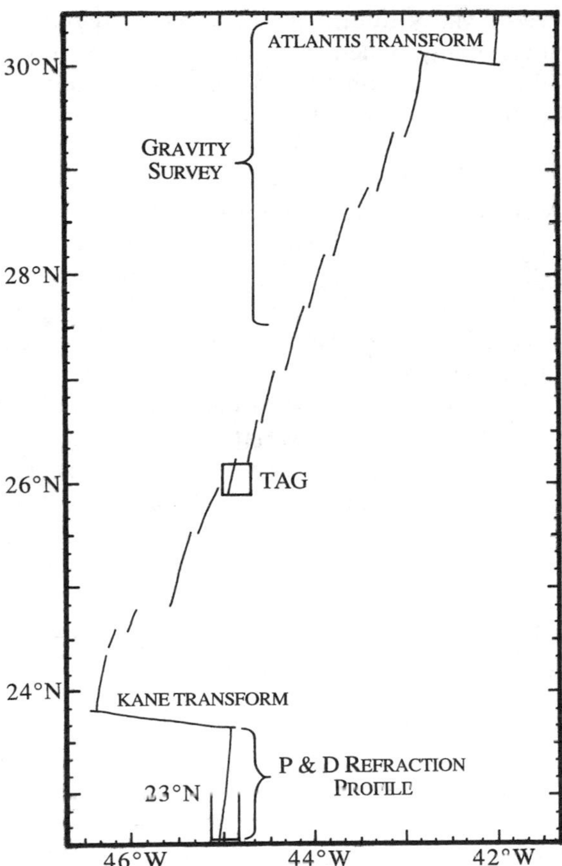

Figure 7 Sites of important geophysical experiments along the axis of the northern Mid-Atlantic Ridge, including the area of the gravity survey of Lin et al (1990), the micro-earthquake and two-dimensional tomography studies at 23°N (Toomey et al 1985, 1988) and 26°N (Kong et al 1992), and the refraction profile of Purdy & Detrick (1986) south of the Kane transform.

Segments are 20–90 km in length. The total along-axis relief within a segment generally increases with increased segment length.

The morphology of the median valley within a segment can be characterized as either narrow and shallow or broad and deep. These characteristics are thought to be thermally controlled, with the former indicating a relatively hot, more recently magmatically active segment and the latter a cooler, more tectonically active one. A single segment may be hour-glass-shaped in map view, with a narrow, shallow rift valley near its center and a broader, deeper valley toward the distal ends, suggesting systematic intra-segment variability in thermal structure. It is not known whether the spectrum of along-axis bathymetry and segmentation observed in surveys to date fully characterize the range of styles of slow-spreading ridge systems.

ALONG-AXIS VARIATIONS: GRAVITY Gravity observations collected along the Mid-Atlantic Ridge provide further constraints on the physical characteristics of segmentation and suggest that magmatic accretion is focused at discrete locations near segment centers (Kuo & Forsyth 1988, Lin et al 1990, Morris & Detrick 1991). The gravity data, corrected for depth and constant crustal thickness to yield mantle Bouguer anomalies, show circular gravity lows centered near the midpoints of morphologically defined segments (Figure 8). The variations in density that accompany cooling of the oceanic lithosphere may account for some component of the cross-axis mantle Bouguer anomaly, but a thermal model consistent with only passive mantle upwelling does not account for the observed along-axis variations. The mantle Bouguer anomalies are more negative for longer ridge segments (Lin et al 1990); see Figure 9. The interpretation of the gravity anomalies is that magmatic accretion is focused near the center of a segment and that the vigor of magmatic activity at the midpoint of a segment increases with segment length.

Two specific hypotheses on the form of focused magmatic accretion have been proposed to explain the gravity anomalies (Lin et al 1990). The first attributes the heterogeneous density structure to along-axis variations in crustal thickness. Since the mantle Bouguer correction is calculated for a fixed crustal thickness, any departures from this structure will appear as an anomaly. Consequently, the mantle Bouguer anomaly pattern could arise from a relatively thick crust near the center of a ridge segment and relatively thinner crust near the segment ends. Alternatively, the along-axis gravity lows may arise from variations in upper-mantle density structure. In this hypothesis, the anomalous density structure is attributed to buoyancy-driven mantle upwelling. There are at least two sources of buoyancy: thermal expansion due to elevated temperature and the compositional buoyancy of depleted, relative to unmelted, mantle material. The two hypotheses are not mutually exclusive but rather constitute simple end-member models for anomalous density structure. Since hotter mantle would normally be expected to give rise to greater melt generation and thus greater crustal thickness, it is likely that the variations in along-axis gravity result from variations in both crustal and mantle structure. For both hypotheses mantle melt production is distinctly three-dimensional and focused near segment centers.

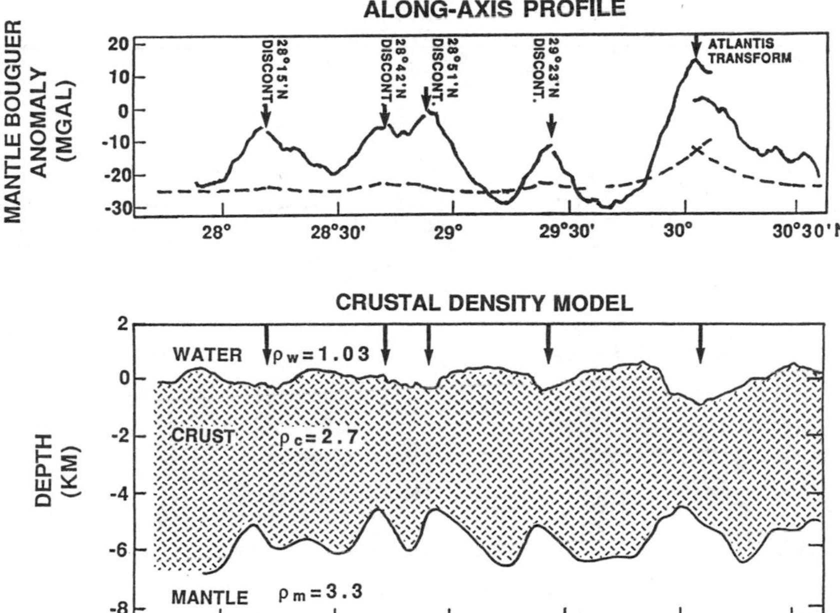

Figure 9 (*Top*) Along-axis profile (Lin et al 1990) of mantle Bouguer gravity anomaly (*solid line*) and the predicted anomaly (*dashed line*) for a model with fixed crustal thickness and the thermal structure given by a passive upwelling model for mantle flow (Phipps Morgan & Forsyth 1988). (*Bottom*) Along-axis bathymetry and crustal thickness for a model in which the entire mantle Bouguer anomaly is due to variations in crustal thickness (Lin et al 1990), the crust in this model is predicted to be thickest near the segment midpoints and thinnest at the segment ends.

ALONG-AXIS VARIATIONS: SEISMIC STRUCTURE AND SEISMICITY The nature of the seismic structure of the crust along the axis of a slow-spreading ridge, and of its variability within and between segments, is well illustrated by a seismic refraction line shot down a 120-km-long section of the axis of the Mid-Atlantic Ridge south of the Kane Fracture Zone (Purdy & Detrick 1986). For most of the length of the profile, the structure is similar to that of normal oceanic crust, with a well-defined Moho transition zone, a mantle P-wave velocity of about 8 km/s, and a normal crustal thickness of 6–7 km. Only a slightly low velocity to layer 3 distinguishes this structure from that of mature oceanic crust. Two significant along-axis structural changes, however, are noteworthy (Figure 10): a 10–15-km-wide zone of low velocities in the lower crust beneath an along-axis high, and a zone of complex structure marking the transition northward to thinner crust near the intersection of the median valley with the Kane transform. On the

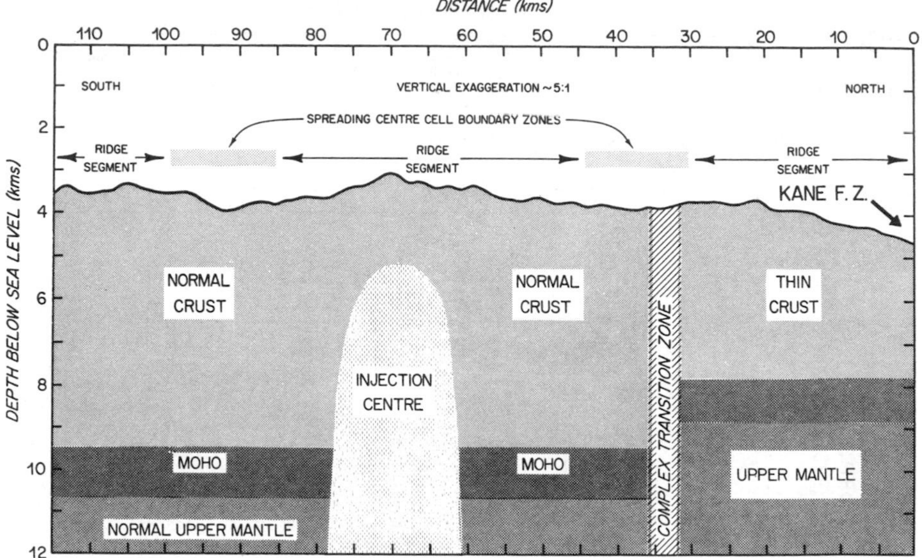

Figure 10 Schematic representation of crustal structure inferred from seismic refraction data along a 120-km-long section of the Mid-Atlantic Ridge axis south of the Kane Fracture Zone (from Purdy & Detrick 1986). On the basis of anomalously low velocities in the lower crust, the axial high near the midpoint of the central segment is interpreted to be the site of most recent injection of (now solidified) magma into the crust.

basis of along-axis bathymetry as well as structure, Purdy & Detrick (1986) interpret these two zones as marking the center and end, respectively, of a spreading center segment. At no location along the 120-km length of ridge was any evidence found for a significant body of molten or partly molten rock. A multichannel seismic reflection profile along the northern portion of the refraction line, including the site of a high-temperature hydrothermal field (ODP Leg 106 Scientific Party 1986), also failed to find evidence for a shallow intracrustal reflector that might be associated with a crustal magma body (Detrick et al 1990).

The distribution of earthquake hypocenters both across and along the median valley has important implications for the thermal and mechanical structure of a ridge segment. According to current models for segmentation, lithosphere near the ends of a segment should be thicker, tectonic stretching should be more important, and faulting should extend deeper than near segment midpoints. From the inversion of teleseismic *P* and *S* waveforms from 50 large earthquakes along slow-spreading ridge systems, Huang & Solomon (1988) showed that the maximum centroid depth, and thus the greatest extent of seismogenic faulting, increased systematically with decreasing spreading rate. This result indicates a strong spreading-rate dependence to the maximum thickness of the mechanically strong lithosphere within the median valley inner floor, where the epicenters of the largest ridge earthquakes are located. Huang & Solomon (1988) and Lin & Bergman (1990) have shown further that there is a clear tendency for the largest earthquakes on slow-spreading ridges to occur near the ends of morphological segments.

Microearthquake experiments with networks of ocean-bottom seismometers and hydrophones permit the delineation of patterns of faulting within a segment in much greater detail than with only teleseismic data. One such microearthquake study, near 23°N, was sited in an along-axis deep at the southern end of a ridge segment approximately 40 km in length (Toomey et al 1985, 1988); the segment is depicted in the center of Figure 10. Earthquakes occurred as deep as 8 km beneath the inner floor, indicating brittle failure throughout the crustal column and probably into the uppermost mantle. Fault-plane solutions for several groups of events indicated normal faulting mechanisms, consistent with lithospheric stretching. In combination with observations of the rate of occurrence and typical fault slip of large earthquakes on this segment (Huang et al 1986), an asymmetry in the inner floor indicated by multibeam bathymetry (Detrick et al 1984), and the nearly normal oceanic crustal structure indicated by seismic refraction (Purdy & Detrick 1986), the microearthquake data imply (Figure 11) that this median valley deep has undergone horizontal extension without axial magmatism for approximately the last 10^4 years (Toomey et al 1988).

A second microearthquake experiment was located near 26°N, at an along-axis high near the midpoint of a ridge segment also approximately 40 km in length (Kong et al 1992). In contrast to the magmatically quiescent, comparatively cooled crust beneath the along-axis deep at 23°N, the axial high of the ridge segment at 26°N is the site of a black-smoker hydrothermal vent field (Rona et al 1986). Within the inner floor, the maximum focal depth of events beneath the region of the along-axis high was less than that of earthquakes beneath the along-axis deep 15 km to the south (Figure 12), consistent with a hotter crust and thinner mechanical lithosphere beneath the high. The fault plane solutions of grouped events indicated diverse source mechanisms and a variable state of stress.

Analysis of the travel-time residuals of *P* waves recorded on seafloor networks also permits tomographic imaging of the local seismic velocity structure of the crust in the vicinity of the network. At 23°N the two-dimensional seismic structure across the median valley inner floor at the site of the axial deep shows that the modest decrease in the layer 3 velocity indicated in the along-axis refraction results (Purdy & Detrick 1986) is

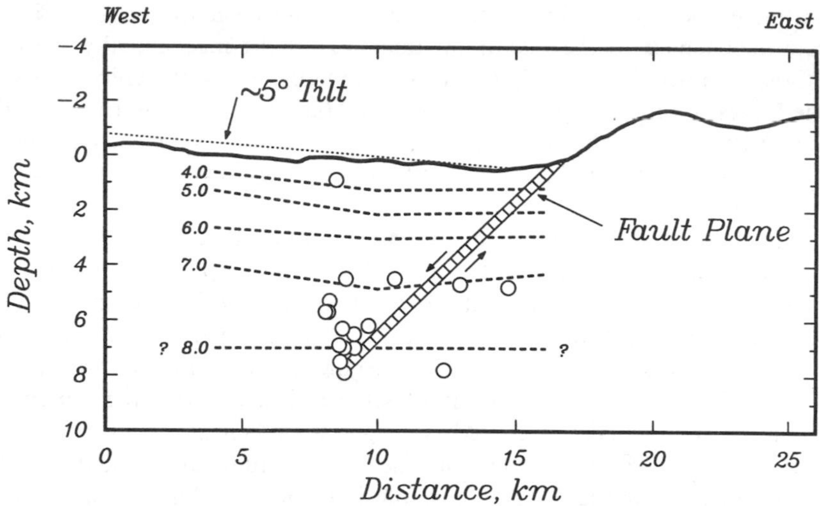

Figure 11 Cross-axis vertical section of the Mid-Atlantic Ridge median valley along-axis deep near 23°N, showing microearthquake hypocenters (*circles*) and contours of *P*-wave velocity (in km/s) obtained from two-dimensional delay-time tomography (Toomey et al 1988). Bathymetry is from the Sea Beam survey of Detrick et al (1984). A conjectural fault plane for recent large ($m_b > 5$) earthquakes, consistent with fault geometry and dimensions inferred from body waveform inversion (Huang et al 1986), is shown as a continuation at depth of the fault making up the steep eastern inner wall of the median valley. If the full rate of plate separation were accommodated by slip on that fault over 10^4 yr, the fault throw and the observed eastward tilt of the inner floor can both be matched.

largely eliminated by an age of a few hundred thousand years (Figure 11). A tomographic inversion for two-dimensional structure along the inner floor axis at 26°N shows anomalously low mid-crustal velocities beneath the along-axis high (Figure 12). Toward the axial deep at the southern end of this ridge segment the seismic structure is comparable to normal oceanic crustal structure, similar to the axial deep near 23°N. This along-axis variation is consistent with the hypothesis that crustal accretion is focused near the center of the ridge segment (Kong et al 1992).

The relationships documented among segment length, axial relief, and mantle Bouguer gravity anomaly minimum (Lin et al 1990) suggest that the ridge segments near 23°N and 26°N are architecturally similar. The two microearthquake experiments, together with teleseismic analysis of large earthquakes and seismic refraction data, may thus begin to characterize the intra-segment seismic structure of a 40-km-long ridge segment on a slowly spreading ridge. Near the segment center the mechanically strong lithosphere is only a few kilometers thick and the tectonic state is

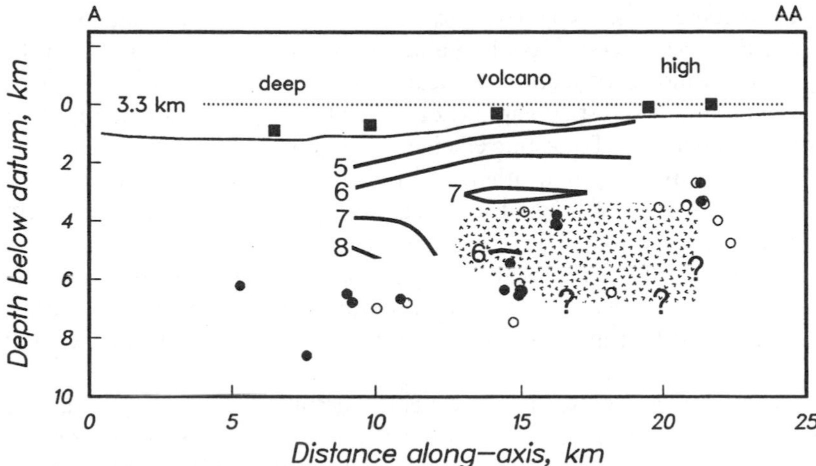

Figure 12 Along-axis vertical section of the Mid-Atlantic ridge median valley near 26°N, showing microearthquake hypocenters (*circles*, filled where the locations were constrained by one or more *S*-wave arrival times) and contours of *P*-wave velocity (in km/s) obtained from two-dimensional delay-time tomography (Kong et al 1992). Filled squares show the locations of ocean-bottom seismic stations. The stippled area shows the location of a region of low velocities imaged in the mid to lower crust beneath the axial high and black-smoker hydrothermal field; the lower and northern limits of this anomaly are uncertain. The low velocities are attributed to elevated temperatures at the site of the most recent crustal injection of (now solidified) magma along this ridge segment.

highly variable, reflecting a mix of extension and thermal stress accompanying rapid cooling (Kong et al 1992), while the segment ends are characterized predominantly by ongoing extension and seismogenic faulting extending to the base of the crust (Toomey et al 1988, Huang & Solomon 1988).

Implications for Crustal Formation

Seismic data on mid-ocean ridge crustal structure indicate that the axial magma chamber departs significantly from the earlier predictions from ophiolites (e.g. the lower panel of Figure 1) that the chamber would be large and nearly steady state. No magma body has been found along slow-spreading ridges despite a concerted search. The axial magma body along fast-spreading rises is only a thin sill, perhaps typically no more than 1 km in width and 100 m or less in thickness (Kent et al 1990). The diagnostic AMC reflector was not seen along nearly 40% of the northern East Pacific Rise axis imaged by multichannel reflection profiling (Detrick et al 1987), indicating that even the thin magma lens is not a steady-state feature at fast spreading rates. Two inferences may be made from these findings. The first is that the process of cooling and solidifying magma delivered to the upper crust along the axial neovolcanic zone must occur on a time scale comparable to (at fast spreading) or faster than (at slow spreading) that of magma delivery. To be this efficient the cooling of the axial crust must occur dominantly by hydrothermal circulation. The second inference is that the transport of magma to the axial crust must be time-dependent and spatially variable. The spatial variability applies equally to the flow regime in the ridge upper mantle responsible for magma generation. The possibility of a large axial magma chamber remains open, but such a feature would be transient and is likely to be a rare event in space and time.

The relationship between the vertical structure of oceanic crust and the evolution of the axial magma chamber also requires modification from the predictions of early ophiolite-based models. Although the upper crustal lithostratigraphy of oceanic crust is similar to the volcanic basalts and sheeted dikes of the ophiolite model (Anderson et al 1982) and such rocks are capable of being emplaced above the modest magma bodies imaged by seismic experiments along fast-spreading rises, the detailed mechanism of emplacement of layer 3 remains to be completely explained. One possibility is that the axial lower crust is episodically filled with a transient magma chamber capable of differentiation of a basal cumulate layer—much like the ophiolite model but with nearly complete solidification—even at fast spreading rates, before a new large batch of magma is again injected into the crust. This possibility would be considerably strengthened were a large magma chamber discovered along at least one ridge segment. A second possibility is that the axial magma chamber never significantly exceeds the dimensions or crustal depth of the magma lenses imaged seismically along the northern East Pacific Rise. Emplacement of the lower 4–5 km of crust must then occur by some form of solid-state flow downward and outward from the magma lens. This scenario raises the question as to how a seismically sharp crust-mantle transition can be consistently formed at the characteristic Moho depth for oceanic crust. The trapping of melt within sills at the base of the crust (Garmany 1989) may play a role in the formation of Moho structure.

Seismic and mechanical structure clearly are segmented in a manner related to other measures of ridge-axis segmentation, but the scales of segmentation within the crust are incompletely defined and the relationship of these scales to causative processes is not established. At the longest scales of segmentation, between major transform faults, the lithosphere is clearly exerting an influence on the pattern of mantle upwelling and melt generation. Along slow-spreading ridges, there appears to be a systematic variation along a segment in the thickness of the strong mechanical layer in a sense consistent with the hypothesis that the delivery of heat and magma from the mantle is concentrated at the segment center. At the 10-km scale of morphological segmentation along fast-spreading rises, the coincidence of intracrustal seismic and thermal segment boundaries with morphological boundaries also suggests control by magmatic processes. Thus the dynamics of the coupled mantle-melt system are of prime importance for some segmentation scales.

RECENT RESULTS ON RIDGE UPPER MANTLE STRUCTURE

The seismic velocity and density of the upper mantle beneath mid-ocean ridges are functions of local temperature and bulk composition. The distribution of these quantities are in turn related to patterns of mantle flow, the extent of mantle melt extraction, and mantle chemical heterogeneity. The seismic velocity of oceanic mantle material is also known to be anisotropic. The sense and magnitude of this anisotropy provide important constraints on the processes of lithospheric spreading and mantle dynamics. Considerable progress has recently been made toward defining the laterally heterogeneous seismic velocity structure of ocean-ridge mantle from tomographic inversions of the phase and group velocities of long-period surface waves and the travel times of long-period body waves. For a review of many of the techniques used in such studies and a summary of recent findings see Romanowicz (1991).

The most prominent lateral variation in structure of the oceanic upper mantle is that associated with the cooling and aging of oceanic lithosphere. Mid-ocean ridge axes are resolved as regions of generally low upper-mantle velocities in global and regional tomographic inversions (Woodhouse & Dziewonski 1984; Grand 1987; Zhang & Tanimoto 1989, 1991; Montagner & Tanimoto 1990), and an age-dependence has been documented for both surface-wave phase and group velocities (Leeds et al 1974; Forsyth 1975, 1977; Yoshii 1975; Mitchell & Yu 1980; Montagner & Jobert 1983; Nishimura & Forsyth 1985, 1988) and the travel times of steeply incident body waves (Duschenes & Solomon 1977, Girardin 1980, Kuo et al 1987, Woodward & Masters 1991, Sheehan & Solomon 1991). Examples of this age dependence are shown in Figures 13 and 14.

An important question is whether the age dependence of seismic velocity is entirely an effect of lithospheric cooling or also contains a contribution from the sublithospheric mantle. Zhang & Tanimoto (1991) find that the variation of Love wave phase velocity versus age shown in Figure 13 can

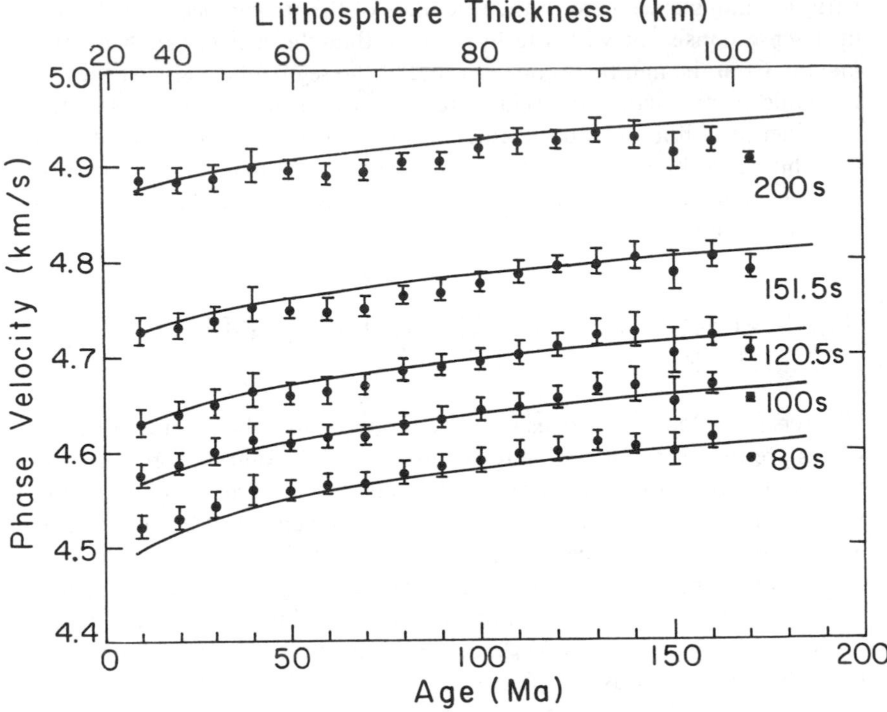

Figure 13 Love wave phase velocity versus seafloor age and wave period in the Pacific Ocean (from Zhang & Tanimoto 1991). Error bars denote one standard deviation. The solid lines are the predicted variations for a model in which a high-velocity lithosphere thickens in proportion to the square root of age.

be explained entirely by the thickening of oceanic lithosphere with age, as long as the thickness is proportional to the square root of seafloor age, even out to ages in excess of 100 Ma. A similar finding has been reported from efforts to model pure-path Rayleigh wave phase velocities (e.g. Leeds et al 1974), although other workers (Yoshii 1975, Forsyth 1977) report that some increase with age in the shear velocity of the underlying asthenosphere is also required. The *PP-P* and *SS-S* differential travel times, plotted at the seafloor age of the surface-reflected phase (*PP* or *SS*) and averaged by age as in Figure 14 (from Woodward & Masters 1991), also show an approximately linear decrease with square root of age out to ages in excess of 100 Ma, although Sheehan & Solomon (1991) note an apparent departure from the linear relation for ages greater than 100 Ma in the North Atlantic. The *SS-S* differential travel time that would result from a

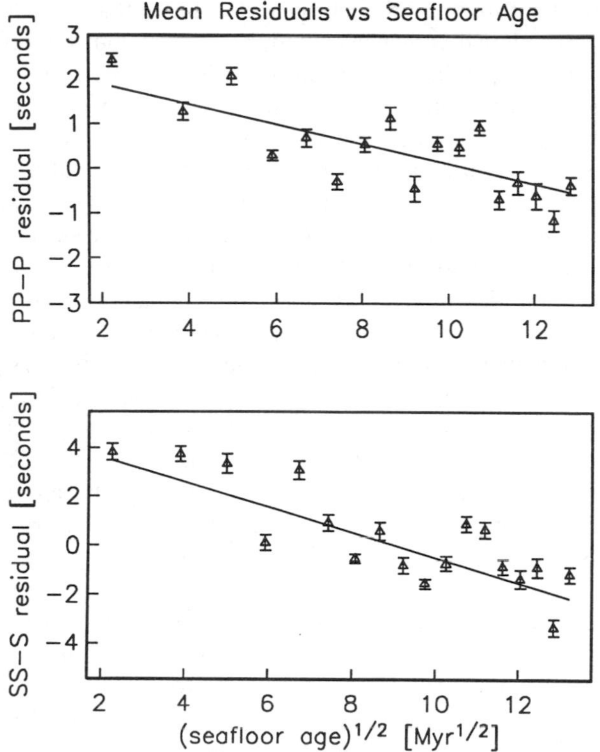

Figure 14 *PP-P* and *SS-S* differential travel-time residuals in oceanic regions versus the square root of age of the seafloor at the bounce point of the surface-reflected phase (*PP* or *SS*) (from Woodward & Masters 1991). Each point shown is an average of about 100 measurements; error bars denote one standard deviation. Most of the variation depicted is due to cooling of the oceanic lithosphere as it spreads from the ridge axis.

standard lithospheric plate cooling model (Parsons & Sclater 1977) and a constant partial derivative of shear wave velocity with temperature (-0.6 m/s K^{-1}) provides a good fit to the data for the North Atlantic for ages less than 100 Ma (Sheehan & Solomon 1991). At the very long wavelengths (typically several thousand kilometers and greater) represented by upper mantle tomographic models, however, there is evidence in the results of inversions of both surface wave or normal mode data and body wave travel times for low-velocity volumes extending as deep as 300–400 km beneath the crestal regions of actively spreading mid-ocean ridges (Woodhouse & Dziewonski 1984, Grand 1987, Zhang & Tanimoto 1989). Most likely these low velocities are a reflection of temperatures that are greater than average for those mantle depths and indicate that mantle upwelling associated with plate divergence extends to at least 300–400 km depth.

If the age-dependent lateral variation in seismic velocity associated with lithospheric cooling is removed from surface wave velocities or differential travel times depicted in map view, then the residual velocities or times are measures of one or more additional components of lateral heterogeneity. Sections of the global ridge system that are extremes in residual Rayleigh wave velocities are the Reykjanes Ridge (anomalously slow, Girardin & Jacoby 1976) and the Southeast Indian Ocean Ridge in the vicinity of the Australian-Antarctic discordance (anomalously fast, Forsyth et al 1987); the axes of these ridge sections are also, respectively, anomalously shallow and deep and likely reflect characteristic mantle temperatures significantly hotter and colder, respectively, than average. Sheehan & Solomon (1991) have documented variations in age-corrected *SS-S* differential travel times at wavelengths of 1000 to 7000 km along the axis of the Mid-Atlantic Ridge; an along-axis variation in the phase velocity of 100-s Love waves (Zhang & Tanimoto 1991) shows a generally similar pattern. The variations in *SS-S* travel-time residuals show strong qualitative correlations with along-axis variations in bathymetry and geoid height (Figure 15) and are likely to be the result of long-wavelength variations in the characteristic temperature and bulk composition of the ridge-axis mantle. Sheehan & Solomon (1991) have formulated a procedure to invert jointly the residual *SS-S* differential travel time, bathymetry, and geoid profiles for lateral variations in upper mantle temperature and Mg # [the molar ratio MgO/(MgO+FeO)]. Temperature and Mg # can be distinguished, in principle, because while an increase in temperature results in a decrease in both seismic velocity and bulk density, an increase in Mg # decreases density but increases seismic wave speed. The resulting models of horizontal variation in mantle temperature and composition along the Mid-Atlantic Ridge (e.g. Figure 16) compare favorably with independent estimates from the composition and geothermometry of peridotites (Michael & Bonatti 1985, Bonatti 1990) and the depth interval of melting implied by the FeO content of mantle-derived basalts (Klein & Langmuir 1989) dredged from along the ridge axis. If the composition of the mantle at several hundred kilometers depth were constant along a mid-ocean ridge, then the characteristic upper mantle temperature should be strongly related to mantle Mg #. A higher temperature should yield a greater extent of mantle melting, and since melt extraction from the mantle is thought to be an efficient process, the residuum should be enriched in MgO relative to less melted mantle (McKenzie 1984). Characteristic temperature and Mg # are not strongly correlated in Figure 16, however; nor are there known to be along-axis variations in crustal thickness that would be

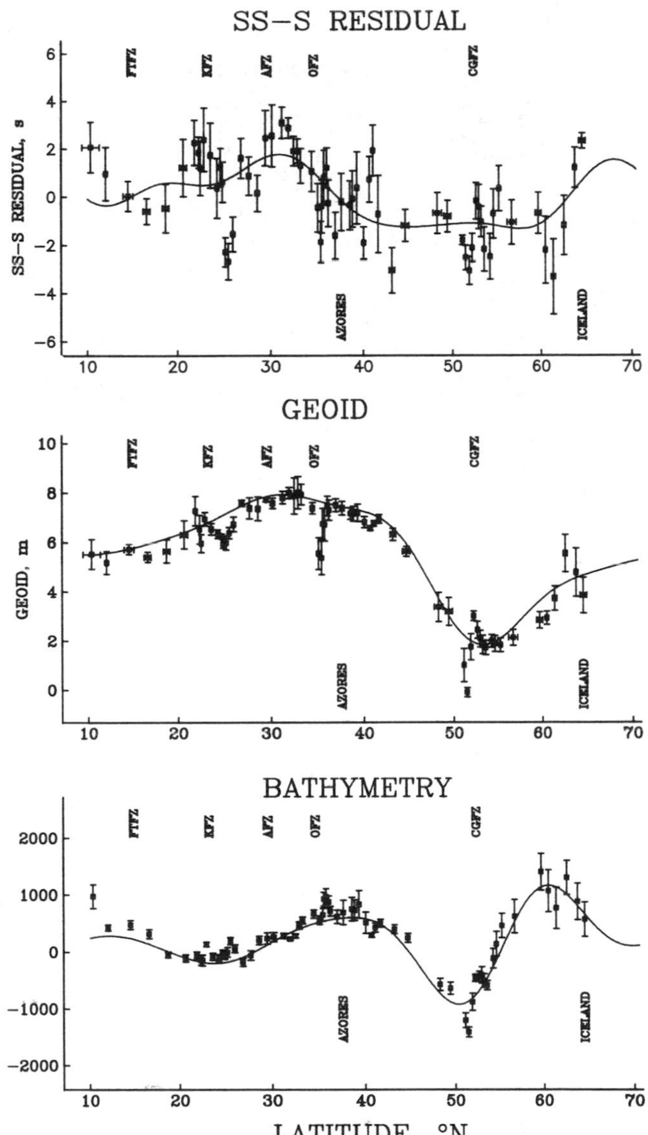

Figure 15 Comparative profiles of age-corrected *SS-S* travel-time residual, bathymetry, and geoid height along the Mid-Atlantic Ridge (from Sheehan & Solomon 1991). All three data types have been high-pass filtered. The points shown are moving averages of 10 adjacent data points, taken for all data types at the *SS*-wave bounce locations; error bars denote standard errors of the means. The solid lines represent filtered versions of the depicted observations, containing only the wavelengths (1400 to 7100 km) used in the inversions for along-axis variations in mantle temperature and composition.

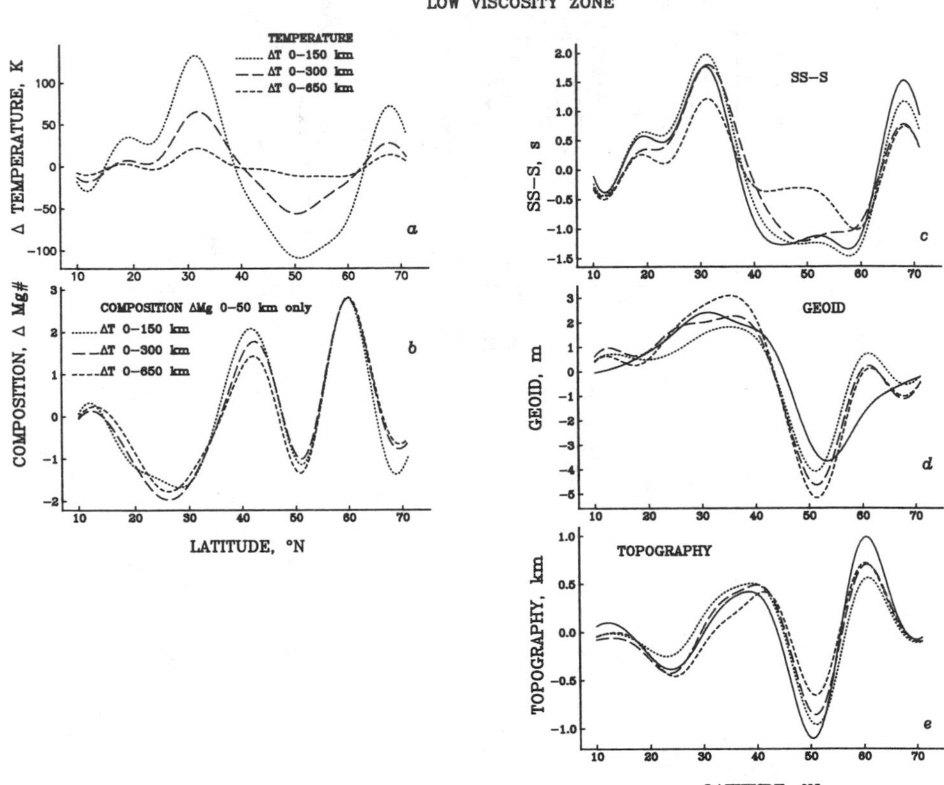

Figure 16 Results of the combined inversion of *SS-S* travel-time residuals, bathymetry, and geoid height for along-axis variations in upper mantle temperature and composition (from Sheehan & Solomon 1991). Composition is parameterized in terms of Mg #. The solutions include the dynamic effects of variations in density; the viscosity structure is taken to consist of a 40-km-thick high-viscosity lid, a low-viscosity zone extending from the base of the lid to a depth of 200 km, and a constant-viscosity underlying mantle. (*a*) Three solutions for along-axis temperature variations: Temperature perturbations are variously constrained to be uniform over 0–150 km depth (*dotted line*), 0–300 km depth (*long dashed line*) or 0–650 km depth (*short dashed line*). (*b*) The corresponding solutions for along-axis composition variations, constrained to occur over the depth interval 0–50 km. (*c*) Observed (*solid line*) and predicted along-axis profiles of *SS-S* travel-time residual. The "observed" profile is the filtered profile of Figure 15. (*d*) Observed and predicted along-axis geoid height profiles. (*e*) Observed and predicted along-axis bathymetry profiles.

expected to result, under the assumption of passive upwelling at ridges (McKenzie 1984), from the along-axis differences in temperature depicted in the figure. Possible explanations for these departures from the expectations of simple models include long-wavelength variations in mantle chemistry unrelated to degree of melt extraction, variations in the volatile content of the melt-generating zones, or a component of dynamic upwelling driven by melting-induced changes in buoyancy (Sheehan & Solomon 1991, Forsyth 1992).

An important element of seismic structure potentially relatable to mantle flow and lithospheric emplacement processes is the anisotropy of seismic velocity in the oceanic upper mantle. As noted above, an azimuthal anisotropy extends throughout most, if not all, of the mantle portion of oceanic lithosphere. In principle, the accumulated strain induced by mantle convective flow should also impart a bulk anisotropy to the sublithospheric mantle, at least that part of the mantle within the stability field of olivine (McKenzie 1979, Ribe 1989). Two forms of anisotropy are potentially indicative of mantle convective flow directions: differences in velocity between horizontally and vertically propagating waves (polarization anisotropy) and variations in velocity with azimuth. Variations in polarization anisotropy can, in principle, distinguish regions of upwelling and downwelling from those areas dominated by horizontal flow, and azimuthal anisotropy indicating a fast direction distinct from the direction of seafloor spreading at the time of formation of the overlying crust (sometimes called the fossil spreading direction) can be a signature of the predominant direction of horizontal asthenospheric flow. Several seismological efforts to detect these forms of anisotropy have been reported. Kuo et al (1987) report that *SS-S* differential travel times in the North Atlantic show a dependence on the azimuth of the *SS* path at the surface-reflection point, with the pattern consistent with a N-S orientation to the olivine *a*-axis in the sublithospheric mantle; they interpreted this pattern as an indication of the orientation of asthenospheric flow in the region (Hager & O'Connell 1979). Neither Woodward & Masters (1991) nor Sheehan & Solomon (1991) found a consistent pattern of significant azimuthal anisotropy in a larger set of *SS-S* times from the North Atlantic, however; Woodward & Masters (1991) resolved an azimuthal dependence of *SS-S* times in the Pacific consistent with the fast direction in the mantle being oriented parallel to the direction of absolute plate motion, but they could not eliminate the possibility that the pattern was an effect of geographic bias in the paths within distinct ranges in azimuth.

Surface waves are sensitive to both polarization and azimuthal anisotropy. Nishimura & Forsyth (1989) inverted Love and Rayleigh wave phase velocities in the Pacific to determine the age and depth dependence of both polarization and azimuthal anisotropy. They found that the ratio of horizontal to vertical shear wave velocity increased from a minimum at young ages (0–4 Ma) to a steady value at ages greater than 20 Ma; they interpreted this result as evidence for dominantly vertical flow beneath the ridge axis. They found that the azimuthal anisotropy, when referenced to the fossil spreading direction, was strongest in regions less than 80 Ma in age, where the effect was a maximum between 25 and 100 km depth but persisted to 200 km depth; since the lithosphere does not extend as deep as 200 km, they interpreted this result as indicating a contribution from asthenospheric flow, which in the young regions of the eastern Pacific is thought to be nearly parallel to the local fossil spreading direction. In regions older than 80 Ma, the azimuthal anisotropy appears to be confined to depths less than 50 km, i.e. less than the thickness of the lithosphere; since this result is, at face value, contrary to both the prevailing model for anisotropy of the mantle lithosphere and evidence from refraction studies (Shimamura 1984), Nishimura & Forsyth (1989) interpreted this result as due to the partial cancellation of lithospheric anisotropy by the differently oriented azimuthal anisotropy of the asthenosphere in the western Pacific.

OUTSTANDING ISSUES AND FUTURE DIRECTIONS

Geophysical experiments designed to characterize the structure of mid-ocean ridges have considerably sharpened the hypotheses for the relationships among plate separation, mantle flow, melt generation, magma transport, and formation of oceanic crust, but these first experiments have, of necessity, been focused on limited areas and scales. Surface wave, normal mode, and long-period body-wave techniques are yielding new results on the oceanic mantle on horizontal scales of 10^3–10^4 km and vertical scales of 10^2 km, while active marine seismic experiments are illuminating the oceanic crust at horizontal scales of 1–10 km and vertical scales of 0.1–1 km. The intermediate horizontal and vertical scales remain important targets of future work. Further, it is well to remember that even the most revealing high-resolution seismic studies are only of local areas, and while some generalizations can be made from regional information such as bathymetry and gravity, most of these generalizations await testing with further, carefully selected high-resolution experiments.

One of the most promising directions for future seismic investigation of ridge processes is to elucidate the structure of the uppermost mantle beneath ridges on a scale sufficient to infer some of the characteristics of mantle flow and melt generation. These topics, heretofore largely in the realm of theoreticians and numerical modelers, are amenable to study through the three-dimensional, anisotropic *P* and *S* wave velocity structure of the ridge upper mantle. The temperature variations associated with upwelling should have direct effects on bulk seismic velocities, and the shearing accompanying flow should yield diagnostic patterns of anisotropy. Beyond the hint from the age dependence of surface wave polarization for ages less than 20 Ma, there is not yet direct seismic confirmation of theoretical expectations that the zone of mantle upwelling and melt generation beneath ridges is of order 100 km in width, much less of various hypotheses advanced to account for the remarkable narrowing by some two orders of magnitude of the zone of the ascending melt into the ~1-km-wide neovolcanic zone (e.g. Phipps Morgan 1987, Scott & Stevenson 1989). What is required experimentally is a passive network of broad-band ocean bottom seismometers situated for one year or more astride a mid-ocean ridge. A combination of body-wave delay time tomography and inversion of surface wave dispersion data can yield two- and three-dimensional images of seismic velocities. Shear wave splitting and azimuthal and polarization anisotropy of surface waves can be used to infer the form of preferred crystal alignment in the mantle and its variation across the zone of upwelling. There is the possibility as well of detecting and characterizing the distribution of zones of melt within the upwelling mantle from the potentially strong effect of any melt fraction on the velocities and attenuation of seismic waves.

The relationship among the various scales of along-axis variability in seismic characteristics and the segmentation of axial processes remains to be explored, particularly the intermediate scales between the 1000-km variation resolvable with surface waves and global tomographic techniques and the 10-km scale addressible with an ESP profile or a crustal tomography experiment. These intermediate scales are those most likely to reflect dynamics of the mantle-melt system. Determination of along-axis variations in the seismic properties of the lowermost crust or uppermost mantle might reveal the nature of variations at this intermediate scale and their connection to other measures of segmentation. Good vertical resolution at these depths will pose a challenge. As with the shallow crust, a combination of seismic techniques probably offers the greatest promise for insight.

The detailed evolution from crustal-level melt to solid oceanic crust, including the important time scales for magma replenishment, solidification, and subsolidus deformation, is not understood. While theoretical, laboratory, petrological, and geochemical analyses of this problem all warrant pursuit, additional seismic experiments addressing this topic can be envisioned. For example, the different states of the magma lenses at 9°

and 13°N indicated by analysis of reflected waves (Harding et al 1989, Vera et al 1990) raise the possibility that systematic mapping of the reflection coefficient of the AMC reflector at near-normal incidence and as a function of range might reveal relationships among magma state, hydrothermal activity, recency of axial volcanism, or other measures of along-axis variability that would help to constrain the controlling processes and their time scales.

At slow spreading rates, the relation between segment length and such measures of mantle upwelling strength as axial depth minimum and mantle Bouguer anomaly magnitude (Lin et al 1990) have not been tested with seismic experiments. Microearthquake experiments designed to map the thickness of the mechanically strong lithosphere and to image the seismic velocity structure in three dimensions with passive seismic tomography should be conducted along slow-spreading ridge segments of a variety of lengths and cross-axis morphology. Because no axial magma body has yet been imaged along slow-spreading ridges, particular attention should be devoted during and after such experiments to any detected volumes of significantly lower than average crustal velocity.

The interpretation of both recently completed and future seismic experiments will benefit substantially from further laboratory measurements of seismic properties of crustal and mantle rocks at a range of pressures, temperatures, and frequencies appropriate to the oceanic crust and upper mantle. Particularly important will be measurements of *P* and *S* wave velocity and attenuation in partially molten mantle material, partially crystalline basaltic melt, and mostly crystalline gabbroic mush. Such measurements are critical to the eventual tracking of melt, by geophysical observation, from initial mantle generation to final crustal solidification.

ACKNOWLEDGMENTS

We thank Joe Cann, Dave Caress, Bob Detrick, Rachel Haymon, Jian Lin, G. M. Purdy, Emilio Vera, Bob Woodward, and Yu-Shen Zhang for copies of figures from their publications. Preparation of this review was supported by the National Science Foundation under grants EAR-9004750, OCE-9000177, and OCE-9106233, the Office of Naval Research under grants N00014-89-J-1257 and N00014-91-J-1216, and the National Aeronautics and Space Administration under grant NAG 5-814.

Literature Cited

Anderson, R. N., Honnorez, J., Becker, K., Adamson, A. C., Alt, J. C., et al. 1982. DSDP hole 504B, the first reference section over 1 km through layer 2 of the oceanic crust. *Nature* 300: 589–94

Bonatti, E. 1990. Not so hot "hot spots" in the oceanic mantle. *Science* 250: 107–11

Bratt, S. R., Purdy, G. M. 1984. Structure and variability of oceanic crust on the flanks of the East Pacific Rise between 11° and 13°N. *J. Geophys. Res.* 89: 6111–25

Bratt, S. R., Solomon, S. C. 1984. Compressional and shear wave structure of the East Pacific Rise at 11°20′N: Constraints from three-component ocean-bottom seismometer data. *J. Geophys. Res.* 89: 6095–110

Bryan, W. B., Moore, J. G. 1977. Compositional variations of young basalts in the Mid-Atlantic Ridge rift valley near lat 36°49′N. *Bull. Geol. Soc. Am.* 88: 556–70

Burnett, M. S., Caress, D. W., Orcutt, J. A. 1989. Tomographic image of the magma chamber at 12°50′N on the East Pacific Rise. *Nature* 339: 206–8

Butler, R. 1985. Anisotropic propagation of P- and S-waves in the western Pacific lithosphere. *Geophys. J. R. Astron. Soc.* 81: 89–101

Cann, J. R. 1974. A model for oceanic crustal structure developed. *Geophys. J. R. Astron. Soc.* 39: 169–87

Caress, D. W., Burnett, M. S., Orcutt, J. A. 1992. Tomographic image of the axial low velocity zone at 12 50′N on the East Pacific Rise. *J. Geophys. Res.* In press

Casey, J. F., Dewey, J. F., Fox, P. J., Karson, J. A., Rosencrantz, E. 1981. Heterogeneous nature of oceanic crust and upper mantle: A perspective from the Bay of Islands ophiolite complex. In *The Sea*, Vol. 7, ed. C. Emiliani, pp. 305–38. New York: Wiley-Interscience

Casey, J. F., Karson, J. A. 1981. Magma chamber profiles from the Bay of Islands ophiolite complex. *Nature* 292: 295–301

Christensen, N. I., Salisbury, M. H. 1975. Structure and constitution of the lower oceanic crust. *Rev. Geophys. Space Phys.* 13: 57–86

Crane, K. 1985. The spacing of rift axis highs: Dependence upon diapiric processes in the underlying asthenosphere? *Earth Planet. Sci. Lett.* 72: 405–14

Detrick, R. S., Buhl, P., Vera, E., Mutter, J., Orcutt, J., Madsen, J., Brocher, T. 1987. Multi-channel seismic imaging of a crustal magma chamber along the East Pacific Rise. *Nature* 326: 35–41

Derick, R. S., Fox, P. J., Kastens, K., Ryan, W. B. F., Mayer, L., Karson, J. A. 1984. Sea Beam survey of the Kane Fracture Zone and the adjacent Mid-Atlantic Ridge rift valley. *Eos, Trans. Am. Geophys. Union* 65: 1006 (Abstr.)

Detrick, R. S., Mutter, J. C., Buhl, P., Kim, I. I. 1990. No evidence from multichannel reflection data for a crustal magma chamber in the MARK area on the Mid-Atlantic Ridge. *Nature* 347: 61–64

Duschenes, J. D., Solomon, S. C. 1977. Shear wave travel-time residuals from oceanic earthquakes and the evolution of oceanic lithosphere. *J. Geophys. Res.* 82: 1985–2000

Forsyth, D. W. 1975. The early structural evolution and anisotropy of the oceanic upper mantle. *Geophys. J. R. Astron. Soc.* 43: 103–62

Forsyth, D. W. 1977. The evolution of the upper mantle beneath mid-ocean ridges. *Tectonophysics* 38: 89–118

Forsyth, D. W. 1992. Geophysical constraints on mantle flow and melt generation beneath mid-ocean ridges. In *Mantle Flow, Melt Generation and Lithospheric Deformation at Mid-Ocean Ridges*, *Geophys. Monogr*. Washington, DC: Am. Geophys. Union. In press

Forsyth, D. W., Ehrenbard, R. L., Chapin, S. 1987. Anomalous upper mantle beneath the Australian-Antarctic discordance. *Earth Planet. Sci. Lett.* 84: 471–78

Fox, P. J., Schreiber, E., Peterson, J. J. 1973. The geology of the oceanic crust: Compressional wave velocities of oceanic rocks. *J. Geophys. Res.* 78: 5155–79

Francheteau, J., Ballard, R. D. 1983. The East Pacific Rise near 21°N, 13°N and 20°S: Inferences for along-strike variability of axial processes of the mid-ocean ridge. *Earth Planet. Sci. Lett.* 64: 93–116

Francis, T. J. G. 1969. Generation of seismic anisotropy along the mid-ocean ridges. *Nature* 221: 162–65

Garmany, J. 1989. Accumulations of melt at the base of young oceanic crust. *Nature* 340: 628–32

Girardin, N. 1980. Travel time residuals of PP waves reflected under oceanic and continental platform regions. *Phys. Earth. Planet. Inter.* 23: 199–206

Girardin, N., Jacoby, W. R. 1976. Rayleigh wave dispersion along Reykjanes Ridge. *Tectonophysics* 55: 155–71

Grand, S. P. 1987. Tomographic inversion for shear velocity beneath the North American plate. *J. Geophys. Res.* 92: 14,065–90

Hager, B. H., O'Connell, R. J. 1979. Kinematic models of large-scale flow in the Earth's mantle. *J. Geophys. Res.* 84: 1031–48

Hale, L. D., Morton, C. J., Sleep, N. H. 1982. Reinterpretation of seismic reflection data over the East Pacific Rise. *J. Geophys. Res.* 87: 7707–17

Harding, A. J., Orcutt, J. A., Kappus, M. E., Vera, E. E., Mutter, J. C., et al. 1989. Structure of young oceanic crust at 13°N on the East Pacific Rise from expanding spread profiles. *J. Geophys. Res.* 94: 12,163–96

Haymon, R. M., Fornari, D. J., Edwards, M. H., Carbotte, S., Wright, D., Macdonald, K. C. 1991. Hydrothermal vent distribution along the East Pacific Rise crest (9°09′–54′N) and its relationship to magmatic and tectonic processes on fast-spreading mid-ocean ridges. *Earth Planet. Sci. Lett.* 104: 513–34

Herron, T. J. 1982. Lava flow layer—East Pacific Rise. *Geophys. Res. Lett.* 9: 17–20

Herron, T. J., Ludwig, W. J., Stoffa, P. L., Kan, T. K., Buhl, P. 1978. Structure of the East Pacific Rise crest from multichannel seismic reflection data. *J. Geophys. Res.* 83: 798–804

Hess, H. H. 1964. Seismic anisotropy of the uppermost mantle under oceans. *Nature* 203: 629–31

Houtz, R., Ewing, J. 1976. Upper crustal structure as a function of plate age. *J. Geophys. Res.* 81: 2490–98

Huang, P. Y., Solomon, S. C. 1988. Centroid depths of mid-ocean ridge earthquakes: Dependence on spreading rate. *J. Geophys. Res.* 93: 13,445–77

Huang, P. Y., Solomon, S. C., Bergman, E. A., Nabelek, J. L. 1986. Focal depths and mechanisms of Mid-Atlantic Ridge earthquakes from body waveform inversion. *J. Geophys. Res.* 91: 579–98

Kent, G. M., Harding, A. J., Orcutt, J. A. 1990. Evidence for a smaller magma chamber beneath the East Pacific Rise at 9°30′N. *Nature* 344: 650–53

Kidd, R. G. W. 1977. A model for the process of formation of the upper oceanic crust. *Geophys. J. R. Astron. Soc.* 50: 149–83

Klein, E. M., Langmuir, C. H. 1989. Local versus global variations in ocean ridge basalt composition: A reply. *J. Geophys. Res.* 94: 4241–52

Kong, L. S. L., Solomon, S. C., Purdy, G. M. 1992. Microearthquake characteristics of a mid-ocean ridge along-axis high. *J. Geophys. Res.* In press

Kuo, B.-Y., Forsyth, D. W. 1988. Gravity anomalies of the ridge-transform system in the South Atlantic between 31 and 34.5°S: Upwelling centers and variations in crustal thickness. *Mar. Geophys. Res.* 10: 205–32

Kuo, B.-Y., Forsyth, D. W., Wysession, M. 1987. Lateral heterogeneity and azimuthal anisotropy in the North Atlantic determined from SS-S differential travel times. *J. Geophys. Res.* 92: 6421–36

Langmuir, C. H., Bender, J. F., Batiza, R. 1986. Petrological and tectonic segmentation of the East Pacific Rise, 5°30′–14°30′N. *Nature* 322: 422–29

Lees, A. R., Knopoff, L., Kausel, E. G. 1974. Variations of upper mantle structure under the Pacific Ocean. *Science* 186: 141–43

Lin, J., Bergman, E. A. 1990. Rift grabens, seismicity, and volcanic segmentation of the Mid-Atlantic Ridge: Kane to Atlantic fracture zones. *Eos, Trans. Am. Geophys. Union* 71: 1572 (Abstr.)

Lin, J., Purdy, G. M., Schouten, H., Sempere, J.-C., Zervas, C. 1990. Evidence from gravity data for focused magmatic accretion along the Mid-Atlantic Ridge. *Nature* 344: 627–32

Macdonald, K. C., Fox, P. J., Perram, L. J., Eisen, M. F., Haymon, R. M., et al. 1988. A new view of the mid-ocean ridge from the behaviour of ridge-axis discontinuities. *Nature* 335: 217–25

Macdonald, K. C., Luyendyk, B. P. 1977. Deep-tow studies of the structure of the Mid-Atlantic Ridge crest near lat 37°N. *Geol. Soc. Am. Bull.* 88: 621–36

McClain, J. S., Orcutt, J. A., Burnett, M. 1985. The East Pacific Rise in cross section: A seismic model. *J. Geophys. Res.* 90: 8627–39

McKenzie, D. 1979. Finite deformation during fluid flow. *Geophys. J. R. Astron. Soc.* 58: 689–715

McKenzie, D. 1984. The generation and compaction of partially molten rock. *J. Petrol.* 25: 713–65

Michael, P. J., Bonatti, E. 1985. Peridotite composition from the North Atlantic: Regional and tectonic variations and implications for partial melting. *Earth Planet. Sci. Lett.* 73: 91–104

Mitchell, B. J., Yu, G.-K. 1980. Surface wave dispersion, regionalized velocity models, and anisotropy of the Pacific crust and upper mantle. *Geophys. J. R. Astron. Soc.* 63: 497–514

Montagner, J. P., Jobert, N. 1983. Variation with age of the deep structure of the Pacific Ocean inferred from long-period Rayleigh wave dispersion. *Geophys. Res. Lett.* 27: 206–22

Montagner, J.-P., Tanimoto, T. 1990. Global anisotropy in the upper mantle inferred from the regionalization of phase velocities. *J. Geophys. Res.* 95: 4797–819

Morris, E., Detrick, R. S. 1991. Three-dimensional analysis of gravity anomalies in the MARK area, Mid-Atlantic Ridge 23°N. *J. Geophys. Res.* 96: 4355–66

Murase, T., McBirney, A. R. 1973. Properties of some common igneous rocks and their melts at high temperatures. *Geol. Soc. Am. Bull.* 84: 3563–92

Mutter, J. C., Barth, G. A., Buhl, P., Detrick, R. S., Orcutt, J., Harding, A. 1988. Magma distribution across ridge-axis discontinuities on the East Pacific Rise from multichannel seismic images. *Nature* 336: 156–58

Nishimura, C. E., Forsyth, D. W. 1985.

Anomalous Love-wave phase velocities in the Pacific: sequential pure-path and-spherical harmonic inversion. *Geophys. J. R. Astron. Soc.* 81: 389–407
Nishimura, C. E., Forsyth, D. W. 1988. Rayleigh wave phase velocities in the Pacific with implications for azimuthal anisotropy and lateral heterogeneities. *Geophys. J. R. Astron Soc.* 94: 479–501
Nishimura, C. E., Forsyth, D. W. 1989. The anisotropic structure of the upper mantle in the Pacific. *Geophys. J. R. Astron. Soc.* 96: 203–29
ODP (Ocean Drilling Program) Leg 106 Scientific Party 1986. Drilling the Snake Pit hydrothermal sulfide deposit on the Mid-Atlantic Ridge, lat 23°22′N. *Geology* 14: 1004–7
Pallister, J. S., Hopson, C. A. 1981. Samail ophiolite plutonic suite: Field relations, phase variation, cryptic variation and layering, and a model of a spreading ridge magma chamber. *J. Geophys. Res.* 86: 2593–644
Parsons, B., Sclater, J. G. 1977. An analysis of the variation of ocean floor bathymetry and heat flow with age. *J. Geophys. Res.* 82: 803–27
Phipps Morgan, J. 1987. Melt migration beneath mid-ocean spreading centers. *Geophys. Res. Lett.* 14: 1238–41
Phipps Morgan, J., Forsyth, D. W. 1988. Three-dimensional flow and temperature perturbations due to a transform offset: effects on oceanic crustal and upper mantle structure. *J. Geophys. Res.* 93: 2955–66
Purdy, G. M. 1982. The variability in seismic structure of layer 2 near the East Pacific Rise at 12°N. *J. Geophys. Res.* 87: 8403–16
Purdy, G. M., Detrick, R. S. 1986. Crustal structure of the Mid-Atlantic Ridge at 23°N from seismic refraction studies. *J. Geophys. Res.* 91: 3739–62
Raitt, R. W. 1963. The crustal rocks. In *The Sea*, Vol. 3, ed. M. N. Hill, pp. 85–102. New York: Wiley-Interscience
Raitt, R. W., Shor, G. G., Francis, T. J. G., Morris, G. B. 1969. Anisotropy of the Pacific upper mantle. *J. Geophys. Res.* 74: 3095–109
Reid, I., Jackson, H. R. 1981. Oceanic spreading rate and crustal thickness. *Mar. Geophys. Res.* 5: 165–72
Ribe, N. M. 1989. Seismic anisotropy and mantle flow. *J. Geophys. Res.* 94: 4213–23
Romanowicz, B. 1991. Seismic tomography of the Earth's mantle. *Annu. Rev. Earth Planet. Sci.* 19: 77–99
Rona, P. A., Klinkhammer, G., Nelson, T. A., Trefry, J. H., Elderfield, H. 1986. Black smokers, massive sulfides and vent biota at the Mid-Atlantic Ridge. *Nature* 321: 33–37
Schouten, H., Klitgord, K. D., Whitehead, J. A. 1985. Segmentation of mid-ocean ridges. *Nature* 317: 225–29
Scott, D. R., Stevenson, D. J. 1989. A self-consistent model of melting, magma migration and buoyancy-driven circulation beneath mid-ocean ridges. *J. Geophys. Res.* 94: 2973–88
Sempere, J.-C., Purdy, G. M., Schouten, H. 1990. Segmentation of the Mid-Atlantic Ridge between 24°N and 30°40′N. *Nature* 344: 427–31
Shearer, P., Orcutt, J. 1985. Anisotropy in the oceanic lithosphere—theory and observations from the Ngendei seismic refraction experiment in the south-west Pacific. *Geophys. J. R. Astron. Soc.* 80: 493–526
Sheehan, A. F., Solomon, S. C. 1991. Joint inversion of shear wave travel time residuals and geoid and depth anomalies for long-wavelength variations in upper mantle temperature and composition along the Mid-Atlantic Ridge. *J. Geophys. Res.* 96: 19,981–20,009
Shimamura, H. 1984. Anisotropy in the oceanic lithosphere of the northwestern Pacific basin. *Geophys. J. R. Astron. Soc.* 76: 253–60
Sleep, N. H. 1975. Formation of oceanic crust: Some thermal constraints. *J. Geophys. Res.* 80: 4037–42
Solomon, S. C., Huang, P. Y., Meinke, L. 1988. The seismic moment budget of slowly spreading ridges. *Nature* 334. 58–61
Spudich, P., Orcutt, J. 1980. A new look at the seismic velocity structure of the oceanic crust. *Rev. Geophys.* 18: 627–45
Toomey, D. R., Purdy, G. M., Solomon, S. C., Wilcock, W. S. D. 1990. The three-dimensional seismic velocity structure of the East Pacific Rise near latitude 9°30′N. *Nature* 347: 639–45
Toomey, D. R., Solomon, S. C., Purdy, G. M. 1988. Microearthquakes beneath the median valley of the Mid-Atlantic ridge near 23°N: Tomography and tectonics. *J. Geophys. Res.* 93: 9093–112
Toomey, D. R., Solomon, S. C., Purdy, G. M., Murray, M. H. 1985. Microearthquakes beneath the median valley of the Mid-Atlantic Ridge near 23°N: Hypocenters and focal mechanisms. *J. Geophys. Res.* 90: 5443–58
Vera, E. E., Mutter, J. C., Buhl, P., Orcutt, J. A., Harding, A. J., et al. 1990. The structure of 0- to 0.2-m.y.-old oceanic crust at 9°N on the East Pacific Rise from expanded spread profiles. *J. Geophys. Res.* 95: 15,529–56

Whitehead, J. A. Jr., Dick, H. J. B., Schouten, H. 1984. A mechanism for magmatic accretion under spreading centres. *Nature* 312: 146–48

Wilcock, W. S. D., Solomon, S. C., Purdy, G. M., Toomey, D. R., Dougherty, M. E. 1991. Seismic attenuation structure of the crust on the axis of the East Pacific Rise at 9°30′N. In *Spring Meeting 1991, Eos, Trans. Am. Geophys. Union* 72, suppl.: 263 (Abstr.)

Woodhouse, J. H., Dziewonski, A. M. 1984. Mapping the upper mantle: Three dimensional modeling of Earth structure by inversion of seismic waveforms. *J. Geophys. Res.* 89: 5953–86

Woodward, R. L., Masters, G. 1991. Global upper mantle structure from long-period differential travel times. *J. Geophys. Res.* 96: 6351–78

Yoshii, T. 1975. Regionality of group velocities of Rayleigh waves in the Pacific and the thickening of the plate. *Earth Planet. Sci. Lett.* 25: 305–12

Zhang, Y.-S., Tanimoto, T. 1989. Three-dimensional modelling of upper mantle structure under the Pacific Ocean and surrounding area. *Geophys. J. Int.* 98: 255–69

Zhang, Y.-S., Tanimoto, T. 1991. Global Love wave phase velocity variation and its significance to plate tectonics. *Phys. Earth Planet. Inter.* 66: 160–202

Annu. Rev. Earth Planet. Sci. 1992. 20:365 88

MIXING IN THE MANTLE

Louise H. Kellogg

Department of Geology, University of California, Davis, California 95616

KEY WORDS: convection, geodynamics

INTRODUCTION

Although it is well known and accepted that convection in the Earth's mantle is the driving force of plate tectonics, little is known about the nature of mantle convection. The planform of convection cells cannot be mapped with assurance, and it is not clear whether the mantle convects in one layer or several layers.

The basalts produced at mid-ocean ridges around the globe are remarkably uniform, both in composition and in the isotopic characteristics that reveal their long-term evolution. This is not surprising, since mantle convection, like all flows, acts to stir the mantle and destroy heterogeneities. But the mantle is not entirely uniform, and oceanic island basalts reveal small and large heterogeneities, some of which have persisted for the entire history of the Earth. Understanding the mixing accomplished by mantle convection helps in the interpretation of mantle geochemistry; likewise the isotope systematics reflect convective processes.

The mid-ocean ridge basalts (MORB) are depleted in trace elements due to the extraction of the crust, which is correspondingly enriched. The isotopic character of MORB indicates that the source region for these basalts has been depleted for some time (several billion years). The oceanic island basalts (OIB), on the other hand, exhibit isotope characteristics ranging from depleted to undepleted to enriched. These heterogeneities can form in myriad ways; for instance by subduction of crustal material, or by preservation of undepleted material from the early Earth.

STRUCTURE OF THE MANTLE

Layered Convection

Schilling (1973) used geochemical observations on the Reykjanes Ridge near Iceland to show that the mantle is heterogeneous and that plume

0084–6597/92/0515–0365$02.00

volcanism produces basalt that is chemically distinct from the neighboring ridge basalts. He postulated a primitive mantle source deep beneath Iceland from which the mantle plume arises.

The phrase "layered mantle convection" has been associated with several models. The first (Figure 1) allows no large-scale exchange of material between the upper and lower mantle (e.g. Allègre & Turcotte 1985). The model of Allègre and Turcotte allows exchange of small amounts of material by entrainment in boundary layer instabilities which grow at 670 km depth. Mobile rare gases such as helium may also move across the boundary by diffusion. If helium is not sufficiently mobile to leak across the discontinuity by diffusion, then models of layered mantle convection must allow "leak" or exchange of material between the layers, to satisfy the isotopic constraints.

An alternative model of layered convection has the upper mantle growing at the expense of the lower mantle due to extraction of the crust (Jacobsen & Wasserburg 1979). In this model, the depleted residue remaining after extraction of the crust from undepleted mantle material is buoyant compared to the lower mantle and remains in the upper regions of the mantle. The exchange of material in this case is in one direction only, and the boundary between the two layers moves deeper into the mantle with time. It is possible that this model applies to the early stages of the Earth, when the continental crust formed rapidly, but that the volume of the upper mantle is now fixed in time.

All of these layered convection models involve reservoirs, namely the upper and lower mantle, which are mechanically isolated from one another to some extent and which are compositionally distinct from one another.

Anderson (1982a,b, 1983) has proposed an alternative model in which the mantle is compositionally stratified. In this model, the upper mantle is divided into a shallow olivine-rich peridotite layer that is enriched in incompatible elements and a depleted eclogite layer that is the source of MORB. The model requires the efficient delamination of the subducted oceanic lithosphere with the basaltic component entering the eclogite layer. An important constraint on mantle processes is that the thickness of the basaltic crust is nearly constant, independent of the spreading rate. Pressure release melting of a fertile mantle beneath an ocean ridge naturally leads to this consequence. However, it appears mechanically impossible to melt a segregated ecologite layer to form an oceanic crust with a thickness that is not dependent on the spreading rate.

A hybrid model was proposed by Silver and coworkers (1988) in which subducted slab can penetrate the lower mantle, but penetration is limited in extent and does not result in the large-scale overturn associated with models of whole-mantle convection. Overturn and the resultant mixing

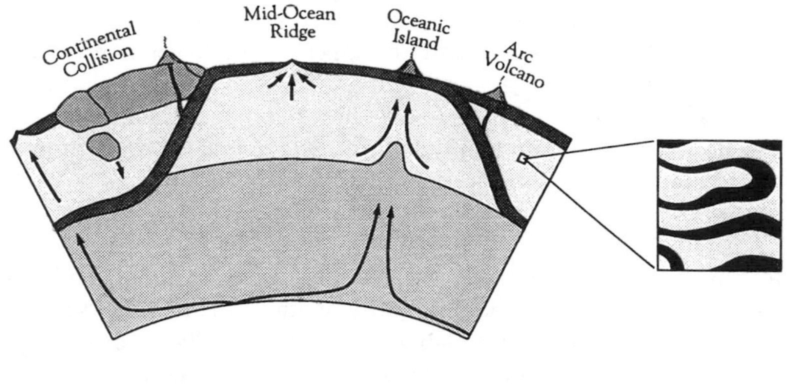

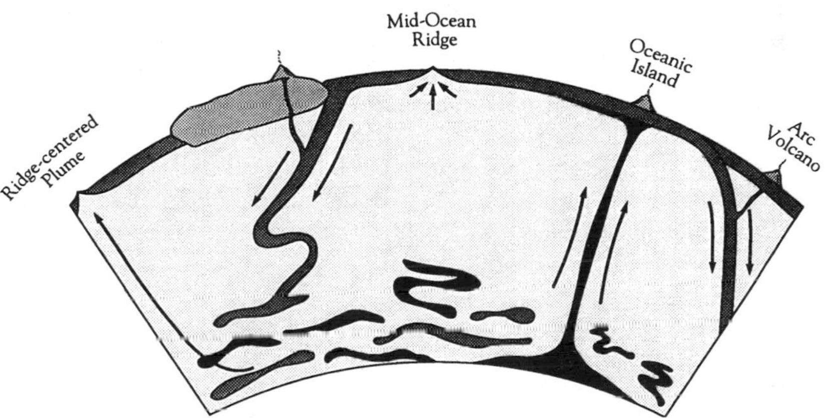

Figure 1 Two possible models of the structure of the mantle. The top figure shows two-layered convection as described by Allègre & Turcotte (1986) and others. The mantle consists of two reservoirs which remain largely isolated through time. The upper mantle provides the source material for MORB; the lower mantle provides the source material for the most primordial of oceanic island basalts. Heterogeneities are introduced by subduction, delamination of continental material, and entrainment of lower mantle material. Heterogeneities are rapidly mixed into the upper mantle by convective stirring. In the process of mixing, they form long, narrow stringers which give the mantle a "marble-cake" structure (inset) as seen in high-temperature peridotites. Variations on the layered mantle picture are described in the text. The lower figure shows whole mantle convection as proposed by Davies (1984) and others. Subducting slabs penetrate into the lower mantle. In this model, mixing is sluggish in the lower mantle due to increased viscosity. Relatively large heterogeneities made up of subducted material and some primordial material persist for long periods of time. Upwelling plumes arising from the core-mantle boundary bring this material to the surface at oceanic islands. Other models than those shown here are also possible.

can be prevented by a density contrast due to a difference in composition between the upper and lower mantle, or by a negative Clapyron slope associated with a phase transition at the upper mantle-lower mantle boundary (Silver et al 1988).

Models by Machetel & Weber (1991) show that phase changes can cause convection to vary intermittently between whole-mantle convection and layered-mantle convection.

Whole-Mantle Convection

Alternatively, the mantle may undergo large-scale overturn—that is, convection in large cells extending the entire depth of the mantle. According to this model, the subducting slab sinks below the 670-km discontinuity to the lower mantle, while upwelling plumes rise from the core-mantle boundary to the surface (Figure 1).

Models of whole-mantle convection must be able to preserve parts of the mantle for long periods of time to satisfy the geochemical requirements that some "primitive" or nearly undifferentiated material exists in the mantle. The geochemical constraints are discussed in a later section.

Davies (1984, 1990b) presented a model of whole-mantle convection that attributes the heterogeneities in MORB and OIB to imperfect mixing in the mantle reservoir (plum pudding model). Loper (1985) has proposed whole-mantle convection but with continuous stratification to provide the heterogeneities associated with the OIB.

ISOTOPIC EVOLUTION OF THE MANTLE

Isotopic Observations

One of the most important sources of information about the Earth's mantle is the chemical characteristics of mantle-derived rocks. Studies of isotope ratios and the distributions of rare earth elements show that normal mid-ocean ridge basalts are nearly uniformly depleted (Gast 1968, Bougault et al 1980, Allègre et al 1980). Normal MORB is defined in this context as the oceanic crust of that portion of the ridge system with bathymetric depths greater than 2 km. This excludes, for example, sections near Iceland and the Azores. The kinematics of plate tectonics cause the mid-ocean ridge system to migrate randomly over the upper mantle. MORB represent a random sample of the upper mantle; the uniformity of normal MORB in terms of rare earth distributions and isotope ratios can be attributed to mixing due to convection in the upper mantle and is evidence that the upper mantle is a nearly uniformly depleted reservoir.

Although normal mid-ocean ridge basalts are nearly uniform in terms of both incompatible element concentrations and isotope ratios, ocean island basalts are not (Gast 1968, Sun & Hanson 1975, Sun 1980). The ocean island basalts represent only a small fraction of the total erupted oceanic basalts and include anomalous segments of the mid-ocean ridge system such as Iceland and the Azores. The Nd, Sr, Pb, and He isotope signatures of OIB and MORB are plotted in Figure 2. On the basis of their isotope signatures, oceanic basalts can be described as falling into five (White 1985) or more groups. Most prominant among these groups is the MORB group, with low $^{87}Sr/^{86}Sr$ ratios and high $^{143}Nd/^{144}Nd$. The MORB group basalts exhibit the strongest correlation between Nd, Sr, and Pb isotopes. White includes Iceland in the MORB group. Other groups are named for the oceanic island whose basalts typify the characteristics of the group. For instance, the Kerguelen group, including Tristan da Cunha and Gough islands, exhibits high $^{87}Sr/^{86}Sr$ ratios, low $^{143}Nd/^{144}Nd$, and low $^{207}Pb/^{204}Pb$ and $^{206}Pb/^{204}Pb$ ratios. Significantly, there is little evidence of mixing between the OIB groups, but MORB appears to have mixed with all the other groups to some degree. White suggests that this results from contamination of the upper mantle (the MORB source region) by plumes.

The isotope signatures of the oceanic island basalts can also be explained as being the result of mixing four (Zindler & Hart 1986) or more components of the mantle. On the basis of Nd, Sr, Pb, and He isotope data Zindler & Hart identify a depleted mantle source, two enriched sources (EM I and EM II), and a source high in radiogenic Pb called HIMU. In this model MORB consists of a mixture of the end-member depleted mantle and small amounts of other components. These authors also propose the existence of a large reservoir of material of some intermediate composition denoted as prevalent mantle (PREMA). This consists of a mixture of depleted mantle, HIMU, enriched mantle, and primitive material. The geochemical data alone cannot distinguish whether PREMA is a single component of distinctive composition or the result of formation of basalts from a mixture of other components.

The concept of a "mantle component" can provide a useful characterization of the source region of an oceanic basalt. A mantle component is identified with a set of idealized chemical and especially isotopic characteristics. A reservoir is identified with a set of samples with similar chemical characteristics which are, as a result of their similarity, assumed to come from the same source region. The concepts of reservoir and component are related and are occasionally used interchangeably, but they are not identical. A reservoir is usually associated with samples originating from a common location. Thus normal mid-ocean ridge basalts, which are much alike in chemical character, can be used to identify a single component (*N*-MORB). Mid-ocean ridge basalts are also usually associated with a single reservoir, namely the shallow mantle.

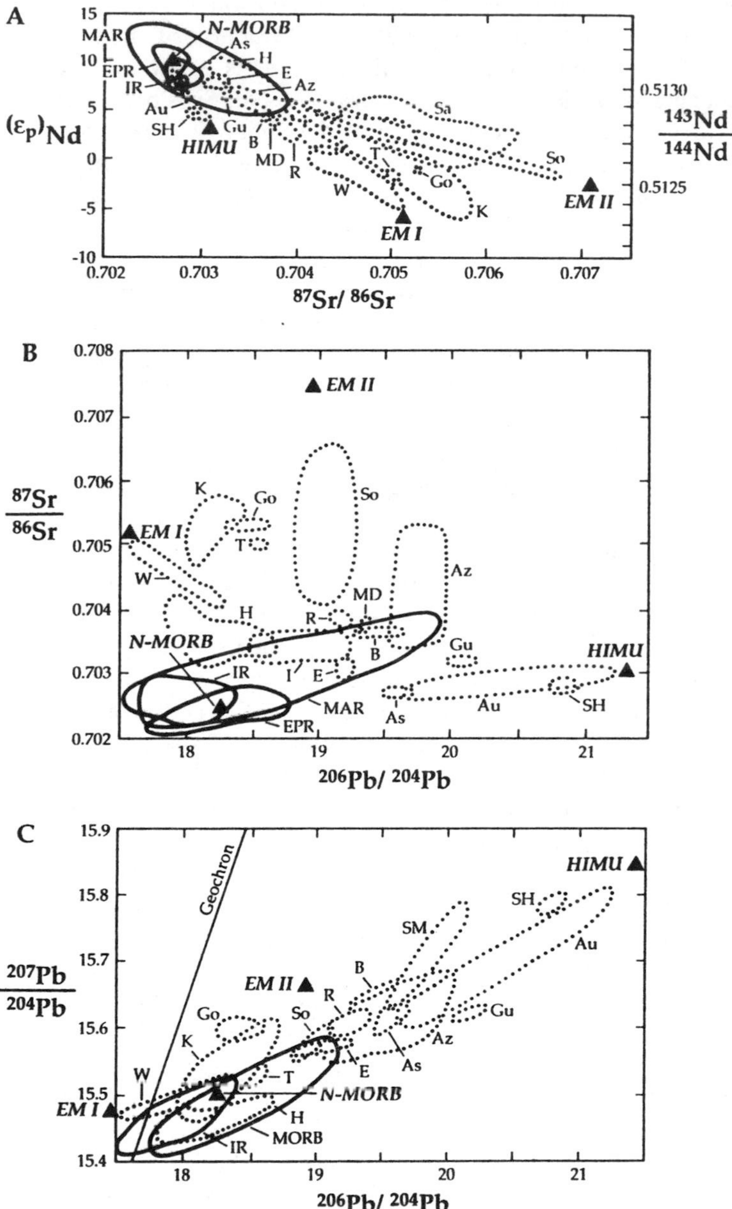
A
MAR
N-MORB
As
H
EPR
E
IR
Az
Au
SH
Gu
HIMU
B
MD
R
Sa
T
Go
So
W
K
EM I
EM II
$(\varepsilon_p)_{Nd}$
$^{143}Nd/^{144}Nd$
0.5130
0.5125
$^{87}Sr/^{86}Sr$
B
EM II
K
Go
So
EM I
T
W
Az
MD
H
R
N-MORB
B
Gu
HIMU
IR
I
E
MAR
As
Au
SH
EPR
$^{87}Sr/^{86}Sr$
$^{206}Pb/^{204}Pb$
C
Geochron
HIMU
SH
SM
Au
B
EM II
R
Go
So
Gu
K
Az
E
As
W
T
N-MORB
EM I
H
MORB
IR
$^{207}Pb/^{204}Pb$
$^{206}Pb/^{204}Pb$

Figure 2

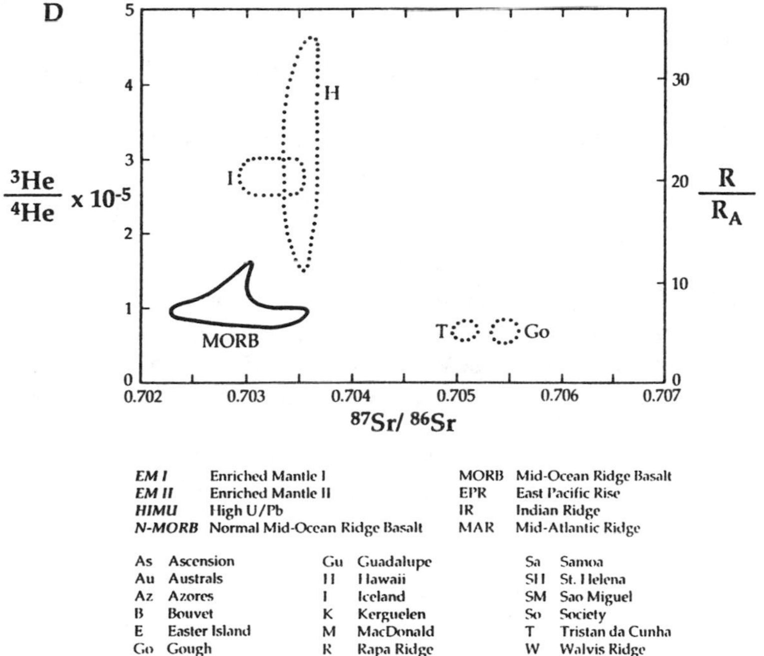

Figure 2 (*a*) Neodymium-strontium correlation for MORB and OIB. (Data collected from the literature.) The fields encircled by solid curves represent MORB; dotted lines represent oceanic island basalts; points represent the mantle components defined by Zindler & Hart (1986). (*b*) Strontium-lead correlation for MORB and OIB. (*c*) Lead-lead correlation for MORB and OIB. (*d*) Helium-strontium correlation for MORB and OIB.

Modeling the effect of exchange of material between reservoirs provides a way to use the isotope signatures of mantle-derived rocks as tracers of mantle convection. However, the isotope signature of a sample provides no information whatsoever about the physical location of the reservoir. The locations of reservoirs must be inferred (with varying degrees of uncertainty) from other information. In addition, some components (for example, the continental crust) do not consist of a reservoir of homogeneous isotopic character, but rather of rocks of diverse isotope signatures (e.g. DePaolo 1980, 1981, 1988). However, isotope studies of the continental crust show that it has been extracted from a mantle source that has been nearly uniform for billions of years (DePaolo 1980). The isotope signature associated with the component or reservoir is usually the mean value (Hart 1988), although occasionally an extremum is used (Hart 1988, Allègre 1982).

Associated with Hawaii and Iceland is a pristine, primitive, or undepleted reservoir which is often identified with the lower mantle beneath the 670-km seismic discontinuity. In the cases of ocean islands such as Tristan, Gough, Kerguelen, St. Helena, the Azores, and the Society Islands, the isotope signatures may result from the presence in the depleted mantle reservoir of incompletely homogenized subducted oceanic crust and entrained sediments (Chase 1981, Hofmann & White 1982, Zindler et al 1982). These isotope signatures probably developed in the continental crust and lithosphere (Allègre & Turcotte 1985), rather than in the depleted mantle reservoir (Hofmann & White 1982), since chemical heterogeneities would not persist long enough in the convecting mantle to develop the observed isotopic variations.

The primary geochemical cycle of the solid earth is directly associated with plate tectonics. The creation of the oceanic crust at oceanic ridges leads to the strong concentration of incompatible elements into the basaltic crust through the partial melting process. Elements are added and removed from the oceanic crust by seawater alteration and the oceanic crust is coated with sediments that are primarily derived from the continents. The subducted oceanic lithosphere is heated and becomes part of the convecting mantle reservoir. Convective mixing stretches and disperses the layered oceanic lithosphere until heterogeneities are reduced to a small scale. The time scale for this process is critical to interpretation of the geochemical observations. This process is discussed in detail in later sections.

At ocean trenches the altered oceanic crust is cycled back into the Earth's interior. Along with the descending plate, some continental material may be recycled into the mantle at subduction zones. Dehydration of the descending plate causes melting of the overlying wedge (Tatsumi et al 1983) at a depth of about 100 km, forming arc volcanos and adding material to the continents. Arc magmas also contain some subducted oceanic crust and sediments (Kay 1980). The complex processes responsible for the formation of the continental crust concentrate incompatible elements into the continents. As a result, the continental crust is a geochemical reservoir that is enriched compared to the bulk earth. These elements have been removed from all or part of the mantle; the mantle, therefore, is a depleted reservoir complementary to the continental crust.

The complementary nature of the continental crust and the MORB source reservoir requires that the material from which the crust has been extracted be mixed back into the MORB source reservoir. The fate of the subducted oceanic lithosphere has long been a subject of controversy.

Dickinson & Luth (1971) suggested that the subducted lithosphere lies at the base of the mantle; Oxburgh & Parmentier (1977) suggested that it underplates the continental lithosphere. However, the chemical evidence is conclusive; a large fraction of the subducted lithosphere is mixed back into the upper mantle reservoir. It is this mixing that has led to the development of the depleted isotopic signature of the MORB source region. If the subducted slab were not substantially mixed into the upper mantle, MORB isotopic signatures would exhibit some of the characteristics of an undepleted reservoir, rather than the depleted signature that is observed (Silver et al 1988).

ISOTOPIC GEODYNAMICS The isotope systems most often used as tracers for mantle convection are listed in Table 1. These decay systems provide tracers for the evolution of the mantle and crust. In effect, the ratio of a radiogenic daughter isotope to the stable isotopes of the same species measures the time-integrated history of the parent/daughter ratios of the sample. A review of the systematics of these systems is provided by Zindler & Hart (1986).

The Nd and Sr data from oceanic basalts, worldwide, are shown in Figure 2*a*. The linear correlation between these led to the conclusion that there were at least two reservoirs in the mantle: a depleted reservoir from which MORB originate, and a complementary enriched reservoir containing some original material that has never been depleted by melt

Table 1 Some radiogenic isotopes used as mantle tracers

Parent		Daughter	Stable isotope	Half-life
^{147}Sm	→	^{143}Nd	^{144}Nd	106×10^9 years
^{87}Rb	→	^{87}Sr	^{86}Sr	48.8×10^9 years
^{238}U	→	^{206}Pb	^{204}Pb	4.47×10^9 years
^{235}U	→	^{207}Pb	^{204}Pb	0.704×10^9years
^{232}Th	→	^{208}Pb	^{204}Pb	14×10^9 years
^{238}U	→	^{4}He	^{3}He	4.47×10^9 years
^{235}U	→	^{4}He	^{3}He	0.704×10^9 years
^{232}Th	→	^{4}He	^{3}He	14×10^9 years

extraction (DePaolo & Wasserburg 1976) or material that has been enriched by the addition of crustal material. Note that these possibilities are not mutually exclusive and that the Nd-Sr correlation plot cannot, by itself, be used to distinguish more than two reservoirs.

Jacobsen & Wasserburg (1979, 1980a, 1981), Wasserburg & DePaolo (1979), O'Nions et al (1979), Allègre et al (1980), DePaolo (1980), Allègre (1982), and Galer et al (1989) have all discussed the implications of crustal extraction for isotope systematics. By treating the MORB source region and the continental crust as complementary reservoirs, Jacobsen & Wasserburg (1979) calculated the fraction of the mantle that must have been depleted to form the crust. The depleted mantle reservoir was associated with the upper mantle. One of the objectives of reservoir studies is to determine the volume of the depleted mantle reservoir. This volume is directly related to the form of mantle convection. If the mantle convects as a whole, then the volume is that of the whole mantle. In the case of layered-mantle convection, plate tectonic processes would deplete the upper mantle relative to the lower mantle and the volume of the depleted reservoir would be that of the upper mantle.

Use of Mixing Studies to Interpret Isotope Systematics

The geochemical reservoir models discussed above are based on exchange of material between hypothetical well-mixed reservoirs, but these models do not indicate how this mixing process takes place. However, any model of mantle dynamics must be able to produce the geochemical observations outlined above and shown in Figure 2. The observation of primary importance is that the mantle source of MORB is very homogeneous. Yet heterogeneities exist in the mantle on a variety of length scales, and although the MORB source region exhibits characteristics that show a geochemical cycle of about 500 Myr based on the Pb isotopes (Galer & O'Nions 1985), some heterogeneities have persisted since the formation of the Earth. The time and length scales of mantle heterogeneity called for by the geochemical observations are summarized below.

PRESERVATION OF PRIMORDIAL MANTLE The discovery of ^{3}He in oceanic waters near mid-ocean ridges (Clarke et al 1969, Mamyrin et al 1969) revealed that the mantle has not been entirely outgassed during 4.5 Byr of convective overturn (Clarke et al 1969, Allègre et al 1983). Helium is extremely volatile, and it is readily lost from the Earth by efficient outgassing associated with melting and the formation of new oceanic crust. In the interior of the Earth, ^{4}He is produced by α-decay of U and Th, while ^{3}He is produced in very small quantities by neutron reactions with Li. Helium from the continents and hydrosphere probably does not enter the mantle in significant amounts by subduction, so virtually all ^{3}He in the Earth's interior has been there since the formation of the Earth. This is consistent with the observed presence of ^{129}Xe (the decay product of extinct ^{129}I) in terrestrial samples, which also indicates that the Earth is incompletely outgassed (Allègre et al 1983, Butler et al 1963, Staudacher & Allègre 1983). Reservoir studies show that the fraction of the mantle that has been outgassed exceeds the fraction that has been depleted by extraction of the continental crust (Allègre et al 1983); if the mantle is layered, this provides further evidence for limited exchange of material between the upper and lower mantles.

SCALE OF MANTLE HETEROGENEITIES

Small Scale of Heterogeneity: Evidence from Variations on Mid-Ocean Ridges Gurnis (1986a) looked at the variations of Sr isotopes in MORB and OIB and determined that the lateral heterogeneities can be distinguished on scales ranging from 1000 km to 1 km. The largest scale is the scale of the Dupal anomaly (Dupre & Allègre 1983, Hart 1984). The smallest scale is probably the minimum scale of mantle heterogeneity visible in MORB; magma mixing is likely to smear out any smaller heterogeneities. Kenyon (1990) observed that the spectrum of heterogeneities observed along ocean ridges drops off below a wavelength of about 1000 km. She calculated that mixing by porous flow dispersion and in magma chambers would require magma chambers about 1000 km long to erase mantle heterogeneities on this scale. She concluded that solid-state processes in the mantle (such as convection) must be responsible for the remainder of the mixing.

Small Scale of Heterogeneity: Evidence from High-Temperature Peridotites High-temperature peridotites such as those found in Beni-Bousera, Rhonda, and Lherz, provide a direct window into the mantle—albeit one that has been influenced by emplacement processes. These massifs are composed of a peridotitic matrix containing embedded layers of pyroxenite with a basaltic composition. The layers make up a few percent by volume of the massif (Boudier & Nicolas 1972, Kornprobst 1969) and their thicknesses range from a few centimeters to a few meters. The Sr, Nd, and Pb isotope ratios of the bands show as much variation on a small spatial scale as do ocean island basalts, worldwide (Polvé & Allègre 1980, Reisberg & Zindler 1986, Zindler et al 1983, Hamelin & Allègre 1988). As the bands do not give a single isochron age (e.g. the age of emplacement of the peridotite), these isotope signatures are thought to represent the original characteristics of the mantle rock. Allègre & Turcotte (1986) argue that the layers are composed of subducted oceanic crust which has been stretched and thinned by convection in the mantle. The persistence of layers with thicknesses of a few centimeters indicate that heterogeneities persist to this scale in the mantle. These heterogeneities will not be reflected in chemical variations in MORB along ridges, since magma mixing will erase heterogeneities on a scale of 1 km (Kenyon 1990).

Large Scale Heterogeneities: The Dupal Anomaly Sr and Pb isotopes in ocean island basalts in the southern hemisphere (especially those from the Indian Ocean) are markedly different from basalts in the northern hemisphere (Dupre & Allègre 1983, Hart 1984). These basalts have more radiogenic (higher) $^{207}Pb/^{204}Pb$ and $^{208}Pb/^{204}Pb$ ratios compared to $^{206}Pb/^{204}Pb$ ratios than samples from the North Atlantic and Pacific oceans. The Sr isotope characteristics of the Indian Ocean basalts are also more radiogenic than those of the North Atlantic and Pacific oceans. The isotopic signature of samples from the Southern Atlantic resembles those of the Indian Ocean, suggesting that this anomalous signature extends over much of the southern hemisphere. Indian Ocean MORB also exhibit this distinctive signature. This difference represents the largest scale of isotopic heterogeneity in the mantle; it is known as the Dupal anomaly. Dupré & Allègre (1983) suggest that this anomaly represents the mixing of material from an isolated lower mantle reservoir with the possible addition of continental crust to the MORB source region (the upper mantle). The Dupal anomaly has persisted for at least 115 million years (Weis et al 1989).

MODELS OF MIXING

The Physics of Mixing

When a blob of dye is introduced into a fluid, it is stretched, thinned, and folded by the shearing action of the flow. This stirring action is one aspect of mixing, and it is what concerns us here. At the same time, diffusion acts to blend the heterogeneity into the surrounding matrix, destroying the chemical or thermal distinctiveness which identifies the heterogeneity. In a highly reactive system, the time scale on which diffusion acts can be of the same order of magnitude or even larger than the time scale on which mechanical stirring acts. This is why, for example, when a hardener is stirred into an epoxy resin, it is a matter of practical concern that the stirring be complete before the epoxy hardens. In the mantle, the diffusion coefficient is about 10^{-15} cm^2/s (Hofmann & Hart 1978); over the course of 10^9 years, diffusion acts to destroy heterogeneities on the scale of centimeters or less (Kellogg & Turcotte 1986/1987) and is ineffective in homogenizing larger-scale heterogeneities. Diffusion is a relatively minor component of mixing in the mantle; mechanical stirring is of much greater interest. The "mixing time" for chemical heterogeneities in the mantle is therefore the time required for stirring to deform heterogeneities to the centimeter scale on which diffusion acts (Kellogg & Turcotte 1990). Another definition of the mixing time would be the time needed to disperse small-scale heterogeneities evenly throughout the flow. These two mixing times need not be the same; weak, large-scale heterogeneities may persist even if the mantle is well-mixed on a local scale (Richter et al 1982).

Mixing by Thermal Convection

A number of papers have considered convective mixing as it applies to the mantle. Richter & Ribe (1979) considered an idealized, two-dimensional model of Rayleigh-Bénard convection and calculated the thinning of heterogeneities with time. Using laboratory and numerical experiments, Richter and coworkers (1982) showed that time-dependent convection is more effective in thinning heterogeneities than steady convection. Olson and coworkers (1984a,b) pointed out the relative importance of pure and simple shear on mixing rates. Numerical calculations of convective stirring (Hoffman & McKenzie 1985) indicate that the thinning is accomplished primarily by pure shear—that is, exponential thinning. Exponential thinning of heterogeneities is a primary characteristic of turbulent, rather than laminar, mixing. In turbulent mixing, heterogeneities may experience continuous stretching accompanied by rapid deformation, or they may experience periods of rapid mixing interspersed with periods of quiescence in which mixing is slow. Thus a distribution of thicknesses of heterogeneities will develop after convective mixing begins. Gurnis (1986a,b), Gurnis & Davies (1986a,b), and Davies (1990a,b) argue that some heterogeneities persist in regions where convective mixing is ineffective.

The rapidity with which heterogeneities are stirred by the mantle flow is quite sensitive to the details of the flow. Christensen (1989) defined a series of mixing regimes into which flows may be categorized, based on the degree of mixing accomplished in the flow. These categories include (*a*) rapid mixing, in which particles are rapidly dispersed and are strained by pure shear; (*b*) slow mixing, in which little dispersion occurs; and (*c*) a hybrid regime in which some regions of the fluid undergo rapid mixing while isolated "islands" of unmixed material remain.

MIXING IN STEADY AND PERIODIC FLOWS These different degrees of stirring can easily be seen in very simple flows. Figure 3 shows the dispersion of particles in a simple flow model created by imposing two modes on a single convection cell and allowing one to vary periodically. The horizontal and vertical velocities are specified by:

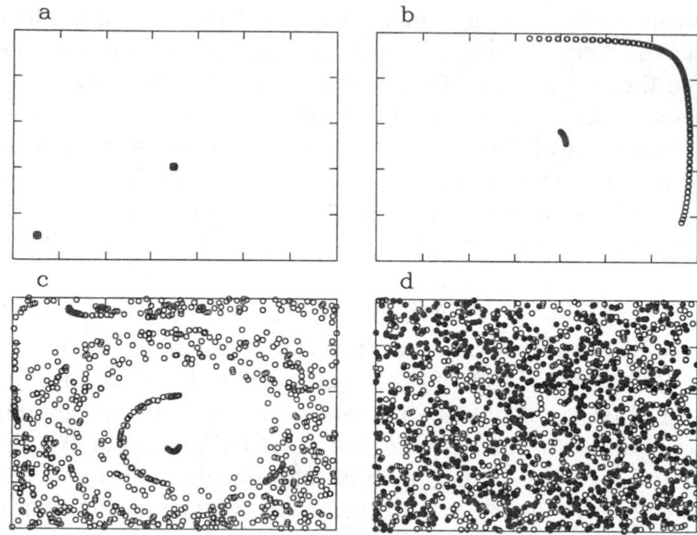

Figure 3 Mixing in a steady and time-periodic two-mode flow in which the secondary mode varies sinusoidally in time. (*a*) Initial placement of passive markers which illustrate the mixing. The two clusters consist of 900 markers each. (*b*) Mixing in a steady flow after about 10 overturns. There is only one mode of convection, consisting of one cell of steady flow in the box. This corresponds to the slow mixing situation described by Christensen (1989). (*c*) Mixing in an unsteady flow. One set of markers (*open circles*) exhibit rapid mixing, but an "island" of unmixed material remains in the center of the flow, and markers in this region are not widely dispersed (*solid circles*). This corresponds to the hybrid mixing regime of Christensen (1989). (*d*) Rapid mixing in a two-mode, periodically-varying flow. The secondary mode has been increased in magnitude. After 10 overturn times, the markers are thoroughly dispersed. This corresponds to the rapid or turbulent mixing regime identified by Christensen (1989).

$$u(x, y, t) = -\frac{2}{\pi}\cos(\pi y)\left[B\sin\left(\frac{\pi x}{\sqrt{2}}\right)+C\sin(\pi\omega t)\sin\left(\frac{2\pi x}{\sqrt{2}}\right)\right] \tag{1}$$

$$v(x, y, t) = \frac{\sqrt{2}}{\pi}\sin(\pi y)\left[B\cos\left(\frac{\pi x}{\sqrt{2}}\right)+2C\sin(\pi\omega t)\cos\left(\frac{2\pi x}{\sqrt{2}}\right)\right]. \tag{2}$$

The first term in the velocity corresponds to a single, steady convection cell in a box of aspect ratio $\sqrt{2}$. The second term corresponds to a two-cell flow in the same box with a sinusoidal variation in time. This simple model illustrates the kinematics of mixing in a two-dimensional flow, in the spirit of the calculations performed by Hoffman & McKenzie (1985), McKenzie (1979), Gurnis (1986a), Gurnis & Davis (1986a), Christensen (1989), and Ottino (1989). Because the flow varies periodically, the style of the flow and the degree of mixing are determined by the relative magnitude of the two cells, C/B, and by the frequency of the periodic time variation, ω. To illustrate the influence of the secondary convection cells on mixing, ω is held constant at a value of 1.0 and B fixed at a value of 6.0. Only C is allowed to vary.

Figure 3 shows the mixing accomplished in this flow for values of C ranging from 0 to 10. Two clusters of 900 particles were placed in the flow (Figure 3*a*). Each cluster is a box of dimensions 0.01 by 0.01 (smaller than the symbols used to plot the markers). When $C = 0$ the flow corresponds to steady state convection with a single cell flowing clockwise. Because the flow is steady, each particle follows a single closed path corresponding to a streamline, and little mixing is accomplished (Figure 3*b*). Similar results are seen when C is small compared to B.

When $C = 5$ the flow occasionally breaks into two cells, and the presence of the additional upwelling or downwelling thermal plume with its corresponding stagnation points enhances the mixing substantially (Figure 3*c*). However, a core of material remains relatively untouched by the secondary mode, since the second plume does not sweep all the way across the cell. Thus, one of the clusters is widely dispersed in the outer regions of the cell (*open circles*), but the second cluster (*closed circles*) remains unmixed in an "island" of nearly stagnant material.

Finally, when C is increased to 10, the flow breaks into two cells, and the plume associated with the second mode sweeps across the cell. This flow rapidly mixes both sets of markers (Figure 3*d*). However, as pointed out by Gurnis (1986b), some small blobs or clusters of particles may still be apparent in such flows. These calculations illustrate the complex range of mixing phenomena that can result from extremely simple flows. Mixing calculations such as these can be used to illustrate the physics of mixing and suggest possible scenarios for the mantle, but they cannot be used to distinguish what actually occurs in the mantle. Only with detailed knowledge of the mantle flow will it be possible to evaluate the mixing time in the mantle.

CHAOTIC MIXING Numerical models of mantle flows indicate that mantle convection is chaotic (Machetel & Yuen 1988, Stewart & Turcotte 1989). Extensive studies of stirring in chaotic flows have been carried out (Aref 1984, Aref & Balachandar 1986, Ottino et al 1988, Solomon & Gollub 1988a,b). Both laboratory and numerical experiments have shown chaotic flows to be extremely effective at mixing; however, these studies have concentrated on the role of periodic time-dependence in producing chaotic mixing rather than on the stirring accomplished by flows with a chaotic time-dependence.

Stewart & Turcotte (1989) determined the route to chaos in a two-dimensional, infinite Prandtl number fluid by taking a Fourier series expansion of the solutions to the governing equations for the flow in which only the coefficients of the series are time-dependent. This establishes a system of ordinary differential equations which can have fixed, periodic, quasi-periodic, or chaotic solutions. These correspond to different attractors in the phase space defined by the coefficients of the terms of the expansion. The flow can then be reconstructed by Fourier synthesis.

This method was used by Saltzman (1962) to study atmospheric dynamics. Lorenz (1963) truncated this series at three modes to obtain the Lorenz equations, which exhibit a well-known "strange attractor" for $\mathrm{Pr} = 20$ (Sparrow 1982). When the viscosity is very large and the thermal diffusivity very small (as, for instance, is the case for the Earth's mantle), it is appropriate to take the limit of infinite Prandtl number, $\mathrm{Pr} = \nu/\kappa \rightarrow \infty$. In the Lorenz equations, the only solution is steady convection in this limit. Stewart & Turcotte (1989) showed that a 12-mode expansion of the infinite Prandtl number Saltzman equations generates chaotic flow at Rayleigh numbers of the order of $\mathrm{Ra} = 5 \times 10^4$. A low number of modes is appropriate for the mantle, since mantle convection will be stabilized by the presence of the stiff plates on the surface.

The resulting flow is spatially simple but results in very rapid mixing. Figure 4 shows a cluster of passive markers being scattered by the flow in a series of "snapshots." The instantaneous streamlines show that the flow tends to be dominated by two cells in the horizontal and one or two in the vertical. Because the flow is time-dependent, these streamlines are not particle paths. The flow is "space-filling" in that any small cluster of particles disperses to fill the box.

Such pictures of the dispersion of heterogeneities are useful in illustrating how the flow mixes, but a quantitative measure of the mixing is necessary to use the results. This can be done by calculating the deformation undergone by passive, elliptical strain markers. Deformation of a marker is characterized by its elongation a/a_0, where a is the length of the semimajor axis and a_0 is the initial length. For infinitesimal markers, deformation is controlled by the local flow field and is obtained by integrating the strain rate tensor along the particle's trajectory (Kellogg & Turcotte 1990).

The rate at which a marker deforms depends on its path, which depends in turn on its initial position. A statistical description of mixing can be obtained by distributing a large number of markers evenly throughout the box, and then tracking their evolution. The distribution of resulting strains is characteristic of the flow. A characteristic of chaotic flows is that strain is accomplished primarily by pure shear (exponential elongation). This is illustrated in Figure 5, in which the median strain attained by 900 strain

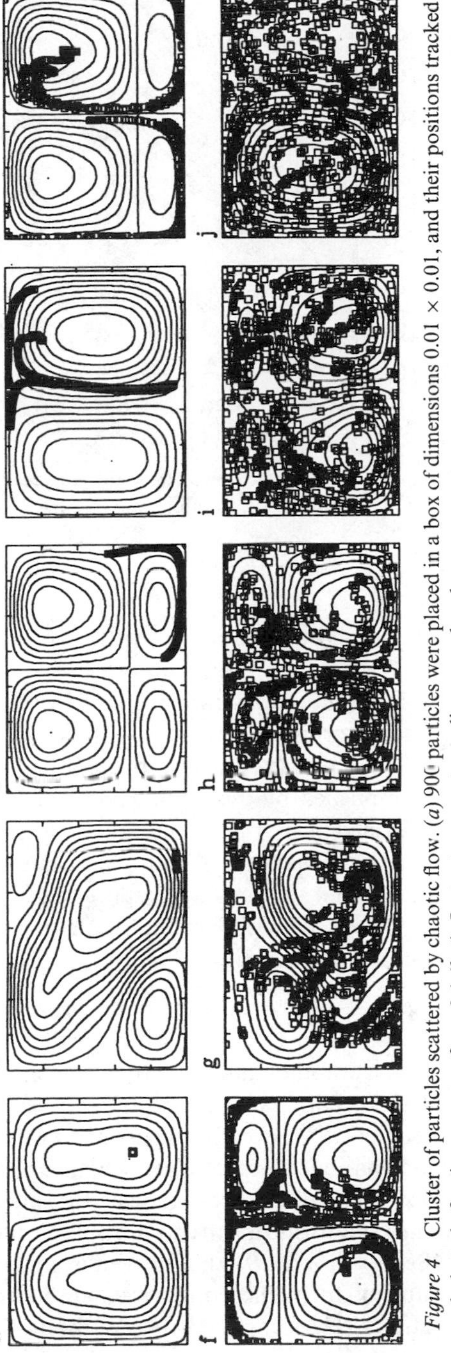

Figure 4 Cluster of particles scattered by chaotic flow. (*a*) 900 particles were placed in a box of dimensions 0.01 × 0.01, and their positions tracked and plotted after increments of $t = 0.1$ (*b–j*). Instantaneous streamlines are also shown.

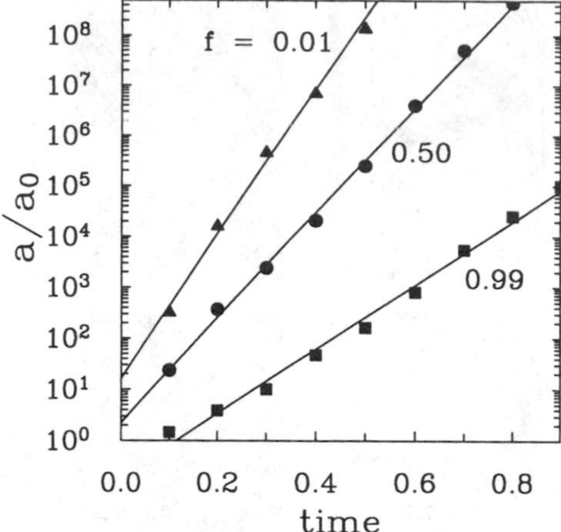

Figure 5 Strain evolution paths in chaotic flow. The strain a/a_0 reached by a fraction of the particles f is plotted as a function of the nondimensional time t for $f = 0.01$, $f = 0.50$, and $f = 0.99$.

markers ($f = 0.50$) is plotted as a function of time. The strain evolution has the form $a = a_0 e^{2\alpha t}$, where α is the strain rate.

Figure 5 also shows the minimum strain attained by 99% of the particles ($f = 0.99$) and the minimum strain attained by the most deformed 1% of the particles ($f = 0.01$). The latter curve provides an estimate of the maximum strain accomplished by the flow. The particles which are least deformed are also deformed primarily by pure shear, although the effective strain rate is low for these markers. From this calculation Kellogg & Stewart (1991) concluded that subducted oceanic crust would be destroyed by convective mixing in less than 1.2 Byr for whole-mantle convection or 500 Myr for layered-mantle convection.

EFFECT OF 3 DIMENSIONS ON MIXING To date, mantle mixing models have been largely restricted to two-dimensional calculations due to the limitations of available computer power. Three-dimensional mixing calculations are still at the preliminary stages. It is to be expected that the addition of another degree of freedom to the problem will lead to increased complexity of phenomena, but it is not clear whether mixing will tend to be more rapid or slower in fully three-dimensional models. In spherical shells convection takes the form of narrow upwelling plumes and sheet-like downwellings (Bercovici et al 1989), and even in a flat layer upwellings and downwellings do not form the continuous sheets imposed by two-dimensional calculations (Houseman 1990). Without the disruption of continuous sheets, particles may be less efficiently scattered in three dimensions (Davies 1990a). On the other hand, even a small amount of pure shear can be so efficient at stretching and dispersing heterogeneities that turbulent mixing may dominate in a relatively stable flow (Christensen 1990, Kellogg & Turcotte 1990).

Gable & O'Connell (1991) show that mixing may be chaotic in a steady three-dimensional flow, while in two dimensions time-dependence is required for efficient mixing. Since mantle convection is certainly time-dependent, it is likely to mix the mantle more efficiently than the steady flows examined by Gable and O'Connell. Much further work is required to resolve this question.

MIXING IN THE MANTLE

Viscosity Variations with Depth

Analysis of the Earth's geoid suggests that the lower mantle is more viscous than the upper mantle (Hager 1984). This observation is borne out by recent studies of post-glacial rebound (Lambeck 1990). It has been suggested that this viscosity jump is due to a phase change at 670 km. Numerical experiments (Gurnis & Hager 1988) show that subducted slabs will penetrate a viscosity boundary if there is no accompanying increase in density of the lower mantle. The slab will bend on encountering the boundary. Davies (1984) suggested that a viscosity increase with depth will profoundly influence the mixing rates in the mantle. Convection will be more sluggish and stirring less efficient than in a mantle with uniform viscosity (Gurnis & Davies 1986b). These authors suggest that primordial material may be preserved in blobs in the lower mantle, but other mixing calculations (Christensen 1990) indicate that turbulent mixing will control the mixing time.

Density Variations in the Mantle

The boundary between the upper and lower mantle is commonly taken to occur at the 670-km seismic discontinuity. Jeanloz & Knittle (1989) found that, in addition to a phase change, there is a change in composition at 670 km making the lower mantle at least 2.6% denser than the upper mantle. As discussed earlier, for a chemical heterogeneity to persist in the mantle requires that convection mixes inefficiently. The density contrast deduced by Jeanloz & Knittle constitutes stable stratification of the mantle. Numerical experiments show that about a 6% density contrast is required to maintain stable layered convection in the mantle.

A number of papers have investigated the structure of convection in a stability layered mantle (e.g. Richter & Johnson 1974, Cserepes &

Rabinowicz 1985, Ellsworth & Schubert 1988). These calculations were done at steady state and allowed at most a slight deflection of the material boundary; they therefore were not designed to address the mixing that takes place between the layers.

Olson (1984) used experiments with glucose solutions to examine mixing by thermal convection in a two-layered flow. Mixing between the layers occurred by entrainment of narrow tendrils of material in convective eddies. By this mechanism it is possible to maintain layered convection with some exchange of material between layers. Material in these experiments is not transported across the boundary in buoyant plumes. It is also possible for entrainment to increase the level of heterogeneity in the mantle by creating a layer of intermediate composition in the transition zone (Olson & Kincaid 1991). Entrainment is also observed in calculations of double-diffusive convection at infinite Prandtl number (Hansen & Yuen 1989, Kellogg 1991). Although these models do not predict large-scale penetration of plumes between the layers, plumes may penetrate the upper mantle as a precursor to large-scale overturn when the density contrast is low (e.g. Olson 1984, Kellogg 1991). This suggests that, if the mantle is layered, the model of Allègre & Turcotte (1985) is more realistic than the model of Jacobsen & Wasserburg (1979).

CONCLUDING REMARKS

Understanding the mixing rate in the mantle is critical to interpreting the geodynamic implications of geochemical observations. Any model of mantle convection must satisfy the constraints provided by isotopic observations. Mid-ocean ridges sample a uniform, depleted reservoir, while oceanic islands show evidence that parts of the mantle contain primordial and enriched material. Some mantle heterogeneities persist for the entire history of the Earth, and some extend to length scales of thousands of kilometers.

Rapid mixing of the entire mantle by convection would disperse these heterogeneities and is incompatible with the observations. Since the mid-ocean ridges randomly sample over the surface of the Earth, it is clear that some sort of compositional stratification is required. The proposed models take two forms. One possibility is that convection does not rapidly homogenize the mantle, so that heterogeneities are preserved for billions of years in the lower mantle. However, many models indicate that convective mixing is rapid, because small components of turbulent mixing will control the mixing time. In this case, it is necessary that the heterogeneities be convectively isolated from each other. This would be accomplished if the mantle is layered with a thermal boundary layer in the interior. Chemical reservoir models suggest that the boundary is located at the 670-km seismic discontinuity.

Models of mixing in the mantle are necessarily incomplete since the pattern of convection is not known. Thus neither model can be definitively ruled out. In particular, it is possible that mixing in three dimensions will differ significantly from mixing in two dimensions. Much further work is needed both to resolve the form of convection in the mantle and to further understand the mechanics of mixing.

ACKNOWLEDGMENTS

This work has benefited from discussions with Donald L. Turcotte, Cheryl A. Stewart, David Spence, Steven Wiggins, Scott King, Geoff Davies, and with participants of the Los Alamos National Lab Mantle Convection Workshop.

Literature Cited

Allègre, C. J. 1982. Chemical geodynamics. *Tectonophysics* 81: 109–32

Allègre, C. J., Brévart, O., Dupré, B., Minster, J.-F. 1980. Isotopic and chemical effects produced in a continuously differentiating convecting earth mantle. *Philos. Trans. R. Soc. London Ser. A* 297: 447–77

Allègre, C. J., Staudacher, T., Sarda, P., Kurz, M. 1983. Constraints on evolution of the earth's mantle from rare gas systematics. *Nature* 303: 762–66

Allègre, C. J., Turcotte, D. L. 1985. Geodynamic mixing in the mesosphere boundary layer and the origin of oceanic islands. *Geophys. Res. Lett.* 12: 207–10

Allègre, C. J., Turcotte, D. L. 1986. Implications of a two-component marble cake mantle. *Nature* 323: 123–27

Anderson, D. L. 1982a. Isotopic evolution of the mantle: the role of magma mixing. *Earth Planet. Sci. Lett.* 57: 1–12

Anderson, D. L. 1982b. Isotopic evolution of the mantle: a model. *Earth Planet. Sci. Lett.* 57: 13–24

Anderson, D. L. 1983. Chemical composition of the mantle. *J. Geophys. Res.* 88: 41–52

Aref, H. 1984. Stirring by chaotic advection. *J. Fluid Mech.* 143: 1–21

Aref, H., Balachandar, S. 1986. Chaotic advection in a Stokes flow. *Phys. Fluids* 29: 3515–21

Bercovici, D., Schubert, G., Glatzmaier, G. A. 1989. 3-dimensional spherical-models of convection in the earth's mantle. *Science* 244: 950–55

Boudier, F., Nicolas, A. 1972. Fusion partielle gabbroique dans la lherzolite de Lanzo (Alpes piémontaises). *Bull. Suisse Mineral. Petrol.* 52: 39–56 (In French)

Bougault, H., Joron, J. L., Treuil, M. 1980. The primordial chrondritic nature and large-scale heterogeneities in the mantle: evidence from high and low partition coefficient elements in oceanic basalts. *Philos. Trans. R. Soc. London Ser. A* 297: 203–13

Butler, W. A., Jeffery, P. M., Reynolds, J. H., Wasserburg, G. J. 1963. Isotopic variations in terrestrial xenon. *J. Geophys. Res.* 96: 3283–91

Chase, C. G. 1981. Oceanic island Pb: two-stage histories and mantle evolution. *Earth Planet. Sci. Lett.* 52: 277–84

Christensen, U. 1989. Mixing by time-dependent convection. *Earth Planet. Sci. Lett.* 95: 382–94

Christensen, U. 1990. Reply to comment by G. F. Davies on "Mixing by time-dependent convection." *Earth Planet. Sci. Lett.* 98: 408–10

Clarke, W. B., Beg, M. A., Craig, H. 1969. Excess ^{3}He in the sea: evidence for terrestrial primordial helium. *Earth Planet. Sci. Lett.* 6: 213–20

Cserepes, L., Rabinowicz, M. 1985/86. Gravity and convection in a two-layer mantle. *Earth Planet. Sci. Lett.* 76: 193–207

Davies, G. F. 1984. Geophysical and isotopic constraints on mantle convection: an interim synthesis. *J. Geophys. Res.* 89: 6017–40

Davies, G. F. 1990a. Comment on "Mixing by time-dependent convection" by U. Christensen. *Earth Planet. Sci. Lett.* 98: 405–7

Davies, G. F. 1990b. Mantle plumes, mantle stirring and hotspot chemistry. *Earth Planet. Sci. Lett.* 99: 94–109

DePaolo, D. J. 1980. Crustal growth and mantle evolution: inferences from models of element transport and Nd and Sr isotopes. *Geochim. Cosmochim. Acta* 44: 1185–96

DePaolo, D. J. 1981. Nd isotopic studies: some new perspectives on earth structure and evolution. *EOS, Trans. Am. Geophys. Union* 62: 137–40

DePaolo, D. J. 1988. Age dependence of the composition of continental crust: evidence from Nd isotopic variations in granitic rocks. *Earth Planet. Sci. Lett.* 90: 263–71

DePaolo, D. J., Wasserburg, G. J. 1976. Nd isotopic variations and petrogenetic models. *Geophys. Res. Lett.* 3: 249–52

Dickinson, W. R., Luth, W. C. 1971. A model for plate tectonic evolution of mantle layers. *Science* 174: 400–4

Dupré, B., Allègre, C. J. 1983. Pb-Sr isotope variation in Indian Ocean basalts and mixing phenomena. *Nature* 303: 142–46

Ellsworth, K., Schubert, G. 1988. Numerical models of thermally and mechanically coupled two-layer convection of highly viscous fluids. *Geophys. J.* 93: 347–63

Gable, C., O'Connell, R. 1991. Chaotic mantle mixing: time dependence is unnecessary. *EOS, Trans. Am. Geophys. Union* 72: 269 (Abstr.)

Galer, S. J. G., Goldstein, S. L., O'Nions, R. K. 1989. Limits on chemical and convective isolation in the earth's interior. *Chem. Geol.* 75: 257–90

Galer, S. J. G., O'Nions, R. K. 1985. Residence time of thorium, uranium and lead in the mantle with implications for mantle convection. *Nature* 316: 778–82

Gast, P. W. 1968. Trace element fractionation and the origin of tholeiitic and alkaline magma types. *Geochim. Cosmochim. Acta* 32: 1057–86

Gurnis, M. 1986a. Quantitative bounds on the size spectrum of isotopic heterogeneity within the mantle. *Nature* 323: 317–20

Gurnis, M. 1986b. Stirring and mixing in the mantle by plate-tectonic flow: large persistent blobs and long tendrils coexist. *Geophys. Res. Lett.* 13: 1464–77

Gurnis, M., Davies, G. F. 1986a. Mixing in numerical models of mantle convection incorporating plate kinematics. *J. Geophys. Res.* 91: 6375–95

Gurnis, M., Davies, G. F. 1986b. The effect of depth-dependent viscosity on convective mixing in the mantle and the possible survival of primitive mantle. *Geophys. Res. Lett.* 13: 541–44

Gurnis, M., Hager, B. H. 1988. Controls of the structure of subducted slabs. *Nature* 335: 317–21

Hager, B. H. 1984. Subducted slabs and the geoid: constraints on mantle rheology and flow. *J. Geophys. Res.* 89: 6003–15

Hamelin, B., Allègre, C. J. 1988. Lead isotope study of orogenic lherzolite massifs. *Earth Planet. Sci. Lett.* 91: 117–31

Hansen, U., Yuen, D. A. 1989. Subcritical double-diffusive convection at infinite Prandtl number. *Geophys. Astrophys. Fluid Dyn.* 47: 199–224

Hart, S. R. 1984. A large-scale isotope anomaly in the Southern Hemisphere mantle. *Nature* 309: 753–57

Hart, S. R. 1988. Heterogeneous mantle domains—Signatures, genesis and mixing chronologies. *Earth Planet. Sci. Lett.* 90: 273–96

Hoffman, N. R. A., McKenzie, D. P. 1985. The destruction of geochemical heterogeneities by differential fluid motions during mantle convection. *Geophys. J. R. Astron. Soc.* 82: 163–206

Hofmann, A. W., Hart, S. R. 1978. An assessment of local and regional isotopic equilibrium in the mantle. *Earth Planet. Sci. Lett.* 38: 44–62

Hofmann, A. W., White, W. M. 1982. Mantle plumes from ancient oceanic crust. *Earth Planet. Sci. Lett.* 57: 421–36

Houseman, G. A. 1990. The thermal structure of mantle plumes—axisymmetric or triple-junction? *Geophys. J. Int.* 102: 15–24

Jacobsen, S. B., Wasserburg, G. J. 1979. The mean age of mantle and crustal reservoirs. *J. Geophys. Res.* 84: 7411–27

Jacobsen, S. B., Wasserburg, G. J. 1980. A two-reservoir recycling model for model-crust evolution. *Proc. Natl. Acad. Sci. USA* 77: 6298–6302

Jacobsen, S. B., Wasserburg, G. J. 1981. Transport models for crust and mantle evolution. *Tectonophysics* 75: 163–79

Jeanloz, R., Knittle, E. 1989. Density and composition of the lower mantle. *Philos. Trans. R. Soc. London Ser. A* 328: 377–89

Kay, R. W. 1980. Volcanic arc magmas: implications of a melting-mixing model for element recycling in the crust–upper mantle system. *J. Geol.* 88: 497–522

Kellogg, L. H. 1991. Interaction of plumes with a compositional boundary at 670 km. *Geophys. Res. Lett.* 18: 865–68

Kellogg, L. H., Stewart, C. A. 1991. Mixing by chaotic convection in an infinite Prandtl number fluid and implications for mantle convection. *Phys. Fluids A* 3: 1374–78

Kellogg, L. H., Turcotte, D. L. 1986/87. Homogenization of the mantle by convective mixing and diffusion. *Earth Planet. Sci. Lett.* 81: 371–78

Kellogg, L. H., Turcotte, D. L. 1990. Mixing and the distribution of heterogeneities in a chaotically convecting mantle. *J. Geophys. Res.* 95: 421–32

Kenyon, P. M. 1990. Trace-element and isotopic effects arising from magma migration beneath midocean ridges. *Earth Planet. Sci. Lett.* 101: 367–78

Kornprobst, J. 1969. Le massif ultrabasique des Beni-Bousera: etude des péridotites de haute température et haute pression et des pyroxénolites avec ou sans grenat qui leur sont associées. *Contrib. Mineral. Petrol.* 23: 283–322 (In French)

Lambeck, K. 1990. Glacial rebound, sea-level change and mantle viscosity. *Q. J. R. Astron. Soc.* 31: 1–30

Lorenz, E. N. 1963. Deterministic nonperiodic flow. *J. Atmos. Sci.* 20: 130–41

Loper, D. E. 1985. A simple model of whole-mantle convection. *J. Geophys. Res.* 90: 1809–36

Machetel, P., Weber, P. 1991. Intermittent layered convection in a model mantle with an endothermic phase-change at 670 km. *Nature* 350: 55–57

Machetel, P., Yuen, D. A. 1988. Infinite Prandtl number spherical-shell convection. In *16th Conf. Math. Geophys.*, ed. M. J. R. Wortel, N. J. Vlaar, G. Nolet, S. A. P. L. Cloetingh, pp. 265–90. Holland: Reidel

Mamyrin, B. A., Tolstikhin, I. N., Anufriev, G. S., Kamanskiy, I. L. 1969. Anomalous isotopic composition of helium in volcanic gases. *Dokl. Akad. Nauk SSSR* 184: 1197–99

McKenzie, D. 1979. Finite deformation during fluid flow. *Geophys. J. R. Astron. Soc.* 58: 689–715

Olson, P. 1984. An experimental approach to thermal convection in a two-layered system. *J. Geophys. Res.* 89: 11,293–301

Olson, P., Kincaid, C. 1992. Layered mantle convection experiments and the structure of the transition zone. *Earth Planet. Sci. Lett.* In press

Olson, P., Yuen, D. A., Balsiger, D. 1984a. Convective mixing and the fine structure of mantle heterogeneity. *Phys. Earth Planet. Inter.* 36: 291–304

Olson, P., Yuen, D. A., Balsiger, D. 1984b. Mixing of passive heterogeneities by mantle convection. *J. Geophys. Res.* 89: 425–36

O'Nions, R. K., Evensen, N. M., Hamilton, P. J. 1979. Geochemical modeling of mantle differentiation and crustal growth. *J. Geophys. Res.* 84: 6091–6101

Ottino, J. M. 1989. *The Kinematics of Mixing: Stretching, Chaos, and Transport.* Cambridge: Cambridge Univ. Press. 364 pp.

Ottino, J. M., Leong, C. W., Rising, H., Swanson, P. D. 1988. Morphological structures produced by mixing in chaotic flows. *Nature* 333: 419–25

Oxburgh, E. R., Parmentier, E. M. 1977. Compositional and density stratification in oceanic lithosphere—causes and consequences. *J. Geol. Soc. London* 133: 343–55

Polvé, M., Allègre, C. J. 1980. Orogenic lherzolite complexes studied by ^{87}Rb-^{87}Sr: a clue to understand the mantle convection processes. *Earth Planet. Sci. Lett.* 51: 71–93

Reisberg, L., Zindler, A. 1986. Extreme isotopic variation in the upper mantle: evidence from Ronda. *Earth Planet. Sci. Lett.* 81: 29–45

Richter, F. M., Daly, S. F., Nataf, H. C. 1982. A parameterized model for the evolution of isotopic heterogeneities in a convecting system. *Earth Planet. Sci. Lett.* 60: 178–94

Richter, F. M., Johnson, C. E. 1974. Stability of a chemically layered mantle. *J. Geophys. Res.* 79: 1635–39

Richter, F. M., Ribe, N. M. 1979. On the importance of advection in determining the local isotopic composition of the mantle. *Earth Planet. Sci. Lett.* 43: 212–22

Saltzman, B. 1962. Finite amplitude free convection as an initial value problem. *J. Atmos Sci.* 19: 329–41

Schilling, J.-G. 1973. Icelandic mantle plume: geochemical evidence along the Reykjanes Ridge. *Nature* 242: 565–71

Silver, P. G., Carlson, R. W., Olson, P. 1988. Deep slabs, geochemical heterogeneity, and the large-scale structure of mantle convection: investigation of an enduring paradox. *Annu. Rev. Earth Planet. Sci.* 16: 477–541

Solomon, T. H., Gollub, J. P. 1988a. Chaotic particle-transport in time-dependent Rayleigh-Benard convection. *Phys. Rev. A* 38: 6280–86

Solomon, T. H., Gollub, J. P. 1988b. Passive transport in steady Rayleigh-Benard convection. *Phys. Fluids* 31: 1372–79

Sparrow, C. 1982. *The Lorenz Equations: Bifurcations, Chaos, and Strange Attractors.* New York: Springer-Verlag

Staudacher, T., Allègre, C. J. 1983. Terrestrial xenology. *Earth Planet. Sci. Lett.* 60: 389–406

Stewart, C. A., Turcotte, D. L. 1989. The route to chaos in thermal-convection at infinite Prandtl number: 1. Some trajectories and bifurcations. *J. Geophys. Res.* 94: 13,707–17

Sun, S. S. 1980. Lead isotopic study of young volcanic rocks from mid ocean ridges, ocean islands, and island arcs. *Phil. Trans. R. Soc. London Ser. A* 297: 409–45

Sun, S. S., Hanson, G. N. 1975. Evolution of the mantle: geochemical evidence from alkali basalt. *Geology* 3: 297–302

Tatsumi, Y., Sakuyama, M., Fukuyama, H., Kushiro, I. 1983. Generation of arc basalt magmas and thermal structure of the mantle wedge in subduction zones. *J. Geophys. Res.* 88: 5815–25

Wasserburg, G. J., DePaolo, D. J. 1979. Models of earth structure inferred from neodymium and strontium isotopic abundances. *Proc. Natl. Acad. Sci. USA* 76: 3594–98

Weis, D., Bassias, Y., Gautier, I., Mennessier, J.-P. 1989. Dupal anomaly in existence 115 ma ago: evidence from isotopic study of the Kerguelen plateau (South Indian Ocean). *Geochim. Cosmochim. Acta* 53: 2125–31

White, W. M. 1985. Sources of oceanic basalts: radiogenic isotopic evidence. *Geology* 13: 115–18

Zindler, A., Hart, S. 1986. Chemical geodynamics. *Annu. Rev. Earth Planet. Sci.* 14: 493–570

Zindler, A., Jagoutz, E., Goldstein, S. 1982. Nd, Sr and Pb isotopic systematics in a three-component mantle: a new perspective. *Nature* 298: 519–23

Zindler, A., Staudigel, H., Hart, S. R., Endres, R., Goldstein, S. 1983. Nd and Sr isotopic study of a mafic layer from Ronda ultramafic complex. *Nature* 304: 226–30

Annu. Rev. Earth Planet. Sci. 1992. 20:389–430

ORIGIN OF NOBLE GASES IN THE TERRESTRIAL PLANETS

Robert O. Pepin

School of Physics and Astronomy, University of Minnesota, Minneapolis, Minnesota 55455

KEY WORDS: planetary atmospheres, sun, solar wind, meteorites, adsorption, hydrodynamic escape

INTRODUCTION

The sources of planetary and meteoritic volatiles, and the mechanisms driving their evolution from primordial to present-day compositions, are not well known. The field bristles with models of one type or another, but none are without problems. Efforts to understand the histories of terrestrial planet atmospheres have concentrated on their noble gases as evolutionary tracers, since these are free of the entanglements of chemical interaction. The elemental and isotopic compositions of these minor atmospheric constituents are rich in clues to the chemical characteristics of their source reservoirs, the physics of their evolution, and the nature of the astrophysical and planetary environments in which it occurred. Most workers agree that atmospheric mass distributions of the nonradiogenic noble gases were probably established very early, through the action of processes operating before, during, or shortly after planetary accretion. But beyond this there is no consensus as yet on the specifics of sources or mechanisms.

The question of past and present volatile inventories on and in the terrestrial planets is intrinsically interesting in the more general context of the evolution of the planets themselves. For each individual body it cross-cuts issues relating to atmospheric origin and compositional history, emergence of a coupled atmosphere-surface system and climate evolution, surface morphology and its record of past geological processing, and planetary rheology, differentiation, and geophysical history. But the volatile problem extends well beyond its relationship to the initial state of a

0084–6597/92/0515–0389$02.00

particular planet and the mechanisms that drove it down its specific evolutionary track. Attempts to decipher planetary histories in the broader context of the evolution of the solar system as a whole are focusing more and more attention on the sources and processing of volatiles in the primordial solar accretion disk, in primitive meteorites, and in the terrestrial planets as a class.

In current models for the history of terrestrial planet atmospheres, primary volatile sources may include the solar nebula, solar wind, comets, one or more known meteorite classes, or other bodies carrying different volatile distributions—with the assumption that these represent materials that either are no longer extant in the solar system or exist but are unsampled. A wide range of possible mechanisms for gain or loss of planetary gases has been proposed. Some of these appear likely to preserve the compositional signatures of the volatile reservoirs on which they act. Such presumably nonfractionating processes include impact-degassing of meteorites or cometary matter accreted by planets, direct gravitational capture of ambient nebular gases, planetary outgassing, and ejection of gases by impact of accreting bodies into preexisting atmospheres. Other mechanisms fractionate either elements alone, or both elements and isotopes. Examples are adsorption of nebular gases or implantation and diffusion of solar wind gases on or in dust grain surfaces prior to accretion, solution of ambient volatiles into the melted surfaces of planets during accretion, partitioning in planetary interiors, and loss of atmospheric constituents from planetesimals or planetary bodies by Jeans or hydrodynamic escape or by nonthermal processes.

We have no model-independent ways to assess the relative importance of these and perhaps other primordial sources and evolutionary processes in establishing the various mass distributions found in present-day planetary atmospheres. Assumptions in specific atmospheric models about the origins and compositions of primary reservoirs, their existence at particular times, the mechanisms responsible for transporting and fractionating their volatiles, and the environments in which such processes operated are judged primarily by their perceived plausibility in the context of other models, conjectures, or ideas concerning the origin and evolution of the solar system. The one hard constraint on these models, which so far they all fail to satisfy to varying degrees, is that they be able to account in detail for the elemental and isotopic compositions of present-day atmospheres.

The key diagnostic volatiles for tracing atmospheric origin and nonbiogenic evolution are the noble gases, nitrogen as N_2, and carbon as CO_2. Signatures of origin and physical processing would be expected to survive most clearly in the chemically inert noble gases. Their record is complex, however, and not readily interpreted. Absolute abundances of noble gases in the atmospheres of Earth, Venus, and Mars and in the carbonaceous chondrites display highly variable depletions with respect to solar abundances. Isotopic patterns in each of these volatile reservoirs are generally distinct from each other and from inferred compositions in the sun and primitive nebula.

PRIMARY SOLAR-SYSTEM VOLATILE SOURCES AND PROCESSES

Gravitational Capture of Nebular Gases

One possible noble gas source for Earth and Venus is direct capture from the nebula. Substantial gravitational condensation of ambient gases requires the growth of protoplanets to appreciable fractions of the present masses of these two planets prior to dissipation of the nebular gas phase. A difficulty that presently confronts this requirement has to do with the relative timing of nebular dissipation versus planetary growth. Current estimated time scales for loss of circumstellar dust and gas, from observation (Strom et al 1988, 1989a,b, 1992; Walter et al 1988; Walter & Barry 1991) and theory (Boss et al 1989), are on the order of $\sim 10^7$ yr or less, compared to $\sim 10^8$ yr for planetary growth to full mass in the standard model of planetary accumulation (Wetherill 1986, 1990a).

Nebular lifetimes inferred from astronomical observation, however, are based solely on evolution of their fine dust component. Measurements of molecular line emissions which could at least set upper limits on the longevity of the gas-phase component in "naked" T-Tauri disks do not yet exist (Strom et al 1992). There is no reason to believe that the disappearance of micrometer-size dust from infrared detectability, say by accretion into larger grains, would necessarily be coincident with gas loss. The "standard model" of planetary accumulation (Wetherill 1986) predicts terrestrial planet growth to roughly 80% of their final masses within $\sim 2 \times 10^7$ yr. So if the observed dust clearing signals the beginning of planet-building, and a significant remnant of gas survived in the inner solar accretion disk for another 2×10^7 yr or so, substantial gravitational capture would have occurred on Earth and Venus. And there are further uncertainties, pertaining to the influence of ambient gas on the dynamics and time scales of the accretion process itself. On the one hand, although they have not been well studied, it is not out of the question that gas drag effects during accumulation could have resulted in full planetary growth within $\approx 10^7$ yr (Wetherill 1990a)—much more commensurate with estimates of nebular lifetimes and thus with condensation of massive primary atmospheres on the largest terrestrial planets. Alternatively, it is possible that *loss* of nebular gas was an important trigger for accumulation of relatively smart ($\sim$Mercury-size) planetary embryos into larger bodies (Wetherill 1990b), foreclosing the possibility of later large-scale capture.

It would appear that current observations and modeling cannot rule firmly for or against gravitational capture of massive solar-composition primary atmospheres on Earth and Venus. If such capture did occur, subsequent evolution from primary to present-day noble gas distributions must have involved processes that fractionated both elements and isotopes. The noble gas elemental abundances in the contemporary atmospheres on both planets shown in Figures 1 and 2 have distinctly nonsolar Ne/Kr ratios, and Earth is further depleted in Ar relative to Kr. There is presently no reason to assume elemental fractionation in the capture process itself, although it may be that this possibility needs more theoretical study. Evidence for isotopic fractionation is discussed in a later section.

Independent of whether or not direct gas capture provided Earth and Venus with dense external primary atmospheres, the process may have played an important role in creating internal volatile reservoirs for later outgassing of secondary atmospheres on the terrestrial planets. Protoplanetary cores, growing through sizes of a few lunar masses in the presence of nebular gas at temperatures and pressures estimated from astronomical observations and from accretion disk models, could have accreted atmospheres by condensation of surrounding gases in their gravitational wells—tenuous atmospheres, to be sure, but still amplified in surface pressure by $\sim$4–6 orders of magnitude above ambient nebular pressures (Pepin 1991). Adsorption and occlusion of these surface gases within growing cores would appear to be a natural consequence of Wetherill's (1990b) new early-stage model of planetary accumulation, in the presence of nebular gas, to $\sim$Mercury-size embryos in $\sim 10^6$ yr. Wetherill speculates that dissipation of nebular gas at about this time may have initiated the next stage of accretion, to essentially final planetary masses by "giant impact" mergers of these embryos. If correct, this would mean that core adsorption would be naturally truncated at $\sim 10^6$ yr or shortly thereafter, but at levels that appear large enough to supply outgassed species on Earth and Mars, and with distribution coefficients for noble gas adsorption on accreting core materials that are not very different from values measured in the laboratory (Pepin 1991).

Volatile-Rich Planetesimals

Another possible source of inner planet noble gases, as well as water, carbon, and nitrogen, is accretion of volatile-rich icy comets scattered inward from the outer solar system. Although noble gas distributions in comets are unknown, solar isotopic compositions would be expected in condensates from the nebula. There is evidence that heavier species (Xe, Kr, and Ar) could have been trapped in ice in approximately solar elemental proportions at very low temperatures, with appreciable abundances per gram of water (Bar-Nun et al 1985, Owen et al 1991). Thus incorporation of a few percent of icy cometary matter into an accreting terrestrial planet could potentially have supplied heavy noble gases of solar composition to its primary atmosphere, in addition to abundant hydrogen which could later have fueled hydrodynamic escape of the atmosphere. This kind of source is particularly attractive in that the low Ne/Ar ratio expected in these ices, for gas occlusion by trapping in amorphous ice (Owen et al 1991) or by clathration (Lunine & Stevenson 1985), would provide a natural explanation for the underabundance of neon in the otherwise solar-like Venus pattern shown in Figures 1 and 2. A cometary carrier of primordial Venus volatiles with just these characteristics has been proposed (Owen 1987, Hunten et al 1988, Owen et al 1991).

A model of partial hydrodynamic escape of Venus's primary atmosphere (Pepin 1991) is described in a later section. An icy planetesimal source supplying both the hydrogen needed for this escape episode and the noble gas abundances initially on Venus would require accretion of ~2–4 weight % of water into the planet with the approximate composition $Xe/H_2O \sim 2 \times 10^{-10}$ g/g, Xe/Kr ~ solar, Ar/Kr ~ 75% of solar, and Ne/Kr ~ 0.75% of solar. Compared to known volatile carriers, the inferred Xe loading is about the same as bulk carbonaceous chondrite (CI) concentrations, or about 15 times smaller than the Xe/H_2O ratio in the CIs. These numbers do not seem intrinsically implausible, but something clearly needs to be known about elemental and isotopic abundances of all these species in comets before the possibility of this kind of source can be assessed in a meaningful way.

Given the nebular temperatures and pressures likely to be found at its radial distance, Mars is too small to have condensed appreciable abundances of ambient gases even in the limiting case of isothermal capture (Hunten 1979, Pepin 1991). Therefore, regardless of the plausibility of gravitational capture as a noble gas source for primary atmospheres on Venus and Earth, some other way is needed to supply Mars. An early inward flux of icy planetesimals would certainly have contributed to all three of the terrestrial planets. There are also some rather convincing arguments that Mars incorporated, in addition, a substantial component of volatile-rich bodies resembling the present-day carbonaceous chondrites. Assuming that the SNC meteorites are in fact samples of Mars (see below), Swindle et al (1986) found that the isotopic composition of Martian atmospheric Xe is just that of mass-fractionated CI-Xe, and Dreibus & Wänke's (1985, 1987, 1989) SNC-based geochemical model of Mars' bulk composition calls for a ~40% mass fraction of volatile-rich, oxidized CI-like material in the planet.

A number of arguments can be made in favor of cometary carriers for inner planet volatiles (Pepin 1991). As noted above, depletion of the Ne/Ar ratio relative to the solar ratio is likely in such ices. This source could have contributed to both Venus and Earth if an initially Venus-like primary atmosphere on Earth was later elementally fractionated in partial mass-dependent ejection by a giant Moon-forming impact, or in some other way. A mixture of noble gases from planetesimal ices and CI meteorites can also be accommodated in models of Martian isotopic evolution.

However this kind of water-noble gas source has potential problems of its own. Cometary ices could have carried other volatiles into primary atmospheres, possibly in large amounts. For instance, material released from Comet Halley [assuming a dust/gas mass ratio of ~ 6 near the upper end of McDonnell et al's (1986) range and taking all H as H_2O and all H_2O in the gas phase] contains ~ 0.2 gC/gH_2O and between ~ 0.02 and less than 0.08 gN/gH_2O (Jessberger et al 1989). Accretion of a carrier of this composition to supply ~ 2–4 weight % of water to Venus would oversupply the present-day atmospheric CO_2 and N_2 inventories by between one and two orders of magnitude. So one would have to postulate an icy source with very much less fractionated C/H and N/H ratios, relative to solar, than Halley, or that all but a few percent of the cometary C and N was retained in the planet during accretion and throughout its subsequent geologic history. Moreover there is a carbon isotope constraint which appears to rule out a strictly Halley-like source. Wyckoff et al (1989) report $^{12}C/^{13}C = 65 \pm 9$ ($\delta^{13}C = 370 \pm 190‰$) in Halley from ground-based spectral measurements, well below the Venus value of 89.3 ± 1.6 (von Zahn et al 1983). However other comets could perhaps satisfy this particular constraint: The average $^{12}C/^{13}C$ in four of them is 103 ± 30 (Wyckoff et al 1989), within range of Venus.

Early Post-Nebular Solar Wind

The compositional characteristics set out above for an icy planetesimal source require occlusion of nebular noble gases with approximately unfractionated elemental ratios for Xe:Kr:Ar but much lower Ne. Thermodynamic modeling suggests that clathrated noble gases, while characterized by low Ne/Ar, do *not* reflect ambient gas-phase composition for Xe/Ar and Kr/Ar but instead are strongly enriched in the heavier species (Lunine & Stevenson 1985). Clathrated gases therefore appear unlikely to serve, and one must appeal to physical adsorption of noble gases on ice which, as yet, has not been extensively studied in the laboratory. First experimental studies in this regime (Owen et al 1991) do indeed indicate nonfractionating Xe, Kr, and Ar adsorption and Ne depletion, but only in a narrow range of very low (<30 K) temperatures.

Noble gases implanted into lunar and meteoritic dust grains by low-energy solar wind irradiation are known to display solar-like elemental abundance ratios, with typical ^{20}Ne loadings in lunar grains of $\approx 10^{-6}$ g/g (e.g. see Pepin et al 1970, Eberhardt et al 1972, Marti et al 1972, Hintenberger et al 1974). Implanted neon (and helium) can be depleted preferentially by subsequent heating, or lost even at comparatively low temperatures by rapid diffusion from certain mineral structures, notably plagioclase (e.g. Frick et al 1988). Accretion of planetesimals containing $\approx$25–40 weight % of solar-wind-irradiated material, loaded to lunar levels in Xe, Kr, and Ar but depleted $\approx$100-fold in ^{20}Ne to a content of $\approx 10^{-8}$ g/g, could account for the absolute noble gas abundances in Venus's atmosphere within the error bars that currently pertain to it.

This general kind of solar-wind source for noble gases on the terrestrial planets has been proposed in various contexts by Wetherill (1981), Donahue et al (1981), and McElroy & Prather (1981). Wetherill's (1981) model specifically addresses the large ^{36}Ar abundance on Venus. He perceived its major difficulty to be that of confining accretion of solar-wind-rich materials largely to just this planet, given the likelihood that gravitational scattering would tend to disperse it throughout the inner solar system. But it is important to note that this may no longer be a fatal objection. Earth could have acquired a compositionally Venus-like primary atmosphere as well, later fractionated by partial ejection, and as noted above this kind of solar-like noble gas component can also be accommodated on Mars (Pepin 1991).

It appears that the possibility of a solar-wind source of this kind for noble gases in the terrestrial planets must be taken seriously. It may turn out on compositional grounds to be more suitable than either gravitational capture or icy planetesimal accretion. Moreover Sasaki's (1991) recent arguments for off-disk penetration of an early and intense solar wind flux into a post-nebular environment rich in fine collisional dust may imply the existence of a hitherto unconsidered ancient reservoir of abundant and heavily loaded carriers of solar-like noble gases.

Adsorption of Nebular Gases

Laboratory studies have shown that noble gases exposed to particular kinds of finely-divided solid materials are adsorbed on or within the surfaces of individual grains. Adsorption is most efficient for various forms of carbon, where it appears to reflect intrinsic structural properties (Frick et al 1979, Niemeyer & Marti 1981, Wacker et al 1985, Zadnik et al 1985, Wacker 1989), but has also been experimentally demonstrated for other minerals (Yang et al 1982, Yang & Anders 1982a,b) and for polymineralic meteorite powder (Fanale & Cannon 1972). Moreover the process occurs naturally in sedimentary materials (Canalas et al 1968, Fanale & Cannon 1971, Phinney 1972, Podosek et al 1980, 1981, Bernatowicz et al 1984). Adsorbed gases on these substrates generally display fractionation patterns, relative to ambient gas-phase abundances, in which heavier species are enriched. For the most part the fractionations are remarkably uniform, considering the wide range of experimental and natural conditions under which they are produced, and are primarily elemental. Although occasional isotopic effects have been reported in natural samples (Phinney 1972), they are not observed in equilibrium adsorption in the laboratory (Bernatowicz & Podosek 1986). Isotopic fractionation does however occur in experimental environments involving ion implantation from plasmas rather than simple adsorption of neutral gases (Frick et al 1979, Dziczkaniec et al 1981, Bernatowicz & Fahey 1986, Bernatowicz & Hagee 1987).

The decline in solar-normalized noble gas abundance ratios from Kr to Ne in planetary atmospheres and meteorites is qualitatively similar to many of the adsorption fractionation patterns seen in the laboratory and in terrestrial sedimentary rocks, as shown in Figure 2. This has led to a widespread belief that adsorption has played a central role in establishing these ratios. Atmospheric Xe, however, is anomalously low in this regard, and the failure of atmospheric Xe/Kr on Earth (and Mars) to follow the steeply-inclined trend defined by Ne/Kr and Ar/Kr triggered a long and unsuccessful search of the terrestrial sedimentary column for the presumably adsorbed "missing" Xe. Moreover adsorption of isotopically solar noble gases cannot by itself generate the isotopic patterns seen in planets, and laboratory estimates of single-stage gas/solid partition coefficients are too low to account for planetary noble gas abundances by adsorption on free-floating dust grains in the open nebula. Nevertheless adsorption from a gas phase could well have been a fundamental mechanism for supplying volatiles to planets, if it occurred from a reservoir well above nebular pressures (for example, from gravitationally captured atmospheres surrounding planetary cores) and, as discussed later, was followed by processes that fractionated isotopes and decoupled the evolutionary history of Xe from that of the lighter noble gases.

THE CURRENT DATA BASE AND ITS IMPLICATIONS FOR EVOLUTIONARY PROCESSING

Present-Day Volatile Inventories

ELEMENTAL ABUNDANCES Noble gases and other species of high volatility were largely lost from the early planetary system when solid matter separated from the gas phase of the solar accretion disk and the disk's gaseous component later dissipated. Planetary atmospheres, meteorites, comets, and the sun are the principal surviving volatile reservoirs. "Solar" or "solar-system" abundances of the elements based on data from the sun's photosphere, the solar wind, meteorites (for nonvolatile species), and galactic sources (Cameron 1982, Anders & Ebihara 1982, Anders & Grevesse 1989) represent our best estimates for the composition of the primordial solar nebula. As shown in Figure 1, noble gases in the atmospheres

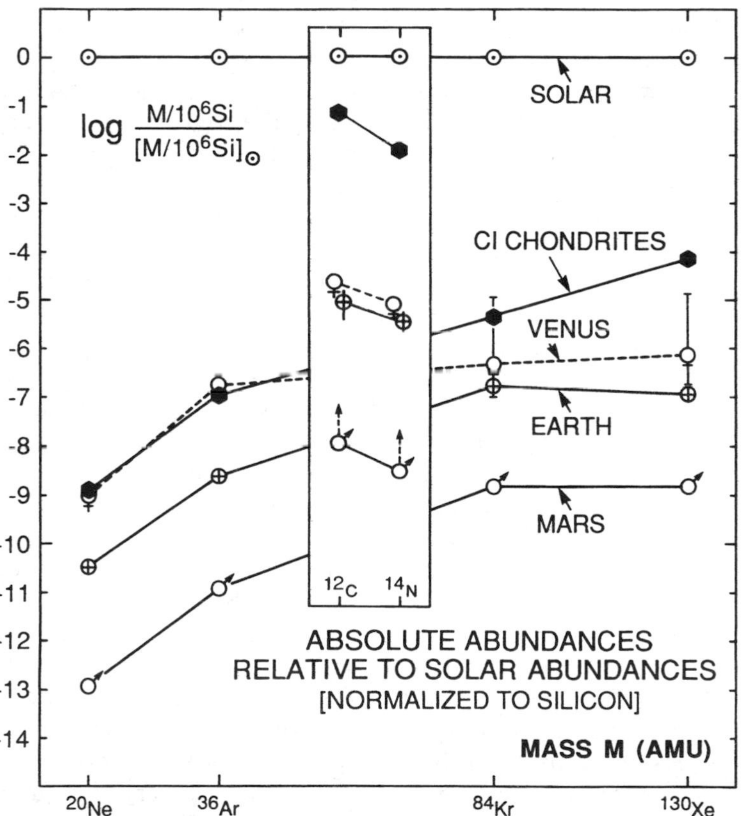

Figure 1 Absolute noble gas, carbon, and nitrogen abundances in planetary atmospheres (atoms per 10^6 planetary Si atoms) and in CI meteorites with respect to the corresponding solar ratios. Data from a compilation by Pepin (1991). "Atmospheric" abundances of terrestrial C and N are summed estimates for all crustal reservoirs from Walker (1977).

of the terrestrial planets and in primitive meteorites are grossly depleted relative to these estimates, by factors ranging from 4–9 orders of magnitude for xenon to 9–13 orders of magnitude for neon. There are no data as yet for comets. Carbon and nitrogen are relatively more abundant, particularly in the meteorites. In addition to large and variable elemental depletions, each of these contemporary volatile reservoirs displays its own set of isotopic signatures.

Helium abundances are not presented in Figure 1. Helium is only weakly bound gravitationally and escapes from the atmospheres of the terrestrial planets. The helium contents and compositions of these atmospheres are thus of little relevance as signatures of their origin and early evolution. However one should note that an isotopically solar-like helium component does exist in the Earth's interior and in meteorites, and the question of its origin—by solar wind implantation, adsorption, or some other process—is of considerable importance.

The planetary data plotted in Figure 1 refer entirely to atmospheric inventories except for terrestrial carbon and nitrogen. A relevant question is whether volatiles once in these atmospheres have been extracted by physical or chemical partitioning into surface materials, and perhaps later sequestered in planetary interiors by crustal cycling. Given the thermal and geologic state of Venus, the answer for that planet is likely to be no (Kaula 1990). On Earth, the hydrosphere, biosphere, and sedimentary column are important subsidiary reservoirs for some atmophilic species. They contain fractions of total "atmospheric" inventories that range from probably minor for the nonradiogenic noble gases—including Xe (Bernatowicz et al 1984, 1985)—to perhaps significant for nitrogen, to dominant for carbon and hydrogen. To the extent that there has been a net transport of volatiles into the upper terrestrial mantle by subduction of sediments over geologic time, the inventory estimates in Figure 1 based on summation over all these "accessible" reservoirs are lower limits. For Mars, theoretical and experimental estimates of the carrying capacity of the cold megaregolith for H_2O and CO_2 are such that, for these species at least, the observable Martian volatile reservoirs (atmosphere, ice caps, and layered deposits) are likely to contain only small fractions of total inventories (Fanale 1976; Fanale & Jakosky 1982; Fanale et al 1982, 1992). In addition, the Martian volatile system is open to loss of molecular species, including N_2, to space. Over the history of the planet, nonthermal escape of at least a few times the present atmospheric nitrogen inventory is required to account for the present-day enrichment in ^{15}N (McElroy et al 1977, Yung et al 1977, Pepin 1991). Martian carbon and nitrogen abundances in Figure 1 are therefore indicated as lower limits.

The data base used here for Mars' atmospheric composition includes both *Viking* spacecraft measurements and analyses of gases trapped in glassy nodules in the SNC meteorite EETA 79001. The geochemical case for origin of the SNC meteorites on Mars, and for trapping of an unfractionated sample of ambient Martian atmosphere in shock-generated melt phases of 79001, has been reviewed by Pepin (1985, 1987), Dreibus & Wänke (1985), Hunten et al (1987), and Pepin & Carr (1992). Detailed numerical comparisons of *Viking* and 79001 determinations for CO_2, N_2, and the noble gases are given in the last two of these reports. All abundances and isotopic compositions common to the two data sets agree within their respective uncertainties. Abundances of the two major atmospheric constituents, CO_2 and N_2, were determined more accurately by *Viking*.

Data from in situ compositional measurements of the Venusian atmosphere by mass spectrometers and gas chromatographs on U.S. and Soviet spacecraft have been comprehensively reviewed and assessed by von Zahn et al (1983). Abundances of C, N, ^{20}Ne, and ^{36}Ar in Figure 1 are calculated from the recommended mixing ratios given in their Table II. There is a profound difference of interpretation in the literature concerning the Kr and Xe concentrations measured by *Pioneer Venus* and by *Venera 11–12*. The von Zahn et al review traces the derivation of the disparate ^{84}Kr mixing ratios from the two sets of instrumental data—their estimate of 0.4 ppm from *Venera*, based on analysis by Istomin et al (1980), versus 0.025 ppm from Donahue et al's (1981) refinement of *Pioneer Venus* data—and sets out the authors' arguments in favor of the former. The contrary case is made by Donahue et al (1981), and their lower value is chosen for Figure 1—in particular because severe terrestrial Kr contamination of the *Venera 11–12* measurements is now thought to be likely, and *Venera 13* and *14* data are in better agreement with the *Pioneer Venus* results (T. Donahue, personal communication).

Venusian Xe estimates are certainly in no better shape—von Zahn et al (1983) suggest that it might not have been detected at all. However, the *Pioneer Venus* ^{132}Xe mixing ratio, reported as an upper limit of 10 ppb by Donahue & Pollack (1983), was further refined by Donahue (1986), where ^{84}Kr/^{132}Xe is given as 12.5 ± 3.5 implying a ^{132}Xe mixing ratio of about 1.9 ppb. This value has been used together with an assumed ^{130}Xe/^{132}Xe isotope ratio of 0.164 to calculate the ^{130}Xe abundance in Figure 1. Kr and Xe concentrations adopted here thus consistently reflect the views of Donahue and his coworkers. They define the strikingly solar-like Ar : Kr : Xe distribution shown for Venus in Figures 1 and 2.

Elemental abundance ratios of noble gases in terrestrial planet atmospheres (from data compiled by Pepin 1991), and in CI carbonaceous chondrites (Mazor et al 1970) and E-chondrites (Crabb & Anders 1981) are plotted normalized to solar ratios in Figure 2. To include all classes of volatile-rich chondrites, relative abundances for CM, CV3, CV4, and CO3 carbonaceous chondrites (Mazor et al 1970, Srinivasan et al 1977, Alaerts et al 1979, Matsuda et al 1980) are also shown. The first three are indis-

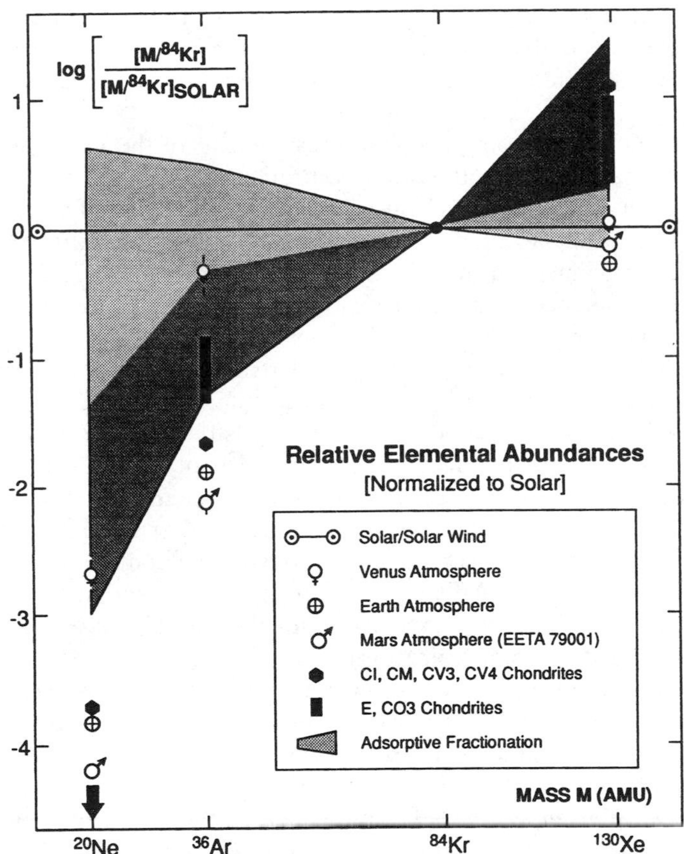

Figure 2 Noble gas $M/^{84}$Kr abundance ratios in terrestrial planet atmospheres and volatile-rich meteorites, plotted with respect to solar relative abundances and compared to the range of elemental fractionations (with respect to ambient gas-phase abundance ratios) determined from laboratory adsorptive experiments and from analyses of natural sedimentary materials. These fractionations generally fall in the darker shaded area of the Figure, except for a number of measurements on carbon black displaying the smaller or reversed patterns within the lighter shading. Data from Pepin (1991), and from references given in the text for CM, CV3, CV4, and CO3 chondrites and for laboratory and sedimentary adsorption studies.

tinguishable from the CI distribution on this scale, and the CO3s fall within the E-chondrite range.

The general pattern of progressively greater relative depletion with decreasing mass for planetary and meteoritic Ar/Kr and Ne/Kr ratios in Figure 2 suggests varying degrees of mass-dependent fractionation of solar-composition nobles gases. (Note that this monotonic pattern with respect to mass does not hold for carbon and nitrogen: Abundances of these elements, relative to ^{84}Kr as in Figure 2, range from about 3x solar for atmospheric nitrogen on Mars to 20,000x solar for carbon in an average E chondrite.) As noted earlier, the similar fractionation patterns in Figure 2 generated in experimental investigations of noble gas adsorption have suggested to many workers that gases adsorbed on nebular dust grains were the dominant contributors to present-day atmospheric and meteoritic volatile inventories. However, as also noted, the adsorptive fractionation process cannot by itself satisfy certain isotopic constraints that are important in evaluating the evolutionary significance of these similarities (and differences) in solar system volatile compositions.

ISOTOPIC COMPOSITIONS Solar, planetary, and meteoritic isotopic data from Pepin (1991) and references therein are plotted in Figures 3–5. Relative isotopic abundances of elements now in the sun, and those once extant in the early solar nebula, are not subject to direct measurement. The best available data on Ne, Ar, and Kr isotopic compositions in the solar wind yield ^{20}Ne/^{22}Ne = 13.7$\pm$0.3, ^{36}Ar/^{38}Ar = 5.6$\pm$0.1, and the range of Kr compositions shown in Figure 4; solar Xe is represented in Figure 5 by a parental Xe composition (U-Xe) derived from meteoritic and terrestrial data (Pepin & Phinney 1978, Pepin 1991). The principal uncertainty for Ne, Ar, and Kr (beyond that imposed by analytic errors) is the general question of whether the wind is an accurate sample of the solar composition and, by extension, that of the early nebula. It is reasonable to suppose that the compositions of the sun and the early nebula are closely related, but the unresolved question for the solar wind data is that of fractionation of the wind with respect to solar photospheric abundances by the processes supplying gases to the corona (Geiss 1982, Geiss & Bochsler 1985, Bochsler et al 1986, von Steiger & Geiss 1989). There is evidence that observed inter-element fractionations arising from variations in first ionization potential (FIP) are not serious for the high-FIP suite of elements considered here (von Steiger & Geiss 1989), but there also are observations that point to element-specific isotope fractionation in solar particle emission. One celebrated example is the large and unexplained discrepancy in the isotopic compositions of Ne measured in the solar wind and in solar flares (Mewaldt et al 1984). Taken at face value, this must mean that either the wind or flare isotopes are fractionated with respect to the true solar ratio.

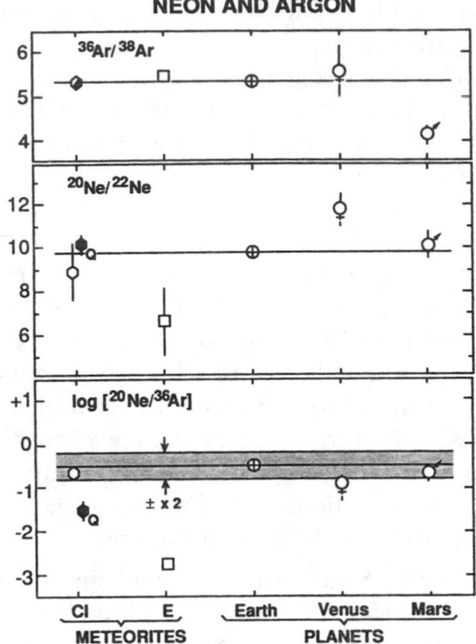

Figure 3 Elemental and isotopic compositions of neon and argon in carbonaceous (CI) and enstatite (E) chondritic meteorites and in terrestrial planet atmospheres. Open CI symbols represent bulk gas contents; see text for definition of "Q"-component gases in the CI meteorites. Data from Pepin (1991) and references therein. Units are atom/atom for isotope ratios, and g/g for the elemental Ne/Ar ratios.

Mars data are from SNC meteorite measurements. Noble gas analyses of EETA 79001 yield most of the isotopic ratios unmeasured or measured with low precision by *Viking* instruments. Of particular interest for Martian atmospheric evolution are the uniquely low $^{36}Ar/^{38}Ar$ ratio of 4.1 ± 0.2 (Figure 3), solar-like Kr (Figure 4), and the signature of mass-fractionated CI meteoritic Xe in the atmospheric Xe composition (Figure 5).

As noted earlier, the Ar/Kr and Xe/Kr elemental ratios for Venus in Figure 2 are not very different from the solar pattern. Ne and Ar isotopic data also suggest a solar affinity for noble gases on Venus, but are too uncertain to establish it rigorously. The $^{20}Ne/^{22}Ne$ ratio from *Pioneer Venus* was originally reported to be 14.3 ± 4.1 (Hoffman et al 1980); the Figure 3 value of 11.8 ± 0.7 is from a later compilation by Donahue (1986), and is seen to be significantly higher than the terrestrial and CI ratios. Thus there appears to be a solar signature in Venusian Ne. It would

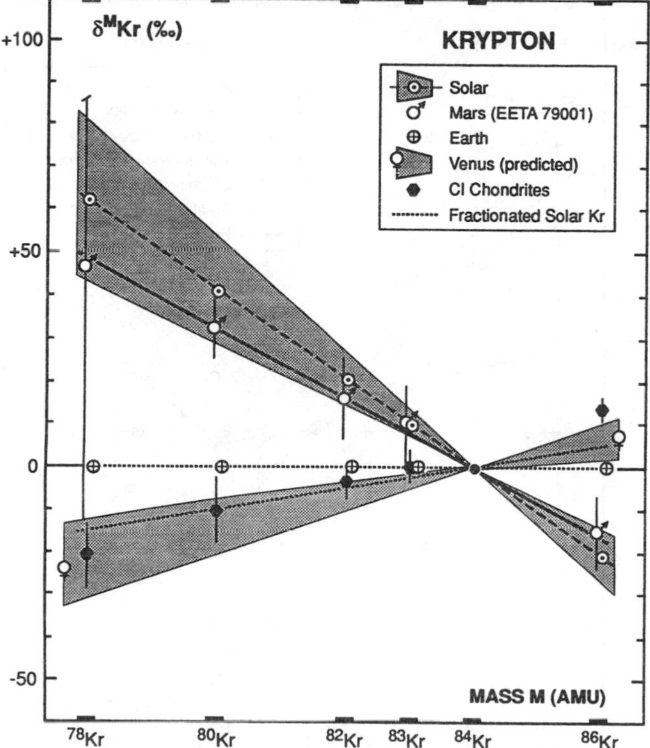

Figure 4 Isotopic compositions of krypton in planetary atmospheres, the CI meteorites, and the solar wind, plotted as deviations in per mil from terrestrial Kr isotope ratios. Data from Pepin (1991) and references therein. Dotted lines show matches to measured Kr isotope ratios on Earth and in the CI chondrites generated by solar Kr fractionations in the planetary and meteoritic hydrodynamic escape models. The upper shaded region is the uncertainty field for solar wind Kr composition; the lower shaded region represents the model-predicted compositional range for present-day Venus Kr. Martian Kr falls within the solar field and is considered to be unfractionated by hydrodynamic loss.

seem, from the nominally high *Pioneer Venus* $^{36}Ar/^{38}Ar$ ratio of 5.6 ± 0.6 (Hoffman et al 1980, Donahue 1986) plotted in Figure 3 that this could also be the case for Ar, but the measurement is too imprecise to be diagnostic. There are no quantitative data on Kr and Xe compositions.

Comparisons and Implications

More than two decades ago, when only the meteoritic and terrestrial patterns in Figure 2 were known, the term "planetary component" was

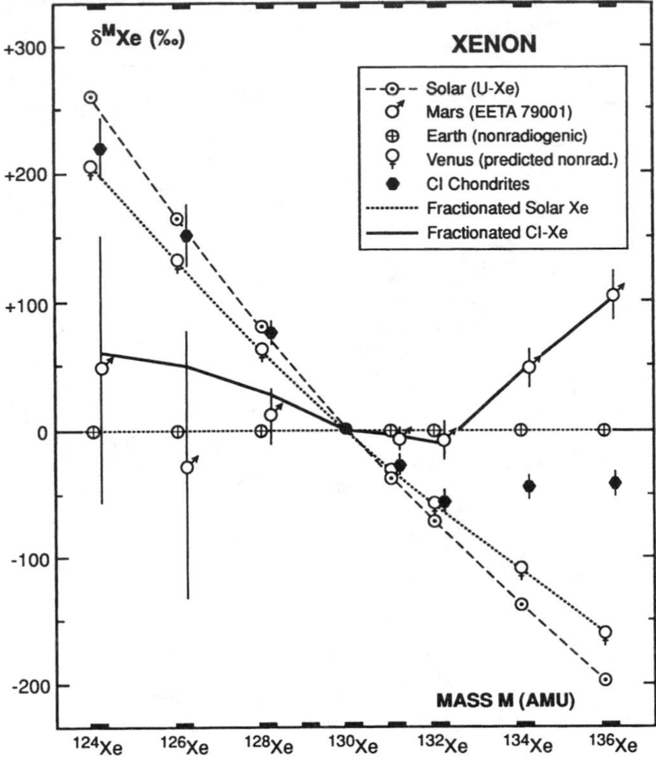

Figure 5 Isotopic compositions of xenon in planetary atmospheres, CI meteorites, and assumed for the primordial nebula (U-Xe), plotted as per mil deviations from nonradiogenic terrestrial Xe isotope ratios. Data from Pepin (1991) and references therein. Matches to nonradiogenic Xe isotope ratios on Earth (*dotted line*) and Mars (*solid line*) are generated respectively by fractionation of solar Xe and CI-Xe in hydrodynamic escape. There are no comparison data for Venus: The dotted curve represents the predicted nonradiogenic Xe composition.

introduced to designate CI-type noble gases, for what seemed then to be good reasons: The CI and Earth relative abundance distributions are virtually identical for Ne : Ar : Kr (Figure 2), the terrestrial $^{20}Ne/^{22}Ne$ and $^{36}Ar/^{38}Ar$ ratios fall within the CI ranges (Figure 3), and the Kr isotopic compositions are rather similar (Figure 4). That the terminology was unfortunate, and misleading, is now clear. It created a wide-spread presumption of a simple genetic relationship between "planetary" and terrestrial noble gases, and thereby established one glaring discrepancy—the obvious CI-Earth differences in Xe/Kr ratio (Figure 2) and in Xe isotopic composition (Figure 5)–as the celebrated "Xe Problem" that somehow had to be solvable within the framework of this presumption. Moreover, "planetary" is a generic term, and its genetic implications could be assumed to apply to terrestrial planet atmospheres other than Earth's. Although it is not clear that a simple relationship between CI and planetary atmospheric noble gases in general can be dismissed on the basis of Figure 2 alone, the "Xe Problem" is now known to occur on at least two planets, and the Ne/Kr and Ar/Kr ratios on Venus are badly out of line.

As noted above, the relative underabundance of Xe in the Earth's atmosphere compared to the CI meteorites initiated a long search for the "missing" terrestrial Xe. Attention focused on crustal volatile reservoirs, and in the particular on gases adsorbed on shales and slates in the sedimentary column (Canalas et al 1968, Fanale & Cannon 1971, Phinney 1972). We now know that the Xe inventories in these reservoirs are much too small to account for the discrepancy (Bernatowicz et al 1984, 1985). Moreover, the "Xe Problem" was always much deeper than just the factor 25 difference in relative Xe/Kr abundances shown in Figure 2. Isotopic distributions of both elements are also involved, and here the data convincingly rule out any simple acquisition of terrestrial gases directly from a CI reservoir.

Implications of observed elemental and isotopic distributions for the origin and evolution of planetary noble gases have been discussed by Pepin (1989a), and for terrestrial noble gases alone, using an earlier and now incomplete data base, by Ozima & Igarashi (1989). A central and much debated question (e.g. see Donahue & Pollack 1983) raised by the Ne-Ar comparisons in Figure 3 is whether the relatively tight grouping of the planetary Ne/Ar ratios necessarily points to a common accretional carrier of these gases, or instead could be a coincidental result of evolution from compositionally different sources by fractionation, mixing, or both. The first view would favor scenarios in which planetary noble gases were acquired by accretion of planetesimals carrying, on average, a single type of volatile mass distribution—perhaps that of bulk CI carbonaceous chondrite material (represented in Figure 3 by the open CI symbol). [The "Q"-component (*filled symbol*) ratio refers to gases sited on CI grain surfaces (Alaerts et al 1979; Wieler et al 1990, 1991; Pepin 1991), and is clearly lower, as is the E-chondrite ratio.] Isotopic considerations, however, severely limit this approach: Venusian $^{20}Ne/^{22}Ne$ is greater than 2σ above the terrestrial value, and Martian $^{36}Ar/^{38}Ar$ is much too low. Simple addition of a common "veneer" material cannot simultaneously accommodate these similar elemental but disparate isotopic compositions, and is therefore unattractive. The data appear to call not only for mechanisms that fractionated isotopes, but also for differing extents of isotopic processing. We are then left with the second alternative: generation of coincidentally rather uniform Ne/Ar ratios as end products of fractionating mechanisms operating to different degrees in the separate post-accretional environments of the individual planets and the meteorite parent bodies.

Isotopic constraints on the nature of evolutionary processing from primordial to present-day compositions, and on the composition of possible primary reservoirs, are even firmer for the two heaviest noble gases. The first-order observation from the krypton and xenon data plotted in Figures 4 and 5 is that the solar, planetary, and meteoritic isotope distributions tend to be different. The difference is subtle, but still significant, for CI-Earth Kr and Mars-Earth Xe. The one exception may be the solar versus Martian Kr compositions shown in Figure 4, which are not resolvable within present measurement uncertainties [indicated for solar Kr (Pepin 1991) by the *shaded band*]. The generally good fits of the data to the simple mass-dependent relationships indicated by the various lines and curves in Figures 4 and 5 point very strongly to the operation of mass-fractionating processes; accretion of a common carrier with one fixed isotopic composition for Kr and Xe (e.g. a CI veneer) is clearly ruled out as a source for these gases on Earth and Mars. The most straightforward scenario for evolution of these planetary and meteoritic mass distributions from a common primordial parent would seem to be the following (Pepin 1989a, 1991): (*a*) For krypton in Figure 4, CI and terrestrial Kr are generated by mass fractionation of initially solar-composition parental reservoirs, leaving the residual gases isotopically heavier. Kr on Mars apparently experienced a much smaller fractionation, or none at all, and consequently is still compositionally close to solar. (*b*) For xenon in Figure 5, solar Xe undergoes severe fractionation into terrestrial Xe, and CI-Xe is derived by mixing unfractionated or slightly fractionated solar Xe with a nucleogenetic $^{131-136}$Xe component carried into the solar accretion disk by infalling dust grains. Xe on Mars differs from that on Earth in that it seems to have evolved from a nonsolar parent. Swindle et al (1986) were the first to note that Martian Xe is most simply interpreted as fractionated CI-Xe. This is illustrated in Figure 5, where the solid line, representing a *calculated* composition of mass-fractionated CI Xe, is seen to be an excellent fit to the measured Martian isotope abundances (excluding ^{129}Xe, where a Mars-specific radiogenic contribution dominates). This genesis for Xe on Mars is fully compatible with Dreibus & Wänke's (1985, 1987, 1989) geochemical modeling, which calls for a large CI accretional component in the planet, and for a massive hydrodynamic blowoff of its primordial atmosphere (see below) which could have implemented the required Xe fractionation.

In terms of the "Xe Problem," it is clear that isotopic considerations destroy the presumption of a CI-Earth noble gas relationship of the simple kind that led to the creation of the problem in the first place. They also point very firmly to the conclusion that mechanisms capable of fractionating isotopes have been at work. For these heavy noble gases, Jeans escape probably played no role since improbably small bodies or high exospheric temperatures are required, but in the proper environment hydrodynamic escape (discussed later) could have been effective. However the measured isotopic patterns impose severe constraints on provenances and processes. Figures 4 and 5 imply a separation of Kr and Xe processing, in locale or time and perhaps also in type. While the terrestrial Xe composition can be generated from a solar-like parent, the fractionation is so large that any associated Kr would have been driven from solar composition to an isotopically much heavier final composition than is observed for terrestrial Kr. On Mars, although the evidence in Figure 5 for derivation of Xe directly from CI-Xe appears firm, note that Martian Kr cannot be generated by fractionation of the isotopically heavier CI composition—instead a solar-like parent is implied. Realistic models for generating planetary volatile distributions will have to accommodate this intriguing decoupling of the evolutionary histories of the two heaviest noble gases.

MODELING APPROACHES TO THE PROBLEM

Modeling the origin and evolution of terrestrial planet atmospheres, in ways that accommodate the data shown in Figures 1–5 in the context of plausible planetary environments and processes, is a classic problem in the planetary sciences. Historically, models of origin based on seemingly straightforward clues in one subset of the extensive data base have encountered inconsistencies in another. A celebrated example of this kind of difficulty involves the Figure 3 neon-argon distributions discussed earlier. The approximate concordance of Ne/Ar elemental ratios on Earth, Venus, and Mars, and in bulk samples of the primitive CI carbonaceous chondrites seems to imply accretion of common parental material as the source of at least these two planetary gases, but isotope systematics argue against this interpretation. Attempts to model the origin of terrestrial planet atmospheres, for Ne-Ar and for other species, have tended in the past to focus on deriving the Figure 2 *elemental* abundance patterns of atmospheric noble gases from primordial meteoritic or nebular reservoirs. Until recently, with a few exceptions, comparatively little attention has been paid to *isotopic* distributions, in particular those of the heavy species krypton and xenon, or to physical environments in which evolutionary processing might plausibly have occurred.

An excellent review of various classes of models, and their successes and failures, has been given by Donahue & Pollack (1983). They divide theories developed up to that time into two general categories of planetary formation, one in "gas-rich" and the other in "gas-poor" environments. The first involves planetary growth to essentially full mass in the presence of nebular gas, leading to gravitational capture of solar-composition primordial atmospheres. It includes Cameron's (1978, 1983) giant gaseous protoplanet hypothesis and the massive, gravitationally captured early terrestrial atmosphere proposed by the Kyoto group (see Hayashi et al 1985 and references therein). Neither of these has been examined rigorously from the point of view of elemental and isotopic evolution of atmospheric constituents during subsequent loss of these protoatmospheres. However Sasaki & Nakazawa (1988) have advanced the Kyoto model considerably in this respect by showing that present-day terrestrial Xe could be a fractionated residue of an early massive atmosphere dissipated by hydrodynamic escape.

Gas-poor models, in which planetary accumulation occurs after loss of the nebula, can be further subdivided into "veneer" and "grain-accretion" theories (Donahue & Pollack 1983). The first of these model classes attributes atmophilic element inventories on one planet or another to late-stage accretion of cometary (Lewis 1974; Owen 1986, 1987; Hunten et al 1988; Owen et al 1991) or meteoritic (Anders & Owen 1977) volatile sources, the second primarily to sorption of nebular gases (Pollack & Black 1979, 1982) or implantation of early solar wind gases (Wetherill 1981, Donahue et al 1981, McElroy & Prather 1981) in precursor materials from which the planets later accreted. Veneer theories are basically mixing models that aim to reproduce planetary volatile inventories by accretion of known (or occasionally hypothetical) meteorite types in specified proportions. Bogard (1988), in a general mixing model approach not tied to any specific view of atmospheric origin, has examined the ability of combinations of known (or suspected) solar system volatile components to generate the noble gas, carbon, and nitrogen abundance patterns observed in planetary atmospheres. Grain-accretion theories constructed to be broadly applicable to all three terrestrial planets (e.g. Pollack & Black 1982) feature nonuniform mixing of host grains for particular kinds of meteoritic and solar wind gases, together with dust-grain carriers of variable amounts of sorbed nebular gases, to form their atmospheres.

Models appealing only to veneers or grain accretion as noble gas sources are intrinsically unable to account for the full range of isotopic variability seen in terrestrial planet atmospheres. The processes of mixing and sorption invoked in these theories do not fractionate isotopes, contrary to the evidence in the compositional patterns shown in Figures 3–5 that such fractionating mechanisms have been at work. For this reason, gas-rich models that assume the presence of dense primordial atmospheres of whatever origin on growing or fully accreted planetary bodies, or on (or in) large preplanetary planetesimals, appear fundamentally more attractive. They offer the potential for isotopic fractionation in the process(es) that subsequently dissipated these atmospheres, and the possibility that the variable distributions of noble gas abundances and isotopes seen in present-day atmospheres may be understood as reflecting different degrees of processing on the individual bodies themselves.

A new generation of evolutionary models is beginning to focus on fractionation during atmospheric escape. In the first of these, Donahue (1986) considered fractionation during classical Jeans escape of pure, solar-composition noble gas atmospheres from large planetesimals. Their atmospheres are assumed to derive from outgassing of nebular gases previously adsorbed on the surfaces of preplanetesimal dust grains, and are subsequently fractionated to different degrees by losses that depend on the rates of planetesimal growth. These bodies later accumulate in various proportions to form the terrestrial planets and their atmospheres. Donahue's model can account reasonably well for relative Ne : Ar : Kr elemental abundances and for Ne isotopic compositions in the atmospheres of Venus, Earth, and Mars. Predicted $^{36}Ar/^{38}Ar$ ratios, however, are much lower than observed, and variations in Kr and Xe elemental and isotopic compositions in different planetary reservoirs cannot be explained since Jeans escape of these heavy species from the parent planetesimals is essentially nil.

These problems are proving to be more tractable in the context of a related thermal loss mechanism: hydrodynamic escape (Hunten et al 1987, 1988, 1989; Zahnle & Kasting 1986; Pepin 1986, 1987, 1989b, 1990, 1991; Sasaki & Nakazawa 1988; Zahnle et al 1990a). Here heavy atmospheric constituents can be lost from large bodies—partly or fully accreted planets. In this process hydrogen-rich primordial atmospheres are assumed to be heated at high altitudes, after the nebula has dissipated, by intense far-ultraviolet radiation from the young sun. Under these conditions thermal escape fluxes of hydrogen can be large enough to exert upward drag forces on heavier atmospheric constituents sufficient to lift them out of the atmosphere, at rates that depend on their mole fractions and masses (Hunten et al 1987, Zahnle & Kasting 1986). Lighter species are entrained and lost with the outflowing hydrogen more readily than heavier ones, leading to mass fractionation of the residual atmosphere. Hydrodynamic escape is particularly attractive for its ability to generate, in an astrophysically plausible environment, large isotopic fractionations of the type displayed by terrestrial xenon with respect to solar Xe (Figure 5)—first observed and attributed to an (unknown) fractionation process 30 years ago (Krummenacher et al 1962)—and by Xe in the atmosphere of Mars relative to Xe in primitive meteorites (Figure 5, Swindle et al 1986).

Escape to space, however, may not be the only mechanism which could have produced such severe fractionations. There are two current models for the origin of planetary Xe from solar Xe—one appealing to hydrodynamic escape (Pepin 1991), where Xe evolution is driven by an escape process operating on the planets themselves, and another originally suggested by Ozima & Nakazawa (1980) (see also Ozima & Igarashi 1989) and recently redeveloped and extended by Zahnle et al (1990b), in which nebular Xe is fractionated into the terrestrial composition by gravitational isotopic separation in large porous planetesimals which are later accreted by the terrestrial planets. This source is supposed by Zahnle et al (1990b) to supply isotopically similar Xe to Earth, Mars, and Venus, and so one difficulty with this theory in its present form is that Earth and Mars in Figure 5 are characterized by significantly different heavy isotope ratios (with the assumption, generally accepted but not proven, that SNC meteorite data do in fact apply to Mars). An isotopic analysis of Xe on Venus, where no data presently exist and where the Xe compositions yielded by the two models (Figure 5) are even more discrepant, should rule between them.

EARLY ASTROPHYSICAL ENVIRONMENTS AND PLANETARY ACCRETION HISTORIES

The comparisons of noble gas mass distributions in terrestrial planet atmospheres, primitive meteorites, and the solar wind (Figures 1–5) strongly suggest that fractionating processes drove the evolution of atmospheres from primordial compositions to their present-day states. The challenge is to deduce what these mechanisms might have been, and to compare the conditions under which they could have acted with current ideas about the nature of the solar system's early astrophysical and planetary environment, and how that environment evolved with time. This approach links the processing of primordial volatiles directly to the observational record and theoretical modeling of conditions and processes in pre-main-sequence stars and their accretion disks.

There is firm observational evidence that short-wavelength luminosities of young solar-type stars in the $\sim 10^6$–3×10^7 yr age range are several orders of magnitude higher than that of the contemporary sun, and that with increasing stellar age these enhanced activities decline with an exponential or power law time dependence (Canuto et al 1982, Zahnle & Walker 1982, Simon et al 1985, Walter et al 1988, Strom et al 1988, Feigelson & Kriss 1989, Walter & Barry 1991). Infrared observations of these young stellar systems also indicate that fine circumstellar dust tends to disappear from inner regions of their accretion disks within $\sim 10^7$ yr or less (Strom et al 1989a,b, 1992), and provide estimates of the earliest times for nebular clearing of dust and gas. As noted above, disappearance of fine dust from IR detectability could be a result of aggregation into larger—but still small—particles, and the actual time scales for gas-phase dissipation in these systems is observationally unconstrained. According to the standard model of planetary accretion, planet-building that began during this epoch would have proceeded rapidly, to ~Mercury-sized embryos within $\sim 10^6$ yr, and arguably in the presence of nebular gas; further growth to full planetary masses during the following $\sim 10^8$ yr by "giant-impact" mergers of these embryos could, among other possibilities, have been triggered by gas loss (Wetherill 1986, 1990a,b). Although time scales for gas dissipation and decay of collisionally-generated dust densities to suitably low transmission opacities are very uncertain and a matter of debate (e.g. Prinn & Fegley 1989), it seems likely that a very large flux of solar coronal and transition-region radiation would have penetrated the midplane of the early solar system to planetary distances within $< 10^8$ yr.

The presence of massive hydrogen-rich primordial atmospheres on the terrestrial planets at this time is conjectural. Although they are a natural consequence of accretion of Earth and Venus to roughly their final masses in the presence of nebular gas, the condition that the gas survived for this length of time may not have been met. Addition of water amounting to a few weight % or less of planetary masses by accretion of icy planetesimals could also have supplied noble gases and abundant hydrogen. However much more work on modeling the probable fluxes of outer-system objects into terrestrial planet space during the first $\sim 10^8$ yr of solar system history is needed to judge whether this source is plausible.

Hydrogen escape fluxes high enough to sweep out and fractionate species as massive as Xe from such atmospheres require energy inputs roughly two to three orders of magnitude greater than presently supplied to terrestrial planetary exospheres by solar extreme-ultraviolet (EUV) radiation. Since early solar EUV radiation could not have penetrated a full gaseous nebula to planetary distances, the applicable time interval of stellar activity is that following transition of the dense, optically thick accretion disks surrounding the classical T-Tauri stars (CTTS) to the tenuous disks associated with "weak-lined" (WTTS; Strom et al 1989a) or "naked" (NTTS; Walter 1986) T-Tauri stars. Interstellar extinction of EUV radiation from the WTTS requires estimation of intensities from measurements at other wavelengths. Among present observational data, soft (~3–60 Å) X-ray fluxes are most likely to be representative of at least the short-wavelength coronal component ($\lambda < 700$ Å) of the EUV spectrum. The combined Walter et al (1988) and Simon et al (1985) data base on X-ray activity has been discussed by Pepin (1989b). Figure 6, a plot of estimated stellar age versus

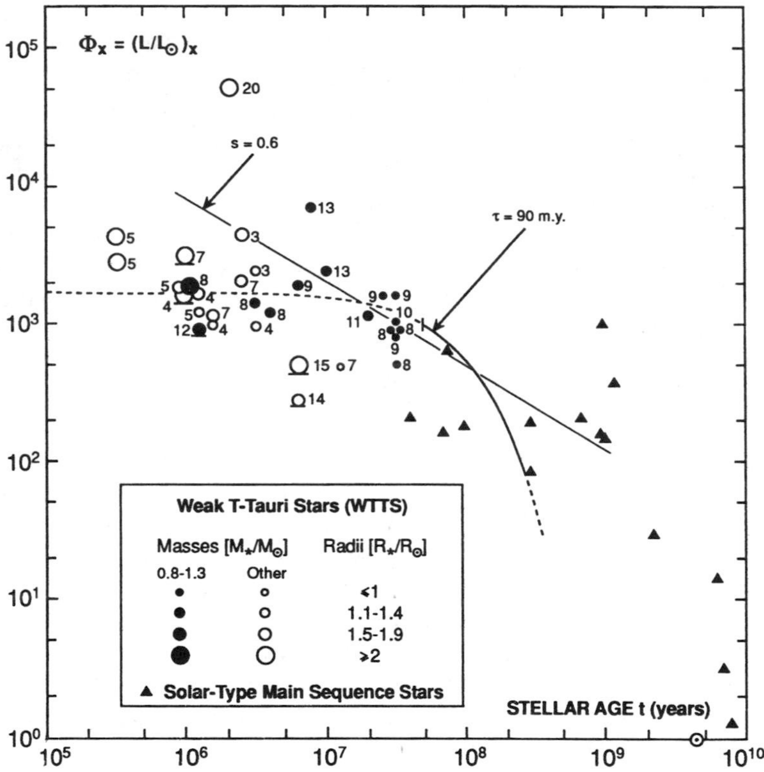

Figure 6 Observed X-ray luminosities L_x, relative to that of the present sun, vs estimated stellar ages for "weak-lined" T-Tauri stars and solar-type main sequence stars (after Pepin 1989b). All WTTS data from Walter et al (1988), MSS data from Simon et al (1985) and Walter & Barry (1991). Numbers adjacent to the plotted WTTS points are stellar mass estimates in tenths of a solar mass. Underlining identifies measurements that yielded only lower limits on luminosity. The plotted exponential, with a mean decay time $\tau = 9 \times 10^7$ yr, represents one way in which soft X-ray (Φ_x) and EUV luminosity enhancements of the young sun could have declined with time. The period of interest for planetary atmospheric evolution is indicated by the solid portion of the plotted curve (Pepin 1991). A $\Phi_x \propto (1/t)^{0.6}$ power law decay (Feigelson & Kriss 1989) is shown for comparison.

$\Phi_x(t) = [L(t)/L_\odot]_x$ (where $L_\odot$ is the X-ray luminosity of the contemporary sun) is an updated version of Pepin's (1989b) Figure 1(*b*) in which Simon et al's (1985) main sequence star (MSS) data have been replotted against Walter & Barry's (1991) more recent estimates of their ages. It is clear that both the WTTS and MSS scatter considerably from a single functional dependence of Φ_x on stellar age. However most of the MSS data for t between $\sim 5 \times 10^7$ and 5×10^8 yr do indicate a decline in Φ_x by a factor of ~ 5 from WTTS levels at $\sim 3 \times 10^7$ yr.

To illustrate the hydrodynamic escape model outlined in the following section, Pepin (1991) assumed that Φ_x declined exponentially with time. The exponential plotted in Figure 6, with a mean decay time τ of 9×10^7 yr, is one possible example of the dependence of Φ_x on t for the young sun. [An alternative time dependence is shown by the straight line, representing a power law flux decline of the type suggested by Feigelson & Kriss (1989) to hold over the entire stellar age range from $\sim 10^6$ to 10^{10} yr.] The period of hydrodynamic loss of planetary atmospheres, beginning by assumption at 5×10^7 yr and terminating at $\leq 4 \times 10^8$ yr, is indicated by the solid portion of the curve. Note that at 5×10^7 yr, $\Phi_x \cong 1000$. This level of enhancement over present-day solar radiation applies strictly only to soft X-ray and short-wavelength EUV radiation from the stellar corona. Current data on radiation longward of the EUV—from transition regions and upper chromospheres of the WTTS and solar-type MSS where additional components of the $\lambda < 1000$ Å EUV radiation also originate—indicate smaller increases (Simon et al 1985, Walter et al 1988, Pepin 1989b). Thus one would expect the total EUV enhancement at 5×10^7 yr to have been somewhat less than 1000.

MODELS OF ATMOSPHERIC EVOLUTION BY HYDRODYNAMIC ESCAPE

This scale of EUV energy deposition into a hydrogen-rich planetary atmosphere would have driven an episode of hydrodynamic escape, initially intense but tapering off as the solar EUV luminosity declined with time. It seems inevitable that some degree of hydrodynamic loss and fractionation of planetary atmospheres would have occurred if the conditions noted above for energy source, hydrogen supply, and midplane transparency were even partially met. Consequently the process is inherently interesting—and perhaps crucially important—for atmospheric evolution in the context of current observations and theories about the astrophysical environment of the early solar system.

The potential power of the hydrodynamic loss mechanism to replicate observed isotopic distributions is well illustrated by Hunten et al's (1987) applications of the process to several specific cases, including the derivation of terrestrial Xe from solar Xe; by Sasaki & Nakazawa's (1988) independent treatment of the terrestrial Xe problem; and by Zahnle et al's (1990a) consideration of Ne and Ar compositions on Earth and Mars. The next step in assessing its more general viability as an actual instrument of volatile evolution is to examine the consequences of hydrodynamic escape for the full range of elemental and isotopic mass distributions found in contemporary planetary atmospheres and volatile-rich meteorites, and to explore the astrophysical and planetary conditions under which the process could reconcile this data base. A first attempt by the author to construct such a model (Pepin 1991) is outlined in the following sections.

Theory

A simple analytic theory of mass fractionation in hydrodynamic escape of gases from planetary atmospheres has been developed by Hunten et al (1987). They considered an atmosphere consisting primarily of a light gaseous species (constituent 1—assumed, in their applications of the theory, to be molecular hydrogen) of mass m_1 and column density N_1 particles cm^{-2}, containing minor amounts of heavier components (constituents 2—for example, primary noble gases) of masses m_2 and inventories N_2. Energy input at high altitudes drives thermal loss of the light constituent with escape flux F_1 particles cm^{-2} sec^{-1}. The escaping gas exerts upward drag forces on heavier species. For a given F_1 the upward drag is sufficient to lift all constituents with masses m_2 less than a critical mass m_c out of the atmosphere. The escape fluxes of these species depend on their mole fractions in the atmosphere and on their masses. The critical or "crossover" mass m_c, the smallest mass m_2 for which the escape flux of constituent 2 is zero, is defined as follows:

$$m_c = m_1 + \frac{kTF_1}{bgX_1} \tag{1}$$

and the escape flux of constituent 2 for $m_2 < m_c$ is then given by

$$F_2 = \frac{X_2}{X_1} F_1 \left(\frac{m_c - m_2}{m_c - m_1} \right), \tag{2}$$

where T is temperature, k is Boltzmann's constant, b is the diffusion parameter (product of diffusion coefficient and total number density) for constituent 2 in constituent 1, X_1 and X_2 are the mole fractions of the major and minor constituents respectively, and g is the gravitational acceleration.

If solar EUV output declined with time over the $\approx 10^6$–10^9 year period of interest for the early history of planetary atmospheres (as seems likely for the young sun), the principal source of upper atmospheric heating likewise diminished. Consequently $F_1(t) = F_1^0 f(t)$, where $f(t)$ is some decreasing function of time and F_1^0 represents the escape flux of constituent 1 at some initial time t_0 when escape began. The corresponding initial crossover mass from Equation (1) is then

$$m_c^0 = m_1 + \frac{kTF_1^0}{bgX_1}, \quad \text{and} \quad F_1^0 = \left(\frac{gX_1}{kT}\right) b(m_c^0 - m_1). \tag{3}$$

Under these conditions the expression for the time rate of change of N_2, the atmospheric inventory of constituent 2, is

$$\frac{dN_2}{N_2} = -\frac{F_1^0}{N_1}\left[f(t) - \frac{m_2 - m_1}{m_c^0 - m_1}\right] dt \tag{4}$$

for any $m_2 < m_c$, if X_1, the mole fraction of constituent 1, is assumed to remain constant and near unity throughout the escape process.

Pepin's (1991) model is applied to planetary atmospheres rich in hydrogen, under one of two different sets of escape conditions. In the first, one assumes that the full inventory N_2 of a minor species is initially in the atmosphere, but hydrogen is replenished from some source at roughly its rate of escape so that the atmospheric hydrogen inventory N_1 is approximately constant throughout the escape episode. The evolution of N_2 from initial time t_0 to time t_2 (when escape of m_2 ceases) is then given in this "constant inventory" model by integration of Equation (4) from t_0 to t_2 with constant N_1, yielding

$$\frac{N_2}{N_2^0} = \exp\left[-\frac{F_1^0}{N_1}\int f(t)\,dt + \frac{F_1^0}{N_1}\left(\frac{m_2 - m_1}{m_c^0 - m_1}\right)(t_2 - t_0)\right], \tag{5}$$

where N_2^0 is the initial inventory of constituent 2. From Equation (4), termination of N_2 loss ($dN_2/N_2 = 0$) at t_2 occurs when

$$f(t_2) = \frac{m_2 - m_1}{m_c^0 - m_1} \tag{6}$$

which yields t_2, given a functional form for $f(t)$—for example, $f(t) = F_1/F_1^0 = \exp[-(t - t_0)/\tau]$ for exponential decay of the solar EUV flux (Figure 6). Over the time interval $t_2 - t_0$, the crossover mass decreases from m_c^0 to m_2 as the hydrogen escape flux declines from F_1^0. All atmospheric species of mass less than m_c^0 are depleted during the escape episode.

As an alternative to the constant hydrogen inventory model, suppose that all of the hydrogen is initially in the atmosphere along with the minor species. With no replenishment, N_1 in this "Rayleigh fractionation" model declines with time along with N_2; $dN_1/dt = -F_1 = -F_1^0 f(t)$, and with $N_1/N_1^0 \equiv g(t)$ Equation (4) becomes

$$\frac{dN_2}{N_2} = \frac{dg(t)}{g(t)} + \frac{F_1^0}{N_1^0}\left(\frac{m_2 - m_1}{m_c^0 - m_1}\right)\frac{dt}{g(t)}, \tag{7}$$

where $g(t) = 1 - (F_1^0/N_1^0)\int f(t)\,dt$ integrated from t_0 to t.

ENERGY-LIMITED ESCAPE FLUX The energy required for escape of a light constituent of mass m_1 from its local gravitational field, at radial distance $r \geq r_s$ from a body of mass M and radius r_s, is Gm_1M/r ergs per particle. If the global mean solar EUV input is ϕ ergs cm^{-2} sec^{-1}, the energy-limited escape flux is

$$F_1 = \frac{r\phi\varepsilon}{Gm_1M} \text{ particles cm}^{-2}\text{ sec}^{-1}, \tag{8}$$

where ε is the fraction of incident EUV flux converted to thermal escape energy of m_1. ϕ is conveniently expressed in terms of the current mean EUV flux at Earth $\phi_\oplus(t_p)$, heliocentric distance R, and the ratio Φ_{EUV} of the flux at past time t to that at present time t_p. We define $\phi(R, t_p) \equiv (R_\oplus/R)^2\phi_\oplus(t_p)$ and $\Phi_{EUV}(t) \equiv \phi(R, t)/\phi(R, t_p)$, yielding $\phi(R, t) = \phi = (R_\oplus/R)^2\phi_\oplus(t_p)\Phi_{EUV}$. Substituting this into Equation (8) and Equation (8) into Equation (1), with $g = GM/r^2$ and M expressed in terms of r_s and density ρ, yields

$$m_c - m_1 = \left(\frac{3}{4\pi G\rho}\right)^2\left(\frac{1}{r_s}\right)^3\left(\frac{R_\oplus}{R}\right)^2\left(\frac{kT}{X_1}\right)\left(\frac{\phi_\oplus(t_p)\varepsilon}{bm_1}\right)\left(\frac{r}{r_s}\right)^3\Phi_{EUV}(t) \tag{9}$$

which relates crossover mass to the EUV flux irradiating a planetary object at heliocentric distance R. $\Phi_{EUV}(t) \propto \exp[-(t - t_r)/\tau]$ for exponential flux decay, and τ is taken to be 9×10^7 yr (Figure 6), assuming that Φ_X and Φ_{EUV} decay together. With known values or estimates for the various bracketed parameters in Equation (9), this expression defines the EUV energy demand of the atmospheric escape model for a particular planet as a function of crossover mass (which must lie above the mass of the heaviest escaping species).

Assumptions

The model rests on three basic postulates:

1. Noble gases in terrestrial planet atmospheres and primitive meteorites, excluding later contributions from radioactive decay and the solar wind, derived from one or both of two primordial sources: "local" volatiles

in the gas phase of the protosolar nebula, and "exotic" volatiles (reviewed by Anders 1987, 1988) sequestered in interstellar grains carried into the accretion disk from presolar molecular cloud environments.

2. Present-day mass distributions of these volatiles evolved from primordial source compositions through fractionation by either or both of two primary processes: gas adsorption and fixation in accreting solid matter or on nebular dust grains, fractionating elements but not isotopes; and hydrodynamic escape of atmospheres from the partially or fully assembled planets and from meteorite parent bodies, fractionating both elements and isotopes.
3. Planetary conditions for hydrodynamic escape were established when primary atmospheres rich in hydrogen or water accumulated around the growing terrestrial planets, and following the dissipation of nebular dust and gas to conditions of low transmission opacity, were exposed to intense solar EUV radiation which declined in strength with time.

Specific assumptions are made within the context of these general postulates. A number of them relate to the noble gas data base. Gases trapped in an SNC meteorite are assumed to be a sample of the Martian atmosphere. Isotopic distributions of neon, argon, and krypton in the gas phase of the protosolar nebula are taken to be similar to those in the solar wind. Nebular xenon is a quasi-theoretical composition (U-Xe) derived from analysis of meteoritic and terrestrial data but only marginally observed directly in meteorites (Niemeyer & Zaikowski 1980). Uncertainties in these estimates of primordial isotopic distributions are countered to some extent by the ability of the model to accommodate different initial compositions.

Serious attempts to derive the compositions of the contemporary atmospheres on Earth and Mars from primitive nebular, planetesimal, or planetary source reservoirs have had to confront, in one way or another, the apparent decoupling of the evolutionary histories of krypton and xenon noted in an earlier section. This implies separate provenances for the two species. In the present model, atmospheric Xe inventories on both planets are considered to be fractionated relicts of their primary atmospheric Xe, while most of the Kr and lighter noble gases are products of planetary outgassing. This effective isolation of the heaviest noble gas from lighter species follows naturally from the processes intrinsic to the model, in all respects but one. Isotope mixing systematics impose a strict upper limit on the allowed level of "contamination" of residual primary Xe by later degassing from the interior. This limit falls well below the amount of Xe that ordinarily would be expected to accompany the outgassed Kr.

Therefore the model's most crucial assumption is that the bulk of this Xe has been retained in the bodies of the planets; this leads to a prediction of geochemical partitioning behavior at high pressures which defines a central experimental test of the model. Xe is known to metallize somewhat above 1 megabar (Goettel et al 1989), and so could have partitioned into metal phases deep in the Earth. But Mars is the more critical case, since interior pressures are, and always were, well below 1 megabar. As yet there are no experimental studies at the relevant Martian pressures of a few hundred kilobars.

The interval needed for dissipation of nebular gas and decay of collisionally-generated dust densities to levels allowing EUV transmission through the inner solar system midplane is taken, more or less arbitrarily, to be 5×10^7 yr. By this time Earth-, Venus-, and Mars-like bodies in the standard model of planetary accretion had grown to greater than 90% of their final masses, and the chances of later giant impacts—which probably would have derailed their atmospheres from their modeled evolutionary tracks in unknown ways—were diminishing (Wetherill 1986, 1990a). Results are not sensitive to this particular choice of initial time t_0 for the beginning of hydrodynamic escape.

During its first 5×10^7 yr of growth to nearly final mass, each planet is assumed to acquire two spatially separated noble gas reservoirs of nebular isotopic composition. The first of these, incorporated deep in the planetary interior, was the source for species later outgassed to the surface. Models of the adsorption process (Pepin 1991) suggest that these interior reservoirs could have originated from gravitationally captured gases, occluded and buried in accreting protoplanetary cores during growth to the "planetary embryo" stage in the presence of the nebular gas phase (Wetherill 1990b). The second was a surface reservoir contained in a hydrogen-rich primary atmosphere coaccreting with the planet, perhaps contributed to all three planets by icy planetesimals and, on Mars, by a substantial additional component of CI-like meteoritic material. When gas and dust opacities declined, intense heating by EUV absorption at high altitudes began to drive the escape of the primary atmosphere.

Results

Integration of Equations (5) or (7) for constant inventory or Rayleigh loss of hydrogen, with EUV and hydrogen escape flux decay described by exponential $f(t)$ with $\tau = 9 \times 10^7$ yr, yields expressions for the time evolution of the inventories N_2 of atmospheric noble gases. Free parameters in these expressions [e.g. the initial and final crossover masses m_c^0 and m_c^f for one species, and the constant (N_1) or initial (N_1^0) hydrogen inventory] are adjusted to yield final (post-escape) fractionated isotope distributions that are in best accord with present-day noble gas isotopic compositions. Values for noble gas abundances N_2^0 initially present in primary or degassed planetary atmospheres at the beginning of their escape episodes are determined from the constraint that final residual inventories from all contributing sources must equal observed abundances in planetary atmospheres. Principal results of applying this escape model to atmospheric evolution on the terrestrial planets are outlined below. Derived isotopic compositions and comparisons with measured compositions are shown in Figures 4 and 5 for krypton and xenon, and in Figure 7 for neon and argon.

EARTH Atmospheric loss and associated elemental and isotopic fractionation of residual atmospheric species, beginning at a crossover mass well above Xe, continued for $t > t_0$ until EUV heating declined to a level where the Xe isotopes could no longer be lifted out of the atmosphere by the drag forces exerted on them by the weakening hydrogen escape flux. At this point the planetary Xe inventory was frozen at its present-day abundance and with an isotopic composition closely matching that of nonradiogenic terrestrial Xe (Figure 5), while escape of lighter constituents continued. By the time of planetary outgassing, presumably triggered by later stages of differentiation, these lighter species from the primary atmosphere had been depleted to levels well below their contemporary atmospheric abundances. Solar-composition Kr, Ar, and Ne degassed from the interior (accompanied by little if any outgassed Xe, in what was noted above to be the central assumption of the model), then entered the atmosphere and were mass-fractionated by continuing hydrodynamic escape to isotopic compositions identical or close to their present-day values (Figures 4 and 7). Elemental fractionation of the source reservoir of outgassed noble gases with respect to solar composition is required to match relative abundances in the Earth's atmosphere. The elemental composition yielded by the model is in reasonable agreement with fractionation patterns produced in laboratory adsorption experiments.

An interesting—and perhaps troublesome—consequence of the requirement to reproduce present-day isotopic compositions is that this stipulation cannot be met for a primary terrestrial atmosphere with the same noble gas elemental abundance pattern as that on Venus (see below). This is contrary to what one would expect if both atmospheres were supplied from the same source (e.g. outer solar system icy planetesimals, solar wind carriers, or gravitational capture). Compared to Venus, the terrestrial Ne/Xe, Ar/Xe, and to a lesser extent Kr/Xe ratios must be depleted by increasingly greater amounts, the lighter the species. One possibility, which needs much more computational attention than it has

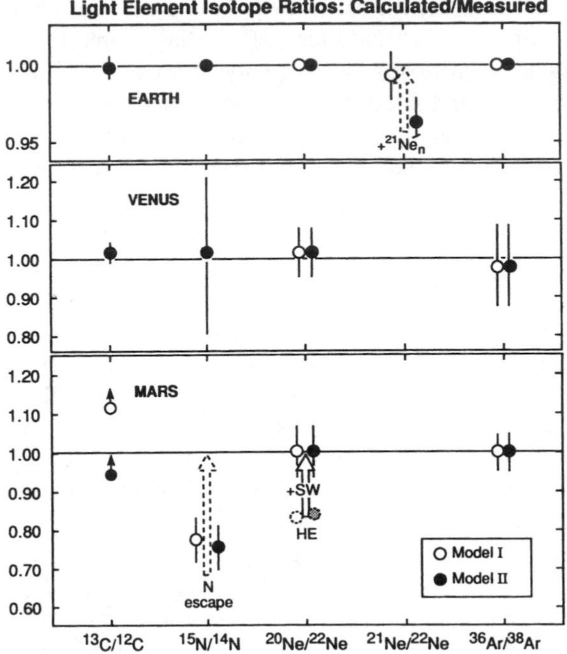

Figure 7 Comparison of measured neon, argon, carbon, and nitrogen isotopic compositions on Earth, Venus, and Mars (data from Pepin 1991 and references therein) with those calculated from hydrodynamic escape models. Filled and open symbols respectively designate results without (Model I) and with (Model II) late-stage accretion of a C-N-rich veneer. Plotted points represent the fractionated compositions left in residual planetary atmospheres at the end of escape, except for Ne on Mars where the dotted "HE" data show the over-fractionation of $^{20}Ne/^{22}Ne$. Upward arrows indicate how some of these residual isotope ratios might later have evolved, on Earth by post-escape degassing of nucleogenetic $^{21}Ne_n$ produced in the crust (Wetherill 1954, Kyser & Rison 1982), and on Mars by nonthermal nitrogen escape (Nier & McElroy 1977, McElroy et al 1977) and capture of solar wind (*SW*) gases.

so far received, is that this fractionation is a signature of partial ejection of an initially Venus-like primary atmosphere during a giant Moon-forming impact on Earth, prior to the beginning of hydrodynamic escape. Loss of some, perhaps most of a preexisting atmosphere would be expected in a large-scale collision, and surviving residues could well be fractionated (Cameron 1983, Hunten et al 1989, Benz & Cameron 1991).

The high rate of exospheric EUV energy input, calculated from Equation

(9), required for escape at the large initial crossover mass needed on Earth (and Venus and Mars as well) is still less than the X-ray enhancements shown in Figure 6 by factors of two or more. Thus the inference that the EUV luminosity of the young sun was sufficient to drive Xe escape rests on direct observations of young solar-type stars at bracketing wavelengths and is reasonably firm. However one should note that this conclusion depends on the assumption of hydrogen-dominated atmospheres. Substantial mixing ratios of species such as CO_2 could have truncated escape at much lower masses, for two reasons (Zahnle et al 1990a): Such molecules are good infrared emitters and thus can radiate away available energy, and the hydrogen escape flux could have been limited by having to diffuse upward through a second abundant and heavier constituent.

VENUS Current data on abundances and isotopic compositions of minor atmospheric constituents on Venus are uncertain—but not to the point of preventing diagnostic comparisons with the other planets. The atmosphere is clearly much richer than Earth's in noble gases, for ^{36}Ar by a factor of ~ 70. There is a pronounced solar-like signature in the elemental Xe/Kr and Ar/Kr ratios and in the Ne isotopes, and nominally in Ar as well, although measurement errors are large. These clues suggest that the noble gas inventories on Venus are dominated by abundant, lightly fractionated remnants of its primary atmosphere. Assuming that it began at the same time as on Earth (and Mars) and was driven by the same solar EUV flux history, an episode of partial hydrodynamic escape of a primary Venus atmosphere that originally was isotopically solar and (except for Ne) elementally near-solar in composition fits the limited isotopic data we have very well, as shown in Figure 7. Contributions by planetary degassing, say at the level experienced by Earth, would constitute only minor fractions of the total noble gas inventories and are not discernable. However, this simple view of Venus's atmospheric evolution does require a much lower initial Ne abundance, by a factor of ~ 100, than would have been present in an atmosphere of purely solar elemental composition. It is partly for this reason that origin of the primary atmosphere from accretion of icy planetesimals carrying noble gases occluded from the nebula at low temperatures—a process likely to discriminate against Ne—is attractive.

MARS An evolution very similar to Earth's, of primary atmospheric and degassed secondary noble gases to present-day compositions by hydrodynamic escape, is indicated for Mars, although derived time scales for outgassing are somewhat longer. The calculated elemental fractionation pattern of degassed solar-like Kr, Ne, and Ar is essentially identical to that deduced for Earth. Retention of Xe within the planet is again required.

However, there are two noteworthy differences between Earth and Mars.

The SNC-derived isotopic composition of contemporary Martian atmospheric Xe (Swindle et al 1986) points to a primordial atmosphere dominated by Xe which was not isotopically solar, as on Earth, but instead resembled that found in the CI carbonaceous chondrites (Figure 5). This observation implies a major CI-like accretional component in Mars. The second difference is the inability of the basic escape model to replicate simultaneously the isotopic compositions of present-day Ar and Ne; any simple escape episode that generates the uniquely low Martian $^{36}Ar/^{38}Ar$ ratio (Figure 3) overfractionates $^{20}Ne/^{22}Ne$, as shown in Figure 7. Since the composition of Ne in the tenuous Martian atmosphere is extraordinarily sensitive to perturbation by external Ne components, this apparent anomaly could be due to post-escape elevation of the $^{20}Ne/^{22}Ne$ ratio by the addition of modest amounts of solar wind gases, via either direct capture or accretion of solar-wind-rich interplanetary dust, over geologic time.

PLANETARY CARBON AND NITROGEN The planetary model outlined above does not address the evolutionary histories of carbon dioxide and nitrogen. One or the other of these two constituents dominates in all three present-day atmospheres. Among several possibilities for origin of the carbon and nitrogen now present in crustal and atmospheric reservoirs on Earth and Venus is one that can be explored by a straightforward extension of the baseline noble gas model: that they were supplied by a late-accreting veneer of C- and N-rich matter toward the end of hydrodynamic escape.

The general argument for derivation of surficial carbon and nitrogen on Earth and Venus from late accretion of C-N-rich veneers is based on the compositionally similar inventories of these elements on the two bodies, in contrast to evidence for markedly dissimilar planetary histories of magmatism, tectonic activity, and outgassing (Kaula 1990). Origin from infall of chemically uniform materials from a common external source therefore appears more straightforward than degassing from internal sources. The present-day elemental and isotopic distributions of C, N, and the noble gases on the two planets set constraints on the abundances and compositions that all of these elements may have in candidate source materials for the veneer. Among extant meteoritic classes, these chemical constraints are remarkably well met by the enstatite (E) chondrites, and only by them. Inclusion in the baseline model of an E-chondrite veneer, accreted more or less simultaneously by the two planets at a time when waning hydrodynamic escape was just capable of fractionating isotopically light E-chondrite nitrogen to its observed composition in Earth's atmosphere, yields satisfactory matches to planetary C/N ratios and isotopic compositions without seriously perturbing the noble gas histories outlined above (Figure 7). To match estimates of the absolute atmospheric and crustal abundances of C and N requires accretion of relatively modest amounts of veneering material: a total of about 3/4 of a lunar mass, with ~75% of this on Venus. The steep falloff with heliocentric distance suggests an inner solar system provenance for the veneering material.

Late accretion of C-N-rich veneers may or may not have contributed to Mars. As noted above, there are geochemical arguments for a large, CI-like Martian accretional component. This type of material would comprise an independent and potent source of both carbon and nitrogen. A minor variant of the baseline model, in which a small fraction of the initial CI volatile inventory is degassed from the planetary interior along with the noble gases, suggests that abundances of C and N emplaced on the surface of Mars could have been considerably greater than current atmospheric inventories. These excesses imply significant sinks, arguably later sequestering of CO_2 in carbonate rock, and losses of N to space.

METEORITIC NOBLE GASES Isotopic distributions of noble gases in meteorites differ in several respects from both primordial solar and present-day planetary compositions. CI carbonaceous chondrite gases were included above as one component of the initial volatile system on Mars without addressing the question of how they themselves might have evolved, prior to planetary accretion, from the primary "local" and "exotic" solar-system gas reservoirs postulated above. Hydrodynamic escape, operating in an environment quite different from the planetary cases, appears to be a promising way to generate the required fractionations. The meteorite model developed by the author (Pepin 1991) assumes the accumulation of CI parent planetesimals, in his example to about four times the mass of Ceres, in a cold enough ambient environment to permit incorporation of thermally unprocessed or lightly processed extrasolar grains carrying adsorbed solar-like gases in grain-mantle ices, and presolar "exotic" gases in grain interiors. Grain-mantle volatiles are assumed to mobilize in higher temperature regions within the planetesimal, and episodically outgas to the cold surface where species of lower volatility condense. The resulting gaseous atmosphere, containing noble gases and a dominant light carrier gas, is transient—i.e. unstable against hydrodynamic blowoff. This type of hydrodynamic escape can be described by a simple analytic formalism. With the assumption that the atmospheric noble gases were initially isotopically solar but elementally depleted in Ne, the model leads to a residual planetesimal atmosphere in which Xe, Kr, Ar, and Ne are fractionated to isotopic compositions in close agreement with the surface-sited "Q"-component in carbonaceous chondrites. Subsequent adsorption of residual atmospheric gases on planetesimal surface grains containing "exotic" nucleogenetic isotopes accounts for the observed siting and for bulk isotopic compositions. Required distribution coefficients for the adsorption process in this environment can be comparable to laboratory values, alleviating a major difficulty with previous approaches. However this model, while promising, rests on a larger number of assumptions than the planetary models and is correspondingly more speculative.

SUMMARY

We have come full circle, in a very real way, with this recent emphasis on fractionation of atmospheric noble gases by escape from gravitational potential wells. This was just the approach taken by Brown (1949) and Suess (1949) in their pioneering attempts to account for the differences between the elemental abundance patterns of solar and terrestrial noble gases. A quote from a review of this early work by Signer & Suess (1963) makes the point: "All but a fraction of about 10^{-7} of the rare gases left the earth's gravitational field during an early stage of evolution by hydrodynamic outflow without undergoing separation. The rest, however, underwent some process by which a fractionation took place, shifting the abundance ratio (of Ne/Xe) by more than a factor of 10^4. A simple explanation for the fractionation of the rare gases is selective loss from a gravitational field during a limited period of time (Suess 1949)." At that time the theory of mass fractionation in hydrodynamic escape had not been developed, and the appeal was to classical Jeans escape which required the fractionating loss to take place from small preplanetary planetesimals. Nevertheless this perception of evolutionary processing has a distinctly modern ring.

The escape models described in the latter part of this review attempt to synthesize approaches taken in previous efforts to understand the origin of volatile distributions in planetary atmospheres and meteorites, and to place them in a context more fully integrated with observational and theoretical inferences about the solar system's early history. The hypothesis of hydrodynamic escape of massive primary atmospheres clearly places the basic planetary model in Donahue & Pollack's (1983) gas-rich class. But direct conceptual antecedents of the adsorbed components in planetary and planetesimal interiors are found in earlier gas-poor grain-accretion theories, and the meteoritic source invoked for carbon and nitrogen on Earth and Venus is obviously just a restricted and specific version of previous veneer scenarios. So these models involve a combination of familiar primordial components and theories, and processes of adsorption and escape that are likewise well known although the fractionating hydrodynamic loss formalism is a relatively new development.

They differ from most earlier approaches, however, in several respects. The full ranges of observed isotopic variations, from primordial to contemporary compositions on and within present-day planetary and meteoritic reservoirs, are taken into account and attributed to two fractionating processes—adsorption and hydrodynamic escape—operating to varying degrees under individual planetary or planetesimal rather than nebular conditions. Final isotopic distributions therefore can be (and are) specific to a particular body. Primordial volatiles are processed in defined astrophysical environments with characteristics that relate directly to, and can eventually be assessed from, observational records of conditions and processes in solar-type pre-main-sequence stars and their accretion disks. The chronology for the various stages of planetary atmospheric evolution is tied directly to time scales for evolution of surface activity on the young sun, via expressions for the time dependence of the solar EUV flux driving hydrodynamic escape from planets.

It is evident from this brief overview that two new approaches have appeared in efforts to model the origin and evolution of planetary noble gases and other volatiles: an emphasis on considering isotopic as well as elemental mass distributions, and a realization that at least part of the processing of atmospheric volatiles could have occurred very early, on the planets themselves and in an energetic and rapidly evolving astrophysical environment. The study of planetary atmospheres therefore relates in very direct ways to the general theme of early planetary history. One may expect that the evolution of atmospheres on the one hand, and that of the sun, its accretion disk, and the planets born from it on the other, will be linked more and more firmly together as these models and their successors are developed.

ACKNOWLEDGMENTS

This work was supported by NASA Grants NAG 9-60 from the Planetary Materials and Geochemistry Program and NAGW-1371 from the Planetary Atmospheres Program.

Literature Cited

Alaerts, L., Lewis, R. S., Anders, E. 1979. Isotopic anomalies of noble gases in meteorites and their origins—IV. C3 (Ornans) carbonaceous chondrites. *Geochim. Cosmochim. Acta* 43: 1421–32

Anders, E. 1987. Local and exotic components of primitive meteorites, and their origin. *Phil. Trans. R. Soc. London Ser. A* 323: 287–304

Anders, E. 1988. Circumstellar material in meteorites: Noble gases, carbon, and nitrogen. In *Meteorites and the Early Solar System*, ed. J. F. Kerridge, M. S. Matthews, pp. 927–55. Tucson: Univ. Ariz. Press

Anders, E., Owen, T. 1977. Mars and Earth: Origin and abundance of volatiles. *Science* 198: 453–65

Anders, E., Ebihara, M. 1982. Solar-system abundances of the elements. *Geochim. Cosmochim. Acta* 46: 2363–80

Anders, E., Grevesse, N. 1989. Abundances of the elements: Meteoritic and solar. *Geochim. Cosmochim. Acta* 53: 197–214

Atreya, S. K., Pollack, J. B., Matthews, M. S., eds. 1989. *Origin and Evolution of Planetary and Satellite Atmospheres*. Tucson: Univ. Ariz. Press. 881 pp.

Bar-Nun, A., Herman, G., Laufer, D., Rappaport, M. L. 1985. Trapping and release of gases by water ice and implications for icy bodies. *Icarus* 63: 317–32

Benz, W., Cameron, A. G. W. 1991. Terrestrial effects of the giant impact. In *Origin of the Earth*, ed. H. E. Newsom, J. H. Jones, pp. 61–67. New York: Oxford Univ. Press

Bernatowicz, T. J., Fahey, A. J. 1986. Xe isotopic fractionation in a cathodeless glow discharge. *Geochim. Cosmochim. Acta* 50: 445–52

Bernatowicz, T. J., Podosek, F. A. 1986. Adsorption and isotopic fractionation of Xe. *Geochim. Cosmochim. Acta* 50: 1503–7

Bernatowicz, T. J., Hagee, B. E. 1987. Isotopic fractionation of Kr and Xe implanted in solids at very low energies. *Geochim. Cosmochim. Acta* 51: 1599–1611

Bernatowicz, T. J., Podosek, F. A., Honda, M., Kramer, F. E. 1984. The atmospheric inventory of xenon and noble gases in shales: The plastic bag experiment. *J. Geophys. Res.* 89: 4597–4611

Bernatowicz, T. J., Kennedy, B. M., Podosek, F. A. 1985. Xe in glacial ice and the atmospheric inventory of noble gases. *Geochim. Cosmochim. Acta* 49: 2561–64

Bochsler, P., Geiss, J., Kunz, S. 1986. Abundances of carbon, oxygen, and neon in the solar wind during the period from August 1978 to June 1982. *Solar Phys.* 103: 177–201

Bogard, D. D. 1988. On the origin of Venus' atmosphere: Possible contributions from simple component mixtures and fractionated solar wind. *Icarus* 74: 3–20

Boss, A. P., Morfill, G. E., Tscharnuter, W. M. 1989. Models of the formation and evolution of the solar nebula. See Atreya et al 1989, pp. 35–72

Brown, H. 1949. Rare gases and formation of the earth's atmosphere. In *The Atmospheres of the Earth and Planets*, ed. G. P. Kuiper, pp. 258–66. Chicago: Univ. Chicago Press

Cameron, A. G. W. 1978. Physics of the primitive solar nebula and of giant gaseous protoplanets. In *Protostars and Planets*, ed. T. Gehrels, pp. 453–87. Tucson: Univ. Ariz. Press

Cameron, A. G. W. 1982. Elemental and nuclidic abundances in the solar system. In *Essays in Nuclear Astrophysics*, ed. C. Barnes, D. Clayton, D. Schramm, pp. 23–43. Cambridge: Cambridge Univ. Press

Cameron, A. G. W. 1983. Origin of the atmospheres of the terrestrial planets. *Icarus* 56: 195–201

Canalas, R., Alexander, E. C. Jr., Manuel, O. K. 1968. Terrestrial abundance of noble gases. *J. Geophys. Res.* 73: 3331–34

Canuto, V. M., Levine, J. S., Augustsson, T. R., Imhoff, C. L. 1982. UV radiation from the young Sun and oxygen and ozone levels in the prebiological palaeoatmosphere. *Nature* 296: 816–20

Crabb, J., Anders, E. 1981. Noble gases in E-chondrites. *Geochim. Cosmochim. Acta* 45: 2443–64

Donahue, T. M. 1986. Fractionation of noble gases by thermal escape from accreting planetesimals. *Icarus* 66: 195–210

Donahue, T. M., Pollack, J. B. 1983. Origin and evolution of the atmosphere of Venus. In *Venus*, ed. D. Hunten, L. Colin, T. Donahue, V. Moroz, pp. 1003–36. Tucson: Univ. Ariz. Press

Donahue, T. M., Hoffman, J. H., Hodges, R. R. Jr. 1981. Krypton and xenon in the atmosphere of Venus. *Geophys. Res. Lett.* 8: 513–16

Dreibus, G., Wänke, H. 1985. Mars, a volatile-rich planet. *Meteoritics* 20: 367–81

Dreibus, G., Wänke, H. 1987. Volatiles on Earth and Mars: A comparison. *Icarus* 71: 225–40

Dreibus, G., Wänke, H. 1989. Supply and loss of volatile constituents during the accretion of terrestrial planets. See Atreya et al 1989, pp. 268–88

Dziczkaniec, M., Lumpkin, G., Donohoe, K., Chang, S. 1981. Plasma synthesis of carbonaceous material with noble gas tracers. *Lunar Planet. Sci. XII*, pp. 246–48. Houston: Lunar Planet. Inst.

Eberhardt, P., Geiss, J., Graf, H., Grögler, N., Mendia, M. D., et al. 1972. Trapped solar wind noble gases in Apollo 12 lunar fines 12001 and Apollo 11 breccia 10046. *Proc. Lunar Sci. Conf., 3rd*, pp. 1821–56. New York: Pergamon

Fanale, F. P. 1976. Martian volatiles: Their degassing history and geochemical fate. *Icarus* 28: 179–202

Fanale, F. P., Cannon, W. A. 1971. Physical adsorption of rare gas on terrigenous sediments. *Earth Planet. Sci. Lett.* 11: 362–68

Fanale, F. P., Cannon, W. A. 1972. Origin of planetary primordial rare gas: The possible role of adsorption. *Geochim. Cosmochim. Acta* 36: 319–28

Fanale, F. P., Jakosky, B. M. 1982. Regolith-atmosphere exchange of water and carbon dioxide on Mars: Effects on atmospheric history and climate change. *Planet. Space Sci.* 30: 819–31

Fanale, F. P., Salvail, J. R., Banerdt, W. B., Saunders, R. S. 1982. Mars: The regolith-atmosphere-cap system and climate change. *Icarus* 50: 381–407

Fanale, F. P., Pollack, J. B., Carr, M. H., Pepin, R. O., Postawko, S. E. 1992. Mars: Epochal climate change and volatile history. In *Mars*, ed. H. H. Kieffer, B. M. Jakosky, M. S. Matthews. Tucson: Univ. Ariz. Press. In press

Feigelson, E. D., Kriss, G. A. 1989. Soft X-ray observations of pre-main-sequence stars in the Chamaeleon dark cloud. *Astrophys. J.* 338: 262–76

Frick, U., Mack, R., Chang, S. 1979. Noble gas trapping and fractionation during synthesis of carbonaceous matter. *Proc. Lunar Planet. Sci. Conf., 10th*, pp. 1961–73. New York: Pergamon

Frick, U., Becker, R. H., Pepin, R. O. 1988. Solar wind record in the lunar regolith: Nitrogen and noble gases. *Proc. Lunar Planet. Sci. Conf., 18th*, pp. 87–120. Cambridge: Univ. Press, Houston: Lunar Planet. Inst.

Geiss, J. 1982. Processes affecting abundances in the solar wind. *Space Sci. Rev.* 33: 201–17

Geiss, J., Bochsler, P. 1985. Ion composition in the solar wind in relation to solar abundances. In *Proc. Int. Conf. on Isotope Ratios in the Solar System, Paris 1984.* Paris: CNES

Goettel, K. A., Eggert, J. H., Silvera, I. F., Moss, W. C. 1989. Optical evidence for the metallization of xenon at 132(5) GPa. *Phys. Rev. Lett.* 62: 665–68

Hayashi, C., Nakazawa, K., Nakagawa, Y. 1985. Formation of the solar system. In *Protostars and Planets II*, ed. D. C. Black, M. S. Matthews, pp. 1100–53. Tucson: Univ. Ariz. Press

Hintenberger, H., Weber, H. W., Schultz, L. 1974. Solar, spallogenic, and radiogenic rare gases in Apollo 17 soils and breccias. *Proc. Lunar Sci. Conf., 5th*, pp. 2005–22. New York: Pergamon

Hoffman, J. H., Hodges, R. R. Jr., Donahue, T. M., McElroy, M. B. 1980. Composition of the Venus lower atmosphere from the Pioneer Venus mass spectrometer. *J. Geophys. Res.* 85: 7882–90

Hunten, D. M. 1979. Capture of Phobos and Deimos by protoatmospheric drag. *Icarus* 37: 113–23

Hunten, D. M., Pepin, R. O., Walker, J. C. G. 1987. Mass fractionation in hydrodynamic escape. *Icarus* 69: 532–49

Hunten, D. M., Pepin, R. O., Owen, T. C. 1988. Planetary Atmospheres. In *Meteorites and the Early Solar System*, ed. J. F. Kerridge, M. S. Matthews, pp. 565–91. Tucson: Univ. Ariz. Press

Hunten, D. M., Donahue, T. M., Walker, J. C. G., Kasting, J. F. 1989. Escape of atmospheres and loss of water. See Atreya et al 1989, pp. 386–422

Istomin, V. G., Grechnev, K. V., Kochnev, V. A. 1980. Mass spectrometry of the lower atmosphere of Venus: Krypton isotopes and other recent results of the Venera 11 and 12 data processing. In *23rd COSPAR Meeting, Budapest*; also Preprint D-298. Moscow: Space Res. Inst., USSR Acad. Sci.

Jessberger, E. K., Kissel, J., Rahe, J. 1989. The composition of comets. See Atreya et al 1989, pp. 167–91

Kaula, W. M. 1990. Venus: A contrast in evolution to Earth. *Science* 247: 1191–96

Krummenacher, D., Merrihue, C. M., Pepin, R. O., Reynolds, J. H. 1962. Meteoritic krypton and barium versus the general isotopic anomalies in xenon. *Geochim. Cosmochim. Acta* 26: 231–49

Kyser, T. K., Rison, W. 1982. Systematics of rare gas isotopes in basic lavas and ultramafic xenoliths. *J. Geophys. Res.* 87: 5611–30

Lewis, J. S. 1974. The temperature gradient in the solar nebula. *Science* 186: 440–43

Lunine, J. I., Stevenson, D. J. 1985. Thermodynamics of clathrate hydrate at low and high pressures with application to the outer solar system. *Astrophys. J. Suppl. Series* 58: 493–531

Marti, K., Wilkening, L. L., Suess, H. E. 1972. Solar rare gases and the abundances of the elements. *Astrophys. J.* 173: 445–50

Matsuda, J., Lewis, R. S., Takahashi, H., Anders, E. 1980. Isotopic anomalies of noble gases in meteorites and their origins—VII. C3V carbonaceous chondrites. *Geochim. Cosmochim. Acta* 44: 1861–74

Mazor, E., Heymann, D., Anders, E. 1970. Noble gases in carbonaceous chondrites. *Geochim. Cosmochim. Acta* 34: 781–820

McDonnell, J. A. M., Kissel, J., Grün, E., Grard, R. J. L., Langevin, Y., et al. 1986. Giotto's dust impact detection system didsy and particulate impact analyser PIA: Interim assessment of the dust distribution and properties within the coma. In *20th ESLAB Symposium on the Exploration of Halley's Comet*, Vol. 2, ed. B. Battrick, E. J., Rolfe, R. Reinhard, pp. 25–38. ESA SP-250

McElroy, M. B., Kong, T. Y., Yung, Y. L. 1977. Photochemistry and evolution of Mars' atmosphere: A Viking perspective. *J. Geophys. Res.* 82: 4379–88

McElroy, M. B., Prather, M. J. 1981. Noble

gases in the terrestrial planets. *Nature* 293: 535–39

Mewaldt, R. A., Spalding, J. D., Stone, E. C. 1984. A high-resolution study of the isotopes of solar flare nuclei. *Astrophys. J.* 280: 892–901

Niemeyer, S., Zaikowski, A. 1980. I-Xe age and trapped Xe components of the Murray (C-2) chondrite. *Earth Planet. Sci. Lett.* 48: 335–47

Niemeyer, S., Marti, K. 1981. Noble gas trapping by laboratory carbon condensates. *Proc. Lunar Planet. Sci. Conf., 12th*, pp. 1177–88. New York: Pergamon

Nier, A. O., McElroy, M. B. 1977. Composition and structure of Mars' upper atmosphere: Results from the neutral mass spectrometers on Viking 1 and 2. *J. Geophys. Res.* 82: 4341–49

Owen, T. 1986. Update of the Anders-Owen model for martian volatiles. In *Workshop on the Evolution of the Martian Atmosphere*, LPI Tech. Report 86-07, ed. M. Carr, P. James, C. Leovy, R. Pepin, pp. 31–32. Houston: Lunar Planet. Inst.

Owen, T. 1987. Could icy impacts reconcile Venus with Earth and Mars? In *Mars: Evolution of its Climate and Atmosphere*, LPI Tech. Report 87-01. ed. V. Baker, M. Carr, F. Fanale, R. Greeley, R. Haberle, C. Leovy, T. Maxwell, p. 91. Houston: Lunar Planet. Inst.

Ozima, M., Nakazawa, K. 1980. Origin of rare gases in the Earth. *Nature* 284: 313–16

Ozima, M., Igarashi, G. 1989. Terrestrial noble gases: Constraints and implications on atmospheric evolution. See Atreya et al 1989, pp. 306–27

Pepin, R. O. 1985. Evidence of martian origins. *Nature* 317: 473–75

Pepin, R. O. 1986. Volatile inventory of Mars—II: Primordial sources and fractionating processes. In *Mars: Evolution of its Climate and Atmosphere*, LPI Tech. Report 87-01, ed. V. Baker, M. Carr, F. Fanale, R. Greeley, R. Haberle, C. Leovy, T. Maxwell, pp. 99–101. Houston: Lunar Planet. Inst.

Pepin, R. O. 1987. Volatile inventories of the terrestrial planets. *Rev. Geophys.* 25: 293–96

Pepin, R. O. 1989a. Atmospheric compositions: Key similarities and differences. See Atreya et al 1989, pp. 291–305

Pepin, R. O. 1989b. On the relationship between early solar activity and the evolution of terrestrial planet atmospheres. In *The Formation and Evolution of Planetary Systems*, ed. H. A. Weaver, L. Danly, pp. 55–74. Cambridge: Cambridge Univ. Press

Pepin, R. O. 1990. A model for noble gas fractionation on a carbonaceous chondrite parent body. *Meteoritics* 25: 397–98

Pepin, R. O. 1991. On the origin and early evolution of terrestrial planet atmospheres and meteoritic volatile. *Icarus* 92: 2–79

Pepin, R. O., Phinney, D. 1978. Components of xenon in the solar system. Unpublished preprint

Pepin, R. O., Carr, M. H. 1992. Major issues and outstanding questions. In *Mars*, ed. H. H. Kieffer, B. M. Jakosky, M. S. Matthews. Tucson: Univ. Ariz. Press. In press

Pepin, R. O., Nyquist, L. E., Phinney, D., Black, D. C. 1970. Rare gases in Apollo 11 lunar material. *Proc. Apollo 11 Lunar Sci. Conf.*, pp. 1435–54. New York: Pergamon

Phinney, D. 1972. ^{36}Ar, Kr, and Xe in terrestrial materials. *Earth Planet. Sci. Lett.* 16: 413–20

Podosek, F. A., Honda, M., Ozima, M. 1980. Sedimentary noble gases. *Geochim. Cosmochim. Acta* 44: 1875–84

Podosek, F. A., Bernatowicz, T. J., Kramer, F. E. 1981. Adsorption of xenon and krypton on shales. *Geochim. Cosmochim. Acta* 45: 2401–15

Pollack, J. B., Black, D. C. 1979. Implications of the gas compositional measurements of Pioneer Venus for the origin of planetary atmospheres. *Science* 205: 56–59

Pollack, J. B., Black, D. C. 1982. Noble gases in planetary atmospheres: Implications for the origin and evolution of atmospheres. *Icarus* 51: 169–98

Prinn, R. G., Fegley, B. Jr. 1989. Solar nebula chemistry: Origin of planetary, satellite, and cometary volatiles. See Atreya et al 1989, pp. 78–136

Sasaki, S. 1991. Off-disk penetration of ancient solar wind. *Icarus* 91: 29–38

Sasaki, S., Nakazawa, K. 1988. Origin of isotopic fractionation of terrestrial Xe: hydrodynamic fractionation during escape of the primordial H_2–He atmosphere. *Earth Planet. Sci. Lett.* 89: 323–34

Signer, P., Suess, H. E. 1963. Rare gases in the sun, in the atmosphere, and in meteorites. In *Earth Science and Meteoritics*, ed. J. Geiss, E. D. Goldberg, pp. 241–72. Amsterdam: North-Holland

Simon, T., Herbig, G., Boesgaard, A. M. 1985. The evolution of chromospheric activity and the spin-down of solar-type stars. *Astrophys. J.* 293: 551–74

Srinivasan, B., Gros, J., Anders, E. 1977. Noble gases in separated meteoritic minerals: Murchison (C2), Ornans (C3), Karoonda (C5), and Abee (E4). *J. Geophys. Res.* 82: 762–78

Strom, S. E., Strom, K. M., Edwards, S. 1988. Energetic winds and circumstellar disks associated with low mass young stellar objects. In *NATO Advanced Study Institute: Galactic and Extragalactic Star Formation*, ed. R. Pudritz, p. 53. Dordrecht: Reidel

Strom, K. M., Strom, S. E., Edwards, S., Cabrit, S., Skrutskie, M. F. 1989a. Circumstellar material associated with solar-type pre-main sequence stars: A possible constraint on the timescales for planet-building. *Astron. J.* 97: 1451–70

Strom, S. E., Edwards, S., Strom, K. M. 1989b. Constraints on the properties and environment of primitive solar nebulae from the astrophysical record provided by young stellar objects. In *The Formation and Evolution of Planetary Systems*, ed. H. A. Weaver, L. Danly, pp. 91–109. Cambridge: Cambridge Univ. Press

Strom, S. E., Edwards, S., Skrutskie, M. F. 1992. Evolutionary timescales for circumstellar disks associated with intermediate and solar-type stars. In *Protostars and Planets III*, ed. E. H. Levy, J. Lunine, M. S. Matthews. Tucson: Univ. Ariz. Press. In press

Suess, H. E. 1949. Die Häufigkeit der Edelgase auf der Erde und im Kosmos. *J. Geology* 57: 600–7

Swindle, T. D., Caffee, M. W., Hohenberg, C. M. 1986. Xenon and other noble gases in shergottites. *Geochim. Cosmochim. Acta* 50: 1001–15

von Steiger, R., Geiss, J. 1989. Supply of fractionated gases to the corona. *Astron. Astrophys.* 225: 222–38

von Zahn, U., Kumar, S., Niemann, H., Prinn, R. 1983. Composition of the Venus atmosphere. In *Venus*, ed. D. Hunten, L. Colin, T. Donahue, V. Moroz, pp. 299–430. Tucson: Univ. Ariz. Press

Wacker, J. F. 1989. Laboratory simulation of meteoritic noble gases. III. Sorption of neon, argon, krypton and xenon on carbon: Elemental fractionation. *Geochim. Cosmochim. Acta* 53: 1421–33

Wacker, J. F., Zadnik, M. G., Anders, E. 1985. Laboratory simulation of meteoritic noble gases. I. Sorption of xenon on carbon: Trapping experiments. *Geochim. Cosmochim. Acta* 49: 1035–48

Walker, J. C. G. 1977. *Evolution of the Atmosphere*, 318 pp. New York: Macmillan

Walter, F. M. 1986. X-ray sources in regions of star formation. I. The naked T Tauri stars. *Astrophys. J.* 306: 573–86

Walter, F. M., Barry, D. C. 1991. Pre- and main sequence evolution of solar activity. In *The Sun in Time*, ed. C. P. Sonett. M. S. Giampapa, M. S. Matthews, pp. 633–57. Tucson: Univ. Ariz. Press

Walter, F. M., Brown, A., Mathieu, R. D., Myers, P. C., Vrba, F. J. 1988. X-ray sources in regions of star formation. III. Naked T Tauri stars associated with the Taurus-Auriga complex. *Astron. J.* 96: 297–325

Wetherill, G. W. 1954. Variations in the isotopic abundances of neon and argon extracted from radioactive minerals. *Phys. Rev.* 96: 679–83

Wetherill, G. W. 1981. Solar wind origin of ^{36}Ar on Venus. *Icarus* 46: 70–80

Wetherill, G. W. 1986. Accumulation of the terrestrial planets and implications concerning lunar origin. In *Origin of the Moon*, ed. W. K. Hartmann, R. J. Phillips, G. J. Taylor, pp. 519–50. Houston: Lunar Planet. Inst.

Wetherill, G. W. 1990a. Formation of the Earth. *Annu. Rev. Earth Planet. Sci.* 18: 205–56

Wetherill, G. W. 1990b. Calculation of mass and velocity distributions of terrestrial and lunar impactors by use of theory of planetary accumulation. In *Abstracts for the International Workshop on Meteorite Impact on the Early Earth*, LPI Contr. No. 746, pp. 54–55. Houston: Lunar Planet. Inst.

Wieler, R., Baur, H., Signer, P., Lewis, R. S. 1990. Noble gases in "Phase Q." Further studies on the Allende and Murchison meteorites. *Meteoritics* 25: 420

Wieler, R., Anders, E., Baur, H., Lewis, R. S., Signer, P. 1991. Noble gases in "Phase Q": Closed-system etching of an Allende residue. *Geochim. Cosmochim. Acta* 55: 1709–22

Wyckoff, S., Lindholm, E., Wehinger, P. A., Peterson, B. A., Zucconi, J.-M., Festou, M. C. 1989. The ^{12}C/^{13}C abundance ratio in comet Halley. *Astrophys. J.* 339: 488–500

Yang, J., Anders, E. 1982a. Sorption of noble gases by solids, with reference to meteorites. II. Chromite and carbon. *Geochim. Cosmochim. Acta* 46: 861–75

Yang, J., Anders, E. 1982b. Sorption of noble gases by solids, with reference to meteorites. III. Sulfides, spinels, and other substances; on the origin of planetary gases. *Geochim. Cosmochim. Acta* 46: 877–92

Yang, J., Lewis, R. S., Anders, E. 1982. Sorption of noble gases by solids, with reference to meteorites. I. Magnetite and carbon. *Geochim. Cosmochim. Acta* 46: 841–60

Yung, Y. L., Strobel, D. F., Kong, T. Y., McElroy, M. B. 1977. Photochemistry of nitrogen in the Martian atmosphere. *Icarus* 30: 26–41

Zadnik, M. G., Wacker, J. F., Lewis, R. S.

1985. Laboratory simulation of meteoritic noble gases. II. Sorption of xenon on carbon: Etching and heating experiments. *Geochim. Cosmochim. Acta* 49: 1049–59

Zahnle, K. J., Walker, J. C. G. 1982. The evolution of solar ultraviolet luminosity. *Rev. Geophys. Space Phys.* 20: 280–92

Zahnle, K. J., Kasting, J. F. 1986. Mass fractionation during transonic hydrodynamic escape and implications for loss of water from Venus and Mars. *Icarus* 68: 462–80

Zahnle, K. J., Kasting, J. F., Pollack, J. B. 1990a. Mass fractionation of noble gases in diffusion-limited hydrodynamic hydrogen escape. *Icarus* 84: 502–27

Zahnle, K. J., Pollack, J. B., Kasting, J. F. 1990b. Xenon fractionation in porous planetesimals. *Geochim. Cosmochim. Acta* 54: 2577–86

Reference added in proof

Owen, T., Bar-Nun, A., Kleinfeld, I. 1991. Noble gases in terrestrial planets: Evidence for cometary impacts? In *Comets in the Post-Halley Era*, Vol. 1, ed. R. L. Newburn Jr., M. Neugebauer, J. Rahe, pp. 429–37. Dordrecht: Kluwer

Annu. Rev. Earth Planet. Sci. 1992. 20:431–68

EVOLUTION OF THE SAN ANDREAS FAULT[1]

Robert E. Powell

U.S. Geological Survey, U.S. Courthouse, Spokane, Washington 99201

Ray J. Weldon II

Department of Geological Sciences, University of Oregon, Eugene, Oregon 97403

KEY WORDS: tectonics, California, structure

INTRODUCTION

The modern San Andreas fault forms an 1100-km transform link between a transform-transform-trench triple junction off Cape Mendocino and a system of seafloor spreading axes and transform faults in the Gulf of California (Figure 1). As a transform fault, the San Andreas is commonly represented as the boundary between the North American and Pacific plates, but actually it is only one fault in a complex system of faults (Figure 2) that absorbs the relative motion between the two plates. The transform fault paradigm satisfactorily explains the existence of the San Andreas fault as a consequence of relative plate motion between the two plates, but because most transform faults are relatively simple oceanic structures, the paradigm provides little insight into the intracontinental evolution of the complex San Andreas system. On the other hand, the geologic record along the San Andreas fault system not only tells of its evolution, but also illuminates how the various parts of the system may have functioned in the evolving plate boundary.

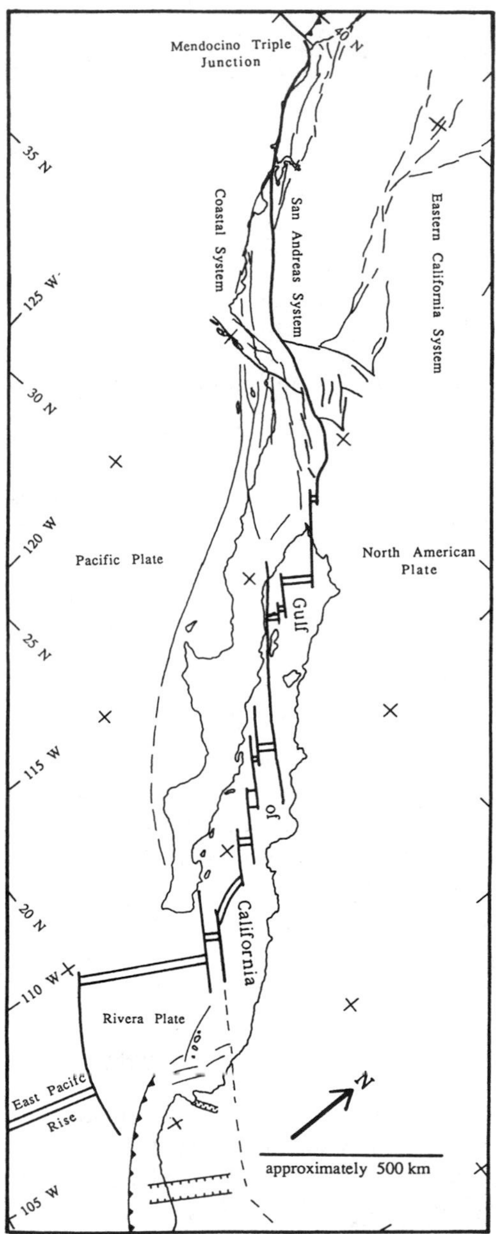

Figure 1 Plate tectonic setting for the modern San Andreas fault. After Humphreys & Weldon (1991).

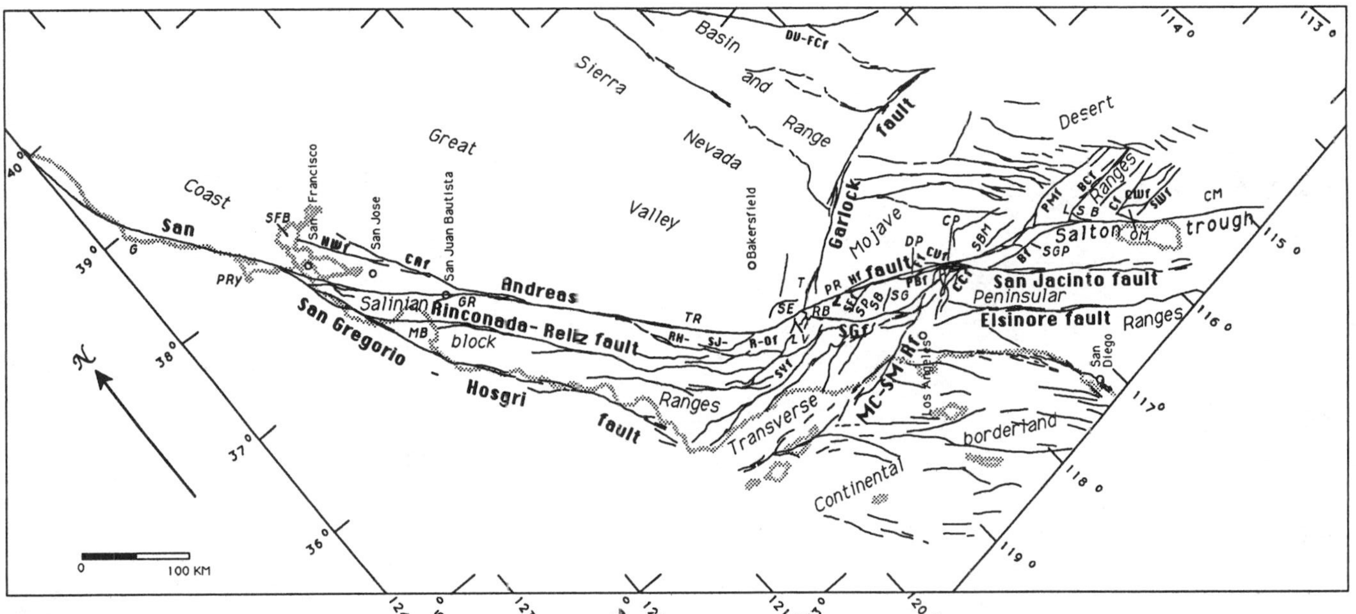

Figure 2 San Andreas fault system in southern and central California (after Powell 1992b). Faults: Bf, Banning fault; BCf, Blue Cut fault; Cf, Chiriaco fault; CAf, Calaveras fault; CCf, Cucamonga fault; CVf, Cajon Valley fault; CWf, Clemens Well fault; DV-FCf, Death Valley–Furnace Creek fault; Ff, Fenner fault; Hf, Hitchbrook fault; HWf, Hayward fault; MC-SM-Rf, Malibu Coast–Santa Monica–Raymond fault; PBf, Punchbowl fault; PMf, Pinto Mountain fault; RH-SJ-R-Of, Red Hills–San Juan–Russell–Ozena fault; SFf, San Francisquito fault; SGf, San Gabriel fault; SWf, Salton Creek fault; SYf, Santa Ynez fault. Physiographic features: *CM*, Chocolate Mountains; *CP*, Cajon Pass; *DP*, Devil's Punchbowl; *G*, Gualala; *GR*, Gabilan Range; *L*, Liebre Mountain; *LSB*, Little San Bernardino Mountains; *LV*, Lockwood Valley; *MB*, Monterey Bay; *OM*, Orocopia Mountains; *PR*, Portal Ridge; *PRy*, Point Reyes; *RB*, Ridge basin; *SB*, Soledad basin; *SBM*, San Bernardino Mountains; *SE*, San Emigdio Mountains; *SFB*, San Francisco Bay; *SG*, San Gabriel Mountains; *SGP*, San Gorgonio Pass; *SP*, Sierra Pelona; *T*, Tehachapi Mountains; *TR*, Temblor Range.

In the past decade, continuing study of displacement, timing of movement, and slip rate on the various faults of the San Andreas system, and of the rate of relative motion between the North American and Pacific plates has yielded a rapidly growing body of evidence that is inconsistent with the widely cited history of the fault system. In this review, we summarize the evidence, published and in press, that is incompatible with the prevailing consensus model and that forms the basis for a new consensus model. We emphasize results from the study of overall displacement, slip rate, and palinspastic restoration of the San Andreas fault system in southern and central California [see papers in Powell et al (1992)]; we also take note of new calculations of the rate of relative motion between the North American and Pacific plates based on seafloor magnetic anomalies in the Gulf of California and on orbital geodetic measurements (DeMets et al 1990, Ward 1990, Ness et al 1991).

The past decade has also seen much research on topics that relate to the history of the San Andreas system. Much recent work has focused on the role of evolving geometric complexities and secondary structures in earthquake segmentation and nucleation (Schwartz & Coppersmith 1984, Sieh et al 1989, Weldon & Springer 1988, Jones 1988), on the strength of the fault in relation to the orientation of regional stress (Zoback et al 1987, Mount & Suppe 1987), on the kinematics of rotated blocks (Luyendyk et al 1985, Hornafius et al 1986, Carter et al 1987), and on the postulated existence of exotic terranes in California (Champion et al 1984, Beck 1986, Butler et al 1991). This research has fostered debate about the age, style, and role of structures associated with the San Andreas system [see papers in Wallace (1990) and Zoback & Lachenbruch (1992)]. While we cannot address the findings of all these studies in this short review, we present a history for the San Andreas system that provides a new perspective and a better foundation from which to address these and other topics.

We begin this review with a history of study of the San Andreas fault system since 1950, where we summarize the findings that culminated in the prevailing model for the fault system's evolution and contemporaneous findings that are incompatible with the model. We then briefly describe the key evidence that provides the impetus and basis for developing a new model for the evolution of the fault system. Finally, we examine the evolving role of the San Andreas fault in the plate boundary from the perspective of this new model.

DEVELOPMENT OF CONCEPTS SINCE 1950

Modern study of the San Andreas fault system began in the early 1950s with the initial documentation of large right-lateral displacement—tens of kilometers on the San Gabriel fault (Crowell 1952) and hundreds of kilometers on the San Andreas fault (Hill & Dibblee 1953). These findings, enhanced by subsequent studies, provided a basis for applying plate tectonic theory to the San Andreas fault system. Combined with the concept that right-lateral shear on the San Andreas system is related to the rifting of Baja California from North America (Wegener 1924, Hamilton 1961, Rusnak et al 1964), Wilson's (1965) proposal that the San Andreas fault is a continental transform quickly led to elaboration of the plate tectonic setting of the San Andreas fault and its role in the Pacific–North American boundary (Larson et al 1968, Moore & Buffington 1968, Atwater 1970, Larson 1972). The transform-fault paradigm gained support as it was tested and debated during the late 1960s and early 1970s (Dickinson & Grantz 1968; Hill 1971, 1974; Dickinson et al 1972; Huffman 1972; Kovach & Nur 1973; Baird et al 1974; Woodburne 1975; Campbell & Yerkes 1976; Matthews 1976).

This scientific colloquy led to additional studies in the late 1970s and, by the early 1980s, four areas of consensus had emerged. (*a*) The Gulf of California has opened 300 km since 4 to 5 Ma at a North American–Pacific plate rate of 5 to 6 cm/yr (Minster & Jordan 1978, Curray & Moore 1984). (*b*) Overall late Cenozoic displacement on the San Andreas fault in central California is ~300 km (Kovach & Nur 1973, Matthews 1976, Nilsen 1984, Ross 1984). (*c*) Late Cenozoic displacement on the San Andreas fault system in southern California is also ~300 km, but that displacement is distributed on three major faults, 240 km on the San Andreas and San Jacinto faults and 60 km on the San Gabriel fault (Crowell 1975, 1981; Ehlig 1981, 1982). (*d*) Late Cenozoic displacement on the San Andreas fault in northern California is ~450 km, which is the sum of displacements of 300 km on the San Andreas fault in central California and up to 150 km on the San Gregorio–Hosgri fault (Graham & Dickinson 1978, Clark et al 1984a).

By 1981, the plate-tectonic paradigm and the geologic history of the San Andreas fault had been integrated into a model that amalgamated the first three consensus items: ~300 km of offset across the San Andreas fault system, divided in southern California into ~240 km across the San Andreas fault proper and ~60 km across the San Gabriel fault, was equated to ~300 km of rifting in the Gulf of California since 5 Ma. Although this consensus transform-fault model incorporates much onland and seafloor data, the model is inconsistent with other key data. Conflicting evidence is found in measurements of (*a*) overall displacement on faults of the San Andreas system in southern California, (*b*) slip rate on faults of the San Andreas system, and (*c*) plate-motion rate between North America and Baja California.

Some measurements of displacement on faults in southern California have proved to be irreconcilable with the widely cited consensus model. Thus, 1981 also saw the elucidation of an alternative model (Powell 1981), in which the San Andreas system has a much longer history than the 4 to 5 m.y. indicated by the consensus model. In this alternative model, ~100 km of right-lateral San Andreas displacement occurred during the early to middle Miocene on a fault that was subsequently disrupted by the San Gabriel fault and the San Andreas fault proper (Figure 3). Remnants of this early San Andreas fault include the Clemens Well, Fenner, and San Francisquito faults in southern California and the San Andreas fault in central and northern California. Displacement on the Clemens Well–Fenner–San Francisquito fault is about 1/3 of the overall displacement on the San Andreas system in southern California.

In the mid-1980s, other investigators documented a displacement of only 150 to 160 km on the San Andreas fault proper in southern California (Matti et al 1985, Frizzell et al 1986, Weldon 1986; see also Dillon 1975). Key evidence for this modest slip is a distinctive Triassic monzogranite that crops out in the San Bernardino Mountains east of the San Andreas fault and on Liebre Mountain west of the fault, but occurs northeast of the Clemens Well–Fenner–San Francisquito fault (Figures 3 and 4). The 150- to 160-km offset of this pluton on the San Andreas proper and the ~100-km displacement on the Clemens Well–Fenner–San Francisquito fault are mutually corroborating in that overall displacement on the San Andreas system as a whole is unchanged. Instead, displacement in southern California is redistributed in space and time within the system by subtracting ~100 km from the San Andreas fault proper and adding ~100 km to the older Clemens Well–Fenner–San Francisquito fault.

Holocene slip rates determined for the San Andreas fault in the 1980s led to further doubts about the consensus model. Sieh & Jahns (1984),

Figure 3 The San Andreas fault zone in California. (*a*) Index map showing the present-day distribution of rocks along the fault. B, Banning block; BL, Ben Lomond block; CH, Chocolate Mountains; CI, Channel Islands; CM, Cargo Muchacho Mountains; CT, Catalina Island; CW, Chuckwalla Mountains; EM, Eagle Mountains; FM, Frazier Mountain; GM, Gila Mountains; GR, Gabilan Range; LM, Liebre Mountain block; LP, La Panza Range; MA, Monte Arido block; MM, Montara Mountain; MP, Mount Pinos; PR, Point Reyes; RC, Rinconada block; SB, San Bernardino Mountains; SG, San Gabriel Mountains; SL, Santa Lucia Range; SLO, San Luis Obispo; SM, Santa Monica Mountains; SY, Santa Ynez Mountains. (*b*) The rocks reconstructed to their position prior to offset across the San Andreas system, ~20 to 22 Ma. Ruled lines show present north, to allow visualization of rotations. The future locations of various faults of the San Andreas system are shown with different symbols. See Powell (1992a) for details of the reconstruction and references for the rotations.

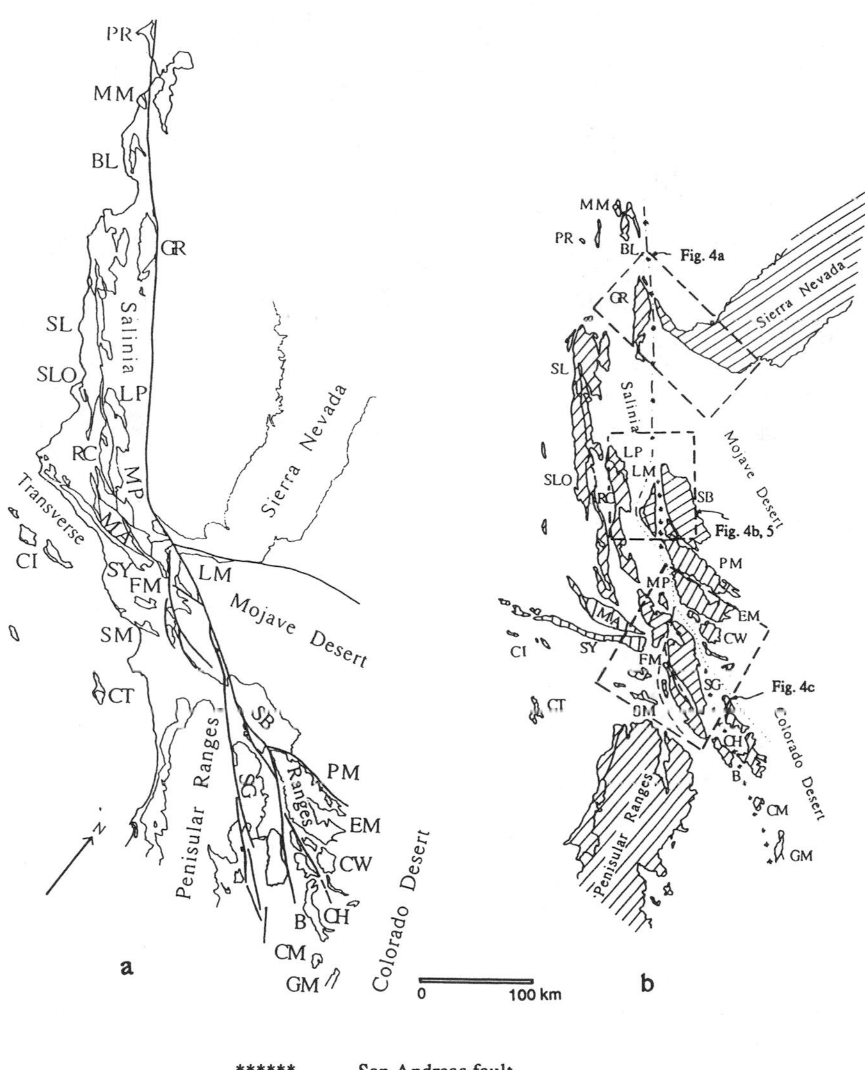

****** San Andreas fault

--------- San Gabriel fault

.... Clemens Well-Fenner-San Francisquito fault

Weldon & Sieh (1985), Perkins et al (1989), and Prentice et al (1991) demonstrated that the fault has slipped at 25 to 35 mm/yr during the Holocene, or about 1/2 to 2/3 the plate rate. Because slower rates have also been established for the San Andreas fault in southern California for the Pleistocene and the Pliocene (Weldon 1986, Weldon et al 1992), the San Andreas fault there would have required 7 to 10 m.y. rather than 4 or 5 m.y. to accumulate the ~240-km offset required by the consensus model. The discrepancy is compounded by the evidence that most of the slip on the San Gabriel fault—which in the model bears 60 of the 300 km of fault displacement that is linked to the opening of the Gulf of California since 4 or 5 Ma—actually occurred between 12 to 10 and 5 Ma (Crowell 1975, Crowell & Link 1982). The Quaternary and Pliocene slip-rate data indicate instead that the San Andreas fault has accumulated a displacement of ~150 km since about 5 Ma, an amount that complements the conclusions drawn from the measurements of overall displacement.

Further conflict between model and data arose when the calculated rate of slip between the Pacific and North American plates was reduced upon reexamination of the magnetic anomalies at the mouth of the Gulf of California (DeMets et al 1987, 1990; Ness et al 1991). The current best estimates range between 46 and 49 mm/yr, rather than the widely cited 56 mm/yr of Minster & Jordan (1978, 1984).

The evidence that the Quaternary and Pliocene slip rate on the San Andreas fault is significantly less than the relative plate motion rate led to discussion of a San Andreas "deficit" and to efforts to find the rest of the plate motion on other faults and to construct quantitative kinematic models for the plate boundary (Bird & Rosenstock 1984, Minster & Jordan 1984, Weldon & Humphreys 1986). Plate boundary deformation not only encompasses a broad zone along the San Andreas fault in the Coast Ranges, Transverse Ranges, Peninsular Ranges, and Salton trough, but also extends well east of the San Andreas in the Mojave Desert and the Basin and Range province and well west of the San Andreas in the western Transverse Ranges, southern Coast Ranges, and continental borderland [see papers and references in Crouch & Bachman (1984), U.S. Geological Survey (1987), Namson & Davis (1988a,b), Dokka & Travis (1990a,b), Lettis et al (1990), Rymer & Ellsworth (1990), Wallace (1990), and a forthcoming Geological Society of America Special Paper on the Seismotectonics of the central California Coast Range]. Structures attributed to shear along the plate boundary include right- and left-slip faults, reverse and normal faults, contractional folds at the surface, and inferred decollements at depth.

In the last five years, space-based geodesy has provided measurements of site-movement along faults of the San Andreas fault system, yielding a present-day plate motion rate of 45 to 50 mm/yr and present-day slip rates that are mostly the same as the Quaternary slip rates (Kroger et al 1987, Ward 1990). These present-day slip rates largely corroborate kinematic models that require the San Andreas fault, as only one fault within a broad, complex plate boundary, to slip at a slower rate than the 60 mm/yr required by the consensus model.

KEY EVIDENCE FOR A NEW MODEL

Displacement on the San Andreas and related late Cenozoic faults is a passive response to the gravitational and thermal driving mechanisms for the fault system. Thus, our understanding of the geologic evolution of the San Andreas fault system depends on our knowledge of displacement and timing of movement on each of the faults in the system and on the kinematic relations among the various faults. Recent efforts to understand the San Andreas system as a whole have emphasized the need to balance the strain budget for the system (Bird & Rosenstock 1984; Weldon & Humphreys 1986; Humphreys & Weldon 1991, 1992) and to maintain mass balance in restoring slip on the system (Powell 1992a). An important consequence of these studies is the recognition that slip rate and displacement measured on the San Andreas fault proper vary along its length depending on the deformation accommodated both in space and time by other faults of the system.

Recent study of the San Andreas fault system has progressed in five complementary ways, leading to reassessment of the magnitude and timing of displacement on various faults of the system. (*a*) Overall displacement has been more completely constrained by reconstruction of paleogeologic domains that existed prior to inception of the fault system. (*b*) Additional incremental displacements have been established by measuring offsets of rocks and structures that formed during the growth of the fault system. (*c*) Timing of fault movement has been further constrained by documenting the ages both of the youngest rocks that are fully displaced and of rocks that have been incrementally displaced. (*d*) The sequence of fault movement has been newly established both by the steps necessary to reconstruct pre–San Andreas paleogeologic domains and by the age of rocks and structures that formed coevally with the fault system. (*e*) Slip rates have been calculated from geologic, geodetic, and seismic measurements of strain.

Magnitude and Timing of Overall Displacements

The regional paleogeologic framework of southern and central California constrains overall reconstruction of rocks along the San Andreas fault system [for additional discussion and references, see Powell (1992a); see also Crowell (1962), Woodburne (1975), Ross (1984), and papers in Crowell (1975), Howell & McDougall (1978), Ernst (1981, 1988), Ingersoll & Woodburne (1982), Ingersoll & Ernst (1987), and Powell et al (1992)]. The oldest part of this framework comprises a northwest-trending belt of cratonic Early and Middle Proterozoic plutonic and metamorphic rocks overlain by metamorphosed Late Proterozoic and Paleozoic strata. This belt is flanked by belts of metamorphosed Mesozoic volcanic and sedimentary rocks, and all the belts were intruded by distinctive suites of Mesozoic plutonic rocks. Fault-bounded terranes were delimited during the late Mesozoic and early Cenozoic: the Pelona, Orocopia, and Rand Schists and related rocks were thrust beneath the cratonic-batholithic terrane; the Franciscan Complex and the Catalina Schist were juxtaposed tectonically outboard of the cratonic-batholithic terrane, and the Coast Range ophiolite and depositionally overlying Upper Jurassic to Lower Cretaceous strata of the lower part of the Great Valley sequence were faulted above the Franciscan-Catalina terrane and, in the southern Coast Ranges and Western Transverse Ranges, against the cratonic-batholithic terrane. Upper Cretaceous through Eocene marine strata crop out along the boundary between the cratonic-batholithic and Franciscan-Catalina terranes in the Coast Ranges, western Transverse Ranges, and Peninsular Ranges. These strata include both those that constitute the upper part of the Great Valley sequence and overlying units and those that depositionally overlap the western margin of the cratonic-batholithic terrane. Along with the Coast Range ophiolite and lower part of the Great Valley sequence, these strata are superposed tectonically over the eastern margin of the Franciscan-Catalina terrane. The Pelona–Orocopia–Rand and Franciscan-Catalina terranes are exposed in antiforms that breach the tectonically superjacent terranes. Upper Oligocene and lower Miocene terrestrial and marine sedimentary and volcanic strata filled basins that formed in the cartonic-batholithic and Coast Range ophiolite–Great Valley terranes and in Upper Cretaceous–Eocene rocks; in the Transverse Ranges and southern Coast Ranges, the Oligocene and Miocene strata were faulted against the Pelona-Orocopia and Franciscan-Catalina terranes along extensional faults associated with the uplift of these metamorphic terranes.

Within the general pattern of these regional paleogeologic terranes, Powell (1992a) has defined four paleogeologic reference domains which consist of specific combinations of rocks and structures from the regional terranes. Three of these reassembled reference domains are shown in Figure 4. The northernmost domain comprises metamorphosed upper Paleozoic or lower Mesozoic strata, Mesozoic gabbroic and granitic plutonic rocks, Pelona Schist and related rocks, Eocene marine strata, and upper Oligocene terrestrial and marine sedimentary and volcanic rocks currently found in the Gabilan Range and vicinity west of the San Andreas fault and in the San Emigdio and Tehachapi Mountains and on Portal Ridge and vicinity east of the San Andreas (Figure 4*a*). This domain is based on well established correlations across the central reach of the San Andreas fault between the Transverse Ranges and San Francisco Bay (Ross 1984; see also Kovach & Nur 1973, Nilsen 1984) and has been displaced by ~295 km along the San Andreas. The youngest rocks in this domain that are displaced ~295 km by the San Andreas fault are the 22- to 24-m.y.-old Pinnacles and Neenach Volcanics (Kovach & Nur 1973, Matthews 1976, Sims 1992).

A second paleogeologic reference domain contains metamorphosed Proterozoic or lower Paleozoic strata, Mesozoic plutonic rocks, and Upper Cretaceous through Eocene marine strata now found in the La Panza Range and Liebre Mountain block west of the San Andreas fault proper and in the San Bernardino Mountains east of the San Andreas (Figure 4*b*). This domain is based in part on recent lithologic correlations (Joseph et al 1982, Frizzell et al 1986) and its reconstruction requires restoring a displacement of 150 to 160 km on the San Andreas fault (Matti et al 1985, Frizzell et al 1986, Matti & Morton 1992, Powell 1992a, Weldon et al 1992) and a combined displacement of 140 to 150 km on the Clemens Well–Fenner–San Francisquito–San Andreas and San Gabriel faults.

A third paleogeologic reference domain consists of Proterozoic gneiss units, Mesozoic plutonic rocks, Eocene marine strata, and upper Oligocene and lower Miocene terrestrial strata currently found west of the San Andreas and San Gabriel faults on Frazier Mountain, Mount Pinos, and in the vicinity; east of the San Andreas and Clemens Well faults in the eastern Orocopia and Chuckwalla Mountains and the vicinity; and between the San Andreas and San Gabriel faults and south of the San Francisquito fault in the Sierra Pelona, Soledad basin, and northern San Gabriel Mountains (Figure 4*c*). Reconstruction of this domain requires a combined displacement of about 305 to 315 km on the Clemens Well–Fenner–San Francisquito, San Gabriel, and San Andreas faults (Powell 1992a). The youngest rocks in this domain that are displaced 305 to 315 km by the San Andreas fault system are: the 20- to 26-m.y.-old volcanic strata (Frizzell & Weigand 1992) of the Plush Ranch Formation of Carman (1964), the Diligencia Formation of Crowell (1975), and the Vasquez Formation; and the ca. 20- to 22-Ma Arikareean(-Hemingfordian?) strata (Woodburne 1975) of the Diligencia Formation and the Tick Canyon Formation of Jahns (1940).

Reconstruction of this third reference domain provides four important constraints on the evolution of the San Andreas fault system (summarized from Powell 1992a). First, it requires that the Clemens Well, Fenner, and

EXPLANATION

SEDIMENTARY AND VOLCANIC ROCKS

Holocene to middle Miocene strata

Early Miocene and Oligocene sedimentary rocks—Includes Plush Ranch (PR), Tick Canyon (TC), Vasquez (V), and Diligencia (D) Formations in Figure 4c

Early Miocene and Oligocene volcanic rocks—Includes Vasquez Formation (V) in Figure 4c, Pinnacles (PV) and Neenach (NV) Volcanics in Figure 4a

Eocene sedimentary rocks

Paleocene sedimentary rocks

Late Cretaceous sedimentary rocks

PLUTONIC AND METAMORPHIC ROCKS

Cretaceous quartz dioritic rocks

Cretaceous gneissic quartz dioritic rocks and Early Cretaceous and older gneiss

Cretaceous and Jurassic granitic rocks, undivided

Cretaceous or Jurassic very coarse-grained monzogranite

Pelona and Orocopia Schists and schist of Sierra de Salinas, undivided

Jurassic and Triassic(?) gabbro and diorite—Jurassic rocks shown in Figure 4a; Triassic(?) rocks shown in Figure 4c

Jurassic to Triassic and late Paleozoic(?) carbonate rocks and schist and quartzite

Triassic monzodioritic rocks—Includes Mount Lowe igneous pluton of Barth & Ehlig (1988)

Paleozoic miogeoclinal strata

Mesozoic and Paleozoic, or Proterozoic metasedimentary and metaigneous suites of Limerock Canyon

Middle Proterozoic anorthosite-syenite complex

Early Proterozoic augen gneiss of Monument Mountain of Powell (1991a)

Early Proterozoic metasedimentary rocks, amphibolite, and laminated granitic orthogneiss—As mapped, also includes Middle Proterozoic gneiss

Early Proterozoic granitic augen gneiss and overlying Early or Middle Proterozoic metasedimentary rocks

San Andreas fault

Punchbowl fault

San Gabriel fault

Clemens Well-Fenner-San Francisquito fault

Figure 4 Detailed reconstructions of key reference domains along the San Andreas fault. After Powell (1992a). (*a*) Gabilan Range–San Emigdio Mountains–Portal Ridge paleogeologic reference domain reassembled by restoring right slip of 295 km on the San Andreas fault proper and left slip of 12 km on the western segment of the Garlock fault. NV, Neenach Volcanics; PV, Pinnacles Volcanics. (*b*) La Panza Range–Liebre Mountain–San Bernardino paleogeologic reference domain reassembled by restoring right slip of 162 km on the San Andreas fault proper, including 42 km on the Punchbowl fault strand; 42 km on the San Gabriel–San Andreas fault; and 110 km on the Clemens Well–Fenner–San Francisquito–San Andreas fault.

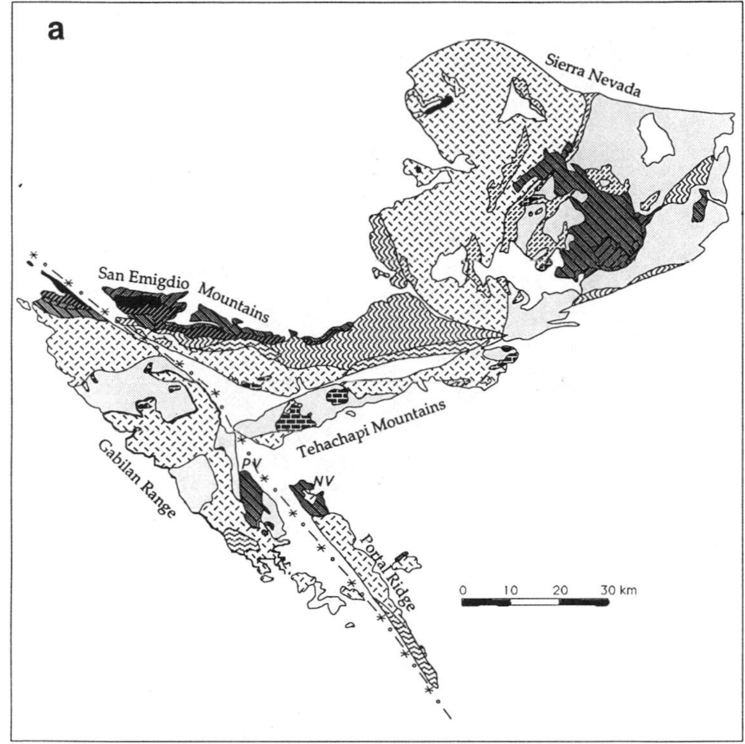

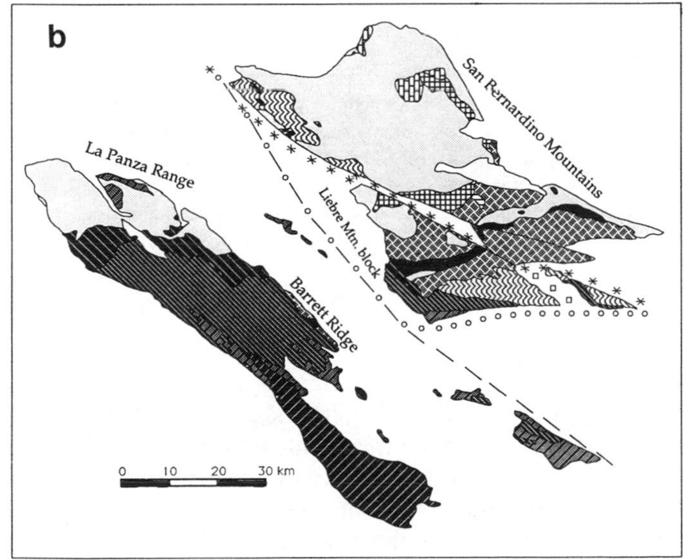

Figure 4—continued

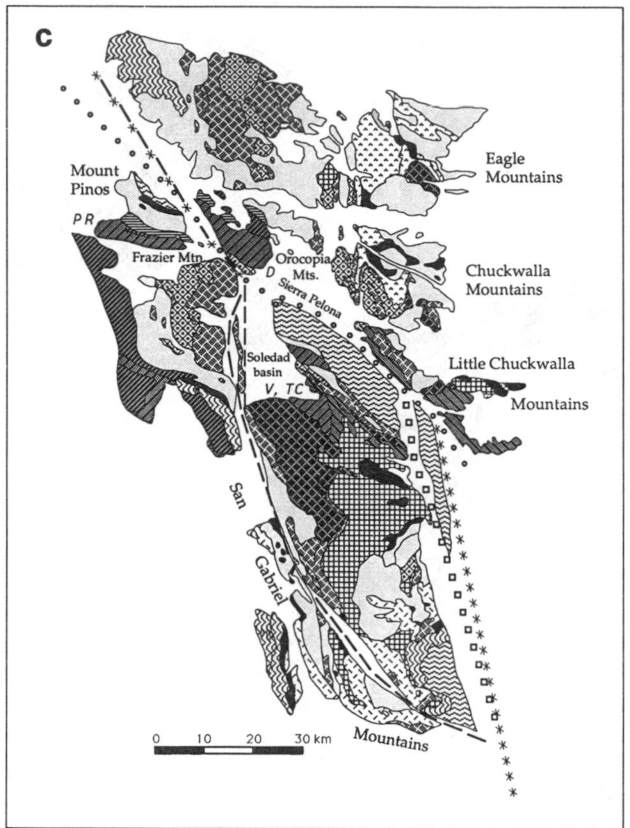

Figure 4c Frazier Mountain–eastern Orocopia Mountains–Soledad basin paleogeologic reference domain reassembled by sequentially restoring left slip of 8 and 11 km on the Salton Creek and Chiriaco faults, respectively, and right slip of 162 km on the San Andreas fault proper, 42 km on the San Gabriel–San Andreas fault, and 110 km on the Clemens Well–Fenner–San Francisquito–San Andreas fault. D, Diligencia Formation of Crowell (1975); PR, Plush Ranch Formation of Carman (1964); TC, Tick Canyon Formation of Jahns (1940); V, Vasquez Formation.

San Francisquito faults were parts of a single, throughgoing strike-slip fault. Second, it indicates that the combined overall displacement on the Clemens Well–Fenner–San Francisquito, San Gabriel, San Andreas, and San Jacinto faults of the San Andreas fault system in southern California is 305 to 315 km. Third, in order to effect the reconstruction, displacement must be restored sequentially, first on the San Jacinto fault and the San Andreas fault proper, then on the San Gabriel fault, and finally on the Clemens Well–Fenner–San Francisquito fault: Any other sequence will not result in the proper reassembly of the rocks of the paleogeologic reference domain. Fourth, in order to allow the Frazier Mountain-Mount Pinos block to be rejuxtaposed against the eastern Orocopia Mountains by restoring slip on the Clemens Well–Fenner–San Francisquito fault, the Frazier Mountain-Mount Pinos block can be restored no more than 42 to 45 km on the San Gabriel fault. Thus, if displacement on the San Andreas fault proper in southern California is ~160 km, then displacement on the Clemens Well–Fenner–San Francisquito fault is about 100 to 110 km.

Simultaneous reconstruction of all the reference domains provides additional constraints (summarized from Powell 1992a). First, while it was active, the Clemens Well–Fenner–San Francisquito fault must have extended northward along the central reach of the San Andreas fault. In other words, during the late early and early middle Miocene, the Clemens Well–Fenner–San Francisquito fault *was* the San Andreas fault in southern California and the Clemens Well–Fenner–San Francisquito–San Andreas fault was the only throughgoing right-lateral strike-slip fault in California. Second, a balanced reconstruction can only be achieved by also restoring displacements on strike-slip faults that were active after about 5 Ma: (*a*) right slip of 24 to 28 km documented on the San Jacinto fault (Sharp 1967, Bartholomew 1970); (*b*) cumulative right slip of 30 to 40 km documented on the northwest-striking faults of the central Mojave Desert (Dokka 1983, Dokka & Travis 1990a, who postulated an additional 40 km of right slip on faults in the eastern Mojave desert); (*c*) left slip of ~12 km postulated for the western segment of the Garlock fault (Powell 1992a), presumably increasing eastward to 56 to 64 km along the central reach of the fault (Smith 1962, Davis & Burchfiel 1973); (*d*) cumulative left slip of 45 to 50 km documented on the east-striking faults of the Transverse Ranges east of the San Andreas fault; and (*e*) as much as ~10 km of left slip on the northeast-trending faults in the southeastern San Gabriel Mountains. If real, the discrepancy in displacement on the western and central parts of the Garlock fault is unresolved. Third, no more than 23 km of displacement on the San Gabriel fault can rejoin the San Andreas fault proper to the southeast and connect with the Salton trough–Gulf of California region (compare Matti & Morton 1992, Weldon et al 1992).

In order to complete a balanced reconstruction of the San Andreas fault system Powell (1992a) suggested that the measured overall displacement of 45 km on the Rinconcada–Reliz fault (Graham 1978) merged northward with the San Gregorio–Hosgri fault, which exhibits an overall displacement of 105 km south of the junction (Nagel & Mullins 1983) and 150 km north of the junction (Clark et al 1984a).

Because paleogeologic patterns in Proterozoic through late Cenozoic rocks in the cratonic-batholithic terrane of southern and central California are fully reassembled by restoration of slip along the faults of the San Andreas system, the balanced reconstruction invalidates displacements that are significantly larger than those used in making the reconstruction (Powell 1992a,b). This restriction applies not only to larger measurements based on cross-fault stratigraphic or lithologic correlations for the faults of the San Andreas system (Ross et al 1973, Graham et al 1989), but also to large displacement postulated on any hypothetical pre-, proto-, or syn-San Andreas faults. Such faults have been postulated to account for perceived differences in overall displacement on the San Andreas north of, versus in and south of, the Transverse Ranges, to accommodate unreconciled plate motion, and to provide a docking structure for rocks that have yielded anomalously low paleomagnetic inclination data (Suppe 1970, Anderson 1971, Dickinson et al 1972, Garfunkel 1973, Beck 1986, Sedlock & Hamilton 1991).

Movement on the Clemens Well–Fenner–San Francisquito–San Andreas fault is bracketed between 20 to 17 and 13 to 12 Ma (Powell 1992a). The older age limit is based on the inferred full displacement of Saucesian (lower Miocene) strata along the central reach of the San Andreas fault (Stanley 1987b), on stratigraphic and sedimentologic evidence in upper Saucesian strata of the Temblor Formation east of the San Andreas fault for a northwest-moving source to the west (O'Day & Sims 1986), and on full displacement of 20- to 26-m.y.-old volcanic rocks and 20- to 22-m.y.-old Arikareean(-Hemingfordian?) (lower lower Miocene) strata in southern California and of 22- to 24-m.y.-old volcanic rocks in central California. The younger age limit is based on the overlap of the Fenner fault by the lower Clarendonian (upper middle Miocene) basal strata of the Punchbowl Formation (Woodburne 1975) and on the overlap of the San Francisquito fault by the Barstovian (middle Miocene) San Francisquito Canyon breccia unit of Sams (1964, Szatai 1961) and the upper Clarendonian (lower upper Miocene) upper part of the Mint Canyon Formation (Woodburne 1975). Based on magnetostratigraphic data, the base of the Punchbowl Formation was deposited at 13 Ma (Liu 1990). Furthermore, clasts in the upper Barstovian lower part of the Mint Canyon Formation that were derived from the north include Pelona Schist from south of the San Francisquito fault and Paleogene marine sandstone from north of the fault (Ehlert 1982), indicating that most movement on the fault was over by the late Barstovian. Evidence for dextral shear in the vicinity of the San Andreas fault between 22 and 26 Ma (Graham et al 1989, Yeats et al 1989) indicates that dextral shear predated the existence of a throughgoing strike-slip fault.

Most of the movement on the San Gabriel, San Gregorio–Hosgri, and Rinconada–Reliz faults occurred between 13 to 11 and 5 Ma (Powell 1992a). The older age limit of 13 to 11 Ma for the San Gabriel fault is indicated by the full displacement of the lower part of the Mint Canyon

Formation, which is late Barstovian to possibly early Clarendonian in age (Woodburne 1975), and the lower two members of the Caliente Formation in Lockwood Valley and the Dry Canyon area of the Cuyama badlands, which range in age from Hemingfordian through early Clarendonian (early and middle Miocene) (Carman 1964, Woodburne 1975). On the basis of fossil and magnetostratigraphic evidence, movement on the San Gabriel fault associated with deposition of the Violin Breccia occurred between about 10 and 6 or 5 Ma (Crowell & Link 1982). After deposition of the Violin Breccia, 1 to 2 km of displacement has occurred on the San Gabriel fault since 3 or 4 Ma. Locally, the San Gabriel fault has been active during the Quaternary (Cotton 1986, Weber 1986). The San Gregorio–Hosgri fault began moving after 13 to 12 Ma and prior to 11 Ma, had accumulated 80 km of displacement by 6 or 7 Ma, and has slipped an additional 70 km since then (Clark et al 1984a). The San Gregorio–Hosgri fault has been active in the Quaternary (Silver & Normark 1978, Weber & Cotton 1981, Hanson & Lettis 1990).

The youngest units that are inferred to have been displaced 150 to 170 km or more on the San Andreas fault include the upper Miocene and Pliocene Imperial Formation in San Gorgonio Pass from the upper Miocene and Pliocene Bouse Formation in the southern Chocolate Mountains and subsurface upper Miocene strata in the Yuma basin (Dillon 1975, Dillon & Ehlig 1992), and the Pliocene Hungry Valley Formation in the Ridge Basin from a crystalline-rock source terrane in the San Bernardino Mountains (Crowell & Link 1982, Matti et al 1985, Frizzell et al 1986, Weldon et al 1992). All of these formations include rocks at least as young as 5 or 6 Ma, indicating that little dextral movement occurred on the Mojave Desert and Salton trough segments of the San Andreas fault prior to about 5 Ma.

During the time the San Andreas fault has been active, most of the deposits which formed along it have been terrigenous sediments, typically limited to small basins. Recently, substantial progress has been made in southern California in characterizing and dating sediments of this type that range in age from late Miocene through Quaternary. For example, there has been the recent recognition and dating of several pull-apart basins along the San Andreas fault that predate the modern San Andreas fault; these include the Mill Creek basin (Sadler et al 1992) and the Devil's Punchbowl (Weldon 1986, Weldon et al 1992). The existence of these basins provides evidence for incipient dextral shear near part of the trace of the modern San Andreas fault prior to 5 Ma and perhaps as early as 13 Ma—that is, during the time that the San Gabriel fault was active.

However, paleogeographic and paleostructural evidence indicates that essentially all of the 150 to 160 km of movement on the Mojave Desert segment of the San Andreas proper in southern California has occurred since about 5 Ma. The Ridge and Santa Ana basins have been displaced ~150 km from their source terranes. The proto–San Bernardino Mountains, which were uplifted on the Squaw Peak thrust from about 10 to 5 Ma, provided the right provenance for most of the sediments deposited in the Ridge basin (south-directed sediment path, Figure 5) (Weldon 1986, Meisling & Weldon 1989, Weldon et al 1992). Moreover, distinctive clasts in sediments in the Santa Ana basin in the San Bernardino Mountains may have been derived from the proto–San Gabriel Mountains (north-directed sediment path, Figure 5) at this time (10 to 5 Ma); alternatively, the clasts may have been shed northeast, orthogonal to the San Andreas since about 5 Ma after the San Andreas had accumulated some offset (Sadler 1992). The Liebre Mountain thrust, which developed as part of the Squaw Peak thrust between 10 and 5 Ma, has since been displaced ~150 km from the Squaw Peak thrust along the San Andreas fault (Weldon 1986, Meisling & Weldon 1989, Weldon et al 1992). The San Gabriel and Cajon Valley faults, which may have been a single fault until about 5 Ma, are now separated by ~150 km along the San Andreas fault (Matti et al 1985, Weldon 1986, Matti & Morton 1992, Weldon et al 1992).

Average Slip Rates

Long-term average slip rates for various faults of the San Andreas system have been calculated from geologic evidence for magnitude and timing of displacement [for additional discussion and references, see Clark et al (1984b), Wesnousky (1986), Brown (1990), and the U.S. Geological Survey (1990)]. Incremental Holocene displacement has been measured at many localities, Pleistocene displacement at a few. Variously well-constrained calculations have yielded different Holocene average dextral slip rates for the various segments of the San Andreas fault proper: between 23 and 35 mm/yr on the Salton trough segment of the San Andreas (Keller et al 1982, Sieh 1986); ~25 mm/yr on the segment of the San Andreas between San Gorgonio and Cajon Passes (Weldon & Sieh 1985, Weldon 1986, Harden & Matti 1989); ~35 mm/yr on the section of the San Andreas bounding the Mojave Desert and Carrizo Plain (Sieh & Jahns 1984, Salyards 1989); between 20 and 25 mm/yr near San Juan Bautista (Perkins et al 1989); at least 9 mm/yr near San Francisco (Brown 1990, Prentice et al 1991); and between 20 and 25 mm/yr north of San Francisco (Prentice et al 1991). Incremental offsets of Pleistocene fanglomerate units have been shown by linking distinctive suites of clasts in fanglomerate units in the Mojave Desert east of the fault to source areas in the San Gabriel Mountains west of the fault that are tapped by major drainages debouching onto the Mojave Desert (Weldon 1986, Weldon et al 1992). These offsets yield a

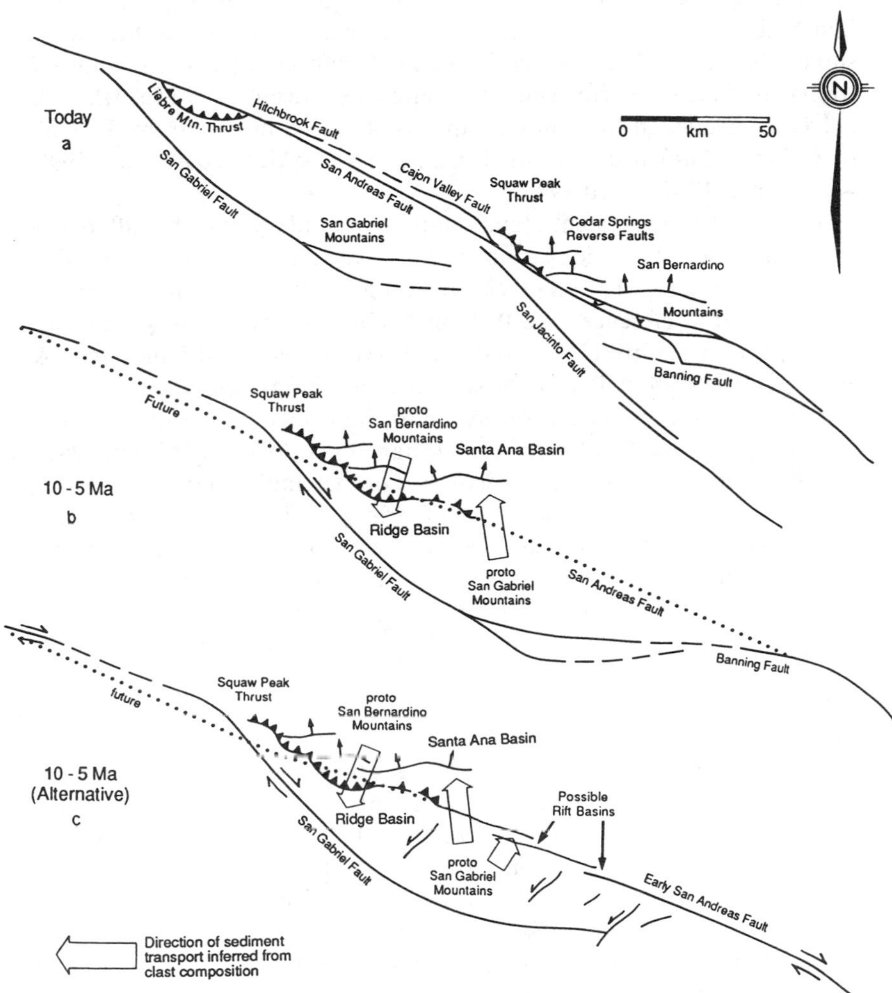

Figure 5 Reconstruction of late Miocene paleogeologic features along the modern San Andreas fault. After Weldon (1986) and Meisling & Weldon (1989). The correlation of rocks, structures, paleogeographic features, and sediment provenance indicates an offset of ~150 km since 5 Ma. Large arrows show direction of sediment transport inferred from clast assemblages. (*a*) Present-day distribution of late Miocene faults and sedimentary basins along the San Andreas fault in the central Transverse Ranges. (*b*) Preferred reconstruction of Squaw Peak–Liebre Mountain thrust fault and late Miocene paleogeography along the San Andreas fault. The San Gabriel fault and Squaw Peak–Liebre Mountain thrust are active. (*c*) Alternative reconstruction of late Miocene faults and paleogeography. The San Gabriel fault, Squaw Peak–Liebre Mountain thrust, and the southern part of the San Andreas fault are active, with pull-apart rift basins developing along the San Andreas.

Pleistocene slip rate of 35 mm/yr for the Mojave Desert segment of the San Andreas fault. Collectively, these data clearly indicate that the rate of slip on the San Andreas proper during the Quaternary has been about 1/2 to 3/4 the Pacific–North American plate rate since 4 or 5 Ma, which is currently calculated to be in the range of 46 to 48 mm/yr from the Gulf of California northward through California (DeMets et al 1987, 1990; Kroger et al 1987; Ward 1990).

The deficit in movement along the plate boundary and the differences in strain rate on the various segments along the San Andreas are approximately balanced by slip rates measured on other structures [for additional discussion and references, see Weldon & Humphreys (1986); Wesnousky (1986); Brown (1990); U.S. Geological Survey (1990); and Humphreys & Weldon (1991, 1992)]. Part of the strain deficit is accommodated on strike-slip faults that intersect the San Andreas fault, including 8 to 17 mm/yr for the right-lateral San Jacinto fault (Sharp 1981, Morton & Matti 1992), 2 to 3 mm/yr for the western reach of the Garlock fault (Clark et al 1984b), 7 mm/yr for the central reach of the Garlock fault during the Holocene and probably since the latest Miocene (Carter 1987), 4 to 12 mm/yr for the right-lateral Calaveras and Hayward faults (Lienkaemper et al 1991), and 5 to 11 mm/yr during the Pleistocene for the right lateral San Gregorio–Hosgri fault (Weber & Cotton 1981, Weber 1983, Hanson & Lettis 1990).

Strain is also accommodated by right-lateral faults that do not intersect the San Andreas, including faults in the north-central Mojave Desert, across which an overall slip rate of 7 mm/yr has been calculated from geodetic measurements (Sauber et al 1986), and the Elsinore fault, for which slip rates in the range 3 to 9 mm/yr have been calculated (Lamar & Rockwell 1986). In addition, Quaternary rates of convergence in the Transverse Ranges range from less than 1 to 6 mm/yr on reverse faults along the south flank of the San Gabriel Mountains (Clark et al 1984b) to 17 to 23 mm/yr distributed chiefly on various reverse faults and contractional folds along the flanks of the Ventura basin, with as much as 10 mm/yr on individual structures (Yeats 1983, U.S. Geological Survey 1987, Rockwell 1988).

In general, average Holocene slip rates for the San Andreas fault system calculated from geologic data agree with modern slip rates calculated from trilateration networks, very long baseline interferometry (VLBI) experiments, and seismic moments [for further discussion and references, see Wesnousky (1986), Kroger et al (1987), Wallace (1990), Ward (1990), and Lisowski et al (1991)].

To date, details about the magnitude and timing of incremental Pliocene displacement on the San Andreas have been inferred only along the Mojave Desert segment (see Weldon et al 1992) where the magnetostratigraphically dated Phelan Peak Formation of Weldon et al (1992) received sediment across the fault from successively more southeastward sources as it was deposited northeast of the San Andreas. These incremental offsets yield a Pliocene slip rate on the San Andreas of 30 to 40 mm/yr—essentially equal to that of the Quaternary (Weldon et al 1992). Other recent calculations yield roughly a 30–40 mm/yr range of Pliocene slip rates in central and northern California—essentially equal to that of the Quaternary rates on the San Andreas in central California and on the San Andreas and Calaveras-Hayward faults in northern California (Fox et al 1985, Sims 1992). Thus, the rate of slip on the San Andreas proper during the Pliocene is also significantly less than the Pacific–North American plate rate, with the deficit balanced by dextral shear distributed over other faults in the system (Weldon & Humphreys 1986).

Uncertainties about the ages of units older than 5 Ma lead to still less well constrained long-term average slip rates for older faults of the San Andreas system. For a displacement of ~45 km in the interval between 12±1 and 5±1 Ma, the range of possible slip rates is 5 to 9 mm/yr for the San Gabriel fault (Powell 1992a) and for the San Andreas fault north of its intersection with the San Gabriel fault (Powell 1992a, Sims 1992). This range of possible rates is 2/5 to 3/4 the rate of 12 mm/yr that can be calculated for a displacement of 60 km accumulating between about 10 and 5 Ma (age range of the Violin Breccia). For a displacement of 100 to 110 km, the range of possible slip rates is 12 to 28 mm/yr for the Clemens Well–Fenner–San Francisquito–San Andreas fault if it was active between 20 to 17 and 13 to 12 Ma (see Powell 1992a).

Time-Displacement History

The history of the San Andreas fault system can be summarized by an offset vs age curve [Figure 6; compare and contrast Hill (1971), Dickinson et al (1972), Graham et al (1989), Sedlock & Hamilton (1991)], from which we draw three conclusions about the time-displacement history of the San Andreas fault. First, the magnitude and timing of movement on the Clemens Well–Fenner–San Francisquito, San Gabriel, San Andreas proper, and San Jacinto faults in and south of the Transverse Ranges are equivalent to the magnitude and timing of movement on the San Andreas fault proper between the Transverse Ranges and San Francisco. Second, there are distinct phases in the time-displacement evolution of the San Andreas fault system: (*a*) ~35 mm/yr since 4 to 5 Ma on the San Andreas fault proper—including the San Jacinto fault south of its junction with the San Andreas and the Calaveras–Hayward fault north of its junction with the San Andreas, (*b*) ~5 to 10 mm/yr between 13 to 11 and 6 to 4 Ma on the San Gabriel–San Andreas fault, and (*c*) ~10 to 30 mm/yr between 20 to 17 and 13 to 12 Ma on the Clemens Well–Fenner–San

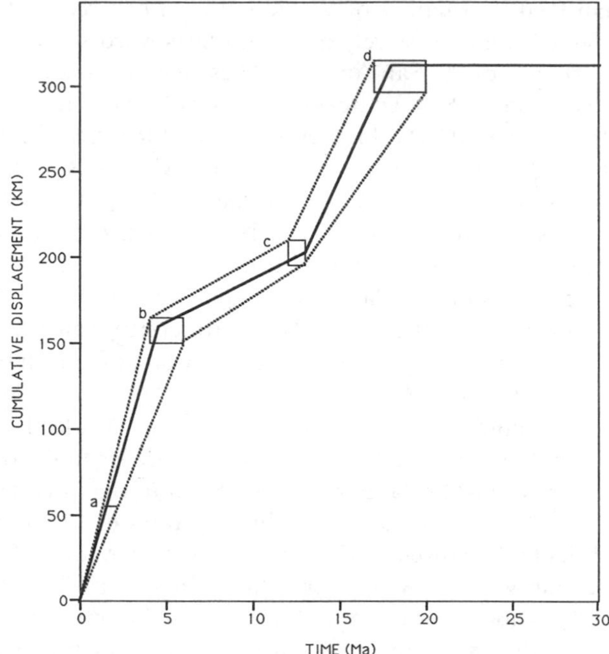

Figure 6 Displacement vs time curve for the San Andreas, San Gabriel, and Clemens Well–Fenner–San Francisquito faults of the San Andreas system. Box *a* is based on a continuous, dated record of offset spanning the Pleistocene, which demonstrates a constant rate of ~35 mm/yr on the San Andreas fault north of the San Gabriel Mountains (Weldon 1986, Weldon et al 1992). Boxes *b* to *d* represent the stratigraphic and structural constraints on magnitude and timing of inception and cessation of successive phases of the fault system, as summarized above. The boxes and curve also incorporate the premises that slip on the San Andreas system was continuous from one phase to next and that slip was transferred completely between phases. The solid curve represents a three-phase history of slip, with 22 mm/yr for the Clemens Well–Fenner–San Francisquito–San Andreas fault between 18 and 13 Ma, 5 mm/yr for the San Gabriel–San Andreas fault between 13 and 5 Ma, and 35 mm/yr for the San Andreas proper since 5 Ma. The uncertainties, however, generate envelopes of possible time-displacement curves.

Francisquito–San Andreas fault. Third, we find no evidence for a gradual or continuous increase in rate with time as hypothesized by early workers (for example, Dickinson et al 1972). The rate increased around 4 to 5 Ma and has changed little.

EVOLVING ROLE IN THE PLATE BOUNDARY

Interaction between the North American and Pacific plates began about 29 Ma when the East Pacific Rise intersected the subduction zone between the Farallon and North American plates at the Pioneer and Mendocino fracture zones (Atwater 1970, 1989). Dextral relative motion between the two plates led to the growth of a transform-fault boundary along which the San Andreas fault formed (Wilson 1965; McKenzie & Morgan 1969; Atwater 1970, 1989; Dickinson & Snyder 1979; Dickinson 1981; Irwin 1990). According to the theoretical geometrical model for transform growth, the transform-ridge-trench triple point that resulted from the intersection of the Mendocino fracture zone and Farallon–North American trench "instantaneously" evolved into a transform-transform-trench triple point and a transform-ridge-trench triple point that migrated apart along a continental-margin transform fault. Currently, however, these plate-boundary triple junctions, located near Cape Mendocino and the mouth of the Gulf of California, are linked by the San Andreas fault and a set of transforms in the gulf, well inland from the expected position along the continental margin.

Most of the models proposed for the plate tectonic evolution of the San Andreas system oversimplify the geologic history of the fault system in a Procrustean effort to "fit" the complex on-land record to that of the seafloor [see Powell (1992b) for discussion and references]. The rich complexity of the on-land record, on the other hand, provides evidence for a temporal and spatial disposition of faulting in the San Andreas system that has yet to be fully utilized in constraining models for the evolution of the plate boundary.

Onshore in California, the earliest phase of dextral shear was transtensional deformation and volcanism between 26 and 22 to 20 Ma (Stanley 1987a, Graham et al 1989, Yeats et al 1989); we find no evidence for a throughgoing strike-slip fault on land during this time interval (Powell 1992a). In southern and central California, the late Oligocene and the early Miocene were characterized by the structural growth of terrestrial and marine sedimentary and volcanic basins in a dextral shear regime. The timing and style of the tectonic transition between this transtensional phase and the emergence of a throughgoing strike-slip fault by late early Miocene (ca. 20 to 18 Ma) warrants further investigation. Either the structures of this early transtensional phase of deformation evolved into a throughgoing strike-slip fault by 20 to 17 Ma, or they provided an inherited framework, part of which the throughgoing fault followed. Relevant to the nature of this transition is a growing body of evidence that calls into question the presumed existence of a transform offshore along the continental margin of central California during this time interval (McCulloch 1987, Tennyson 1989).

Between 20 to 17 Ma and 13 to 12 Ma, the Clemens Well–Fenner–San Francisquito–San Andreas fault accumulated 100 to 110 km of displacement (Powell 1981, 1992a; Sims 1992) and major sinistral deformation occurred along the southern boundary of the Transverse Ranges either as an ancestral strike-slip fault or as a zone of sinistral deflection along the trend of the modern Malibu Coast, Santa Monica, Raymond, Cucamonga, and Banning faults (Campbell & Yerkes 1976, Silver 1982, Powell 1992a). Coeval dextral slip of ~180 km has been postulated along the East Santa Cruz Basin fault in the southern California continental borderland (Howell et al 1974, Howell 1976a) and several tens of kilometers of right-oblique shear have been documented along Walker lane east of the Sierra Nevada (Bohannon 1979, Stewart 1983, Dilles 1989). Within the context of the transform fault paradigm, displacement on the Clemens Well–Fenner–San Francisquito–San Andreas fault probably connected northwestward with the Mendocino triple junction, but its disposition to the southeast is problematic. Hypothetical scenarios for the southeastward disposition of displacement are shown in Figure 7 (summarized from Powell 1992a). In one scenario, dextral displacement on the Clemens Well–Fenner–San Francisquito–San Andreas fault stepped right to the continental margin to connect with the southward-migrating southern triple junction along a fault such as the East Santa Cruz Basin fault (Figure 7*a*). The required large right step would have taken place by means of the well-documented zone of coeval sinistral and extensional deformation along the southern boundary of the Transverse Ranges west of the San Andreas (Campbell & Yerkes 1976). In this scenario, favored by Powell (1992a), right-lateral displacement is transformed into left-oblique extensional tectonism across an extensional wedge perhaps analogous to that between the Death Valley and Garlock faults. In this scenario, extreme extension near the corners of the "Z" would provide a tectonic regime for exhuming, on coeval detachment faults, the deep-seated rocks of the Catalina Schist in the Los Angeles basin area and of Pelona and Orocopia Schists in the restored Sierra Pelona–Chocolate Mountains area.

In another scenario, displacement on the southeastward extension of the Clemens Well fault stepped right across the Chocolate Mountains to a hypothetical fault—near the current trace of the Salton trough segment of the San Andreas proper—which connected southward with the gulf region (Figure 7*b*). At present, there is a lack of evidence for the required displacement in the Salton trough (Powell 1992a), and for widespread extensional volcanism and tectonism in the Gulf of California region prior to about 12 to 14 Ma (Withjack & Jamison 1986, Lonsdale 1989, Stock & Hodges 1989, Humphreys & Weldon 1992)—although the evidence from some localities does not preclude extensional faulting having started prior to 14 Ma (Henry 1989). In the absence of strong supporting evidence, however, it seems unlikely that the Clemens Well–Fenner–San Francisquito–San Andreas fault extended into the gulf region.

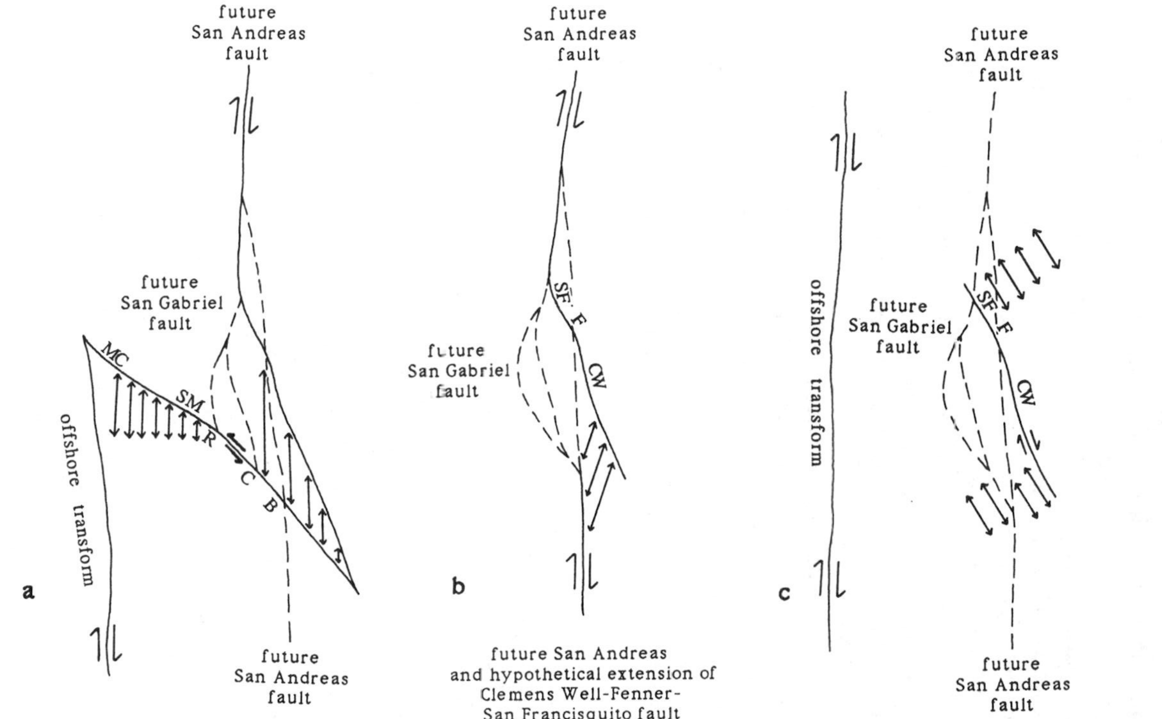

Figure 7 Models for the Clemens Well–Fenner–San Francisquito–San Andreas fault (after Powell 1992a). To the north, the dextral Clemens Well–Fenner–San Francisquito–San Andreas fault probably connected with the Mendocino triple junction. To the south, the disposition of displacement on the Clemens Well–Fenner–San Francisquito–San Andreas can be modeled in three ways: (*a*) right slip stepped west to the continental margin to connect with the southward migrating southern triple junction via an offshore transform fault—possibly the East Santa Cruz Basin fault of Howell et al (1974); (*b*) right slip stepped west to the area of the future Salton trough and Gulf of California; or (*c*) right slip was accommodated by extensional faulting in southeasternmost California and southwesternmost Arizona. B, Banning fault; C, Cucamonga fault; CW, Clemens Well fault; F, Fenner fault; MC, Malibu Coast fault; R, Raymond fault; SF, San Francisquito fault; SM, Santa Monica fault. Double-tipped arrows show extensional domains required by each model.

In a third scenario, the southeastward projection of the Clemens Well–Fenner–San Francisquito–San Andreas fault can be modeled as an intracontinental transform fault which accommodated differential displacement rather than connecting to the southern triple junction (Figure 7*c*). This tectonic setting is partly analogous to that proposed for the Garlock fault or the Las Vegas shear (Davis & Burchfiel 1973, Wernicke et al 1982), and this scenario requires that right-lateral displacement be transformed into Basin and Range type extensional faulting in southeasternmost California and southwesternmost Arizona. The viability of this scenario depends on the as yet undocumented existence of a regional extensional domain that projected south to southwest from the Clemens Well–Fenner–San Francisquito–San Andreas fault.

These possibilities provide testable alternatives in the timing and displacement across structures in southern California and the Gulf of California region, and more work should resolve the evolution of the tectonic setting of the fault zone. Whatever tectonic role the Clemens Well–Fenner–San Francisquito–San Andreas fault played, it accounts for ~100 km of displacement usually erroneously distributed to the modern San Andreas fault; removing this extra displacement provides a later history that is much more reasonable.

Between 13 to 11 and 6 to 4 Ma, the principal active right-lateral faults of the San Andreas system were the San Gabriel, San Gregorio–Hosgri, and Rinconada–Reliz faults and the San Andreas fault northwest of its intersection with the San Gabriel fault. The San Gabriel fault accumulated a displacement of 42 to 45 km that passed northward onto the central reach of the San Andreas fault, the Rinconada–Reliz fault accumulated a displacement of as much as ~45 km that probably passed northward onto the San Gregorio fault, and the San Gregorio–Hosgri fault accumulated a displacement of as much as 150 km that passed northward onto the northern reach of the San Andreas fault. To the north, these faults probably connected with the Mendocino triple junction via the San Andreas. To the south, where these dextral faults intersect the sinistral and reverse faults of the western Transverse Ranges, the disposition of displacements is controversial. The San Gregorio–Hosgri and Rinconada–Reliz faults abut the western Transverse Ranges: We find no evidence that these faults extend through or east of the western Transverse Ranges, and their displacements either are absorbed by an east-trending domain of left-oblique extension in the western Transverse ranges or are stepped right to the continental margin across such a domain.

The San Gabriel fault splits southward into several branches in the western and southern San Gabriel Mountains (Figure 3), where the disposition of its displacement can be modeled in different ways as shown in

Figure 8 [summarized from Powell (1992a), Weldon et al (1992)]. Typically, dextral slip on the San Gabriel fault is linked kinematically to right-oblique extensional tectonism associated with the proto–Gulf of California (Crowell 1981, Lonsdale 1989, Stock & Hodges 1989, Humphreys & Weldon 1991). However, although the displacement of 22 to 23 km on the San Gabriel fault in the San Gabriel Mountains (Ehlig 1975) probably extends west through San Gorgonio Pass (Matti & Morton 1992), where it connects with the Salton trough–Gulf of California region (Figure 8*a*), disposition of the remainder of the displacement on the San Gabriel fault is less clear. Southeastward projection of large dextral displacement along the south flank of the San Gabriel Mountains is not supported by disrupted paleogeologic patterns and is disallowed by the continuous distribution of the 15-m.y.-old Glendora Volcanics across the projected fault path. One alternative possibility is that dextral shear stepped west to the continental margin, causing left-oblique extension in the Los Angeles basin area (Figure 8*b*). Another possibility is that dextral motion stepped east to the Punchbowl–San Andreas fault, causing a transpressional step (Figure 8*c*) that could have uplifted the proto–Transverse Ranges in the late Miocene (Meisling & Weldon 1989, Weldon et al 1992). The best solution may be a combination of all three (Figure 8*d*), in which about half of the displacement stepped west in parallel with that on the San Gregorio–Hosgri and Rinconada–Reliz faults, a small amount stepped east where it was associated with the early growth of pull-apart basins along the Punchbowl–San Andreas fault, and up to half connected with the proto–Gulf of California.

In the central Transverse Ranges, the transition from the Clemens Well–Fenner–San Francisquito fault to the San Gabriel fault, as the principal strike-slip fault of the San Andreas system, was accompanied by an abrupt change in the structural and stratigraphic character of contemporaneous basins along the San Andreas fault (Weldon et al 1992). Basins contemporaneous with the Clemens Well–Fenner–San Francisquito fault were characterized by throughgoing drainage systems, far-traveled clast assemblages, and relatively broad distribution of facies; basins contemporaneous with the San Gabriel fault, on the other hand, were characterized by local relief and clast sources, and by abrupt facies changes. Moreover, the growth of the San Gabriel fault coincided with uplift and erosion of the ancestral San Bernardino Mountains and central San Gabriel Mountains (Figure 5).

Other faults that may have been active between 10 and 5 Ma include the Punchbowl fault, the Garlock fault, and some of the right-lateral faults of the north-central Mojave Desert. The pull-apart origin proposed for the Punchbowl and Mill Creek Formations implies that early right-oblique

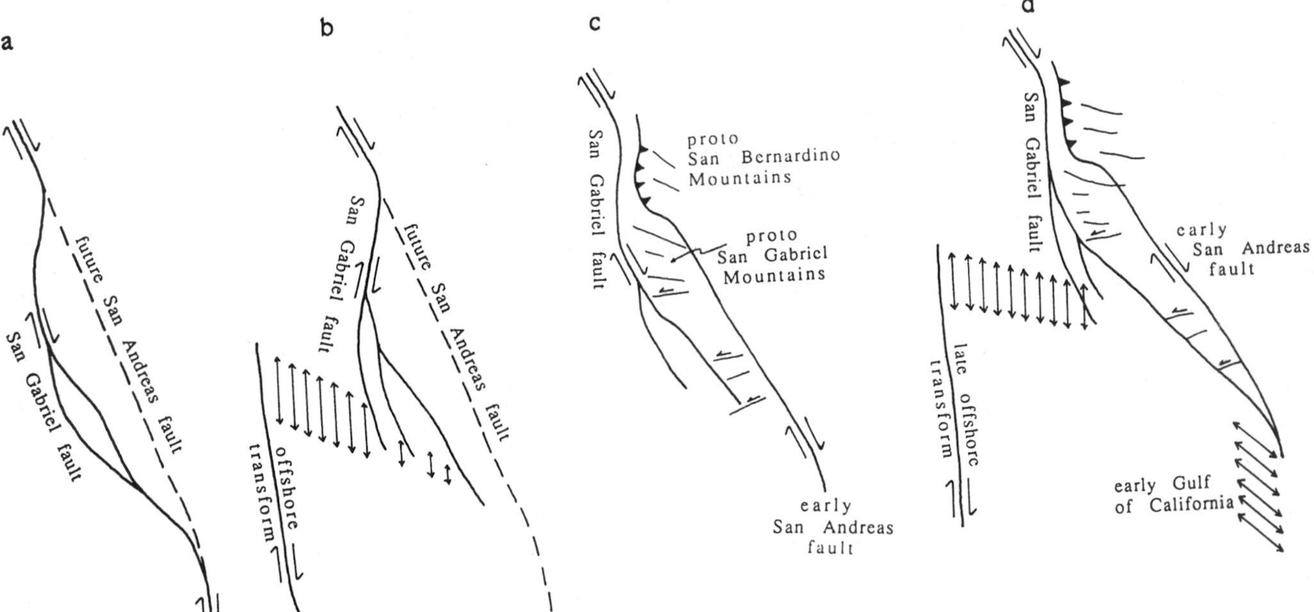

Figure 8 Models for the San Gabriel–San Andreas fault. After Powell (1992a) and Weldon et al (1992). To the north, the dextral San Gabriel–San Andreas fault connected with the Mendocino triple junction. To the south, disposition of displacement is problematic and can be modeled in four ways: (*a*) right slip merged eastward with the future trace of the San Andreas fault; (*b*) right slip stepped west to the continental margin to connect with the southward migrating southern triple junction; (*c*) right slip stepped east to the incipient Punchbowl–San Andreas fault; or (*d*) a combination of (*a*) to (*c*). Double-tipped arrows show extensional domains required by each model: Extensional domains between the San Gabriel fault and the offshore transforms in (*b*) and (*d*) are left-oblique; extensional domain in early Gulf of California region in (*d*) is right-oblique.

shear occurred on the Punchbowl fault or near the trace of the San Andreas fault in the vicinity of the pull-apart basins. As inferred from stratigraphic and paleomagnetic evidence, the Garlock fault may have begun to move at 9 or 10 Ma (Carter 1987, Loomis & Burbank 1988) and some of the right-lateral faults in the Mojave Desert may have begun to move between 10 and 6 Ma (Dokka & Travis 1990a).

Since 4 to 6 Ma, the major active right-lateral fault has been the San Andreas fault proper, which has accumulated a displacement of 150 to 160 km. The San Andreas proper in and south of the Transverse Ranges formed as the link between the San Andreas fault north of its junction with the San Gabriel fault and the opening Gulf of California. The rapid acceleration of slip rate to 35 mm/yr at 4 to 5 Ma signals the establishment (or reestablishment if the Clemens Well–Fenner–San Francisquito fault is considered part of the San Andreas) of the San Andreas fault as the dominant fault in the plate boundary. The jump in the locus of seafloor spreading into the Gulf of California at this time (Atwater 1970, 1989; Lonsdale 1989; Stock & Hodges 1989) is unquestionably related kinematically to acceleration of slip on the San Andreas fault. The apparently constant slip rate since indicates that the basic kinematics of the plate boundary seen today (Figure 9) have persisted for the past 4 to 5 Ma. Humphreys & Weldon (1992) speculated that there was a change in the coastal fault system around 2 Ma; slip that now crosses the northern Baja Peninsula (trans-peninsular faults, Figure 9)—and is kinematically related to the high rate of Quaternary deformation in the western Transverse Ranges and Los Angeles basin region—may have extended farther down the coast and connected to the plate boundary in or south of southern Baja. The magnitude of deformation extending to the San Andreas system from the gulf does not appear to have changed at this time.

Since its inception, the San Andreas fault proper has slipped at less than the plate rate of 48 mm/yr. In southern and central California between Cajon Pass and the Gabilan Range, the San Andreas fault has slipped at 35 mm/yr, whereas north and south of that segment the San Andreas fault proper has slipped at 20 to 25 mm/yr. The plate motion deficit is being accommodated by at least three mechanisms that operated coevally with the Pliocene and Quaternary San Andreas fault (Minster & Jordan 1978; Fox et al 1985; Weldon & Humphreys 1986; U.S. Geological Survey 1990; Ward 1990; Humphreys & Weldon 1991, 1992; Lisowski et al 1991): (*a*) A subordinate system of right- and left-lateral strike-slip faults developed including, from south to north, dextral faults in the Peninsular Ranges and possibly in the continental borderland, sinistral faults in the Transverse Ranges, dextral faults in the Mojave Desert, the sinistral Garlock fault, dextral faults in the Death Valley area, and the dextral Calaveras–Hayward

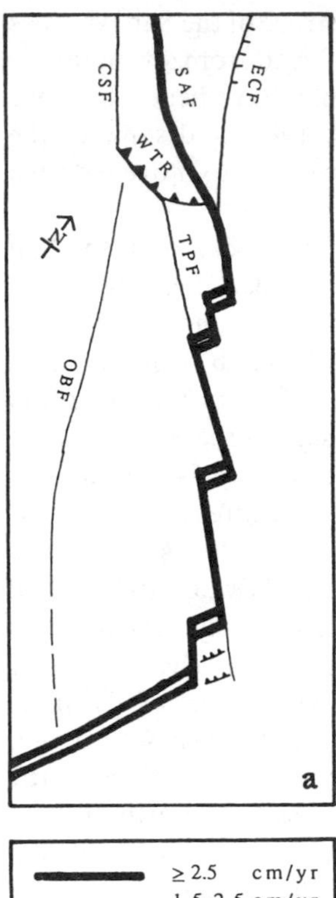

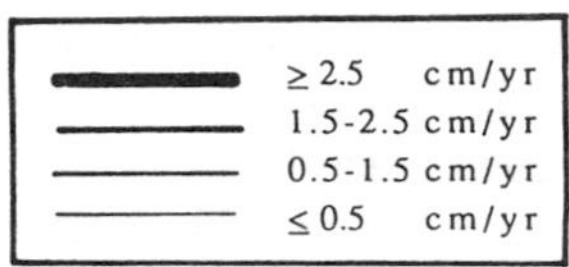

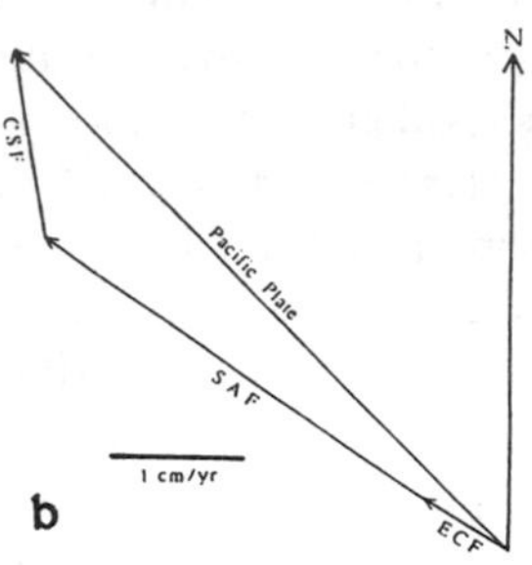

Figure 9 Kinematic relations for the modern San Andreas fault modeled from actual plate boundary structures (Figure 1): (*a*) sketch of the plate boundary fault system (from Humphreys & Weldon 1991); (*b*) vector boundary showing how the slip rates on the San Andreas fault, coastal faults, and eastern California faults sum to the plate total (after Humphreys & Weldon 1992). CSF, coastal system faults; ECF, eastern California faults; OBF, outer borderland faults; SAF, San Andreas fault system; TPF, trans-peninsular faults; WTR, western Transverse Ranges deformation.

fault. (*b*) Oblique contractional structures have developed in the Transverse Ranges (≤ 23 mm/yr) and in the southern Coast Ranges (≤ 5 mm/yr). (*c*) The Basin and Range province underwent right-oblique extension (~8 mm/yr).

In Figure 9, the sketch and vector diagram (modeled from the map in Figure 1) of the faults of the San Andreas system highlight a three-fold division of the plate boundary in California since 4 to 5 Ma. The San Andreas fault provides the most direct connection between the two triple junctions and is most active, but much deformation also occurs on subparallel zones to the east and west (Figures 1 and 9*a*). The San Andreas system and the coastal fault system, particularly near the Transverse Ranges, trend more westerly than the motion across them, so they are transpressional. The eastern California system trends more northerly and is transtensional. The vector diagram (Figure 9*b*) is the sum of the three zones at the latitude of the Transverse Ranges, and shows how the three zones sum to the plate boundary total.

Prior to 4 to 5 Ma, the San Gabriel–San Andreas and Clemens Well–Fenner–San Francisquito–San Andreas faults constituted even smaller fractions of published estimates for total plate motion (Engebretson et al 1985, Stock & Molnar 1988, Atwater 1989) than the modern San Andreas does. The distribution of the rest of the plate motion between about 20 and 5 Ma is not well established. Although it has been shown that Walker lane and perhaps the northern Mojave Desert accommodated right-oblique shear between 20 and 5 Ma (Wernicke et al 1982; Stewart 1983; Dilles 1989; Dokka & Travis 1990a,b), these displacements have yet to be integrated with those of the San Andreas system to balance the strain budget for the pre–5 Ma plate boundary. Inasmuch as the geometrical model for transform growth predicts the existence of a transform fault along the actual edge of the North American plate, one might expect to find an early dextral fault of the San Andreas transform system along the boundary between continental and oceanic crust offshore from southern and central California (Dickinson 1981). Although such a fault would have accommodated part of the plate motion, no transform fault has been identified along the continental margin. On the contrary, the boundary between oceanic and continental crust is interpreted as an inactive east-dipping low-angle fault, along which oceanic crust was subducted beneath the west edge of the North American plate prior to overlap by undeformed Miocene strata (McCulloch 1987, Irwin 1990).

CONCLUSIONS

The San Andreas fault system constitutes a largely intracontinental plate boundary between the North American and Pacific plates. The San Andreas fault itself, with the largest displacement and highest slip rate of any fault in the system, is the principal link between the Mendocino junction and the seafloor system of spreading axes and transforms in the Gulf of California.

Existing transform fault models for the evolution of the San Andreas system are inconsistent with many findings on displacement, timing of movement, and slip rate for the faults of the San Andreas system. Palinspastic reconstruction of paleogeologic features that evolved during and prior to the evolution of the fault system shows that throughgoing intracontinental faults of the San Andreas system emerged at 20 to 17 Ma and evolved in at least three distinct phases. In the earliest phase, between 20 to 17 Ma and 13 to 12 Ma, the Clemens Well–Fenner–San Francisquito–San Andreas fault accumulated a displacement of 100 to 110 km at an average rate in the range of 10 to 30 mm/yr.

In the middle phase, between 13 to 12 Ma and 6 to 4 Ma, the San Gabriel, Rinconada–Reliz, and San Gregorio–Hosgri faults developed southwest of the Clemens Well–Fenner–San Francisquito–San Andreas fault with a more northerly strike than that fault. Northwest of their junctions with the earlier fault, displacement on the new faults passed onto the San Andreas segment of the earlier fault; southeast of its junction with the San Gabriel fault, the Clemens Well–Fenner–San Francisquito segment of the earlier fault was abandoned. The San Gabriel fault accumulated a displacement of 42 to 45 km at a rate in the range of 5 to 10 mm/yr.

In the latest evolutionary phase of the San Andreas system, the modern San Andreas fault emerged at 5 to 4 Ma as a link between the nascent Gulf of California and the Mendocino triple junction. The San Gabriel fault was largely abandoned, but the Rinconada–Reliz and San Gregorio–Hosgri faults continued to accumulate displacement. The slip-rate on the San Andreas fault (20 to 35 mm/yr) during this phase consistently has been less than the plate rate, resulting in the growth of other structures to help accommodate the full plate movement. These other structures include a subordinate system of conjugate strike-slip faults, oblique contractional structures in the Transverse and Coast Ranges, and oblique extensional structures in the Basin and Range province.

Although it provides an appealingly simple plate tectonic framework for the existence of the San Andreas fault system, assigning a transform role to the San Andreas system raises questions about why the San Andreas fault system developed intracontinentally rather than along the boundary between oceanic and continental crust. By considering the transform fault paradigm for the Pacific–North American plate boundary from the perspective of this three-phase evolution for the San Andreas fault system, we can look for ways to further our understanding of why the San Andreas system evolved as a transform.

ACKNOWLEDGMENTS

In studying the evolution and tectonics of the San Andreas fault system, we have benefitted greatly from numerous discussions with Jon Matti and Gene Humphreys. We thank Brian Atwater, Kevin Burke, and Doug Morton for their thorough and thoughtful reviews, and Keith Dodson and David Couzens for their editorial patience. We also thank John Stimac for drafting some of the figures. Most of the work was funded internally or externally by the U.S. Geological Survey.

Literature Cited

Anderson, D. L. 1971. The San Andreas fault. *Sci. Am.* 225: 52–68

Atwater, T. 1970. Implications of plate tectonics for the Cenozoic tectonic evolution of western North America. *Geol. Soc. Am. Bull.* 81: 3513–35

Atwater, T. 1989. Plate tectonic history of the northeast Pacific and western North America. See Winterer et al 1989, pp. 21–72

Baird, A. K., Morton, D. M., Woodford, A. O., Baird, K. W. 1974. Transverse Ranges province: A unique structural-petrochemical belt across the San Andreas fault system. *Geol. Soc. Am. Bull.* 85: 163–74

Barth, A. P., Ehlig, P. L. 1988. Geochemistry and petrogenesis of the marginal zone of the Mount Lowe intrusion, central San Gabriel Mountains, California. *Contrib. Mineral. Petrol.* 100: 192–204

Bartholomew, M. J. 1970. San Jacinto fault zone in the northern Imperial Valley, California. *Geol. Soc. Am. Bull.* 81: 3161–66

Beck, M. E. Jr. 1986. Model for late Mesozoic-early Tertiary tectonics of coastal California and western Mexico and speculations on the origin of the San Andreas fault. *Tectonics* 5: 49–64

Bird, P., Rosenstock, R. W. 1984. Kinematics of present crust and mantle flow in southern California. *Geol. Soc. Am. Bull.* 95: 946–57

Bohannon, R. G. 1979. Strike-slip faults of the Lake Mead region of southern Nevada. In *Cenozoic Paleogeography of the Western United States; Pacific Coast Paleogeography Symposium 3*, ed. J. M. Armentrout, M. R. Cole, H. TerBest, Jr., pp. 129–39. Los Angeles: Pac. Sect., Soc. Econ. Paleontol. Mineral. 335 pp.

Brown, R. D. Jr. 1990. Quaternary deformation. See Wallace 1990, pp. 82–113

Butler, R. F., Dickinson, W. R., Gehrels, G. E. 1991. Paleomagnetism of coastal California and Baja California: Alternatives to large-scale northward transport. *Tectonics* 10: 561–76

Campbell, R. H., Yerkes, R. F. 1976. Cenozoic evolution of the Los Angeles basin area—relation to plate tectonics. See Howell 1976b, pp. 541–58

Carman, M. F. Jr. 1964. Geology of the Lockwood Valley area, Kern and Ventura Counties, California. *Calif. Div. Mines Geol. Spec. Rep. 81.* 62 pp.

Carter, B. A. 1987. Quaternary fault-line features of the central Garlock fault, Kern County, California. In *Geological Society of America Centennial Field Guide Volume 1, Cordilleran Section*, ed. M. L. Hill, pp. 133–35. Boulder, Colo.: Geol. Soc. Am. 490 pp.

Carter, J. N., Luyendyk, B. P., Terres, R. R. 1987. Neogene clockwise tectonic rotation of the eastern Transverse Ranges, California, suggested by paleomagnetic vectors. *Geol. Soc. Am. Bull.* 98: 199–206

Champion, D. E., Howell, D. G., Gromme, C. S. 1984. Paleomagnetic and geologic data indicating 2500 km of northward displacement for the Salinian and related terranes, California. *J. Geophys. Res.* 89: 7736–52

Clark, J. C., Brabb, E. E., Greene, H. G., Ross, D. C. 1984a. Geology of Point Reyes peninsula and implications for San Gregorio fault history. See Crouch & Bachman 1984, pp. 67–85

Clark, M. M., Harms, K. K., Lienkaemper, J. J., Harwood, D. S., Lajoie, K. R. et al. 1984b. Preliminary slip-rate table and map of late-Quaternary faults of California. *US Geol. Surv. Open-File Rep. 84-106.* 13 pp. Scale 1:1,000,000

Cotton, W. R. 1986. Holocene paleoseismology of the San Gabriel fault, Saugus/Castaic area, Los Angeles County, California. See Ehlig 1986, pp. 33–41

Crouch, J. K., Bachman, S. B., eds. 1984. *Tectonics and Sedimentation along the California Margin.* Los Angeles: Pac. Sect., Soc. Econ. Paleontol. & Mineral., vol. 38. 188 pp.

Crowell, J. C. 1952. Probable large lateral displacement on San Gabriel fault, southern California. *Am. Assoc. Petrol. Geol. Bull.* 36: 2026–35

Crowell, J. C. 1962. Displacement along the San Andreas fault, California. *Geol. Soc. Am. Spec. Pap. 71.* 61 pp.

Crowell, J. C., ed. 1975. San Andreas fault in southern California. *Calif. Div. Mines & Geol. Spec. Rep. 118.* 272 pp.

Crowell, J. C. 1981. An outline of the tectonic history of southeastern California. See Ernst 1981, pp. 583–600

Crowell, J. C., Link, M. H., eds. 1982. *Geologic History of Ridge Basin, Southern California.* Los Angeles: Pac. Sect., Soc. Econ. Paleontol. & Mineral. 304 pp.

Curray, J. R., Moore, D. G. 1984. Geologic history of the mouth of the Gulf of California. See Crouch & Bachman 1984, pp. 17–35

Davis, G. A., Burchfiel, B. C. 1973. Garlock fault: An intracontinental transform structure, southern California. *Geol. Soc. Am. Bull.* 84: 1407–22

DeMets, C., Gordon, R. G., Stein, S., Argus, D. F. 1987. A revised estimate of Pacific-

North America motion and implications for western North America plate boundary zone tectonics. *Geophys. Res. Lett.* 14: 911–14

DeMets, C., Gordon, R. G., Argus, D. F., Stein, S. 1990. Current plate motions. *Geophys. J. Int.* 101: 425–78

Dickinson, W. R. 1981. Plate tectonics and the continental margin of California. See Ernst 1981, pp. 1–28

Dickinson, W. R., Cowan, D. S., Schweickert, R. A. 1972. Discussion of "Test of new global tectonics." *Am. Assoc. Petrol. Geol. Bull.* 56: 375–84

Dickinson, W. R., Grantz, A., eds. 1968. *Proc. Conf. Geol. Probl. San Andreas Fault System.* Stanford, Calif.: Stanford Univ. Pub. Geol. Sci., vol. 11. 374 pp.

Dickinson, W. R., Snyder, W. S. 1979. Geometry of triple junctions related to the San Andreas transform. *J. Geophys. Res.* 84: 561–72

Dilles, J. H. 1989. Late Cenozoic normal and strike-slip faults, northern Wassuk Range, Nevada. *Geol. Soc. Am. Abstr. with Programs* 21: 73

Dillon, J. T. 1975. *Geology of the Chocolate and Cargo Muchacho Mountains, southeasternmost California.* PhD thesis. Univ. Calif., Santa Barbara. 405 pp.

Dillon, J. T., Ehlig, P. L. 1992. Displacement on the southern San Andreas fault. See Powell et al 1992. In press

Dokka, R. K. 1983. Displacements on late Cenozoic strike-slip faults of the central Mojave Desert, California. *Geology* 11: 305–8

Dokka, R. K., Travis, C. J. 1990a. Late Cenozoic strike-slip faulting in the Mojave Desert, California. *Tectonics* 9: 311–40

Dokka, R. K., Travis, C. J. 1990b. Role of the eastern California shear zone in accommodating Pacific-North American plate motion. *Geophys. Res. Lett.* 17: 1323–26

Ehlert, K. W. 1982. Basin analysis of the Miocene Mint Canyon Formation, southern California. See Ingersoll & Woodburne 1982, pp. 51–64

Ehlig, P. L. 1975. Basement rocks of the San Gabriel Mountains, south of the San Andreas fault, southern California. See Crowell 1975, pp. 177–86

Ehlig, P. L. 1981. Origin and tectonic history of the basement terrane of the San Gabriel Mountains, central Transverse Ranges. See Ernst 1981, pp. 253–83

Ehlig, P. L. 1982. The Vincent thrust: Its nature, paleogeographic reconstruction across the San Andreas fault and bearing on the evolution of the Transverse Ranges. See Fife & Minch 1982, pp. 370–79

Ehlig, P. L., compiler 1986. *Neotectonics and Faulting in Southern California.* Geol. Soc. Am., Cordilleran Sect. Annu. Meet., 82nd, Guidebook and Vol. 208 pp.

Engebretson, D. C., Cox, A., Gordon, R. G. 1985. Relative motions between oceanic and continental plates in the Pacific basin. *Geol. Soc. Am. Spec. Pap. 206.* 59 pp.

Ernst, W. G., ed. 1981. *The Geotectonic Development of California; Rubey Volume I.* Englewood Cliffs, NJ: Prentice-Hall. 706 pp.

Ernst, W. G., ed. 1988. *Metamorphism and Crustal Evolution of the Western United States; Rubey Volume VII.* Englewood Cliffs, NJ: Prentice-Hall. 1153 pp.

Fife, D. L., Minch, J. A., eds. 1982. *Geology and Mineral Wealth of the California Transverse Ranges; Mason Hill Volume.* Santa Ana, Calif.: South Coast Geol. Soc. Annu. Symp. & Guidebook 10. 699 pp.

Fox, K. F. Jr., Fleck, R. J., Curtis, G. H., Meyer, C. E. 1985. Implications of the northwestwardly younger age of the volcanic rocks of west-central California. *Geol. Soc. Am. Bull.* 96: 647–54

Frizzell, V. A. Jr., Mattinson, J. M., Matti, J. C. 1986. Distinctive Triassic megaporphyritic monzogranite: Evidence for only 160 km offset along the San Andreas fault, southern California. *J. Geophys. Res.* 91: 14,080–88

Frizzell, V. A. Jr., Weigand, P. W. 1992. Whole-rock K-Ar ages and geochemical data from middle Cenozoic volcanic rocks, southern California—A test of correlations across the San Andreas fault. See Powell et al 1992. In press

Garfunkel, Z. 1973. History of the San Andreas fault as a plate boundary. *Geol. Soc. Am. Bull.* 84: 2035–42

Graham, S. A. 1978. Role of Salinian block in evolution of San Andreas fault system, California. *Am. Assoc. Petrol. Geol. Bull.* 62: 2214–31

Graham, S. A. Dickinson, W. R. 1978. Evidence for 115 kilometers of right slip on the San Gregorio-Hosgri fault trend. *Science* 199: 179–81

Graham, S. A., Stanley, R. G., Bent, J. V., Carter, J. B. 1989. Oligocene and Miocene paleogeography of central California and displacement along the San Andreas fault. *Geol. Soc. Am. Bull.* 101: 711–30

Hamilton, W. 1961. Origin of the Gulf of California. *Geol. Soc. Am. Bull.* 72: 1307–18

Hanson, K. L., Lettis, W. R. 1990. Pleistocene slip rates for the San Simeon fault zone based on offset marine terraces and displaced drainages. See Lettis et al 1990, pp. 191–224

Harden, J. W., Matti, J. C. 1989. Holocene and late Pleistocene slip rates on the San

Andreas fault in Yucaipa, California, using displaced alluvial-fan deposits and soil chronology. *Geol. Soc. Am. Bull.* 101: 1107–17

Henry, C. D. 1989. Late Cenozoic Basin and Range structure in western Mexico adjacent to the Gulf of California. *Geol. Soc. Am. Bull.* 101: 1147–56

Hill, M. L. 1971. A test of new global tectonics: Comparison of northeast Pacific and California structures. *Am. Assoc. Petrol. Geol. Bull.* 55: 3–9

Hill, M. L. 1974. Is the San Andreas a transform fault? *Geology* 2: 535–36

Hill, M. L., Dibblee, T. W. Jr. 1953. San Andreas, Garlock, and Big Pine faults; A study of the character, history, and tectonic significance of their displacements. *Geol. Soc. Am. Bull.* 64: 443–58

Hornafius, J. S., Luyendyk, B. P., Terres, R. R., Kamerling, M. J. 1986. Timing and extent of Neogene tectonic rotation in the western Transverse Ranges, California. *Geol. Soc. Am. Bull.* 97: 1476–87

Howell, D. G., Stuart, C. J., Platt, J. P., Hill, D. J. 1974. Possible strike-slip faulting in the southern California borderland. *Geology* 2: 93–98

Howell, D. G. 1976a. A model to accommodate 1000 kilometres of right-slip, Neogene, displacement in the southern California area. See Howell 1976b, pp. 530–40

Howell, D. G., ed. 1976b. *Aspects of the Geologic History of the California Continental Borderland.* Pac. Sect., Am. Assoc. Petrol. Geol. Misc. Publ. 24. 561 pp.

Howell, D. G., McDougall, K. A., eds. 1978. *Mesozoic Paleogeography of the Western United States; Pacific Coast Paleogeography Symp. 2.* Los Angeles: Pac. Sect., Soc. Econ. Paleontol. & Mineral. 573 pp.

Huffman, O. F. 1972. Lateral displacement of upper Miocene rocks and the Neogene history of offset along the San Andreas fault in central California. *Geol. Soc. Am. Bull.* 82: 2913–46

Humphreys, E. D., Weldon, R. J. II 1991. Kinematic constraints on the rifting of Baja California. In *The Gulf and Peninsular Province of the Californias*, ed. J. P. Dauphin, B. R. T. Simoneit, pp. 217–29. Tulsa: Am. Assoc. Petrol. Geol. Mem. 47. 848 pp.

Humphreys, E. D., Weldon, R. J. II 1992. Deformation across southern California: A local determination of Pacific-North America relative plate motion. *J. Geophys. Res.* Submitted

Ingersoll, R. V., Ernst, W. G., eds. 1987. *Cenozoic Basin Development of Coastal California; Rubey Volume VI.* Englewood Cliffs, NJ: Prentice-Hall. 496 pp.

Ingersoll, R. V., Woodburne, M. O., eds. 1982. *Cenozoic Nonmarine Deposits of California and Arizona.* Los Angeles: Pac. Sect., Soc. Econ. Paleontol. & Mineral. 122 pp.

Irwin, W. P. 1990. Geology and plate-tectonic development. See Wallace 1990, pp. 60–80

Jahns, R. H. 1940. Stratigraphy of the easternmost Ventura basin, California, with a description of a new lower Miocene mammalian fauna from the Tick Canyon Formation. *Carnegie Inst. Wash. Publ.* 514: 145–94

Jones, L. M. 1988. Focal mechanisms and the state of stress on the San Andreas fault in southern California. *J. Geophys. Res.* 93: 8869–91

Joseph, S. E., Davis, T. E., Ehlig, P. L. 1982. Strontium isotopic correlation of the La Panza Range granitic rocks with similar rocks in the central and eastern Transverse Ranges. See Fife & Minch 1982, pp. 310–20

Keller, E. A., Bonkowski, M. S., Korsch, R. J., Shlemon, R. J. 1982. Tectonic geomorphology of the San Andreas fault zone in the southern Indio Hills, Coachella Valley, California. *Geol. Soc. Am. Bull.* 93: 46–56

Kovach, R. L., Nur, A., eds. 1973. *Proc. Conf. Tectonic Probl. San Andreas Fault System.* Stanford, Calif.: Stanford Univ. Publ. Geol. Sci., vol. 13. 494 pp.

Kroger, P. M., Lyzenga, G. A., Wallace, K. S., Davidson, J. M. 1987. Tectonic motion in the western United States inferred from very long baseline interferometry measurements, 1980–1986. *J. Geophys. Res.* 92: 14,151–63

Lamar, D. L., Rockwell, T. K. 1986. An overview of the tectonics of the Elsinore fault zone. See Ehlig 1986, pp. 149–58

Larson, R. L. 1972. Bathymetry, magnetic anomalies, and plate tectonic history of the mouth of the Gulf of California. *Geol. Soc. Am. Bull.* 83: 3345–59

Larson, R. L., Menard, H. W., Smith, S. M. 1968. Gulf of California: A result of ocean-floor spreading and transform faulting. *Science* 161: 781–84

Lettis, W. R., Hanson, K. L., Kelson, K. I., Wesling, J. R., eds. 1990. *Neotectonics of South-Central Coastal California.* Friends of the Pleistocene, Pacific Cell, Field Trip Guidebook. San Francisco: Geomatrix. 383 pp.

Lienkaemper, J. J., Borchardt, G., Lisowski, M. 1991. Historic creep rate and potential for seismic slip along the Hayward fault, California. *J. Geophys. Res.* 96: 18,261–83

Lisowski, M., Savage, J. C., Prescott, W. H. 1991. The velocity field along the San Andreas fault in central and southern California. *J. Geophys. Res.* 96: 8369–89

Liu, W. 1990. *Paleomagnetism of Miocene sedimentary rocks in the Transverse Ranges: The implications for tectonic history*. PhD thesis. Calif. Inst. Technol., Pasadena

Lonsdale, P. 1989. Geology and tectonic history of the Gulf of California. See Winterer et al 1989, pp. 499–521

Loomis, D. P., Burbank, D. W. 1988. The stratigraphic evolution of the El Paso basin, southern California: Implications for the Miocene development of the Garlock fault and uplift of the Sierra Nevada. *Geol. Soc. Am. Bull.* 100: 12–28

Luyendyk, B. P., Kamerling, M. J., Terres, R. R., Hornafius, J. S. 1985. Simple shear of southern California during Neogene time suggested by paleomagnetic declinations. *J. Geophys. Res.* 90: 12,454–66

Matthews, V. III 1976. Correlation of Pinnacles and Neenach Volcanic Formations and their bearing on San Andreas fault problem. *Am. Assoc. Petrol. Geol. Bull.* 60: 2128–41

Matti, J. C., Morton, D. M. 1992. Paleogeographic evolution of the San Andreas fault in southern California; A reconstruction based on a new cross-fault correlation. See Powell et al 1992. In press

Matti, J. C., Morton, D. M., Cox, B. F. 1985. Distribution and geologic relations of fault systems in the vicinity of the central Transverse Ranges, southern California. *US Geol. Surv. Open-File Rep. 85-365*. 27 pp. Scale 1:250,000

McCulloch, D. S. 1987. Regional geology and hydrocarbon potential of offshore central California. In *Geology and Resource Potential of the Continental Margin of Western North America and Adjacent Ocean Basins—Beaufort Sea to Baja California*, ed. D. W. Scholl, A. Grantz, J. G. Vedder, pp. 353–401. Houston: Circum-Pacific Counc. Energy Miner. Resour. Earth Sci. Ser., vol. 6. 799 pp.

McKenzie, D. P., Morgan, W. J. 1969. Evolution of triple junctions. *Nature* 224: 125–33

Meisling, K. E., Weldon, R. J. 1989. Late Cenozoic tectonics of the northwestern San Bernardino Mountains, southern California. *Geol. Soc. Am. Bull.* 101. 106–28

Minster, J. B., Jordan, T. H. 1978. Present-day plate motions. *J. Geophys. Res.* 83: 5331–54

Minster, J. B., Jordan, T. H. 1984. Vector constraints on Quaternary deformation of the western United States east and west of the San Andreas fault. See Crouch & Bachman 1984, pp. 1–16

Moore, D. G., Buffington, E. C. 1968. Transform faulting and growth of the Gulf of California since the late Pliocene. *Science* 161: 1238–41

Morton, D. M., Matti, J. C. 1992. Extension and contraction within an evolving divergent strike-slip fault complex: The San Andreas and San Jacinto fault zones at their convergence in southern California. See Powell et al 1992. In press

Mount, V. S., Suppe, J. 1987. State of stress near the San Andreas fault: Implications for wrench tectonics. *Geology* 15: 1143–46

Nagel, D. K., Mullins, H. T. 1983. Late Cenozoic offset and uplift along the San Gregorio fault zone: Central California continental margin. In *Tectonics and Sedimentation along Faults of the San Andreas System*, ed. D. W. Andersen, M. J. Rymer, pp. 91–103. Los Angeles: Pac. Sect., Soc. Econ. Paleontol. & Mineral. 110 pp.

Namson, J., Davis, T. 1988a. Structural transect of the western Transverse Ranges, California: Implications for lithospheric kinematics and seismic risk evaluation. *Geology* 16: 675–79

Namson, J. S., Davis, T. L. 1988b. Seismically active fold and thrust belt in the San Joaquin Valley, central California. *Geol. Soc. Am. Bull.* 100: 257–73

Ness, G. E., Lyle, M. W., Couch, R. W. 1991. Marine magnetic anomalies and oceanic crustal isochrons of the gulf and peninsular province of the Californias. In *The Gulf and Peninsular Province of the Californias*, ed. J. P. Dauphin, B. R. T. Simoneit, pp. 47–69. Tulsa: Am. Assoc. Petrol. Geol. Mem. 47. 848 pp.

Nilsen, T. H. 1984. Offset along the San Andreas fault of Eocene strata from the San Juan Bautista area and western San Emigdio Mountains, California. *Geol. Soc. Am. Bull.* 95: 599–609

O'Day, P. A., Sims, J. D. 1986. Sandstone composition and paleogeography of the Temblor Formation, central California—Evidence for early to middle Miocene right-lateral displacement on the San Andreas fault system. *Geol. Soc. Am. Abstr. with Programs* 18: 165

Perkins, J. A., Sims, J. D., Sturgess, S. S. 1989. Late Holocene movement along the San Andreas fault at Melendy Ranch: Implications for the distribution of fault slip in central California. *J. Geophys. Res.* 94: 10,217–30

Powell, R. E. 1981. *Geology of the crystalline basement complex, eastern Transverse Ranges, southern California: Constraints on regional tectonic interpretation*. PhD thesis. Calif. Inst. Technol., Pasadena. 441 pp.

Powell, R. E. 1992a. Balanced palinspastic reconstruction of pre-late Cenozoic paleogeology, southern California; Geologic and kinematic evolution of the San

Andreas fault system. See Powell et al 1992. In press

Powell, R. E. 1992b. Foreword. See Powell et al 1992. In press

Powell, R. E., Weldon, R. J. II, Matti, J. C., eds. 1992. The San Andreas fault system; Displacement, palinspastic reconstruction, and geologic evolution. *Geol. Soc. Am. Mem. 178*. In press

Prentice, C., Niemi, T. M., Hall, N. T. 1991. Quaternary tectonics of the nothern San Andreas fault, San Francisco Peninsula, Point Reyes, and Point Arena, California. In *Geologic Excursions in Northern California: San Francisco to the Sierra Nevada*, ed. D. Sloan, D. L. Wagner, pp. 25–34. Sacramento: Calif. Div. Mines Geol. Spec. Publ. 109

Rockwell, T. 1988. Neotectonics of the San Cayetano fault, Transverse Ranges, California. *Geol. Soc. Am. Bull.* 100: 500–13

Ross, D. C. 1984. Possible correlations of basement rocks across the San Andreas, San Gregorio-Hosgri, and Rinconada-Reliz-King City faults, California. *US Geol. Surv. Prof. Pap. 1317*. 37 pp.

Ross, D. C., Wentworth, C. M., McKee, E. H. 1973. Cretaceous mafic conglomerate near Gualala offset 350 miles by San Andreas fault from oceanic crustal source near Eagle Rest Peak, California. *US Geol. Surv. J. Res.* 1: 45–52

Rusnak, G. A., Fisher, R. L., Shepard, F. P. 1964. Bathymetry and faults of Gulf of California. In *Marine Geology of the Gulf of California*, ed. T. H. Van Andel, G. G. Shor, Jr., pp. 59–75. Tulsa: Am. Assoc. Petrol. Geol. Mem. 3. 408 pp.

Rymer, M. J., Ellsworth, W. L., eds. 1990. The Coalinga, California, earthquake of May 2, 1983. *US Geol. Surv. Prof. Pap. 1487*. 417 pp.

Sadler, P. M. 1992. Pelona Schist clast assemblages in the Mio-Pliocene Santa Ana Sandstone, central San Bernardino Mountains, southern California. See Powell et al 1992. In press

Sadler, P. M., Demirer, A., West, D., Hillenbrand, J. M. 1992. The Mill Creek basin and fault strands in the San Andreas fault zone south of the San Bernardino Mountains. See Powell et al 1992. In press

Salyards, S. L. 1989. *Dating and characterizing late Holocene earthquakes using paleomagnetics*. PhD thesis. Calif. Inst. Technol., Pasadena. 217 pp.

Sams, R. H. 1964. *Geology of the Charlie Canyon area, northwest Los Angeles County, California*. MA thesis. Univ. Calif., Los Angeles. 101 pp.

Sauber, J., Thatcher, W., Solomon, S. C. 1986. Geodetic measurement of deformation in the central Mojave Desert, California. *J. Geophys. Res.* 91: 12,683–93

Schwartz, D. P., Coppersmith, K. J. 1984. Fault behavior and characteristic earthquakes: Examples from the Wasatch and San Andreas fault zones. *J. Geophys. Res.* 89: 5681–98

Sedlock, R. L., Hamilton, D. H. 1991. Late Cenozoic tectonic evolution of southwestern California. *J. Geophys. Res.* 96: 2325–51

Sharp, R. V. 1967. San Jacinto fault zone in the Peninsular Ranges of southern California. *Geol. Soc. Am. Bull.* 78: 705–29

Sharp, R. V. 1981. Variable rates of late Quaternary strike slip on the San Jacinto fault zone, southern California. *J. Geophys. Res.* 86: 1754–62

Sieh, K. 1986. Slip rate across the San Andreas fault and prehistoric earthquakes at Indio, California. *EOS, Trans. Am. Geophys. Union* 67: 1200 (Abstr.)

Sieh, K. E., Jahns, R. H. 1984. Holocene activity of the San Andreas fault at Wallace Creek, California. *Geol. Soc. Am. Bull.* 95: 883–96

Sieh, K., Stuiver, M., Brillinger, D. 1989. A more precise chronology of earthquakes produced by the San Andreas fault in southern California. *J. Geophys. Res.* 94: 603–23

Silver, E. A., Normark, W. R., eds. 1978. San Gregorio-Hosgri fault zone, California. *Calif. Div. Mines Geol. Spec. Rep. 137*. 56 pp.

Silver, L. T. 1982. Evidence and a model for west-directed early to mid-Cenozoic basement overthrusting in southern California. *Geol. Soc. Am. Abstr. with Programs* 14: 617

Sims, J. D. 1992. Chronology of displacement on San Andreas fault in central California: Evidence from reversed positions of exotic rock bodies near Parkfield, California. See Powell et al 1992. In press

Smith, G. I. 1962. Large lateral displacement on Garlock fault, California, as measured from offset dike swarm. *Am. Assoc. Petrol. Geol. Bull.* 46: 85–104

Stanley, R. G. 1987a. Implications of the northwestwardly younger age of the volcanic rocks of west-central California: Alternative interpretation. *Geol. Soc. Am. Bull.* 98: 612–14

Stanley, R. G. 1987b. New estimates of displacement along the San Andreas fault in central California based on paleobathymetry and paleogeography. *Geology* 15: 171–74

Stewart, J. H. 1983. Extensional tectonics in the Death Valley area, California: Transport of the Panamint Range structural block 80 km northwestward. *Geology* 11: 153–57

Stock, J. M., Hodges, K. V. 1989. Pre-Pliocene extension around the Gulf of California and the transfer of Baja California to the Pacific plate. *Tectonics* 8: 99–115

Stock, J., Molnar, P. 1988. Uncertainties and implications of the Late Cretaceous and Tertiary position of North America relative to the Farallon, Kula, and Pacific plates. *Tectonics* 7: 1339–84

Suppe, J. 1970. Offset of late Mesozoic basement terrains by the San Andreas fault system. *Geol. Soc. Am. Bull.* 81: 3253–57

Szatai, J. E. 1961. *The geology of parts of the Redrock Mountain, Warm Spring, Violin Canyon, and Red Mountain quadrangles, Los Angeles County, California.* PhD thesis. Univ. So. Calif., Los Angeles

Tennyson, M. E. 1989. Pre-transform early Miocene extension in western California. *Geology* 17: 792–96

U.S. Geological Survey 1987. Recent reverse faulting in the Transverse Ranges, California. *US Geol. Surv. Prof. Pap. 1339.* 203 pp.

U.S. Geological Survey 1990. Probabilities of large earthquakes in the San Francisco Bay region, California. *US Geol. Surv. Circ. 1503.* 51 pp.

Wallace, R. E., ed. 1990. The San Andreas fault system, California. *US Geol. Surv. Prof. Pap. 1515.* 283 pp.

Ward, S. N. 1990. Pacific-North America plate motions: New results from very long baseline interferometry. *J. Geophys. Res.* 95: 21,965–81

Weber, F. H. Jr. 1986. Geologic relationships along the San Gabriel fault between Castaic and the San Andreas fault, Kern, Los Angeles, and Ventura Counties, California. See Ehlig 1986, pp. 109–22

Weber, G. E. 1983. Geological investigation of the marine terraces of the San Simeon region and Pleistocene activity on the San Simeon fault zone, San Luis Obispo County, California. *Technical report to US Geol. Surv. under Contract 14-08-00001-18230.* 66 pp.

Weber, G. E., Cotton, W. R. 1981. Geologic investigation of recurrence intervals and recency of faulting along the San Gregorio fault zone, San Mateo County, California. *US Geol. Surv. Open-File Rep. 81-263.* 99 pp.

Wegener, A. 1924. *The Origin of Continents and Oceans.* Transl. J. G. A. Skerl. New York: Dutton. 212 pp. 3rd ed. (From German)

Weldon, R. J. II 1986. *The late Cenozoic geology of Cajon Pass; Implications for tectonics and sedimentation along the San Andreas fault.* PhD thesis. Calif. Inst. Technol., Pasadena. 400 pp.

Weldon, R., Humphreys, E. 1986. A kinematic model of southern California. *Tectonics* 5: 33–48

Weldon, R. J., Meisling, K. E., Alexander, J. 1992. A speculative history of the San Andreas fault system in the central Transverse Ranges. See Powell et al 1992. In press

Weldon, R. J. II, Sieh, K. E. 1985. Holocene rate of slip and tentative recurrence interval for large earthquakes on the San Andreas fault, Cajon Pass, southern California. *Geol. Soc. Am. Bull.* 96: 793–812

Weldon, R. J. II, Springer, J. E. 1988. Active faulting near the Cajon Pass well, southern California; Implications for the stress orientation near the San Andreas fault. *Geophys. Res. Lett.* 15: 993–96

Wernicke, B., Spencer, J. E., Burchfiel, B. C., Guth, P. L. 1982. Magnitude of crustal extension in the southern Great Basin. *Geology* 10: 499–502

Wesnousky, S. G. 1986. Earthquakes, Quaternary faults, and seismic hazard in California. *J. Geophys. Res.* 91: 12,587–631

Wilson, J. T. 1965. A new class of faults and their bearing on continental drift. *Nature* 207: 343–47

Winterer, E. L., Hussong, D. M., Decker, R. W., eds. 1989. *The Eastern Pacific Ocean and Hawaii.* Boulder, Colo.: Geol. Soc. Am., The Geology of North America, vol. N. 563 pp.

Withjack, M. O., Jamison, W. R. 1986. Deformation produced by oblique rifting. *Tectonophysics* 126: 99–124

Woodburne, M. O. 1975. Cenozoic stratigraphy of the Transverse Ranges and adjacent areas, southern California. *Geol. Soc. Am. Spec. Pap. 162.* 91 pp.

Yeats, R. S. 1983. Large-scale Quaternary detachments in Ventura basin, southern California. *J. Geophys. Res.* 88: 569–83

Yeats, R. S., Calhoun, J. A., Nevins, B. B., Schwing, H. F., Spitz, H. M. 1989. Russell fault: An early strike-slip fault of California Coast Ranges. *Am. Assoc. Petrol. Geol. Bull.* 73: 1089–1102

Zoback, M. D., Lachenbruch, A. H. 1992. Introduction to papers on the Cajon Pass scientific drilling project. *J. Geophys. Res.* In press

Zoback, M. D., Zoback, M. L., Mount, V. S., Suppe, J., Eaton, J. P., et al. 1987. New evidence on the state of stress of the San Andreas fault system. *Science* 238: 1105–11

Annu. Rev. Earth Planet. Sci. 1992. 20:469–500

NUTATIONS OF THE EARTH[1]

P. M. Mathews

Department of Theoretical Physics, University of Madras, Madras 600 025, India and Harvard-Smithsonian Center for Astrophysics, Cambridge, Massachusetts 02138

I. I. Shapiro

Harvard-Smithsonian Center for Astrophysics, Cambridge, Massachusetts 02138

KEY WORDS: Earth rotation, wobbles, normal modes, free core nutations

EARTH ROTATION: GENERAL FEATURES

The figure of the Earth is very nearly that of an oblate spheroid. Its moments of inertia about principal axes in the equatorial plane differ from each other only by about two parts in 10^5, and may be taken to have a common value A, for the purposes of nutation theory. The moment of inertia C about the third principal axis, in the direction of the poles, is larger by about one part in 300. The dynamical ellipticity, defined as $e \equiv (C-A)/A$, has the dominant role in determining the Earth's nutations. The flattening of the Earth's surface, i.e. its geometrical ellipticity, is defined as the fractional difference between the equatorial axis and the polar axis, and is also about 1/300. This equatorial bulge and the associated difference between C and A are almost wholly accounted for by centrifugal forces associated with a rotation of the Earth about its polar axis at a mean rate of one cycle per sidereal day (1 cpsd), equivalent to an angular velocity Ω_0 which is 7.29212×10^{-5} rad sec^{-1} at present. We will find it convenient to express all angular velocities in units of Ω_0. In the following, an angular velocity $f\Omega_0$, or the equivalent frequency f cpsd, will often be referred to simply as "frequency f," though f itself is nondimensional. The corresponding period is K/f solar days, where $K \approx 365.24/366.24$ is the ratio of the length of a sidereal day to that of a solar day.

The rotation of the Earth is not entirely steady. The rotation rate fluctuates about the mean value Ω_0 which itself is decreasing slowly at the rate of about 7×10^{-22} rad sec^{-2} (see e.g. Lambeck 1980). Further, the direction in inertial space of the Earth's figure axis, which we denote by the unit vector $\mathbf{i}_3$, undergoes both precession and nutation. (We shall use the term "figure axis" or "symmetry axis" for the instantaneous direction of the polar axis of the spheroidal figure of the Earth, ignoring the deviations from such a figure due to deformations; reference will be made later to the considerations involved in a careful definition of such an axis for deformable Earth models.) Precession is the motion of $\mathbf{i}_3$ around the pole of the ecliptic, the normal to the Earth's orbital plane, at a nearly constant rate and constant angle of $\sim 23.5°$ to this normal; the full circuit takes about 26,000 years. Nutation consists of a fluctuating motion of $\mathbf{i}_3$ in space about the smooth precessional path. The observed nutation is in fact the motion in space of the direction of the figure axis, not that of the axis of rotation or of the angular momentum vector; see, for instance, Ooe & Sasao (1974), Kinoshita et al (1979), and Wahr (1981c). It is convenient to visualize the motions of the unit vector $\mathbf{i}_3$ with the precessional component removed; the residual (nutational) motions of $\mathbf{i}_3$ would then be around a mean direction $\mathbf{I}_3$ which is fixed in space. References elsewhere in this review to $\mathbf{I}_3$ as a space-fixed axis must be understood in this sense. The nutations cause angular separations of up to nearly 10 arc seconds between $\mathbf{I}_3$ and the instantaneous $\mathbf{i}_3$. Nutation is necessarily accompanied by wobble, i.e. a separation of the direction of the instantaneous angular velocity vector $\mathbf{\Omega}$ of the solid mantle of the Earth from that of the figure axis $\mathbf{i}_3$, together with a motion of these two vectors around each other. Nutation-cum-wobble occurs whenever the axis of rotation is inclined to the symmetry axis. When this happens in the absence of any external influences, one has free nutation-wobble. Forced nutation and precession of the Earth are caused by the torque associated with the gravitational potential ϕ_e due to sources external to the Earth, namely, the Sun and the Moon and, to a marginal extent, also the planets. Since ϕ_e is also responsible for the tides, it is referred to as the tidal potential.

Nutation was discovered, and so named, by James Bradley. By contrast with his discovery of aberration which was serendipitous, Bradley's inference of the cause of the variations in stellar positions which he observed was grounded in the basic physics of "the Moon's action upon the equatorial parts of the Earth"; and he set out on a 19-year program of observations which culminated, in 1748, with the confirmation that these variations were indeed caused by changes, associated with the motion of the nodes of the lunar orbit with a period close to 19 years, in the Moon's gravitational torque on the Earth. For details of the lunisolar potential and its spectral components, the reader may consult Melchior (1978). Treatments of the various aspects of Earth rotation, including nutation, may be found in the books by Munk & MacDonald (1960), Jeffreys (1976), Lambeck (1980), and Moritz & Mueller (1987), and in the recent reviews by Wahr (1988a), Kinoshita & Sasao (1989), and Wahr & de Vries (1990). The surveys by Herring (1991) and Masters (1991) of recent work in geodesy and seismology, respectively, also contain much useful information.

GRAVITATIONAL PERTURBATIONS

The tidal potential $\phi_e(\mathbf{r}, t)$ at a point $\mathbf{r}$ at a time t is made up primarily of the second degree ($n = 2$) terms in the expansion of the lunisolar gravitational potential in terms of spherical harmonics $Y_n^m(\theta, \varphi) \equiv P_n^m(\theta)e^{-im\varphi}$ defined relative to a coordinate system that has its origin at the geocenter and its polar axis along $\mathbf{I}_3$, which is the mean direction of both the rotation and the figure axes of the Earth. Terms of degrees $n > 2$ in the potential are too small to be of interest at present in nutation theory. The potential $\phi_e(\mathbf{r}, t)$ at any space-fixed point in the region of the Earth has periodicities associated with the apparent mean motions of the Sun and the Moon around the Earth and with the slow changes in the parameters of the lunar orbit. The most important of these periods $T_\alpha^{(S)}$, in a space-fixed frame, range in magnitude from about 9 days to about 19 years; the corresponding frequencies, in cpsd, are $\sigma_\alpha^{(S)} \equiv K/T_\alpha^{(S)}$. Consider a particular spectral component of the degree 2 potential ϕ_e, characterized by the time dependence $e^{i\sigma^{(S)}\Omega_0 t}$ where $\sigma^{(S)}$ is any of the frequencies $\sigma_\alpha^{(S)}$. The component evidently has the phase factor $\exp[i(\sigma^{(S)}\Omega_0 t - m\varphi^{(S)})]$, where $\varphi^{(S)}$ is the azimuthal angle of the point being considered, relative to a space-fixed frame with polar axis $\mathbf{I}_3$. Therefore this part of the potential, for positive m, moves in space in the prograde or retrograde sense around $\mathbf{I}_3$, i.e. in the direction of increasing or decreasing values of $\varphi^{(S)}$, depending on whether the sign of $\sigma^{(S)}$ is positive or negative. Conversely, prograde or retrograde components of the gravitational perturbation will be characterized by positive or negative signs, respectively, for $\sigma_\alpha^{(S)}$ and $T_\alpha^{(S)}$. (Negative values of m appear automatically in the corresponding complex conjugate terms in the potential, and need not be separately considered.) A fixed point on the rotating Earth has $\varphi^{(S)} = \varphi_0 + \Omega_0 t$; the phase factor at an Earth-fixed point is therefore $\exp[i(\sigma^{(S)} - m)\Omega_0 t - m\varphi_0)]$. The frequency of this potential term, as experienced on the rotating Earth, is thus $(\sigma^{(S)} - m)$ cpsd; and the body and ocean tide components arising from this term are also of the same frequency. Since the magnitudes of the $\sigma^{(S)}$ are of the order of 0.1 or less, the tidal frequencies corresponding to a fixed m and various values of $\sigma^{(S)}$ cluster within a narrow band; thus one has the long period, diurnal, and semidiurnal tidal bands for $m = 0$, 1, and 2, respectively. A particular tidal frequency is identified as prograde if the sign of the corresponding $\sigma^{(S)}$ is positive and as retrograde if the sign is negative. Along with each prograde frequency $\sigma_\alpha^{(S)}$, the corresponding retrograde frequency $-\sigma_\alpha^{(S)}$ is also present in the tidal potential; their amplitudes are in general unequal.

On account of the spheroidal shape of the Earth and of the fact that the orbits of the Sun and the Moon do not lie in the Earth's equatorial plane, a torque about some equatorial axis is produced on the Earth by the tesseral ($m = 1$) part of the degree 2 tidal potential; the zonal ($m = 0$) and sectorial ($m = 2$) parts do not give rise to any torques on a spheroidal Earth. The tesseral potential has the form

$$\phi_{\mathrm{e}}^{(m=1)}(\mathbf{r}, t) = -\Omega_0^2[\phi_{\mathrm{a}}(t)xz + \phi_{\mathrm{b}}(t)yz] \tag{1}$$

with

$$\phi_{\mathrm{a}} = \sum_s \Phi_s \cos\varphi_s, \qquad \phi_{\mathrm{b}} = \sum_s \Phi_s \sin\varphi_s, \tag{2}$$

where $\Phi_s = (3GM_s \sin 2\theta_s/2d_s^3\Omega_0^2)$, and the summation is over the sources of the gravitational field (Sun, Moon, and planets). The direction and the distance of the center of the source from the center of the Earth are denoted by (θ_s, φ_s) and d_s, respectively, and the mass of the source by M_s. Now, $GM_s/d_s^3 \approx \Omega_{\mathrm{E}}^2$ or $\Omega_{\mathrm{M}}^2 M_{\mathrm{M}}/M_{\mathrm{E}}$, according as the source s is the Sun or the Moon. (The subscripts E and M refer to the Earth and the Moon, respectively, and Ω_{M} and Ω_{E} are the mean angular velocities of orbital motion of the respective bodies.) It is trivial then to verify that $\Phi_s < 10^{-5}$ in either case, and hence that ϕ_{a} and ϕ_{b} are $O(10^{-5})$.

The torque on the Earth due to the tesseral potential (1) is

$$\mathbf{\Gamma} = -\Omega_0^2 eA\,\mathbf{i}_3 \times \boldsymbol{\phi} \tag{3}$$

where $\boldsymbol{\phi}$ is the vector with components $(\phi_{\mathrm{a}}, \phi_{\mathrm{b}}, 0)$. Relative to the rotating Earth, the frequency of a typical spectral component of this $m = 1$ torque is $\sigma \equiv (\sigma^{(S)} - 1)$ cpsd.

PRECESSION, NUTATION, AND WOBBLE

The gyroscopic response of the Earth to that Fourier component of the torque which is of zero frequency in inertial space ($\sigma^{(S)} = 0$, implying $\sigma = -1$) is precession. The rate of precession is proportional to

$H \equiv (C-A)/C$; it does not depend on any other feature of the Earth. Astronomical observations permit an accurate determination of H. From the IAU 1976 system of astronomical constants, Kinoshita et al (1979) find that $H = 0.00327399$; the corresponding value of the Earth's dynamical ellipticity is $e \equiv (C-A)/A \equiv H/(1-H) = 0.00328475$. Analyses of currently available LLR (lunar laser ranging) and VLBI (very long baseline interferometry) data yield a value for e which is about 6 parts in 10^5 lower (Williams et al 1991, Herring et al 1991).

Forced nutation is the response of the Earth to the time-varying part of the lunisolar torque, manifested as a motion of the figure axis $\mathbf{i}_3$ of the Earth around the space-fixed $\mathbf{I}_3$. It is traditional, in the astronomy literature, to represent this motion in terms of the angular displacements of $\mathbf{i}_3$ from $\mathbf{I}_3$, or equivalently, of the components of the vector $\boldsymbol{\eta} \equiv (\mathbf{i}_3-\mathbf{I}_3)$, in two orthogonal directions: the nutations in obliquity ($\Delta\varepsilon$) and longitude ($\Delta\psi$). The coefficients of the nutations in obliquity and longitude at the angular frequency $\nu\Omega_0$ of a Fourier component of the lunisolar gravitational potential are defined in terms of the spectral expansions:

$$\Delta\varepsilon = \sum_{\nu} [\Delta\varepsilon_{\nu \mathrm{I}} \cos \nu\Omega_0 t + \Delta\varepsilon_{\nu \mathrm{O}} \sin \nu\Omega_0 t], \tag{4a}$$

$$\Delta\psi = \sum_{\nu} [\Delta\psi_{\nu \mathrm{I}} \sin \nu\Omega_0 t + \Delta\psi_{\nu \mathrm{O}} \cos \nu\Omega_0 t]. \tag{4b}$$

$\Delta\varepsilon_{\nu\mathrm{I}}$ and $\Delta\varepsilon_{\nu\mathrm{O}}$ are, respectively, the in-phase and out-of-phase coefficients of the nutation in obliquity, and $\Delta\psi_{\nu\mathrm{I}}$ and $\Delta\psi_{\nu\mathrm{O}}$, of the nutation in longitude. (The out-of-phase coefficients arise from dissipative processes; they are very much smaller than the corresponding in-phase ones.) The frequencies ν are linear combinations, with nonnegative integer coefficients, of fundamental frequencies defined through the time rates of change of Brown's fundamental arguments l, l', F, D, and Ω (see, for example, Woolard 1953). Most, but not all, of the important ν are positive; for any particular ν, $|\nu|$ and $-|\nu|$ are, respectively, prograde and retrograde members of the set of tidal frequencies $\sigma_\alpha^{(S)}$ referred to earlier.

The in-phase parts of the spectral components of $\Delta\varepsilon$ and $\Delta\psi$ corresponding to any particular ν constitutes an elliptical nutation; the out-of-phase parts constitute another such nutation. Each of these paired elliptical nutations may be resolved into two circular components, one of which represents a uniform rotation of the figure axis $\mathbf{i}_3$ around $\mathbf{I}_3$ in the prograde sense, and the other, in the retrograde sense. The combination of the two prograde circular nutations, characterized by the frequency $\sigma^{(S)} = |\nu|$, results in one with a complex amplitude $\eta^{(\mathrm{pro})}$; the composite retrograde nutation has $\sigma^{(S)} = -|\nu|$ and a complex amplitude $\eta^{(\mathrm{ret})}$. These amplitudes are related to the coefficients of nutation in obliquity and longitude by

$$\eta^{(\mathrm{pro})} = -\frac{1}{2}\left(\Delta\varepsilon_{\nu\mathrm{I}} - \frac{\nu}{|\nu|}\Delta\psi_{\nu\mathrm{I}}\sin\varepsilon_0\right) + \frac{i}{2}\frac{\nu}{|\nu|}\left(\Delta\varepsilon_{\nu\mathrm{O}} + \frac{\nu}{|\nu|}\Delta\psi_{\nu\mathrm{O}}\sin\varepsilon_0\right), \quad (5a)$$

$$\eta^{(\mathrm{ret})} = -\frac{1}{2}\left(\Delta\varepsilon_{\nu\mathrm{I}} + \frac{\nu}{|\nu|}\Delta\psi_{\nu\mathrm{I}}\sin\varepsilon_0\right) - \frac{i}{2}\frac{\nu}{|\nu|}\left(\Delta\varepsilon_{\nu\mathrm{O}} - \frac{\nu}{|\nu|}\Delta\psi_{\nu\mathrm{O}}\sin\varepsilon_0\right), \quad (5b)$$

where ε_0 is the mean obliquity. The amplitudes a^+ and a^- defined in the literature (e.g. Herring et al 1991) are to be identified as $a^+ = \eta^{(\mathrm{ret})}$, $a^- = \eta^{(\mathrm{pro})}$ for $\nu > 0$ and $a^+ = \eta^{(\mathrm{pro})}$, $a^- = \eta^{(\mathrm{ret})}$ for $\nu < 0$.

A circular nutation having a frequency $\sigma^{(S)}$ as seen from inertial space is necessarily accompanied by a circular wobble with frequency $\sigma = \sigma^{(S)} - 1$ relative to the rotating Earth; in other words, the part transverse to $\mathbf{i}_3$ of the vector $\mathbf{m} \equiv (\mathbf{\Omega}/\Omega_0 - \mathbf{i}_3)$ has a spectral component $\mathbf{m}(\sigma)$ which executes a circular motion around $\mathbf{i}_3$ with frequency σ. The combination $\tilde{m}(\sigma) \equiv m_1(\sigma) + im_2(\sigma)$ is the complex wobble amplitude at frequency σ. It is simply related to the complex amplitude $\tilde{\eta}(\sigma)$ of the corresponding nutation which is defined in terms of the nutation vector $\boldsymbol{\eta} \equiv \mathbf{i}_3 - \mathbf{I}_3$ and has frequency $\sigma^{(S)} \equiv (1+\sigma)$ in inertial space:

$$\tilde{\eta}(\sigma) = -\frac{\tilde{m}(\sigma)}{(1+\sigma)}. \quad (6)$$

This kinematic relation (see, for example, Ooe & Sasao 1974) is independent of the properties of the Earth, and holds whether the nutation-wobble is forced or free. However, the nutational and wobble responses to a given perturbation do, individually, depend on the Earth's structure and material properties. The study of nutations is therefore of interest, not only for obtaining accurate information about Earth rotation, but also as a means of gaining knowledge of the Earth itself. Because the nutational normal modes, which cause resonances in the forced nutations, have definite prograde ($\sigma^{(S)} > 0$) or retrograde ($\sigma^{(S)} < 0$) character, it is especially useful, in the geophysical context, to express the nutation-wobbles in circular component form. The most important of the circular nutations, for geophysical purposes, are the fortnightly, semiannual, annual, and principal nutations; they correspond to motions of the $\mathbf{i}_3$ axis in space with periods of ± 13.7, ± 186.2, ± 365.3, and ± 6798.4 solar days, respectively, with the sign indicating the prograde or retrograde nature of the nutation. A list of a large number of nutation periods together with frequencies of the corresponding tides may be found in Melchior (1978).

It is an elementary result, following directly from Euler's theory of the rotation of a symmetric top, that a *rigid* spheroidal Earth model (indicated by a subscript R in the following) would have a free wobble mode of frequency e cpsd, e being the dynamical ellipticity of the Earth model, and that forcing by the part $Y_2^1\tilde{\phi}(\sigma)e^{i\sigma\Omega_0 t}$ of the potential (1) would produce forced wobble and nutation with amplitudes

$$\tilde{m}_R(\sigma) = \frac{e}{e-\sigma}\tilde{\phi}(\sigma), \qquad \tilde{\eta}_R(\sigma) = -\frac{e}{(1+\sigma)(e-\sigma)}\tilde{\phi}(\sigma). \tag{7}$$

Computation of the rigid-Earth amplitudes to an accuracy of 0.1 milliarcsecond (mas), improving upon the earlier standard results of Woolard (1953), was done by Kinoshita et al (1979). The greatly extended calculations needed to enhance the accuracy to 0.01 mas have since been carried out by Zhu & Groten (1989) and Kinoshita & Souchay (1990). The rigid-Earth results are fundamental to studies on more realistic Earth models, the amplitudes of which are expressed in terms of a normalized amplitude $\tilde{\eta}_N(\sigma)$ computed for the model of interest:

$$\tilde{\eta}_N(\sigma) = \frac{\tilde{\eta}(\sigma)}{\tilde{\eta}_R(\sigma)} = \frac{\tilde{m}(\sigma)}{\tilde{m}_R(\sigma)}, \qquad \tilde{\eta}(\sigma) = \tilde{\eta}_N(\sigma)\tilde{\eta}_R(\sigma). \tag{8}$$

All information regarding the structure and material properties of the Earth is contained in the normalized amplitude $\tilde{\eta}_N$, which is independent of $\tilde{\phi}$, and hence free of any error in the computation of the tidal potential. For the calculation of $\tilde{\eta}(\sigma)$, it is not necessary to know $\tilde{\phi}(\sigma)$ explicitly, since it is already incorporated in the model-independent factor $\tilde{\eta}_R$ which is tabulated in the works just cited.

NORMAL MODES

The amplitude of the nutational response to a torque of frequency σ cpsd may, in general, be decomposed into a linear combination of resonance terms of the form $A_n/(\sigma-\sigma_n)$ where the σ_n are the eigenfrequencies associated with normal modes of the Earth. If any one or more of the eigenfrequencies lie within the diurnal tidal band, resonant enhancement of the amplitudes of nutations at frequencies σ close to any such σ_n is to be expected. It is important, therefore, to understand the normal modes of free motions of the Earth. A theoretical treatment of the normal modes of the rotating ellipsoidal Earth has been given by Dahlen (1968) and Dahlen & Smith (1975); numerical results on normal modes and eigenfrequencies are given in an illuminating article by Smith (1977).

The only normal mode of a rigid axially symmetric body is its Eulerian free wobble. The free wobble consists of a coning of the angular velocity vector $\mathbf{\Omega}$ about the symmetry axis of the body, wherein the component parallel to $\mathbf{i}_3$ has a constant value Ω_0, while the transverse part of $\mathbf{\Omega}$ rotates around $\mathbf{i}_3$ with angular frequency $e\Omega_0$. If the Earth were rigid, its Eulerian wobble period should be $1/e \approx 305$ days. But when the free wobble was discovered, its period, as determined by Chandler in 1891, was found to be about 14 months. The deformability of the Earth (Love 1909, Larmor 1909) and the existence of a fluid core (Hough 1895, Poincaré 1910) were invoked to explain the difference. A careful accounting by Smith & Dahlen (1981) showed a significant contribution from the pole tide produced in the oceans by the variations in pole position due to the wobble, and a residual part which was attributed to anelasticity in the mantle and used to deduce parameter values in models of anelasticity. The amplitude of the Chandler wobble is somewhat variable, but has a mean of about one fifth of an arcsecond.

The nonspherical (spheroidal) shape of the core-mantle boundary (CMB) has the consequence that any rotational motion of the fluid core relative to the mantle, with the core's rotation axis inclined to the symmetry axis of the CMB, causes imbalance of fluid pressure on the boundary, and a resultant torque which tends to bring the two axes into alignment. This inertial, or pressure, coupling between the core and the mantle gives rise to a second wobble normal mode (Hough 1895, Poincaré 1910). This mode is the nearly diurnal free wobble (NDFW), with an eigenfrequency σ_{NDFW} very close to, and on the retrograde side of, -1 cpsd. It is also referred to as the free core nutation (FCN). It is characterized by wobble motions of the mantle and the fluid core in opposite directions, with the latter having an amplitude about 4000 times as large as the former (see, for instance, Mathews et al 1991b). The location of the NDFW eigenfrequency within the diurnal tidal band provokes resonant enhancement of the wobble-nutation response of the Earth to lunisolar perturbations. Various tidal parameters such as the Love numbers h, k, and l also exhibit this resonance. This fact has been invoked, starting with the early work of Warburton & Goodkind (1978), for the estimation of the frequency of the NDFW from Earth-tide measurements. Neuberg et al (1987) estimate from the frequency dependence of the gravimetric factor $\delta \equiv (1+h-3k/2)$, measured at several stations in central Europe, that the period $(1+\sigma_{\mathrm{NDFW}})^{-1}$ of the FCN is -434 ± 7 sidereal days. This value differs considerably from Wahr's (1981c) theoretical value of -460 days but is close to the result of the first determination from VLBI data by Herring et al (1986). Current data (Herring et al 1991) are consistent with the period of -430 days estimated theoretically by Mathews et al (1991b). The FCN frequency is thus well established from the Earth's response to tidal forcing, but the *free* nutation itself has remained rather elusive.

Though an amplitude as high as 18 mas has been claimed from astrometric observations (Hefty 1989), the latest analysis of VLBI data yields an amplitude of only 0.26 ± 0.04 mas (Herring et al 1991), well within the upper bounds of 3 mas and 2 mas, respectively, estimated earlier from LLR data by Wahr & Larden (1983) and Dickey et al (1983), and also below the upper bound of 0.6 mas set by Herring et al (1988) from VLBI data.

Recent theoretical work (Mathews et al 1991a,b; de Vries & Wahr 1991) has shown that the presence of a solid inner core (SIC) gives rise to another nearly diurnal wobble mode which had not revealed itself in Wahr's (1981c) study of eigenfrequencies by numerical methods. This wobble and the associated nutation are prograde. Mathews et al (1991b) found the relation between the wobble motions of the fluid outer core and the mantle, in this mode, to be very nearly the same as in the already known retrograde FCN, and have used the term "retrograde FCN" (RFCN) for the latter and "prograde FCN" (PFCN) for the new mode; de Vries & Wahr (1991) refer to the new mode as the "free inner core nutation" or FICN. The two FCN modes differ markedly in the relative motions of the fluid outer core and the solid inner core: The wobble amplitudes of the two are comparable and of the same sign in the RFCN, while they are of opposite signs in the PFCN with the inner core amplitude being 300 times larger. Mathews et al (1991a,b) show also that there exists yet another mode, the inner core wobble (ICW), which would reduce to the free wobble of the SIC if the forces between the SIC and the rest of the Earth could be turned off. It is predominantly a rotation of the figure axis of the inner core relative to the mantle, the motions of the various axes of rotation being smaller in amplitude by several orders of magnitude.

The kinematic factor $1/(1+\sigma)$ in Equation (6) is often interpreted as representing a resonance associated with a normal mode with a characteristic frequency of exactly -1 cpsd, independent of any properties of the Earth. This pseudo-resonance, which occurs at zero frequency in inertial space, appears only in the nutations and not in the wobbles; it is in fact a manifestation of the precessional motion. Since it involves only a continuously evolving tilt of the Earth as a whole with no associated internal movements, it is called the "tilt-over" mode (TOM). The strength of this resonance is proportional to $e/(1+e)$ and to the combination $(GM_E/a^3\Omega_0^2)$, where a is the mean radius of the Earth; it is independent of all other Earth parameters.

Besides the above-listed modes which produce resonant effects, large or small, in the forced nutations, there exist other normal modes (in fact, infinite classes of them) which have essentially no wobble component. The elastic vibration modes, which leave their imprint on seismograms and have periods less than one hour, are the best known; they have characteristic frequencies that are high enough to make the associated resonances unimportant in the nearly diurnal wobbles. Normal modes of the fluid outer core are much less well understood despite a number of studies on their frequency spectra (e.g. Crossley 1975, 1980; Shen & Mansinha 1976; Smith 1977; Friedlander 1987). A search for signals associated with core modes in gravity records (Cummins et al 1991) has not resulted in any positive detection. These modes are believed to have no detectable influence on nutations.

FORCED NUTATIONS: THEORETICAL CONSIDERATIONS FOR IDEALIZED EARTH MODELS

Theoretical computations of nutation amplitudes are based on oceanless and perfectly elastic Earth models characterized by hydrostatic equilibrium density and elastic parameter profiles deduced from seismological data. After the pioneering work of Jeffreys & Vicente (1957a,b) employing such models, theoretical treatments of the nutation problem have followed either of two broad approaches. One of these, typified by the works of Smith (1977) and Wahr (1981a,b,c), seeks to determine the complete displacement field $\mathbf{u}(\mathbf{r})$ produced by the perturbations, i.e. the displacements of the material points throughout the Earth from their positions $\mathbf{r}$ in an unperturbed configuration. This configuration is taken to correspond to hydrostatic equilibrium in a state of uniform rotation with angular velocity Ω_0 about $\mathbf{I}_3$, and is characterized by time-independent functions $\rho_0(\mathbf{r})$, $\phi_{g0}(\mathbf{r})$, $\lambda_0(\mathbf{r})$, and $\mu_0(\mathbf{r})$ which have spheroidal symmetry about $\mathbf{I}_3$ and represent, respectively, the density, the geopotential, and the two Lamé parameters describing the elastic properties, of the Earth. The field $\mathbf{u}(\mathbf{r})$ is subject to the momentum balance equation which governs the motion of material elements of the Earth under elastic and gravitational forces, including external perturbations. Relative to a coordinate system rotating with the constant angular velocity $\mathbf{\Omega}_0 \equiv \Omega_0 \mathbf{I}_3$, this equation, linearized in the perturbations (Dahlen 1972, Smith 1974), may be written in the form

$$\frac{\partial^2 \mathbf{u}}{\partial t^2} + 2\Omega_0 \times \frac{\partial \mathbf{u}}{\partial t} - \frac{1}{\rho_0}\nabla \cdot [\mathscr{T}] - (\nabla \cdot \mathbf{u})\nabla\phi_{g0} - \nabla[\phi' + \mathbf{u}\cdot\nabla\phi_{g0}] = -\nabla\phi_e, \tag{9}$$

wherein $[\mathscr{T}]$ is the elastic stress tensor and ϕ' is the change in the geopotential caused by the perturbations. The components of $[\mathscr{T}]$ are

$$T_{ij} = \lambda_0 \delta_{ij} + \mu_0 \left(\frac{\partial u_i}{\partial x_j} + \frac{\partial u_j}{\partial x_i} \right), \tag{10}$$

and ϕ' is related through Poisson's equation to the density perturbations associated with the displacements, which may, in turn, be expressed in terms of $\mathbf{u}(\mathbf{r})$:

$$\nabla^2 \phi' = -4\pi G \nabla \cdot (\rho_0 \mathbf{u}). \tag{11}$$

The field $\mathbf{u}(\mathbf{r})$ is the solution, subject to appropriate boundary conditions, of the pair of partial differential equations (9) and (11). A unique decomposition of the field $\mathbf{u}(\mathbf{r})$ at any given depth in the Earth into a deformational part and a part representing rigid rotation of a spherical shell at that depth is possible (see, for instance, Smith 1977 and Mathews et al 1991a). Specifically, the degree 1 toroidal terms in an expansion of the displacement field in terms of vector spherical harmonics constitute the rotational part, which primarily represents nutation, but also includes a very small torsional deformation between different radial layers of the Earth. The deformational part of $\mathbf{u}(\mathbf{r})$, excluding the torsional part just referred to, is composed of all the vector spherical harmonic terms of degree higher than 1; it contains all the information about tidal phenomena.

The second approach, developed originally by Molodensky (1961) in his treatment of the nutations of a stratified and deformable Earth with a fluid core, has relied on the quasistatic spherical approximation (neglecting ellipticity and inertial force terms) in computing the deformability parameters of the core and the mantle from the tidal deformation equations. The equations of angular momentum balance, which govern rotational motions of the mantle and the core, involve these parameters as well as others which can be evaluated directly as integrals involving the functions in terms of which the Earth model is specified. The nutation amplitudes are obtained by solving these equations, which generalize the rigid-Earth equation of Euler, for the components transverse to $\mathbf{i}_3$ of the angular velocity vector $\mathbf{\Omega}$ of the mantle. This method has been greatly simplified and placed on a more sound footing in the formulation of Sasao et al (1980), and further developed and extended by Mathews et al (1991a) to take account of independent motions of the inner core.

In employing the first approach to study the nutations and deformations caused by perturbations at any tidal frequency, Wahr (1981a,b,c) developed a resonance expansion for the displacement field $\mathbf{u}(\mathbf{r})$ in terms of the normal modes $\mathbf{u}_k(\mathbf{r})$, from which resonance formulae for the amplitude of forced nutation and for tidal parameters such as Love numbers also follow. For computational purposes, however, he took the contributions from the elastic vibrational (seismic) modes to the resonance expansions to be effectively independent of frequency within the diurnal tidal band, the width of the band being rather small compared to the distance of the band from the nearest seismic eigenfrequency. The resonance expression for $\tilde{\eta}(\sigma)$ could then be replaced, for any σ within the band, by $\tilde{\eta}(\sigma_0)$ at a particular frequency σ_0 which Wahr took to be that of the O_1 tide, plus the difference between the contributions to $\tilde{\eta}(\sigma)$ and $\tilde{\eta}(\sigma_0)$ from the rotational eigenmodes, namely the tilt-over mode, and the NDFW and Chandler wobble modes. The problem was thus reduced to one of computing $\tilde{\eta}(\sigma_0)$ and the parameters in the three retained resonance terms. The process of solving the basic field equations for this purpose is based on an expansion of $\mathbf{u}(\mathbf{r})$ in terms of spheroidal functions $\tau_n^m(\mathbf{r}) \equiv [U_n^m(r)\hat{r}Y_n^m(\theta,\varphi)+V_n^m(r)r\nabla Y_n^m(\theta,\varphi)]$ and toroidal functions $W_n^m(r)[-\mathbf{r}\times\nabla Y_n^m(\theta,\varphi)]$, wherein scalar functions U_n^m, V_n^m, and W_n^m of the radial variable appear as coefficients of vector spherical harmonics. The expansion enables a reduction of the field equations to an infinite set of ordinary first-order differential equations for the radial functions. Such a reduction, retaining terms up to the first order in deviations from sphericity of the Earth, was carried out by Smith (1974). With a degree 2 tesseral perturbation, the expansion of $\mathbf{u}(\mathbf{r})$ has the form $\mathbf{u}(\mathbf{r}) = (\boldsymbol{\tau}_1^1+\boldsymbol{\sigma}_2^1+\boldsymbol{\tau}_3^1+\cdots)$, with toroidal $(\boldsymbol{\tau}_n^1)$ and spheroidal $(\boldsymbol{\sigma}_n^1)$ terms alternating in the series (Dahlen 1968, Smith 1974). The admixture of higher degree $(n>2)$ terms with $\boldsymbol{\tau}_1^1$ (rotational displacement) and $\boldsymbol{\sigma}_2^1$ (the deformational response of a spherical nonrotating Earth) is due to Coriolis forces and ellipticity, which also cause the radial equations for all n to be mutually coupled. Solutions for chosen Earth models were obtained numerically by Smith (1977) and Wahr (1981b,c), after truncating this infinite coupled system to a manageable finite size by restricting the terms in the spherical harmonic expansion to $\boldsymbol{\tau}_1^1$, $\boldsymbol{\sigma}_2^1$, and $\boldsymbol{\tau}_3^1$ only. The eigenfrequencies and resonance strength parameters as well as the constant $\tilde{\eta}(\sigma_0)$ in Wahr's (1981c) resonance expansion for $\tilde{\eta}(\sigma)$ were obtained from numerical solutions, computed in this manner, for the eigenfunctions $\mathbf{u}_k(\mathbf{r})$ and for the displacement field $\mathbf{u}(\mathbf{r})$ produced by tidal forcing. He converted this expansion, after adjusting the coefficient of the tilt-over mode term to conform to the value of e determined from precession, into a resonance formula for $\tilde{\eta}_N(\sigma)$, the nutation amplitude normalized relative to that for a rigid-Earth model having the same ellipticity e. In our notation, the formula is

$$\tilde{\eta}_N(\sigma) \equiv \frac{\tilde{\eta}(\sigma)}{\tilde{\eta}_R(\sigma)} = 1-(\sigma+1)(\sigma-e) \times \left\{B_0-(\sigma-\sigma_0)\left[\frac{B_1}{\sigma-\sigma_1}+\frac{B_2}{\sigma-\sigma_2}-\frac{B_0'}{\sigma-e}\right]\right\}, \quad (12)$$

wherein σ_1 and σ_2 are, respectively, the frequencies of the Chandler wobble and the NDFW. The following values were obtained by Wahr (1981c) for the various numerical constants, from numerical computations for Earth model 1066A of Gilbert & Dziewonski (1975), with σ_0 chosen as the frequency of the O_1 tide: $B_0 = 0.416$, $B_1 = 0.810$, $B_2 = 0.665$, $B_0' = 1.06$, $e = 3.28 \times 10^{-3}$, $\sigma_0 = -0.927$, $\sigma_1 = 2.478 \times 10^{-3}$, and $\sigma_2 = -1.0021714$. (Signs of all the frequencies are reversed, to conform to our convention.) The coefficients in the IAU 1980 nutation series were obtained from formula (12), using the above constants and with values of $\tilde{\eta}_R(\sigma)$ taken from Kinoshita et al (1979).

For a study of nutations, the angular momentum approach is much simpler than that based on determination of the displacement field. The central role in the former is played by the so-called Liouville equation

$$\frac{d\mathbf{H}}{dt} + \mathbf{\Omega} \times \mathbf{H} = \mathbf{\Gamma} \tag{13}$$

which governs the evolution, as seen from a reference frame rotating with angular velocity $\mathbf{\Omega}$, of the angular momentum $\mathbf{H}$ of a body subject to a torque $\mathbf{\Gamma}$. In general, $\mathbf{H} = [\mathscr{C}]\cdot\mathbf{\Omega}+\mathbf{h}$, wherein $[\mathscr{C}]$ is the instantaneous moment of inertia tensor of the body, $[\mathscr{C}]\cdot\mathbf{\Omega}$ represents the angular momentum which the body would have if it were rotating rigidly with the angular velocity $\mathbf{\Omega}$ of the chosen reference frame, and $\mathbf{h}$ is the angular momentum arising from motions of the body or parts of it relative to the reference frame.

For a rigid-Earth model, a system of axes $(\mathbf{i}_1, \mathbf{i}_2, \mathbf{i}_3)$, with $\mathbf{i}_3$ along the body symmetry axis and the other mutually orthogonal axes in the equatorial plane, is a natural choice. With reference to such a frame whose angular velocity $\mathbf{\Omega}$ characterizes the rotation of the body itself, the inertia tensor $[\mathscr{C}] = [\mathscr{C}_0]$, where $[\mathscr{C}_0]$ is diagonal with time-independent elements A, A, C; and the relative angular momentum $\mathbf{h}$ vanishes. Equation (13) then reduces to Euler's equations for the components of $\mathbf{\Omega}$.

However, the Earth is not a rigid body. It is not only elastic but also has a fluid outer core enclosing a solid inner core. Nevertheless, theoretical treatments in the literature assume (not always in specific terms, and often implicitly) that it is still possible to define an "**i**-system" of coordinate axes which satisfies the following requirements: (*a*) It provides a "crust-fixed" reference frame—i.e. its rotation is the same as that of the Earth's surface layer from which astronomical observations are made; (*b*) it is a "body-fixed" principal axis system, in the sense that the inertia tensor $[\mathscr{C}]$ relative to this system, with the contributions from the irrotational part of the displacement field removed, remains constant and equal to $[\mathscr{C}_0]$ at all times;

(*c*) it is a Tisserand mean axis system, defined as one with respect to which $\mathbf{h} = \mathbf{0}$; or at least a Tisserand mean mantle system wherein the contribution to $\mathbf{h}$ from the mantle vanishes. These requirements are not strictly mutually compatible, because of the radial dependence of the rotational part of the displacements in a deformable Earth. Nevertheless, a choice of $\mathbf{\Omega}$ and an $\mathbf{i}$-system can indeed be made so as to meet the desired requirements to the degree of accuracy that is currently needed for nutation theory, as has been verified in the work of Mathews et al (1991a).

With $\mathbf{\Omega}$ and the coordinate system thus chosen, one writes

$$[\mathscr{C}] = [\mathscr{C}_0] + [\delta\mathscr{C}], \qquad \mathbf{\Omega} = \Omega_0 \mathbf{i}_3 + \boldsymbol{\omega} \equiv \Omega_0(\mathbf{i}_3 + \mathbf{m}), \tag{14}$$

where $[\delta\mathscr{C}]/A$ and $\mathbf{m}$ represent the small fractional deviations from the unperturbed state, defined as one of rotation with angular velocity Ω_0 about $\mathbf{i}_3$. The small magnitude of the gravitational perturbations, as indicated above Equation (3), has the consequence that theoretical results with accuracy better than that of current observations of nutations can be obtained from a linearized form of Equation (13), linear in $[\delta\mathscr{C}]$, $\mathbf{m}$, and $\boldsymbol{\phi}$. However, terms of order $\varepsilon|\mathbf{m}|$ need to be retained, ε being the typical magnitude of the ellipticity ($\varepsilon \approx 1/300$). This degree of approximation has been referred to by Mathews et al (1991a) as $O(m\varepsilon)$.

To the first order in $[\delta\mathscr{C}]$ and $\mathbf{m}$, the equation of motion (13) reduces to

$$\Omega_0\left\{[\mathscr{C}_0]\cdot\frac{d\mathbf{m}}{dt} + \frac{d}{dt}[\delta\mathscr{C}]\cdot\mathbf{i}_3\right\} + \frac{d\mathbf{h}}{dt} + \Omega_0^2\mathbf{i}_3 \times \{[\mathscr{C}_0]\cdot\mathbf{m} + [\delta\mathscr{C}]\cdot\mathbf{i}_3\}$$
$$+ \Omega_0^2\mathbf{i}_3 \times ([\mathscr{C}_0]\cdot\mathbf{i}_3) = -\Omega_0^2 Ae\,\mathbf{i}_3 \times \boldsymbol{\phi}, \tag{15}$$

where $\mathbf{h}$ arises from motions of the fluid outer core and the solid inner core relative to the $\mathbf{i}$-system. Only the first two components of this vector equation are needed to describe the wobble, represented by m_1 and m_2. They may be combined into one equation for the complex amplitude $\tilde{m} = m_1 + im_2$ on using the explicit form of $[\mathscr{C}_0]$:

$$\left(\frac{d}{dt} - ie\Omega_0\right)\tilde{m} + \frac{1}{\Omega_0 A}\frac{d\tilde{h}}{dt} + \frac{1}{A}\left(\frac{d}{dt} + i\Omega_0\right)\tilde{c}_3 + \frac{i}{A}\tilde{h} = -ie\Omega_0\tilde{\phi}, \tag{16}$$

where $\tilde{h} = h_1 + ih_2$, $\tilde{\phi} = \phi_1 + i\phi_2$, and $\tilde{c}_3 = c_{13} + ic_{23}$, c_{13}, and c_{23} being elements of the tensor $[\delta\mathscr{C}]$.

In the simple case of a wholly solid but elastic model Earth, the terms in $\tilde{h}$ drop out in consequence of the definition of the $\mathbf{i}$-system; and the deformations represented by $[\delta\mathscr{C}]$ are attributable to the gravitational perturbation and the incremental centrifugal potential associated with the difference between $\mathbf{\Omega}$ and $\mathbf{\Omega}_0$. In the linear approximation, which is sufficient, one has then

$$\tilde{c}_3/A = \kappa(\tilde{m}-\tilde{\phi}). \tag{17}$$

This equation defines the compliance κ which represents the deformability of the Earth as a whole. The value of the degree 2 Love number k far from any resonances is related to κ by $k = 3GA\kappa/a^5\Omega_0^2$. Appearance of the combination $(\tilde{m}-\tilde{\phi})$ in (17) is due to the incremental centrifugal potential having precisely the same form as $\phi_e^{(m=1)}$ of Equation (1) except for the replacements $\phi_a \rightarrow -m_1$, $\phi_b \rightarrow -m_2$. Introducing (17) into (16) and restricting attention to a single Fourier component with the time dependence $e^{i\sigma\Omega_0 t}$ (i.e. with a frequency of σ cpsd), one obtains the equation

$$[\sigma+(1+\sigma)\kappa-e]\tilde{m} = [(1+\sigma)\kappa-e]\tilde{\phi}. \tag{18}$$

The solution of this equation in the absence of external excitation ($\tilde{\phi} = 0$) yields the natural frequency $\sigma_1 = (e-\kappa)/(1+\kappa)$ of the only free wobble mode which exists; it is lower than that of the rigid Earth ($\sigma_R \equiv e$). Under forcing by $\tilde{\phi}$, the wobble $\tilde{m}$ produced is obtained trivially from (18). It exhibits a resonance at $\sigma = \sigma_1$.

The presence of a fluid core necessitates separate consideration of its angular momentum. An elegant formulation of the torque equation for an ellipsoidally stratified and compressible fluid core, starting from the hydrodynamical equation, is due to Sasao et al (1980). Central to their treatment was the observation that the core flow which accompanies a periodic nutational motion of the mantle is equivalent to a "rigid" rotation of the fluid as a whole, apart from boundary effects whose contribution to the flow field is smaller than the rotational part of the flow by a factor of the order of 1/400, the flattening of the core-mantle boundary. This rotation can be characterized by an angular velocity vector $\mathbf{\Omega}_f$, such that the angular momentum $\mathbf{H}_f$ of the fluid is just $[\mathscr{C}_f]\cdot\mathbf{\Omega}_f$, where $[\mathscr{C}_f]$ is the inertia tensor of the fluid core alone. Sasao et al (1980) showed that the "boundary flow" does not contribute to the angular momentum balance of the fluid core to $O(m\varepsilon)$, and hence that $\mathbf{H}_f$ satisfies the equation

$$\frac{d\mathbf{H}_f}{dt} - \boldsymbol{\omega}_f \times \mathbf{H}_f = 0, \tag{19}$$

where $\boldsymbol{\omega}_f$ characterizes the differential rotation of the fluid core relative to the mantle:

$$\boldsymbol{\omega}_{\rm f} \equiv \Omega_0 \mathbf{m}_{\rm f} \equiv \boldsymbol{\Omega}_{\rm f} - \boldsymbol{\Omega}. \tag{20}$$

An equation identical in form to (19) had been obtained by Poincaré (1910) for an incompressible and homogeneous core from the Helmholtz vorticity theorem. The derivation by Sasao et al (1980) showed it to be valid also for a stratified and compressible core, to $O(m\varepsilon)$.

Linearization of Equation (19), followed by the introduction of complex quantities $\tilde{m}_{\rm f}$ and $\tilde{c}_3^{\rm f}$ which are analogous to $\tilde{m}$ and $\tilde{c}_3$ but pertain to the fluid core alone, leads to an equation involving $\tilde{m}_{\rm f}$, $\tilde{c}_3^{\rm f}$, and $\tilde{m}$. As regards the torque equation (13) for the whole Earth, its linearized form (16) now gets a nonvanishing contribution from $\tilde{h}$ because of the rotation of the core relative to the mantle: $\mathbf{h} = [\mathscr{C}_{\rm f}] \cdot \boldsymbol{\omega}_{\rm f}$. Furthermore, both $\tilde{c}_3$ and $\tilde{c}_3^{\rm f}$ will now involve $\tilde{m}_{\rm f}$ as well as $(\tilde{m} - \tilde{\phi})$. As a result, there are now four compliances, defined through the equations

$$\tilde{c}_3 = A[\kappa(\tilde{m} - \tilde{\phi}) + \xi \tilde{m}_{\rm f}],$$
$$\tilde{c}_3^{\rm f} = A_{\rm f}[\gamma(\tilde{m} - \tilde{\phi}) + \beta \tilde{m}_{\rm f}], \tag{21}$$

by Sasao et al (1980), who also proved a reciprocity relation, $A\xi = A_{\rm f}\gamma$, connecting two of the compliances. On introducing Equations (21) into the linearized versions of the angular momentum equations for the whole Earth and for the fluid core, one obtains the following coupled equations for $\tilde{m}$ and $\tilde{m}_{\rm f}$:

$$[\sigma + (1+\sigma)\kappa - e]\tilde{m} + (1+\sigma)(\xi + A_{\rm f}/A)\tilde{m}_{\rm f} = [(1+\sigma)\kappa - e]\tilde{\phi}, \tag{22a}$$

$$\sigma(1+\gamma)\tilde{m} + [(1+\sigma) + \beta\sigma + e_{\rm f}]\tilde{m}_{\rm f} = \gamma\sigma\tilde{\phi}. \tag{22b}$$

With $\tilde{\phi} = 0$, these equations determine the eigenfrequencies of the free wobble modes. To the first order in small quantities, they are

$$\sigma_1 = (A/A_m)(e - \kappa), \qquad \sigma_2 = -1 - (A/A_m)(e_{\rm f} - \beta). \tag{23}$$

These correspond to the Chandler wobble and the nearly diurnal free wobble (NDFW), respectively.

Mathews et al (1991a) extended the treatment of Sasao et al (1980) by (*a*) giving an independent proof of the validity of (19) even in the presence of the inner core, (*b*) modifying the detailed structure of $\mathbf{H}$ and $\mathbf{H}_{\rm f}$ to allow for independent orientation and independent rotation of the inner core, and (*c*) adding two more equations of motion: one governing the dynamics of the angular velocity vector $\boldsymbol{\Omega}_{\rm s}$ of the inner core, and the other affirming that the motion of the symmetry axis $\mathbf{i}'_3$ of the solid inner core is governed by the rotation $\boldsymbol{\Omega}_{\rm s}$. Equations (22) then get replaced by a set of four simultaneous linear equations in the dynamical variables $\tilde{m}$, $\tilde{m}_{\rm f}$, $\tilde{m}_{\rm s}$

and $\tilde{n}_s$, where $\tilde{m}_s$ and $\tilde{n}_s$ are the complex amplitudes associated with $\mathbf{m}_s \equiv (\mathbf{\Omega}_s - \mathbf{\Omega})/\Omega_0$ and $\mathbf{n}_s \equiv \mathbf{i}'_3 - \mathbf{i}_3$, respectively, and where the subscript s refers to the solid inner core. This set of equations, or the equivalent matrix equation, leads to four normal modes, of which the PFCN and ICW are new. The expressions given in Equation (23) for the frequencies of the Chandler wobble and the RFCN get modified only to the extent of a replacement of A/A_m in σ_2 by $(A-A_s)/A_m$. The PFCN frequency is $-1+(1+A_s/A_m)X$, where X is controlled primarily by the ellipticity e_s of the inner core, though it depends also on other Earth parameters. In the Earth models studies, $X \approx 0.8e_s$, and the ICW frequency is $\approx 0.2e_s$ cpsd. The frequency dependence of the amplitudes of forced nutation follows from the structure of the matrix equation, and has been expressed by Mathews et al (1991a) in the form of a resonance formula for the normalized amplitude $\tilde{\eta}_N$:

$$\tilde{\eta}_N(\sigma) = R + R'(1+\sigma) + \sum_{\alpha=1}^{4} \frac{R_\alpha}{\sigma - \sigma_\alpha}, \tag{24}$$

wherein the eigenfrequencies σ_α and resonance coefficients R_α refer to the four eigenmodes: Chandler wobble, RFCN, PFCN, and ICW. Nonrigidity of the Earth is responsible for the deviation of $\tilde{\eta}_N(\sigma)$ from unity. Besides influencing the resonance terms which result from the existence of the fluid outer and solid inner cores, elasticity of the Earth also makes a nonresonant contribution to $\tilde{\eta}_N(\sigma)$; it is represented by $(R-1)$ together with the R' term which takes account of its small variation with frequency across the diurnal band.

A valuable feature of the angular momentum approach is that the nutation problem in the frequency domain is reduced to the solution of a matrix equation wherein those aspects of the Earth's structural and material properties which affect the nutations are consolidated into a set of parameters appearing in the matrix elements. The set consists of the dynamical ellipticities and the ratios of the moments of inertia of the different regions, the compliances, the fluid density at the inner core boundary, and another parameter which measures the strength of the torque on the inner core when out of alignment with the mantle. When the inner core is explicitly taken into account, the expressions for $\tilde{c}_3$ and $\tilde{c}_3^f$ contain additional terms proportional to $\tilde{m}_s$; and a similar expression for $\tilde{c}_3^s$ also enters the picture. The number of compliances is then nine in all. There exist reciprocity relations among them (Mathews et al 1991a) which are equivalent to symmetry of the matrix of coefficients of $(\tilde{m}-\tilde{\phi})$, $\tilde{m}_f$, and $\tilde{m}_s$ in the expressions for $\tilde{c}_3$, $\tilde{c}_3^f$, and $\tilde{c}_3^s$. The magnitudes of the compliances are $O(\varepsilon)$ or smaller: $\kappa \approx 1.0 \times 10^{-3}$, $\gamma \approx 2.0 \times 10^{-3}$, and $\beta \approx 0.6 \times 10^{-3}$ are the largest; the others range in value down to about 10^{-9}. Mathews et al (1991b) have tabulated the values of all the relevant parameters for Earth model PREM of Dziewonski & Anderson (1981) as well as for 1066A.

Because the dependence of the matrix equation on the Earth model parameters is explicit, it is easy to determine the sensitivity of the eigenfrequencies and nutation amplitudes to any small changes in the value of any of these parameters. This feature greatly facilitates the interpretation of the nutation data in geophysical terms. The fact that quite accurate analytical solutions are obtainable, such as the expressions (23) for the eigenfrequencies and their generalizations taking explicit account of the inner core, is especially valuable; it was used to advantage by Gwinn et al (1986). Tables of sensitivities of R, R', R_α, and σ_α, as well as of the nutation amplitudes at important tidal frequencies, are given by Mathews et al (1991b).

Both the displacement field and angular momentum approaches retain in the equations of motion only terms up to the first order in the ellipticity. The former approach employs the additional approximation, already mentioned, of truncating the spherical harmonic expansion of the displacement field. It has, however, been rather widely believed (see, for example, Kinoshita & Sasao 1989 and Wahr & de Vries 1990) that the angular momentum approach is inherently less accurate than the displacement field method. One reason for this belief lies in the quasistatic spherical approximation used in evaluating the compliances. This approximation is not a necessary part of the method, however. The effects of the Earth's rotation and ellipticity on deformations, hitherto neglected, have been computed (Buffett et al 1991); they contribute only $O(\varepsilon)$ corrections to the compliances, as expected, and the changes caused by them in the theoretical nutation amplitudes are less than the standard deviations of the corresponding amplitudes inferred from VLBI observations. A second reason has to do with the fact that only the purely rotational part $\boldsymbol{\omega}_f \times \mathbf{r}$ of the flow field in the fluid core plays a role in the angular momentum balance equations, giving rise to the impression that a severe truncation, retaining only the $\boldsymbol{\tau}_1^1$ part of the field, has been performed. However, the noninvolvement of the residual part of the flow field is merely a consequence of the negligible magnitude of its contribution to the angular momentum. The fact that it is of a higher order of smallness than $O(m\varepsilon)$ was shown by Sasao et al (1980) on the basis of the hydrodynamical equation of fluid flow in the core; the proof has been sharpened and extended by Mathews et al (1991a). It is nevertheless true that the residual flow caused by the nonsphericity of the boundaries contributes to deformations; it is properly accounted for when ellipticity and rotation corrections are added (Buffett et al 1991) to the compliances. It seems safe to conclude that with the inclusion of these corrections, the Liouville equation method would be no less accurate than the displacement field approach.

DEVIATIONS FROM IDEALIZED MODELS

Hydrostatic equilibrium Earth models yield values for the dynamic ellipticity e that are too low by over 1%, compared to the value determined from the precession rate. The process of computing nutation amplitudes from such Earth models must therefore include an adjustment to correct for this difference. Wahr (1981c) makes this adjustment by introducing the astronomically determined value of e in the TOM resonance term in his nutation formula, while leaving all other parameters in the formula to be determined from Earth models. Recent work (Zhu et al 1990; Mathews et al 1991a,b) has shown that this adjustment is incomplete and not entirely adequate. The equations resulting from the angular momentum approach involve e as an explicit parameter; in making numerical computations, Mathews et al (1991b) simply use the precession-determined value for this parameter, instead of that pertaining to the Earth model. They find that the difference between the two procedures causes a difference of over 0.55 mas in the computed values of the prograde semiannual amplitude, which is about 14 times the present standard deviation in the VLBI estimate of this amplitude.

The observed nutations include the effects of ocean tides, mantle anelasticity, and noninertial forces (viscous, magnetic) which affect the motion of the fluid outer core relative to the mantle and the solid inner core, and various other effects such as those due to lateral inhomogeneities in the Earth. Some of these mechanisms cause the response of the Earth to be slightly out of phase with the gravitational torque; their contributions to the amplitudes of circular nutations have, therefore, an imaginary part. Both the real and imaginary parts of such contributions need to be taken into account while comparing theoretical and observational results.

Ocean tidal movements in the diurnal band of frequencies are influenced by the FCN resonance, and cause deformations of the Earth which affect the nutations. Wahr & Sasao (1981) have made theoretical estimates of the contributions from ocean tides to nutation amplitudes, using as inputs the RFCN eigenfrequency from Wahr (1981c) and tide heights from contemporary models (Schwiderski 1980, Parke & Handershott 1979). Comparison of these models with the determinations from satellite observations (e.g. Marsh et al 1990) suggests that tide height estimates now are good to about 15%. However, modifications to the Wahr & Sasao results need to be made to take account of the difference of the RFCN eigenfrequency, as currently estimated, from Wahr's value.

Dissipative mechanisms associated with anelasticity in the mantle influence the dynamics of the Earth through their effects on the elastic moduli. Seismological studies indicate that dissipation in compressional movements is very small compared with that in shear deformations (see, for instance, Karato & Spetzler 1990). It is usual therefore to confine attention to the corrections from anelasticity to the rigidity modulus inferred from seismic studies. These corrections, which are frequency-dependent and include an imaginary part, vary with depth in the mantle. The most widely used class of models for anelasticity assumes a power-law frequency dependence for the mantle Q-factor: $Q(\omega) = Q(\omega_m)(\omega/\omega_m)^{\alpha}$. The fractional change in the rigidity modulus corresponding to this behavior is given (see, for instance, Wahr & Bergen 1986) by the complex quantity

$$\frac{\delta\mu}{\mu_0} = \frac{1}{Q(\omega_m)}\left\{\cot\left(\frac{\alpha\pi}{2}\right)\left[1-\left(\frac{\omega_m}{\omega}\right)^{\alpha}\right]+i\left(\frac{\omega_m}{\omega}\right)^{\alpha}\right\}. \tag{25}$$

Different models differ in the values assigned to $Q(\omega_m)$ in different parts of the mantle and to the reference frequency ω_m. Smith & Dahlen (1981), using their own estimate of the Q of the Chandler wobble together with the Q for the seismic frequency band, estimated the value of α that would be appropriate for the intervening frequency region which includes diurnal frequencies. They found the most plausible value of α to be 0.15 for the QMU model of Sailor & Dziewonski (1978) and 0.09 for the β model (Sipkin & Jordan 1980). A survey of various estimates of the frequency dependence of Q may be found in Karato & Spetzler (1990). Anelasticity contributions to nutation amplitudes were estimated by Wahr & Bergen (1986) on the basis of the models considered by Smith & Dahlen (1981). Computations of nutation amplitudes including the effects of anelasticity were made by Dehant (1988, 1990), by using a modified form of the procedure of Wahr (1981b,c) to solve displacement field equations wherein complex functions of the radial variable r characteristic of various anelastic Earth models were employed for the rigidity modulus. Differences among estimates obtained from the different computations and for the various models are large enough to raise the possibility of discrimination among them on the basis of nutation observations: Anelastic contributions estimated by Wahr & Bergen on the basis of the β model are about 70% higher than those based on the QMU model, while the values obtained by Dehant (1990) are 15% to 200% higher.

Relative motions between the fluid outer core and the solid regions across its outer and inner boundaries are impeded by viscous drag as well as by electromagnetic forces arising from the motion of the flux lines of the frozen-in core magnetic field through the inner core and any conducting region in the lowermost mantle. Estimates of the resulting torques, based on Rochester's (1976) calculations, lead to negligible contributions to nutations. However, it has been suggested recently (Buffett 1991) that these estimates may be too low.

ESTIMATION OF NUTATION AMPLITUDES: THE VLBI METHOD

The most accurate estimates available now for the nutation amplitudes are obtained from analyses of VLBI group-delay data. The group delay for a pair of radio telescopes is the difference between the transit times of an electromagnetic signal from a given radio source (usually a quasar) to the two telescopes; it is given by $\tau_g = (\mathbf{B}\cdot\mathbf{s}/c)+\Delta\tau$, where **B** is the instantaneous baseline vector connecting the two telescope locations, **s** is the unit vector in the direction from the barycenter of the solar system to the radio source and $\Delta\tau$ comprises the effects of aberration, relativistic corrections to the travel times, delays due to the decreased speed of propagation of the signals in the atmosphere and ionosphere, clock "synchronization" errors, and other instrumental delays. The radio signals from a single source, after amplification, heterodyning and "clipping," are simultaneously recorded on magnetic tape, along with a time base generated by a local hydrogen maser clock, at each of a number of radio telescopes belonging to the VLBI network (see, e.g. Clark et al 1985). After a recording period of 100–300 sec, referred to as an "epoch" of observation, the telescopes switch simultaneously to another source, and this process continues usually for a period of 24–26 hours. The correlation function $\langle F_i(t)F_j(t-\tau)\rangle$ between the recordings made at telescopes i and j is expected to be a maximum when τ is equal to the group delay $(\tau_g)_{ij}$ for this pair of telescopes. This basic property is exploited with the aid of various sophisticated techniques to determine the group delay for any given pair of telescopes, the actual correlation being done after shipping the magnetic tapes to a common correlator facility. From each one-day experiment, one thus determines, for each of the epochs, a set of group delays corresponding to the various pairs of telescopes in the VLBI network. The aggregate of such numbers determined over a period of years from multisite experiments (performed once every five days or more often since 1984, and less frequently earlier) constitutes the raw data for nutation studies.

Relative to a space-fixed reference frame wherein **s** remains invariant, the vector **B** varies with time because the baseline rotates with the Earth, and also because changes in the baseline take place due to tidal deformations, tectonic plate movements, etc. Earth rotation is decomposed into uniform diurnal rotation, precession, nutations due to tidal forcing, and wobbles unrelated to these forced nutations. Using a priori expressions for these elements of the rotation (e.g. the IAU 1980 series for the nutation) and models for the other processes which contribute to the group delays, the expected group delay at each of the epochs of observation during a day-long experiment is computed for each of the telescope pairs involved. The set of residual differences between the observed group delays from a one-day session and the corresponding expected values is used to estimate, among other parameters, the deviations in obliquity and longitude of the Earth's pole from its expected (a priori) position for that day. The VLBI data from several years of observations thus provide two time series, with values for each day of observation, of corrections to the nutations in obliquity and longitude. Corrections to the a priori values of the coefficients $\Delta\varepsilon_{\nu I}$, $\Delta\psi_{\nu I}$, $\Delta\varepsilon_{\nu O}$, and $\Delta\psi_{\nu O}$ at the various tidal frequencies are obtained by spectral analysis. The corresponding corrections to the amplitudes of circular nutations then follow from Equations (5). The standard deviations of these estimates are now only 0.04 mas for periods up to 430 days and 1.0 mas for longer periods.

OBSERVATIONAL RESULTS AND THEORETICAL INTERPRETATION

With the attainment of submilliarcsecond accuracies in the determination of nutation amplitudes from VLBI data, the need for modifications to some features of standard Earth models became apparent. Herring et al (1986) found a difference of almost 2 mas between the real part of the VLBI-determined amplitude of the retrograde annual nutation and the theoretical value (Wahr 1981c) adopted in the IAU 1980 nutation series. Noting the influence of the FCN resonance on this amplitude, they observed that the discrepancy could be eliminated by changing the period of the FCN from Wahr's figure of 460 days to 433 days. More recently, Zhu et al (1990) found $\sigma_2 = -1.0023064$, equivalent to an FCN period of 433.5 days, from a three-parameter fit of the Wahr formula (12) to the VLBI-determined fortnightly, semiannual, annual, and 18.6 year nutation amplitudes after applying corrections for ocean tides. (Values recommended by them for the other parameters, on the basis of this fit, are: $B_0 = 0.4708$, $B_1 = 0.8121$, $B_2 = 0.6754$, $B_3 = 1.068$, $\sigma_1 = 0.0022995$, $\sigma_3 = 0.003285$.) Gwinn et al (1986) interpreted the modified period, through Equation (23), in terms of a higher value of the ellipticity e_f of the fluid outer core than in hydrostatic-equilibrium Earth models. From the higher e_f, they inferred an extra flattening of the core-mantle boundary relative to hydrostatic equilibrium, equivalent to an increase of about 0.5 ± 0.1 km in the difference between the equatorial and polar radii of this boundary. Mathews et al (1991b) obtain, from their theoretical computation, a value of about 430 m. Presence of a low density layer at the top of the fluid core, or lateral heterogeneities in the mantle or the core, could affect these estimates, but no clear indication of any need for revision has emerged from studies of this question (Yoder & Ivins 1988a, Wahr & de Vries 1989). Estimates of nonhydrostatic flattening of the CMB made by global seismic travel-time tomography (Creager & Jordan 1986, Morelli & Dziewonski 1987, Doornbos & Hilton 1989) have been at variance with each other and with the estimate from nutations (see also Masters 1991). Rodgers & Wahr (1990) have concluded from travel-time inversions using different ISC data sets and also synthetic data that reliable inference of CMB topography is not possible from available data. Gundmundsson & Clayton (1991) conclude, from a test study, that an uncertainty of 5 to 10 km exists in maps of the CMB based on seismic tomography inferred from seismic travel-time inversion, and that inhomogeneities in the upper mantle play a significant role in obscuring deep Earth structure. It appears therefore that the best information available at present about the flattening of the CMB comes from nutations.

Modification of the fluid core ellipticity e_f as described above still leaves discrepancies between theory and observation in amplitudes other than that for the retrograde annual nutations, even after including corrections for anelastic and ocean tidal effects. The largest one, in the prograde semiannual amplitude, appeared to be worsened by the anelasticity corrections (Wahr & Bergen 1986). With a magnitude close to 1 mas, this discrepancy was worrisome (Herring 1988, Wahr 1988b, Wahr & de Vries 1990). However, the situation has changed recently. Analysis of VLBI data by Herring et al (1991) yielded new estimates for nutation amplitudes, and also a correction of -3.2 ± 1.3. mas/yr to the precession constant, equivalent to a decrease of about 6 parts in 10^5 in the value of the Earth's ellipticity e. Their results agree, within the root-sum-square standard deviations, with estimates of the 18.6 year nutation from optical astrometry (Capitaine et al 1988, Li 1989), and of the precession constant (estimated correction: -2.7 ± 0.4 mas/yr) and the 18.6 year and annual nutations from lunar laser ranging data (Williams et al 1991). The theoretical results of Mathews et al (1991b) for the real parts of the amplitudes of circular nutations and the observational estimates of Herring et al (1991) are in good agreement, consistent with the statistical uncertainties, at all tidal frequencies except one. The exception is in the prograde fortnightly amplitude, where there is a discrepancy of 0.24 mas; this too is found to disappear (Herring 1991 and personal communication) when VLBI data analysis is done anew with the addition of semidiurnal UT1 variations to the set of quantities to be estimated, which hitherto included, along with nutation amplitudes and other parameters, only one UT1 variation for each day of observation. [This change in the VLBI estimate of the fortnightly nutation amplitude should also affect the parameter values determined by Zhu et al (1990), referred to above.] Besides the dynamical effects of the solid inner core, the following aspects of the treatment of Mathews et al (1991a,b) have significant roles in leading to the new theoretical values: (*a*) use of the new rigid-Earth nutation series of Zhu & Groten (1989), modified to take account of the above-mentioned correction to the precession constant; (*b*) use of the value of e determined from the precession constant, instead of the Earth model value of e, in the final equations of the theory; (*c*) corrections for ocean tidal effects computed following Wahr & Sasao (1981), and corrections for anelasticity based on the Wahr & Bergen (1986) theory, using the QMU model with $\alpha = 0.15$ (corrections based on other models being either inconsistent with VLBI data, or leading to poorer agreement); and (*d*) adjustment of the ellipticity e_f of the fluid outer core to match the observed retrograde annual amplitude with the corresponding theoretical value, the latter having been computed using PREM values for all the Earth model parameters except e and e_f, and corrected for ocean tidal and anelastic effects. The values of the parameters in the resonance formula (24) when the Earth model is characterized by these modified PREM parameters (Mathews et al 1991b) are shown in Table 1. [They have a weak dependence on the observed retrograde annual amplitude, since it was used to determine the modified e_f. If the estimate of this amplitude were to change by one standard deviation (0.04 mas), the parameters R_2 and $(1+\sigma_2)$ would need to be modified by 0.08%, while the remaining parameters would remain practically unaffected.] The nutation amplitudes obtained from the theory of Mathews et al (1991a,b) are shown in Table 2, along with the corrections applied. Table 3 presents a comparison, with the IAU 1980 values, of these results as well as of the observational estimates of Herring et al (1991) and the results of the data fitting done by Zhu et al (1990).

Significant imaginary components, exceeding 3 times the standard deviation of 0.04 mas, are found in the VLBI estimates by Herring et al (1991) of the retrograde and prograde annual amplitudes (0.32 and 0.15 mas, respectively) and of the prograde semiannual amplitude (−0.47 mas). The imaginary parts of the contributions to these amplitudes from anelasticity and ocean tides, computed using the same models as above, account for the last of these, leaving a residual of only 0.04 mas; but discrepancies of 0.39 and 0.13 mas, respectively, remain in the first two amplitudes. Buffett (1991) has presented a possible explanation of these discrepancies, which

Table 1 Normal mode frequencies and resonance coefficients for modified PREM[a]

	σ_α	T_α	R_α
CW	0.0025312	394.0	-5.7992×10^{-4}
RFCN	−1.0023203	−429.8	-1.1976×10^{-4}
PFCN	−0.9979031	475.6	5.1098×10^{-7}
ICW	0.0004139	2409	-4.5447×10^{-8}
	$R = 1.05128$	$R' = -0.28037$	

[a] The periods T_α shown are in solar days; they are relative to a space-fixed frame for RFCN and PFCN, and relative to an Earth-fixed frame for CW and ICW. The frequencies σ_α are in cpsd. See text for definitions and abbreviations. The tabulated values differ slightly from those given by Mathews et al (1991b), in consequence of an error correction which changes the value of the ellipticity of the Earth to $e = 0.00328454$. This error had resulted from the use of the wrong sign while applying the correction inferred from VLBI data by Herring et al (1991) to the precession constant. The changes in the nutation amplitudes resulting from the error correction are tiny, the largest being 0.005 mas.

does not significantly affect the agreement elsewhere, in terms of electromagnetic torques between the fluid core and the mantle; this explanation requires that there be a highly conducting layer at the bottom of the mantle, and that the energy in the magnetic field at the core-mantle boundary be about five times the conventional estimates. These estimates are based on surface measurements of the magnetic field which take into account spherical harmonic components of the field up to order 10. Buffett suggests that the extra energy could be contributed by a large number of spherical harmonic components of degree higher than 10 and that the high inverse-power dependence of such components on radius would make them undetectable at the Earth's surface.

The presence of dissipative processes has the consequence that at least some of the compliances cease to be real, and correspondingly, the parameters in resonance formulae like (12) and (24) become complex. Dissipative torques acting between the core and the mantle [considered, for instance, by Sasao et al (1977) and Buffett (1991)] make β complex. Yoder & Ivins (1988b) sought to represent tidal effects by adding complex corrections to κ, ξ, γ, and β of the oceanless Earth. However, a clear theoretical basis for determination of the complex contributions to the compliances from the various dissipative processes needs to be developed, and uncertainties in the modeling of such processes need to be narrowed down, before a single

Table 2 New theoretical results for nutation amplitudes[a]

	Nutation amplitudes		Corrections		
Tidal period	Rigid Earth	Modified PREM	Ocean tides	Anelas-ticity	Corrected amplitudes
−182.6	−22.59	−24.66	0.06	0.04	−24.56
−365.3	−24.89	−33.47	0.18	0.29	−33.00
−6798.4	−8050.98	−8023.40	−0.96	−0.34	−8024.70
6798.4	-1177.03	−1180.57	0.12	0.05	−1180.40
365.3	25.03	25.74	−0.02	−0.01	25.71
182.6	−530.75	−549.37	0.60	0.28	−548.49
13.7	−91.51	−94.24	0.03	0.06	−94.15

[a] Tidal periods are in solar days, and nutation amplitudes and corrections in mas. The rigid-Earth amplitudes are from Zhu et al (1989) as modified by Herring et al (1991) to correspond to the new value of the Earth's ellipticity, $e = 0.00328454$. The third column is for an oceanless, dissipationless Earth, with the above value for e and with PREM values for all other parameters except e_f which is adjusted as explained in the text. The remaining columns are from Mathews et al (1991b) and Herring et al (1991).

formula which accurately represents both the real and imaginary parts of the nutation amplitudes can be arrived at. The complex eigenfrequency of the NDFW has been estimated in recent years both from VLBI and tidal gravimetric observations. While the two approaches give concurring results for the real part, as noted in an earlier section, they lead to widely different estimates for the imaginary part, or equivalently, for the Q and the decay time of the NDFW resonance. The imaginary part σ_{im}, in cpsd, of the eigenfrequency as estimated by Gwinn et al (1986) from VLBI data was $(2.3 \pm 1.1) \times 10^{-5}$; a more recent estimate (T. A. Herring, personal communication) is $(1.7 \pm 0.1) \times 10^{-5}$. Since the decay time τ, in sidereal days, is

$$\tau = \frac{1}{2\pi\sigma_{\mathrm{im}}} = \frac{2Q}{2\pi\sigma_{\mathrm{r}}}, \tag{26}$$

where $\sigma_{\mathrm{r}} = |\sigma_2| \approx 1 + 1/430$ is the magnitude of the real part of the eigenfrequency, the decay times corresponding to these estimates are ≈ 6900 and 9400 days, respectively. However, analyses of tidal gravimetric data by Neuberg et al (1987) and Richter & Zürn (1988) yield $Q \approx 2800$ ($\tau \approx 890$

Table 3 Comparison of recent results with IAU 1980 nutation amplitudes[a]

Tidal period	Amplitude IAU 1980	Differences from IAU 1980 VLBI (H91)	Theory (M91; H91)	Data fit (Z90)
−182.6	−24.53	−0.05	−0.03	−0.09
−365.3	−31.06	−1.94	−1.94	−1.84
−6798.4	−8022.06	−3.11	−2.65	−2.71
6798.4	−1180.45	−0.28	0.05	−0.09
365.3	25.66	0.04	0.05	0.05
182.6	−549.07	0.52	0.58	0.49
13.7	−94.08	−0.32	−0.07	−0.34

[a] Tidal periods are in solar days and nutation amplitudes and differences in mas. IAU 1980 values are for an oceanless dissipationless Earth (Model 1066A) with rigid-Earth amplitudes and e from Kinoshita et al (1979). H91: Herring et al (1991); M91: Mathews et al (1991b), reproduced in the last column of Table 2; Z90: Zhu et al (1990). The IAU values have been subtracted from the amplitudes given in these papers to obtain the differences tabulated. The standard deviations of the VLBI estimates shown above are 1.0 mas for the ± 6798.4 day periods and 0.04 mas for all others. The observed amplitudes at all the periods shown in this table were used in the data fit in Z90.

days) and $Q = 3120 \pm 320$ ($\tau \approx 990$ days), respectively; a more recent result (Neuberg et al 1990) is $Q = 2800 \pm 50$. The apparent persistent difference between the estimates obtained by the two methods is yet to be explained. An understanding of the mechanisms responsible for the damping of the NDFW has to await the resolution of this problem.

DISCUSSION AND OUTLOOK

The accuracy with which Earth orientation can be determined as a function of time has increased tremendously over the past decade, due to advances in the technology of VLBI and in techniques of data analysis, and also due to the expanding volume of data over a lengthening time span. Of the factors which limit the accuracy of the determination, one of the most important is the atmospheric contribution to group delays, especially the part due to water vapor which is the hardest to estimate reliably (see e.g. Elgered et al 1991). In the case of the 18.6 year nutations, the accuracy of the estimated amplitudes is limited by the time span for which quality data are available. Presently this span covers only about half the period, which makes it difficult to separate this long-period nutation from the precessional motion of the pole. Reduction of the uncertainty in the 18.6 (and 9.3) year nutation amplitudes to the same level as in others will happen with the passage of years; the concomitant increase in accuracy of the precession constant will greatly enhance our ability to extract geophysical information from the observed nutations.

Uncertainties in the estimated contributions to the observed amplitudes from anelasticity, ocean tides, and other effects constitute a serious limitation on the extent to which idealized-Earth parameters may be constrained by the observations. Such contributions have to be removed from the VLBI-estimated amplitudes to get the equivalent values for an idealized Earth, before comparing with the theoretical idealized-Earth results. In this process, the uncertainties in the estimates of these contributions combine with the estimated errors in the VLBI estimates to widen the limits on the "observed" idealized-Earth amplitudes and make it correspondingly more difficult to get tight constraints on the Earth parameters on which the amplitudes depend. Estimates of ocean tidal quantities are now available from satellite tracking data. Spatial and temporal changes in the Earth's gravity field associated with ocean and body tides cause perturbations of satellite orbits; and estimations of the amplitudes and phases of various components of the ocean tide form part of the analysis of satellite data. The amplitudes found from the GEM-T2 analysis (Marsh et al 1990) for the O_1, P_1, and K_1 tides carry standard deviations of 6 to 15% and differ from the corresponding estimates of Schwiderski (1980) by 10 to 20%; uncertainties in the phase angles range from about 3° to 8°. The perturbations of the gravity field include also the effects of mantle anelasticity equivalent to an increase of 1 to 2% in the Love number k, according to estimates by Wahr & Bergen (1986) and Dehant (1987). But these effects are very much smaller than those of the ocean tides, and uncertainties in the latter make it doubtful whether useful information about anelasticity can be obtained from satellite tracking data (see, for instance, Zschau 1986). There seems to have been no effort so far to confront with experiment the theoretical estimates, by the same authors, of anelasticity corrections of 1 to 2% to the h Love number. The effect of anelasticity on the tidal gravimetric factor $\delta \equiv (1+h-3k/2)$ is too small to be identified. Though the values of δ obtained by Baker et al (1989) from measurements at a number of European stations differ from the theoretical values of Dehant (1987) for an anelastic Earth only by about 0.2%, the contributions from anelasticity to δ according to varying estimates by Wahr & Bergen (1986), Dehant (1987), and Dehant & Zschau

(1989) are still smaller: They lie in the range of 0.03 to 0.08%, except for values up to 0.15% when the β model of Sipkin & Jordan (1980) is used. In regard to ocean tides, however, substantial reduction in the uncertainties in our knowledge of their characteristics may be expected with the rapidly increasing quantity of satellite tracking data and improvements in their accuracy. With developments also in the theoretical modeling of ocean tidal movements, including that of currents (see, for instance, Brosche et al 1989), it should be possible to estimate ocean tide corrections to nutation amplitudes with greater accuracy than at present. Such a development, leading to more stringent limits on the observed amplitudes corrected for ocean tide effects, is likely to result in effective constraints on models of anelasticity: It appears, in fact, that we may already be on the threshold of being able to attain some degree of discrimination among various models of anelasticity (Herring et al 1991) from comparison of the theoretical and the observational results on nutations. In any case, it seems safe to predict that significant information on Earth structure and dynamics, including some difficult to obtain by other means, will emerge from studies on nutations in the next few years.

ACKNOWLEDGMENTS

We thank T. A. Herring and J. L. Davis for carefully reading the review and for helpful comments. Part of this review was written while one of us (P.M.M.) was visiting at the Indian Institute of Science, Bangalore; he wishes to thank J. Pasupathy and the Centre for Theoretical Studies at the Institute for their hospitality. This work was supported in part by grant EAR-89-05560 from the National Science Foundation.

Literature Cited

Babcock, A. K., Wilkins, G. A., eds. 1988. *The Earth's Rotation and Reference Frames for Geodesy and Geodynamics, Proc. IAU Symp. No. 128*. Dordrecht: Kluwer

Baker, T. F., Edge, R. J., Jeffries, G. 1989. European tidal gravity: an improved agreement between observations and models. *Geophys. Res. Lett.* 16: 1109–12

Brosche, P., Seiler, U., Sündermann, J., Wünsch, J. 1989. Periodic changes in Earth's rotation due to oceanic tides. *Astron. Astrophys.* 220: 318–20

Buffett, B. A. 1991. Constraints on magnetic energy and mantle conductivity from the forced nutations of the Earth. *J. Geophys. Res.* Submitted

Buffett, B. A., Mathews, P. M., Herring, T. A., Shapiro, I. I. 1991. Forced nutations of the Earth: influence of inner core dynamics 4. Elastic deformation. *J. Geophys. Res.* Submitted

Capitaine, N., Li, Z. X., Nie, S. Z. 1988. Determination of the principal nutation from improved BIH astrometric data. *Astron. Astrophys.* 202: 306–8

Clark, T. A., Corey, B. E., Davis, J. L., Elgered, G., Herring, T. A., et al. 1985. Precision geodesy using the Mark-III very-long-baseline interferometer system. *IEEE Trans. Geosci. Remote Sens.* GE-23: 438–49

Creager, K. C., Jordan, T. H. 1986. Aspherical structure of the core-mantle boundary from PKP travel times. *Geophys. Res. Lett.* 13: 1497–1500

Crossley, D. J. 1975. Core undertones with rotation. *Geophys. J. R. Astron. Soc.* 42: 477–88

Crossley, D. J. 1980. Simple core undertones. *Geophys. J. R. Astron. Soc.* 60: 129–61

Cummins, P., Wahr, J. M., Agnew, D. C., Tamura, Y. 1991. Constraining core undertones using stacked IDA gravity records. *Geophys. J. Int.* 106: 189–98

Dahlen, F. A. 1968. The normal modes of a rotating elliptical Earth. *Geophys. J. R. Astron. Soc.* 16: 329–67

Dahlen, F. A. 1972. Elastic dislocation theory for a self-gravitating elastic configuration with an initial static stress field. *Geophys. J. R. Astron. Soc.* 28: 357–83

Dahlen, F. A., Smith, M. L. 1975. The influence of rotation on the free oscillations of the Earth. *Philos. Trans. R. Soc. Ser. A* 279: 583–624

de Vries, D., Wahr, J. M. 1991. The effects of the solid inner core and nonhydrostatic structure on the Earth's forced nutations and Earth tides. *J. Geophys. Res.* 96: 8275–93

Dehant, V. 1987. Tidal parameters for an inelastic Earth. *Phys. Earth Planet. Inter.* 49: 97–116

Dehant, V. 1988. Nutations and inelasticity of the Earth. See Babcock & Wilkins 1988, pp. 323–29

Dehant, V. 1990. On the nutations of a more realistic Earth model. *Geophys. J. Int.* 100: 477–83

Dehant, V., Zschau, J. 1989. The effect of mantle inelasticity on tidal gravity: a comparison between the spherical and the elliptical Earth model. *Geophys. J.* 97: 549–55

Dickey, J. O., Williams, J. G., Newhall, X. X., Yoder, C. F. 1983. Geophysical applications of lunar laser ranging. *Proc. Int. Union Geodesy Geophys./Int. Assoc. Geodesy Symp., Hamburg*, pp. 509–21. Columbus: Dept. Geodetic Sci., Ohio State Univ.

Doornbos, D. J., Hilton, T. 1989. Models of the core-mantle boundary and the travel times of internally reflected core phases. *J. Geophys. Res.* 94. 15,741–51

Dziewonski, A. M., Anderson, D. L. 1981. Preliminary reference Earth model. *Phys. Earth Planet. Inter.* 25: 297–356

Elgered, G., Davis, J. L., Herring, T. A., Shapiro, I. I. 1991. Geodesy by radio interferometry: water vapor radiometry for estimation of the wet delay. *J. Geophys. Res.* 96: 6541–55

Friedlander, S. 1987. Internal waves in a rotating stratified spherical shell. *Geophys. J. R. Astron. Soc.* 89: 637–55

Gilbert, F., Dziewonski, A. M. 1975. An application of normal mode theory to the retrieval of structural parameters and source mechanisms from seismic spectra. *Philos. Trans. R. Soc. Ser. A* 278: 187–269

Gudmundsson, O., Clayton, R. W. 1991. A 2-D synthetic study of global travel time tomography. *Geophys. J. Int.* 106: 53–65

Gwinn, C. R., Herring, T. A., Shapiro, I. I. 1986. Geodesy by radio interferometry: studies of the forced nutations of the Earth 2. Interpretation. *J. Geophys. Res.* 91: 4755–65

Hefty, J. 1989. Periodic fluctuations at frequency of the nearly diurnal free wobble in Ondřejov PZT observations. *Stud. Geophys. Geod.* 33: 117–32

Herring, T. A. 1988. VLBI studies in the nutations of the Earth. See Reid & Moran 1988, pp. 371–75

Herring, T. A. 1991. The rotation of the Earth. *Rev. Geophys. Suppl., US Natl. Rep. IUGG 1987–90*, pp. 172–75. Washington, DC: Am. Geophys. Union

Herring, T. A., Buffett, B. A., Mathews, P. M., Shapiro, I. I. 1991. Forced nutations of the Earth: influence of inner core dynamics. 3. Very long interferometry data analysis. *J. Geophys. Res.* 96: 8259–73

Herring, T. A., Gwinn, C. R., Buffett, B. A., Shapiro, I. I. 1988. Bound on the amplitude of the Earth's free core-nutation. See Babcock & Wilkins 1988, pp. 293–300

Herring, T. A., Gwinn, C. R., Shapiro, I. I. 1986. Geodesy by radio interferometry: studies of forced nutations of the Earth 1. Data analysis. *J. Geophys. Res.* 91: 4745–54 (Correction: 1986. *J. Geophys. Res.* 91: 14,165)

Hough, S. S. 1895. The oscillations of a rotating ellipsoidal shell containing fluid. *Philos. Trans. R. Soc. Ser. A* 186: 469–506

Jeffreys, H. 1976. *The Earth.* Cambridge: Cambridge Univ. Press. 6th ed.

Jeffreys, H., Vincente, R. O. 1957a. The theory of nutation and the variation of latitude. *Mon. Not. R. Astron. Soc.* 117: 142–61

Jeffreys, H., Vincente, R. O. 1957b. The theory of nutation and the variation of latitude: the Roche model core. *Mon. Not. R. Astron. Soc.* 117: 162–73

Karato, S., Spetzler, H. A. 1990. Defect microdynamics in minerals and solid-state mechanisms of seismic wave attenuation and velocity dispersion in the mantle. *Rev. Geophys.* 28: 399–421

Kinoshita, H., Nakajima, K., Kubo, Y., Nakagawa, I., Sasao, T., Yokoyama, K. 1979. Note on nutation in ephemerides. *Publ. Int. Latit. Obs. Mizusawa* 12: 71–108

Kinoshita, H., Sasao, T. 1989. Theoretical aspects of the Earth rotation. In *Reference Frames in Astronomy and Geophysics*, ed. J. Kovalevsky, I. I. Müller, B. Kolaczek, pp. 173–211. Dordrecht: Kluwer

Kinoshita, H., Souchay, J. 1990. The theory of the nutation for the rigid Earth model at the second order. *Celest. Mech.* 48: 187–265

Lambeck, K. 1980. *The Earth's Variable Rotation: Geophysical Causes and Consequences*. Cambridge: Cambridge Univ. Press

Larmor, J. 1909. The relation of the Earth's free precessional nutation to its resistance against tidal deformation. *Proc. R. Soc. London Ser. A* 82: 89–96

Li, Z. X. 1989. A determination of the errors in the 1980 IAU principal nutation constants. *Chin. Astron. Astrophys.* 13: 307–12

Love, A. E. H. 1909. The yielding of the Earth to disturbing forces. *Proc. R. Soc. London Ser. A* 82: 73–88

Marsh, J. G., Lerch, F. J., Putney, B. H., Felsentreger, T. L., Sanchez, B. V., et al. 1990. The GEM-T2 gravitational model. *J. Geophys. Res.* 95: 22043–71

Masters, T. G. 1991. Structure of the Earth: mantle and core. *Rev. Geophys. Suppl., US Natl. Rep. IUGG 1987–90*, pp. 671–79. Washington, DC: Am. Geophys. Union

Mathews, P. M., Buffett, B. A., Herring, T. A., Shapiro, I. I. 1991a. Forced nutations of the Earth: influence of inner core dynamics 1. Theory. *J. Geophys. Res.* 96: 8219–42

Mathews, P. M., Buffett, B. A., Herring, T. A., Shapiro, I. I. 1991b. Forced nutations of the Earth: influence of inner core dynamics 2. Numerical results and comparisons. *J. Geophys. Res.* 96: 8243–57

McCarthy, D. D., Carter, W. E., eds. 1990. *Variations in Earth rotation, AGU Monogr. No. 59*. Washington, DC: Am. Geophys. Union

Melchior, P. 1978. *The Tides of the Planet Earth*. Oxford: Pergamon

Molodensky, M. S. 1961. The theory of nutation and diurnal Earth tides. *Commun. Obs. R. Belg.* 188: 25–56

Morelli, A., Dziewonski, A. M. 1987. Topography of the core-mantle boundary. *Nature* 325: 678–83

Munk, W. H., MacDonald, G. J. F. 1960. *The Rotation of the Earth*. Cambridge: Cambridge Univ. Press

Neuberg, J., Hinderer, J., Zürn, W. 1987. Stacking gravity tide observations in Central Europe for the retrieval of the complex eigenfrequency of the nearly diurnal free wobble. *Geophys. J. R. Astron. Soc.* 91: 853–68

Neuberg, J., Hinderer, J., Zürn, W. 1990. On the complex eigenfrequency of the "nearly diurnal free wobble" and its geophysical interpretation. See McCarthy & Carter 1990, pp. 11–16

Ooe, M., Sasao, T. 1974. Polar wobble, sway and astronomical latitude-longitude observations. *Publ. Int. Latit. Obs. Mizusawa* 9: 223–33

Parke, M. E., Hendershott, M. C. 1979. M_2, S_2, K_1 models of the global ocean tide on the elastic Earth. *Mar. Geod.* 3: 379–408

Poincaré, H. 1910. Sur la précession des corps deformables. *Bull. Astron.* 27: 321–56

Reid, M. J., Moran, J. M., eds. 1988. *The Impact of VLBI on Astrophysics and Geophysics, Proc. IAU Symp. No. 129*. Dordrecht: Kluwer

Richter, B., Zürn, W. 1988. Chandler effect and nearly diurnal free wobble as determined from observations with a superconducting gravimeter. See Babcock & Wilkins 1988, pp. 309–15

Rochester, M. G. 1976. The secular decrease in obliquity due to dissipative core-mantle coupling. *Geophys. J. R. Astron. Soc.* 46: 109–26

Rodgers, A., Wahr, J. M. 1990. CMB topography from ISC travel times and synthetic tests of noise tolerance. *Eos Trans. Am. Geophys. Union* 71: 1465

Sailor, R. V., Dziewonski, A. M. 1978. Measurements and interpretation of normal mode attenuation. *Geophys. J. R. Astron. Soc.* 53: 559–81

Sasao, T., Okamoto, I., Sakai, S. 1977. Dissipative core-mantle coupling and nutational motion of the Earth. *Publ. Astron. Soc. Jpn.* 29: 83–105

Sasao, T., Okubo, S., Saito, M. 1980. A simple theory on the dynamical effects of a stratified fluid core upon nutational motion of the Earth. In *Proc. IAU Symp. No. 78*, ed. E. P. Federov, M. L. Smith, P. L. Bender, pp. 165–83. Hingham: Reidel

Schwiderski, E. W. 1980. On charting global ocean tides. *Rev. Geophys.* 18: 243–68

Shen, P.-Y., Mansinha, L. 1976. Oscillation, nutation and wobble of an elliptical Earth with liquid outer core. *Geophys. J. R. Astron. Soc.* 46: 467–96

Sipkin, S. A., Jordan, T. H. 1980. Frequency dependence of Q_{ScS}. *Bull. Seismol. Soc. Am.* 70: 1071–1102

Smith, M. L. 1974. The scalar equations of infinitesimal elastic-gravitational motion for a rotating, slightly elliptical Earth. *Geophys. J. R. Astron. Soc.* 37: 491–526

Smith, M. L. 1977. Wobble and nutation of the Earth. *Geophys. J. R. Astron. Soc.* 50: 103–40

Smith, M. L., Dahlen, F. A. 1981. The period

and *Q* of the Chandler wobble. *Geophys. J. R. Astron. Soc.* 64: 223–81

Wahr, J. M. 1981a. A normal mode expansion for the forced response of a rotating Earth. *Geophys. J. R. Astron. Soc.* 64: 651–75

Wahr, J. M. 1981b. Body tides on an elliptical rotating, elastic and oceanless Earth. *Geophys. J. R. Astron. Soc.* 64: 677–703

Wahr, J. M. 1981c. The forced nutations of an elliptical, rotating, elastic and oceanless Earth. *Geophys. J. R. Astron. Soc.* 64: 705–27

Wahr, J. M. 1988a. The Earth's rotation. *Annu. Rev. Earth Planet. Sci.* 16: 231–49

Wahr, J. M. 1988b. The theory of the Earth's orientation, with some new results for nutations. See Reid & Moran 1988, pp. 381–90

Wahr, J. M., Bergen, Z. 1986. The effects of mantle anelasticity on nutations, Earth tides and tidal variations in rotation rate. *Geophys. J. R. Astron. Soc.* 87: 633–68

Wahr, J. M., de Vries, D. 1989. The possibility of lateral structure inside the core and its implications for nutation and Earth tide observations. *Geophys. J. Int.* 99: 511–19

Wahr, J. M., de Vries, D. 1990. The Earth's forced nutations: geophysical implications. See McCarthy & Carter 1990, pp. 79–84

Wahr, J. M., Larden, D. R. 1983. An analysis of lunar laser ranging data for the Earth's free core nutation. In *Proc. Ninth Int. Symp. Earth Tides*, ed. J. T. Kuo, pp. 547–53. Stuttgart: Schweizerbart

Wahr, J. M., Sasao, T. 1981. A diurnal resonance in the ocean tide and the Earth's load response due to the resonant free "core nutation." *Geophys. J. R. Astron. Soc.* 64: 747–65

Warburton, R. J., Goodkind, J. M. 1978. Detailed gravity tide spectrum between one and four cycles per day. *Geophys. J. R. Astron. Soc.* 52: 117–36

Williams, J. G., Newhall, X. X., Dickey, J. O. 1991. Lunisolar precession—determination from lunar laser ranges. *Astron. Astrophys.* 241: L9–12

Woolard, E. W. 1953. Theory of the rotation of the Earth around its center of mass. *Astron. Pap. Am. Ephem.* 15: 3–165

Yoder, C. F., Ivins, E. R. 1988a. On the ellipticity of the core-mantle boundary from Earth's nutations and gravity. See Babcock & Wilkins 1988, pp. 317–22

Yoder, C. F., Ivins, E. R. 1988b. Improved analytic nutation model. See Reid & Moran 1988, pp. 379–80

Zhu, S. Y., Groten, E. 1989. Various aspects of numerical determination of nutation constants. I. Improvement of rigid-Earth nutation. *Astron. J.* 98: 1104–11

Zhu, S. Y., Groten, E., Reigber, Ch. 1990. Various aspects of numerical determination of nutation constants. II. Improved nutation series for a deformable Earth. *Astron. J.* 99: 1024–44

Zschau, J. 1986. Tidal friction in the solid Earth: constraints from the Chandler wobble period. In *Space Geodesy and Geodynamics*, ed. A. J. Anderson, A. Cazenave, pp. 315–44. London: Academic

NOTE ADDED IN PROOF

In a paper that appeared after this review was completed, McCarthy & Luzum (1991) have presented estimates of nutation amplitudes from different data sets and comparisons with values calculated from a number of theoretical treatments.

References added in proof

McCarthy, D. D., Luzum, B. J. 1991. Observation of lunisolar and free core nutation. *Astron. J.* 102: 1889–95

Moritz, H., Mueller, I. I. 1987. *Earth Rotation. Theory and Observation.* New York: Unger

Annu. Rev. Earth Planet. Sci. 1992. 20:501–26

ANTARCTICA; A TALE OF TWO SUPERCONTINENTS?

Ian W. D. Dalziel

Institute for Geophysics, University of Texas at Austin, 8701 Mopac Boulevard, Austin, Texas 78759-8345

KEY WORDS: plate tectonics, Gondwana, continental reconstruction

INTRODUCTION

Dog Sleds to Satellites

On March 29, 1912, Captain Robert Falcon Scott and his companions, Dr. Edward A. Wilson and Lieutenant H. R. Bowers died in their tent during a blizzard on the Ross Ice Shelf in Antarctica from cold, exhaustion, and lack of nourishment. They were a mere 18 km from a supply depot they had cached months earlier, before their journey to the South Pole. They died disappointed men, having learned on their arrival at the pole that Roald Amundsen and his Norwegian party had beaten them to the long-sought goal by one month. During their return journey, despite full awareness of the seriousness of their situation, Scott and his men had taken time to "geologize" in the Transantarctic Mountains (Scott 1913). They man-hauled 16 kg of rock samples collected there towards civilization until they succumbed. The specimens were retrieved by the party that found their frozen remains the next summer.

The accurate geologic positioning of Antarctica as the "keystone" of a southern supercontinent by Alex Du Toit in his reconstruction of 1937, following the early concepts of Taylor (1910) and Wegener (1912), would not have been on firm ground without the collections made on Scott's expeditions, and on the expedition of Ernest H. Shackleton that pioneered the route through the Transantarctic Mountains to the Polar Plateau in 1908 (Shackleton 1909). Specimens of the *Glossopteris* flora firmly linked Antarctica to the other southern continents, though full acceptance of the hypothesis of a Gondwana supercontinent by the scientific community

0084–6597/92/0515–0501$02.00

had to await acquisition of marine geophysical data from the southern oceans and its interpretation in terms of seafloor spreading (Norton & Sclater 1979). Satellite altimetry data acquired in the late 1980s now clearly show fracture zones representing the "flow lines" along which the other southern continents and India moved away from the Antarctic keystone during fragmentation of the Gondwana supercontinent (Royer et al 1990, see Figure 1), spectacular confirmation of Wegener's concept of a continental "Polflucht" (flight from the poles) and of Du Toit's geologic correlations.

The scientists on the first expeditions to the continental interior also collected rocks that link Antarctica to a supercontinent that may have predated Gondwana. The global marine transgression and accompanying explosion of life that marked the dawn of the Phanerozoic Eon (c. 550 Ma) are widely believed to have followed fragmentation of a Neoproterozoic[1] supercontinent, with Cambrian marine transgressive sequences on several continents being deposited on subsiding continental margins (Bond et al 1984). The effects of early Paleozoic compression and magmatism, and the present-day ice cap, have combined to obscure the possible late Precambrian–earliest Cambrian rift history in Antarctica. However, Frank Wild collected isolated archaeocyathans—Cambrian reef organisms—on Shackleton's 1908–9 expedition (Taylor 1914), and Scott's journal records Edward Wilson's discovery of "a specimen of limestone with archeocyathus" on February 8, 1912 as his party studied a moraine near Mount Buckley in the Transantarctic Mountains on their return from the pole (Scott 1913). Archeocyathan-bearing Lower Cambrian limestones have now been found to occur along the Transantarctic Mountains from the Weddell Sea to the Ross Sea (Rowell & Rees 1989, 1991). Recently it has been proposed that these deposits reflect marine transgression which followed the separation of Antarctica from North America during supercontinental fragmentation in the late Precambrian, and hence the birth of the Pacific Ocean basin (Moores 1991, Dalziel 1991). The Cambrian limestone collected on Scott's last expedition illustrates that Antarctic geology may be a tale of two supercontinents: Gondwana and a late Precambrian predecessor.

I use a series of global paleogeographic maps to review our understanding of the development of the Antarctic continent and plate in a framework of both the established Gondwana supercontinent and the hypothetical older supercontinent.

[1] The International Union of Geological Sciences has formally approved division of the Proterozoic Eon into the Paleo-, Meso-, and Neoproterozoic eras at 1.6 and 1.0 Ga (Plumb, K. A., *Episodes*, v. 14, No. 2, pp. 139–40, June 1991).

Antarctica Today

Antarctica is divided physiographically and geologically by the Transantarctic Mountains which extend for 4500 km from the Ross Sea to the Weddell Sea (Figure 1). The crust of East (or Greater) Antarctica was amalgamated from Archean nuclei during the Mesoproterozoic (1.6–1.0 Ga) and formed a major segment of the Precambrian craton of the Gondwana supercontinent from its amalgamation in latest Precambrian ($\sim$600 Ma) until the birth of the southern oceans in the late Mesozoic ($\sim$150 Ma). West (or Lesser) Antarctica is part of the circum-Pacific mobile belt.

The earliest tectonism clearly associated with the Transantarctic (i.e. proto-Pacific) margin of the East Antarctic craton was rifting during the Neoproterozoic. Marine transgression of the margin took place in the Early Cambrian. By the Middle to Late Cambrian (c. 520–500 Ma) it was undergoing compression, along-strike displacements, and calc-alkaline magmatism in an event known as the Ross orogeny. This event, called the Delamerian orogeny in eastern Australia, involved subduction beneath the craton. Mid-Paleozoic tectonism (c. 500–350 Ma) involved further compression, strike-slip displacements, accretion of terranes, and continental margin arc magmatism. In the late Paleozoic and early Mesozoic (c. 350–200 Ma) the convergent margin moved oceanward as a result of accretion. The sedimentary strata of that age now exposed in the Transantarctic Mountains—the classic Gondwana sequence that covers the craton and lower-middle Paleozoic orogens, thereby uniting the southern continents stratigraphically represent intracratonic basinal strata.

West Antarctica consists of four geologically distinct terranes (Dalziel & Elliot 1982) separated by subglacial depressions (Jankowski & Drewry 1981, Figure 1). The Ellsworth-Whitmore Mountains block (EWM) is a displaced segment of the craton margin (Schopf 1969). The Antarctic Peninsula (AP), Thurston Island (TI), and Marie Byrd Land (MBL) blocks, together with the New Zealand microcontinent are all Paleozoic to Mesozoic fore-arc and magmatic-arc terranes developed along the sector of the Pacific margin of the Gondwana craton extending between southern South America and eastern Australia. They were displaced after the Gondwanian orogeny in the early Mesozoic when the sedimentary cover of the Gondwana craton was folded and thrust cratonward along the Pacific margin. Uplift and widespread bimodal magmatism also preceded the final fragmentation of Gondwana which started with separation of East and West Gondwana by seafloor spreading in the Somali and Mozambique basins. The timing of breakaway of the major continents from Antarctica is well documented by marine data (Lawver et al 1991, Figure 1), and the

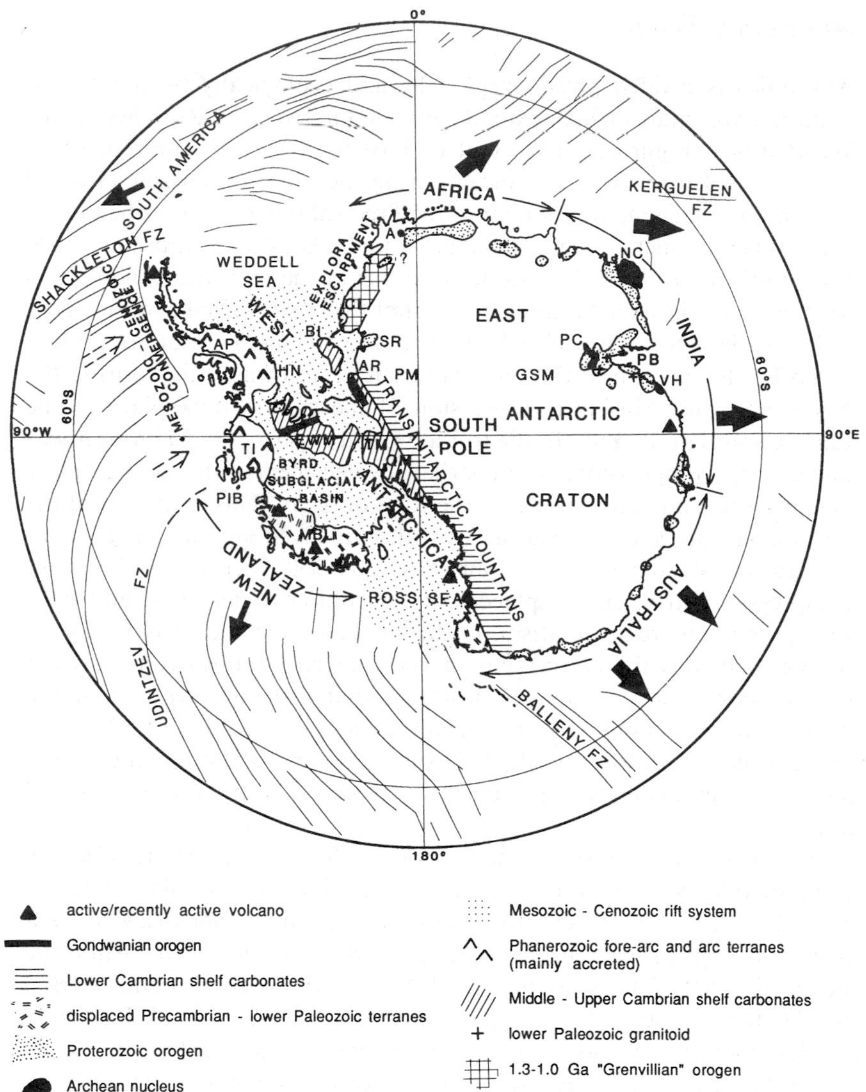

Figure 1 Simplified tectonic map of the Antarctic continent and surrounding ocean basins (from Dalziel 1991, Tingey 1991, Royer et al 1990). Bold arrows show directions taken by the major fragments of the Gondwana supercontinent during fragmentation. A—Annandagstoppane; AP—Antarctic Peninsula; AR—Argentine Range; BI—Berkner Island; CL—Coats Land nunataks (Bertrab, Littlewood, and Moltke); EWM—Ellsworth-Whitmore Mountains block; GSM—Gamburtzev Subglacial Mountains; H—Heimefrontfjella; HN—Haag Nunataks; MBL—Marie Byrd Land; NC—Napier Complex; NVL—North Victoria Land; PB—Prydz Bay; PC—Prince Charles Mountains; PIB—Pine Island Bay; PM—Pensacola Mountains; QM—Queen Maud Mountains; SR—Shackleton Range; TI—Thurston Island; TM—Thiel Mountains; VH—Vestfold Hills.

motions of the West Antarctic terranes relative to each other and to the East Antarctic craton are constrained by paleomagnetic data (Grunow et al 1991).

The Cenozoic history of the Antarctic continent has been dominated by extension and alkaline volcanism in the Ross embayment and along the Pacific margin of West Antarctica (Le Masurier & Thomson 1990), by block uplift of the Transantarctic Mountains (Fitzgerald 1989), and by the development and fluctuation of the ice cap (Barrett 1989).

A LATE PRECAMBRIAN SUPERCONTINENT?

The East Antarctic craton is almost entirely covered by ice; bedrock is only exposed ($<1\%$) along the coastal fringe and in the Transantarctic Mountains. Nonetheless four Archean nuclei are known (Figure 1). They are separated by Proterozoic metamorphic belts, the youngest of these reflecting crustal growth and suturing of the older continental nuclei during the late Mesoproterozoic (1.3–1.0 Ga) (Tingey 1991, Figure 1).

Reconstruction

Neoproterozoic rifting along the Pacific margins of both East Antarctica and the Australian craton is marked by thick craton-derived turbiditic sequences with associated bimodal igneous rocks. These sequences were deformed in the latest Precambrian to earliest Cambrian and are unconformably overlain by Cambrian marine strata (e.g. Schmidt et al 1965, Rowell et al 1992). This history is remarkably similar to that of other major continental margins interpreted as late Precambrian-Early Cambrian rifts (Bond et al 1984). It suggests that East Antarctica–Australia, restored to their configuration prior to the Mesozoic-Cenozoic development of the southern oceans, separated from some other Neoproterozoic continent along what is now the Transantarctic Mountains front and its continuation into the region of the Flinders Ranges of South Australia. If correct, this concept adds $\sim 25\%$ to the length of latest Precambrian to Early Cambrian rifted margins globally, thus making the general case for a Neoproterozoic supercontinent more compelling.

Moores (1991) has recently put forward evidence that the cratons of North America and East Antarctica–Australia were joined during the late Precambrian. Dalziel (1991) has presented a computer reconstruction showing not only that this is geometrically possible, but also that the boundary separating the 1.3–1.0 Ga Grenville province from the older radiometric provinces of the Laurentian cratonic nucleus may have continued into East Antarctica along the eastern margin of the Weddell Sea

(Figure 2). Although the paleomagnetic data from uppermost Precambrian and Cambrian rocks are still confusing, this reconstruction is quite compatible with widely accepted poles for the Gondwana supercontinent and North America from the Late Cambrian onward (Van der Voo et al 1984, Dalziel 1991; Figures 3 and 4). Thus, the similarities of the North America and East Antarctica–Australia cratons, the piercing points provided by the boundaries of the Grenville-age (1.3–1.0 Ga) belts, and the similar rifting and transgression histories (Moores 1991, Dalziel 1991) make the former juxtaposition of these 4500 km long margins an attractive possibility. Furthermore, there are no obvious alternative conjugates for either of them.

Dalziel (1991) and Hoffman (1991) have broadened the hypothesis of Moores (1991), reviving the suggestion that the eastern margin of Laurentia may have rifted from western South America (Bond et al 1984, Hartnady 1986). According to this scheme North America broke out of a Neoproterozoic supercontinent (Figure 2) to form the Pacific and Iapetus ocean basins while the Gondwana supercontinent was still amalgamating (Figure 3). The juxtaposition of eastern North America and western South America was originally suggested on the basis of minimum required displacement between breakup of a supercontinent at the end of the Precambrian and established early to middle Paleozoic reconstructions (Bond et al 1984). It is supported by near-simultaneous initiation of thermal subsidence in the mid-Atlantic region of North America and in northwestern Argentina, as well as by the presence in South America of the Cambrian Olenellid trilobite fauna characteristic of North America, and by Late Cambrian paleomagnetic data (Figure 3). The exact tectonic setting of the Cambrian strata in South America is unclear due to a complex history of late Precambrian rifting, late Precambrian and early Paleozoic terrane collision (Ramos 1988), and the superimposed Andean orogenesis. However, the Upper Proterozoic to Lower Cambrian Puncoviscana Formation of northwestern Argentina is a turbidite sequence derived from the craton which is interpreted as a passive margin fan (Aceñolaza et al 1988). It is very similar to the Patuxent Formation of the Transantarctic Mountains (Jezek et al 1985), so that North America is bordered in the reconstruction by continental margins with Neoproterozoic rift-related sequences and Cambrian transgressive sequences comparable to those along its Appalachian, Cordilleran, and Gulf of Mexico margins (Figure 2). The faunal similarities between the Appalachian and Andean margins, as well as the southwestern margins of North America and Africa, imply shoreline colonization routes and/or tectonic displacement of terranes along a mutual boundary prior to complete separation during the early to mid-Paleozoic (Hartnady 1986, Figure 4).

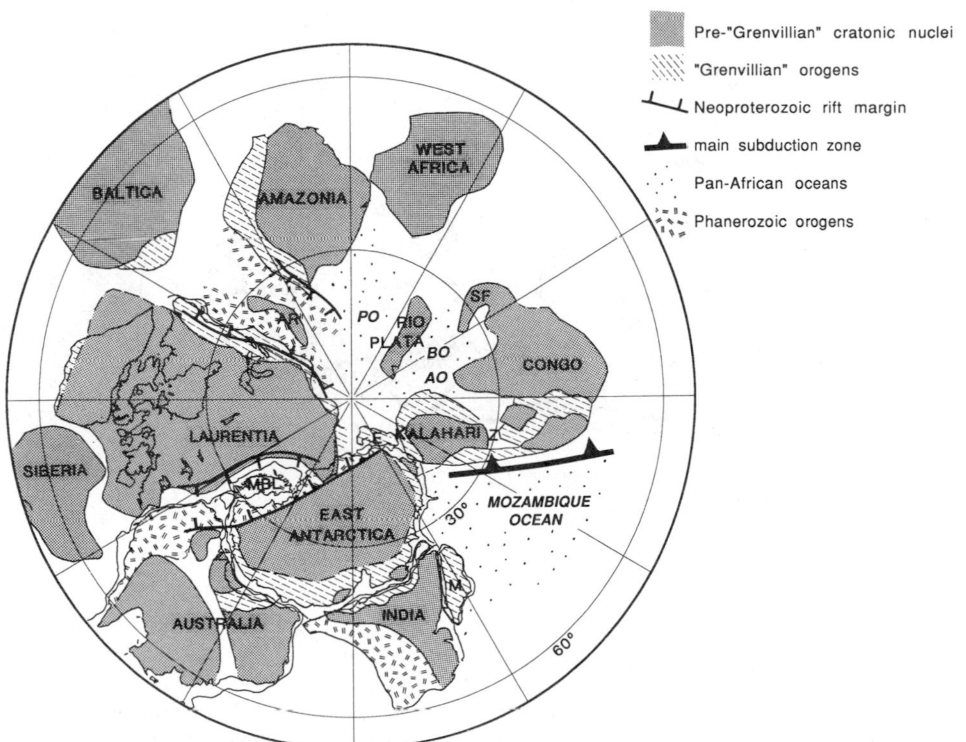

Figure 2 Hypothetical reconstruction of a late Precambrian supercontinent (c. 750 Ma) based on the North America–East Antarctica/Australia relationship suggested by Moores (1991) and amplified by Dalziel (1991). The pole approximates a generalized Euler pole for the fragmentation of the supercontinent and the amalgamation of Gondwana (Figure 3). The small circles about the pole at 30° and 60° are for scale only. Terranes of western North America and West Antarctica have been removed with the exception of Marie Byrd Land and the Ellsworth-Whitmore Mountains block which are shown in restored positions relative to the East Antarctic craton. The relationship of eastern North America and western South America is that of Dalziel (1991) and Hoffman (1991), and was first suggested by Bond et al (1984). Accreted terranes of the Patagonian region have been removed. Positioning of Baltica and Siberia follows Hoffman (1991). Position of the Kalahari craton adjoining East Antarctica is after Groenwald et al (1991), and relationship of the Sao Francisco–Congo and Kalahari cratons is after Hartnady et al (1985) and Hanson et al (1988). Location of ocean basins follows Hartnady et al (1985), Ramos (1988), Porada (1989), Dalziel (1991), and Hoffman (1991). AO—Adomaster Ocean; AR—Arequipa massif; BO—"Brazilide Ocean"; E—Ellsworth-Whitmore Mountains block; K—Kalahari craton; M—Madagascar; MBL—Marie Byrd Land; PO—"Pampean Ocean"; SF—Sao Francisco craton; Z—Zambesi belt.

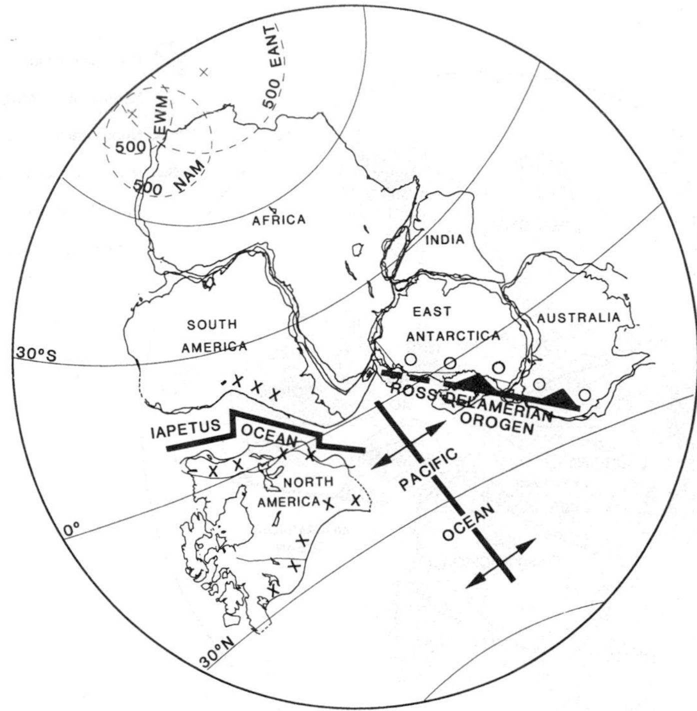

Figure 3 Paleomagnetically controlled reconstruction of the Pacific and southern (i.e. South America–North America) Iaptus oceans on final amalgamation of the Gondwana supercontinent at the end of the Cambrian (c. 500 Ma). The paleomagnetic poles and circles of confidence (alpha 95) are as described in Dalziel (1991, Figure 2). Crosses and circles respectively denote the North American and East Antarctic/Australian trilobite faunal realms of Jell (1974).

Eastern Gondwana (East Antarctica–Australia–India) was assembled by the end of the "Grenvillian" orogenesis at 1.0 Ga. The Kalahari craton of southern Africa was attached to the Archean rocks of the Annandagstoppane in East Antarctica (Groenwald et al 1991), and the Zambezi belt between the eastern extremities of the Kalahari and Congo–Sao Francisco cratons were sutured by ~0.8 Ga (Hanson et al 1988). Therefore, if the juxtapositions of East Antarctica–Australia with western North America and eastern North America with western South America are accepted, a major supercontinent existed at that time (Figure 2). Western Gondwana (South America–Africa), however, has several late Precambrian sutures (Hartnady et al 1985, Ramos 1988, Porada 1989). It is along the boundaries between the Brazilian–West African, Rio de la Plata, and Sao Francisco–

Congo–Kalahari cratons, and particularly along the Mozambique belt between West Gondwana and East Gondwana (where there are ophiolitic and eclogitic fragments, and blueschists), that ocean basins must have closed during the final amalgamation of the "daughter" Gondwana supercontinent during the break-out of Laurasia from the "parent" Neoproterozoic supercontinent. The computer reconstruction for c. 0.75 Ga that is presented in Figure 2, while hypothetical, permits ocean-continent relations to be studied within an accurate global framework.

Fragmentation

Critical issues for the hypothesis of a late Precambrian supercontinent are: 1. the timing of rift-drift transition along the East Antarctic–Australian, western Laurentian, eastern Laurentian, and western South American margins; 2. the timing of final collision along the sutures within West Gondwana and the Mozambique belt; and 3. the length of the Cambrian Period. The geometry of the model shown in Figure 2 requires consumption of ocean basins with a total width of approximately 8500 km. If the Cambrian is significantly shorter than the 60 million years (570–510 Ma) suggested by Harland et al (1990), and there are indications that this is the case (see discussion by McElhinny 1991), the changes required in the models of Dalziel (1991) and Hoffman (1991) would be difficult to envisage as happening between the Precambrian-Cambrian boundary (the approximate time of rift-drift transition according to the thermal subsidence curves of Bond et al 1984) and the Cambrian-Ordovician boundary (widely accepted as the youngest possible time for final consolidation of Gondwana; Figures 2 and 3). They could require unlikely seafloor spreading and convergence rates. Moreover, although "Pan African" radiometric ages as young as ~500 Ma are widespread in Gondwana, in many cases they appear to represent a thermal event post-dating final suturing of that supercontinent (see Stern et al 1992).

Hence the Neoproterozoic rift systems shown on Figure 2 are vital. Do they actually reflect the formation of oceanic lithosphere significantly earlier than the Precambrian-Cambrian boundary? If so there would have been more time to amalgamate Gondwana by continental collision along the Western Gondwana and Mozambique sutures by 650–600 Ma as required by the geochronologic data (Hartnady 1991, Dalziel 1992, Stern et al 1992), and to generate the embryonic Pacific and Iapetus ocean basins which were 9000 km and 1700 km wide respectively by the end of the Cambrian as shown in Figure 3.

Along the Transantarctic Mountains, there are igneous rocks associated with the Upper Proterozoic turbidites of the Patuxent Formation and its equivalents, the Goldie, La Gorce, and Duncan formations of the

Beardmore Group (Stump et al 1986). Geochemical study of mafic pillow lavas interbedded with the Goldie Formation yielded a Sm-Nd mineral isochron of age of 762 ± 24 Ma with a low Sm/Nd ratio and an εNd value of $+6.85$, suggesting to Borg et al (1990) that the rocks formed in an oceanic rather than a continental setting. Data recently acquired as part of the Joint US-UK West Antarctic Tectonics Project from mafic rocks in the Patuxent Formation demonstrate that these are significantly more evolved and do not fall within the range predicted for mid-ocean-ridge basalts or intra-oceanic arc magmas. While their isotopic compositions are similar to those of hot-spot related volcanics such as those of Bouvet Island, and to primitive subduction-related lavas, they could also have been derived from enriched mantle of the sub-continental lithosphere (Storey et al 1991). Their geologic setting, provenance, and association with felsic volcanic rocks supports an intracontinental rift setting. Hence the rocks of the Beardmore Group appear to represent tectonic environments ranging from an intracontinental rift to at least a narrow ocean basin. They were all deformed in the latest Precambrian or Early Cambrian prior to deposition of Middle Cambrian marine strata (Rowell et al 1992). The rocks of the Patuxent Formation, for example, are folded by two sets of tight, upright folds with sub-horizontal hinge lines, and also by late-stage, steep and variably plunging folds.

The Neoproterozoic Beardmore Group along the Pacific margin of East Antarctica–Australia and its counterparts in North America, the Belt-Purcell and Windermere sequences, cut across the fabric of the older cratonic crystalline basement. At the very least, therefore, these rocks appear to represent rifts that weakened the continental crust of a Neoproterozoic supercontinent and controlled the final breakup. Some workers believe the Windermere rocks represent a passive continental margin sequence (Ross 1991). The equivalent rocks in eastern North America and western South America (the Chilhowee-Catoctin and Puncoviscana sequences respectively) also controlled final rifting, but differ in that they themselves follow a Mesoproterozoic "Grenvillian" orogenic belt (Figure 2).

The geochemical data from the central Transantarctic Mountains indicates that some of the Neoproterozoic basins along the East Antarctica–Australia margin were partly floored by oceanic lithosphere. The absence of subduction-related magmatism associated with the Beardmore deformational event, however, suggests that the Beardmore Ocean itself was narrow (Borg et al 1990), perhaps like that between Madagascar and Africa today. The deformation may be of no more significance than the late compression in the Keeweenawan rift (Van Schmus et al 1987), or may reflect significant transpression between two major continents as they first started to move apart and may even have involved thrusting of Archean amphibolite over Proterozoic crystalline rocks in the Miller Range (Goodge et al 1991). Motion and deformation of this type, like the late Mesozoic interaction between South America and Africa in the equatorial Atlantic, would be expected if North America and East Antarctica–Australia moved as suggested in Figures 2 and 3 between the latest Precambrian and the end of the Cambrian. Deformation in the Puncoviscana Formation of Argentina (Aceñolaza et al 1988) could likewise reflect transpressional North America–South America relative motion.

At present, therefore, the supercontinental model of Figure 2 seems reasonable for the early to mid-Neoproterozoic ($\sim$0.8–0.7 Ga), with amalgamation of the "daughter" Gondwana supercontinent before the end of the Precambrian. This does, however, require a separate explanation of the thermal subsidence in the latest Precambrian to earliest Cambrian that initiated the Cordilleran miogeocline along the western margin of Laurentia. Reliable paleomagnetic data for the major continents during the latest Precambrian to Middle Cambrian are urgently needed to test these intriguing ideas.

THE GONDWANA SUPERCONTINENT

Early to Mid-Paleozoic

From the Cambrian until the birth of the southern oceans in the Middle Jurassic, Antarctica formed the core of the Gondwana Supercontinent. During that interval of approximately 350 million years, Gondwana appears to have moved significantly with respect to the spin axis of the earth (Figure 4). Although the paleomagnetic data are not unequivocal, they are supported by geologic evidence of two distinct glaciations. North Africa was positioned over the South Pole during the Late Cambrian and Ordovician (c. 500–450 Ma) when the Saharan area underwent glaciation. Motion over the pole may have put southern South America there by the Middle Silurian (c. 425 Ma), the drift then reversing so that central Africa was in a polar position in the Late Devonian (375 Ma). The drift direction apparently reversed again in the late Paleozoic, the supercontinent moving across the pole so that present day southern South America, southern Africa, India, East Antarctica, and Australia were affected by the Permo-Carboniferous glaciation (300–250 Ma). By the beginning of the Mesozoic (245 Ma) the South Pole was in the Pacific Ocean basin off West Antarctica.

The motion of a continent relative to the spin axis of the earth as derived from paleomagnetic data alone does not yield information regarding relative plate motions along its boundaries. However, major episodes of compression, subduction-related magmatism, and tectonic accretion along the

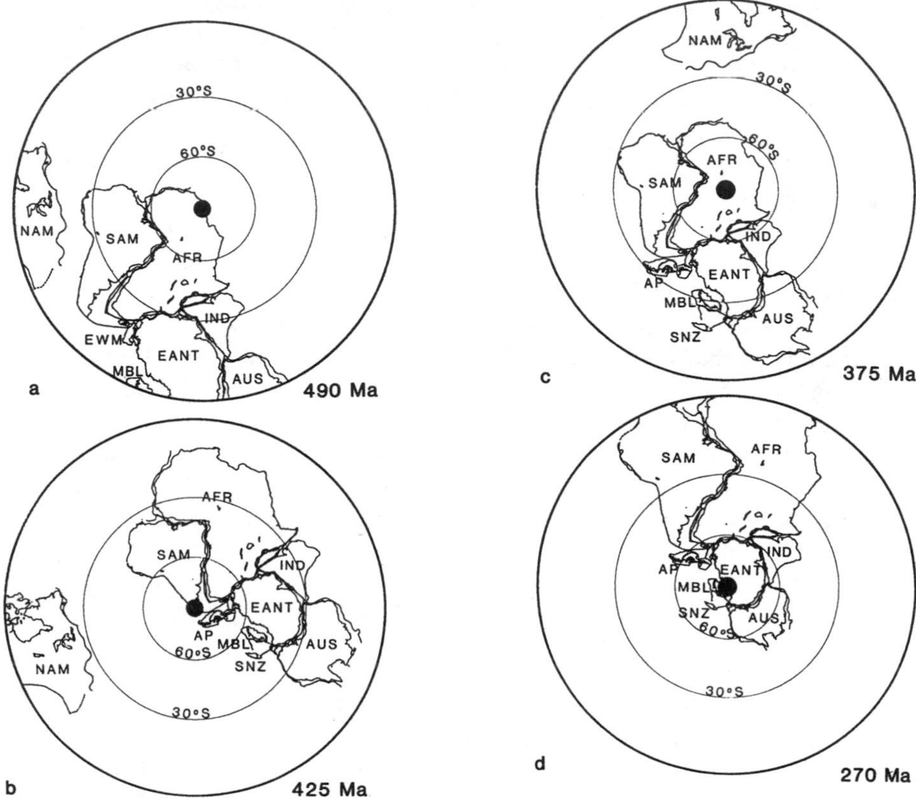

Figure 4 Drift of the Gondwana supercontinent and Laurentia relative to the south geographic pole (*filled circle*) from the Ordovician until the Permian, based on paleomagnetic data described in Dalziel (1991, Figure 2).

Pacific margin of North America and South America during the late Mesozoic and Cenozoic (150 Ma to present), when motions are also constrained by marine geophysical data, coincide with periods when those continents were converging with the floor of the Pacific Ocean (e.g. Dalziel 1986). A general correlation between the drift history of Gondwana between the Cambrian and the Middle Jurassic (500–150 Ma) and the tectonic development of its Pacific margin may reflect the same type of process (Coney 1990, Coney et al 1990).

The earliest period of tectonism along the Pacific margin of East Antarctica–Australia following the Neoproterozoic rifting and deformation recorded by the Beardmore Group, is the Cambrian-Ordovician Ross-Delamerian event (c. 500 Ma). The exact nature and timing of the

Ross-Delamerian event are uncertain because of inadequate data and the difficulty of relating stratigraphic and geochronometric data in the Cambrian. The event appears to have involved both strike-slip and compressional deformation as well as subduction-related magmatism above a cratonward-dipping slab (Rowell & Rees 1989, Borg et al 1990). Strike-slip deformation may have started as early as the Middle Cambrian (Rowell et al 1992).

The differences between the Antarctica-Australia and North America trilobite faunas in the Cambrian indicate that there was a substantial oceanic barrier between the two margins by the Middle Cambrian (Jell 1974). This is in keeping with the reconstruction shown in Figure 3 which suggests the Pacific Ocean was ~9000 km wide by the end of the Cambrian, making transformation of the rifted East Antarctica–Australia margin into a zone of convergence by that time quite plausible. On the other hand, the southern Iapetus between eastern North America and western South America remained narrow (~1700 km). Destruction of the East Antarctica–Australia late Precambrian–Early Cambrian rifted margin by the end of the Cambrian explains the limited development of the early Paleozoic shelf sequence. Outer parts of the East Antarctica–Australia margin (Figure 1) may have been displaced laterally (Rowell & Rees 1989, Rowell et al 1992).

Portions of the displaced margins such as the Pensacola and Queen Maud mountains form a narrow exterior zone of the craton along the Transantarctic Mountains (Figure 1). Others may have been accreted in North Victoria Land and southeastern Australia where the geology is extremely complex (Bradshaw et al 1985). Although opinions on details differ widely, there is general agreement that a collage of mixed continental, oceanic island arc, Andean-type magmatic arc, and deep marine sedimentary rocks was tectonically emplaced against the Ross orogen of North Victoria Land by the latest Devonian to Permian (c. 375–250 Ma; see Borg & De Paolo 1991). Subduction beneath this composite terrane prior to docking appears to have been opposed to that of the Ross orogen, as the chemistry of the Devonian Admiralty intrusives has the opposite polarity to that of the Cambro-Ordovician intrusives along the craton margin (Borg et al 1987, 1990).

Late Paleozoic to Early Mesozoic

Convergence dominated the history of the entire Pacific margin of Gondwana during the Paleozoic (Coney 1990), and the basement of the southern Andes and Antarctic Peninsula exposes subduction-related magmatic arc rocks of mid- to late Paleozoic age (Pankhurst 1990, Milne & Millar 1991). There was massive fore-arc accretion along these margins during the late

Paleozoic to early Mesozoic, including accretion of exotic seamounts with caps of fusulinid limestone (Dalziel 1982, Figure 5). Ramos (1986) has suggested that these rocks were accreted to the margin of the Gondwana supercontinent as a single entity—the Patagonia terrane, the collision generating the early Mesozoic Gondwanide fold belt of the Sierra de la Ventana in Argentina, the Cape fold belt in southern Africa, and the Ellsworth and Pensacola Mountains of Antarctica (Figure 5). There are no reliable paleomagnetic data directly indicating that the Patagonia terrane is allochthonous. However, deformation within the African craton, and the rotation of the Falkland Islands (FI) and EWM terranes followed Gondwanide folding and occurred prior to the appearance of seafloor within the Gondwana craton (Daly et al 1991, Dalziel & Grunow 1991). This intraplate deformation and the localized Gondwanide folding may have resulted from a collisional event. Offset of the subduction zones along the Antarctic Peninsula and NZ margins in the Triassic reconstruction of Grunow et al (1991) (Figure 5) could reflect closure of a marginal basin floored by stretched continental crust, there being no convincing evidence of a suture.

During the Devonian to Early Jurassic, the Gondwana craton-cover strata were deposited in the interior of the Gondwana supercontinent. The Beacon Supergroup of the Transantarctic Mountains represents this sequence in Antarctica. The Beacon was deposited in a series of basins behind the convergent Pacific margin, and records both the movement of the supercontinent relative to the South Pole (Figure 4), and events along the Pacific margin. Most prominent are the deposits resulting from Permo-Carboniferous glaciation when Antarctica was located over the South Pole, and the development of a foredeep basin in the Pensacola and Ellsworth Mountains region during the Permian in response to loading of the continental margin during the Gondwanide orogeny (Collinson 1991, Barrett 1991; Figure 5). It is the Permian strata that carry the *Glossopteris* flora found by the first geologists to visit the continent.

THE ANTARCTIC CONTINENT AND PLATE

Initial Fragmentation of Gondwana

In reviewing our understanding of the development of the present-day Antarctic continent and plate from the breakup of the Gondwana supercontinent to the present day, I use a series of maps (Figure 5–10) recently developed by Grunow et al (1991). These use data from the oceanic basins to constrain the major continents, and paleomagnetic data together with geologic information to position microcontinents and other displaced terranes. Displacement of the FI and EWM blocks following the Gondwanide folding was the first tectonism associated with the fragmentation of the

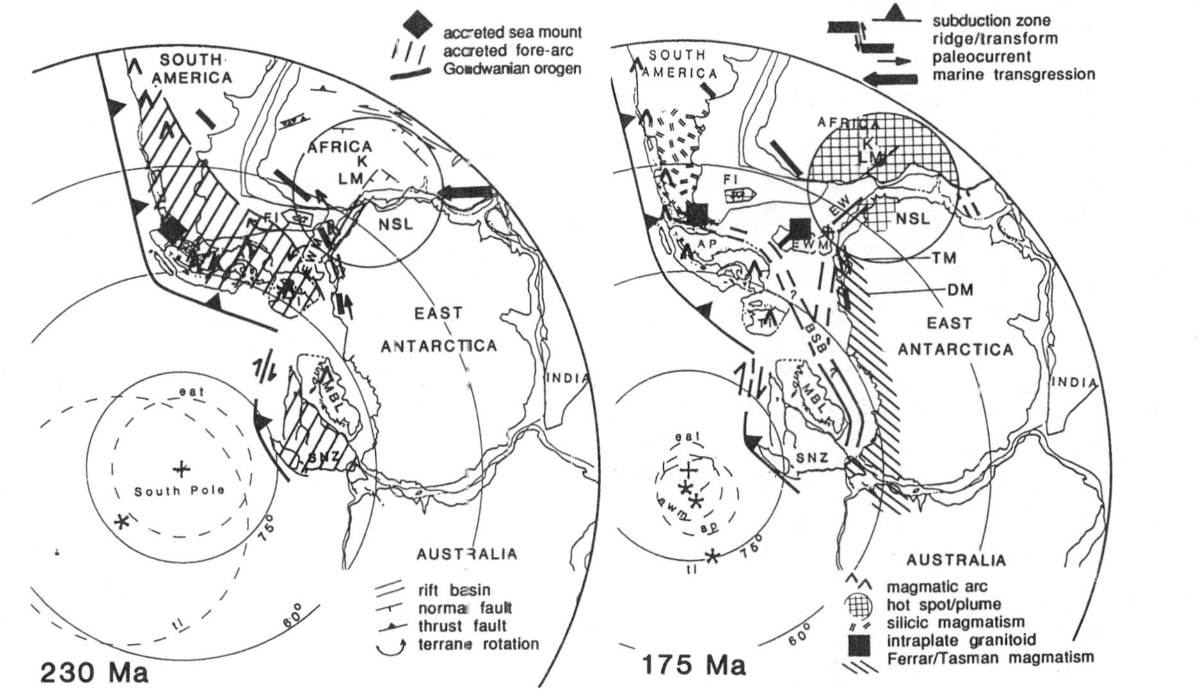

Figure 5 (*Left*) Gondwana during the Late Triassic (230 Ma) after the Gondwanide orogeny; based on the reconstruction of Grunow et al (1991), intraplate deformation in Africa after Daly et al (1991), and reconstruction of the Gondwanide orogen after Dalziel & Grunow (1991). Arc magmatism from Pankhurst (1990). The positioning of the incipient Karoo/Bouvet plume (*circle*) follows that of White & McKenzie (1989). AP—Antarctic Peninsula block; EWM—Ellsworth-Whitmore Mountains block; FI—Falkland Islands terrane (shape schematic); K—Karoo basin; LM—Lebombo monocline; MBL—Marie Byrd Land; NSL—Neuschwabenland; SNZ—south New Zealand (South Island, Campbell Plateau, and Chatham Rise); TI—Thurston Island block. Dashed circles are confidence circles about paleomagnetic poles (*asterisks* and *crosses*) of Grunow et al (1991) and are identified by letters as above. The reconstructions in Figures 5 through 10 are all polar stereographic projections centered on the paleomagnetic south pole (*cross*) determined for East Antarctica (*eat*).

Figure 6 (*Right*) Gondwana in the Middle Jurassic (175 Ma) after the rotation of the Falkland Island and Ellsworth-Whitmore Mountains blocks; based on the reconstruction of Grunow et al (1991). The location of the magmatic provinces follows Cox (1978), Dalziel et al (1987), White & McKenzie (1989), and Storey & Alabaster (1991). Additional abbreviations: BSB—Byrd Subglacial Basin (including Bentley Subglacial Trough); DM—Dufek Massif; EW—Explora wedge; TM—Theron Mountains.

Gondwana supercontinent. Initial counterclockwise rotation of the EWM block accompanied opening of a northeast-trending rift flanked by inward-dipping reflectors along the eastern side of the Weddell Sea (Kristoffersen & Hinz 1991). The northern end of this rift appears to have been the Explora wedge of seaward-dipping reflectors on the Antarctic side, and the Lebombo monocline on the African side (Figures 5 and 6, see Martin & Hartnady 1986). The latter is an east-facing flexure in the Karoo plateau lavas, and the former is interpreted to be a thick volcanic sequence (Hinz & Krause 1982). Karoo volcanism had two main peaks, at 193 ± 5 and 178 ± 5 Ma (Cox 1988), and the Explora wedge is interpreted to be Middle Jurassic based on data from ODP Leg 113 (Doyle et al 1991). The layered mafic Dufek massif at the southern end of the failed rift in Antarctica is part of the Ferrar mafic igneous province which has been dated at 179 ± 7 Ma (Kyle et al 1981), and intraplate granites cutting the folded Paleozoic strata of the EWM block average 173 ± 3 Ma (Millar & Pankhurst 1987). White & McKenzie (1989) have postulated that the Karoo magmatism, and that of the Weddell Sea margin, resulted from the initial expression of a mantle plume now represented by the Bouvet Island hot-spot, and Brewer (1989) has identified Jurassic dikes in the Theron Mountains as representing the limit of magmatism related to the suggested Karoo-Bouvet plume along the Transantarctic margin (Figure 6). However, the persistence of the Ferrar diabases and related silicic volcanic rocks along this entire margin into Tasmania and South Australia (Elliot 1991), and the kinematics of related faulting (Wilson 1991), demonstrate Middle Jurassic rifting within Gondwana far beyond the immediate influence of the proposed plume. This extension could have been controlled by stresses generated above the Pacific margin subduction zone (Cox 1978, Dalziel et al 1987).

Initial seafloor spreading between the Somali and Mozambique basins during the Late Jurassic (c. 165 Ma) truncated the Explora-Lebombo rift by right-lateral shearing between Africa and East Antarctica (Lawver et al 1991, Figure 7). This separation was accompanied by widespread extension and silicic magmatism in southern South America, leading to the development of the ophiolitic floor of the Rocas Verdes basin along the Andean margin (Dalziel 1981). Again the influence of a Pacific margin extensional stress regime is apparent. The Larsen Harbour Complex on South Georgia has been dated using U-Pb/zircon geochronometry at 150 Ma (Mukasa et al 1988). Storey & Alabaster (1991) have recently reviewed the geochemistry of the widespread magmatism associated in time and space with the initial fragmentation of Gondwana. They argue that during the initial stages of breakup, the Pacific margin magmas of the Antarctic Peninsula and southernmost South America (magmatic arc of Figures 6 and 7), and the contemporaneous Ferrar-Tasman suite, were both derived from a lithospheric mantle source enriched by the long-lived Paleozoic to early Mesozoic subduction of sediments. Lower to Middle Jurassic continental magnesian andesites comparable to those of Baja California and the Setouchi volcanic belt in Japan occur along the Pacific margin of the Antarctic Peninsula and on South Georgia, and they suggest that the high geothermal gradient indicated by the composition of these rocks was generated by subduction of an active spreading ridge. However, propagation of the main spreading ridge separating East and West Gondwana, through the embryonic Weddell Sea, into an extensional regime beyond the Antarctic Peninsula/Andean arc as shown by Grunow et al (1991, Figure 7) would also provide a thermal regime for magma generation of this type. Separation of the Antarctic Peninsula from southern South America at this time was very slow, and small relative motions between the EWM, TI, and AP blocks resulted in localized deformation such as the Late Jurassic to Early Cretaceous folding event at the base of the AP known as the Palmer Land orogeny (Grunow et al 1991).

Isolation of Antarctica

During the Early Cretaceous the South Atlantic started to open, turning the two plate East Antarctica–Africa system into the three plate South America–East Antarctica–Africa system (Lawver et al 1991), and in the model of Grunow et al (1991) the EWM block joined the Antarctic Peninsula and TI blocks to move essentially as one entity (Weddellia) during the opening of the Weddell Sea. Weddellia appears to have rotated clockwise along a transform fault between Berkner Island and the EWM block (Figure 8). The precise timing of the opening of the Weddell Sea remains to be resolved, but the ~30° clockwise rotation of Weddellia required by the paleomagnetic data from the TI block appears to have occurred between ~125 and ~110 Ma, and to have been transferred to relative motion across the magmatically active Pacific margin subduction zone along the boundary between the TI and MBL/New Zealand blocks—present-day Pine Island Bay (Grunow et al 1991). This almost exactly reversed the motion that generated the Gondwanide fold belt (compare Figures 5 and 8).

With the rapid opening of the South Atlantic Ocean basin during the Cretaceous normal interval, opening of the Weddell Sea was completed, and with the probable exception of some East Antarctica–MBL motion, the present-day shape of Antarctica was completed (Figure 9). Mid-Cretaceous increase in the convergence rate along the Andean margin and westward motion of the South American plate in a mantle reference frame (Dalziel 1986) resulted in collapse and inversion of the Rocas Verdes basin

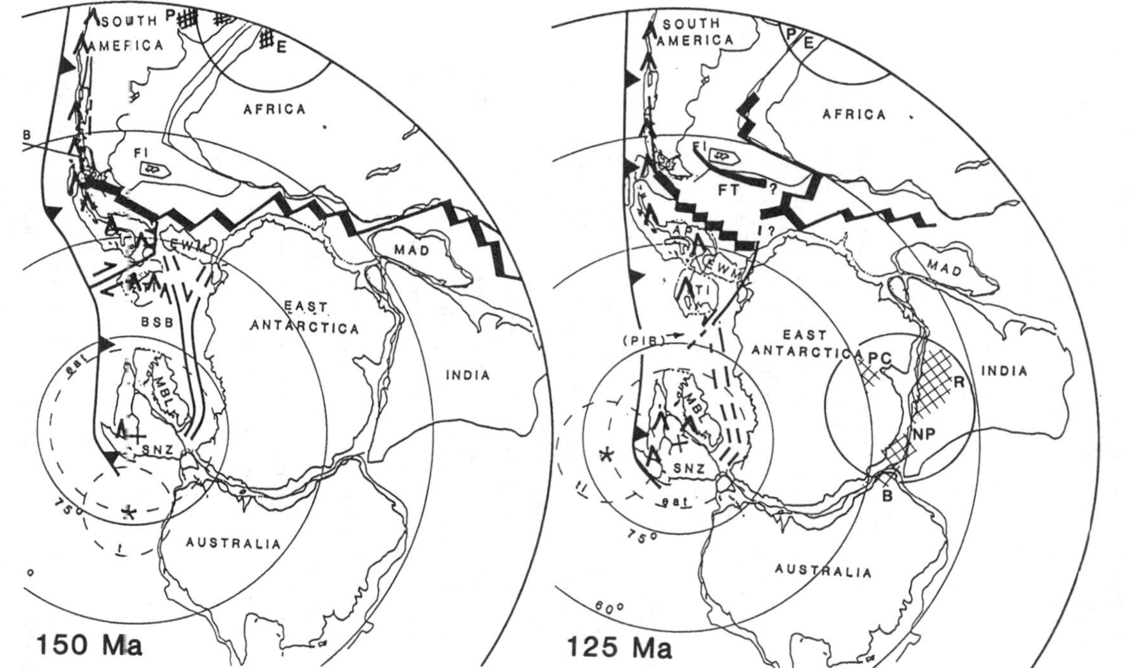

Figure 7 (*Left*) Gondwana in the latest Jurassic (150 Ma) after the initial separation of East and West Gondwana; based on the reconstruction of Grunow et al (1991). Opening of the Rocas Verdes basin is based on Mukasa et al (1988), and the location of the incipient Parana/Tristan da Cunha plume follows White & McKenzie (1989). Legend as in Figures 5 and 6. Additional abbreviations: E—Etendeka volcanics; P—Parana volcanics; RVB—Rocas Verdes basin.

Figure 8 (*Right*) Gondwana in the Early Cretaceous (125 Ma) immediately after initial opening of the South Atlantic Ocean basin; based on the reconstruction of Grunow et al (1991). Location of the incipient Kerguelen/Broken Ridge plume follows Royer & Coffin (1992). Additional abbreviations: B—Bunbury basalts; K—Kerguelen/Broken Ridge plume; FT—Falkland Trough; NP—Naturaliste Plateau; PC—Prince Charles Mountains; R—Rajmahal traps.

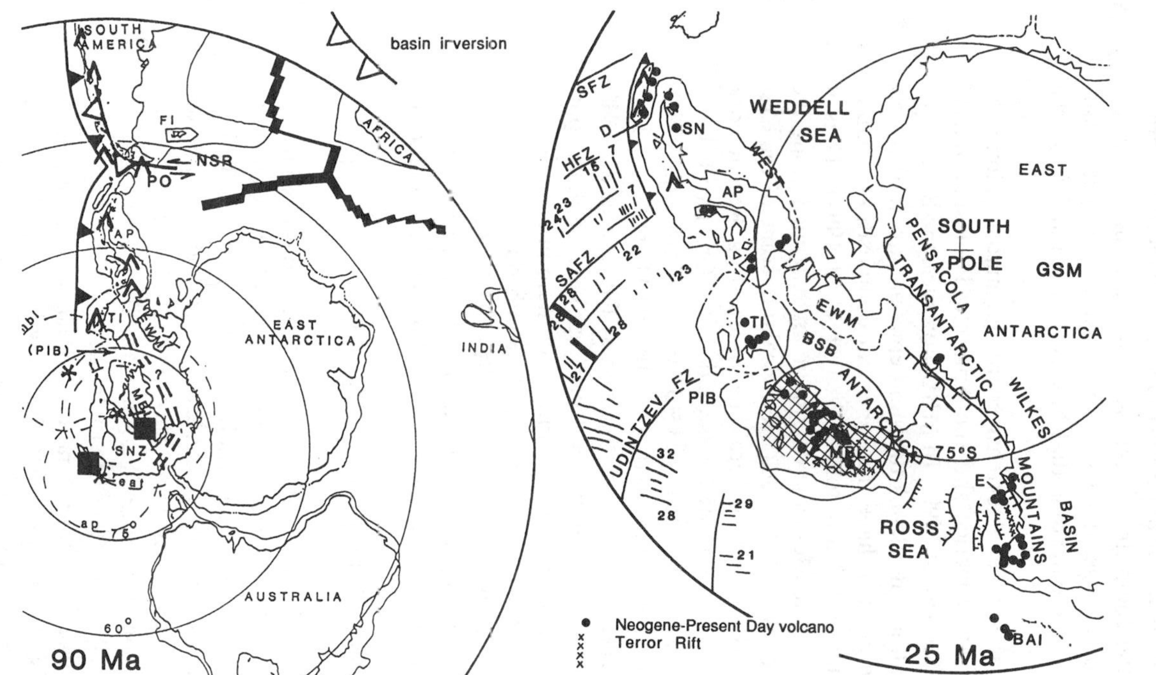

Figure 9 (*Left*) The Gondwana continents bordering the Pacific in the mid-Cretaceous (90 Ma) during formation of the Patagonian orocline and initial separation of South America and the Antarctic Peninsula; based on the reconstruction of Grunow et al (1991). The inversion of the Rocas Verdes basin to initiate Andean orogenesis follows Dalziel (1981), and the MBL/SNZ granitoids are based on Tulloch & Kimbrough (1989), Weaver et al (1991), Kimbrough (1991), and Richard et al (1991). Legend as in Figure 5. Additional abbreviations: NSR—North Scotia Ridge; PIB—Pine Island Bay; PO—Patagonian orocline.

Figure 10 (*Right*) Antarctica during the latest Paleogene (25 Ma) immediately following the opening of Drake Passage. Magnetic anomaly identification in the southern Pacific Ocean basin follows Mayes et al (1990). Location of Cenozoic volcanics (including Neogene and Present-Day deposits) is from LeMasurier & Thomson (1990), uplift of the Transantarctic Mountains from Fitzgerald et al (1989), and subsidence of the Pensacola-Wilkes basin from Stern & ten Brink (1989). Additional abbreviations: B—Balleny Islands; D—Deception Island; E—Mount Erebus; HFZ—Hero Fracture Zone; JRI—James Ross Island; SAFZ—South Anvers Fracture Zone; SFZ—Shackleton Fracture Zone; SN—Seal Nunataks.

and the development of the Patagonian orocline by counterclockwise rotation of the Andean magmatic arc in Tierra del Fuego (Cunningham et al 1991). This accompanied the initiation of left-lateral motion of South America relative to the completed Antarctic plate along the site of the present-day North Scotia Ridge, and isolated West Antarctica tectonically from South America at the same time that East Antarctica and Australia finally separated.

The Cretaceous was a time of extensive arc magmatism along the Andean margin of South America and the Antarctic Peninsula and TI blocks of Antarctica (Pankhurst 1990). In MBL, however, the Byrd Coast granites of that age, although calc-alkaline and epizonal, are mostly peraluminous, and some of them have trace element compositions characteristic of A-type anorogenic granitoids (Weaver et al 1991). Thus it appears that although a late Mesozoic magmatic arc extended along the entire Pacific margin of West Antarctica/New Zealand, some of the Cretaceous granitoids of MBL reflect a different tectonic setting. Recent work in the Fosdick Mountains area of MBL has yielded mid-Cretaceous U-Pb zircon ages on Byrd Coast granites (e.g. 103 ± 3 Ma, see Kimbrough 1991), while older work yielded Rb-Sr mineral dates of 102–92 Ma reflecting Late Cretaceous uplift of deep-seated migmatitic rocks (Richard et al 1991). These ages are similar to those determined from high-grade metamorphic rocks from the South Island of New Zealand which were interpreted as a tectonically denuded core complex (Tulloch & Kimbrough 1989), and suggest mid–Late Cretaceous extension and uplift in MBL and New Zealand immediately preceding separation of the latter from West Antarctica along the fracture zones revealed by the satellite altimetry data (Figure 1).

ICE, TECTONISM, AND THE FUTURE

Ice now covers the Antarctic continent. Marine seismic and side-scan data, supported by drilling results, show that the ice cap formerly extended to the edge of the continental shelf. Reports in the 1960s that mid-Tertiary volcanics rest on a glaciated surface were greeted with extreme skepticism. Drilling in Prydz Bay (Figure 1) has now shown that the history of glaciation extends back through the entire Oligocene, and there are indications of Eocene glaciation even on the northern fringes of West Antarctica. There is therefore a possibility that glaciation in East Antarctica extends back even into the Cretaceous, and that the growth and decline of ice sheets has been much more rapid and frequent than formerly believed (Webb 1990).

Development and fluctuation of the Antarctic ice sheet accompanied major tectonic changes even after the basic geography of the continent was established in the mid-Cretaceous. Immediately following rifting of New Zealand from MBL at about 90 Ma (Figure 9), subduction ceased sequentially eastward along the margin of West Antarctica from Pine Island Bay to the Hero Fracture Zone (Mayes et al 1990, Figure 10). Uplift of the Transantarctic Mountains along the western side of the Ross Sea embayment, initiated in the Mesozoic (Fitzgerald & Stump 1991), was renewed at about 60 Ma (Fitzgerald 1989). A bimodal alkaline volcanic province developed by the Oligocene, and spread from the Transantarctic Mountains to the tip of the Antarctic Peninsula (LeMasurier & Thomson 1990, Figure 10). Uplift of the Transantarctic Mountains, and related flexural development of the Pensacolas-Wilkes Subglacial Basin, has recently been modeled as the result of heating of the free craton edge (Stern & ten Brink 1989)—which originated in the Neoproterozoic—during extension in the Ross Sea embayment, notably the volcanically active Terror Rift (Behrendt & Cooper 1991, Figure 10). This extension may be related to final movement of MBL with respect to East Antarctica. Faulting associated with uplift of the Transantarctic Mountains indicates a component of right-lateral displacement (Wilson 1991).

The cause of the extension and volcanicity is still unclear. Antarctica has become almost stationary with respect to the South Pole since the Late Cretaceous (Figure 9), and it is therefore tempting to suggest that it is related to a mantle plume as Kyle et al (1991) have recently done for Marie Byrd Land (Figure 10). From Pine Island Bay to the tip of the Antarctic Peninsula, however, the volcanicity appears to be related to a continental margin extensional regime which developed after the cessation of subduction, and the small isolated Gaussberg alkaline volcano at $\sim 90°E$ on the margin of the East Antarctic craton (LeMasurier & Thomson 1990) occurs at the end of the Kerguelen Plateau. The Cenozoic volcanic and extensional regime does not, therefore, appear to have one simple tectonic explanation.

A relationship between environmental changes and tectonism has been recognized for many years. Opening of gateways allowing deep ocean circulation around the entire Antarctic continent and uplift of the Transantarctic Mountains both occurred during the Cenozoic and have been linked to climatic deterioration and development of the Antarctic ice cap by several authors. The Prydz Bay drilling results indicate, however, that significant glaciation preceded development of a complete deep water pathway for the circum-polar current. Recently Behrendt & Cooper (1991) have suggested that the start of the latest cold period which changed Antarctic glaciation from temperate to polar (2.5 Ma) was related to rapid (~ 1 km/m.y.) uplift of the Transantarctic Mountains.

This important topic leads to the issue of future studies of the Antarctic continent. Although major advances have been made in the study of both tectonic and environmental changes through time, too little is known of their relationship. The continental interior may hold significant clues. The thickest part of the West Antarctic ice sheet is grounded in the Byrd Subglacial Basin (Figure 1) which reaches near-oceanic depths, yet the age and tectonic nature of this basin is unknown. It formed part of a seaway between the Indo-Atlantic and Pacific Oceans until glaciation, and may have had an important influence on oceanic circulation. The East Antarctic ice sheet covers the Gamburtzev Subglacial Mountains which have a relief of 3–4 km and come close to the surface, yet they are totally unknown geologically. The drainage basins of this ice sheet date back to the Permian, possibly to the Proterozoic. Debris in moraines along the Transantarctic Mountains indicate that they contain Neogene marine strata and suggest rapid and frequent fluctuations in the extent of the ice cover which are not obvious in sediment cores from the deep ocean basins (Webb 1990). The tantalizing indications of tectonic control of climatic change from the fringes of the Antarctic continent demand that we put more effort into studies of the rocks beneath the ice and submerged around the continent in this unique and critical global laboratory.

ACKNOWLEDGMENTS

My work in the Antarctic over the past twenty years has been supported by the Division of Polar Programs of the National Science Foundation. Funds from DPP 89-17127 facilitated the preparation of this review. All the reconstructions were made with the help of Lisa Gahagan through the Paleoceanographic Mapping Project (now PLATES) of the Institute for Geophysics, University of Texas at Austin. Anne Grunow's paleomagnetic data, generated in Dennis Kent's laboratory on material she and I collected as part of a joint US-UK project, form the basis of the Mesozoic reconstructions. I appreciate being able to use the maps we developed in the course of this work prior to publication.

Many colleagues around the world who provided me with preprints of articles in press have helped make this review as up-to-date as possible. I apologize that space severely limits the number of citations. Constructive comments from Kevin Burke, Mike Coffin, Dickson Cunningham, Olav Eldholm, Anne Grunow, Richard Hanson, Larry Lawver, and Terry Wilson greatly improved the manuscript.

This paper is the University of Texas Institute for Geophysics contribution number 889.

Literature Cited

Aceñolaza, F. G., Miller, H., Toselli, A. J. 1988. The Puncoviscana Formation (Late Precambrian-Early Cambrian)—Sedimentology, tectonometamorphic history and age of the oldest rocks of NW Argentina. In *Lecture Notes in Earth Sciences 17*, ed. H. Balhlburg, Ch. Breitkreuz, P. Giese, pp. 25–37. Heidelberg: Springer-Verlag

Barrett, P. J., ed. 1989. Antarctic Cenozoic history from the CIROS-1 drillhole, McMurdo Sound. *DSIR Bull.* 245: 5–6. Wellington: DSIR

Barrett, P. J. 1991. The Devonian to Triassic Beacon Supergroup of the Transantarctic Mountains and correlatives in other parts of Antarctica. See Tingey 1991

Behrendt, J. C., Cooper, A. 1991. Evidence of rapid Cenozoic uplift of the shoulder escarpment of the West Antarctic rift system and a speculation on possible climate forcing. *Geology* 19: 315–19

Bond, G. C., Kominz, M., Devlin, W. 1991. Rapid global-scale events during the Precambrian-Cambrian transition: consequence of rapid dispersal of a late Proterozoic supercontinent? *Geol. Soc. Am. Abstr. with Programs* 23: A112 (Abstr.)

Bond, G. C., Nickeson, P. A., Kominz, M. A. 1984. Breakup of a supercontinent between 625 Ma and 555 Ma: New evidence and implications for continental histories. *Earth Planet. Sci. Lett.* 70: 325–45

Borg, S. G., DePaolo, D. J. 1991. A tectonic model of the Antarctic Gondwana margin with implications for southeastern Australia: Isotopic and geochemical evidence. *Tectonophysics* 196: 339–58

Borg, S. G., DePaolo, D. J., Smith, B. M. 1990. Isotopic structure and tectonics of the Central Transantarctic Mountains. *J. Geophys. Res.* 95: 6647–67

Borg, S. G., Stump, E., Chappel, B. W., McCulloch, M. T., Wyborn, D., et al. 1987. Granitoids of northern Victoria Land, Antarctica: Implications of chemical and isotopic variations to regional crustal structure and tectonics. *Am. J. Sci.* 287: 127–69

Bradshaw, J. D., Weaver, S. D., Laird, M. G. 1985. Suspect terranes and Cambrian tectonics in northern Victoria Land, Antarctica. In *Tectonostratigraphic Terranes of the Circum-Pacific Region*, ed. D. G. Howell, pp. 467–79. Houston: Earth Sci. Ser. No. 1

Brewer, T. S. 1989. Mesozoic dolerites from Whichaway Nunataks. *Antarct. Sci.* 1: 151–55

Collinson, J. W. 1991. The palaeo-Pacific margin as seen from East Antarctica. See Thompson et al 1991, pp. 199–204

Coney, P. J. 1990. Terranes, tectonics and the Pacific Rim. *Aust. Inst. Mining Metall., Pacific Rim* 90: 19–30

Coney, P. J., Edwards, A., Hine, R., Morrison, F., Windrim, D. 1990. The regional tectonics of the Tasman orogenic system, eastern Australia. *J. Struct. Geol.* 12: 514–44

Cox, K. G. 1978. Flood basalts, subduction and the break-up of Gondwanaland. *Nature* 274: 47–49

Cox, K. G. 1988. The Karoo Province. In *Continental Flood Basalts*, ed. J. D. Macdougall, pp. 239–72. Dordrecht: Kluwer

Cunningham, W. D., Klepeis, K. A., Gose, W. A., Dalziel, I. W. D. 1991. The Patagonian orocline: New paleomagnetic data from the Andean magmatic arc in Tierra del Fuego. *J. Geophys. Res.* 96: 16,061–68

Daly, M. C., Lawrence, S. R., Kimun'a, D., Binga, M. 1991. Late Palaeozoic deformation in central Africa: a result of distant collision? *Nature* 350: 605–7

Dalziel, I. W. D. 1981. Back-arc extension in the southern Andes, a review and critical reappraisal. *Philos. Trans. R. Soc. London Ser. A* 300: 319–35

Dalziel, I. W. D. 1982. Pre-Jurassic history of the Scotia Arc region. In *Antarctic Geoscience*, ed. C. Craddock, pp. 111–26. Madison: Univ. Wis. Press

Dalziel, I. W. D. 1986. Collision and cordilleran orogenesis, an Andean perspective. In *Collision Tectonics*, ed. M. P. Coward, A. C. Ries, Geol. Soc. London Spec. Publ. No. 19, pp. 389–404

Dalziel, I. W. D. 1991. Pacific margin of Larentia and East Antarctica/Australia as a conjugate rift pair: Evidence and implications for an Eocambrian supercontinent. *Geology* 19: 598–601

Dalziel, I. W. D. 1992. Reply to comment on "Pacific margins of Laurentia and East Antarctica-Australia as a conjugate rift pair: Evidence and implications for an Eocambrian supercontinent" by Ian W. D. Dalziel (Geology, V. 19, pp. 598–601), by Robert J. Stern, Mohamed Sultan and Mohamed Gamal-Salam. *Geology*. In press

Dalziel, I. W. D., Elliot, D. H. 1982. West Antarctica: Problem child of Gondwanaland. *Tectonics* 1: 3–19

Dalziel, I. W. D., Grunow, A. M. 1992. Late Gondwanide tectonic rotations within Gondwanaland: Causes and consequences. *Tectonics*. In press

Dalziel, I. W. D., Storey, B. C., Garrett, S. W., Grunow, A. M., Herrod, L. D. B., et al. 1987. Extensional tectonics and the fragmentation of Gondwanaland. In *Continental Extensional Tectonics*, ed. J. F. Dewey, M. P. Coward, P. Hancock, Geol. Soc. London Spec. Publ. 28, pp. 433–41

Doyle, P., Crame, J. A., Thomson, M. R. A. 1991. Late Jurassic-Early Cretaceous macrofossils from Leg 113, Hole 692B, eastern Weddell Sea. In *Proc. ODP, Sci. Results, 113*, ed. P. F. Barker, J. P. Kennett et al, pp. 443–48

Du Toit, A. L. 1937. *Our Wandering Continents*. Edinburgh: Oliver & Boyd. 366 pp.

Elliot, D. H. 1991. Triassic-Early Cretaceous evolution of Antarctica. In *Geological Evolution of Antarctica*, ed. M. R. A. Thomson, J. A. Crame, J. W. Thomson, pp. 541–48. Cambridge: Cambridge Univ. Press

Fitzgerald, P. G. 1989. Uplift and formation of the Transantarctic Mountains application of apatite fission trace analysis to tectonic problems. *28th Int'l. Geo. Congress*, Washington, DC. Abstr. p. 491

Fitzgerald, P. G., Stump, E. 1991. Late Cretaceous and Cenozoic uplift of the Transantarctic mountains: Implications for regional tectonics. *Geol. Soc. Am. Abstr. with Programs* 23: A305 (Abstr.)

Goodge, J. W., Borg, S. G., Smith, B. K., Bennett, V. C. 1991. Tectonic significance of Proterozoic ductile shortening and translation along the Antarctic margin of Gondwana. *Earth Planet. Sci. Lett.* 102: 58–70

Groenwald, P. B., Grantham, G. H., Watkeys, M. K. 1991. Geological evidence for a Proterozoic to Mesozoic link between southeastern Africa and Dronning Maud Land, Antarctica. *J. Geol. Soc. London*. 148: 1115–24

Grunow, A. M., Kent, D. V., Dalziel, I. W. D. 1991. New paleomagnetic data from Thurston Island: implications for the tectonics of West Antarctica and the Weddell Sea. *J. Geophys. Res.* 96: 17,935–54

Hanson, R. E., Wilson, T. J., Brueckner, H. K., Onstott, T. C., Wardlaw, M. S., et al. 1988. Reconnaissance geochronology, tectonothermal evolution, and regional significance of the middle Proterozoic Choma-Kalomo block, southern Zambia. *Precambrian Res.* 42: 39–61

Harland, W. B., Armstrong, R. L., Cox, A. V., Craig, L. E., Smith, A. G., Smith, D. G. 1990. *A Geologic Time Scale 1989*. Cambridge: Cambridge Univ. Press. 263 pp.

Hartnady, C. J. H. 1986. Was North America ("Laurentia") part of south-western Gondwanaland during the late Proterozoic era? *Suid-Afr. Tydskr. Wet.* 82: 251–54

Hartnady, C. J. H. 1991. About turn for supercontinents. *Nature* 352: 476–78

Hinz, K., Krause, W. 1982. The continental margin of Queen Maud Land, Antarctica: Seismic sequences, structural elements, and geological development. *Geol. Jahrb.* E23: 17–41

Hoffman, P. F. 1991. Did the breakout of Laurentia turn Gondwanaland inside out? *Science* 252: 1409–12

Jankowski, E. J., Drewry, D. J. 1981. The structure of West Antarctica from geophysical studies. *Nature* 291: 17–21

Jell, P. A. 1974. Faunal provinces and possible planetary reconstruction of the middle Cambrian. *J. Geol.* 82: 319–50

Jezek, P., Willner, A. P., Aceñolaza, F. G., Miller, H. 1985. The Puncoviscana trough—a large basin of Late Precambrian to Early Cambrian age on the Pacific edge of the Brazilian shield. *Geol. Rundsch.* 74/3: 573–84

Kimbrough, D. L., Richard, S. M. 1991. *Geochronology of basement rocks in the Fosdick Mountain region of West Antarctica and correlation with Cretaceous extensional terrains of southern New Zealand. 6th Int. Symp. Antarc. Earth Sci., Ranzan, Japan*, pp. 313–16 (Abstr.)

Kristoffersen, Y., Hinz, K. 1991. Evolution of the Gondwana plate boundary in the Weddell Sea area. See Thomson et al 1991, pp. 225–30

Kyle, P. R., Elliot, D. H., Sutter, J. F. 1981. Jurassic Ferrar Supergroup tholeiites from the Transantarctic Mountains, Antarctica, and their relationship to the initial fragmentation of Gondwana. In *Gondwana Five*, ed. M. M. Cresswell, P. Vella, pp. 283–87. Rotterdam: A. A. Balkema

Kyle, P. R., McIntosh, W. C., Panter, K., Smellie, J. 1991. *Is volcanism in Marie Byrd Land related to a mantle plume? Gondwana 8*, Hobart, Australia (Abstr.)

Lawver, L. A., Royer, J.-Y., Sandwell, D. T., Scotese, C. R. 1991. Evolution of the Antarctic continental margins. See Thomson et al 1991, pp. 533–40

LeMasurier, W., Thomson, J. W., eds. 1990. Volcanoes of the Antarctic plate and Southern Ocean. *Antarctic Research Series, 48*. Washington, DC: Am. Geophys. Union

Martin, A. K., Hartnady, C. J. H. 1986. Plate tectonic development of the Southwest Indian Ocean: A revised reconstruction of East Antarctica and Africa. *J. Geophys. Res.* 91: 4767–86

Mayes, C., Sandwell, D. T., Lawver, L. A. 1990. Tectonic history and new isochron chart of the South Pacific. *J. Geophys. Res.* 95: 8543–67

McElhinny, M. 1991. A review of "A geologic time scale 1989" by Harland, W. B., et al eds., 1990. Cambridge University Press. 263 pp. *Nature* 349: 26

Millar, I. L., Pankhurst, R. J. 1987. Rb-Sr Geochronology of the region between the Antarctic Peninsula and the Transantarctic Mountains: Haag nunataks and Mesozoic granitoids. In *Gondwana Six: Structure, Tectonics, and Geophysics*, ed. G. D. McKenzie, pp. 151–60. Washington, DC: Am. Geophys. Union

Milne, A. J., Millar, I. L. 1991. Mid-Palaeozoic basement in eastern Graham Land and its relation to the Pacific margin of Gondwana. See Thomson et al 1991, pp. 335–40

Minister, J. B., Jordan, T. H. 1978. Present-day plate motions. *J. Geophys. Res.* 83: 5331–54

Moores, E. M. 1991. The Southwest U.S.–East Antarctic (SWEAT) connection: A hypothesis. *Geology* 19: 425–28

Mukasa, S. B., Dalziel, I. W. D., Brueckner, H. K. 1988. Zircon U-Pb Constraints on the Kinematic Evolution of the Northern Scotia Arc. *GSA Annu. Meet. Denver, CO, 1988, Abstr. with Programs*, p. A12

Norton, I. O., Sclater, J. G. 1979. A model for the evolution of the Indian Ocean and the breakup of Gondwanaland. *J. Geophys. Res.* 84: 6803–30

Pankhurst, R. J. 1990. The Paleozoic and Andean magmatic arcs of West Antarctica and southern South America. In *Plutonism from Antarctica to Alaska*, ed. S. M. Kay, C. W. Rapela, pp. 1–7. Geol. Soc. Am. Spec. Pap. 241

Porada, H. 1989. Pan-African rifting and orogenesis in southern to equatorial Africa and eastern Brazil. *Precambrian Res.* 44: 103–36

Ramos, V. A. 1986. Discussion of "Tectonostratigraphy, as applied to analysis of South African Phanerozoic basins," by H. de la R. Winter. *Trans. Geol. Soc. South Afr.* 89: 427–29

Ramos, V. A. 1988. Late Proterozoic-early Paleozoic of South America—a collisional history. *Episodes* 11: 168–74

Rees, M. N., Pratt, B. R., Rowell, A. J. 1989. Early Cambrian reefs, reef complexes, and associated lithofacies of the Shackleton Limestone, Transantarctic Mountains. *Sedimentology* 36: 341–61

Richard, S. M., Smith, C. H., Luyenduyk, B. P., Kimbrough, D. L. 1991. *Geological and geophysical investigations in the Northern Ford ranges, Marie Byrd Land, West Antarctica: implications for a zone of Late Mesozoic, deep crustal migmatization and flow in the Fosdick metamorphic complex. 6th Int. Symp. Ant. Earth Sci., Ranzan, Japan*, pp. 501–2 (Abstr.)

Ross, G. M. 1991. Tectonic setting of the Windermere Supergroup: the continental perspective. *Geology* 19: 1125–28

Rowell, A. J., Rees, M. N. 1989. Early Paleozoic history of the upper Beardmore Glacier area: implications for a major Antarctic structural boundary within the Transantarctic Mountains. *Antarct. Sci.* 1: 249–60

Rowell, A. J., Rees, M. N. 1991. Setting and significance of the Shackleton Limestone, central Transantarctic Mountains. See Thomson et al 1991, pp. 171–76

Rowell, A. J., Rees, M. N., Evans, K. R. 1992. Evidence of major Middle Cambrian deformation in the Ross orogen Antarctica. *Geology*. In press

Royer, J.-Y., Coffin, M. F. 1992. Jurassic to Eocene plate tectonic reconstructions in the Kerguelen Plateau region. *Proc. ODP Sci. Results, 120*. In press

Royer, J.-Y., Gahagan, L. M., Lawver, L. A., Mayes, C. L., Nürnberg, D., et al. 1990. A tectonic chart for the Southern Ocean derived from Geosat Altimetry data. In *Antarctica as an Exploration Frontier—hydrocarbon potential, geology and hazards*, ed. A. Grantz, L. Johnson, J. F. Sweeney. AAPG Studies in Geology 31: 89–100

Schmidt, D. L., Williams, P. L., Nelson, W. J., Ege, J. R. 1965. Upper Precambrian and Palaeozoic stratigraphy and structure of the Neptune Range of Antarctica. *US Geol. Surv. Prof. Pap. 525D*, pp. 112–19

Schopf, J. M. 1969. Ellsworth Mountains: position in West Antarctica due to sea floor spreading. *Science* 164: 63–66

Scott, R. F. 1913. *Scott's Last Expedition*, ed. L. Huxley. New York: Dodd & Mead

Shackleton, E. H. 1909. *The Heart of the Antarctic*. London: Heinemann. 419 pp.

Stern, R. J., Sultan, M., Abdel-Salam, M. G. 1992. Discussion of the paper "Pacific margins of Laurentia and East Antarctica-Australia as a conjugate rift pair: Evidence and implications for an Eocambrian supercontinent," by I. W. D. Dalziel (Geology, v. 19, pp. 598–601). *Geology*. In press

Stern, T. A., ten Brink, U. S. 1989. Flexural uplift of the Transantarctic Mountains. *J. Geophys. Res.* 94: 10,315–30

Storey, B. C., Alabaster, T. 1991. Tectonomagmatic controls on Gondwana break-up models: evidence from the proto-Pacific margin of Antarctica. *Tectonics*. In press

Storey, B. C., Macdonald, D. I. M., Millar, I. L., Pankhurst, R. J., Alabaster, T., et al.

1991. *Upper Proterozoic rift-related basin and bimodal volcanic sequences in the Pensacola Mountains, Antarctica: precursor to Cambrian supercontinent breakup? Gondwana 8*, Hobart, Australia (Abstr.)

Stump, E., Smit, J. H., Self, S. 1986. Timing of events during the late Proterozoic Beardmore Orogeny, Antarctica: geological evidence from the La Gorce Mountains. *Geol. Soc. Am. Bull.* 97: 953–65

Taylor, F. B. 1910. Bearing of the Tertiary mountain belt on the origin of the Earth's plan. *Bull. Geol. Soc. Am.* 21: 179–226

Taylor, G. G. 1914. Short notes on palaeontology. In *Glaciology, physiography, stratigraphy, and tectonic geology of south Victoria Land*, ed. T. W. E. David, R. E. Priestley, British Antarctic Expedition 1907–9. Reports on the Scientific Investigations. *Geology* 1: 1–319

Thomson, M. R. A., Crame, J. A., Thomson, J. W., eds. 1991. *Geologic Evolution of Antarctica*. Cambridge: Cambridge Univ. Press

Tingey, R. J. 1991. *The Geology of Antarctica*. Oxford: Oxford Univ. Press. In press

Tulloch, A. J., Kimbrough, D. L. 1989. The Paparoa metamorphic core complex, New Zealand: Cretaceous extension associated with fragmentation of the Pacific margin of Gondwana. *Tectonics* 8: 1217–34

Van der Voo, R., McCabe, C., Scotese, C. 1984. Was Laurentia part of an Eocambrian supercontinent? In *Plate reconstructions from Paleozoic paleomagnetism*, ed. R. Van der Voo, C. Scotese, N. Bonhommet, pp. 131–36. Am. Geophys. Union Geodyn. Ser. 2

Van Schmus, W. R., Bickford, M. E., Zietz, I. 1987. Early and middle Proterozoic provinces in the central United States. In *Proterozoic Lithospheric Evolution*, ed. A. Kroner, pp. 43–68. Am. Geophys. Union Geodyn. Ser. 17

Weaver, S. D., Bradshaw, J. D., Adams, C. J. 1991. Granitoids of the Ford Ranges, Marie Byrd Land, Antarctica. See Thomson et al 1991, pp. 345–52

Webb, P. N. 1990. The Cenozoic history of Antarctica and its global impact. *Antarct. Sci.* 1: 3–21

Wegener, A. 1912. Die Entstehung der Kontinente. *Petermanns Mitteilungen*, pp. 185–95, 253–56, 305–9

White, R., McKenzie, D. 1989. Magmatism at rift zones: the generation of volcanic continental margins and flood basalts. *J. Geophys. Res.* 94: 7685–7729

Wilson, T. J. 1991. Mesozoic and Cenozoic structural patterns in the Transantarctic Mountains, southern Victoria Land. *Antarct. J. U.S.* In press

Annu. Rev. Earth Planet. Sci. 1992. 20:527–52

CALCULATION OF ELASTIC PROPERTIES FROM THERMODYNAMIC EQUATION OF STATE PRINCIPLES

Craig R. Bina

Department of Geological Sciences, Northwestern University, Evanston, Illinois 60208

George R. Helffrich

Department of Terrestrial Magnetism, Carnegie Institution of Washington, Washington, DC 20015

KEY WORDS: seismic velocities, mantle structure, mantle composition, elastic moduli, finite strain theory, Mie-Grüneisen theory, Debye theory, lattice dynamics

INTRODUCTION

The analysis of seismic velocities at depth within the Earth has a long history and attracts an increasingly broad spectrum of researchers. Almost as soon as seismic travel times were available, Williamson & Adams (1923) used them to obtain velocity profiles and check their consistency with the densities that would result from self-compression of a uniform material. Birch (1939) addressed the same question of homogeneity and composition by incorporating the effect of pressure in his extrapolation of the shallow Earth's density and wave speeds to mantle depths. This traditional use of seismological data to investigate mantle properties and phenomena continues since seismic waves (in a broad sense that includes normal modes as well) constitute our primary probe of Earth structure. For example, core-diffracted *P* and *S* waves (Wysession & Okal 1988, 1989) carry

0084–6597/92/0515–0527$02.00

information concerning the properties of the D″ layer at the core-mantle boundary (Bullen 1949, Lay & Helmberger 1983, Lay 1986) that can be related to temperature and composition (Wysession 1991). Slabs of subducted oceanic lithosphere influence the travel times and waveforms of seismic waves (Oliver & Isacks 1967; Toksöz et al 1971; Sleep 1973; Engdahl et al 1977; Jordan 1977; Suyehiro & Sacks 1979; Creager & Jordan 1984, 1986; Silver & Chan 1986; Vidale 1987; Fischer et al 1988; Cormier 1989) which can be related to their thermal structure and composition. Seismic waves reflected and refracted at the slab-mantle interface also provide constraints on the properties of this interface that can be interpreted by computing seismic velocities in slab mineralogies (Helffrich et al 1989). In a similar fashion, the Earth's velocity structure and discontinuities provide information concerning the composition of the mantle (Birch 1952, Lees et al 1983, Anderson & Bass 1984, Bass & Anderson 1984, Bina & Wood 1984, Weidner 1985, Bina & Wood 1987, Akaogi et al 1989, Duffy & Anderson 1989). Finally, the issue of the composition of the lower mantle can be addressed by comparing computed densities of the candidate mineralogies with density profiles of the lower mantle (Watt & O'Connell 1978, Jackson 1983, Anderson 1987, Chopelas & Boehler 1989, Bina & Silver 1990, Fei et al 1991).

The variety of approaches involved in these studies demonstrates the need for a review of the methodology used to compute elastic wave speeds, in the spirit of an earlier review by Anderson et al (1968). Three broad areas bear on this topic: thermodynamic analysis, continuum mechanics, and solid state physics. Experiments provide the raw data and some important rules of thumb that make it possible to compute elastic velocities inside the Earth. Our intent is to combine theory and data into a practical method that will facilitate calculations of this type and to indicate some of the failings and alternative approaches that may be pursued.

The elastic properties of the Earth's interior—i.e. density and elastic wave velocities—are functions of composition, mineralogy, pressure, and temperature. The dependence of the elastic properties on these factors has been measured for numerous minerals over a range of conditions. Compositional variations traditionally have been treated through empirical systematics based on the mineral structure and the mean atomic weight of a mineral (e.g. Birch 1961; Anderson & Nafe 1965; Anderson 1967, 1987). On the other hand, temperature and pressure effects have been investigated through laboratory measurements (e.g. Christensen 1984). These results have given rise to the view that pressure and temperature derivatives of wave speeds, assumed to be constant and applied to speeds measured at room conditions, yield elastic wave speeds under mantle conditions. This process requires extensive extrapolations, by factors of up to ~20 in temperature and 100 in pressure. In general, however, these derivatives are not constant throughout the extrapolation interval, and this assumption is likely to lead to incorrect results. When large extrapolations away from measured properties are necessary, they should be guided by a theoretical framework that controls the functional form. The purpose of this paper is to review the relations that provide a conceptual basis for bridging the gap between experimental and mantle conditions. We deal first with the relevant thermodynamic relations and then briefly with the continuum mechanical and lattice dynamical theories that bear on elastic properties at mantle conditions.

BASIC PARAMETERS

Thermodynamic relations link the responses of substances to changes in their ambient conditions. A number of equally valid formalisms may be invoked to describe these relations (see Appendix), but we live in an environment that exposes us to the responses of substances to changes in temperature and composition at constant pressure. Not only do these conditions shape our intuition, they affect the relative difficulty of certain types of measurements. This, in turn, affects the type of experimental information available for computing elastic properties.

In the laboratory, the most easily controlled parameters are pressure, temperature, and bulk composition. For this reason the Gibbs potential G is often used to characterize such systems. In this formulation the natural variables are P (pressure), T (temperature), and the N_i (compositional quantities), and the equations of state (see Appendix and Table A1) are

$$S = S(T, P, \{N_i\}), \quad V = V(T, P, \{N_i\}), \quad \mu_i = \mu_i(T, P, \{N_i\}), \tag{1}$$

where S, V, and μ_i are the entropy, volume, and chemical potentials, respectively.

As elastic waves pass through a substance, they deform it and may perturb its volume. These volume perturbations comprise one of the links between elasticity and thermodynamic analysis because through them many thermodynamic constraints may be brought to bear on seismic wave propagation. Another important factor is that elastic waves travel faster than heat diffuses so the thermodynamic system may be viewed as adiabatic (closed to heat transfer). The deformation, moreover, may often be viewed as a constant-entropy (isentropic) process, since any such deformation is essentially reversible in the absence of significant elastic attenuation. Since any entropy changes are negligible or poorly characterized, we focus on the effect of volume perturbations.

An obvious, measurable property of a substance is its volume. Its temperature dependence is given by the *volume coefficient of thermal expansion* α, where

$$\alpha \equiv \frac{1}{V}\left(\frac{\partial V}{\partial T}\right)_P = \left(\frac{\partial \ln V}{\partial T}\right)_P, \tag{2}$$

a measure of the volume increase upon heating. α is generally positive but may be zero or negative for some substances, and is O (10^{-5} K^{-1}) for most minerals at room conditions. Similarly, the pressure dependence of the volume is given by a measure of the volume decrease upon compression, the *isothermal compressibility* β_T, where

$$\beta_T \equiv \frac{-1}{V}\left(\frac{\partial V}{\partial P}\right)_T = -\left(\frac{\partial \ln V}{\partial P}\right)_T, \tag{3}$$

or by its reciprocal, the *isothermal bulk modulus* or *isothermal incompressibility* defined by

$$K_T \equiv \frac{1}{\beta_T} = -V\left(\frac{\partial P}{\partial V}\right)_T, \tag{4}$$

an O (10^8 bars) quantity. (Seismologists regrettably use α and β for the *P*- and *S*-wave speeds; here these will be designated v_P and v_S.)

The entropy S is a less intuitive property closely related [from the first of Equations (1)] to a directly measurable quantity, the *isobaric heat capacity* or *specific heat at constant pressure* C_P, defined by

$$C_P \equiv T\left(\frac{\partial S}{\partial T}\right)_P = \left(\frac{đq}{dT}\right)_P, \tag{5}$$

a measure of the heat $đq$ absorbed by one mole of the substance per unit temperature change.

The parameters defined in Equations (2)–(5) above are also functions of pressure and temperature and thus possess their own temperature and pressure derivatives. Of these, the pressure derivative of K_T

$$K_T' \equiv \left(\frac{\partial K_T}{\partial P}\right)_T = -\left(\frac{\partial \ln K_T}{\partial \ln V}\right)_T \tag{6}$$

figures prominently in finite strain equations. Another useful quantity is the temperature dependence of K_T, given by the *isothermal Anderson-Grüneisen ratio* (Barron 1979) or the *isothermal second Grüneisen parameter* δ_T (Anderson et al 1968):

$$\delta_T \equiv \frac{-1}{\alpha K_T}\left(\frac{\partial K_T}{\partial T}\right)_P = -\left(\frac{\partial \ln K_T}{\partial \ln V}\right)_P. \tag{7}$$

Since S and V are state functions, their mixed partial derivatives are equal

$$\frac{\partial^2 S}{\partial P\,\partial T} = \frac{\partial^2 S}{\partial T\,\partial P}, \qquad \frac{\partial^2 V}{\partial P\,\partial T} = \frac{\partial^2 V}{\partial T\,\partial P}, \tag{8}$$

leading to additional relations for the pressure dependence of α and C_P

$$\left(\frac{\partial \alpha}{\partial P}\right)_T = \frac{1}{K_T^2}\left(\frac{\partial K_T}{\partial T}\right)_P = \frac{-\alpha\delta_T}{K_T}, \tag{9}$$

$$\left(\frac{\partial C_P}{\partial P}\right)_T = -TV\left[\alpha^2 + \left(\frac{\partial \alpha}{\partial T}\right)_P\right]. \tag{10}$$

Elastic waves are assumed to propagate adiabatically/isentropically, yet the relations given thus far are isothermal or isobaric. Various expressions relate these to isentropic conditions. Temperature and pressure under such conditions are related by the *adiabatic gradient*

$$\left(\frac{\partial T}{\partial P}\right)_S = \frac{TV\alpha}{C_P}. \tag{11}$$

The defined quantities that follow are analogous to the isothermal relations: the *adiabatic compressibility* β_S and the *adiabatic bulk modulus* or *adiabatic incompressibility* K_S

$$K_S \equiv \frac{1}{\beta_S} \equiv -V\left(\frac{\partial P}{\partial V}\right)_S, \tag{12}$$

the *isochoric heat capacity* or *specific heat at constant volume* C_V

$$C_V \equiv T\left(\frac{\partial S}{\partial T}\right)_V = \left(\frac{dq}{dT}\right)_V, \tag{13}$$

and the isothermal pressure derivative of K_S

$$K_S' \equiv \left(\frac{\partial K_S}{\partial P}\right)_T. \tag{14}$$

The temperature dependence of K_S is given by the *adiabatic Anderson-Grüneisen ratio* or the *adiabatic second Grüneisen parameter* δ_S:

$$\delta_S \equiv \frac{-1}{\alpha K_S}\left(\frac{\partial K_S}{\partial T}\right)_P = \frac{-1}{\alpha}\left(\frac{\partial \ln K_S}{\partial T}\right)_P. \tag{15}$$

The adiabatic and isothermal parameters are related via the *thermal Grüneisen ratio* or *first Grüneisen parameter* γ_{th}:

$$\gamma_{th} \equiv \frac{\alpha K_T V}{C_V} = \frac{\alpha K_S V}{C_P} = \frac{\alpha K_T V}{C_P - T\alpha^2 K_T V}. \tag{16}$$

Implicit in γ_{th} is a deep connection between thermodynamic properties and the atomic structure of solids. Lattice dynamical theories relate macroscopic solid properties to the effects of a lattice of oscillating atoms interacting through their mutual interatomic forces (Born & Huang 1954, Leibfried & Ludwig 1961). Grüneisen-like parameters characterize the volume dependence of the lattice's oscillator frequencies and are important parts of the Mie-Grüneisen equations of state which figure prominently in shock wave studies (Courant & Friedrichs 1948, McQueen et al 1967, Rigden et al 1988, Anderson & Zou 1989, Jeanloz 1989) and in thermal equations of state in general (Jeanloz & Knittle 1989). γ_{th} is an averaged measure of such volume dependence of frequency, and characterizes anharmonic lattice behavior (Slater 1939, Ashcroft & Mermin 1976). It also functions to interconvert adiabatically defined quantities and isothermally defined ones through

$$\frac{K_S}{K_T} = \frac{C_P}{C_V} = 1 + T\alpha\gamma_{th}. \tag{17}$$

HIGH PRESSURE PARAMETERIZATION

The pressures of the Earth's interior lead to considerable material compressions. Continuum mechanical theory provides the link with thermodynamic properties by first postulating the existence of a strain energy function and then identifying it with one of the thermodynamic potentials (see Appendix). Under constant temperature conditions the natural choice is the Helmholz free energy F because $P = -(\partial F/\partial V)_T$, where ∂V clearly relates to the strain imposed by material compression. If instead constant entropy conditions are imposed, the choice is the internal energy U since $P = -(\partial U/\partial V)_S$. Either of these imposed conditions leads to identical equations in which the appropriate modulus, either adiabatic or isothermal, must be used. The choice of F as the strain energy function leads to the well-known Birch-Murnaghan equation that employs K_T. Application of the thermodynamics of F to the analysis of silicate melting relationships at high pressures is demonstrated by Stixrude & Bukowinski (1990).

In these functional relations, volume (strain) depends upon pressure in a rather awkward implicit fashion. Expanding the elastic potential energy arising from an applied hydrostatic finite strain in a Taylor series in strain derivatives and truncating to fourth order in strain leads to (Davies 1973, 1974; Davies & Dziewonski 1975)

$$P = -\left(\frac{\partial F}{\partial V}\right)_T = 3K_{T0}f(1+2f)^{5/2}\left\{1-2\xi f + \frac{f^2}{6}[9K_{T0}K''_{T0}+4\xi(4-3K'_{T0})+5(3K'_{T0}-5)]\right\}. \quad (18)$$

Here

$$\xi \equiv \frac{3}{4}(4-K'_{T0}), \qquad f \equiv \frac{1}{2}\left[\left(\frac{V_0}{V(P)}\right)^{2/3}-1\right], \quad (19)$$

where V_0, K_{T0}, K'_{T0}, and K''_{T0} are all evaluated at zero pressure. A third-order truncation yields the Birch-Murnaghan equation (Birch 1938, 1939, 1952, 1978). Use of this fourth-order expansion may be considered optimistic since the higher derivatives of the bulk modulus are difficult to reliably measure (Bell et al 1987, Jeanloz 1981) and mantle strains are adequately modeled to third order [but not strains in the core or those encountered in shock wave studies (Jeanloz 1989)]. We include it to show the strain order at which these derivatives become significant and for consistency with the pressure dependence of the elastic moduli that follow.

Similar methods yield the pressure dependence of the individual isotropic elastic moduli. K_T follows from differentiation of (18), but to ensure a consistent truncation at fourth order we use the Davies & Dziewonski (1975) formulation

$$K_T(f) = K_{T0}(1+2f)^{5/2}\left\{1-f(5-3K'_{T0}) + \frac{f^2}{2}[9K_{T0}K''_{T0}+(3K'_{T0}-7)(3K'_{T0}-5)]\right\}. \quad (20)$$

The pressure dependence of the shear modulus μ is derived with greater difficulty but may be cast in a similar form (Birch 1939; Davies 1973, 1974; Davies & Dziewonski 1975):

$$\mu(f) = \mu_0(1+2f)^{5/2}\left\{1-f(5-3\mu'_0K_{T0}/\mu_0) + \frac{f^2}{2}[9\mu''_0K^2_{T0}/\mu_0+9\mu'_0K_{T0}/\mu_0(K'_{T0}-4)+35]\right\}. \quad (21)$$

Bracketed terms which are second order in f involve the higher-order pressure derivatives $K''_T = (\partial^2 K_T/\partial P^2)_T$ and $\mu'' = (\partial^2\mu/\partial P^2)_T$. Their inclusion is not warranted by compression data available to date, but they are given here to show where these higher-order strain parameters enter the expressions. K''_{T0} and μ''_0 appear to be of order K_{T0}^{-1} and μ_0^{-1} (Davies & Dziewonski 1975, Cohen 1987, Isaak et al 1990, Yoneda 1990) and so make $O(1)$ contributions to the bracketed f^2 terms in (20) and (21). In the absence of reliable K''_{T0} and μ''_0 values, one may approximate them as K_{T0}^{-1} and μ_0^{-1} or ignore f^2 terms completely (Duffy & Anderson 1989, Gwanmesia et al 1990).

THERMAL PARAMETERIZATION

Some important approximations greatly simplify elastic property calculations. The first concerns the pressure dependence of γ_{th}. To a good approximation (McQueen et al 1967)

$$\frac{\partial}{\partial P}\left(\frac{\gamma_{th}}{V}\right)_T \approx 0, \tag{22}$$

implying that

$$(\gamma_{th}/\gamma_{th0}) = (V/V_0)^q \tag{23}$$

with $q = 1$. Departures from this approximation are known (e.g. Jeanloz & Ahrens 1980a,b) and the approximation has been refined (Irvine & Stacey 1975), but the error introduced by this approximation when used to calculate seismic velocities is small. The exponent q is related to other thermodynamic properties (Bassett et al 1968, Anderson 1984, Anderson & Yamamoto 1987) through

$$q = \delta_T + 1 - K'_{T0} - \frac{T}{\alpha K_T V}\left(\frac{\partial(\alpha K_T)}{\partial T}\right)_V. \tag{24}$$

If q is constant, a relation among these is implied, leading to the second important approximation. Anderson (1984) has drawn attention to the fact that αK_T is essentially constant. While his conclusions are based on metal and alkali halide measurements, they also hold for olivine (Isaak et al 1989), and we assume this to be true for other mantle materials. The last term of (24) is therefore zero, and along with the $q = 1$ approximation this leads to

$$\delta_T = K'_{T0}. \tag{25}$$

This is a good rule of thumb and provides a consistency check between compression measurements (yielding K_{T0} and K'_{T0}) and α through (7).

δ_T is an expression of the temperature behavior of the elastic constants. Quasi-harmonic lattice dynamics theory predicts a linear dependence of the elastic constants on temperature in the high-temperature limit (Born & Huang 1954; Leibfried & Ludwig 1961; Davies 1973, 1974; Garber & Granato 1975a,b), implying that $(\partial K_T/\partial T)_P$ is constant [as well as $(\partial \mu/\partial T)_P$]. In this same limit, δ_T in many materials is observationally constant (Anderson et al 1968, Anderson & Goto 1989, Isaak et al 1989). Since αK_T appears constant as well as $(\partial K_T/\partial T)_P$, δ_T should also be constant by (7). We thus have a consistent picture of mantle material behavior within the framework of the two approximations.

Recent high-quality elastic constant measurements agree with the linear temperature derivatives predicted by theory (Anderson & Goto 1989, Goto et al 1989, Isaak et al 1989). What of the temperature dependence of K_T' and μ'? Relation (25) and the constancy of δ_T at high temperatures imply that the temperature dependence of K_T' is weak. More generally, Davies (1973) notes that the thermal contributions to F are $O(f^{-2})$ where f is the strain. Since K_T' and K_T'' appear in third- and fourth-order strain terms, their temperature dependence is negligible compared to the temperature dependence of V (with a first-order dependence on strain) and K_T (where the dependence is second order). The T dependence of μ' and μ'' can be neglected by a similar argument. Lattice dynamical simulation supports this argument (Hemley et al 1989) by showing less than 10% variation in K_T' in $MgSiO_3$ perovskite over a 300–2000 K temperature range.

Integration of (7) gives the temperature dependence of K_T:

$$\ln \frac{K_T(T)}{K_T(T_{ref})} = -\int_{T_{ref}}^{T} \alpha \delta_T \, d\hat{T}. \tag{26}$$

The large body of $\alpha(T)$ measurements at room pressure strongly recommends 1-bar evaluation of this integral. δ_T is only weakly temperature dependent and can be taken as constant, but the high-temperature value [or its approximation (25)] should be used if possible. Another useful simplification is that α is generally linear with T at high temperature so the integral can be analytically transformed to a quadratic polynomial in T.

The adiabatic moduli are needed to compute elastic wave speeds while K_T, not K_S, is the modulus measured by static compression. Shear deformations preserve volume, so the adiabatic and isothermal shear moduli are identical. Given K_T, (17) gives K_S but the conversion must be done at the conditions of interest if (18) and (20) are used since the up-P path they imply is isothermal. This introduces the need for the pressure dependence of α.

α at pressure may be obtained by straightforward integration of (9)

$$\ln \frac{\alpha(P)}{\alpha(P_{\text{ref}})} = -\int_{P_{\text{ref}}}^{P} \frac{\delta_T}{K_T} d\hat{P} = \int_{P_{\text{ref}}}^{P} \delta_T d\ln V, \tag{27}$$

but this requires characterization of δ_T at pressure. Some simple dependences are now becoming clear (Chopelas & Boehler 1989, Anderson et al 1990), suggesting that

$$(\partial \ln \alpha/\partial \ln V)_T = \delta_T. \tag{28}$$

If borne out by further experimental and theoretical work, this relation will greatly simplify calculation of $\alpha(P)$. Relations (28) and (27) together imply δ_T is independent of V.

A more involved α pressure dependence can be obtained as follows. Rearranging (16) following Jeanloz & Ahrens (1980b), differentiating and invoking approximation (22), we obtain

$$\left(\frac{\partial \alpha}{\partial P}\right)_T = \frac{1}{K_T}\left(\frac{\gamma_{\text{th}}}{V}\right)\left[\left(\frac{\partial C_V}{\partial P}\right)_T - \frac{C_V}{K_T}\left(\frac{\partial K_T}{\partial P}\right)_T\right]. \tag{29}$$

Moving the focus to the pressure dependence of C_V, note that Debye theory parameterizes C_V as a function of T and a characteristic temperature θ_D summarizing the lattice vibration frequency distribution (Slater 1939; Ashcroft & Mermin 1976; Kieffer 1979, 1982). The vibrational Grüneisen parameter γ_{vib} is related to the averaged frequency dependence of lattice vibrations on the lattice volume

$$\gamma_{\text{vib}} = -\frac{d\ln \theta_D}{d\ln V}, \tag{30}$$

and is one member of an entire family of Grüneisen-type parameters of differing theoretical heritages (Quareni & Mulariga 1988). Since $C_V = C_V(T, \theta_D)$, and $\gamma_{\text{th}} = \gamma_{\text{vib}}$ in the quasi-harmonic approximation, we can use the volume dependence of γ_{th} to parameterize the pressure dependence of C_V following Kieffer (1982). Writing

$$\left(\frac{\partial C_V}{\partial P}\right)_T = \left(\frac{\partial C_V}{\partial \theta_D}\right)_T\left(\frac{\partial \theta_D}{\partial P}\right)_T = -\left(\frac{\partial C_V}{\partial \theta_D}\right)_T\left(\frac{\gamma_{\text{th}}}{V}\right)\theta_D\left(\frac{\partial V}{\partial P}\right)_T$$

$$= \left(\frac{\partial C_V}{\partial \theta_D}\right)_T\left(\frac{\gamma_{\text{th}}}{V}\right)\theta_D \frac{V}{K_T}, \tag{31}$$

we note that each of the terms above is known: $\partial C_V/\partial \theta_D$ is calculable from its definition (Slater 1939)

$$C_V(T, \theta_D) = 9nR\left(\frac{T}{\theta_D}\right)^3 \int_0^{\theta_D/T} \frac{x^4 e^x}{(e^x - 1)^2} dx \tag{32}$$

(where n is the number of atoms per molecular formula unit); (18) and (20) give V and K_T; and γ_{th}/V is approximately constant (22). The terms involving θ_D explicitly read:

$$\theta_D(P) = \theta_D \exp\left\{\frac{\gamma_{th}}{V}\bigg|_{V=V_{ref}} [V_{ref} - V(P)]\right\},$$

$$\frac{1}{9nR}\left(\frac{\partial C_V}{\partial \theta_D}\right)_T = \frac{\theta_D}{T^2} \frac{e^{\theta_D/T}}{(e^{\theta_D/T} - 1)^2} - \frac{3}{\theta_D}\left(\frac{T}{\theta_D}\right)^3 \int_0^{\theta_D/T} \frac{x^4 e^x}{(e^x - 1)^2} dx. \tag{33}$$

Rearranging (29) and (31) and eliminating C_V with (16) yields a differential equation for $\alpha(P)$

$$\left(\frac{\partial \alpha}{\partial P}\right)_T + \alpha \frac{K_T'}{K_T} - \frac{\theta_D(P)}{K_T}\left(\frac{\gamma_{th}}{V}\right)^2 \left(\frac{\partial C_V}{\partial \theta_D}\right)_T \frac{V}{K_T} = 0 \tag{34}$$

that readily yields to numerical solution as an initial value problem (e.g. Press et al 1987). Note that (34) is equivalent, to first order, to substituting (25) into (27).

We show in Figure 1 the approximate α pressure dependence for MgO.

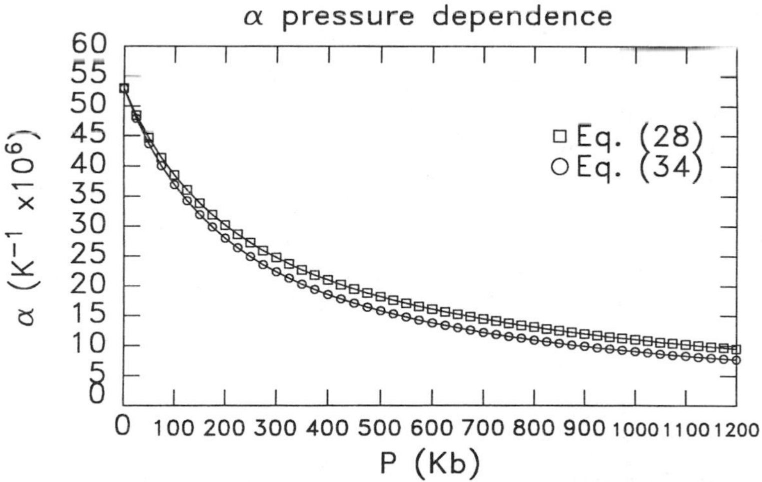

Figure 1 Pressure dependence of thermal expansion α in MgO at 1500°C. Equation (28) denotes method of Anderson et al (1990) where $(\partial \ln \alpha/\partial \ln V)_T = \delta_T$. Equation (34) denotes method whereby $(\partial \alpha/\partial P)_T$ is obtained via the pressure dependence of the heat capacity. Data for MgO are from Fei et al (1991), from which $\delta_{T_{(1200K)}} = 4.47$ was computed, and Sumino & Anderson (1984), whence $\gamma_{th} = 1.52$.

The P dependence for each formulation is similar, and the α values given by (28) and (34) differ by about 20% at core-mantle boundary pressures. This leads to about a 0.5% difference in $1+T\alpha\gamma_{th}$ which can be considered negligible when compared to the errors in the various moduli measurements themselves.

VELOCITY CALCULATIONS

The density ρ and bulk sound velocity v_ϕ of a material are functions of the volume and the adiabatic bulk modulus

$$\rho \equiv \frac{M}{V}, \qquad v_\phi^2 \equiv \frac{K_S}{\rho}, \tag{35}$$

where M is the molar mass of the material. The isotropic P- and S-wave velocities, v_P and v_S, are additionally functions of the shear modulus

$$v_P^2 \equiv \frac{K_S + \frac{4}{3}\mu}{\rho}, \qquad v_S^2 \equiv \frac{\mu}{\rho}. \tag{36}$$

Given P and T conditions where v_P and v_S are desired, we may choose any P-T path along which to evaluate them since V, K_S, and μ are all state functions. The choice is one of convenience. Volumetric thermal expansivity and heat capacity data are widely available for minerals at room pressure (Skinner 1966; Touloukian 1967; Helgeson et al 1978; Robie et al 1979; Taylor 1986; Fei & Saxena 1986; Berman 1988; Fei et al 1990, 1991) but are scarce at high pressure. This suggests that the preferred route to the high P-T state is a single step up-temperature $T_{ref} \rightarrow T$ at room pressure P_{ref} followed by a second step up-pressure $P_{ref} \rightarrow P$ at high temperature T. Since most volume measurements are made at room conditions and thermal expansion measurements are made at room pressure, 298 K and 1 bar is a natural reference state. For simplicity, the 1 bar pressure reference may be taken as zero bars for solids.

To evaluate the volume at P_{ref} and T, integrate (2)

$$\ln \frac{V(T)}{V(T_{ref})} = \int_{T_{ref}}^{T} \alpha d\hat{T}. \tag{37}$$

Similarly, K_T may be evaluated at T from its value at T_{ref} by (26), and $C_P(T)$ may be obtained from data fitted to polynomials. In integrating the polynomial formulation for $\alpha(T)$ in (26) and (37), it is helpful to note that α increases nearly linearly with T above intermediate temperatures (Jeanloz & Ahrens 1980b, Saxena 1989). This simple behavior suggests a linear parameterization of α from a high-temperature reference state such as 1000 K.

To evaluate $V(P, T)$ we insert $V(P_{ref}, T)$, $K_T(P_{ref}, T)$, $K'_T(P_{ref})$, and P into Equation (18). This equation is implicit for V as a function of P, so iterative numerical techniques must be used to find V at a corresponding P; the Newton-Raphson method converges rapidly on the solution (e.g. Press et al 1987). $K_T(P, T)$ is obtained from Equation (20) using the f value from (18). Then $\alpha(P, T)$ is obtained from $\alpha(P_{ref}, T)$ using (28) or (34), at which point $K_S(P, T)$ can be computed from $K_T(P, T)$ via (17).

To obtain $\mu(P, T)$ from its value at the reference state, we first account for the temperature dependence, linear in the high temperature limit, via

$$\mu(T) = \mu(T_{ref}) + (d\mu/dT)\cdot(T - T_{ref}), \tag{38}$$

and we than account for the pressure dependence via (21). Finally, we insert values for V, K_S, and μ at P and T into (35) and (36) to obtain ρ, v_ϕ, v_P, and v_S.

EXAMPLE

As an example of the use of this method, we calculate the velocity difference between the olivine α and β phases at the mantle conditions of the 400-km discontinuity. The pressure and temperature here are approximately 133 kb and 1500°C (Bina & Wood 1987). To calculate these values we use the elastic data in Table 1. Mantle olivines are taken to be of $(Mg_{0.9}Fe_{0.1})_2SiO_4$ composition (Jeanloz & Thompson 1983), so their properties must be interpolated from the tabulated end-member values. The standard practice with elastic properties is to average on the mole fractions of the components in the phase (Weidner et al 1984). At 133 kb and 1500°C, values for ρ, v_P, v_S, and v_ϕ in $(Mg_{0.9}Fe_{0.1})_2SiO_4$ are 3.491 g cm^{-3}, 8.92 km s^{-1}, 4.73 km s^{-1}, and 7.05 km s^{-1} for the α phase and 3.693 g cm^{-3}, 9.62 km s^{-1}, 5.17 km s^{-1}, and 7.55 km s^{-1} for the β phase. Comparing with the mantle Δv_ϕ value of 4.9% from Bina & Wood (1987), the computed Δv_ϕ value suggests that the mantle is ~70% olivine by volume (Figure 2). The bulk sound velocity v_ϕ provides the most reliable comparison of the three velocities since it is computed entirely from volumetric material properties that are better constrained by compression measurements. The shear modulus influences both v_P and v_S and is known only from elastic constant or sound speed measurements. The mutually contradictory results from comparison of v_P and v_S illustrate this difference. Comparing the mantle 4.5–6% Δv_P and 4.6% Δv_S values reported in Bina & Wood (1987) with those calculated above leads to a ~55–75% mantle olivine content based on matching Δv_P yet a ~50% olivine content based on Δv_S. Thus, a

Table 1 Elastic data used in calculations

Component	ρ_{300K} g/cm^3	α_{300K} $\times 10^6\ K^{-1}$	$d\alpha/dT$ $\times 10^9\ K^{-2}$	K_T Mbar	dK_T/dP	μ Mbar	$d\mu/dP$	$-d\mu/dT$ bar K^{-1}	γ_{th}	$\delta_{T(1200K)}$
Mg_2SiO_4 (α)	3.366	26.11	15	1.278	5.1	0.810	1.82	309	1.2	3.96
Fe_2SiO_4 (α)	4.404	26.20	12	1.359	5.2	0.510	0.62	130	1.1	5.45
Mg_2SiO_4 (β)	3.474	20.63	17	1.726	4.73	1.140	1.82	228	1.1	4.27
Fe_2SiO_4 (β)	4.424	22.1	11	1.859	4.73	0.730	0.62	146	1.1	5.51
$Mg_2Si_2O_6$[a] (Opx, Cpx)	3.194	16.12	84	1.054	2.36	0.750	2.36	180	1.1	4.48
$Fe_2Si_2O_6$[a] (Opx, Cpx)	3.998	29.96	26	1.030	2.36	0.750	2.36	180	1.1	4.46
$NaAlSi_2O_6$[a] (Cpx)	3.347	24.72	0	1.417	2.40	0.840	2.40	84	1.3	7.33
$CaMgSi_2O_6$[a] (Opx, Cpx)	3.276	33.27	0	1.118	2.30	0.670	2.30	154	1.1	5.94
$Py_{73}Alm_{16}Uv_6$[a] (Garnet)	3.711	19.20	8	1.700	4.46	0.927	1.47	87	1.0	6.76

300 K reference temperature for all properties. Opx, orthopyroxene; Cpx, clinopyroxene. Except as noted, source of elastic data is Bina & Wood (1987).

[a] Helffrich et al (1989), $d\mu/dP$ values corrected.

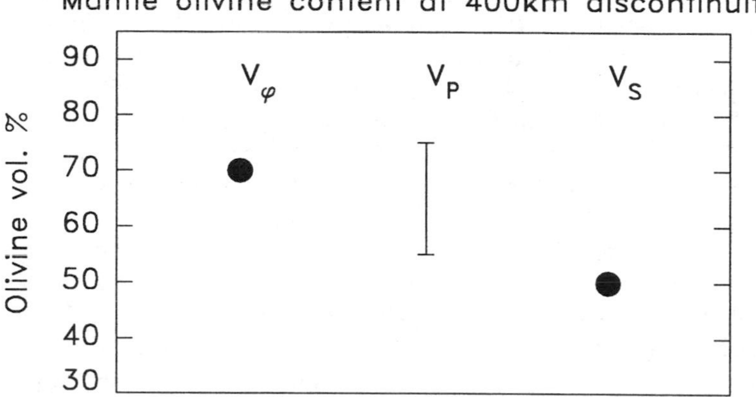

Figure 2 Olivine content at 400 km depth in the mantle computed from velocity contrast between α and β $(Mg_{0.9}Fe_{0.1})_2SiO_4$ olivine. When compared with the seismologically observed velocity jumps in bulk sound velocity v_ϕ, *P* velocity v_P, and *S* velocity v_S, different estimates of mantle olivine content are obtained.

pyrolitic upper mantle composition is suggested by both Δv_ϕ and the upper Δv_P value but not by either Δv_S or the lower Δv_P value. One possible explanation for this systematic variation in compositional estimates as a function of velocity parameterization would be overestimation of $(\mu_\beta - \mu_\alpha)_{P,T}$.

MINERAL AGGREGATES

Four methods have been advanced for averaging elastic moduli in isotropic mineral aggregates. Two of these are the familiar Reuss and Voigt bounds which are derived by assuming either stress continuity or strain (displacement) continuity, respectively, at individual mineral interfaces. There is nothing to particularly favor either assumption, so the third method is to average the Reuss and Voigt bounds; this Hill average constitutes the well-known Voigt-Reuss-Hill (VRH) approximation. Finally, the Hashin-Shtrikman (HS) bounds seek to define the properties of the composite solid by a variational technique that incorporates shape effects, unlike the other averaging schemes. If only the elastic moduli and volumetric proportions of its constituents are known, the HS bounds appear to provide the most restrictive bounds for an isotropic composite (Watt & O'Connell 1978). These methods are all based upon the assumption of homogeneous, isotropic mineral aggregates and thus are not directly applicable to studies of seismic anisotropy (e.g. Silver & Chan 1988).

All of these methods are described by Watt et al (1976), who also give cxamples of the averaging methods applied to sample mixtures. When all four bounds are computed for the same aggregate, the Reuss and Voigt bounds place broad upper and lower limits on the material properties. The HS bounds define similar but narrower limits outside of which the VRH average may occasionally lie—probably indicative of the lack of theoretical basis for the VRH average. Although the VRH average is easier to calculate, the HS bounds are not difficult to compute and should be the preferred averaging method. When the upper and lower HS bounds themselves are averaged, they give an approximate average like VRH.

All of these methods require the determination of the volumetric proportion v_i of each phase i of the N phases present in the aggregate:

$$v_i \equiv \frac{X_i V_i}{\sum_{j=1}^{N} X_j V_j}, \tag{39}$$

where X_i is the mole fraction of phase i in the assemblage. Formulae for the Reuss-, Voigt-, and Hill-averaged moduli (M_{R}, M_{V}, and M_{H}), where $\mathbf{M}$ represents either K or μ, are:

$$M_{\mathrm{R}} = \left[\sum_{i=1}^{N} v_i/\mathbf{M}_i\right]^{-1}, \quad M_{\mathrm{V}} = \sum_{i=1}^{N} v_i\mathbf{M}_i, \quad M_{\mathrm{H}} = \frac{M_{\mathrm{R}}+M_{\mathrm{V}}}{2}. \tag{40}$$

Those for the HS moduli $K_{\mathrm{HS}\pm}$ and $\mu_{\mathrm{HS}\pm}$ are as follows:

$$K_{\mathrm{HS}-} = K_1+[A_1/(1+\alpha_1 A_1)], \quad K_{\mathrm{HS}+} = K_N+[A_N/(1+\alpha_N A_N)],$$

$$K_{\overline{\mathrm{HS}}} = \frac{K_{\mathrm{HS}+}+K_{\mathrm{HS}-}}{2},$$

$$\alpha_1 \equiv -3/(3K_1+4\mu_1), \qquad \alpha_N \equiv -3/(3K_N+4\mu_N),$$

$$A_1 \equiv \sum_{i=2}^{N} \frac{v_i}{(K_i-K_1)^{-1}-\alpha_1}, \qquad A_N \equiv \sum_{i=1}^{N-1} \frac{v_i}{(K_i-K_N)^{-1}-\alpha_N}, \tag{41a}$$

$$\mu_{\mathrm{HS}-} = \mu_1+\frac{1}{2}[B_1/(1+\beta_1 B_1)], \quad \mu_{\mathrm{HS}+} = \mu_N+\frac{1}{2}[B_N/(1+\beta_N B_N)],$$

$$\mu_{\overline{\mathrm{HS}}} = \frac{\mu_{\mathrm{HS}+}+\mu_{\mathrm{HS}-}}{2},$$

$$\beta_1 \equiv -\frac{3(K_1+2\mu_1)}{5\mu_1(3K_1+4\mu_1)}, \qquad \beta_N \equiv -\frac{3(K_N+2\mu_N)}{5\mu_N(3K_N+4\mu_N)},$$

$$B_1 \equiv \sum_{i=2}^{N} \frac{v_i}{[2(\mu_i-\mu_1)]^{-1}-\beta_1}, \qquad B_N \equiv \sum_{i=1}^{N-1} \frac{v_i}{[2(\mu_i-\mu_N)]^{-1}-\beta_N}. \tag{41b}$$

The HS formulae require sorting of the moduli: K_1 and μ_1 represent the smallest values and K_N and μ_N the largest. [See Watt et al (1976) for shape-dependent formulae for HS bounds, useful for modeling the properties of fluid-solid aggregates (e.g. Helffrich et al 1989).]

For a sample calculation using this method, we will compute the temperature dependence of velocities in a pyrolitic bulk composition at 96 kb and 1000°C. This is a useful exercise to show the magnitude of the P and S velocity increase anticipated within a slab of subducted lithosphere. A temperature decrease of about 500°C from the slab-mantle interface temperature of 1000°C is expected from thermal modeling (Helffrich et al 1989). Table 2 lists the mineral densities, volumetric proportions, and velocities at 1500°C, 1000°C, and 500°C, computed from the elastic data of Table 1. Note that the Hashin-Shtrikman bounds are much tighter than the Voigt and Reuss bounds, though the VRH and HS averages are identical.

The temperature dependence based on the difference between the 1500°C and 1000°C values is -0.56 m s^{-1} K^{-1} for P and -0.34 m s^{-1} K^{-1} for S. Between 1000°C and 500°C the P figure drops to -0.44 m s^{-1} K^{-1} but the temperature dependence of S is the same [see (38)]. The temperature

Table 2 Volumetric proportions of minerals in pyrolitic mantle composition at 96 kb[a]

	1500°C					1000°C					500°C				
Phase	vol%	ρ	v_P	v_S	v_ϕ	vol%	ρ	v_P	v_S	v_ϕ	vol%	ρ	v_P	v_S	v_ϕ
Cpx	**3.6**	**3.419**	**8.36**	**4.79**	**6.27**	**3.6**	**3.470**	**8.66**	**4.96**	**6.50**	**3.6**	**3.513**	**8.92**	**5.12**	**6.69**
Opx	**22.3**	**3.329**	**7.96**	**4.51**	**6.03**	**22.1**	**3.415**	**8.31**	**4.73**	**6.26**	**22.0**	**3.482**	**8.60**	**4.95**	**6.43**
Garnet	**15.0**	**3.810**	**8.95**	**4.94**	**6.90**	**15.1**	**3.845**	**9.16**	**5.04**	**7.08**	**15.2**	**3.877**	**9.36**	**5.13**	**7.24**
Olivine	**59.1**	**3.439**	**8.49**	**4.60**	**6.62**	**59.2**	**3.488**	**8.75**	**4.78**	**6.79**	**59.2**	**3.531**	**8.99**	**4.95**	**6.94**
bulk ρ		**3.469**					**3.525**					**3.572**			
Averages —															
HS			**8.42**	**4.64**	**6.50**			**8.70**	**4.81**	**6.70**			**8.95**	**4.98**	**6.85**
VRH			**8.43**	**4.64**	**6.51**			**8.71**	**4.81**	**6.70**			**8.95**	**4.98**	**6.85**
Bounds —															
HS-			**8.42**	**4.64**	**6.50**			**8.70**	**4.81**	**6.69**			**8.95**	**4.98**	**6.85**
HS+			**8.43**	**4.64**	**6.51**			**8.70**	**4.81**	**6.70**			**8.95**	**4.98**	**6.85**
Voigt			**8.46**	**4.65**	**6.53**			**8.73**	**4.82**	**6.72**			**8.97**	**4.99**	**6.87**
Reuss			**8.40**	**4.63**	**6.48**			**8.68**	**4.81**	**6.68**			**8.93**	**4.98**	**6.84**

[a] Elastic data given in Table 1. Zero pressure volumetric proportions and compositions of phases: 3.5 vol. % Cpx ($Jd_{43}Di_{27}Fs_4En_{26}$), 22.1 vol. % Opx ($En_{60.6}Di_{34.5}Fe_{5.1}$), 15.0 vol. % garnet ($Py_{73}Alm_{16}$), 59.0 vol. % olivine ($Fo_{90}Fa_{10}$) (Helffrich et al 1989).

dependence for P is at the low end or lower than the range used in some seismological studies (-0.5 to -0.9 m s^{-1} K^{-1}) (e.g. Jacob 1972, Sleep 1973, Creager & Jordan 1984) but is compatible with others (e.g. Creager & Jordan 1986). These low temperature derivatives suggest that the 5–10% Δv_P changes implied by seismic waves reflected and converted at the slab-mantle interface are not entirely due to thermal effects (Helffrich et al 1989).

CONCLUSIONS

Fundamental thermodynamic relations prescribe a method for the calculation of densities and seismic velocities at high pressures and temperatures. The mineralogical data required for these calculations continue to improve. Two challenges to the mineral physics and seismological communities are presented by the current state of affairs, however.

The seismological parameter that can be computed with most confidence from current mineralogical data is the bulk sound velocity v_ϕ. This is simply due to the quality and availability of static compression data that give K_T (and hence K_S). Seismologically, however, v_ϕ is a relatively poorly determined property of the Earth, since it is not measured directly but is derived from v_P and v_S profiles. These are separately solved for (e.g. Jeffreys & Bullen 1988, Kennett & Engdahl 1991) and may have different systematic biases leading to v_ϕ profiles of uncertain accuracy. On the other hand, density profiles are obtained independently of body wave travel times but are directly related to v_ϕ since in the Earth $v_\phi^2 = (\partial P/\partial \rho)_S$ (Bullen 1963), so some consistency checking is feasible. Nonetheless, direct inversion for v_ϕ would supply the mineral physics community with a useful datum that could be compared with models of the Earth's constitution.

Seismology already provides very useful data for mineral physics that have not been fully exploited to date: notably the v_S profiles, which depend upon the depth variation of a single elastic modulus, μ, in addition to the density. Laboratory determination of μ requires direct elastic constant or elastic velocity measurements (e.g. Gwanmesia et al 1990, Weidner & Carleton 1977, Anderson & Goto 1989) that are technically more demanding than compression measurements. The biggest gap in our knowledge of μ is the matter of its pressure dependence. A concerted experimental and computational modeling effort (e.g. Cohen 1987, Isaak et al 1990) could make v_S a much more valuable tool for research into Earth composition.

For the moment, the set of material properties at hand must be used, and it is much more complete than in 1968 (Anderson et al 1968). Room pressure values of $\alpha(T)$, $C_P(T)$, K, K', and μ are available for virtually all known mantle minerals. This permits v_ϕ and ρ calculations to be made with some confidence. Perhaps the coming quarter-century will bring a further expansion in materials property knowledge that will improve our understanding of the Earth's interior as much as the first seismic wave travel time tables did for Adams and Williamson.

ACKNOWLEDGMENTS

C. Bina acknowledges the hospitality of Tokyo University's Geophysical Institute and the Carnegie Institution of Washington where portions of this review were prepared. Thanks also to G. Helffrich's colleagues at DTM for tolerating his closeted work habits during the final weeks of work on this manuscript and to Lars Stixrude for helpful comments.

APPENDIX: THERMODYNAMIC POTENTIALS AND FUNDAMENTAL EQUATIONS

The thermodynamic state of a system can be described by a number of parameters (cf Callen 1960). The entropy S, internal energy U, volume V, and number of moles N_i of each component i of the system are called *extensive parameters*; they scale linearly with the size of the system (i.e. amount of matter) under consideration. Entropy is a homogeneous first-order function of the other extensive parameters:

$$S = S(U, V, \{N_i\}) \tag{A1}$$

and is called a *thermodynamic potential* by analogy with the potentials of physics. The differential quantities of the potential are its measurable properties. Equation (A1) is the fundamental equation of the thermo dynamic potential and constitutes a complete thermodynamic description of a system. The thermodynamic potential is a state function; that is, the potential difference between any two states is independent of the path taken between those states. Given this fundamental equation, we may write the total differential of the entropy:

$$dS = \left(\frac{\partial S}{\partial U}\right)_{V,\{N_i\}} dU + \left(\frac{\partial S}{\partial V}\right)_{U,\{N_i\}} dV + \sum_i \left(\frac{\partial S}{\partial N_i}\right)_{U,V,\{N_{j\neq i}\}} dN_i. \tag{A2}$$

We now define three additional parameters for describing the thermodynamic state of a system:

$$P \equiv -\left(\frac{\partial U}{\partial V}\right)_{S,\{N_i\}}, \quad T \equiv \left(\frac{\partial U}{\partial S}\right)_{V,\{N_i\}}, \quad \mu_i \equiv \left(\frac{\partial U}{\partial N_i}\right)_{S,V,\{N_{j\neq i}\}}, \tag{A3}$$

where the pressure P, temperature T, and chemical potential μ_i of component i are called *intensive parameters* and do not scale with the size of the system. Upon inserting these definitions (A3) into the total differential

(A2), we obtain the following expression for the total differential of the entropy:

$$dS = \frac{1}{T}dU + \frac{P}{T}dV - \sum_i \frac{\mu_i}{T}dN_i. \tag{A4}$$

The significance of this result is that when the state of a system changes, the changes in the extensive parameters describing that system are not all independent. This provides a constraint on the mutual variation of the extensive parameters of the system, in this case volume, entropy, and composition.

Since the fundamental equation (A1) is a first-order homogeneous form, we have

$$S = U\left(\frac{\partial S}{\partial U}\right)_{V,\{N_i\}} + V\left(\frac{\partial S}{\partial V}\right)_{U,\{N_i\}} + \sum_i N_i\left(\frac{\partial S}{\partial N_i}\right)_{U,V,\{N_{j\neq i}\}}$$

$$= \frac{1}{T}U + \frac{P}{T}V - \sum_i \frac{\mu_i}{T}N_i, \tag{A5}$$

which is called the *Euler equation* for the potential S. Upon differentiating Equation (A5) and subtracting our expression (A4) for the total differential of S, we obtain

$$0 = Ud\left(\frac{1}{T}\right) + Vd\left(\frac{P}{T}\right) - \sum_i N_i d\left(\frac{\mu_i}{T}\right), \tag{A6}$$

which is called the *Gibbs-Duhem relation* for the potential S. While the total differential (A4) expresses the relationships between changes in the extensive parameters describing a system, the Gibbs-Duhem relation (A6) expresses the relationships between changes in the intensive parameters of the system.

In the fundamental equation (A1) for the thermodynamic potential S, we have written the potential as a homogeneous first-order function of certain thermodynamic parameters. Hence, the additional parameters (A3) introduced as coefficients in the total differential (A4) are homogeneous zero-order functions of these same thermodynamic parameters:

$$T = T(U, V, \{N_i\}), \quad P = P(U, V, \{N_i\}), \quad \mu_i = \mu_i(U, V, \{N_i\}). \tag{A7}$$

The functions (A7) are called the *equations of state*, and the arguments are called the natural variables of the functions. Knowledge of all three equations of state is equivalent, through the Euler equation, to complete knowledge of the fundamental equation of a system. While the equations of state may be written as functions of other parameters, the natural variables have optimum informational content and are the only ones to permit a complete thermodynamic description of a system (Callen 1960).

Finally, the state of *thermodynamic equilibrium* is given by the *entropy maximum principle*:

$$[dS(U, V, \{N_i\})]_U = 0, \qquad [d^2S]_U < 0. \tag{A8}$$

For example, solution of the first of Equations (A8) for equilibrium between two multi-component phases α and β yields

$$T_\alpha = T_\beta, \quad P_\alpha = P_\beta, \quad \mu_i^\alpha = \mu_i^\beta, \tag{A9}$$

which are the conditions of thermal, mechanical, and chemical equilibrium, respectively. Solution of the second of Equations (A8) yields the thermodynamic *stability conditions*:

$$\frac{dT}{đq} > 0, \qquad \left(\frac{\partial P}{\partial V}\right)_T < 0, \tag{A10}$$

where $đq$ represents the molar heat absorbed by a substance.

In the above discussion, and in Equations (A1)–(A8), we have treated the entropy as the thermodynamic potential, with the internal energy, volume, and mole numbers as the natural variables. However, we may recast the entire discussion from an *entropy representation* into an *energy representation*:

$$S = S(U, V, \{N_i\}) \rightarrow U = U(S, V, \{N_i\}). \tag{A11}$$

The internal energy U is now the thermodynamic potential, and the total differential becomes

$$dU = TdS - PdV + \sum_i \mu_i dN_i. \tag{A12}$$

The equilibrium state, for example, is now given by the *energy minimum principle*:

$$[dU(S, V, \{N_i\})]_S = 0, \qquad [d^2U]_S > 0. \tag{A13}$$

However, for both of the thermodynamic potentials S and U, the natural variables are all extensive parameters. For a particular problem, it may be useful to have a potential whose natural variables include intensive parameters. We may define numerous such thermodynamic potentials through the technique of partial Legendre transformation. For example, we may define the Helmholtz potential F as the quantity $U-TS$ so that the fundamental equation becomes

Table A1 Summary of some common thermodynamic potentials[a]

Summary of some common thermodynamic potentials
Entropy: $S = S(U,V,\{N_i\})$
$dS \equiv \left[\frac{\partial S}{\partial U}\right]_{V,\{N_i\}} dU + \left[\frac{\partial S}{\partial V}\right]_{U,\{N_i\}} dV + \sum_i \left[\frac{\partial S}{\partial N_i}\right]_{U,V,\{N_{j\neq i}\}} dN_i = \left[\frac{1}{T}\right] dU + \left[\frac{P}{T}\right] dV - \sum_i \left[\frac{\mu_i}{T}\right] dN_i$
$S = \left[\frac{1}{T}\right] U + \left[\frac{P}{T}\right] V - \sum_i \left[\frac{\mu_i}{T}\right] N_i \ , \quad 0 = U d\left[\frac{1}{T}\right] + V d\left[\frac{P}{T}\right] - \sum_i N_i d\left[\frac{\mu_i}{T}\right]$
$T = T(U,V,\{N_i\}) \ , \quad P = P(U,V,\{N_i\}) \ , \quad \mu_i = \mu_i(U,V,\{N_i\})$
Equilibrium maximizes S at constant U.
Internal Energy: $U = U(S,V,\{N_i\})$
$dU \equiv \left[\frac{\partial U}{\partial S}\right]_{V,\{N_i\}} dS + \left[\frac{\partial U}{\partial V}\right]_{S,\{N_i\}} dV + \sum_i \left[\frac{\partial U}{\partial N_i}\right]_{S,V,\{N_{j\neq i}\}} dN_i = TdS - PdV + \sum_i \mu_i dN_i$
$U = TS - PV + \sum_i \mu_i N_i \ , \quad 0 = SdT - VdP + \sum_i N_i d\mu_i$
$T = T(S,V,\{N_i\}) \ , \quad P = P(S,V,\{N_i\}) \ , \quad \mu_i = \mu_i(S,V,\{N_i\})$
Equilibrium minimizes U at constant S.
Helmholtz Potential: $F \equiv U - TS = F(T,V,\{N_i\})$
$dF \equiv \left[\frac{\partial F}{\partial T}\right]_{V,\{N_i\}} dT + \left[\frac{\partial F}{\partial V}\right]_{T,\{N_i\}} dV + \sum_i \left[\frac{\partial F}{\partial N_i}\right]_{T,V,\{N_{j\neq i}\}} dN_i = -SdT - PdV + \sum_i \mu_i dN_i$
$F = -PV + \sum_i \mu_i N_i \ , \quad 0 = SdT - VdP + \sum_i N_i d\mu_i$
$S = S(T,V,\{N_i\}) \ , \quad P = P(T,V,\{N_i\}) \ , \quad \mu_i = \mu_i(T,V,\{N_i\})$
Equilibrium minimizes F at constant *T*.
Enthalpy: $H \equiv U + PV = H(S,P,\{N_i\})$
$dH \equiv \left[\frac{\partial H}{\partial S}\right]_{P,\{N_i\}} dS + \left[\frac{\partial H}{\partial P}\right]_{S,\{N_i\}} dP + \sum_i \left[\frac{\partial H}{\partial N_i}\right]_{S,P,\{N_{j\neq i}\}} dN_i = TdS + VdP + \sum_i \mu_i dN_i$
$H = TS + \sum_i \mu_i N_i \ , \quad 0 = SdT - VdP + \sum_i N_i d\mu_i$
$T = T(S,P,\{N_i\}) \ , \quad V = V(S,P,\{N_i\}) \ , \quad \mu_i = \mu_i(S,P,\{N_i\})$
Equilibrium minimizes H at constant *P*.
Gibbs Potential: $G \equiv U - TS + PV = H - TS = G(T,P,\{N_i\})$
$dG \equiv \left[\frac{\partial G}{\partial T}\right]_{P,\{N_i\}} dT + \left[\frac{\partial G}{\partial P}\right]_{T,\{N_i\}} dP + \sum_i \left[\frac{\partial G}{\partial N_i}\right]_{T,P,\{N_{j\neq i}\}} dN_i = -SdT + VdP + \sum_i \mu_i dN_i$
$G = \sum_i \mu_i N_i \ , \quad 0 = SdT - VdP + \sum_i N_i d\mu_i$
$S = S(T,P,\{N_i\}) \ , \quad V = V(T,P,\{N_i\}) \ , \quad \mu_i = \mu_i(T,P,\{N_i\})$
Equilibrium minimizes G at constant *P* and *T*.

[a] For each potential: line 1 gives name, definition (if applicable), and fundamental equation; line 2 gives total differential; line 3 gives Euler equation and Gibbs-Duhem relation; line 4 gives three equations of state associated with fundamental equation; line 5 gives equilibrium extremum principle.

$$U = U(S, V, \{N_i\}) \rightarrow F = F(T, V, \{N_i\}). \tag{A14}$$

Note that this transformation has replaced the entropy S with the temperature T as a natural variable. Similarly, we may define the enthalpy H which replaces V with P, or we may define the Gibbs potential G which replaces both S and V with T and P. Table A1 summarizes the name, fundamental equation, total differential, Euler equation, Gibbs-Duhem relation, equations of state, and equilibrium extremum principle for each of these common thermodynamic potentials.

Literature Cited

Akaogi, M., Ito, E., Navrotsky, A. 1989. Olivine-modified spinel-spinel transitions in the system $Mg_2SiO_4-Fe_2SiO_4$: Calorimetric measurements, thermochemical calculation, and geophysical application. *J. Geophys. Res.* 94: 15,671–85

Anderson, D. L. 1967. A seismic equation of state. *Geophys. J. R. Astron. Soc.* 13: 9–30

Anderson, D. L. 1987. A seismic equation of state II. Shear properties and thermodynamics of the lower mantle. *Earth Planet. Sci. Lett.* 45: 307–23

Anderson, D. L., Bass, J. 1984. Mineralogy and composition of the upper mantle. *Geophys. Res. Lett.* 11: 637–40

Anderson, O. L. 1984. A universal thermal equation-of-state. *J. Geodyn.* 1: 185–214

Anderson, O. L., Goto, T. 1989. Measurement of elastic constants of mantle-related minerals at temperatures up to 1800 K. *Phys. Earth Planet. Inter.* 55: 241–53

Anderson, O. L., Nafe, J. E. 1965. The bulk modulus-volume relationship for oxide compounds and related geophysical problems. *J. Geophys. Res.* 70: 3951–63

Anderson, O. L., Yamamoto, S. 1987. The interrelationship of thermodynamic properties obtained by the piston-cylinder high pressure experiments and RPR high temperature experiments for NaCl. In *Geophys. Monogr. 39, High-Pressure Research in Mineral Physics*, ed. M. H. Manghnani, Y. Syono, pp. 289–98. Washington: AGU

Anderson, O. L., Zou, K. 1989. Formulation of the thermodynamic functions for mantle minerals: MgO as an example. *Phys. Chem. Miner.* 16: 642–48

Anderson, O. L., Chopelas, A., Boehler, R. 1990. Thermal expansivity versus pressure at constant temperature: A re-examination. *Geophys. Res. Lett.* 17: 685–88

Anderson, O. L., Schreiber, E., Lieberman, R. C., Soga, N. 1968. Some elastic constant data on minerals relevant to geophysics. *Rev. Geophys.* 6: 491–524

Ashcroft, N. W., Mermin, N. D. 1976. *Solid State Physics*. Hong Kong: HRW Int.

Barron, T. H. K. 1979. A note on the Anderson-Grüneisen functions. *J. Phys. C* 12: L155–59

Bass, J., Anderson, D. L. 1984. Composition of the upper mantle: Geophysical tests of two petrological models. *Geophys. Res. Lett.* 11: 237–40

Bassett, W. A., Takahashi, T., Mao, H., Weaver, J. S. 1968. Pressure induced phase transformation of NaCl. *J. Appl. Phys.* 39: 319–25

Bell, P. M., Mao, H. K., Xu, J. A. 1987. Error analysis and parameter-fitting in equations of state for mantle minerals. In *Geophys. Monogr. 39, High-Pressure Research in Mineral Physics*, ed. M. H. Manghnani, Y. Syono, pp. 447–54. Washington: AGU

Berman, R. G. 1988. Internally-consistent thermodynamic data for minerals in the system Na_2O-K_2O-CaO-MgO-FeO-Fe_2O_3 - Al_2O_3 - SiO_2 - TiO_2 - H_2O - CO_2. *J. Petrol.* 29: 445–522

Bina, C. R., Silver, P. G. 1990. Constraints on lower mantle composition and temperature from density and bulk sound velocity profiles. *Geophys. Res. Lett.* 17: 1153–56

Bina, C. R., Wood, B. J. 1984. The eclogite to garnetite transition—Experimental and thermodynamic constraints. *Geophys. Res. Lett.* 11: 955–58

Bina, C. R., Wood, B. J. 1987. The olivine-spinel transitions: Experimental and thermodynamic constraints and implications for the nature of the 400 km seismic discontinuity. *J. Geophys. Res.* 92: 4853–66

Birch, F. 1938. The effect of pressure upon the elastic parameters for isotropic solids, according to Murnaghan's theory of finite strain. *J. Appl. Phys.* 9: 279–88

Birch, F. 1939. The variation of seismic velocities within a simplified earth model, in accordance with the theory of finite strain. *Bull. Seismol. Soc. Am.* 29: 463–79

Birch, F. 1952. Elasticity and constitution of the Earth's interior. *J. Geophys. Res.* 57: 227–86

Birch, F. 1961. The velocity of compressional waves in rocks to 10 kilobars, part 2. *J. Geophys. Res.* 66: 2199–2224

Birch, F. 1978. Finite strain isotherm and velocities for single-crystal and polycrystalline NaCl at high pressures and 300°K. *J. Geophys. Res.* 83: 1257–68

Born, M., Huang, K. 1954. *Dynamical Theory of Crystal Lattices*. Oxford: Clarendon

Bullen, K. E. 1949. Compressibility-pressure hypothesis and the Earth's interior. *Mon. Not. R. Astron. Soc.* 5: 355–68

Bullen, K. E. 1963. *Introduction to the Theory of Seismology*. Cambridge: Cambridge Univ. Press

Callen, H. B. 1960. *Thermodynamics*. New York: Wiley

Chopelas, A., Boehler, R. 1989. Thermal expansion at very high pressure, systematics, and a case for a chemically homogeneous mantle. *Geophys. Res. Lett.* 16: 1347–50

Christensen, N. I. 1984. Seismic velocities. In *Handbook of Physical Properties of Rocks*, vol. 2, ed. R. S. Carmichael, pp. 1–345. Boca Raton, Fla: CRC

Cohen, R. 1987. Elasticity and equation of state of $MgSiO_3$ perovskite. *Geophys. Res. Lett.* 14: 1053–56

Cormier, V. F. 1989. Slab diffraction of *S* waves. *J. Geophys. Res.* 94: 3006–24

Courant, R., Friedrichs, O. 1948. *Supersonic Flow and Shock Waves*. Heidelberg: Springer-Verlag

Creager, K. C., Jordan, T. H. 1984. Slab penetration into the lower mantle. *J. Geophys. Res.* 89: 3031–49

Creager, K. C., Jordan, T. H. 1986. Slab penetration into the lower mantle beneath the Mariana and other island arcs of the northwest Pacific. *J. Geophys. Res.* 91: 3573–89

Davies, G. 1973. Quasi harmonic finite strain equations of state of solids. *J. Phys. Chem. Solids* 34: 1417–29

Davies, G. 1974. Effective elastic moduli under hydrostatic stress—I. Quasi-harmonic theory. *J. Phys. Chem. Solids* 35: 1513–20

Davies, G., Dziewonski, A. 1975. Homogeneity and constitution of the earth's lower mantle and outer core. *Phys. Earth Planet. Inter.* 10: 336–43

Duffy, T. S., Anderson, D. L. 1989. Seismic velocities in mantle minerals and the mineralogy of the upper mantle. *J. Geophys. Res* 94: 1895–1912

Engdahl, E. R., Sleep, N. H., Lin, M. T. 1977. Plate effects in North Pacific subduction zones. *Tectonophysics* 37: 95–116

Fei, Y., Saxena, S. K. 1986. A thermochemical data base for phase equilibria in the system Fe-Mg-Si-O at high pressure and temperature. *Phys. Chem. Miner.* 13: 311–24

Fei, Y., Mao, Ho-K., Bysen, B. O. 1991. Experimental determination of element partitioning and calculation of phase relations in the MgO-FeO-SiO_2 system at high pressure and high temperature. *J. Geophys. Res.* 96: 2157–69

Fei, Y., Saxena, S. K., Navrotsky, A. 1990. Internally consistent thermodynamic data and equilibrium phase relations for compounds in the system MgO-SiO_2 at high pressure and high temperature. *J. Geophys. Res.* 95: 6915–28

Fischer, K. M., Jordan, T. H., Creager, K. C. 1988. Seismic constraints on the morphology of deep slabs. *J. Geophys. Res.* 93: 4773–84

Garber, J. A., Granato, A. V. 1975a. Theory of the temperature dependence of second-order elastic constants in cubic materials. *Phys. Rev. B* 11: 3990–97

Garber, J. A., Granato, A. V. 1975b. Fourth-order elastic constants and the temperature dependence of second-order elastic constants in cubic materials. *Phys. Rev. B* 11: 3998–4007

Goto, J., Anderson, O. L., Ohno, I., Yamamoto, S. 1989. Elastic constants of corundum up to 1825 K. *J. Geophys. Res.* 94: 7588–7602

Gwanmesia, G. D., Rigden, S., Jackson, I., Liebermann, R. C. 1990. Pressure dependence of elastic wave velocity for β-Mg_2SiO_4 and the composition of the earth's mantle. *Science* 250: 794–97

Helffrich, G. R., Stein, S., Wood, B. J. 1989. Subduction zone thermal structure and mineralogy and their relationship to seismic wave reflections and conversions at the slab/mantle interface. *J. Geophys. Res.* 94: 753–63

Helgelson, H., Delany, J. M., Nesbitt, H. W., Bird, D. K. 1978. Summary and critique of the thermodynamic properties of the rock-forming minerals. *Am. J. Sci.* 278-A: 1–229

Hemley, R. J., Cohen, R. E., Yeganeh-Haeri, A., Mao, H. K., Weidner, D. J., Ito, E. 1989. Raman spectroscopy and lattice dynamics of $MgSiO_3$-perovskite at high pressure. In *Geophys. Monogr. 45, Perovskite: A Structure of Great Interst to Geophysics and Materials Science*, ed. A.

Navrotsky, D. J., Weidner, pp. 35–44. Washington: AGU

Irvine, R. D., Stacey, F. D. 1975. Pressure dependence of the thermal Grüneisen parameter, with application to the Earth's lower mantle and outer core. *Phys. Earth Planet. Inter.* 11: 157–65

Isaak, D. G., Anderson, O. L., Goto, T. 1989. Elasticity of single-crystal forsterite measured to 1700 K. *J. Geophys. Res.* 94: 5895–5906

Isaak, D. G., Cohen, R. E., Mehl, M. J. 1990. Calculated elastic and thermal properties of MgO at high pressures and temperatures. *J. Geophys. Res.* 95: 7055–67

Jacob, K. 1972. Global tectonic implications of anomalous seismic *P* travel times from the nuclear explosion Longshot. *J. Geophys. Res.* 77: 2556–73

Jackson, I. 1983. Some geophysical constraints on the chemical composition of the Earth's lower mantle. *Earth Planet. Sci. Lett.* 62: 91–103

Jeanloz, R. 1981. Finite-strain equation of state for high-pressure phases. *Geophys. Res. Lett.* 8: 1219–22

Jeanloz, R. 1989. Shock wave equation of state and finite strain theory. *J. Geophys. Res.* 94: 5873–86

Jeanloz, R., Ahrens, T. J. 1980a. Equations of state of FeO and CaO. *Geophys. J. R. Astron. Soc.* 62: 505–28

Jeanloz, R., Ahrens, T. J. 1980b. Anorthite: Thermal equation of state to high pressures. *Geophys. J. R. Astron. Soc.* 62: 529–49

Jeanloz, R., Knittle, E. 1989. Density and composition of the lower mantle. *Philos. Trans. R. Soc. London Ser. A* 328: 377–89

Jeanloz, R., Thompson, A. B. 1983. Phase transitions and mantle discontinuities. *Rev. Geophys. Space Phys.* 21: 51–74

Jeffreys, H., Bullen, K. E. 1988. *Seismological Tables*. Cambridge: Br. Assoc. Seismol. Invest. Comm.

Jordan, T. H. 1977. Lithospheric slab penetration into the lower mantle beneath the Sea of Okhotsk. *J. Geophys.* 43: 473–96

Kennett, B. L. N., Engdahl, E. R. 1991. Traveltimes for global earthquake location and phase identification. *Geophys. J. Int.* 1991: 429–65

Kieffer, S. W. 1979. Thermodynamics and lattice vibrations of minerals: 1. Mineral heat capacities and their relationships to simple lattice vibrational models. *Rev. Geophys. Space Phys.* 17: 1–19

Kieffer, S. W. 1982. Thermodynamics and lattice vibrations of minerals: 5. Application to phase equilibria, isotopic fractionation, and high-pressure thermodynamic properties. *Rev. Geophys. Space Phys.* 20: 827–49

Lay, T. 1986. Evidence for a lower mantle shear velocity discontinuity in *S* and *sS* phases. *Geophys. Res. Lett.* 13: 1493–96

Lay, T., Helmberger, D. 1983. The shear wave velocity gradient at the base of the mantle. *J. Geophys. Res.* 88: 8160–70

Lees, A. C., Bukowinski, M. S. T., Jeanloz, R. 1983. Reflection properties of phase transition and compositional change models of the 670 km discontinuity. *J. Geophys. Res.* 88: 8145–59

Leibfried, G., Ludwig, W. 1961. Theory of anharmonic effects in crystals. *Solid State Phys.* 12: 275–444

McQueen, R. G., Marsh, S. P., Fritz, J. N. 1967. Hugoniot equations of state of twelve rocks. *J. Geophys. Res.* 72: 4999–5036

Oliver, J., Isacks, B. 1967. Deep earthquake zones, anomalous structures in the upper mantle, and the lithosphere. *J. Geophys. Res.* 72: 4259–75

Press, W. H., Flannery, B. P., Teukolsky, S. A., Vetterling, W. T. 1987. *Numerical Recipes*. Cambridge, UK: Cambridge Univ. Press

Quareni, F., Mulariga, F. 1988. The validity of the common approximate expressions for the Grüneisen parameters. *Geophys. J.* 93: 505–19

Rigden, S. M., Ahrens, T. J., Stolper, E. M. 1988. Shock compression of molten silicate: Results for a model basaltic composition. *J. Geophys. Res.* 93: 367–82

Robie, R. A., Hemingway, B. S., Fisher, J. R. 1979. Thermodynamic properties of minerals and related substances at 298.15 K and 1 bar (10^5 pascals) pressure and at high temperatures. *US Geol. Surv. Bull. 1452*, 456 pp.

Saxena, S. 1989. Assessment of bulk modulus, thermal expansion and heat capacity of minerals. *Geochim. Cosmochim. Acta.* 53: 785–89

Silver, P. G., Chan, W. W. 1986. Observations of body wave multipathing from broadband seismograms: Evidence for lower mantle slab penetration beneath the Sea of Okhotsk. *J. Geophys. Res.* 91: 13,787–802

Silver, P. G., Chan, W. W. 1988. Implications for continental structure and evolution from seismic anisotropy. *Nature* 335: 34–39

Skinner, B. J. 1966. Thermal expansion. *Mem. Geol. Soc. Am.* 97: 75–96

Slater, J. C. 1939. *Introduction to Chemical Physics*. New York: McGraw-Hill

Sleep, N. H. 1973. Teleseismic *P*-wave transmission through slabs. *Bull. Seismol. Soc. Am.* 63: 1349–73

Stixrude, L., Bukowinski, M. S. T. 1990.

Fundamental thermodynamic relations and silicate melting with implications for the constitution of D. *J. Geophys. Res.* 95: 19,311–25

Sumino, Y., Anderson, O. 1984. Elastic constants of minerals. In *Handbook of Physical Properties of Rocks*, vol. 3, ed. R. S. Carmichael, pp. 39–138. Boca Raton, Fla: CRC

Suyehiro, K., Sacks, I. S. 1979. *P*- and *S*-wave velocity anomalies associated with the subducting lithosphere determined from travel-time residuals in the Japan region. *Bull. Seismol. Soc. Am.* 69: 97–114

Taylor, D. 1986. Thermal expansion data. *J. Br. Ceramic Soc.* 85: 111–14

Toksöz, M. N., Minear, J. W., Julian, B. R. 1971. Temperature field and geophysical effects of a downgoing slab. *J. Geophys. Res.* 76: 1113–38

Touloukian, Y. S., ed. 1967. *Thermophysical Properties of High Temperature Solid Materials*. New York: MacMillan

Vidale, J. E. 1987. Waveform effects of a high velocity subducted slab. *Geophys. Res. Lett.* 14: 542–45

Watt, J. P., O'Connell, J. 1978. Mixed-oxide and perovskite-structure model mantles from 700–1200 km. *Geophys. J. R. Astron. Soc.* 54: 601–30

Watt, J. P., Davies, G. F., O'Connell, J. 1976. The elastic properties of composite minerals. *Rev. Geophys. Space Phys.* 14: 541–63

Weidner, D. J. 1985. A mineral physics test of a pyrolite mantle. *Geophys. Res. Lett.* 12: 417–20

Weidner, D. J., Carleton, H. R. 1977. Elasticity of coesite. *J. Geophys. Res.* 82: 414–26

Weidner, D. J., Sawamoto, H., Sasaki, S., Kumazawa, M. 1984. Single-crystal elastic properties of the spinel phase of Mg_2SiO_4. *J. Geophys. Res.* 89: 7852–60

Williamson, E. D., Adams, L. H. 1923. Density distribution in the Earth. *J. Wash. Acad. Sci.* 13: 413–32

Wysession, M. E. 1991. *Diffracted seismic waves and the dynamics of the core/mantle boundary*. PhD thesis. Northwestern Univ.

Wysession, M. E., Okal, E. A. 1988. Evidence for lateral heterogeneity at the core-mantle boundary from the slowness of diffracted *S* profiles. In *Geophys. Monogr. 46, Structure and Dynamics of the Earth's Deep Interior*, ed. D. E. Smylie, R. Hide, pp. 55–63. Washington: AGU

Wysession, M. E., Okal, E. A. 1989. Regional analysis of D″ velocities from the ray parameters of diffracted *P* profiles. *Geophys. Res. Lett.* 16: 1417–20

Yoneda, A. 1990. Pressure derivatives of elastic constants of single crystal MgO and $MgAl_2O_4$. *J. Phys. Earth* 38: 19–55

Annu. Rev. Earth Planet. Sci. 1992. 20:553–600

SILICATE PEROVSKITE

Russell J. Hemley and Ronald E. Cohen

Geophysical Laboratory and Center for High Pressure Research, Carnegie Institution of Washington, Washington, DC 20015

KEY WORDS: mineralogy, Earth's lower mantle, high pressure, ferromagnesian silicate, calcium silicate, physical properties of minerals

INTRODUCTION

High-pressure experiments have established that the silicate minerals that comprise the Earth's crust and upper mantle transform to assemblages dominated by minerals having perovskite-type structures at pressures corresponding to those of the lower mantle. This region of the Earth's interior extends from 670 km to 2900 km depth and occupies approximately 60% of the volume of the planet. The observation of the high-pressure stability of silicate perovskite has opened up the possibility of understanding the mineralogy of this vast region of the planet's interior. Indeed, characterization of the chemical and physical properties of a range of silicate perovskites during the past fifteen years has provided new constraints on the composition and structure of the mantle, new insights into core-mantle interaction and coupling, and new bounds on mantle and core formation and long-term evolution.

Significant progress in understanding the properties of silicate perovskites has been made in the past five years. A number of obstacles have been encountered in the study of these high-pressure phases, owing to both the small quantities in which they can be synthesized in the laboratory and their intrinsic instability at low (e.g. ambient) pressure conditions. The recent progress in investigations of silicate perovskites has been made possible in large measure by advances in high-pressure techniques, such as diamond-cell methods including synchrotron radiation and laser techniques, multi-anvil press technology which has made possible the preparation of single crystals and petrologic studies, and advances in theoretical calculations such as new accurate electronic-structure methods. As a result

0084–6597/92/0515–0553$02.00

of these advances, the implications for the Earth on a global scale—many of which were speculations based on, at best, educated guesses just a few years ago—are now grounded on experimental data supported by a growing theoretical understanding. The emerging body of data on these materials also provides a firm basis for ruling out alternative models. On the other hand, a number of outstanding issues remain, and these have given rise to some recent controversies regarding the properties of perovskite and their geophysical implications.

Several reviews of various aspects of silicate perovskites and our current understanding of their implications for the Earth and planetary interiors have appeared during the last few years (Liu & Bassett 1986, Navrotsky & Weidner 1989, Hazen 1988, Anderson 1989). In this paper we review the mineralogy and properties of this class of geophysically and geochemically important materials, emphasizing recent experimental data, theoretical insights, and current problems. This is a field of active research, and we present this review as a progress report rather than as review of a settled subject. In the next few years, we expect many of the issues raised here to be solved by further experiments under *P-T* conditions of the lower mantle and by development and application of refined theoretical methods.

CRYSTAL CHEMISTRY

The perovskite structure for a stoichiometric compound with a composition ABX_3 consists of corner sharing anion X octahedra containing B cations with larger A cations between the octahedra. In the perovskite-type silicates, tetravalent silicon occupies the smaller B site, with a variety of other (typically divalent) cations (Mg, Fe, Ca) in the A site. The ideal, high symmetry structure (aristotype) is shown in Figure 1*a*. A variety of distortions from this ideal structure are possible in general (Glazer 1972, 1975; Aleksandrov 1976, 1978; Salje 1989). These include various combinations of coupled rotations and cation displacements (off-centering). Some octahedral-tilting transitions can be understood in terms of condensation of phonons at the boundary of the Brillouin zone, and other transitions are by necessity first-order phase transitions (see below). Glazer (1972) discussed the classification of distorted perovskites that involve nearly rigid rotations of the oxygen octahedra. In Glazer's notation, the tilts in the *Pbnm* structure are of type ($a^-a^-c^+$), which means that the octahedra are tilted along an axis parallel to *c* with the same direction of tilts in adjacent planes along *c* (c^+), and are tilted by a different rotation angle along *a* and *b* with opposite directions of tilts in adjacent planes (Figure 1*b*). Aleksandrov (1976, 1978) used a different notation which emphasizes the relationship of the c^+ type tilt to an *M*-point phonon

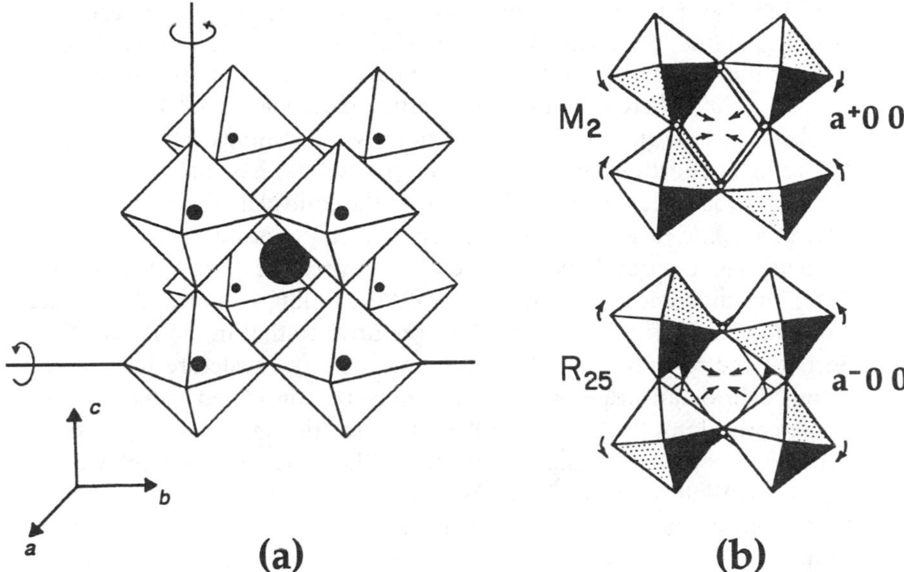

Figure 1 (*a*) Ideal perovskite structure for ABX_3 compounds, illustrating the corner-shared BX_6 octahedra. Two rotation axes leading to tilts are shown with the arrows. (*b*) Combinations of tilts giving rise to the *R*- and *M*-point instabilities (modified from Wolf & Jeanloz 1985).

instability and the a^- type tilts to an *R*-point instability, as discussed below.

The structure of $(Mg,Fe)SiO_3$ perovskite (over the *P-T* range under which it has been studied experimentally) is orthorhombic, with space group *Pbnm*. This structure was proposed on the basis of powder x-ray diffraction on $MgSiO_3$ by Yagi et al (1977, 1978a) and Ito & Matsui (1978), who studied the material quenched to room conditions. This structure was confirmed by single crystal x-ray diffraction techniques by Horiuchi et al (1987). A significant recent advance is the development of techniques for growth of single crystals (Ito & Weidner 1986) for x-ray structure determination and other techniques requiring high-quality single crystals (see below). The single-crystal structure of Horiuchi et al (1987) is very close to that proposed by Yagi et al (1977, 1978a) and Ito & Matsui (1978). The larger, pseudo-tetragonal unit cell suggested by early transmission electron microscopy and diffraction measurements (Madon et al 1980) has been reinterpreted in terms of a high degree of twinning of crystals with the *Pbnm* orthorhombic structure (Madon et al 1989).

In $MgSiO_3$ the tilt angle around c is 11.2°, and the angle of tilt around a and b is about 16.7°. The unit cell in $MgSiO_3$ is rotated approximately 45° from that in cubic perovskite with the new a- and b-axes approximately equal to $\sqrt{2}a'$, where a' is the cubic lattice parameter, and the c-axis is doubled. Thus the unit cell size is quadrupled with 20 atoms in the primitive unit cell. The orthorhombic $MgSiO_3$ perovskite has the axial ratio $a:b:c = 0.968:1:1.394$, which differs from the equivalent ratio of the ideal cubic perovskite, 1:1:1.414. In contrast, $CaSiO_3$ crystallizes in the high-symmetry cubic perovskite structure at high pressure (Liu & Ringwood 1975). This difference in the structures of (Mg,Fe)SiO_3 and $CaSiO_3$ is due to the small size of the ion in the A site relative to that in the B site. The effects of the ratio of the cation ionic radii on the structure have been examined in detail in perovskites that may be considered analogues to silicate perovskites (Sasaki et al 1983). This relationship is generally supported by theoretical calculations based on the ionic model (Hemley et al 1985, 1987; Wolf & Jeanloz 1985; Wolf & Bukowinski 1985, 1987). These calculations predict that the cubic $CaSiO_3$ should transform to the lower symmetry orthorhombic perovskite structure at high compressions, although this has not yet been observed experimentally. Thus, despite the variety of structures observed in perovskites (e.g. Glazer 1972), only two structure-types (cubic *Pm3m* and orthorhombic *Pbnm*) appear to be mineralogically significant for lower mantle silicates. Others have been predicted theoretically and are likely to occur on crystal chemical grounds for other compositions (see below).

Iron enters the orthorhombic $MgSiO_3$ perovskite up to a mole fraction of ~20% (see below). The effect on the molar volume is shown in Figure 2, which summarizes early powder diffraction results and the results of more recent crystal structure refinements by the Reitveld powder method (Parise et al 1990) and single-crystal techniques (Kudoh et al 1990) using synchrotron radiation. Iron substitution affects the structure by causing a decrease in the degree of distortion. Aluminum is also soluble in the orthorhombic perovskite, with a maximum solubility of 25% (Weng et al 1982), and results in a significantly larger increase in the unit-cell volume. Although substitution of both Fe and Al expands the structure, the former results in a decrease in distortion from cubic (smaller octahedral tilts) whereas the latter causes the distortion to increase. Weng et al (1982) propose that the increase in distortion on addition of Al arises from equal distribution of the trivalent cation on both the A and B sites [i.e. (Mg,Al)(Si,Al)O_3], which results in inefficient packing. Preliminary evidence for the effect of simultaneous Fe and Al substitution has been reported recently (O'Neill et al 1991).

The coordination of iron and cation site occupancy in (Mg,Fe)SiO_3

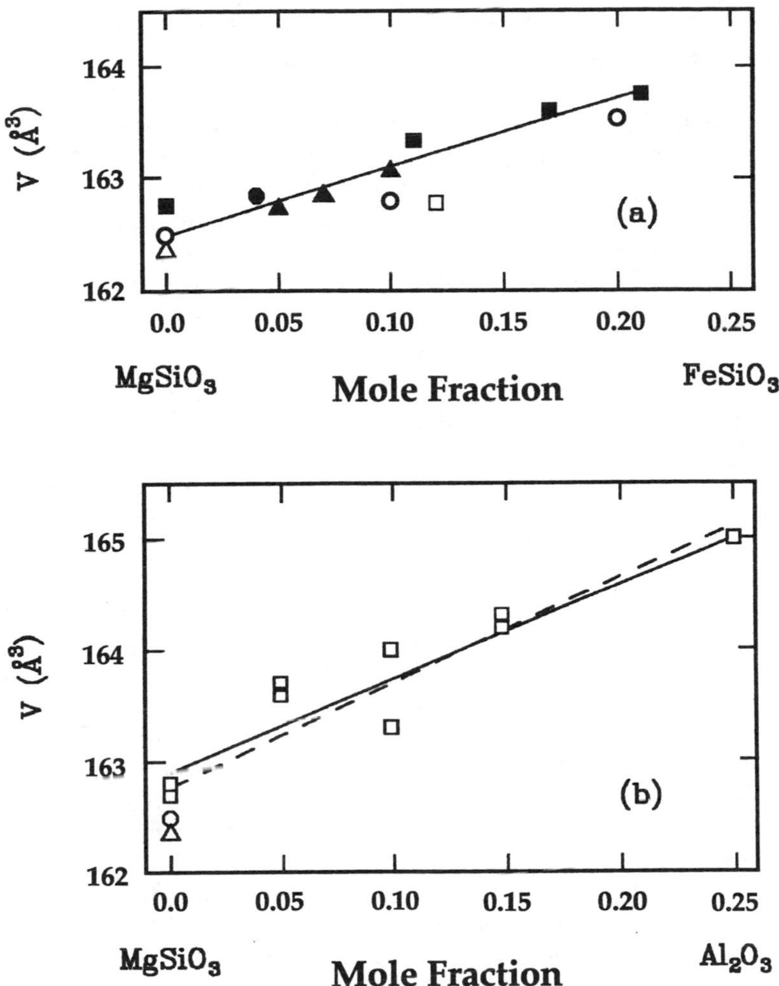

Figure 2 (*a*) Unit-cell volume of the Mg-rich orthorhombic perovskite as a function of mole fraction of $FeSiO_3$ component. ■, Yagi et al (1978b); △, Ito & Matsui (1978); ▲, Ito & Yamada (1982); □, Knittle & Jeanloz (1987a); ●, Kudoh et al (1990); ○, Mao et al (1991). The solid line is the fit $V(\text{Å}^3) = 162.48(7)+6.2(7)X_{Fe}$ reported by Kudoh et al (1990). (*b*) Unit-cell volume as a function of mole fraction of Al_2O_3. □, Weng et al (1982); △, Ito & Matsui (1978); ○, Mao et al (1991). The solid line is the fit $V(\text{Å}^3) = 162.9+8.4X_{Al}$ reported by Weng et al (1982). The dashed line is a fit that includes the more recent zero-pressure data: $V(\text{Å}^3) = 162.9(1)+9.4(9)X_{Al}$.

perovskite has been the subject of much recent discussion. On the basis of EXAFS (extended x-ray absorption fine structure) measurements on $(Mg,Fe)SiO_3$ samples synthesized in a diamond-cell, Jackson et al (1987) propose that Fe substitutes for Si, with an associated vacancy and occupation of the 12-fold A site by Si. Evidencc for two different Fe sites from Mössbauer measurements of diamond-cell samples of $(Mg,Fe)SiO_3$ perovskite was interpreted in terms of mixing of Fe on the A and B sites (K. Weng et al, unpublished). In addition to its crystal chemical significance, such behavior would give rise to an important entropic stabilization of perovskite and perhaps unusual geochemical properties at high temperatures. More recent measurements have been interpreted as evidence for Fe occupying the A site alone (Jeanloz et al 1991). On the other hand, the single-crystal x-ray diffraction data of Kudoh et al (1990) and powder x-ray diffraction and XANES (x-ray absorption near-edge spectroscopy) measurements of Parise et al (1990) on perovskite samples synthesized in multi-anvil devices show no evidence for anomalous site occupancy. Recent NMR measurements on ^{29}Si-enriched $MgSiO_3$ perovskite, also synthesized in a multi-anvil device, indicate highly ordered Si in octahedral coordination with no Si in the A site (Kirkpatrick et al 1991). Unfortunately, similar tests on iron-bearing samples are not possible because Fe forms paramagnetic impurities which overwhelm the NMR signal. Differences in samples resulting from differing preparation procedures—including *P-T* range, approach to equilibrium, impurities, fO_2, point defects, and Fe^{+3} content—could be responsible (Hirsch & Shankland 1991). The observation of octahedral Fe in such samples would be expected if a small amount of glass or magnesiowüstite is present (see below), since Fe partitions into melt and the oxide relative to perovskite (Bell et al 1979, Heinz & Jeanloz 1987). Although there is now good agreement on many of the measured physical properties of perovskite, differences in the behavior of samples synthesized in diamond-cell and multi-anvil devices are evident and not fully understood.

Knowledge of the effects of pressure and temperature on the crystal structure is essential for understanding the properties of the material under mantle conditions. This includes possible displacive transitions, which may give rise to anomalous changes in physical properties with important consequences for the interpretation of geophysical data (see below). Pressure effects on the structure of Mg-rich silicate perovskites have been determined under hydrostatic conditions by both single-crystal and powder x-ray diffraction (Yagi et al 1982; Kudoh et al 1987; Ross & Hazen 1989; Mao et al 1989b, 1991). Figure 3 summarizes the effects of *P* and *T* on the axial (lattice-parameter) ratios c/a and b/a for $(Mg,Fe)SiO_3$ perovskites. The experimental data show that the orthorhombic

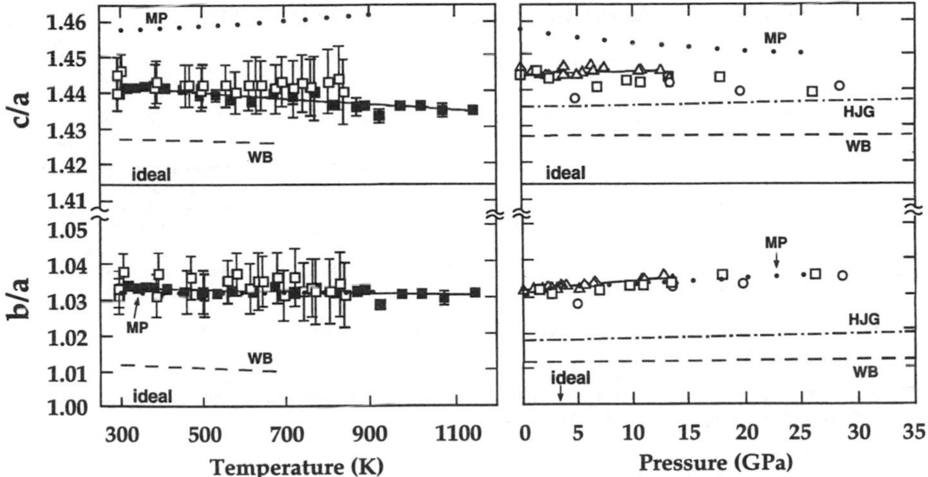

Figure 3 Effect of pressure and temperature on the unit-cell axial ratios for orthorhombic (Mg,Fe)SiO_3. Experimental high-temperature ($\gtrsim$300 K) are included: □, Knittle et al (1986); ■, Wang et al (1991a). The high-pressure data are hydrostatic results: △, Ross & Hazen (1990); ○, □, Mao et al (1991). The predictions of nonempirical theoretical ionic model calculations are shown by the dot-dashed line (*HJG*—Hemley et al 1987) and the dashed line (*WB*—Wolf & Bukowinski 1987). The dotted line is the result of the molecular dynamics simulation using potentials fit to experimental data (*MP*—Matsui & Price 1991, and to be published).

(Mg,Fe)SiO_3 perovskite is elastically anisotropic with the *b*-axis having the smallest linear compressibility, β_b, approximately 25% smaller than β_a or β_c (Table 1). The *a* and *c* compressibilities are similar, with *c* slightly more compressible. With increasing pressure, the difference between *a* and *b* increases and the structure becomes more distorted, consistent with crystal chemical arguments (O'Keeffe et al 1979) and theoretical calculations (Hemley et al 1987), although the changes are relatively small. These changes also indicate that the material becomes more elastically anisotropic with pressure (Meade & Jeanloz 1991). The temperature changes are also relatively small over the range of existing measurements, and the axial ratios remain far from their ideal values (Knittle et al 1986, Ross & Hazen 1989, Parise et al 1990). Nevertheless, recent measurements to ~1200 K do show a decrease in c/a and b/a (Wang et al 1991a). The temperature effects largely mirror the pressure effects over the range of the measurements, suggesting that the structural changes are largely controlled by volumetric terms (e.g. Hemley et al 1987). Perovskites containing open-shell cations may not exhibit the same *P-T* effects on structure because of the higher degree of directional bonding (crystal-field effects) in such compounds (e.g. $SrZrO_3$, Andrault & Poirier 1991).

Table 1 Zero-pressure bulk modulus and linear compressibilities of $(Mg_{1-x}Fe_x)SiO_3$ perovskite (298 K)

K_0 GPa	K_0'	β_{a0} TPa^{-1}	β_{b0} TPa^{-1}	β_{c0} TPa^{-1}	P_{max} GPa	P media[a]	Sample[b]	x	Reference[c]
246	---	1.31	1.20	1.56	0	---	SC	0	Yeganeh-Haeri et al (1989a,b)
247	4	1.41	1.07	1.57	10	M-E-W	SC	0	Kudoh et al (1987)
254	4	1.30	1.04	1.24	13	M-E, Ne	SC	0	Ross and Hazen (1989)
258	4	1.58	1.19	1.10	7	M-E	P	0	Yagi et al (1982)
266	3.9	---	---	---	112	none	P	0.12	Knittle and Jeanloz (1987a)
261	4	1.34	1.08	1.43	30	Ne	P	0, 0.1, 0.2	Mao et al (1991)
---	---	1.29	1.03	1.31	7	NaCl	P	0	Wang et al (1991a)

[a] M-E: methanol-ethanol mixture; M-E-W: methanol-ethanol-water mixture.

[b] SC: single crystal; P: powder.

[c] Adiabatic bulk modulus and compressibilities were measured using Brillouin scattering at zero pressure by Yeganeh-Haeri et al (1989a,b). Isothermal moduli and compressibilities were determined from fitting *P-V-a-b-c* data measured by x-ray diffraction techniques in the remaining studies.

BONDING AND ENERGETICS

Considerable insight into the crystal chemistry, bonding, and basis of physical properties has been obtained from calculations using theoretical and computational methods. Calculations with nonempirical theoretical models based on first principle methods have been particularly useful. These calculations correctly predict the *M*- and *R*-point instabilities that lead to the *Pbnm* structure for $MgSiO_3$ (Hemley et al 1985, 1987; Wolf & Jeanloz 1985; Wolf & Bukowinski 1987). One of the goals of these techniques is an accurate prediction of the density, equation of state, and structural properties of the crystal from a single, unified model. Several sets of calculations based on the Gordon-Kim (electron-gas) model have been performed. Hemley et al (1985, 1987) used a modified electron-gas approach to examine principally the structural distortions at zero pressure and at high compression. Wolf & Jeanloz (1985) and Wolf & Bukowinski (1987) applied an earlier version of the electron-gas model and found good agreement for the bulk modulus. Cohen (1987) used the potential induced breathing (PIB) model, which includes dynamical charge relaxation effects and different assumptions within the Gordon-Kim approximation, good agreement was obtained for the bulk modulus and predictions for the single crystal elastic moduli were made. A variety of calculations based on empirical potentials have also been performed (e.g. Miyamoto & Takeda 1984, Wall et al 1986, Matsui et al 1987, Matsui 1988, Price et al 1989, Wright & Price 1989, Reynard & Price 1990, Matsui & Price 1991). These calculations have been useful for predictions of structural properties, defect energies, and high *P-T* behavior (see below). One important detail that is very difficult to describe accurately with various potential models is the exact degree of distortion from the cubic structure and its variation with pressure. This is because the distortion depends critically on a delicate interplay between the Si-O and Mg-O interactions, as well as the O-O interaction.

All of the above calculations assumed an ionic description of the bonding in silicate perovskite and their general qualitative success indicates that $MgSiO_3$ is likely to be very ionic in comparison to lower pressure silicate minerals having tetrahedrally coordinated Si. Electronic structure calculations on cubic $MgSiO_3$ perovskite have been performed to determine the degree of covalent relative to ionic bonding and to obtain the electronic band structure (Cohen et al 1989) (Figures 4 and 5). These calculations, which appear to represent the best description to date of the bonding in silicate perovskites, show that the Mg is nearly a perfectly spherical Mg^{2+} ion. There is some charge transfer back to Si from O and the charges are less than the nominal Si^{4+} and O^{2-}. The valence charge on Si has both *s* and *d* character. There is also a small covalent bond charge in between the Si and O that contains about 0.1 electrons. Although $MgSiO_3$ is very ionic, it is not described particularly well by a rigid ion model. In fact, the oxygen ion density changes with pressure. If this "breathing" of the oxygen ions is not included, it is difficult to describe accurately both the elastic properties and the equation of state (pressure-volume relations).

Three relatively strong predictions of theoretical calculations are that

1. The distortion (tilts) should increase with pressure, and the energy difference between the equilibrium distorted structure and the undistorted cubic structure increases with pressure;
2. the distorted phase is significantly denser than the undistorted structure at a given pressure; and
3. the cubic structure has a higher bulk modulus than the orthorhombic structure.

The effect of Fe substitution, which is expected to lower the band gap and introduce crystal field stabilization and additional directional bonding, has not yet been explored in these calculations. Theoretical calculations on other iron-bearing oxides, simplified models, and available experimental data suggest that small amounts of Fe soluble in silicate perovskite do not greatly affect its cohesive and elastic properties (e.g. Wright & Price 1989). On the other hand, as discussed below, Fe significantly affects the electrical conductivity and other transport properties, as might be expected from its open-shell character.

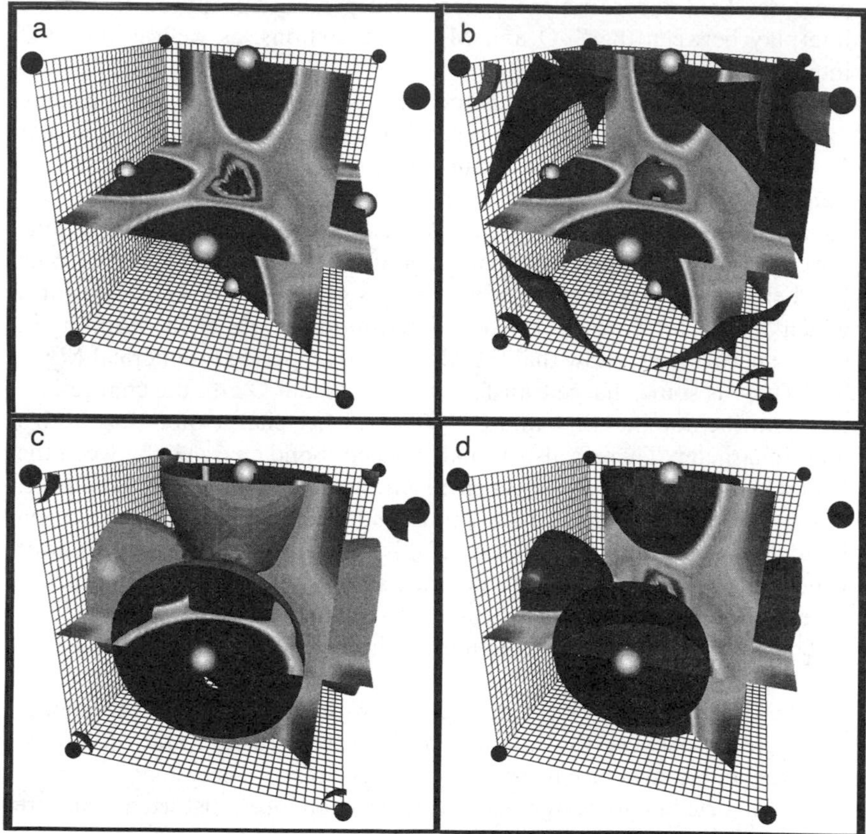

Figure 4 Calculated valence charge density for cubic $MgSiO_3$ perovskite using the LAPW method. The valence charge density is the electron density associated with chemical bonding. In $MgSiO_3$ perovskite the valence charge density is very ionic (i.e. mostly O^{2-}) but there is some hybridization with the Si *d* and *s* states. (*a*) Two slice planes along (*001*) and (*110*). The magnesium ions are at the corners, the oxygen ions are at the centers of the faces, and silicon is at the center. (*b*) Low-density isosurface at a density of 0.014 electrons (e^-)/bohr^3 (1 bohr = 0.529 Å). Note the density around the Mg ions at the corners and the interesting cubic box of density around the Si at the center. (*c*) Isosurface at 0.05 e^-/bohr^3. This surface indicates there is excess charge in the triangles formed between two oxygens and the Si, as illustrated by the "necking" of the surfaces. (*d*) Contour level of 0.1 e^-/bohr^3. The principal contribution to the higher density surfaces is from the oxygen, though there is also a small region of spherical charge around the Si, which consists primarily of Si *s* states.

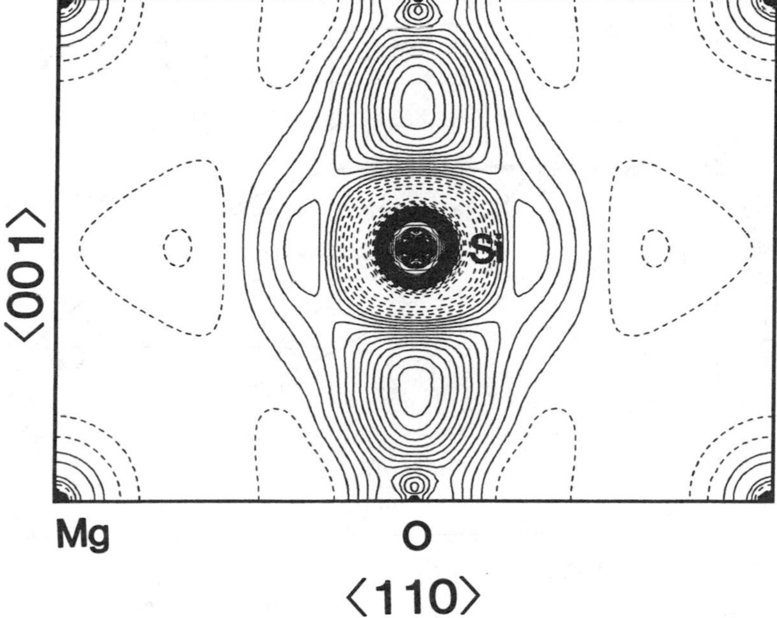

Figure 5 Difference between self-consistent LAPW charge density and overlapping spherical ions (Mg^{2+}, Si^{4+}, O^{2-}) for cubic $MgSiO_3$ along the *(110)* plane ($P = 155$ GPa). The contour interval is 0.005 $e^-/bohr^3$. Negative contours are dashed (from Cohen et al 1989).

STABILITY AND PHASE RELATIONS

The possibility that silicates may adopt a perovskite-type structure at high pressures was suggested by Ringwood (1962) on the basis of crystal chemical systematics. A variety of studies of high-pressure silicate perovskite analogues were carried out by Ringwood's group in the late 1960s (see Reid & Ringwood 1975). The formation of a perovskite-structured silicate was first demonstrated by Liu (1974, 1975a,b) in ferromagnesian silicates using the diamond-anvil cell. Subsequent to the discovery of $(Mg,Fe)SiO_3$ perovskite, considerable effort was directed toward understanding its stability field as a function of composition (Yagi et al 1977, Ito 1977). The phase relations in the pseudo binary systems Mg_2SiO_4-Fe_2SiO_4 and $MgSiO_3$-$FeSiO_3$ are shown in Figure 6, and the effect of pressure on the MgO-FeO-SiO_2 ternary diagram is presented in Figure 7. The composition range in which $(Mg,Fe)_2SiO_4$ spinel dissociates into perovskite and magnesiowüstite $(Mg,Fe)O$ extends over the range

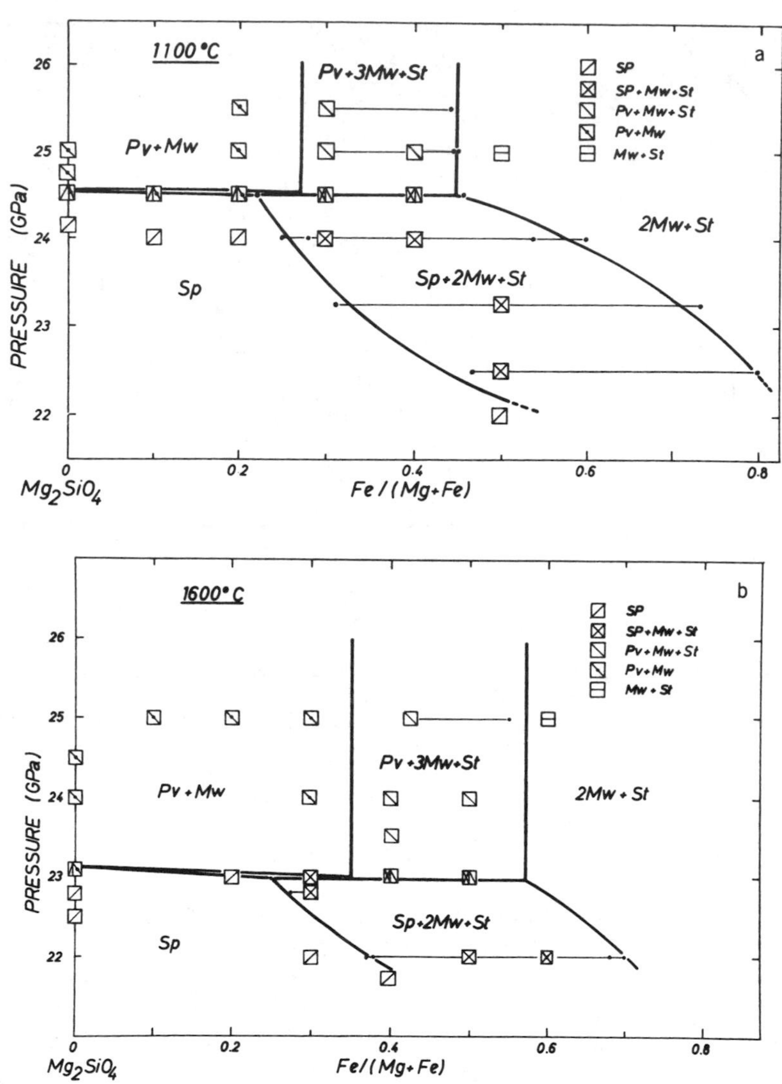

Figure 6 Pseudobinary diagrams in the system Mg_2SiO_4-Fe_2SiO_4 at 1373 K and 1873 K. Sp, spinel; Mw, magnesiowüstite; St, stishovite, Pv, perovskite (from Ito & Takahashi 1989).

Fe/(Mg + Fe) = 0 to 0.22; it thus covers the range of likely mantle compositions (Ringwood 1975, Anderson 1989). Yagi et al (1979c) performed phase equilibrium studies on laser-heated diamond-cell samples quenched from high pressures and temperatures. The experiments showed that the

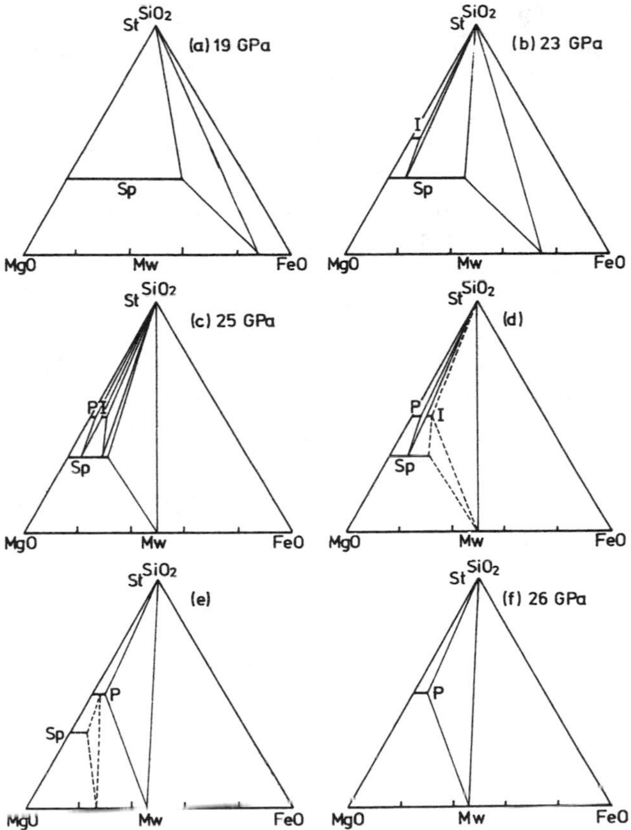

Figure 7 Ternary diagrams in the system FeO-MgO-SiO_2 as a function of pressure from 19 GPa to 26 GPa at 1373 K. Sp, spinel; Mw, magnesiowüstite; St, stishovite; I, ilmenite; Pv, perovskite (from Ito 1984).

three-phase loop of spinel, magnesiowüstite, and stishovite is very narrow in pressure, in fact narrower than the resolution of the experiments at that time. This result has been confirmed in more detailed phase equilibria studies with a large-volume multi-anvil press (Ito & Takahashi 1989), in which phase relations topologically close to those of the earlier studies were also found. There is now strong evidence that the dissociation reaction occurs within 0.2 GPa at ~1870 K, which corresponds to less than 4–6 km depth within the Earth (Wood 1990, Fei et al 1991b, Kuskov & Panferov 1991). The pressure of the disproportionation reaction and its sharpness has important implications for the interpretation of the 670-km discontinuity, which can be explained as the result of the phase transition without the requirement of a change in chemical composition at this depth (see below).

Despite the emerging consensus on certain aspects of the phase relations in this system, several outstanding problems remain. In the diamond-cell experiments, the Pv+3Mw+St three-phase field (Figure 7) extends to higher FeO content than is indicated by the multi-anvil experiments (Ito & Takahashi 1989). The latter workers also find that the three-phase triangle moves toward more FeO-rich compositions with increasing temperature. However, attempts to model this behavior based on available thermochemical data show the opposite effect of temperature (Fei et al 1991b). The difference in maximum solubility of Fe in perovskite determined in diamond-cell and multi-anvil experiments is not yet resolved; possible differences in ferric iron content need to be examined (Fei et al 1991b, Hirsch & Shankland 1991). Guyot et al (1989) have obtained evidence for a pressure effect on the partitioning of iron between perovskite and magnesiowüstite. Preliminary measurements of the spinel to perovskite transition carried out in situ in a CO_2-laser-heated diamond-cell suggest that the *P-T* boundary could be up to 300°C higher than that determined in the multi-anvil quench experiments (Boehler & Chopelas 1991, 1992).

Another important problem in the phase diagram is the negative *P-T* (Clapeyron) slope of perovskite-forming transitions, which has important geophysical implications because if dP/dT is sufficiently negative it could inhibit mixing of the upper and lower mantle. Phase equilibrium and calorimetric data now indicate that the slope of the Il → Pv and Sp → Pv+Mw transitions is negative (Ito & Yamada 1982; Ito & Takahashi 1987a, 1989; Ito et al 1990). The value of dP/dT is approximately -0.004 GPa/K (Ito et al 1990), which is close to the value needed to preclude two-layer convection according to the calculations of Christensen & Yuen (1984). The value reported by Ito et al (1990) requires information on the relative thermal expansivities of the phases at high *P* and *T*. The results are most consistent with the phase equilibrium data of Ito & Takahashi (1989) if a zero-pressure thermal expansion coefficient $\alpha = 2.2 \times 10^{-5}\,K^{-1}$ is assumed (see below). Recent simulations of mantle convection show that intermittent mixing of the upper and lower mantle is possible if dP/dT has an intermediate value of -0.002 GPa/K (Machetel & Weber 1991).

A major question with regard to the presence of silicate perovskite throughout the lower mantle is the pressure range of its stability. Diamond-cell studies of quenched material by Mao et al (1977) indicated that $(Mg,Fe)SiO_3$ perovskite is stable to at least 60 GPa. It was these experiments that first gave rise to the experimentally-based proposal that

$(Mg,Fe)SiO_3$ perovskite may be the most abundant mineral in the Earth. More recently, Knittle & Jeanloz (1987a) performed in situ x-ray diffraction measurements of $(Mg,Fe)SiO_3$ and found evidence for its stability to pressures above 100 GPa. The range for stability of the perovskite as a function of *P-T-X* has not been studied in detail at very high pressures (e.g. >50 GPa), however. The stability of the perovskite phase relative to simple oxides is derived from its 2% higher density relative to the oxide assemblage under ambient conditions, but a possible perovskite breakdown under very high *P-T* conditions (e.g. D'' region) needs to be examined (even in the absence of iron) because the relative densities under these conditions are poorly known (see below). Fei et al (1991a) have examined available thermochemical data for $(Mg,Fe)SiO_3$ and delineated the conditions under which it may break down to simple oxides at high *P* and *T*. On the basis of calculations of the melting curve, Stixrude & Bukowinski (1990) propose that $MgSiO_3$ breaks down to simple oxides at the *P-T* conditions of D'' and thus may be responsible for seismic structure at the base of the lower mantle, although later calculations find a greater stability for the perovskite phase (Stixrude & Bukowinski 1992). These proposals need to be examined by direct experimental investigation. Little is known about the chemistry of the perovskites at deep mantle conditions (including the D'' region), as detailed phase equilibrium studies need to be performed under these ultrahigh pressures. Nevertheless, there is evidence for new pressure-induced reactions under these conditions, for example, reactions between perovskite and molten iron above 70 GPa (Knittle & Jeanloz 1989b, 1991).

Liu & Ringwood (1975) reported the synthesis of cubic $CaSiO_3$ perovskite at 16 GPa, which was found to be nonquenchable. More recent work has established that the calcium perovskite forms at significantly lower pressure than does the magnesium perovskite (11–13 GPa) (Mao et al 1989a, Chen et al 1989, Tarrida & Richet 1989, Tamai & Yagi 1989, Irifune et al 1989). Recent in situ x-ray diffraction measurements have shown that the perovskite phase of $CaSiO_3$ is stable and retains its cubic structure to at least the pressure of the core-mantle boundary (Mao et al 1989a). Similar results have been obtained by Tarrida & Richet (1989) and Yagi et al (1989) to 94 GPa and above 100 GPa, respectively. The wide stability field of the cubic form is in general agreement with theoretical predictions (Hemley et al 1985, 1987; Wolf & Bukowinski 1987). The sharpness of diffraction lines rules out the possibility of tetragonal or orthorhombic distortion of more than 0.5%. The predicted pressure-induced distortion to a lower symmetry perovskite structure must take place at still higher pressures. Like the magnesium-rich perovskite, the temperature range of stability of the calcium perovskite has not been determined in detail experimentally, and analysis of available equations of state shows that free energy of the oxide assemblage is very close to that of perovskite at very high pressures (Mao et al 1989a, Fei et al 1991a). During release of pressure, the perovskite persists metastably at pressures close to ambient. Near zero pressure, however, the phase could be preserved for only a few hours, although small quantities have been observed recently to persist over longer time periods (Kanzaki et al 1991). The observation of amorphization on decompression may be related to theoretical predictions that $CaSiO_3$ perovskite becomes dynamically unstable on expansion of the structure (Hemley & Cohen, unpublished).

Several experimental studies indicate very limited solubility between $(Mg,Fe)SiO_3$ and $CaSiO_3$-perovskite (Liu 1977; Mao et al 1977; Weng et al 1982; Irifune & Ringwood 1987a,b; Tamai & Yagi 1989, Irifune et al 1989). Diopside ($CaMgSi_2O_6$) breaks down to form orthorhombic Mg-rich and cubic Ca-rich perovskites above 24 GPa rather than a single perovskite phase. For example, Irifune et al (1989) report $<2\%$ solubility of $CaSiO_3$ in $MgSiO_3$ and 2–5% solubility of $MgSiO_3$ in $CaSiO_3$. By contrast, Liu (1987) reported complete solid solution between $CaSiO_3$ and $CaMgSi_2O_6$ in the cubic perovskite structure, and evidence for cubic perovskite with a nearly ideal diopside composition formed in garnet peridotite has been reported (Ito & Takahashi 1987a, Takahashi & Ito 1987). The differences could be associated with problems with metastability. Kim et al (1991) report formation of Ca-perovskite from hedenbergite ($CaFeSi_2O_6$) at 19–26 GPa. Silicate perovskites containing heavier alkaline earths Sr and Ba are predicted theoretically to be dynamically stable but only at high pressure (Hemley & Cohen, unpublished); the calculations indicate that they become dynamically unstable on decompression because of the large cation in the A site and therefore may be unquenchable from high pressures and temperature.

There have been limited reconnaissance studies of the phase relations involving a wider range of compositions (cf. Liu & Bassett 1986). A very important component is Al_2O_3. As discussed above, Weng et al (1982) showed that Al_2O_3 dissolves in magnesium perovskite up to 25 mole % instead of forming another phase such as garnet. Recently, a new high-pressure rhombohedral perovskite phase with composition $Ca_2AlSiO_{5.5}$ was reported by Fitz Gerald & Ringwood (1991). Solubility of alkali elements (e.g. Na) in silicate perovskite has been reported and the resulting phases generally appear to be nonquenchable. For example, Liu (1980, 1987) report evidence for the formation of a nonquenchable, cubic perovskite phase of omphacite composition. Experimental studies of the partitioning of transition metals between Mg-perovskite and metallic iron at high pressures have begun (Ohtani et al 1992; see also Knittle & Jeanloz

1989b, 1991). The solubility of hydrogen and rare gases in perovskite at high pressure is of great importance but remains to be determined.

PRESSURE-VOLUME RELATIONS

Measurement of the *P-V* equation of state is essential for geophysical models because such data in principle provide a determination of density ρ and bulk modulus K_S (or K_T), or bulk sound velocity $V_B = K_S/\rho$ as functions of pressure. The isothermal and adiabatic bulk moduli are related as: $K_S = K_T(1+\alpha\gamma T)$, where α is the volume thermal expansion coefficient and γ is the Grüneisen parameter. One of the advances in research on silicate perovskites during the past five years has been the direct measurement of the *P-V* relations for both $(Mg,Fe)SiO_3$ and $CaSiO_3$ perovskite over most of the pressure range of the lower mantle (> 100 GPa). Figure 8 summarizes these results. A comparison of recent equation of state parameters for $(Mg,Fe)SiO_3$ is given in Table 1.

Yagi et al (1979b, 1982) first measured the *P-V* equation of state of $MgSiO_3$ by static compression under hydrostatic conditions using x-ray diffraction; they reported a value of the zero-pressure bulk modulus K_{0T} of 258(±20) GPa ($dK_{0T}/dP \equiv K'_{0T} = 4$ assumed). The maximum pressure in this early study was 8 GPa—well below the stability field of the perovskite. Knittle & Jeanloz (1987a) measured the equation of state of $(Mg_{0.9}Fe_{0.1})SiO_3$ perovskite to pressures in excess of 100 GPa by x-ray diffraction of laser-heated samples (without a pressure medium). The x-ray diffraction patterns were indexed as orthorhombic perovskite. A third-order Birch Murnaghan equation-of-state fit to the data gave zero-pressure parameters $K_{0T} = 266(\pm 6)$ GPa and $K'_{0T} = 3.9(\pm 4)$. Subsequent static compression measurements have been carried out using single-crystal x-ray diffraction under hydrostatic conditions but at lower pressures. Kudoh et al (1987) report a lower value of $K_{0T} = 247$ GPa whereas Ross & Hazen (1990) find $K_{0T} = 254(\pm 13)$ GPa. Recent high-resolution powder diffraction measurements using synchrotron radiation to 30 GPa by Mao et al (1991) give $K_{0T} = 261(\pm 4)$. In each of these studies $K'_{0T} = 4$ was assumed. Yeganeh-Haeri et al (1989a,b) report a value for the adiabatic bulk modulus K_{0S} of 246(±1) GPa on the basis of single-crystal Brillouin scattering spectroscopy carried out at zero pressure; the value for K_{0T} calculated from this result is 243(±1) GPa (see below). Recent results indicate that, within experimental uncertainties, the bulk moduli of $Mg_{1-x}Fe_xSiO_3$ perovskites, with $x = 0$, 0.1, and 0.2, are independent of x (Mao et al 1989b, 1991). Therefore, the density of perovskite at high pressure increases with x in proportion to the density dependence of x at zero pressure (Figure 2).

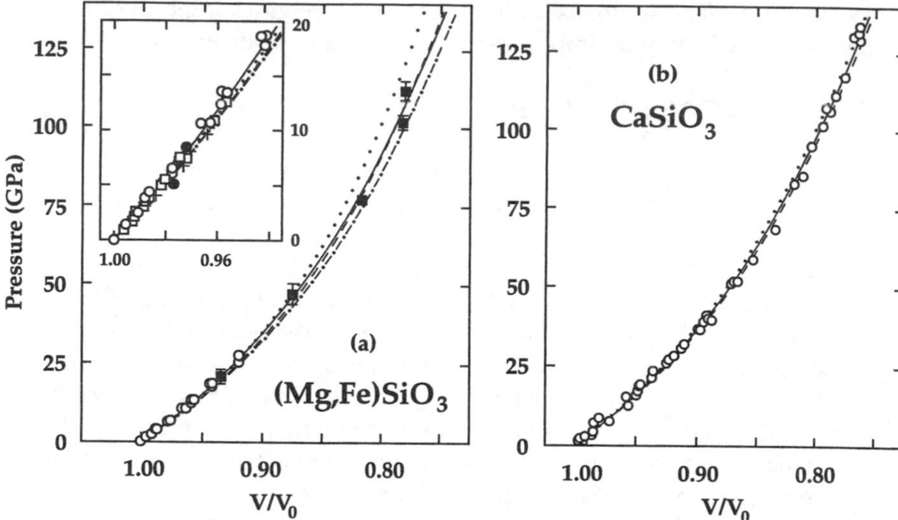

Figure 8 Static compression data and *P-V* equations of state for silicate perovskites (room temperature). The experimental points are from high-pressure x-ray diffraction measurements and the curves represent least-squares fits to different data sets using Birch-Murnaghan finite-strain equations of state (see text and Table 1). (*a*) $(Mg,Fe)SiO_3$-peroskite: ●, Yagi et al (1982); +, Kudoh et al (1987); ■, Knittle & Jeanloz (1987); □, Ross & Hazen (1990); ○, Mao et al (1991). (*Solid line*) $K_{0T} = 266(\pm 6)$ GPa and $K'_{0T} = 3.9(\pm 0.4)$ (Knittle & Jeanloz 1987a), and $K_{0T} = 261(\pm 4)$ GPa and $K'_{0T} = 4$ assumed (Mao et al 1991); these curves are indistinguishable on this scale; (*dotted dashed line*) $K_{0T} = 246$ GPa and $K'_{0T} = 4$ assumed (Kudoh et al 1987; see also Yeganeh-Haeri et al 1989a,b; Chopelas & Boehler 1989); (*dashed line*) $K_{0T} = 246$ GPa and $K'_{0T} = 4.5$ (e.g. Stixrude & Bukowinski 1992); (*dotted line*) $K_{0T} = 246$ GPa and $K'_{0T} = 5.5$ (see Mao et al 1991). The unit-cell volumes are given in Figure 2. (*b*) $CaSiO_3$ perovskite: ○, Mao et al (1989a). (*Solid line*) $V_0 = 45.31(\pm 0.08)$ Å^3 (per unit cell), $K_{0T} = 281(\pm 6)$ GPa, $K'_{0T} = 4$ assumed (Mao et al 1989a); (*dashed line*) $V_0 = 45.60(\pm 0.10)$ Å^3, $K_{0T} = 275(\pm 15)$ GPa, $K'_{0T} = 4$ assumed (Tarrida & Richet 1989); (*dotted line*) $V_0 = 45.58(\pm 0.07)$ Å^3, $K_{0T} = 288(\pm 13)$ GPa, $K'_{0T} = 4$ assumed (Yagi et al 1989). Agreement among the three studies is very good; for clarity only the data points from Mao et al (1989a) are shown.

Although these differences in the bulk moduli determinations are perhaps typical of inter-laboratory comparisons, particularly for incompressible materials and different techniques, they nevertheless have important geophysical implications because they can give rise to significant differences in extrapolated values for the density (and bulk sound velocity) of perovskites at lower-mantle pressures. Some of these apparent discrepancies can be removed by recognizing the trade-off between K_{0T} and K'_{0T} in equation-of-state fits to the data (Mao et al 1991). For example, to reconcile the low bulk modulus obtained from Brillouin scattering (Yeganeh-Haeri et al 1989a,b) with the high-resolution diffraction data of Mao et al (1991), a relatively high value for $K'_{0T} \sim 5.5$ must be assumed (Mao et al 1991). The high K'_{0T} however, is inconsistent with the higher pressure diffraction data of Knittle & Jeanloz (1987a) (see Hemley et al 1992) (Figure 8a). A consistent set of pressure-volume relations is therefore only available for pressures corresponding to the top of the lower mantle (23 GPa to perhaps 50 GPa). At higher pressures, the differences become more pronounced, which adds uncertainty to constraints on mineralogy and composition of the lower mantle based on extrapolated equations of state, even in the absence of the thermal effects described below.

Comparison of recent measurements of the linear compressibilities of (Mg,Fe)SiO_3 perovskite are listed in Table 1. In general, the present conclusion that the *b*-axis is the least compressible is consistent with the zero-pressure Brillouin scattering study of $MgSiO_3$ perovskite by Yeganeh-Haeri et al (1989a,b) and single-crystal x-ray diffraction studies of $MgSiO_3$ perovskite by Kudoh et al (1987) and Ross & Hazen (1989). The remaining differences among these results may be explained as uncertainties associated with the smaller pressure range of study for the single-crystal data, although structural differences between the (Mg,Fe)SiO_3 synthesized in diamond-anvil cells and in large-volume apparati cannot be ruled out. It is possible, for example, that this is associated with different amounts of ferric iron (Fe^{+3}) formed due to differences in oxygen fugacity in the two types of experiments (Fei et al 1991b). Nonhydrostatic pressure has a major effect on the relative compressibilities of lattice parameters. In addition, without the resolution to separate the orthorhombic splitting of the equivalent cubic diffraction peaks, relative compressibilities of lattice parameters could not be accurately determined.

The room temperature *P-V* equation of state of $CaSiO_3$ has been measured by x-ray diffraction in diamond-cells to pressures in the 100 GPa range by three groups (Mao et al 1989a, Yagi et al 1989, Tarrida & Richet 1989). The results are shown in Figure 8*b*. The agreement is comparable to that observed for (Mg,Fe)SiO_3 equation of state studies. The similar values for the bulk moduli of the two perovskites has lead to the proposal that $CaSiO_3$ perovskite could be an invisible component in the lower mantle (Mao et al 1989a, Tarrida & Richet 1989).

THERMAL EXPANSION

Because of the high temperatures prevailing in the lower mantle (>1700 K, e.g. Jeanloz & Morris 1986), determination of the thermal expansivity of silicate perovskite is essential for estimation of the density (and bulk modulus) of the material within the Earth. The critical role of the thermal expansivity of the material for controlling chemical layering and the style of convection within the mantle was pointed out by Jackson (1983) and Jeanloz & Thompson (1983). Figure 9 summarizes the results of thermal expansion studies of $(Mg,Fe)SiO_3$ perovskite at zero pressure. Knittle et al (1986) measured the thermal expansion from 300–870 K at zero pressure of polycrystalline $(Mg_{0.9}Fe_{0.1})SiO_3$ perovskite synthesized in a diamond-cell and obtained a value for the volume thermal expansion coefficient α exceeding 4×10^{-5} K^{-1} above 500 K. Ross & Hazen (1989) measured the thermal expansion at low temperatures (77–400 K), and more recently Parise et al (1990) performed powder diffraction measurements from 10 to 433 K. These studies have provided accurate determination of the thermal expansivity, with $\alpha = 1.9$–2.1×10^{-5} K^{-1} at 300 K, but they are

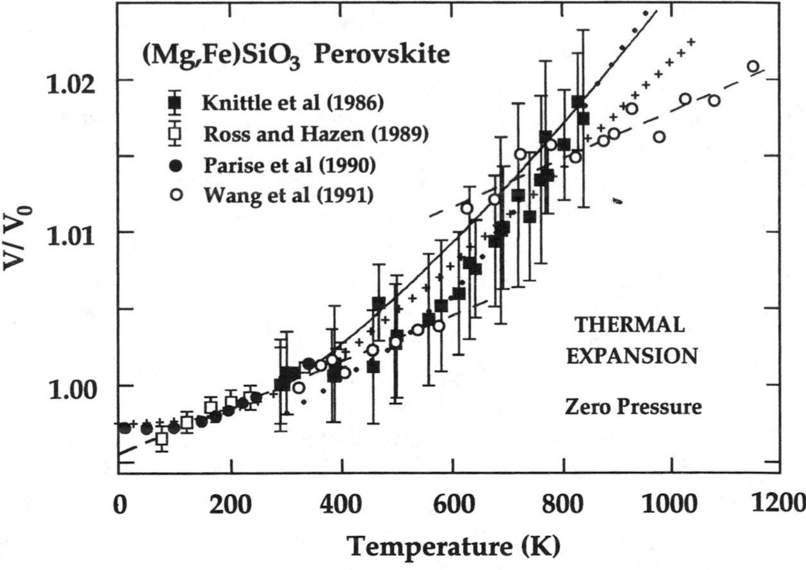

Figure 9 Relative volume of $(Mg,Fe)SiO_3$ perovskite as a function of temperature at zero-pressure (V_0 is the room-temperature volume). Knittle et al (1986) studied $(Mg_{0.9}Fe_{0.1})SiO_3$ composition; Ross & Hazen (1989), Parise et al (1990), and Wang et al (1991a) examined the $MgSiO_3$ endmember. The lines are fits to the experimental data: (*dotted line*) Knittle et al (1986), Grüneisen-Suzuki model fit to the high-temperature data, with $V_0 = 162.25$ Å^3, $K_0 = 260$ GPa, $K'_{0T} = 4$, $\Theta_D = 825$ K, and $\gamma_D = 2.20$; (*solid line*) Mao et al (1991), obtained from fitting high-*P*-*T* data, with $\alpha = 0.301 \times 10^{-4} + 1.500 \times 10^{-8}\ T - 1.139\ T^{-2}$ (zero pressure); (+ + + +) Stixrude & Bukowinski (1990), Mie-Grüneisen-Debye model fit to zero-pressure thermal expansivity and phase equilibria data, $V_0 = 162.25(\pm 0.05)$ Å^3, $\Theta_D = 1016(\pm 6)$ K, and $\gamma_D = 1.73(\pm 0.14)$; (*dashed line*) Wang et al (1991a), linear fit showing proposed phase transition at 600 K.

limited to relatively low temperatures. The results are compatible with the higher thermal expansion found by Knittle et al (1986) if one assumes a temperature dependent α, as found for other silicates. There has been concern, however, since these measurements of thermal expansivity (as well as other properties) were made well outside the perovskite stability field (Navrotsky 1989, Hill & Jackson 1990) where the phase is very unstable. Wang et al (1991a) have reported that $MgSiO_3$ perovskite undergoes a metastable phase transition at 600 K and 7.1 GPa, which is responsible for the high thermal expansivity observed at low pressures (Figure 7); this is discussed in further detail below. Very recently, Wang et al (1991b) reported low thermal expansivities which they attribute to partial conversion to an amorphous phase at zero pressure. Hence this topic continues to be a subject of debate.

A variety of models have been used to examine this question. The Grüneisen parameter can be used to evaluate the thermal expansivity (or thermal pressure) using the thermodynamic Grüneisen parameter defined as $\gamma_{th} = \alpha K_T V/C_V$. Constraints on γ_{th} for silicate perovskites have been obtained from high-pressure vibrational spectra described below (Williams et al 1987, Hemley et al 1989) and from direct fits to high-temperature expansivity data (Knittle et al 1986, Jeanloz & Knittle 1989, Stixrude & Bukowinski 1990, Hemley et al 1992). Simple thermal models such as Einstein and Debye approximations have been used, including the formalism developed by Suzuki (1975) based on Grüneisen theory (Knittle et al 1986). These models involve a parameterization of the thermal pressure with a single characteristic frequency, together with a Grüneisen parameter γ_D, defined as $\gamma_D = -d\ln\Theta_D/d\ln V$, and possibly higher-order terms, where Θ_D is an effective Debye temperature, which is equivalent to the Einstein temperature ($\Theta_D = \sqrt{5/3}\Theta_E$) for $T > \Theta_D$. One of the problems is that γ_{th} is not determined directly in these studies: The γ_D and that determined by spectroscopic data need not be the same as the thermodynamic Grüneisen parameter γ_{th}. Nevertheless, both the spectroscopic data and independent theoretical predictions indicate that γ_{th} is 1.7–2.0 (Williams et al 1987, Hemley et al 1989, Hemley 1991), which is higher than is typically found for other materials (silicates and oxides), consistent with the γ_D determined from the thermal expansivity data (Knittle et al 1986). Estimates of Θ_D range from 725 to 1200 K (Knittle et al 1986; Wolf & Bukowinski 1987; Hemley et al 1987, 1989; Jeanloz & Knittle 1989; Stixrude & Bukowinski 1990); the range in estimates is due in part to the fact that the vibrational density of states is not well represented by a Debye model. Nevertheless, similar fits to the high-temperature thermal expansivity can be obtained over this range because of the trade-off in parameters (Hemley et al 1992). The problem has also been examined by the use of theoretical calculations with nonempirical (ab initio) models (Hemley et al 1987, 1989; Cohen 1987; Wolf & Bukowinski 1987) and empirical approaches (Matsui 1988). The recent lattice dynamical models are not inconsistent with the experimentally observed expansivity, including its temperature dependence, but it should be noted that lattice dynamical predictions of thermal expansion are in general highly sensitive to the interatomic potentials.

The crucial property for comparison of the density of silicate perovskites and other possible lower mantle phases with seismological density profiles is the density at the relevant P-T conditions, which requires a knowledge of the variation in α with pressure as well as temperature. The variation in α with pressure is characterized by the Anderson-Grüneisen parameter $\delta_{T,S}$, which is defined as $\delta_{T,S} = -1/(\alpha K_{T,S})(dK_{T,S}/dT)$. Assuming $\delta_{T,S}$ is independent of P and T, it can be shown that $\delta_{T,S} = (d\ln\alpha/d\ln V)_P$. Constraints on this quantity have been obtained by the use of systematics developed at lower P and T for other materials (Chopelas & Boehler 1989, Anderson et al 1990, Bina & Helffrich 1992). Alternatively, theoretical models provide useful predictions of this quantity (Wolf & Bukowinski 1987, Hemley et al 1987, Isaak et al 1990). Jeanloz & Knittle (1989) developed empirical equations of state for both $(Mg,Fe)SiO_3$ and (Mg,Fe)O based on an anharmonic Einstein model, predicting the densities of both phases under lower-mantle pressures and temperatures.

Recently, the pressure dependence of α has been determined from direct measurements of the density of $(Mg,Fe)SiO_3$ at high pressures and temperatures by the use of x-ray diffraction techniques (Mao et al 1991). These results are summarized in Figure 10. The results support a high value for δ_T ($\sim$6.5) although measurements are not yet possible at lower-mantle temperatures (about 1700 K). Notably, these results are also consistent with the zero-pressure values for α reported by Knittle et al (1986). Because at the highest pressures the measurements were performed within the stability field of the perovskite, no problems with formation of an amorphous phase were encountered under these conditions. Since a decrease in δ_T is expected above the Debye temperature (e.g. Anderson et al 1990), it is important that these measurements be extended to higher temperatures (i.e. above 900 K). Extension of the measurements to higher pressures would also provide the direct determination of the pressure dependence of δ_T, which could be significant under deep lower mantle conditions (Isaak et al 1990).

SHEAR MODULUS

The shear modulus is an even more difficult quantity to measure than either the bulk modulus and equation of state, particularly at high pressures, and

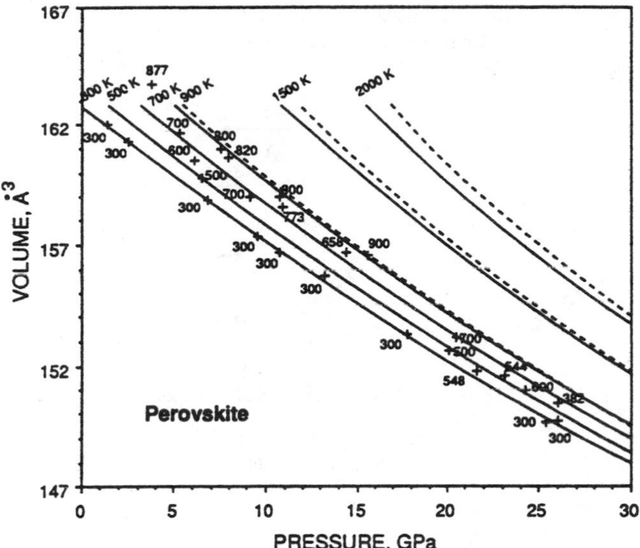

Figure 10 P-V-T relations for $(Mg,Fe)SiO_3$ perovskite. The curves were fits to the high P-T volume measurements (*points*) with $K_{0T} = 261$ GPa, $K'_{0T} = 4$. $\delta_T = 7.4$ (*dashed line*) and $\delta_T = 6.5$ (*solid line*) (from Mao et al 1991).

little was known about this quantity until recently. The shear modulus μ is important because primary seismological data give shear and longitudinal wave velocities (V_S and V_P), both of which depend on the shear modulus; i.e. $V_S = (G/\rho)^{1/2}$ and $V_P = [(K_S + 4/3G)/\rho]^{1/2}$. Predictions of G for silicate perovskite ranging from 140 to 190 GPa were made on the basis of crystal chemical systematics and theoretical models (e.g. Jeanloz & Thompson 1983, Cohen 1987, Matsui et al 1987). The high value of 190 GPa for $MgSiO_3$ perovskite under ambient conditions predicted theoretically using the PIB model (Cohen 1987) was subsequently confirmed by Brillouin scattering measurements of Yeganeh-Haeri et al (1989a,b), who reported a value of 184.2($\pm$4.0) GPa (both Voight-Reuss-Hill bounds). Table 2 lists the complete set of single-crystal elastic moduli from these two studies. One of the striking features of these results is that the value of G under room conditions is equivalent to that of the lower mantle at 1000 km depth (e.g. Dziewonski & Anderson 1981). If the lower mantle is predominantly perovskite, this result indicates that the pressure and temperature effects nearly completely offset each other at this depth (40 GPa and 1900–2300 K).

Table 2 Elastic constants (in GPa) for $MgSiO_3$ perovskite at zero pressure

	Theory[a]			Exp[b]
	c_{ij} Static	c_{ij} 298K	$\partial c_{ij}/\partial P$ 298K	c_{ij} 295 K
c_{11}	548	531	6.1	520
c_{12}	54	44	3.2	114
c_{22}	551	531	5.6	510
c_{13}	153	143	3.0	118
c_{23}	175	166	3.0	139
c_{33}	441	425	6.3	437
c_{44}	241	237	1.9	181
c_{55}	253	249	1.4	202
c_{66}	139	136	1.7	176
K	256	249	4.1	245
G	196	192	1.7	184

[a] Potential-induced breathing model (Cohen 1987).

[b] Brillouin scattering spectroscopy (Yeganeh-Haeri et al 1989a,b).

Although measurements of G under these P-T conditions are not yet possible, theoretical calculations can be used to assess the relative contributions of pressure and temperature. Given a model for the interatomic forces, the calculations of the pressure derivatives of the elastic moduli are straightforward and have been performed for $MgSiO_3$ perovskite (Cohen 1987). The calculation of the temperature derivatives (thermal elastic constants) are more difficult for low-symmetry structures where there are Raman-active modes, since the Raman modes couple with the elastic constants. At a pressure corresponding to 1071 km depth, $G = 255$ GPa (at 300 K), which indicates that $dG/dT \approx 0.04$ GPa/K to be consistent with geophysical data assuming a perovskite-dominated lower mantle (cf Yeganeh-Haeri et al 1989a). It should be pointed out that the effect of iron content on the shear modulus has not been determined and could be significant.

MELTING

Information on the melting curve and melting relations of silicate perovskites is central to a number of problems in solid-Earth geophysics. Melting provides the most important mechanism for chemical differentiation in the Earth; hence, perovskite melting figures prominently in models of the early evolution of the planet. Because of the absence of widespread melting in the lower mantle, the determination of the solidus of silicate perovskite in principle provides an upper bound on the temperature distribution through a large fraction of the planet. In addition, measurement of the fusion curve provides bounds on the density of the liquid at high pressures and the extent to which ultrabasic liquids can become neutrally buoyant relative to surrounding crystalline phases at depth. Although virtually nothing was known about the melting of silicate perovskites just five years ago, there is now considerable information as a result of a series of pioneering experiments at the forefront of high *P-T* technology.

The results are shown in Figure 11. Heinz & Jeanloz (1987) reported diamond-cell measurements of the melting curve of $(Mg,Fe)SiO_3$ to 60 GPa and concluded that melting is independent of pressure above ~30 GPa (T_m ~ 3000 K). This curve was significantly lower than predictions based on empirical models and extrapolations of low-pressure data (Ohtani 1983, Poirier 1989). Subsequent measurements by Bassett et al (1988)

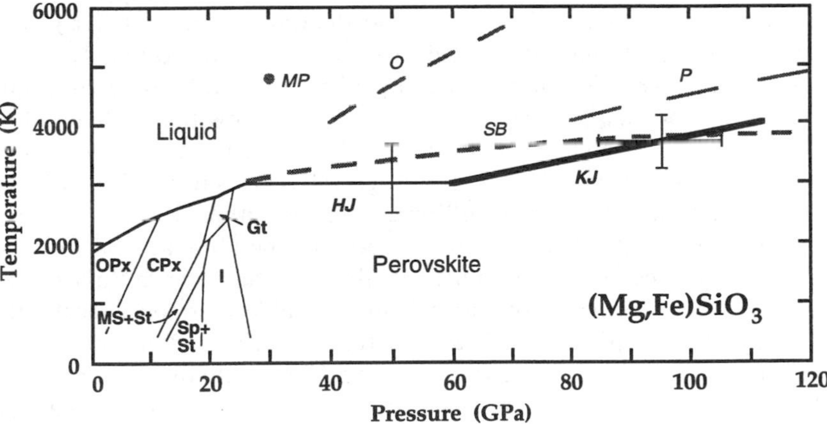

Figure 11 Melting curve for $(Mg,Fe)SiO_3$. The solid line (*HJ*) is the experimental result of Heinz & Jeanloz (1987) for $(Mg_{0.9}Fe_{0.1})SiO_3$ perovskite obtained to 60(±5) GPa, and the bold line at higher pressures (*KJ*) is the result of Knittle & Jeanloz (1989a) determined experimentally to 96(±10) GPa. The error bars give a measure of the estimated uncertainties in the measurements at the highest pressures (see original references for details). The dashed curves are calculated results for $MgSiO_3$ perovskite using a variety of methods: O—Ohtani (1983), P—Poirier (1989), SB—Stixrude & Bukowinski (1990), MP—Matsui & Price (1991). The lower pressure phase relations for $MgSiO_3$ composition are also shown (Kato & Kumazawa 1985).

support the Heinz & Jeanloz result. A higher pressure study by Knittle & Jeanloz (1989) indicates there may be a weak increase in T_m with pressure above 50 GPa. Ito & Katsura (1992) studied the melting of $MgSiO_3$ using a multi-anvil press and found T_m of 2870 K with a slope of 30 K/GPa, but the observations were limited to 21–25 GPa. Further measurements are certainly required. It is not known, for example, if the melting is congruent at very high pressures (Knittle & Jeanloz 1989a, Kubicki & Hemley 1989). Williams (1990) melted $(Mg_{0.88}Fe_{0.12})SiO_4$ in a laser-heated diamond-cell at 50–55 GPa. Infrared measurements of the quenched glasses show evidence for enrichment in SiO_2 with Mw as the liquidus phase. Ito & Katsura (1992) find eutectic melting between Mw and Pv, infer that eutectic composition shifts toward pyroxene composition with pressure, and suggest incongruent melting at deep mantle pressures.

A decrease in the melting slope with pressure indicates that the volume change on fusion approaches zero, which would require significant changes in liquid structure to take place over this pressure range. The extent to which such a change in liquid structure for $MgSiO_3$ compositions is associated with coordination changes in the liquid at these pressures has been a subject of continued discussion (Heinz & Jeanloz 1987; Knittle & Jeanloz 1989b; Jeanloz 1990; Miller et al 1991a,b). Early molecular dynamics simulations predicted a coordination change at very high compressions, although the actual pressures were not accurately determined (Matsui & Kawamura 1980, Matsui et al 1982). More recent simulations also predict this transition, but they also show that high densities can be reached without changes in Si coordination changes, i.e. by repacking of SiO_4 tetrahedra (Kubicki & Lasaga 1991). This result also seems to be consistent with high-pressure Raman studies, although these measurements have been limited to glasses at room temperature and may not be direct analogues of melts (Kubicki et al 1992). Stixrude & Bukowinski (1990) calculated the fusion curve from available thermochemical data (mostly at lower pressures). They find that a leveling off and possibly a maximum in the melting curve is compatible with the current data. However, the result is extremely sensitive to key parameters (such as the compressibility of the melt) which have not been measured at mantle conditions (Hemley & Kubicki 1991), so this result remains speculative. Recent molecular dynamics calculations of Matsui & Price (1991) predict a significantly higher melting curve, with no maximum under lower mantle conditions, which may be taken as evidence against neutral buoyancy of $MgSiO_3$ melts in the lower mantle. On the other hand, on the basis of recent shock-wave studies of komatiite liquid to 36 GPa, Miller et al (1991a,b) suggest that liquidus perovskite is neutrally buoyant at 70 GPa.

VIBRATIONAL DYNAMICS

The vibrational properties of perovskite are important for understanding the nature of possible distortions and for identifying phase transitions in these materials. In this regard, vibrational Raman scattering is particularly relevant because the activity of the first-order Raman spectrum is strongly dependent on the distortions of the crystal from the high-symmetry cubic form. This arises from the fact that all of the Raman-active vibrations in the distorted (e.g. orthorhombic) perovskite structures are derived from modes that occur at the edge of the Brillouin zone in the cubic form (cubic perovskite has no Raman-active modes). Certain of these modes are associated with displacive transitions between low and high-symmetry perovskite structures; that is, they may be soft modes, vibrations whose frequencies decrease and eventually vanish as a function of either pressure or temperature at the transition (cf. McMillan 1988).

There are 20 atoms in the primitive unit cell in *Pbnm* perovskite; thus there are 60 zone center modes including the 3 translational modes. The optic modes decompose as $7A_g+7B_{1g}+5B_{2g}+5B_{3g}+8A_u+7B_{1u}+9B_{2u}+9B_{3u}$. There are 24 Raman-active modes (A_g, B_{1g}, B_{2g}, and B_{3g}) and 25 infrared-active modes (B_{1u}, B_{2u}, and B_{3u}). The symmetry vibrations for orthorhombic *Pbnm* perovskite are shown in Figure 12.

The infrared spectrum of $MgSiO_3$ perovskite, first measured by Weng et al (1983) at zero pressure shows vibrations characteristic of octahedrally coordinated Si. Williams et al (1987) reported measurements of the pressure dependence of the infrared spectrum to 27 GPa as well as the Raman spectrum at zero pressure. Hemley et al (1989) measured the pressure dependence of the Raman spectrum of single crystals of $MgSiO_3$, and recently Chopelas & Boehler (1992) measured spectra of polycrystalline samples. The pressure dependences of the vibrational modes documented in these studies are shown in Figure 13. Additional preliminary data on the pressure dependence of the far-infrared spectrum have been reported (Hofmeister et al 1987, see also Hemley et al 1989). No soft-mode behavior was observed in any of the high-pressure measurements. Chopelas & Boehler (1992) report a decrease in intensity and a small change in pressure shifts of two modes at 37 GPa, which they have interpreted as evidence for a phase transition. However, no transition has yet been observed by x-ray diffraction in this pressure range, as discussed above. In general there is considerable coupling of the polyhedral vibrations in the perovskite. This strong coupling of the modes is associated with characteristic face-sharing of the component polyhedra (SiO_6 octahedra and MgO_{12} distorted dodecahedra). The observed vibrational modes have not yet been assigned

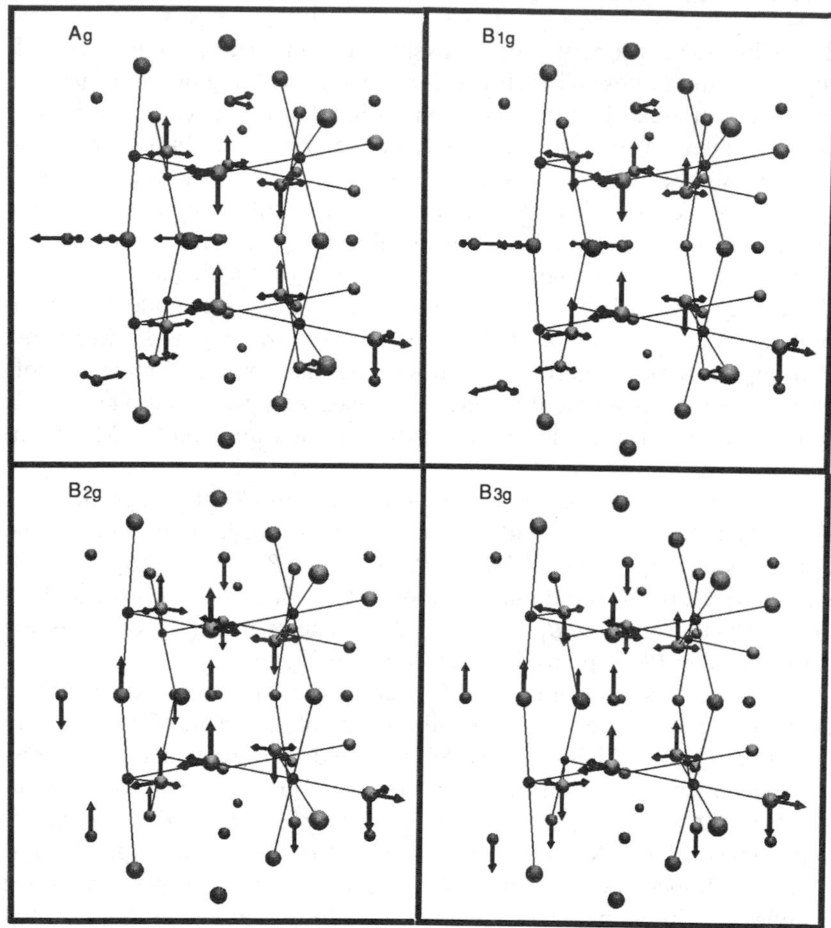

Figure 12 (*a*) Raman-active symmetry modes for *Pbnm* perovskite. The view is along (*110*) in the orthorhombic cell, or (*100*) in cubic coordinates. The *c*-axis is vertical. Si-O "bonds" are indicated by solid lines. The arrows indicate the atomic displacements for each type of symmetry mode. Only the minimum number of displacement vectors are shown, and the displacements are copied by unit-cell translations. The actual vibrational modes are linear combinations of the motions shown. Thus for A_g there are 7 modes, which are linear combinations of O(1) and Mg displacements in the *x*- and *y*-directions, and O(2) displacements in the *x*-, *y*-, and *z*-directions. Only the A_g symmetry modes correspond to possible tilt transitions. (*b*) Odd-symmetry modes for *Pbnm* perovskite. The A_u modes are inactive, and the B_{1u}, B_{2u}, and B_{3u} are infrared active. All of these modes break inversion symmetry, but only the *B* modes generate a dipole moment that can interact with infrared radiation.

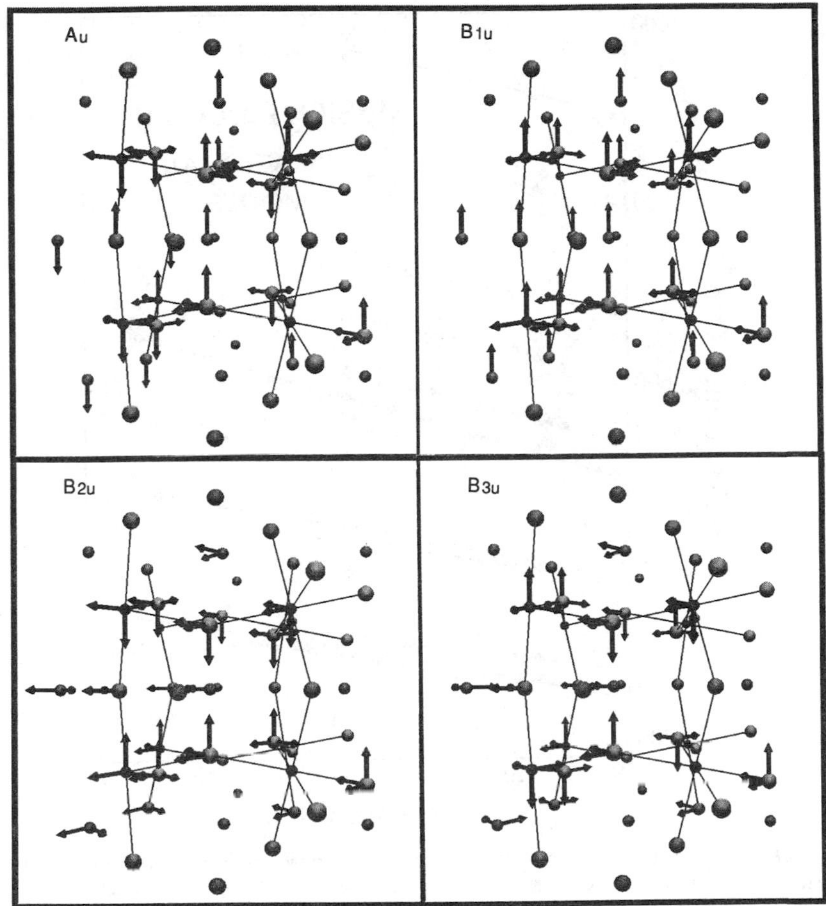

Figure 12b

to specific symmetry species, although it is likely that the strongest Raman bands have A_g symmetry.

Mode-Grüneisen parameters γ_i can be obtained from least-squares fits of the high-pressure mid-infrared and Raman data (Williams et al 1987, Hemley et al 1989, Chopelas & Boehler 1992). This information is important for assessing the microscopic origin of thermodynamic properties such as thermal expansivity and entropy. The γ_{0i} (values at zero pressure) are comparatively large, which indicates that the material has appreciable anharmonicity, at least at low pressure. The results can be used to calculate the Grüneisen parameter via the approximation, $\gamma_{th} \approx \langle\gamma_i\rangle = \Sigma\gamma_i c_i/\Sigma c_i$, where c_i is the heat capacity function for an Einstein oscillator and the

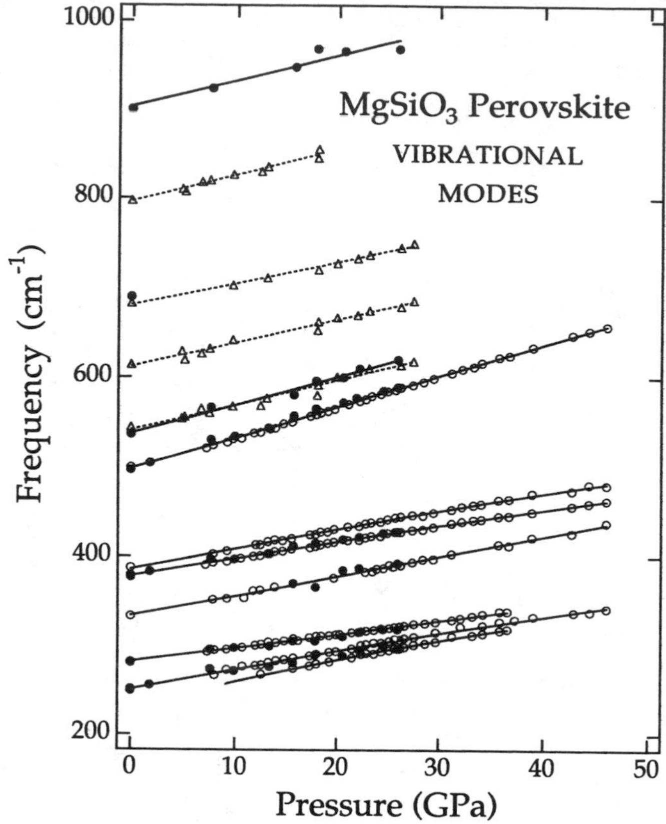

Figure 13 Pressure dependence of the vibrational frequencies for $MgSiO_3$ perovskite (room temperature). Raman modes: ●, Hemley et al (1989); ○, Chopelas & Boehler (1992). Infrared modes: △, Williams et al (1987). The solid and dashed lines are least-squares fits to the pressure shifts for the Raman and infrared modes, respectively.

sum includes in principle all vibrational modes i (see Hemley et al 1989). The sum over the available data gives $\langle\gamma_i\rangle = 1.80\text{–}2.03$ ($q = 0-2$) at 300 K and $\langle\gamma_i\rangle = 1.67\text{–}1.87$ at high temperature which is close to previous results (Hemley et al 1989, Williams et al 1989). Despite the wealth of new information on the vibrational properties of this material at high pressures, accurate calculation of γ_{th} from spectroscopic data is still problematic because the pressure shifts for most of the modes have not been measured and there may be inadequate sampling of the vibrational density of states.

An additional application of vibrational spectroscopic measurements of silicate perovskite has been their use in characterizing the materials at low pressures. This has been particularly important because of the ease with which samples transform to glassy phases at low pressures (e.g. zero pressure) when subjected to heating, laser illumination, and grinding or polishing (Hemley et al 1989, Yeganeh-Haeri et al 1989b, Wolf et al 1990). The glass phase, which is formed prior to crystallization of enstatite (Knittle & Jeanloz 1987b), is readily observed by vibrational spectroscopy and may not be apparent in standard x-ray diffraction or x-ray spectroscopic measurements. There is evidence that the extent of amorphization depends on both sample history and the method of preparation. Amorphization also complicates high-temperature Raman measurements on silicate perovskites, which are important for studying possible temperature-induced soft-mode behavior. Preliminary Raman spectra as a function of temperature up to the point of vitrification at zero pressure have been reported for $MgSiO_3$ (Wolf et al 1990, Durben & Wolf 1991). The frequencies ν decrease with temperature, but no anomalous soft-mode behavior (defined as $\nu \rightarrow 0$) has yet been observed. Because of amorphization, such measurements need to be performed at high pressure, within the stability field of the perovskite.

HIGH-TEMPERATURE PHASE TRANSITIONS

The question of high-temperature phase transitions in distorted perovskites is closely related to the vibrational dynamics as observed in Raman spectroscopy because many of the possible transitions would involve soft-mode behavior with some mode frequencies vanishing (or becoming small) at the phase transition. Continuous transitions in perovskite, if they did occur in the Earth, would be particularly important if they involve soft Raman modes because they would then lead to elastic or acoustic anomalies in shear moduli at the phase transition. Thus, for example, a soft-mode transition from an orthorhombic phase to a tetragonal phase would involve probable softening of c_{11}–c_{12}, and transition from the orthorhombic phase to a monoclinic phase would involve softening of a shear mode (e.g. c_{66}). Also, the average shear modulus could show non-monotonic behavior as a function of pressure and/or temperature.

There are a number of possible phase transitions in perovskite with increasing temperature that involve loss of the various octahedral tilts, culminating with the cubic structure, possibly at temperatures above the melting point. Aleksandrov (1976, 1978) discussed some possibilities for phase transitions from *Pbnm*, but his analysis assumed that symmetry will always be increased with increasing temperature—which is not necessarily the case. The subset of transitions he considered that refer to *Pbnm* is shown by the square blocks in Figure 14. The structures in the circles are for possible intermediate phases which tend to have lower symmetry but lead from one tilt system to another in cases where Aleksandrov considered only discontinuous phase transitions possible. Thus if $(Mg,Fe)SiO_3$ does transform at high temperatures to another distorted perovskite [$q^-a^-c^+$ using Glazer's (1975) notation], it is most likely to transform to tetragonal $D_{4h}^5(00c^+)$, orthorhombic $D_{2h}^{28}(a^-a^-0)$, monoclinic $C_{2h}^2(a^-b^-c^+)$, or to orthorhombic $D_{2h}^{17}(0a^-c^+)$.

Glazer's + and − type tilts can be related to M_2 and R_{25} instabilities. In other words, the atomic displacements are modulated by $\cos[-q \cdot r]$, where q is given by $(0.5, 0.5, 0)$ (in units of $2\pi/a$) for the M-point and $(0.5, 0.5, 0.5)$ for the R-point and r is the lattice vector (relative to the cubic lattice). Figure 1 shows the M_2- and R_{25}-type rotations. There is one three-dimensionally degenerate R-point in the cubic Brillouin zone and there are three distinct M-points. Thus the rigid tilts in perovskite can be represented

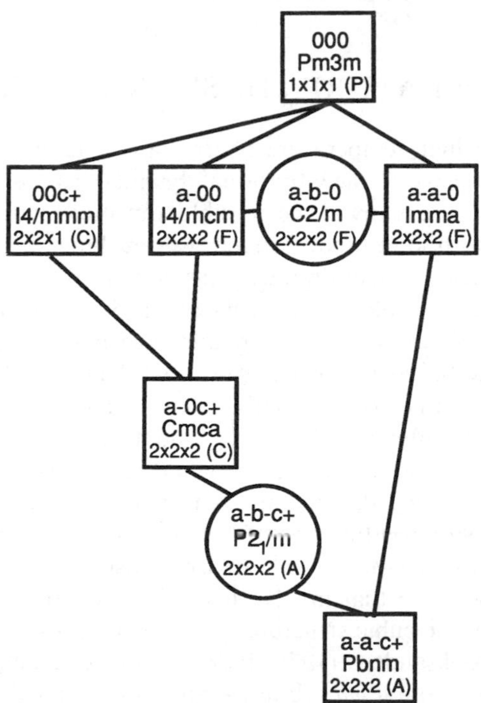

Figure 14 Possible structural phase transitions in perovskite from cubic *Pm3m* to lower symmetry perovskite structures.

as coordinates in a six-dimensional space, only a subset of which can be described by Glazer's notation. The R_{25} and M_2 modes are folded into the zone center in *Pbnm* and become Raman-active modes. One or more of these modes might be expected to display soft-mode behavior near a continuous high-temperature phase transition. However, many of the Raman modes are not observed since they are too weak (Williams et al 1987, Hemley et al 1989, Chopelas & Boehler 1992), and no such soft-mode behavior has yet been observed in $MgSiO_3$ perovskite.

It is crucial to investigate which modes in *Pbnm* perovskite correspond to the R_{25} and M_2 modes in cubic perovskite and whether these modes are expected to go soft at various phase transitions. A symmetry analysis of *Pbnm* perovskite was performed using the method of Boyer (1974). By examining the symmetry modes, we find that only A_g symmetry modes are compatible with tilt distortions that can lead to continuous phase transitions at high temperatures. There are two A_g modes compatible with M_2 symmetry and two A_g modes compatible with R_{25} symmetry. Thus lack of soft-mode behavior in the A_g modes is sufficient to rule out soft-mode behavior in a *Pbnm* perovskite. A transition to *Imma* would likely be a soft-mode transition. A soft-mode transition to *I4/mmm* would require two R_{25}-type modes going soft simultaneously. A direct transition to *Cmcm* would be first-order, though *Cmcm* could be reached continuously through a monoclinic $P2_1/m$ structure. The latter path also would not involve soft-mode behavior. In general, a second-order, continuous phase transition does not require soft-mode behavior. An example is the transitions in $La_{2-x}Ba_xCuO_4$ (Pickett et al 1991). The phase transition could have order-disorder character, where above the phase transition temperature T_c there would be uncorrelated local tilts which become correlated as temperature decreases until a phase transition occurs, at which point the tilts are all in phase. Order-disorder transitions can couple with elastic constants and cause elastic anomalies as well as soft-mode transitions.

A variety of calculations have considered the relative stability of cubic and orthorhombic $MgSiO_3$. Wolf & Bukowinski (1987) and Bukowinski & Wolf (1988) predicted that the series of transitions orthorhombic → tetragonal → cubic in $MgSiO_3$ occurs in the mantle with increasing temperature. These calculations, however, significantly underestimate the distortion of the orthorhombic form in comparison to experiment (Figure 3), and hence its stability relative to cubic. Also, the phase transition temperature was not calculated directly but was estimated by extrapolating the results of quasi-harmonic calculations assuming critical behavior in the tilts as functions of temperature; however, quasi-harmonic calculations do not include the fluctuations that cause critical behavior. It is necessary to perform anharmonic calculations with interatomic potentials that more accurately reproduce the room *P-T* structure (e.g. Hemley et al 1987). Using the PIB model Cohen (1987) found that the energy difference (expressed as temperature) between cubic and orthorhombic $MgSiO_3$ rose rapidly by 60 K GPa^{-1}. Though the energy difference is 6700 K at 100 GPa, the results are sensitive to small changes in the interatomic potentials. Matsui & Price (1991), using their empirical rigid ion potential, predict that the transition occurs just below their theoretically predicted T_m, which is very high (Figure 11). They found a continuous transition from orthorhombic to cubic, which is not allowed in a soft-mode transition, and indicates order-disorder character.

Wang et al (1990, 1992) have interpreted the extensive twinning observed in electron microscopic studies of quenched samples of $(Mg,Fe)SiO_3$ perovskite as evidence for a structural transformation at high *P* and *T*. The observation of twins (Wang et al 1990) does not in itself indicate that the twins are formed by cooling through a phase transition, but the more recent observations (Wang et al 1992) that crystals quenched from above 1870 K at 26 GPa contain more twins than samples annealed below 1570 K is suggestive of a phase transition. They suggest that the transition is to the tetragonal *P4/mbm* ($00c^+$) or *I4/mcm* (a^-00) structures. Wang et al (1992) suggest that the kink observed in the melting curve of perovskite by Knittle & Jeanloz (1989a) arises from the intersection with the same perovskite structural phase transition they observe at lower pressures (Figure 11). If correct, this proposal would imply that seismic velocity (or density) anomalies arising from this transition should occur at a depth corresponding to ~50 GPa in the lower mantle. A transition from orthorhombic to tetragonal should correspond to a softening of the shear modulus $c_{11}-c_{12}$ at the transition. Although these results are quite compelling, the interpretations are complicated by sample decomposition effects under the electron beam, the possibility that samples annealed at higher temperatures have coarser twins (and thus are easier to observe), greater equilibrium density of twins at higher temperature, and compositional changes in the various runs. Wang et al (1992) also studied analogue perovskites $CaTiO_3$, $CaGeO_3$, $MnGeO_3$, $LaGaO_3$, and $SmAlO_3$ which have *Pbnm* structure at ambient conditions and observed similar twinning textures for samples quenched from below and above a phase transition. Although this provides further support for the proposal for a phase transition in $(Mg,Fe)SiO_3$, it should be pointed out that intermediate plagioclase feldspars contain ubiquitous polysynthetic twins which are not the result of any phase transitions.

Evidence for another phase transition has been reported by Wang et al (1991a) on the basis of direct x-ray diffraction measurements of $MgSiO_3$ perovskite as a function of temperature to 1200 K at 7.1 GPa. On the basis of shifts of the *113* diffraction peak, a temperature-induced phase transition to another orthorhombic structure at 600 K was proposed (see Figure 7). They concluded that the thermal expansion coefficient of the perovskite is significantly smaller than the previous measurements and further that the structure of the *Pbnm* orthorhombic perovskite may not be representative of the mineral in the Earth. However, the magnitude of the volume discontinuity initially observed at the proposed phase transition is reported to be smaller in subsequent studies (Weidner et al 1991) and may not be statistically significant. Other diffraction experiments carried out in this pressure range also do not support the results of the Wang et al study (Mao et al 1991, Y. Fei, personal communication). The discrepancy could be associated with texturing effects in the former samples, which can introduce errors in the determination of peak positions in powder diffraction patterns. In view of the complications arising from studies of perovskite at lower pressures (e.g. amorphization), the issue also reinforces the need for careful measurements on the material within its stability field ($P > 23$ GPa).

The observation of ferroelectricity in perovskites has given rise to the suggestion that silicate perovskites might undergo similar ferroelectric transitions, perhaps at lower mantle conditions (Litov & Anderson 1978, Anderson 1988). The charge distribution of the end-member $MgSiO_3$ is largely ionic and does not reveal the type of Ti *d*-orbital hybridization found for $BaTiO_3$ that may be responsible for ferroelectric behavior in the latter material (Cohen & Krakauer 1990). This does not rule out possible ferroelectric-type behavior on substitution of other cations in silicate perovskite.

TRANSPORT PROPERTIES

Other important physical properties are the transport properties of silicate perovskite, such as electrical conductivity, thermal conductivity, and rheology. These properties are needed for modeling convection, thermal history, and the Earth's magnetic field, but they are very difficult to study because of the instability of perovskite under ambient conditions and technical difficulties associated with measurement of transport properties under extreme conditions of temperature and pressure. Despite these problems, significant progress has been made, and the first direct measurements have been reported.

The electrical conductivity of mantle phases strongly controls the strength of the magnetic field at the Earth's surface. Measurements of the electrical conductivity of silicate perovskite at high P and T has therefore been of great interest. These challenging experiments have been at the limit of high-pressure techniques, and the results are accordingly controversial. Li & Jeanloz (1987, 1990) reported measurements of electrical conductivity of $(Mg,Fe)SiO_3$ at high P and T using laser-heated diamond-cell techniques. Their results indicated that the conductivity of the Mg end-member is several orders of magnitude lower than geophysical observations for the lower mantle (Li & Jeanloz 1987). Subsequent measurements indicated a large effect from Fe, but the conductivity of $(Mg_{0.9}Fe_{0.1})SiO_3$ perovskite is still lower than geophysical data for the lower mantle (Li & Jeanloz 1990, Li et al 1991). Peyronneau & Poirier (1989) measured the conductivity of both perovskite and perovskite-magnesiowüstite assemblages at high pressures, but using resistance heating (to 673 K); extrapolating these results to lower-mantle temperatures gives a value for the conductivity in good agreement with geomagnetic models. The measurements are extremely sensitive to contamination by water (Li & Jeanloz 1991, Poirier & Peyronneau 1991). Both laboratories now obtain agreement for the conductivity of high iron-content samples containing 20% Fe/(Fe+Mg), which consist of perovskite and magnesiowüstite mixtures, although there is still a discrepancy for iron compositions. These results are not consistent with the dominant conduction mechanism for perovskite being ionic (O'Keeffe & Bovin 1979, Poirier et al 1983, Kapusta & Guillope 1988, Wall & Price 1989), although very recent theoretical calculations predict that ionic conductivity could be important very close to melting (Matsui & Price 1991). It should be noted that the evidence for ionic conductivity in fluoride analogues (O'Keeffe et al 1979) has been attributed to surface effects (Andersen et al 1985). Finally, these experiments also indicate that the conductivity of (Mg,Fe)O is significantly higher than that of $(Mg,Fe)SiO_3$ perovskite, and that this mineral may dominate the conductivity of the lower mantle (Li & Jeanloz 1990, Wood & Nell 1991).

Little is known about thermal conductivity, κ, at high pressures and temperatures except for model extrapolations from low-pressure measurements (Kieffer 1976, Roufosse & Jeanloz 1983, Brown 1986). Until recently, no measurements were available for silicate perovskite. Osako & Ito (1991) measured the thermal diffusivity $a = C_p\rho\kappa$ of a large polycrystalline sample of $MgSiO_3$ between 160 and 340 K and fit the results to $1/a = A+BT$ where $A = (-6.2 \pm 1.4) \times 10^4$ m^{-2} s and $B = (2.15 \pm 0.05) \times 10^3$ m^2 s K^{-1}. They obtain $\kappa = 5.1$ W m^{-1} K^{-1} under ambient conditions which is similar to other silicates. Extrapolating to lower mantle conditions they estimated $\kappa = 3.0$ W m^{-1} K^{-1} at 670 km depth at 1900 K and 12 W m^{-1} K^{-1} at the top of the D'' layer at 2500 K. They suggest that the thermal conductivity would be high enough to prevent the D'' layer at the base of the mantle from being a thermal boundary layer.

An understanding of the rheological properties of silicate perovskite at high pressures and temperatures is crucial for detailed modeling of convection in the Earth. This has been addressed in several theoretical studies of ion migration in $MgSiO_3$ perovskite (Wall et al 1986, Price et al 1989, Miyamoto 1988). This is a very difficult problem experimentally because laboratory time scales are many orders of magnitude shorter than flow time scales in the Earth. Nevertheless, measurements of the plasticity of silicate perovskites are beginning. The plasticity of single crystals of $MgSiO_3$ perovskite has been studied at 300 K and found to be higher than that of olivine and enstatite (Karato et al 1990). Estimates of the shear strength of $(Mg,Fe)SiO_3$ and magnesiowüstite have been obtained from measurements and analyses of the pressure distribution supported by polycrystalline samples in the diamond-cell (Meade & Jeanloz 1990). The results indicated a maximum shear stress of approximately 7 GPa under lower-mantle pressures and room temperature, compared with 8.5 GPa for $(Mg,Fe)_2SiO_4$ spinel, and the authors speculated that the lower mantle may be more ductile than the transition zone. Ito & Sato (1991) argue that the absence of seismicity below 700 km arises from superplasticity associated with the fine-grain texture of perovskite and magnesiowüstite assemblages in the lower mantle.

Under the high temperatures and low strain rates of the lower mantle, yield stress itself is not a pertinent quantity, rather the viscosity or strain rate versus stress is important. The latter are essentially unknown for minerals under lower mantle conditions and in fact may be governed by magnesiowüstite rather than perovskite. In order to address the problem of temperature dependence of the rheology of perovskite, Beauchesne & Poirier (1989, 1990) have studied high-temperature creep in $BaTiO_3$, $KTaO_3$, $KNbO_3$, $RbCaF_3$, and $KZnF_3$ perovskites and found that each perovskite behaved qualitatively differently; thus it was impossible to draw general conclusions about perovskite deformation at high temperatures. The results were fit to the form: $\dot{\varepsilon} = A\sigma^n \exp(-Q/RT)$, where $\dot{\varepsilon}$ is the strain rate, σ is the shear stress, R is the gas constant, T is temperature, Q is the activation energy, and n is the stress exponent. They found that n varies from 1 to 3.7, ln A varies from -11 to -62 (σ is in Pa), and Q varies from 116 to 469 kJ/mole in the three oxide perovskites they studied. This would imply a drop in strength (or increase in strain rates) of many orders of magnitude between room temperature and mantle temperatures. In summary, little remains known about deformation of perovskite under mantle conditions, but progress has been made at low temperatures and high pressures and, on other perovskites, at high temperatures and zero pressure.

DISCUSSION

The increasing number of measurements of the physical and chemical properties of silicate perovskites has provided an essential database for unifying mineralogical, geochemical, and seismological models of the Earth's lower mantle (Silver et al 1988, Anderson 1989). There is now a fair amount of agreement concerning the structural properties and equation of state, including the bulk modulus and thermal expansivity. The measurements provide new insight into interpretation of the 670-km discontinuity, seismologically determined density profiles, the D'' region, and the core-mantle boundary. However, it must also be said that a complete and accurate description of the physical properties of these phases at mantle conditions is not yet in hand, as is evident from the number of controversies that need to be resolved. Thus, one of the central questions in solid-earth geophysics—the composition of the lower mantle, and whether it is chemically distinct and convecting separately from that of the upper mantle—continues to be a matter of active debate. A detailed review of this topic, including mineral physics data for both silicate perovskites and oxide phases and analyses of different seismological models is presented by Bina & Hemley (to be published). Here we point out key constraints from measurements on silicate perovskites.

The phase equilibrium study of Ito & Takahashi (1989) has reaffirmed the coincidence of $(Mg,Fe)_2SiO_4$ silicate spinel to perovskite plus magnesiowüstite transition with the 670-km discontinuity, including the sharpness of the transition, in agreement with earlier diamond-cell studies. They argue that the 670-km discontinuity is therefore compatible with the occurrence of an isochemical transformation in this system and with a homogeneous peridotitic composition for the upper and lower mantle. In principle, the P-T slopes of the perovskite-forming reactions provide another constraint, but the measured values are not sufficiently negative, or well enough determined experimentally, to rule out mixing of the upper and lower mantle (Ito et al 1990). Currently, the most reliable constraint on the chemical composition is provided by density measurements on candidate upper-mantle compositions at lower-mantle pressures. Jeanloz & Knittle (1989) have calculated densities for perovskite and magnesiowüstite at lower mantle conditions from equation of state models determined in part from experimental data. They conclude that for a range of predicted geotherms upper mantle compositions are incompatible with seismic data (Dziewonski & Anderson 1981) in the lower mantle; specifically, an iron enrichment (bulk iron component of 15%) is required, with the intrinsic density of the lower mantle 2.6% (and possibly up to 5%) higher than the upper mantle. Their results are dependent on as-yet unmeasured parameters at high P-T conditions (Chopelas & Boehler 1989, Bina & Silver 1990, Bukowinski & Wolf 1990). The high-P-T x-ray diffraction results (Mao et al 1991) support the equations of state used in this analysis but indicate there is a significant decrease in thermal expansion that is not included in these calculations. Hemley et al (1992) have shown that including this effect decreases the need for iron enrichment at the top of the lower mantle (with no thermal boundary layer), although silica enrichment is then required (Liu 1979, Bass & Anderson 1984, Kuskov & Panfcrov 1991). Further analysis of the density profile for the entire lower mantle, where the density is better constrained seismologically, as well as for the seismic parameter, is given by Bina & Hemley (1992). A change in chemical composition should give rise to a thermal boundary layer, and two separately convecting systems in the upper and lower mantle. If so, iron enrichment then becomes a plausible means by which to offset the effect of higher temperatures (thermal expansion) in order to be consistent with the seismic density profiles (Jeanloz & Knittle 1989). A compositional boundary, if present, may not occur at the identical depth as the phase transition to perovskite-bearing assemblages (Jeanloz 1991).

As described above, transitions to higher symmetry forms have been predicted to occur with increasing temperature in $(Mg,Fe)SiO_3$ perovskite, and these transitions could complicate such inferences for the composition of the lower mantle (Wang et al 1992). There is currently no direct experimental evidence for these transitions, although similar transitions are documented for isostructural perovskite materials. If perovskite is cubic under lower mantle P-T conditions, its bulk modulus is expected to increase as a result of the loss of degrees of freedom for compression. Theoretical calculations indicate that the bulk modulus could increase by $\sim 5\%$ on going from orthorhombic to cubic (Cohen et al 1989). Such antiferroelastic transitions, which may be soft-mode driven, may give rise to unusual shear softening as the transitions are approached. Yeganeh-Haeri et al (1989a) have proposed that transitions of this type may be responsible for the high values of $d\ln V_s/d\ln V_p$ (≈ 2–3) observed in studies of lateral heterogeneity in the lower mantle (e.g. Dziewonski & Woodhouse 1987). Recent calculations show, however, that such behavior can arise from the effect of temperature on the shear modulus relative to the bulk modulus at high P and T in simple materials such as MgO (Isaak et al 1990, 1992; Agnon & Bukowinski 1990; Anderson et al 1992), which do not exhibit these transitions.

The $CaSiO_3$ perovskite exists in the lower mantle as a major separate phase with an abundance subordinate only to ferromagnesian silicate perovskite and, most probably, to magnesiowüstite. The weight percentage of $CaSiO_3$ perovskite is 6.2% in the pyrolite model, 7.0% in a recent model based on solar abundances, and 12% in the piclogite model. Since $CaSiO_3$ also has a higher bulk modulus than $(Mg,Fe)SiO_3$ (by ~3%) (Mao et al 1991), an appreciable enrichment of Ca in the lower mantle could contribute to increasing K_S. Such a comparison could be used to bound the Ca abundance in the lower mantle, since sufficient constraints cannot be obtained from the density alone (see Mao et al 1989a). These results will also depend on the thermal expansivity of the calcium perovskite at mantle conditions, which has not been measured. Ita & Stixrude (1991) have suggested that the $CaSiO_3$ perovskite transition could be responsible for seismic structure near 520 km (i.e. in the transition zone) (Shearer 1990). Recent studies have indicated that partitioning of rare earth elements in $CaSiO_3$ perovskite is significantly higher than that in $(Mg,Fe)SiO_3$ perovskite (Ito & Takahashi 1987b; Kato et al 1988a,b). If so, cubic $CaSiO_3$ perovskite could represent a large reservoir of rare earth elements in the mantle. However, the effects of disequilibrium in such experiments have been questioned (Walker & Agee 1989).

OUTLOOK

Progress on the study of silicate perovskite during the past few years may be characterized as moving from the study of analogue materials, to studies carried out at ambient (room) conditions, and finally to the in situ study at high pressures. In situ studies of silicate perovskites under combined high pressure and high temperatures are in their infancy (Mao et al 1991), and few have yet been performed under lower mantle *P-T* conditions. Future work will likely be characterized by in situ measurements under these extreme conditions, including petrologic studies of whole rock samples at deep mantle conditions (e.g. O'Neill & Jeanloz 1990). As in the past, accurate ultrahigh-*P-T* measurements will require continued development of both high-pressure technology and analytical techniques. With these developments it is probable that further progress will also be made in the study of transport properties, which could provide important additional constraints on mantle mineralogy. These may be controlled by the defect chemistry of silicate perovskites at high temperatures and pressures, about which little is currently known.

If our current understanding of the phase equilibria under lower mantle conditions is correct, then the study of the properties of $(Mg,Fe)SiO_3$ and $CaSiO_3$ perovskite, along with magnesiowüstite and stishovite, should fully describe the physical properties of the lower mantle. It is important to recognize that our understanding of the deep earth may be biased by the properties of those materials that can be quenched to low (ambient) conditions or otherwise easily studied in the laboratory. Evidence for new nonquenchable aluminous, calcic, and magnesium silicate phases which are stable under lower mantle conditions has been reported (Guyot et al 1989, Ito 1989, Kubicki & Hemley 1989) and others have been predicted (e.g. Finger & Hazen 1991), and new classes of chemical reactions occurring at deep mantle conditions are indeed likely (e.g. Knittle & Jeanloz 1989b, 1991a). It is likely to take many years to fully describe the elastic, rheological, and transport properties of these materials and the deep rocks that they constitute. In so doing, new insights into properties of silicates and oxides will be gained, new high-*P-T* technology and computational techniques will be developed, and a better understanding of the Earth and its history will be achieved.

ACKNOWLEDGMENTS

We are grateful to the following for their help in the preparation of this review: Y. Fei, L. W. Finger, R. M. Hazen, E. Ito, R. Jeanloz, H. K. Mao, C. E. Meade, A. Navrotsky, C. T. Prewitt, G. D. Price, L. Stixrude, Y. Wang, Q. Williams, and G. H. Wolf. Computations were performed on the Cray 2 at the NCSA under the auspices of the NSF. NSF grants EAR-8904080, EAR-8920239, and EAR-8916754 supported this work.

Literature Cited

Agnon, A., Bukowinski, M. S. T. 1990. δ_s at high pressure and $d\ln V_s/d\ln V_p$ in the lower mantle. *Geophys. Res. Lett.* 17: 1149–52

Aleksandrov, K. S. 1976. The sequences of structural phase transitions in perovskites. *Ferroelectrics* 14: 801–5

Aleksandrov, K. S. 1978. Mechanisms of the ferroelectric and structural phase transitions, structural distortions in perovskites. *Ferroelectrics* 20: 61–67

Andersen, N. H., Kjems, J. K., Hayes, W. 1985. Ionic conductivity of the perovskites $NaMgF_3$, $KMgF_3$, and $KZnF_3$ at high temperatures. *Solid State Ionics* 17: 143–45

Anderson, D. L. 1989. *Theory of the Earth.* Boston: Blackwell. 366 pp.

Anderson, O. L. 1988. A ferroelectric transition in the lower mantle? *Eos Trans. Am. Geophys. Union* 69: 1451 (Abstr.)

Anderson, O. L., Chopelas, A., Boehler, R. 1990. Thermal expansivity versus pressure at constant temperature: A re-examination. *Geophys. Res. Lett.* 17: 685–88

Anderson, O. L., Isaak, D., Oda, H. 1992. High-temperature elastic constant data on minerals relevant to geophysics. *Rev. Geophys.* In press

Andrault, D., Poirier, J. P. 1991. Evolution of the distortion of perovskites under pressure: an EXAFS study of $BaZrO_3$, $SrZrO_3$, and $CaGeO_3$. *Phys. Chem. Miner.* 18: 91–105

Bass, J. D., Anderson, D. L. 1984. Composition of the upper mantle: geophysical tests of two petrological models. *Geophys. Res. Lett.* 11: 237–40

Bassett, W. A., Weathers, M. S., Huang, E., Onomichi, M. 1988. Melting curve of $Mg_{0.88}Fe_{0.12}SiO_3$ up to 700 kbar. *Eos Trans. Am. Geophys. Union* 69: 1451 (Abstr.)

Beauchesne, S., Poirier, J. P. 1989. Creep of barium titanate perovskite: a contribution to a systematic approach to the viscosity of the lower mantle. *Phys. Earth Planet. Inter.* 55: 187–99

Beauchesne, S., Poirier, J. P. 1990. In search of a systematics for the viscosity of perovskites: creep of potassium tantalate and niobate. *Phys. Earth Planet. Inter.* 61: 182–98

Bell, P. M., Yagi, T., Mao, H. K. 1979. Iron-magnesium distribution coefficients between spinel [$(Mg,Fe)_2SiO_4$], magnesiowüstite [(Mg,Fe)O], and perovskite [$(Mg,Fe)SiO_3$]. *Carnegie Inst. Washington Yearb.* 78: 618–21

Bina, C. R., Helffrich, G. R. 1992. Calculation of elastic properties from thermodynamic equation of state principles. *Annu. Rev. Earth Planet. Sci.* 20: 527–52

Bina, C. R., Silver, P. 1990. Constraints on lower mantle composition and temperature from density and bulk sound velocity profiles. *Geophys. Res. Lett.* 17: 1153–56

Boehler, R., Chopelas, A. 1991. A new approach to laser heating in high pressure mineral physics. *Geophys. Res. Lett.* 18: 1147–50

Boehler, R., Chopelas, A. 1992. Phase transitions in a 500 kbar–3000 K gas apparatus. See Syono & Manghnani 1992. In press

Boyer, L. L. 1974. Computerized group theory for lattice dynamical problems. *J. Comput. Physics* 16: 167–85

Brown, J. M. 1986. Interpretation of the *D″* zone at the base of the mantle: dependence on assumed values of thermal conductivity. *Geophys. Res. Lett.* 13: 1509–12

Bukowinski, M. S. T., Wolf, G. H. 1988. Equation of state and possible critical phase transitions in $MgSiO_3$ perovskite at lower-mantle conditions. In *Advances in Physical Geochemistry, Vol. 7 Structural and Magnetic Phase Transitions*, ed. S. Ghose, J. M. D. Coey, E. Salje, pp. 91–112. New York: Springer-Verlag

Bukowinski, M. S. T., Wolf, G. H. 1990. Thermodynamically consistent decompression: implications for lower mantle composition. *J. Geophys. Res.* 95: 12,583–93

Chen, L. C., Mao, H. K., Hemley, R. J. 1989. Compression and polymorphism of $CaSiO_3$ at high pressures and temperatures. *Annu. Rep. Dir. Geophys. Lab.* 1988–1989: 94–98

Chopelas, A., Boehler, R. 1989. Thermal expansion at very high pressure, systematics, and a case for a chemically homogeneous mantle. *Geophys. Res. Lett.* 16: 1347–50

Chopelas, A., Boehler, R. 1992. Raman spectroscopy of high pressure $MgSiO_3$ phases synthesized in a CO_2 laser heated diamond anvil cell: perovskite and clinopyroxene. See Syono & Manghnani 1992. In press

Choudhury, N., Chaplot, S. L., Rao, K. R., Ghose, S. 1988. Lattice dynamics of $MgSiO_3$ perovskite. *Pramana J. Phys.* 30: 423–28

Christensen, U. R., Yuen, D. A. 1984. The interaction of a subducting lithosphere with a chemical or phase boundary. *J. Geophys. Res.* 89: 4389–4402

Cohen, R. E. 1987. Elasticity and equation of state of $MgSiO_3$ perovskite. *Geophys. Res. Lett.* 14: 1053–56

Cohen, R. E., Boyer, L. L., Mehl, M. J., Pickett, W. E., Krakauer, H. 1989. Electronic structure and total energy calculations for oxide perovskites and superconductors. See Navrotsky & Weidner 1989, pp. 55–66

Cohen, R. E., Krakauer, H. 1990. Lattice dynamics and origin of ferroelectricity in $BaTiO_3$: Linearized-augmented-plane-wave total-energy calculations. *Phys. Rev. B* 42: 6416–23

Durben, D. J., Wolf, G. H. 1991. Thermally-induced vitrification and internal stress in magnesium silicate ($MgSiO_3$) perovskite. *Eos Trans. Am. Geophys. Union* 72(44): 464 (Abstr.)

Dziewonski, A. M., Anderson, D. L. 1981. Preliminary reference earth model. *Phys. Earth Planet. Inter.* 25: 297–356

Dziewonski, A. M., Woodhouse, J. H. 1987. Global images of the Earth's interior. *Science* 236: 37–48

Fei, Y., Mao, H. K., Hemley, R. J., Shu, J. 1991a. Simultaneous high-P-T diffraction measurements of $(Fe,Mg)SiO_3$-perovskite and (Fe,Mg)O magnesiowüstite: implications for lower mantle composition. *Annu. Rep. Dir. Geophys. Lab.* 1990–1991: 107–14

Fei, Y., Mao, H. K., Mysen, B. O. 1991b. Experimental determination of element partitioning and calculation of phase relations in the MgO-FeO-SiO_2 system at high pressure and high temperature. *J. Geophys. Res.* 96: 2157–69

Finger, L. W., Hazen, R. M. 1991. Crystal chemistry of six-coordinated silicon: a key to understanding the Earth's deep interior. *Acta Crystallogr.* B47: 561–80

Fitz Gerald, J. D., Ringwood, A. E. 1991. High pressure rhombohedral perovskite phase $Ca_2AlSiO_{5.5}$. *Phys. Chem. Miner.* 18: 40–46

Glazer, A. M. 1972. The classification of tilted octahedra in perovskites. *Acta Crystallogr. Ser. B* 28: 3384–92

Glazer, A. M. 1975. Simple ways of determining perovskite structures. *Acta Crystallogr. Ser. A* 31: 756–62

Guyot, F., Madon, M., Peyronneau, J., Poirier, J. P. 1989. X-ray microanalysis of high-pressure/high-temperature phases synthesized from natural olivine in a diamond-anvil cell. *Earth Planet. Sci. Lett.* 90: 52–64

Hazen, R. M. 1988. Perovskites. *Scientific American*, June 1988: 74–81

Heinz, D. L. 1991. Split decision on the mantle. *Nature* 351: 346–47

Heinz, D. L., Jeanloz, R. 1987. Measurement of the melting curve of $Mg_{0.1}Fe_{0.9}SiO_3$ perovskite at lower mantle conditions and its geophysical implications. *J. Geophys. Res.* 92: 11,437–44

Hemley, R. J. 1991. Spectroscopic models for thermodynamic properties of minerals at high pressures and temperatures. *EOS Trans. Am. Geophys. Union* 72(12): 282 (Abstr.)

Hemley, R. J., Cohen, R. E., Yeganeh-Haeri, A., Mao, H. K., Weidner, D. J., Ito, E. 1989. Raman spectroscopy and lattice dynamics of $MgSiO_3$ perovskite at high pressure. See Navrotsky & Weidner 1989, pp. 35–44

Hemley, R. J., Jackson, M. D., Gordon, R. G. 1985. Lattice dynamics and equations of state of high-pressure mineral phases studied with electron-gas theory. *Eos Trans. Am. Geophys. Union* 66: 357 (Abstr.)

Hemley, R. J., Jackson, M. D., Gordon, R. G. 1987. Theoretical study of the structure, lattice dynamics, and equations of state of perovskite-type $MgSiO_3$ and $CaSiO_3$. *Phys. Chem. Miner.* 14: 2–12

Hemley, R. J., Kubicki, J. D. 1991. Deep mantle melting. *Nature* 349: 283–84

Hemley, R. J., Stixrude, L., Fei, Y., Mao, H. K. 1992. Constraints on lower mantle composition from *P-V-T* measurements of $(Fe,Mg)SiO_3$ perovskite and (Fe,Mg)O. See Syono & Manghnani 1992. In press

Hill, R. J., Jackson, I. 1990. The thermal expansion of $ScAlO_3$—a silicate perovskite analogue. *Phys. Chem. Miner.* 17: 89–96

Hirsch, L. M., Shankland, T. J. 1991. Point defects in $(Mg,Fe)SiO_3$ perovskite. *Geophys. Res. Lett.* 18: 1305–8

Hofmeister, A. M., Williams, Q., Jeanloz, R. 1987. Thermodynamic and elastic properties of $MgSiO_3$ perovskite from far-IR spectra at pressure. *Eos Trans. Am. Geophys. Union* 68: 1469 (Abstr.)

Horiuchi, H., Ito, E., Weidner, D. J. 1987. Perovskite-type $MgSiO_3$: single crystal x-ray diffraction study. *Am. Mineral.* 72: 357–60

Irifune, T., Ringwood, A. E. 1987a. Phase transformations in primitive morb and pyrolite compositions to 25 GPa and some geophysical implications. See Manghnani & Syono 1987, pp. 231–42

Irifune, T., Ringwood, A. E. 1987b. Phase transformations in a harzburgite composition to 26 GPa: implications for dynamical behaviour of the subducting slab. *Earth Planet. Sci. Lett.* 86: 365–76

Irifune, T., Susaki, J., Yagi, T., Sawamoto, H. 1989. Phase transformations in diopside $CaMgSi_2O_6$ at pressures up to 25 GPa. *Geophys. Res. Lett.* 16: 187–90

Isaak, D. G., Anderson, O. L., Cohen, R. E. 1992. The relationship between shear and compressional velocities at high pressures: reconciliation of seismic tomography and mineral physics. *Geophys. Res. Lett.* Submitted

Isaak, D. G., Cohen, R. E., Mehl, M. J. 1990. Calculated elastic and thermal properties of MgO at high pressures and temperatures. *J. Geophys. Res.* 95: 7055–67

Ita, J., Stixrude, L. 1991. Petrology, elasticity and composition of the mantle transition zone. *J. Geophys. Res.* In press

Ito, E. 1977. The absence of oxide mixture in high-pressure phases of Mg-silicates. *Geophys. Res. Lett.* 4: 72–74

Ito, E. 1984. Ultra-high pressure phase relations of the system $MgO-FeO-SiO_2$ and their geophysical implications. In *Materials Science of the Earth's Interior*, ed. I. Sunagawa, pp. 387–94. Tokyo: Terra Scientific

Ito, E. 1989. Stability relations of silicate perovskite under subsolidus conditions. See Navrotsky & Weidner 1989, pp. 27–32

Ito, E., Akaogi, M., Topor, L., Navrotsky, A. 1990. Negative pressure-temperature slopes for reactions forming $MgSiO_3$ perovskite from calorimetry. *Science* 249: 1275–78

Ito, E., Katsura, T. 1992. Melting of ferromagnesian silicates under the lower mantle conditions. See Syono & Manghnani 1992

Ito, E., Matsui, Y. 1978. Synthesis and crystal-chemical characterization of $MgSiO_3$ perovskite. *Earth Planet. Sci. Lett.* 38: 443–50

Ito, E., Matsui, Y. 1979. High-pressure transformations in silicates, germanates, and titanates with ABO_3 stoichiometry. *Phys. Chem. Miner.* 4: 265–73

Ito, E., Sato, H. 1991. Aseismicity in the lower mantle by superplasticity of the descending slab. *Nature* 351: 140–41

Ito, E., Takahashi, E. 1987a. Ultrahigh-pressure phase transformations and the constitution of the deep mantle. See Manghnani & Syono 1987, pp. 221–29

Ito, E., Takahashi, E. 1987b. Melting of peridotite at uppermost lower-mantle conditions. *Nature* 328: 514–16

Ito, E., Takahashi, E. 1989. Postspinel transformations in the system Mg_2SiO_4-Fe_2SiO_4 and some geophysical implications. *J. Geophys. Res.* 94: 10,637–46

Ito, E., Takahashi, E., Matsui, Y. 1984. The mineralogy and chemistry of the lower mantle: an implication of the ultrahigh-pressure phase relations in the system $MgO-FeO-SiO_2$. *Earth Planet. Sci. Lett.* 67: 238–48

Ito, E., Weidner, D. J. 1986. Crystal growth of $MgSiO_3$ perovskite. *Geophys. Res. Lett.* 13: 464–66

Ito, E., Yamada, H. 1982. Stability relations

of silicate spinels, ilmenites and perovskite. In *High Pressure Research in Geophysics*, ed. S. Akimoto, M. H. Manghnani, pp. 405–19. Tokyo: Academic

Jackson, I. 1983. Some geophysical constraints on the chemical composition of the Earth's lower mantle. *Earth Planet. Sci. Lett.* 62: 91–103

Jackson, W. E., Knittle, E., Brown, G. E., Jeanloz, R. 1987. Partitioning of Fe within high-pressure silicate perovskite: evidence for unusual geochemistry in the lower mantle. *Geophys. Res. Lett.* 14: 224–26

Jeanloz, R. 1990. Thermodynamics and evolution of the Earth's interior: high-pressure melting of silicate perovskite as an example. *Proc. Gibbs Symp., Yale Univ., May 15–17, 1989*. Am. Math. Soc., pp. 211–26

Jeanloz, R. 1991. Effects of phase transitions and possible compositional changes in the seismological structure near 650 km depth. *Geophys. Res. Lett.* 18: 1743–46

Jeanloz, R., Knittle, E. 1989. Density and composition of the lower mantle. *Philos. Trans. R. Soc. London Ser. A* 328: 377–89

Jeanloz, R., Morris, S. 1986. Temperature distribution in the crust and mantle. *Annu. Rev. Earth Planet. Sci.* 14: 377–415

Jeanloz, R., O'Neill, B., Pasternak, M. P., Taylor, R. D., Bohlen, S. R. 1991. Mössbauer spectroscopy of $Mg_{0.1}Fe_{0.9}SiO_3$ perovskite. *Eos Trans. Am. Geophys. Union* 72(44): 464 (Abstr.)

Jeanloz, R., Thompson, A. B. 1983. Phase transitions and mantle discontinuities. *Rev. Geophys. Space Phys.* 21: 51–74

Kanzaki, M., Stebbins, J. F., Xue, X. 1991. Characterization of quenched high pressure phases in the $CaSiO_3$ system by XRD and ^{29}Si NMR. *Geophys. Res. Lett.* 18: 463–66

Kapusta, B., Guillope, M. 1988. High ionic diffusivity in the perovskite $MgSiO_3$: a molecular-dynamics study. *Phil. Mag. A* 58: 809–16

Karato, S., Fujino, K., Ito, E. 1990. Plasticity of $MgSiO_3$ perovskite: the results of microhardness tests on single crystals. *Geophys. Res. Lett.* 17: 13–16

Kato, T., Kumazawa, M. 1985. Garnet phase of $MgSiO_3$ filling the pyroxene-ilmenite gap at very high temperature. *Nature* 316: 803–5

Kato, T., Ringwood, A. E., Irifune, T. 1988a. Experimental determination of element partitioning between silicate perovskites, garnets, and liquids: constraints on early differentiation of the mantle. *Earth Planet. Sci. Lett.* 89: 123–45

Kato, T., Ringwood, A. E., Irifune, T. 1988b. Constraints on element partitioning coefficients between $MgSiO_3$ perovskite and liquid determined by direct measurements. *Earth Planet. Sci. Lett.* 90: 65–68

Kieffer, S. W. 1976. Lattice thermal conductivity within the earth and considerations of a relationship between the pressure dependence of the thermal diffusivity and the volume dependence of the Grüneisen parameter. *J. Geophys. Res.* 81: 3025–30

Kim, Y.-H., Ming, L. C., Manghnani, M. H., Ko, J. 1991. Phase transformation studies on a synthetic hedenbergite up to 26 GPa at 1200°C. *Phys. Chem. Miner.* 17: 540–44

Kirkpatrick, R. J., Howell, D., Phillips, B. L., Cong, X.-D., Ito, E., Navrotsky, A. 1991. MAS NMR spectroscopic study of $Mg^{29}SiO_3$ with the perovskite structure. *Am. Mineral.* 76: 673–76

Knittle, E., Closman, C., Li, G., Wang, X., Bridges, F. G. 1991. New x-ray absorption measurements on the site distribution of iron in silicate perovskite. *Eos Trans. Am. Geophys. Union* 72(44): 464 (Abstr.)

Knittle, E., Jeanloz, R. 1987a. Synthesis and equation of state of $(Mg,Fe)SiO_3$ perovskite to over 100 gigapascals. *Science* 235: 668–70

Knittle, E., Jeanloz, R. 1987b. The activation energy of the back transformation of silicate perovskite to enstatite. See Manghnani & Syono 1987, pp. 243–50

Knittle, E., Jeanloz, R. 1989a. Melting curve of $(Mg,Fe)SiO_3$ perovskite to 96 GPa: evidence for a structural transition in lower mantle melts. *Geophys. Res. Lett.* 16: 421–24

Knittle, E., Jeanloz, R. 1989b. Simulating the core-mantle boundary: An experimental study of high-pressure reactions between silicates and liquid ion. *Geophys. Res. Lett.* 16: 609–12

Knittle, E., Jeanloz, R. 1991. Earth's core-mantle boundary: Results of experiments at high pressures and temperatures. *Science* 251: 1438–43

Knittle, E., Jeanloz, R., Smith, G. L. 1986. The thermal expansion of silicate perovskite and stratification of the Earth's mantle. *Nature* 319: 214–16

Kubicki, J. D., Hemley, R. J. 1989. Spectroscopic evidence for a new high-pressure magnesium silicate phase. *Annu. Rep. Dir. Geophys. Lab.* 1988–1989: 91–94

Kubicki, J. D., Hemley, R. J., Hofmeister, A. M. 1992. Raman and infrared study of pressure-induced structural changes in $MgSiO_3$, $CaMgSi_2O_6$, and $CaSiO_3$ glasses. *Am. Mineral.* In press

Kubicki, J. D., Lasaga, A. C. 1991. Molecular dynamics simulations of pressure and temperature effects on $MgSiO_3$ and Mg_2SiO_4 melts and glasses. *Phys. Chem. Miner.* 17: 661–73

Kudoh, Y., Ito, E., Takeda, H. 1987. Effect of pressure on the crystal structure of perovskite-type $MgSiO_3$. *Phys. Chem. Miner.* 14: 350–54

Kudoh, Y., Prewitt, C. T., Finger, L. W., Darovskikh, A., Ito, E. 1990. Effect of iron on the crystal structure of (Mg,Fe)SiO_3 perovskite. *Geophys. Res. Lett.* 17: 1481–84

Kuskov, O. L., Panferov, A. B. 1991. Phase diagrams of the FeO-MgO-SiO_2 system and the structure of the mantle discontinuities. *Phys. Chem. Miner.* 17: 642–53

Leinenweber, K., Navrotsky, A. 1988. A transferable interatomic potential for crystalline phases in the system MgO-SiO_2. *Phys. Chem. Miner.* 15: 588–96

Li, X., Jeanloz, R. 1987. Electrical conductivity of (Mg,Fe)SiO_3 perovskite and a perovskite-dominated assemblage at lower mantle conditions. *Geophys. Res. Lett.* 14: 1075–78

Li, X., Jeanloz, R. 1990. Laboratory studies of the electrical conductivity of silicate perovskites at high pressures and temperatures. *J. Geophys. Res.* 95: 5067–78

Li, X., Jeanloz, R. 1991. Phases and electrical conductivity of a hydrous silicate assemblage at lower-mantle conditions. *Nature* 350: 332–34

Li, X., Ming, L.-C., Manghnani, M. H., Wang, Y., Jeanloz, R. 1991. High-pressure experimental studies of the electrical conductivity of $(Mg_{0.9}Fe_{0.1})SiO_3$ perovskite using both cubic-anvil large-volume apparatus and laser-heated diamond-anvil cell. *Eos Trans. Am. Geophys. Union* 72(44): 529 (Abstr.)

Litov, E., Anderson, O. L. 1978. The possibility of ferroelectric-like phenomena in the mantle of the Earth. *J. Geophys. Res.* 83: 1692–98

Liu, L.-G. 1974. Silicate perovskite from phase transformations of pyrope-garnet at high pressure and temperature. *Geophys. Res. Lett.* 1: 277–80

Liu, L.-G. 1975. Post-oxide phases of forsterite and enstatite. *Geophys. Res. Lett.* 2: 417–19

Liu, L.-G. 1976. Orthorhombic perovskite phases observed in olivine, pyroxene and garnet at high pressures and temperatures. *Phys. Earth Planet. Inter.* 11: 289–98

Liu, L.-G. 1977. The system enstatite-pyrope at high pressures and temperatures and the mineralogy of the earth's mantle. *Earth Planet. Sci. Lett.* 36: 237–45

Liu, L.-G. 1979. On the 650-km discontinuity. *Earth Planet. Sci. Lett.* 42: 202–8

Liu, L.-G. 1980. Phase relations in the system diopside-jadeite at high pressure and high temperatures. *Earth Planet. Sci. Lett.* 47: 398–402

Liu, L.-G. 1987. New silicate perovskites. *Geophys. Res. Lett.* 14: 1079–82

Liu, L.-G., Bassett, W. A. 1986. *Elements, Oxides, Silicates.* New York: Oxford Univ. Press. 250 pp.

Liu, L.-G., Ringwood, A. E. 1975. Synthesis of a perovskite-type polymorph of $CaSiO_3$. *Earth Planet. Sci. Lett.* 28: 209–11

Liu, X., Wang, Y., Liebermann, R. C., Maniar, P. D., Navrotsky, A. 1991. Phase transition in $CaGeO_3$ perovskite: Evidence from x-ray powder diffraction thermal expansion, and heat capacity. *Phys. Chem. Miner.* In press

Machetel, P., Weber, P. 1991. Intermittent layered convection in a model mantle with an endothermic phase change at 670 km. *Nature* 350: 55–57

Madon, M., Bell, P. M., Mao, H. K., Poirier, J. P. 1980. Transmission electron diffraction and microscopy of synthetic high pressure $MgSiO_3$ phase with perovskite structure. *Geophys. Res. Lett.* 7: 629–32

Madon, M., Guyot, F., Poirier, J. P. 1989. Electron microscopy of high-pressure phases synthesized from natural olivine in diamond anvil cell. *Phys. Chem. Miner.* 16: 320–30

Manghnani, M. H., Syono, Y., eds. 1987. *High Pressure Research in Mineral Physics.* Tokyo: Terra Scientific, Washington, DC: Am. Geophys. Union

Mao, H. K., Chen, L. C., Hemley, R. J., Jephcoat, A. P., Wu, Y., Bassett, W. A. 1989a. Stability and equation of state of $CaSiO_3$ perovskite to 134 GPa. *J. Geophys. Res.* 94: 17,889–94

Mao, H. K., Hemley, R. J., Fei, Y., Shu, J. F., Chen, L. C., et al. 1991. Effect of pressure, temperature, and composition on lattice parameters and density of (Fe,Mg)SiO_3-perovskites to 30 GPa. *J. Geophys. Res.* 96: 8069–79

Mao, H. K., Hemley, R. J., Shu, J., Chen, L., Jephcoat, A. P., et al. 1989b. The effect of pressure, temperature, and composition on lattice parameters and density of (Fe,Mg)SiO_3-perovskites to 30 GPa. *Annu. Rep. Dir. Geophys. Lab.* 1988–1989: 82–89

Mao, H. K., Yagi, T., Bell, P. M. 1977. Mineralogy of the earth's deep mantle: quenching experiments on mineral compositions at high pressures and temperature. *Carnegie Inst. Washington Yearb.* 76: 502–4

Matsui, M. 1988. Molecular dynamics study

of $MgSiO_3$ perovskite. *Phys. Chem. Miner.* 16: 234–38

Matsui, M., Akaogi, M., Matsumoto, T. 1987. Computational model of the structural and elastic properties of the ilmenite and perovskite phases of $MgSiO_3$. *Phys. Chem. Miner.* 14: 101–6

Matsui, M., Price, G. D. 1991. Simulation of the pre-melting behavior of $MgSiO_3$ perovskite at high pressures and temperatures. *Nature* 351: 735–37

Matsui, Y., Kawamura, K. 1980. Instantaneous structure of an $MgSiO_3$ melt simulated by molecular dynamics. *Nature* 285: 648–49

Matsui, Y., Kawamura, K., Syono, Y. 1982. Molecular dynamics calculations applied to silicate systems: molten and vitreous $MgSiO_3$ and Mg_2SiO_4 under low and high pressures. In *High Pressure Research in Geophysics*, ed. S. Akimoto, M. H. Manghnani, pp. 511–24. Tokyo: Academic

McMillan, P. F. 1989. Raman spectroscopy in mineralogy and geochemistry. *Annu. Rev. Earth Planet. Sci.* 17: 255–83

Meade, C., Jeanloz, R. 1990. The strength of mantle silicates at high pressures and room temperature: implications for the viscosity of the mantle. *Nature* 348: 533–35

Meade, C., Jeanloz, R. 1991. Linear finite strain theory for anisotropic materials. *Eos Trans. Am. Geophys. Union* 72(12): 282 (Abstr.)

Miller, G. H., Stolper, E. M., Ahrens, T. J. 1991a. The equation of state of a molten komatiite. 1. Shock wave compression to 36 GPa. *J. Geophys. Res.* 96: 11,831–48

Miller, G. H., Stolper, E. M., Ahrens, T. J. 1991b. The equation of state of a molten komatiite. 2. Application to komatiite petrogenesis and the hadean mantle. *J. Geophys. Res.* 96: 11,849–64

Miyamoto, M. 1988. Ion migration in $MgSiO_3$-perovskite and olivine by molecular dynamics calculations. *Phys. Chem. Miner.* 15: 601–4

Miyamoto, M., Takeda, H. 1984. An attempt to simulate high pressure structures of Mg-silicates by an energy minimization method. *Am. Mineral.* 69: 711–18

Navrotsky, A., Wiedner, D. J., eds. 1989. *Perovskite: A Structure of Great Interest to Geophysics and Materials Science.* Washington, DC: Am. Geophys. Union. 146 pp.

Navrotsky, A. 1989. Thermochemistry of perovskites. See Navrotsky & Weidner 1989, pp. 67–80

Ohtani, E. 1983. Melting temperature distribution and fractionation in the lower mantle. *Phys. Earth Planet. Inter.* 33: 12–25

Ohtani, E., Kato, T., Ito, E. 1992. Transition metal partitioning between lower mantle and core materials at 27 GPa. *Geophys. Res. Lett.* 18: 85–88

O'Keeffe, M., Bovin, J.-O. 1979. Solid electrolyte behavior of $NaMgF_3$. Geophysical implications. *Science* 206: 599–600

O'Keeffe, M., Hyde, B. G., Bovin, J.-O. 1979. Contribution to the crystal chemistry of orthorhombic perovskites: $MgSiO_3$ and $NaMgF_3$. *Phys. Chem. Miner.* 4: 299–305

O'Neill, B., Brown, D., Jeanloz, R. 1991. Effect of simultaneous Fe and Al substitution on the volume of magnesium-silicate perovskite. *Eos Trans. Am. Geophys. Union* 72(44): 464 (Abstr.)

O'Neill, B., Jeanloz, R. 1990. Experimental petrology of the lower mantle: a natural peridotite taken to 54 GPa. *Geophys. Res. Lett.* 17: 1477–80

Osako, M., Ito, E. 1991. Thermal diffusivity of $MgSiO_3$ perovskite. *Geophys. Res. Lett.* 18: 239–42

Parise, J. B., Wang, Y., Yeganeh-Haeri, A., Cox, D. E., Fei, Y. 1990. Crystal structure and thermal expansion of $(Mg,Fe)SiO_3$ perovskite. *Geophys. Res. Lett.* 17: 2089–92

Peyronneau, J., Poirier, J. P. 1989. Electrical conductivity of the earth's lower mantle. *Nature* 342: 537–39

Pickett, W. E., Cohen, R. E., Krakauer, H. 1991. Lattice instabilities, isotope effect, and high-T_c superconductivity in $La_{2-x}Ba_xCuO_4$. *Phys. Rev. Lett.* 67: 228–31

Poirier, J. P. 1989. Lindemann Law and the melting temperature of perovskites. *Phys. Earth Planet. Inter.* 54: 364–69

Poirier, J. P., Peyronneau, J. 1992. Experimental determination of the electrical conductivity of the material of the Earth's lower mantle. See Syono & Manghnani 1992

Poirier, J. P., Peyronneau, J., Gesland, J. Y., Brebec, G. 1983. Viscosity and conductivity of the lower mantle, an experimental study on a $MgSiO_3$ perovskite analog. *Phys. Earth Planet. Inter.* 32: 273–87

Price, G. D., Wall, A., Parker, S. C. 1989. The properties and behavior of mantle minerals: a computer-simulation approach. *Philos. Trans. R. Soc. London Ser. A* 328: 391–407

Reid, A. F., Ringwood, A. E. 1975. High-pressure modification of $ScAlO_3$ and some geophysical implications. *J. Geophys. Res.* 80: 3363–70

Reynard, B., Price, G. D. 1990. Thermal expansion of mantle minerals at high pressures—a theoretical study. *Geophys. Res. Lett.* 17: 689–92

Richet, P., Mao, H. K., Bell, P. M. 1989.

Bulk moduli of magnesiowüstites from static compression measurements. *J. Geophys. Res.* 94: 3037–45
Ringwood, A. E. 1975. *Composition and Petrology of the Earth's Mantle*. New York: McGraw-Hill. 618 pp.
Ringwood, A. E. 1962. Mineralogical constitution of the deep mantle. *J. Geophys. Res.* 67: 4005–10
Ross, N. L., Hazen, R. M. 1989. Single crystal x-ray diffraction study of $MgSiO_3$ perovskite from 77 to 400 K. *Phys. Chem. Miner.* 16: 415–20
Ross, N. L., Hazen, R. M. 1990. High-pressure crystal chemistry of $MgSiO_3$ perovskite. *Phys. Chem. Miner.* 17: 228–37
Roufosse, M. C., Jeanloz, R. 1983. Thermal conductivity of minerals at high pressure: the effect of phase transitions. *J. Geophys. Res.* 88: 7399–7409
Salje, E. 1989. Characteristics of perovskite-related materials. *Philos. Trans. R. Soc. London Ser. A* 328: 409–16
Sasaki, S., Prewitt, C. T., Lieberman, R. C. 1983. The crystal structure of $CaGeO_3$ and the crystal chemistry of the $GdFeO_3$ type perovskites. *Am. Mineral.* 68: 1189–98
Shearer, P. M. 1990. Seismic imaging of upper-mantle structure with new evidence for a 520-km discontinuity. *Nature* 344: 121–26
Silver, P. G., Carlson, R. W., Olsen, P. 1988. Deep slabs, geochemical heterogeneity, and the large-scale structure of mantle convection: Investigation of an enduring paradox. *Annu. Rev. Earth Planet. Sci.* 16: 477–541
Stixrude, L., Bukowinski, M. S. T. 1990. Fundamental thermodynamic relations and silicate melting with implications for the constitution of *D″*. *J. Geophys. Res.* 95: 19,311–25
Stixrude, L., Bukowinski, M. S. T. 1992. Stability of $(Mg,Fe)SiO_3$ perovskite and the structure of the lowermost mantle. *Geophys. Res. Lett.* In review
Suzuki, I. 1975. Thermal expansion of periclase and olivine, and their anharmonic properties. *J. Phys. Earth* 23: 145–59
Syono, Y., Manghnani, M. H., eds. 1992. *High Pressure Research in Mineral Physics: Applications to Earth and Planetary Sciences*. Tokyo: Terra Scientific. In press
Takahashi, E., Ito, E. 1987. Mineralogy of mantle peridotite along a model geotherm up to 700 km depth. See Manghnani & Syono 1987, pp. 427–37
Tamai, H., Yagi, T. 1989. High-pressure and high-temperature phase relations in $CaSiO_3$ and $CaMgSi_2O_6$ and elasticity of perovskite-type $CaSiO_3$. *Phys. Earth Planet. Inter.* 54: 370–77
Tarrida, M., Richet, P. 1989. Equation of state of $CaSiO_3$ perovskite to 96 GPa. *Geophys. Res. Lett.* 16: 1351–54
Walker, D., Agee, C. 1989. Partitioning "equilibrium," temperature gradients, and constraints on Earth differentiation. *Earth Planet. Sci. Lett.* 96: 49–60
Wall, A., Price, G. D. 1989. Electrical conductivity of the lower mantle: a molecular dynamics simulation of $MgSiO_3$ perovskite. *Phys. Earth Planet. Inter.* 58: 192–204
Wall, A., Price, G. D., Parker, S. C. 1986. A computer simulation of the structure and elastic properties of $MgSiO_3$ perovskite. *Mineral Mag.* 50: 693–707
Wang, Y., Guyot, F., Liebermann, R. C. 1992. Electron microscopic evidence of structural phase transitions in $Mg_{1-x}Fe_x SiO_3$ perovskite and implications for the lower mantle. *J. Geophys. Res.* In press
Wang, Y., Guyot, F., Yeganeh-Haeri, A., Liebermann, R. C. 1990. Twinning in $MgSiO_3$ perovskite. *Science* 248: 468–71
Wang, Y., Weidner, D. J., Liebermann, R. C., Liu, X., Ko, J., et al. 1991a. Phase transition and thermal expansion of $MgSiO_3$ perovskite. *Science* 251: 410–13
Wang, Y., Zhao, Y., Weidner, D. J., Liebermann, R. C., Parise, J. B., Cox, D. E. 1991b. High temperature behavior of $(Mg,Fe)SiO_3$ perovskite at 1 bar. *Eos Trans. Am. Geophys. Union* 72(44): 464 (Abstr.)
Weidner, D. J., Wang, Y., Liebermann, R. C., Vaughan, M. T., Leinenweber, K., et al. 1991. More on high-pressure, high-temperature equation of state of $MgSiO_3$ perovskite. *Eos Trans. Am. Geophys. Union* 72(44): 435 (Abstr.)
Weng, K., Mao, H. K., Bell, P. M. 1982. Lattice parameters of the perovskite phases in the system $MgSiO_3$-$CaSiO_3$-Al_2O_3. *Carnegie Inst. Washington Yearb.* 81: 273–77
Weng, K., Xu, J., Mao, H. K., Bell, P. M. 1983. Preliminary Fourier-transform infrared spectral data on the SiO_6^{8-} octahedral group in silicate-perovskite. *Carnegie Inst. Washington Yearb.* 82: 355–56
Williams, Q. 1990. Molten $(Mg_{0.88}Fe_{0.12})_2SiO_4$ at lower mantle conditions: melting products and structure of quenched glasses. *Geophys. Res. Lett.* 17: 635–38
Williams, Q., Jeanloz, R., McMillan, P. 1987. Vibrational spectrum of $MgSiO_3$ perovskite: zero-pressure Raman and mid-infrared spectra to 27 GPa. *J. Geophys. Res.* 92: 8116–28
Williams, Q., Knittle, E., Jeanloz, R. 1989. Geophysical and crystal chemical sig-

nificance of (Mg,Fe)SiO_3 perovskite. See Navrotsky & Weidner 1989, pp. 1–12

Wolf, G., Bukowinski, M. 1985. Ab initio structural and thermoelastic properties of orthorhombic $MgSiO_3$ perovskite. *Geophys. Res. Lett.* 12: 809–12

Wolf, G., Bukowinski, M. S. T. 1987. Theoretical study of the structural and thermodynamic properties of $MgSiO_3$ and $CaSiO_3$ perovskites: implication for lower mantle composition. See Manghnani & Syono 1987, pp. 313–31

Wolf, G. H., Durben, D. J., McMillan, P. F. 1990. Raman spectroscopic study of the vibrational properties and reversion of $CaGeO_3$ and $MgSiO_3$ perovskites as a function of temperature. *Eos Trans. Am. Geophys. Union* 71: 1667 (Abstr.)

Wolf, G. H., Jeanloz, R. 1985. Lattice dynamics and structural distortions of $CaSiO_3$ and $MgSiO_3$ perovskites. *Geophys. Res. Lett.* 12: 413–16

Wood, B. J. 1989. Mineralogical phase change at the 670-km discontinuity. *Nature* 341: 278

Wood, B. J. 1990. Postspinel transformations and the width of the 670-km discontinuity: A comment on "Postspinel transformations in the system Mg_2SiO_4-Fe_2SiO_4 and some geophysical implications" by E. Ito and E. Takahashi. *J. Geophys. Res.* 95: 12681–85

Wood, B. J., Nell, J. 1991. High-temperature electrical conductivity of the lower mantle phase (Mg,Fe)O. *Nature* 351: 309–11

Wright, K., Price, G. D. 1989. Computer simulation of iron in magnesium silicate perovskite. *Geophys. Res. Lett.* 16: 1399–1402

Yagi, T., Bell, P. M., Mao, H. K. 1979c. Phase relations in the system MgO-FeO-SiO_2 between 150 and 700 kbar at 1000°C. *Carnegie Inst. Washington Yearb.* 78: 614–16

Yagi, T., Kusanaga, S., Tsuchida, Y., Fukai, Y. 1989. Isothermal compression and stability of perovskite-type $CaSiO_3$. *Proc. Jpn. Acad. Ser. B* 65: 129–32

Yagi, T., Mao, H. K., Bell, P. M. 1977. Crystal structure of $MgSiO_3$ perovskite. *Carnegie Inst. Washington Yearb.* 76: 516–19

Yagi, T., Mao, H. K., Bell, P. M. 1978a. Structure and crystal chemistry of perovskite-type $MgSiO_3$. *Phys. Chem. Miner.* 3: 97–110

Yagi, T., Mao, H. K., Bell, P. M. 1978b. Effect of iron on the stability and unit-cell parameters of ferromagnesian silicate perovskite. *Carnegie Inst. Washington Yearb.* 77: 837–41

Yagi, T., Mao, H. K., Bell, P. M. 1979a. Lattice parameters and specific volume for the perovskite phase of orthopyroxene composition, (Mg,Fe)SiO_3. *Carnegie Inst. Washington Yearb.* 78: 612–13

Yagi, T., Mao, H. K., Bell, P. M. 1979b. Hydrostatic compression of $MgSiO_3$ of perovskite structure. *Carnegie Inst. Washington Yearb.* 78: 613–14

Yagi, T., Mao, H. K., Bell, P. M. 1982. Hydrostatic compression perovskite-type $MgSiO_3$. In *Advances in Physical Geochemistry*, Vol. 2, ed. S. K. Saxena, pp. 317–25. New York: Springer-Verlag

Yeganeh-Haeri, A., Weidner, D. J., Ito, E. 1989a. Elasticity of $MgSiO_3$ in the perovskite structure. *Science* 248: 787–89

Yeganeh-Haeri, A., Weidner, D. J., Ito, E. 1989b. Single crystal elastic moduli of magnesium metasilicate perovskite. See Navrotsky & Weidner 1989, pp. 13–25

SUBJECT INDEX

D

E

CUMULATIVE INDEXES

CONTRIBUTING AUTHORS, VOLUMES 1–20

CHAPTER TITLES, VOLUMES 1–20

GEOPHYSICS AND PLANETARY SCIENCE

OCEANOGRAPHY, METEOROLOGY, AND PALEOCLIMATOLOGY

PALEONTOLOGY, STRATIGRAPHY, AND SEDIMENTOLOGY